Handbook of Experimental Pharmacology

Volume 135/II

Springer
Berlin
Heidelberg
New York
Barcelona
Hong Kong
London
Milan
Paris
Singapore
Tokyo

Estrogens and Antiestrogens II

Pharmacology and Clinical Application of Estrogens and Antiestrogens

Contributors

E. Anderson, A.S. Bhatnagar, H. Blode, C. Christiansen, M. Dören, M. Dowsett, K.H. Fritzemeier, R. Gallo, A.R. Genazzani, J.-Å. Gustafsson, Ch. Hegele-Hartung, V.W. Henderson, A. Howell, S.M. Hyder, E.V. Jensen, S.R.D. Johnston, S.K. Jonas, H.L. Jørgensen, H. Kuhl, W. Kuhnz, G. Kuiper, S. Mäkelä, E.F. Mammen, H.H.D. Meyer, W.R. Miller, N. Navizadeh, M. Oettel, F. Petraglia, A. Purohit, M.J. Reed, L. Sahlin, G. Samsioe, B. von Schoultz, R. Sitruk-Ware, G. Söderqvist, A. Spinetti, G.M. Stancel, F.Z. Stanczyk, M. Stomati, A.E. Wakeling, M. Warner, R.Q. Wharton, B. Winding, H. Zimmermann

Editors

M. Oettel and E. Schillinger

Springer

Professor Dr. med.vet.habil. MICHAEL OETTEL
Leiter Forschung und Entwicklung
Jenapharm GmbH & Co. KG
Otto-Schott-Str. 15
D-07745 Jena
GERMANY

Dr. rer.nat. EKKEHARD SCHILLINGER
Leiter Forschung
Fertilitätskontrolle und Hormontherapie
Schering Aktiengesellschaft
Müllerstr. 170–178
D-13342 Berlin
GERMANY

With 85 Figures and 62 Tables

ISBN-13: 978-3-642-64261-6 e-ISBN-13: 978-3-642-60107-1
DOI: 10.1007/978-3-642-60107-1
Library of Congress Cataloging-in-Publication Data

Estrogens and antiestrogens / editors, Michael Oettel and Ekkehard Schillinger. p. cm. – (Handbook of experimental pharmacology; 135) Includes bibliographical references and index. Contents: I. Physiology and mechanisms of estrogens and antiestrogens / contributors, S.E. Alves ... [et al.]. – II. Pharmacology and clinical application of estrogens and antiestrogens / contributors, E. Anderson ... [et al].
ISBN 978-3-642-64261-6 1. Estrogen–Physiological effect. 2. Estrogen–Therapeutic use. 3. Estrogen–Inhibitors–Physiological effect. 4. Estrogen–Inhibitors–Therapeutic use. I. Oettel, Michael, 1939– . II. Series. [DNLM: 1. Estrogens–pharmacology. 2. Estrogen Antagonists–pharmacology. 3. Estrogen Antagonists–therapeutic use. 4. Estrogens–metabolism. W1 HA51L v. 135 1999 / WP 522 E807 1999] QP905.H3 vol. 135 [QP572.E857] 615′.1 s–dc21 [615′.366] DNLM/DLC for Library of Congress 99-15299 CIP

Softcover reprint of the hardcover 1st edition 1999

Cover design: *design & production* GmbH, Heidelberg

Typesetting: Best-set Typesetter Ltd., Hong Kong

Production Editor: Angélique Gcouta

SPIN: 10714172 27/3020 – 5 4 3 2 1 0 – Printed on acid-free paper

Preface

For many years, Springer has been publishing an impressive series of textbooks of pharmacology which have set standards in medical science. Surprisingly, an extensive overview of the current state of the art in research on estrogens and antiestrogens was still lacking. The present two volumes on estrogens and antiestrogens provide a comprehensive review of a field of research in which remarkable progress has been made over the past few years. New insights into the mechanisms of steroid hormone action resulted in a tremendous number of publications from which new principles of preventive and therapeutic applications of estrogens and antiestrogens emerged. Although various electronic data bases provide easy access to this copious information, there was a clear necessity for a monograph-style textbook which assesses and summarizes current knowledge in this rapidly expanding field of research. It should be noted, however, that, due to this dynamic development, it is barely possible to comprehensively update every aspect of basic and clinical knowledge on estrogens and antiestrogens. Thus, the intention of the editors was to provide the reader with an overview of the "classic" and most recently explored areas of research and stimulate future interests in basic and applied endocrinology.

Estrogens were among the first steroid hormones described in the scientific literature. Since they were first isolated, since the chemical, synthesic and pharmacological characterization of naturally occurring estrogens and, later on, of orally active derivatives, estrogen research has produced continuously hallmark results in reproductive endocrinology worldwide. Development of oral contraceptives consisting of estrogens and progestins has revolutionized not only fertility regulation but also society itself. The introduction of hormone replacement therapy and an emerging awareness of its benefits has dramatically improved the quality of life of postmenopausal women. The discovery of antiestrogens and their implication for the treatment of hormone-dependent cancers has saved millions of lives. At present, an explosion of knowledge in the field of molecular biology has a tremendous influence on basic research, which addresses cellular and molecular mechanisms of action of estrogens and antiestrogens. Estrogen research has gained substantial momentum from the discovery of different estrogen receptor isoforms (as exemplified in the chapter by Sven-Ake Gustafsson's group in Vol. I), identification of co-activators and co-repressors which interact with estrogen receptor-mediated transcription, and new insights into the physiological importance of estrogens gathered from studies on transgenic animals. Increasing understanding of basic mechanisms of estrogen action has paved the way for the discovery of a new class of selective ligands of the estrogen receptor, the so-called selective estrogen receptor modulators (SERMs). These compounds may open new avenues in the prevention and treatment of diseases. The growing importance of phytoestrogens has also been recognized. Recently, evidence has accumulated in support of the view that the biological role of estrogens is not confined to the female organism, but also has significant physiological and clinical effect on the male. The central nervous system and bone are constantly gaining importance as crucial targets of estrogen action. Finally, the issues of environmental pollution by estrogen-like compounds and metabolites and its biological consequences have become a source of increasing concern.

The contributors to this book are renowned experts in the corresponding fields of research. The editors thus focussed their effort on providing a balanced representation of basic and clinical research from academia and the pharmaceutical industry, while deliberately abstaining from any major interference with the content and style of individual chapters. This might have resulted in duplication of certain aspects; however, one should consider that even research can be contemplated from different points of view.

It was a rewarding endeavor to closely interact with the authors in the course of the editorial process, and we are indebted to them for their cooperation. We extend our thanks to Doris Walker from Springer for her patience and guidance through various stages of the development of this book. Many thanks also to Hans Herken – the editor-in-chief of the whole series – for his constant interest, his moderate but inevitable pressure on the editors and finally, his patience with us. The editorial assistance of Horst Wagner is gratefully acknowledged. We also appreciate the support of many staff members at Springer, Schering, and Jenapharm.

M. OETTEL and E. SCHILLINGER
Jena and Berlin, July 1999

List of Contributors

ANDERSON, E., Royal Marsden NHS Trust, Fulham Road, London SW3 6JJ, United Kingdom

BHATNAGAR, A.S., Novartis Pharma AG, K-125.15.16, CH-4002 Basel, Switzerland

BLODE, H., Research Laboratories, Pharmakokinetik, Schering AG, D-13342 Berlin, Germany

CHRISTIANSEN, C., Center for Clinical and Basic Research, Ballerup Byvej 222, DK-2750 Ballerup, Denmark

DÖREN, M., King's College Hospital, Department of Family Planning, Denmark Hill, London SE5 9RS, United Kingdom

DOWSETT, M., Academic Department of Biochemistry, Christie Hospital NHS Trust, Manchester M20 4BX, United Kingdom

FRITZEMEIER, K.H., Research Laboratories of Schering AG, Müllerstr. 178, D-13342 Berlin, Germany

GALLO, R., Department of Reproductive Medicine and Child Development, Division of Gynecology and Obstetrics, University of Pisa, Pisa, Italy

GENAZZANI, Department of Reproductive Medicine and Child Development, Division of Gynecology and Obstetrics, University of Pisa, Pisa, Italy

GUSTAFSSON, J.-Å., Department of Medical Nutrition, Karolinska Institute, NOVUM, S-14186 Huddinge, Sweden

HEGELE-HARTUNG, Ch., Research Laboratories of Schering AG, Müllerstr. 178, D-13342 Berlin, Germany

HENDERSON, V.W., Department of Neurology, University of Southern California, Los Angeles, CA 90033, USA

HOWELL, A., CRC Department of Medical Oncology (University of Manchester), Christie Hospital NHS Trust, Manchester M20 4BX, United Kingdom

HYDER, S.M., Department of Integrative Biology, Pharmacology and Physiology, The University of Texas Medical School at Houston, P.O. Box 20108, Houston, TX 77225, USA

JENSEN, E.V., Department of Medical Nutrition, Karolinska Institutet NOVUM, S-141 86 Huddinge, Sweden

JOHNSTON, S.R.D., Department of Medicine, The Royal Marsden NHS Trust, Fulham Road, London SW3 6JJ, United Kingdom

JONAS, S.K., Department of Surgery, Imperial College School of Medicine, Chelsea and Westminster Hospital, 369, Fulham Road, London SW10 9NH, United Kingdom

JØRGENSEN, H.L., Medicon, Gydevang 24–26, P.O. Box 220, DK-3450 Allerød, Denmark

KUHL, H., Klinik für gynäkologische Endokrinologie, Universitäts-Frauenklinik, Theodor-Stern-Kai 7, D-60590 Frankfurt/Main, Germany

KUHNZ, W., Research Laboratories, Pharmakokinetik, Schering AG, D-13342 Berlin, Germany

KUIPER, G., Department of Medical Nutrition and Department of Biosciences, Karolinska Institute, NOVUM, S-14186 Huddinge, Sweden

MÄKELÄ, S., University of Turku, Institute of Biomedicine, Department of Anatomy, Kiinamyllynkatu 10, FIN-20520 Turku, Finland

MAMMEN, E.F., Department of Obstetrics and Gynecology, Wayne State University, School of Medicine, C.S. Mott Center, 275 East Hancock Ave., Detroit, MI 48201, USA
Mailing address: Route 1, 1553-H2 Michibay Drive, Manistique, MI 49854, USA

Meyer, H.H.D., Forschungszentrum für Milch und Lebensmittel Weihenstephan, Institut für Physiologie, Technische Universität München, D-85350 Freising-Weihenstephan, Germany

Miller, W.R., Breast Unit Research Group, Western General Hospital, Edinburgh EH4 2XU, United Kingdom

Navizadeh, N., University of Southern California School of Medicine, 1240 North Mission Road, Room IM2, Los Angeles, CA 90033, USA

Oettel, M., Jenapharm GmbH & Co. KG, Forschung und Entwicklung, Otto-Schott-Str. 15, D-07745 Jena, Germany

Petraglia, F., Department of Surgical Science, Chair of Obstetrics and Gynecology, University of Udine, Udine, Italy

Purohit, A., Endocrinology and Metabolic Medicine, Imperial College of Science, Technology and Medicine, St. Mary's Hospital, London W2 1NY, United Kingdom

Reed, M.J., Endocrinology and Metabolic Medicine, Imperial College of Science, Technology and Medicine, St. Mary's Hospital, London W2 1NY, United Kingdom

Sahlin, L., Division for Reproductive Endocrinology, Department of Woman and Child Health, Karolinska Hospital L5:01, S-171 76 Stockholm, Sweden

Samsioe, G., Department of Obstetrics and Gynecology, Lund University Hospital, S-221 85 Lund, Sweden

Schoultz, B. von, Division for Obstetrics and Gynecology, Department of Woman and Child Health, Karolinska Hospital, P.O. Box 140, S-171 76 Stockholm, Sweden

Sitruk-Ware, R., Reproductive Endocrinology, Department of Endocrinology, Saint-Antoine Hospital, Paris, France *Mailing address:* Exelgyn Labs, 6 rue Christophe Colomb, F-75008 Paris, France

Söderqvist, G., Department of Woman and Child Health, Division for Obstetrics and Gynecology, Karolinska Hospital, S-171 76 Stockholm Sweden

Spinetti, A., Department of Reproductive Medicine and Child Development, Division of Gynecology and Obstetrics, University of Pisa, Pisa, Italy

Stancel, G.M., Department of Integrative Biology, Pharmacology and Physiology, The University of Texas Medical School at Houston, P.O. Box 207087, Houston, TX 77225, USA

Stanczyk, F.Z., University of Southern California School of Medicine, Department of Obstetrics and Gynecology, 1240 North Mission Road, Los Angeles, CA 90033, USA

Stomati, M., Department of Reproductive Medicine and Child Development, Division of Gynecology and Obstetrics, University of Pisa, Pisa, Italy

Wakeling, A.E., Cancer and Infection Research Department, Zeneca Pharmaceuticals, Alderley Park, Macclesfield, Cheshire SK10 4TG, United Kingdom

Warner, M., Department of Medical Nutrition and Department of Biosciences, Karolinska Institute, NOVUM, S-14186 Huddinge, Sweden

Wharton, R.Q., Department of Surgery, Imperial College School of Medicine, Chelsea and Westminster Hospital, 369, Fulham Road, London SW10 9NH, United Kingdom

Winding, B., Center for Clinical and Basic Research, Ballerup Byvej 222, DK-2750 Ballerup, Denmark

Zimmermann, H., Medical Research Department, Jenapharm GmbH & Co. KG, Otto-Schott-Str. 15, D-07745 Jena, Germany

Contents

Part 5: Pharmacology of Estrogens and Antiestrogens

Part 7: Clinical Application and Potential of Estrogens and Antiestrogens

CHAPTER 38

Hormonal Contraception

CHAPTER 45

Estrogens and Antiestrogens in the Male

Part 9: Estrogens, Antiestrogens, and the Environment

CHAPTER 48

Environmental Estrogens
S. MÄKELÄ, S.M. HYDER, and G.M. STANCEL. With 7 Figures 613

Contents of Companion Volume 135/I

Part 5
Pharmacology of Estrogens and Antiestrogens

CHAPTER 21

In Vitro and In Vivo Models to Characterise Estrogens and Antiestrogens

K.-H. FRITZEMEIER and C. HEGELE-HARTUNG

A. General Aspects

In the following chapter, we describe in vitro and in vivo assays used to measure estrogenic and antiestrogenic activities of natural and synthetic hormones (JORDAN et al. 1985; GORILL and MARSHALL 1986), phytoestrogens (MIKSICEK 1992; MIKSICEK 1994; KNIGHT and EDEN 1995; ANDERSON 1997) and potentially estrogenic environmental chemicals (WHITE et al. 1994; DODGE et al. 1996; RAMAMOORTHY et al. 1997; ROUTLEDGE and SUMPTER 1997). The spectrum of in vitro assays, reflecting the sequence of steps leading from binding of the ligand to the receptor, through gene activation, to a biological response, covers estrogen-receptor (ER) binding experiments, assays to measure the binding of the liganded ER to DNA, transactivation assays and other cell-based assays able to predict some of the tissue-specific effects of estrogens.

In addition to assays focused on ER-dependent actions of estrogenic chemicals, test systems allowing for the determination of non-genomic (specifically antioxidative) activities of estrogens are briefly reviewed. Other non-genomic actions of estrogens are not discussed in this article but have been reviewed recently by others (REVELLI et al. 1998).

The in vivo assays include estrogen bioassays performed in rodents and primates. Estrogen and antiestrogen effects on female reproductive organs are measured, and changes in the vagina, uterus and pituitary–ovarian axis are evaluated. An overview of representative estrogens and antiestrogens is given by Table 11 at the end of this chapter. Before describing methods, we will give a brief overview on the physiological targets of estrogens, on classical ligands of the ER and on the basic mechanisms of estrogen action.

I. Physiological Targets of Estrogens

Estrogens control a diversity of physiological functions in different tissues (NORMAN and LITWACK 1987; SUTHERLAND et al. 1988). In mammals, the organs of the female reproductive tract, including the uterus, vagina, ovary (JENNINGS and CREASMAN 1997), mammary gland (DICKSON and LIPPMAN 1987; LIPPMAN and DICKSON 1987; PRICHARD 1997; SATYASWAROOP 1997) pituitary and hypothalamus, are major targets of estrogens. However, estrogens also influence non-reproductive organs and systems. In the liver, estrogens cause alterations in the production of plasmatic proteins and in lipid metabolism (VON SCHOULTZ et al. 1989; STEINGOLD et al. 1991; KRATTENMACHER et al. 1994). In addition, evidence has accumulated over the last two decades, demonstrating that estrogens exhibit an important regulatory impact on bone (TURNER 1997), the urogenital tract (KELLEHER and CARDOZO 1997), the cardiovascular system (LOBO 1990; FOEGH 1992; SARREL 1992; CLARKSON and ANTHONY 1997; NATHAN and CHAUDHURI 1997) and on brain functions (BIRGE 1997; HALBREICH 1997; HENDERSON 1997; MCEWEN et al. 1997b; SHERWIN 1997a,b). Furthermore, it is well documented that estrogens are important regulators of physiological function not only in females, but also in males (WEST and BRENNER 1990; COOKE et al. 1991; GRECO et al. 1993; LUBAHN et al. 1993; MORISHIMA et al. 1995; KORACH et al. 1996; HESS et al. 1997).

Beyond their physiological actions, estrogens also play a role in certain pathophysiological processes. Thus, estrogens are involved in the initiation and progression of breast (DICKSON and LIPPMAN 1987; LIPPMAN and DICKSON 1989; JENNINGS and CREASMAN 1997; PRICHARD 1997; SATYASWAROOP 1997); endometrial (ZIEL 1982; PIKE et al. 1997), ovarian (RODRIGUEZ et al. 1995; CLINTON and HUA 1997) and prostate cancers (HABENICHT et al. 1987; NATIV et al. 1997).

In oviparous animals, such as reptiles, amphibia and birds, estrogens play a fundamental role in the control of egg production. In these species, estrogens control growth and development of the oviduct, the synthesis of egg-white proteins in the ovary (SCHIMKE et al. 1975) and of egg-yolk proteins in the liver (CLEMENS 1974). Because of the high gene expression, their promotors have become valuable tools for analysing the molecular mechanism of estrogen action (YAMAMOTO 1985) and have provided the basis for designing reporter gene assays (KLEIN-HITPASS et al. 1986).

II. ER Ligands

The evolution of knowledge about estrogens and their actions started in the 1930s with the isolation of the hormonal metabolites, oestrone and oestriol (oestrone: Doisy and Butenandt; oestriol: Marrian), which were found afterwards to be the metabolites of oestradiol (for review, BUTENANDT 1933; LAPIERE 1978; JORDAN et al. 1985). The search for other natural estrogens resulted in the purification of equine estrogens (for review, GIRARD et al. 1932a,b; BHAVNANI 1988) and in the detection of phytoestrogens. Besides the natural hormones, synthetic compounds were found to exhibit estrogenic effects. The estrogenic activity of stilbene derivatives was detected in the 1930s by DODDS et al. (1939). 4,4′-dihydroxy-diethyl stilbene (diethylstilboestrol, DES), the first synthetic estrogen to be used as a therapeutic agent, was described by DODDS et al. (1938). Today, the most commonly used synthetic estrogen, specifically in oral contraceptives, is ethinyl oestradiol (EE), which was first synthesised by INHOFFEN et al. (1938).

Tamoxifen, which is structurally closely related to DES, was described as an antiestrogen by Harper and Walpole (1967) (Wakeling 1985 for review). This drug exhibits residual estrogenic activity. In 1987, the first pure antiestrogen, ZM 164,384 (Wakeling and Bowler 1987), and in 1991 the more potent analog ZM 182,780 (Wakeling et al. 1991) were described. Recent findings that the antiestrogen raloxifen, a mixed agonist/antagonist like tamoxifen (Fuchs-Young et al. 1995), exhibits estrogen-like protective effects on bone without causing uterine stimulation (Bryant et al. 1996) has opened the door for designing new selective estrogens (Kauffman and Bryant 1995; von Angerer 1995; Tonetti and Jordan 1996).

III. The Receptors – Mediators of Hormone Action, Targets for Antihormones and Modulators of Gene Expression

Estrogens exhibit most of their physiological effects through a nuclear receptor protein, the ER. Until recently, it had been assumed that there was only one ER. However, in 1996, a second ER was detected (Kuiper et al. 1996a; Mosselman et al. 1996; Tremblay et al. 1997) that shares a high degree of homology with the "classic" counterpart, particularly in the DNA- and ligand-binding domains. The newly detected receptor was called ERβ and the "classic" subtype was renamed ERα. The two ERs are encoded by two distinct genes. The fact that the two receptor subtypes exhibit different patterns of expression in tissues and cells suggests that they may exhibit distinct biological functions (Kuiper et al. 1996b; Shughrue et al. 1996; Byers et al. 1997; Iafrati et al. 1997; Li et al. 1997; Onoe et al. 1997; Saunders et al. 1997; Shughrue et al. 1997a).

The ERs, as members of the steroid hormone receptor superfamily, are soluble nuclear localised proteins which act as ligand-activated transcription factors; binding of an agonist causes a conformational change in the receptor protein, the formation of a receptor dimer and binding of the dimer to a specific nucleotide sequence in the target-gene promotor. The corresponding conserved DNA sequence is called the estrogen response element (ERE). Binding of the receptor dimer to the ERE causes the recruitment of essential transcription factors, the formation of a transcription initiation complex and the transcription of estrogen-regulated genes (Tsai and O'Malley 1994; Beato et al. 1995).

Two domains of the ER molecules are essential for interaction with other transcription factors and therefore for the ERs' transcriptional activity. These two domains are called activation functions, AF1 and AF2 (Berry et al. 1990). The ability of these regions to contribute to ER transcriptional activity varies with the cell and promotor examined. In some contexts, individual activation domains are the major determinants but, in most cases, AF1 and AF2 synergise to stimulate transactivation of gene expression.

Agonists of the ER, like natural hormones or the synthetic estrogens EE and diethylstilboestrol, cause a conformational change of the hormone-binding domain which allows both AF1 and AF2 to be active (BEEKMAN et al. 1993). Pure antagonists of the ER, exemplified by the two "pure" antiestrogens, ZM164,384 (WAKELING 1985) and ZM182,780 (WAKELING et al. 1991), are unable to activate the ER in nearly all instances and efficiently antagonise ER function (for review: VON ANGERER 1995b; PARCZYK and SCHNEIDER 1996). In contrast to the pure antiestrogens, mixed agonists/antagonists, such as hydroxy tamoxifen and raloxifen, inhibit ER activity in a selective manner and may even cause activation of genes under certain conditions (BERRY et al. 1990; RAMKUR and ADLER 1995; VON ANGERER 1995a; FAN et al. 1996). This type of compound causes a conformational change in the ligand-binding domain of the ER that is distinct from estrogens and that allows activation of gene expression via AF1. In contrast, AF2 is inhibited by this type of compound. Whether AF1-dependent activation of a gene occurs depends on the target-gene promotor and the cellular background.

Recent studies have identified some of the proteins that coactivate the ER. SRC-1 (ONATE et al. 1995) and CBP (KAMEI et al. 1996; TORCHIA et al. 1997) were shown to interact with steroid-hormone receptors and to enhance their transcriptional activity. The repressor proteins, SMRT and N-CoR, are assumed to repress trancriptional activity of the unliganded receptor and to be released upon binding of the ligand (HORLEIN et al. 1995; CHEN and EVANS 1996; HEERY et al. 1997).

A number of experiments carried out in tissue culture and animal models have shown that the ER can be activated by growth-factor-mediated signal-transduction pathways (IGNAR-TROWBRIDGE et al. 1992). The ligand-independent pathways for activation of steroid receptors have been reviewed elsewhere (O'MALLEY et al. 1995; DE CUPIS and FAVONI 1997; WEIGEL and ZHANG 1998). The stimulating effects of growth factors such as epidermal growth factor (EGF) and insulin-like growth factor (IGF-1) (KATZENELLENBOGEN and NORMAN 1990; ARONICA and KATZENELLENBOGEN 1993) require intact AF-1 function. As mentioned, antiestrogens such as tamoxifen do not block AF1 and are therefore not effective in blocking growth factor effects. For this reason, the search for antiestrogens is focused on "pure" antiestrogens that block both activation functions (PARCZYK and SCHNEIDER 1996).

One striking observation regarding pure antiestrogens is that they reduce ER protein levels in target tissues and cells by increasing ER turnover (DAUVAIS et al. 1992). This ER destabilisation has been observed in several species and tissues, including cancer tissue (MCCLELLAND et al. 1996). The effect is unique to pure antiestrogens; it is not observed following estrogen depletion or treatment with partial agonists such as tamoxifen (which, in contrast, increase ER levels) (KIANG et al. 1989; NOGUCHI et al. 1993; JIN et al. 1995).

Besides using the classic mode of gene activation, which precludes binding of the ER to an ERE, the ER was shown to cause activation of specific promoters (via the transcription factor AP-1) without physically interacting with the DNA (WEBB et al. 1995). Furthermore, the agonist-activated ER may act as a repressor of gene expression. Thus, the ER was shown to interfere with cytokine-induced activity of the transcription factor nuclear factor kappa B (NF-κB) (POTTRATZ et al. 1993; STEIN and YANG 1995; GALIEN et al. 1996; RAY et al. 1997). This mechanism is discussed as being partly responsible for the bone-protective activity of estrogens (JILKA et al. 1992; POLI et al. 1994). Estrogens cause downmodulation of the NF-κB-activated cyctokine interleukin (IL)-6 in bone and are thus assumed to inhibit the recruitment of osteoclasts.

B. In Vitro Methods to Characterise Estrogens and Antiestrogens

I. Receptor-Binding Assay

JENSEN and JACOBSON (1960) were the first to describe that, following administration of [^{3}H]-oestradiol to female rats, the radioactive ligand was specifically retained within tissues known to be responsive to estrogens. A few years later, the ER was determined to be the primary mediator of oestradiol's biological action (JENSEN and DE SOMBRE 1973; JENSEN et al. 1974; GORSKI and GANNON 1976). Binding of a potentially estrogenic or antiestrogenic substance to the ER is a prerequisite for the compound to exhibit – or interfere with – estrogen-like activity.

1. Principle

The first step in evaluating new drugs or environmental chemicals is determinating their binding affinity to the ER. Generally, the binding affinity is determined using a competitive binding assay. In principal, any estrogen target tissue or cell can be used as the source of ER. Receptor-binding studies are generally performed with a high-speed centrifugal fraction referred to as cytosol preparation (100,000 g supernatant) prepared from crude extracts of tissue (most commonly from rat, rabbit, calf, lamb or human uterus) or cells expressing the ER, e.g. MCF-7 mammary carcinoma cells. For screening purposes, ERs may be produced in transfected cells, e.g. in Chinese hamster ovary (COS) cells transfected with an ER expression vector or, using a baculovirus expression system, in SF9 insect cells (SUMMERS and SMITH 1987).

To perform the competition assay, the ER-containing cytosol is incubated in the presence of a radioactive ligand (predominantly [^{3}H]-oestradiol) and various concentrations of the competitor compound. The incubation may be done at 4°C or 25°C. Different incubation times are used in different laboratories, and range from 1–24 h. If comparing receptor-binding data from different laboratories, the incubation temperatures and times have to be taken into account. The incubation temperature influences on- and off-rate of binding of both reference and unknown compounds to the ER. Therefore, relative affinities determined at different temperatures or after different incubation times may differ considerably.

Test compounds are dissolved in ethanol, dimethyl formamide (DMF) (KATZENELLENBOGEN et al. 1973; WAKELING and BOWLER 1987) or dimethyl sulfoxide (DMSO), depending on their solubility and are then transferred to the incubation buffer. The final solvent concentration during incubation with the receptor must not exceed 3% (v/v). After incubation, unbound compounds are removed by adsorbtion to dextran-coated charcoal (DCC) (or by hydroxyl apatite or ion-exchange chromatography). Centrifuging removes the DCC and aliquots of the supernatants are withdrawn and counted for radioactivity in a scintillation spectrometer. Non-specific binding is determined by parallel incubation with a 500-fold excess of unlabelled oestradiol (E2).

To determine the relative affinity of compounds, the displacement of [^{3}H]-E2 with unlabeled compound is plotted as percentage amount versus log molar concentration of the competitor compound. The activity of a compound is generally expressed as the relative binding affinity (RBA), expressed as a percentage, which is defined as follows:

$$\text{RBA} = \text{IC}_{50}\ \text{ligand}/\text{IC}_{50}\ \text{test compound} \times 100$$
$$(\text{IC} = \text{inhibitory concentration})$$

Sometimes, the RBA is given as competition factor (CF), defined as IC_{50} test compound/IC_{50} ligand (LUEBKE et al. 1976).

Several studies have demonstrated that ERs from different species and tissues exhibit similar specificity and (high) affinity for oestradiol. Table 1 presents a survey of the literature describing ER binding in various tissues and species. A comparative study that examines estrogen and antiestrogen binding with ERs from different organs and tissues (including those from rat, rabbit and humans) was performed by BERGINK et al. (1983). The influence of different substituents on the affinity of oestradiol derivatives was studied by KASPAR and WITZEL (1985), using rat uterus cytosolic ERs.

A comparative study of the binding affinity of selected compounds for ERα and ERβ was performed by KUIPER et al. (1996b). Though the two receptor subtypes exhibit high homology within the hormone-binding domain, differences in the binding affinity of some ER-ligands were found (KUIPER et al. 1996a, 1996b). Interestingly, some phytoestrogens exhibit higher affinity for ERβ than ERα; coumestrol, genistein, β-zearalenol and apigenin are better ligands of ERβ than ERα. Furthermore, an appoximately threefold preference in favor of ERβ was found for the physiological steroidal metabolite androstendiol.

2. Binding of Estrogens and Antiestrogens to Insect-Cell-Produced Human ERα (hERα)

In a comparative study, we determinded the affinity of selected estrogens and antiestrogens to ERα. ER was produced in SF9 insect cells Table 2. The ERα

Table 1. Dissociation constant (K_D) for binding of oestradiol to the estrogen receptor isolated from different organs and different species

Species	Organ	KD (nmol/l)	Reference
Human	Uterus	0.1–0.4	Van der Walt et al. 1986; Gibbons et al. 1979
	Endometrium	0.143	Punnonen and Lukola 1982
	Myometrium	0.1	Bergink et al. 1984
	Ovary	0.07	Lukola and Punnonen 1983
	Vagina	0.144–0.4	Punnonen et al. 1982; Gibbons et al. 1979
	Breast	0.2	Geier et al. 1979
	Prostate	0.1	Ekman et al. 1983
Rat	Uterus	0.122–0.15	Carlson and Gorski 1980; Saiduddin et al. 1977; Ginsburg et al. 1977
	Ovary	0.413–0.58	Saiduddin et al. 1977, Kudolo et al. 1984
	Brain	0.2–0.6	Clark et al. 1982
	Hypothalamus	0.13	Ginsburg et al. 1977
	Cortex	0.14	Ginsburg et al. 1977
	Amygdala	0.13	Ginsburg et al. 1977
	Pituitary	0.1–0.33	Clark et al. 1982; Ginsburg et al. 1977
Rabbit	Uterus	0.31–0.4	Batra et al. 1989; Tong et al. 1983
	Vagina	0.51	Batra et al. 1989
	Kidney	0.4	Barrack et al. 1980
Monkey	Uterus	0.315–0.32	Potgieter et al. 1985; Klein et al. 1985
	Endometrium	0.29	Batra et al. 1989
Mouse	Uterus	1.4	Horigome et al. 1987

expression vector was provided by Prof. P. Chambon, Strasbourg, France. The SF9-cell-produced receptor exhibits high binding affinity for E2 as documented by the dissociation constant (K_d) of E2 for ERα (K_d = 0.5 nmol/l) (Boemer, Berlin, unpublished results).

a) Preparation of ER-Containing Cytosol

Insect cells infected with ERα-encoding baculovirus vectors were harvested by centrifugation for 10 min at 150 *g*, 48 h after infection. The resultant cell pellet was suspended in buffer [0.02 mmol/l TRIS-buffer, pH 7.5, 0.5 mmol/l EDTA, 2 mmol/l dithiothreitol, 20% (v/v) glycerol, 20 mmol/l molybdate, 0.3 mmol/l phenylmethanesulfonic acid fluoride, 0.3 mmol/l aprotinin, 1 mol/l pepstatin, 10 mol/l leupeptin]. The cell suspension was treated by repeated thawing and freezing (at least four times), and was afterwards transferred to an ice bath. The homogenisation was done by repeated suction and expellation through a pipette and, subsequently, through a cannula (Braun, Melsungen, Germany, Sterican, 0.45 × 23 m, 26 G × 7/8). The suspension was

Table 2. Binding affinity of estrogens and antiestrogens to insect cell produced human estrogen receptor (hER)α

Compound	RBA
Oestradiol	100
17α-ethinyl oestradiol	111
cyclodiol	100
cyclotriol	40
Oestriol	7.1
Oestrone	6.7
17α-Oestradiol	11.8
17α-Equilin	18.6
Androstenediol	0.85
Androstenedione	<0.1
Coumestrol	18
Genistein	4
β-Zearalenol	0.6
Apigenin	0.04
Raloxifen	62.5
RU 39,411	73.0
OH-tamoxifen	75.2
CP 336,156	166
ZM 182,780	42

then centrifuged at 105,000 g for 90 min, and the supernatant was used in competition experiments as ER-containing cytosol.

b) Competition Experiment

10 μl [^{3}H]-oestradiol [[2,3,6,7–3H]-oestradiol [oestra-1,3,5(10)-triene-3,17β-diol; specific activity 2 TBq/mmol, (Amersham, Buchler, Braunschweig, Germany)] to give a final concentration of 5×10^{-9} M and either 10 μl unlabeled E2 for the standard curve or 10 μl of the test substance in appropriate concentrations were added to 20 μl cytosol. The steroids were dissolved in DMF; the final concentration did not exceed 3%. The samples were incubated for 120 min at 22°C. After incubation, unbound compounds were adsorbed by incubation with 0.5 ml of a suspension of DCC in buffer [0.01 mol/l TRIS-buffer, pH 7.5, containing 1.5 mmol/l EDTA and 10% (v/v) glycerol] for 10 min at 4°C. After centrifugation for 5 min at 15,000 g, an aliquot of the supernatant was withdrawn and counted for radioactivity. The RBA was determined as described.

c) Results

The results of the binding studies are listed in Table 2.

II. Ligand Effects on DNA Binding of the ER

The classical mode of gene activation by the ER precludes binding of the receptor complex to the ERE in the target-gene promotor (Beato et al. 1990;

CARSON-JURICA et al. 1990). Binding of the ER complex leads to its interaction with basal transcription factors; these interactions are assumed to stabilise the pre-initiation complex at the promotor. The ability of the ER to bind to DNA is influenced by the ligand (KUMAR and CHAMBON 1988). The influence of the ligand on the formation and the stability of the ER–ERE complex is assumed to have predictive value for the ligand's biological activity (CHESKIS et al. 1997). The "pure" antiestrogen ZM182,780 was found to reduce DNA binding of the ER (METZGER et al. 1995).

Different methods have been applied to study the interaction of the ER with DNA. In "band-shift" (also called "gel-shift" or "gel-retardation") experiments, a radioactively labelled oligonuleotide corresponding to an ERE is incubated with ER-containing cells (ARCKBUCKLE et al. 1992; DAUVAIS et al. 1992; CHRISTMAN et al. 1995) or tissue extracts (CURTIS and KORACH 1991; SABBAH et al. 1991; FURLOW et al. 1993); the mixture is then submitted to polyacrylamide gel electrophoresis. The "free" oligonucleotide moves faster during electrophoresis than the ER–ERE complex which is retarded due to its increased size.

Kinetics of the interaction between the ER and the DNA response element cannot be evaluated using the "band-shift" technique. As such, it is difficult to perform comparative studies that examine the influence of different ligands on ER DNA binding. Another disadvantage of the method is that the formation of the receptor–DNA complex in vitro and its stability during gel electrophoresis is influenced considerably by the incubation conditions (salt concentration, temperature, receptor concentration). The susceptibility of this method to minor changes in procedure may partly explain why controversial findings have been reported both for DNA binding of the unliganded (as opposed to the ligand-occupied) receptor and for the effects of pure antiestrogens on the ER–DNA interaction (PHAM et al. 1991; FURLOW et al. 1993).

The recently developed BIACORE technology allows for a quantitative investigation of macromolecular interactions (KARLSSON et al. 1991). The BIACORE system is a biosensor instrument. Its detection principle relies on the optical phenomenon of surface plasmon resonance (SPR). Changes in the refractive index of the solution close to the surface of a sensor chip loaded with one of the interaction partners, e.g. the ERE, are measured. The refractory index is directly correlated with the concentration of solute (the ER) in the layer that is injected over the surface in a controlled flow. Any change in the surface concentration that results from the interaction of the ERE and the ER is detected as an SPR signal, expressed in resonance units (RU). The continuous display of RU as a function of time – referred to as a sensogram – provides a complete record of the association and dissociation of ERE and ER.

CHESKIS et al. (1997) have applied real-time interaction analysis to investigate the kinetics of human ER binding to DNA in the absence and presence of 17β-oestradiol, 17α-EE, analogs of tamoxifen, raloxifen, and ZM182,780.

The authors showed that ligand binding influenced the kinetics of hER interaction with specific DNA. Oestradiol induced the rapid formation of a relatively unstable ER–ERE complex, and binding of ZM182,780 led to slow formation of a relatively stable receptor–DNA complex (the K_d is almost two orders of magnitude lower). Binding of oestradiol accelerated the frequency of receptor–DNA complex formation more than 50-fold, compared with unliganded ER, and more than 1000-fold compared with ERs liganded with ZM182,780. The authors hypothesise that a correlation exists between the rate of gene transcription and the frequency of receptor–DNA complex formation.

III. Transactivation Assays for Detection of Estrogenic and Antiestrogenic Activity

1. Principle

The ERs function by modulating the transcription of target genes. The classical mode of gene activation by the ER begins up on binding of the ligand with the formation of a receptor dimer (homodimer or heterodimer composed of ERα and/or ERβ) (COWLEY et al. 1997). The dimer then binds to a specific nucleotide sequence in the promotor of target genes, the ERE (KUMAR and CHAMBON 1988). The classical ERE is a palindromic hexanucleotide, first identified within the vitellogenin A2 promotor (KLEIN-HITPASS et al. 1986; KLEIN-HITPASS et al. 1988) of *Xenopus laevis*. Several other response elements have been identified that confer estrogen responsiveness to the respective gene. Table 3 gives an overview of estrogen-responsive genes that have different promotor elements confering estrogen sensitivity. A systematic search for novel response elements using a random oligonucleotide library was performed by DANA et al. (1994) using a yeast expression system.

Transactivation assays for the characterisation of compounds are based on the ability of the ER to cause gene activation in a ligand-dependent way. Cell lines are co-transfected with an ER-expression vector and a plasmid that contains an ERE placed in front of a reporter gene [often: chloramphenicol acetyltransferase (GORMAN et al. 1982) or luciferase (DE WET et al. 1987; BRASIER et al. 1989)]. Transfection of the cells can occur transiently. Alternatively, stably transfected cell lines have been created, e.g. MVLN (DEMIRPENCE et al. 1993) and LeC-9 cells (MAYR et al. 1992).

Cells expressing high levels of ER endogenously, e.g. MCF-7 cells (SOMASEKHAR and GORSKI 1988), or after stable transfection with ER, e.g. HTB 96 osteosarcoma cells (WATTS et al. 1989) or Fe33 rat hepatoma cells (KALING et al. 1990), may also be used to establish transactivation assays. These cells are transfected with the reporter gene only, which is activated via the endogenously produced receptor.

Reporter assays with different estrogen-sensitive promotors (see Table 3) have been described. Four of the most frequently used promotors are:

Table 3. Estrogen responsive genes and the corresponding promotor elements conferring estrogen sensitivity

Estrogen responsive gene	Estrogen-responsive promotor element	Reference
Xenopus vitellogenin A2	Palindromic sequence 5′GGTCACAGTGACC-3′ (classical ERE)	Klein-hitpass et al. 1988
Chinook salmon gonadotropin II beta subunit	Three coupled variations of the classical ERE	Liu et al. 1995
Rainbow trout ER gene	ERE with one base exchange at +242 to 254	Le crean et al. 1995
Chicken ovalbumin	Half-palindromic ERE	Kato et al. 1992; Tora et al. 1988
Rat calbindin-D-9k (rCaBP9k)	Imperfect palindromic ERE	Darwish et al. 1991
Mouse calbindin-D-9k (mCaBP9k)	Multiple imperfect half-palindromic EREs	Gill and Christakos 1995
Rat prolactin	ERE located between –1530 to 1950	Somasekhar and Gorski 1988
Human c-fos	Imperfect palindromic ERE (CGGCAGCGTGACC) coupled with an AP-1 binding site	Weisz and Rosales 1990
Rabbit progesterone receptor	Modified ERE (GGTCACATGACT) within first exon	Savouret et al. 1991
Rat progesterone receptor	Multiple distinct regions including 4 imperfect palindromic EREs	Kraus et al. 1994
Human progesterone receptor	Two estrogen inducible promotors, but no consensus palindromic ERE	Kastner et al. 1990
Human complement 3 (C 3) gene	Three synergising EREs, one resembling, two displaying no homology to the classic ERE	Norris et al. 1996
Lactoferrin	Overlapping retinoic acid and estrogen response element	Liu and Teng 1992; Lee et al. 1995
PS2	Imperfect palindromic sequence	Berry et al. 1989; Mori et al. 1990
Cathepsin D	SP1 and ERE-half-site in the –199 to –165 promotor region	Wang et al. 1997; Augereau et al. 1994
Heat shock protein 27 (HSP 27)	SP1 and ERE-half-site	Porter et al. 1996
Retinoic acid receptor alpha (RAR alpha)	Imperfect half-palindromic ERE and SP1-site	Rishi et al. 1995
Rat oxytocin	Composed element of imperfect ERE and DR-0-type direct repeat	Adan et al. 1993
Rat angiotensinogen	Half-palindromic ERE	Feldmer et al. 1991
Alkaline phosphatase	Not determined	Albert et al. 1990

The vitellogenin promotor (KLEIN-HITPASS et al. 1986) needs AF2 of the ER to be active. Therefore, only "pure" agonists are able to stimulate this promotor.

The pS2 (BERRY et al. 1989), the rabbit progesterone receptor (rPR) (SAVOURET et al. 1991) and the C3 (NORRIS et al. 1996) promotors were shown to be activated by OH-tamoxifen, a mixed agonist/antagonist of the ER that activates the ER's AF1.

2. Transactivation Assay with the Vitellogenin A2-ERE-tk-CAT Reporter Gene in HeLa Cells

a) *Comparative Study of the Agonistic Potency of Four Selected Estrogens*

HeLa cells (human cervix carcinoma cells, American Type Culture Collection) were transiently transfected with an expression vector for the hERα (GREEN et al. 1988; TORA et al. 1989) and with a reporter gene consisting of the vitellogenin A2 promotor put in front of the chloramphenicol acetyl transferase (CAT) reporter gene (BURCH et al. 1988). Cells were transfected using the calcium phosphate co-precipitation technique as described for HeLa cells by BOQUEL et al. (1989).

To reduce the level of the basal reporter gene activity, the test on estrogenic activity was performed in the presence of the antiestrogen ZM182,780 (1 nmol/l). One hour after transfection of the cells, oestradiol or test compound solubilised in 10% ethanol/Tris-buffer and the antiestrogen (1 nmol/l) were added to the transfected cells in concentrations ranging from 10 pmol/l to 1 μmol/l. The ethanol concentration in the medium of the cultured cells was 1%. Forty-eight hours after addition of compounds, cells were harvested, lysed and CAT protein levels were determined using a CAT/enzyme-linked immunosorbent assay (ELISA) kit (Boehringer, Mannheim, Germany).

In dose–response curves, the activity of compounds is given as percentage of the maximal induction of CAT protein by E2. The ED_{50} of E2 in this assay was found to be 0.3 nmol/l. The three synthetic compounds exhibit slightly higher potency (Fig. 1A).

b) *Comparative Study of the Antagonistic Potency of Selected Antiestrogens*

The assay described was used to evaluate the antagonistic potency of selected antiestrogens. The transfected cells were treated with 0.1 nmol/l E2 to stimulate the reporter gene activity; increasing concentrations of antiestrogen were co-administered. The resultant dose–response curves are shown in Fig. 1B. The IC_{50} values of the antiestrogens tested are listed in Table 4.

3. Transactivation Assay with MVLN Cells

MCF-7 cells stably transfected with a vitellogenin-ERE-luciferase reporter gene have been established by PONS and NICOLAS (DEMIRPENCE et al. 1993). In these cells, referred to as MVLN cells, the reporter gene is driven by the

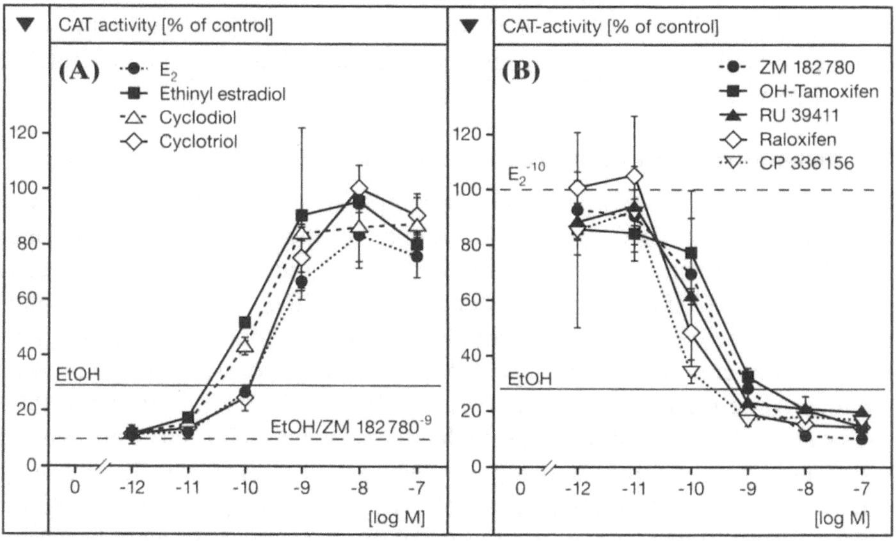

Fig. 1A,B. Agonistic activity of estrogens (A) and antagonistic activity of antiestrogens (B) measured by transactivation assay with a vitellogenin-estrogen response element (ERE)-tk-chloramphenicol acetyl transferase (CAT) reporter gene. **A** HeLa (human cervix carcinoma) cells were transiently cotransfected with the human estrogen receptor (ER) expression vector, HEGO (TORA et al. 1989) and with a VitA2-ERE-tk-CAT reporter gene (KLEIN-HITPASS et al. 1986). To reduce the level of reporter gene activity produced by the unliganded ER, the assay was performed in the presence of ZM182,780 (1 nmol/l). The estrogens oestradiol (●), ethinyl oestradiol (■), cyclodiol (Δ) and cyclotriol (◇) were tested in the concentration range 0.001 nmol/l–100 nmol/l. **B** The same transactivation assay as described in (A) was used to quantity the potency of selected antiestrogens. The CAT-reporter gene was stimulated by E_2 (0.1 nmol/l) and increasing concentrations (range 0.001–100 nmol/l) of antiestrogens were added to inhibit the E2 effect. The ED_{50} of the different antiestrogens is summarised in Table 4

Table 4. Antiestrogenic activity of selected antiestrogens in the vitellogenin reporter gene assay

Antiestrogen	IC50 (inhibition of E2-stimulated reporter gene activity) [nmol/l]
ZM 182,780	0.3
RU 39411	0.2
OH-tamoxifen	0.45
Raloxifen	0.1
CP 336156	0.06

endogenous ER that is expressed at high levels in the mammary carcinoma cell line. Rapid investigation of many compounds for estrogenic and anti-estrogenic activity is possible because the cell-based assay can be performed using a 96-well format.

4. Comparative Study of the Transactivational Activity of ERα and ERβ

MOSSELMAN et al. (1996) performed a comparative study of the transactivational activity of ERα and ERβ. Details of the experiment have been described (MOSSELMAN et al. 1996). Briefly, Chinese hamster ovary cells, CHO K1 and CCL61, were transfected with hERα and hERβ expression vectors, respectively, together with an estrogen-responsive reporter gene (MOSSELMAN et al. 1996). The reporter expression vector was based on the rat oxytocin gene regulatory region position –363/+16 (IVELL and RICHTER 1984; ADAN et al. 1993) linked to the firefly luciferase gene. CHO cells were seeded, transfected with receptor and reporter genes and treated with hormone 5h after transfection with lipofectin reagent (Life Technologies, Palo Alto, Calif., USA). Twenty-four hours after addition of the hormone, cells were harvested and reporter gene activity was determined.

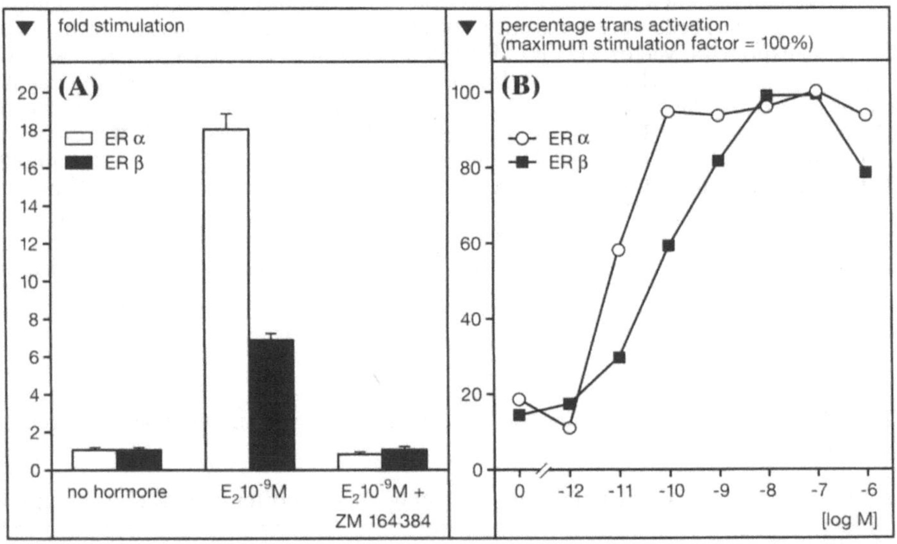

Fig. 2A,B. Comparison of estrogen receptor (ER) α- and ERβ-stimulated transactivation (with permission from MOSSELMANN et al. 1996). **A** Effect of 17β-oestradiol (E2) alone or in combination with the antiestrogen ZM164,384 on the transcriptional activity of ERα (*open bars*) and ERβ (*black bars*) using an estrogen response element (ERE)-based β-galactosidase reporter gene. Indicated is the fold induction of the luciferase activity compared with cells receiving no hormone. All luciferase activities were determined in triplicate and were normalised for differences in transfection efficiency by measuring β-galactosidase activity in the same lysate. **B** Concentration-dependent transactivation of ERα and ERβ by 17β-oestradiol. Transactivation at 10^{-7}M was arbitrarily set at 100%

As documented in Fig. 2A, the efficiency of ERβ with respect to E2-stimulated activation of the oxytocin reporter gene is lower than that of ERα. A comparison of the dose–response curves for oestradiol stimulation of the reporter gene reveals that about tenfold higher E2 concentrations are needed to cause half-maximal activation by ERβ than by ERα (Fig. 2B). Similar results have been obtained by others (Tremblay et al. 1997).

Currently, it is not possible to decide whether the difference in induction of the reporter gene by ERβ and ERα is due to a weaker activation function of ERβ or if other factors (affinity of the ligand to ERβ, affinity of ERβ to the responsive element, promotor context, cellular background) are responsible for the difference in activity.

The antiestrogens OH-tamoxifen and ZM182,780 were tested for their antagonistic potency in an ERα- and ERβ-dependent transactivation assay, respectively (Tremblay et al. 1997). The two antiestrogens were effective in inhibiting both ERα and ERβ (Fig. 3).

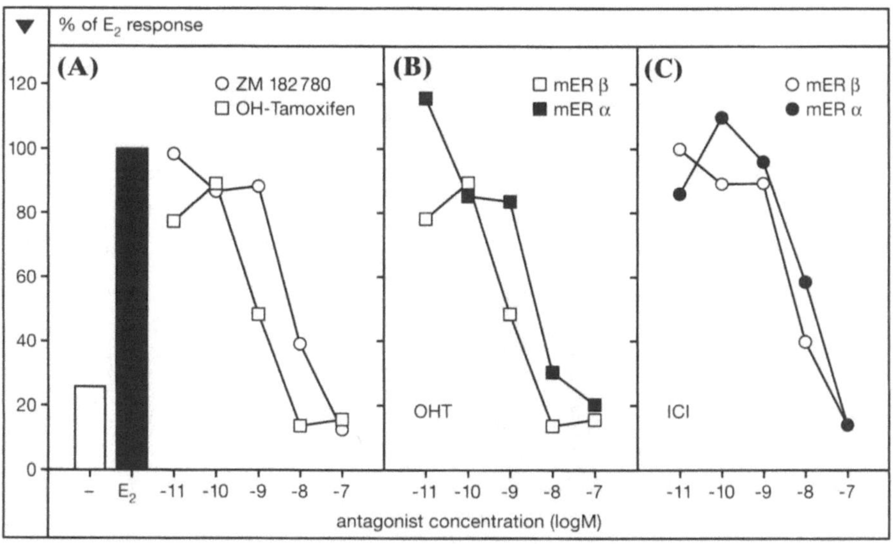

Fig. 3A–C. Effect of antagonists on estrogen receptor (ER) β-mediated transactivation (with permission from Tremblay et al. 1997). Cos-1 cells cotransfected with pCMX-mERβ or -mERα and VitA2-estrogen response element (ERE)-tk-Luc plasmids were incubated for 12 h in the presence or absence of 10 nmol/l E2 and different concentrations of the antagonists indicated. **A** Dose–response of OH-tamoxifen and ZM182,780 on the E2-induced (10 nmol/l) reporter gene in mERβ-expressing Cos-1 cells. The maximal induction by E2 alone was arbitrarily set at 100%. The untreated mERβ basal level is also shown. Compounds within a panel are differentiated by *open squares* (OH-tamoxifen) and *circles* (ZM182,780). **B** Comparative panel of the dose responses to OH-tamoxifen in the presence of 10 nm E2 between mERβ (*open squares*) and mERα (*filled squares*). **C** Comparative panel of the dose responses to ZM182,780 in the presence of 10 nmol E2 between mERβ (*open circles*) and mERα (*filled circles*)

5. Transactivation Assay with an rPR-ERE-tk-CAT Reporter Gene

The rPR-ERE-tk-CAT reporter gene (SAVOURET et al. 1991) is used to detect and quantify the agonistic activity of mixed agonists/antagonists. Hela cells were transfected with an ER expression vector (HEGO), hERα (GREEN et al. 1988; TORA et al. 1989) and the reporter gene rPR-ERE-tk-CAT, using the calcium phosphate co-precipitation technique described for HeLa cells (BOQUEL et al. 1989).

To quantify the agonistic activity of antiestrogens by this reporter-gene assay, the compounds were tested in the absence of E2. As shown in Fig. 4A, the pure antiestrogen ZM182,780 exhibited no agonistic activity but, rather, reduced the basal reporter gene activity. In contrast, OH-tamoxifen (1 μmol/l) and RU 39,411 (1 μmol/l) caused activation of the reporter gene by 83% and 70% compared with the reference E2 (0,1 μmol/l) whose effect was set at 100%.

The antiestrogens were tested for antagonistic activity in concentrations of 1 nmol/l–1 μmol/l (Fig. 4B). ZM182,780 fully antagonised the E2-stimulated reporter gene activity (Fig. 4; PARCZYK et al. 1995). The E2- (1 nmol/l) stimu-

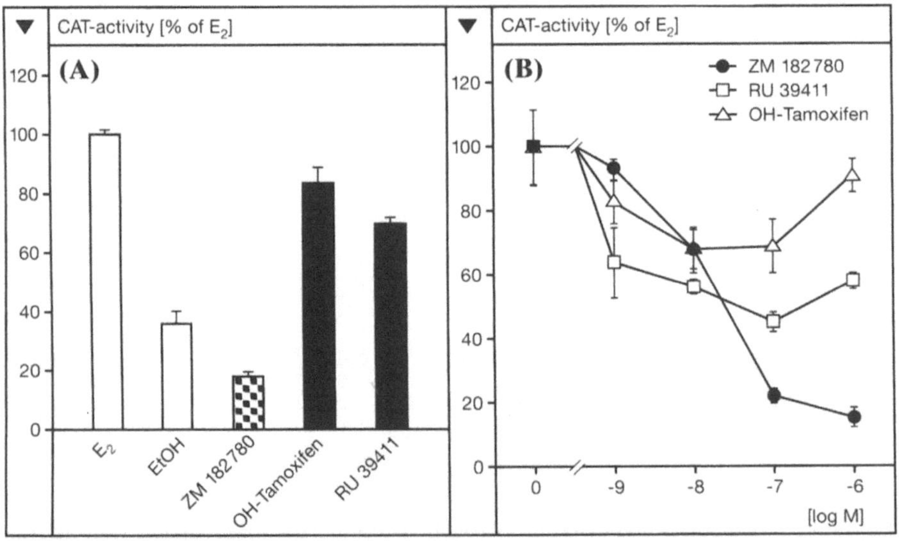

Fig. 4A,B. Transactivation assay to characterise mixed agonists/antagonists of the estrogen receptor (ER). **A** Effect of 17β-oestradiol (E2; 100 nmol/l) and different antiestrogens (1 μmol/l) on the rabbit progesterone receptor (rPR)-estrogen response element (ERE)-tk-chloramphenicol acetyl transferase (CAT) reporter gene activity. The maximal induction by E2 was arbitrarily set at 100%. **B** Different antiestrogens were tested for antagonistic activity in the concentration range 1 nmol/l to 1 μmol/l in the presence of 1 nmol/l 17β-E2. The induction of the reporter gene by E2 alone was arbitrarily set at 100%. ZM182,780 exhibits a dose–response curve typical for a 'pure' antiestrogen. The mixed agonists/antagonists RU39,411 and OH-tamoxifen exhibit partial antagonistic activity lated reporter gene activity was set to 100%. In contrast to the pure antiestrogen, the partial agonist/antagonists RU39,411 (GOTTARDIS et al. 1989) and OH-tamoxifen only partially inhibited the E2-induced reporter gene activity.

IV. Transactivation Assay in Yeast as a Model System to Characterise Estrogenic Compounds

1. Principle

Evidence accumulated within the last decade demonstrates that the mechanisms of transcriptional activation that operate in yeast and mammalian cells are similar. It could be proven that yeast transcription factors are active in mammalian cells, and vice versa. Thus, WEBSTER and coworkers (1988) demonstrated that the yeast transcription factor Gal 4, when tethered to DNA via the ER–DNA-binding domain, activated ER-responsive target genes in mammalian cells. METZGER and colleagues (1988) were the first to prove that the ER, when expressed in Saccharomyces cerevisiae, could function as a ligand-dependent transcription factor in yeast (METZGER et al. 1988).

The use of yeast as a model system to study steroid-hormone receptors and as a screening system to search for steroid-receptor ligands was reviewed recently by WAGNER and MCDONNELL (1996). We therefore confine this discussion to a yeast transactivation assay that was used to characterise phyto- and mycoestrogens (M. HUSEMANN, Berlin, unpublished observations). The yeast system has also been proven to be useful for evaluating the estrogenic activity of alkylphenolic compounds and other chemicals (GAIDO et al. 1997a; GAIDO et al. 1997b; ROUTLEDGE and SUMPTER 1997).

2. Determination of the Agonistic Activity of Selected Phyto- and Mycoestrogens by Transactivation Assay in Yeast

The principle of the assay is illustrated in Fig. 5. Yeast cells are stably transfected with two plasmids, one carrying the ER gene under the control of a Cu^{2+}-inducible promotor, the other carrying an estrogen-reponsive reporter gene (ERE-ERE-CYC-lacZ). Estrogen-inducible β-galactosidase (encoded by the *lacZ* gene) activity is measured by adding a chromogenic substrate.

The recombinant yeast cultures are treated with different concentrations of test compound or reference and 24 h after the addition of compounds, β-galactosidase activity is measured by determining the conversion of the chromogenic substrate ortho-nitrophenyl-β-D-thio-galactoside spectrophotometrically.

The Yeast Strains

Saccharomyces cerevisiae BJ3505 (MATa, pep4:HIS3, prb1-δ1.6R, HIS3, lys2–208, trp1-δ101, ura3–52, gal2, can1) was obtained from the Yeast Genetic Stock Center, Berkeley, USA. Recombinant yeast cells were grown

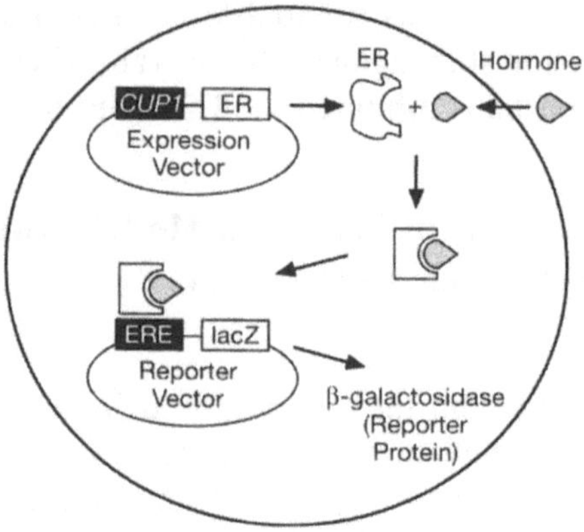

Fig. 5. (With permission from WAGNER and MCDONNELL 1996). Assay of estrogen receptor (*ER*) transcriptional activity in yeast using a β-galactosidase assay. Yeast cells are transformed with two plasmids: one of the plasmids is an expression vector encoding the human estrogen receptor under the control of the yeast metallothionein promoter (*CUP1*) and the other plasmid is a reporter vector that encodes the *E. coli* β-galactosidase (*lacZ*) gene under the control of an estrogen reporter element (*ERE*)-containing *CYC1* promoter. When the ER binds hormone, it undergoes a conformational change, allowing the activated receptor to bind the ERE within the reporter plasmid. The hormone-activated receptor increases the transcription of the *lacZ* reporter gene. β-Galactosidase is quantified using a colorimetric enzyme activity assay

in tryptophan and uracil-deficient minimal medium containing 0.67% yeast nitrogen base (without amino acids) (Difco), 2% glucose, 0.072% amino acid mixture (BIO101), pH 7.0.

Vector Plasmids

The 2-μm-based expression vector YEPubstu was obtained from R. KÖLLING (University of Düsseldorf, Germany). It contains a copper-inducible yeast promoter (*CUP1*) that directs the expression of the yeast ubiquitin gene, a downstream multiple-cloning site for in-frame fusion of the expression cassette to the ubiquitin gene, a *CYC1* transcriptional terminator, and a tryptophan-auxotrophy complementing marker gene (*TRP1*) for selection on a tryptophan-deficient minimal medium. The multi-copy reporter plasmid pLGδ-312 was obtained from L. GUARENTE (MIT, Boston). It contains a *CYC1* promoter upstream of the *E. coli-lacZ* reporter gene and a *URA3*-marker gene for selection on uracil-deficient minimal medium.

Recombinant Yeast Strains

For construction of the ER expression plasmid, a 1.8-kb *Eco*RI complementary DNA (cDNA) fragment coding for the complete human ER (HEGO) was obtained from PIERRE CHAMBON, Strasbourg, France. This fragment was recloned into pSELECT-1 (Promega, Madison, USA). The 5′-terminus preceding the start codon was modified, using the site-directed Mutagenesis Kit (Promega), to introduce, in-frame, a segment coding for the six C-terminal amino acids of ubiquitin, including an *Afl*II site. The modified fragment was digested with *Afl*II plus *Sac*I and recloned into the *Afl*II-*Sac*I digested yeast vector YEPubstu downstream of the *CUP1* promoter in frame with the yeast ubiquitin gene.

The estrogen-responsive reporter plasmid was constructed by substituting the 0.13-kb *Sma*I–*Xho*I fragment of pLGδ-312 (comprising the upstream activating sequences of the *CYC1* promoter) with one copy of a double-stranded synthetic *Sma*I–*Xho*I fragment containing dimeric EREs (5′-CCCGGGAG-GTCACAGTGACCTAGCTACGGATCCAGGTCACAGTGACCTAGC-TAGCTCGAG-3′) derived from the *Xenopus laevis* vitellogenin A2 gene promoter. Yeast cells were successively transfected with reporter and receptor expression plasmids by electroporation using a BIORAD Gene Pulser (Munich, Germany). Recombinant cells were selected on tryptophan and uracil-deficient minimal medium containing 1 M sorbitol and confirmed by plasmid rescue as described by ROBZYK and KASSIR (1992).

Transactivation Assay

Selective minimal medium (10 ml of 1%) was inoculated with an overnight culture of recombinant yeast cells and grown at 300 C with shaking to an optical denity at 600 nm (OD_{600}) of 0.1 (early log phase). Receptor expression was induced by adding 100 μM $CuSO_4$ before transferring 90-μl aliquots into each well of a 96-well microtiter plate already containing 10 μl of test compound dissolved in 10–20% DMSO. The plates were sealed with self-adhesive tape to minimise evaporation and then shaken overnight at 300 C. Cell growth was measured by determining the OD_{595} with the help of a BIORAD ELISA-reader.

To determine the β-galactosidase activity in whole cells, the modified method of MILLER (1972) was adapted to the 96-well microtiter format. Briefly, 10-μl aliquots were transferred to a new plate and mixed with 100 μl of *lacZ* assay buffer (Z buffer: 60 mM Na_2HPO_4, 40 mM NaH_2PO_4, 10 mM KCl, 1 mM $MgSO_4$, 50 mM 2-Mercapto-ethanol, 0.01% SDS, pH 7.0). After incubation for 1 h at room temperature, 25 μl of the chromogenic β-galactosidase substrate ortho-nitrophenyl-β-D-thio-galactoside (4 mg/ml in 0.1 M K-PO_4, pH 7.0) was added and incubated for one more hour. The reaction was stopped by adding 50 μl of a 1-M Na_2CO_3 solution before measuring the OD_{415} with the help of an ELISA reader. β-galactosidase activity was calculated as (OD_{415} 1000)/(δt OD_{595}) (modified Miller-Units). All samples were tested in duplicate. The assay was applied to determine the agonistic activity of selected phyto- and mycoestrogens. The result of the study is shown by Fig. 6.

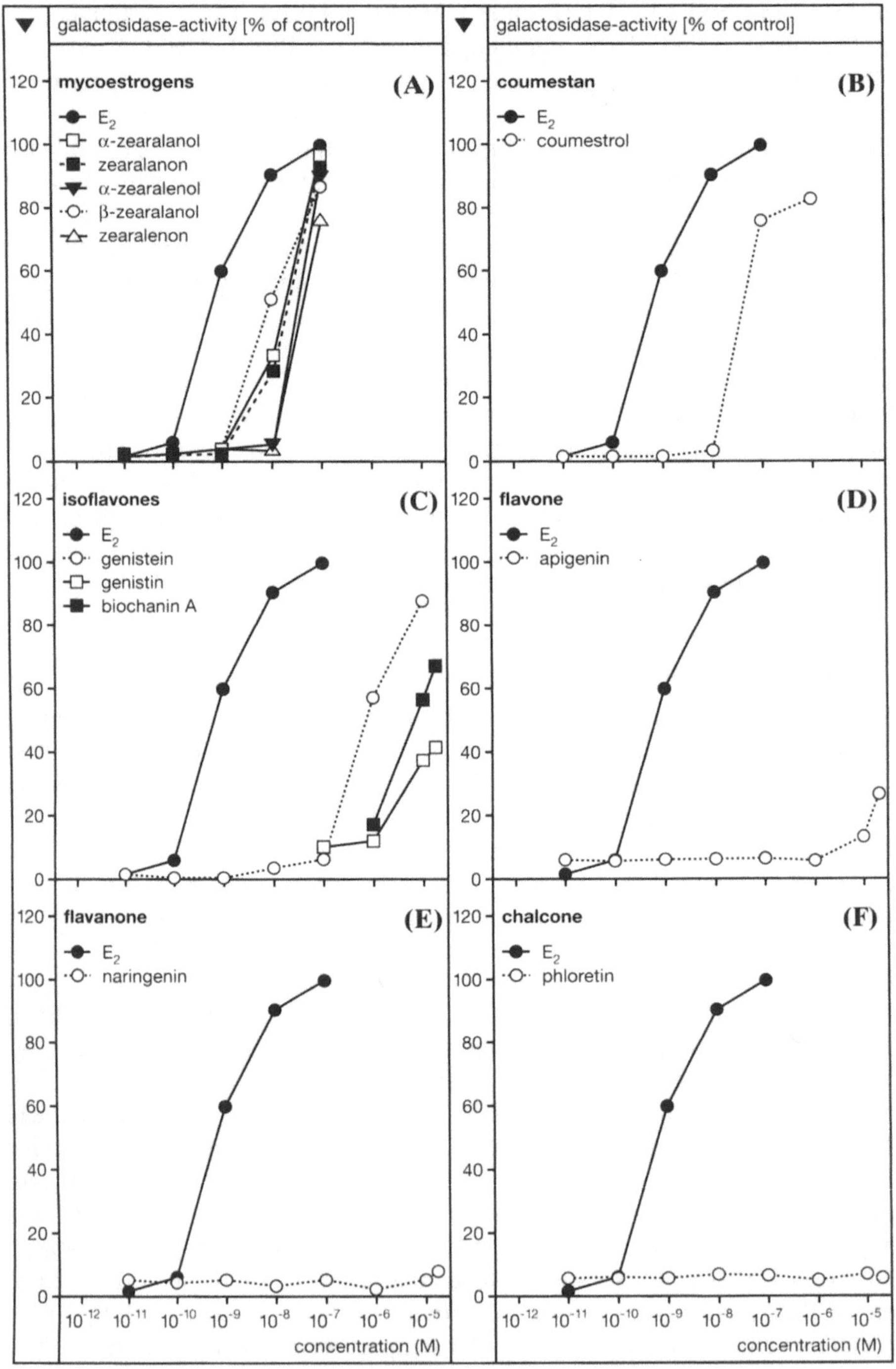

galactosidase-activity [% of control]
mycoestrogens
(A)
E2
α-zearalanol
zearalanon
α-zearalenol
β-zearalanol
zearalenon
coumestan
(B)
E2
coumestrol
isoflavones
(C)
E2
genistein
genistin
biochanin A
flavone
(D)
E2
apigenin
flavanone
(E)
E2
naringenin
chalcone
(F)
E2
phloretin
0
20
40
60
80
100
120
10^-12 10^-11 10^-10 10^-9 10^-8 10^-7 10^-6 10^-5
concentration (M)

V. Reporter Assays Based on "Non-Classical" Mechanisms of Gene Activation by the ER

1. ER Activation of Genes Through AP-1

Besides using the classical mode of gene activation, which precludes binding of the ER to an ERE, the ER was shown to cause activation of specific promoters (via the transcription factor AP-1) without physically interacting with the DNA (WEBB et al. 1995). The collagenase promotor, which contains an AP-1 response element (binding site for the FOS/JUN transcription factor complex (ANGEL and KARIN 1991), was activated by ERα, irrespective of being loaded with estrogens or antiestrogens in a cell-type-specific manner. Thus, tamoxifen activated a collagenase promotor construct in Ishikawa cells, whereas it exhibited little or no effect in the mammary carcinoma cell lines MDA 453, MCF-7 and ZR75 (WEBB et al. 1995). This non-classical type of gene activation by the ER is assumed to play a role for some of the estrogen-like effects of tamoxifen.

Recently an "inverse pharmacology" of estrogens and antiestrogens on the collagenase promotor was reported if the compounds acted through ERβ (PEACH et al. 1997). Antiestrogens activated the collagenase promotor, whereas pure agonists had no effect in the absence of antihormone and inhibited the stimulatory effect of antiestrogens. So far, however, in vivo correlates for these findings are missing, and it is not possible to judge the biological significance of the in vitro observation.

2. TGF-β3 Induction Through a Non-Classical ERE

In a reporter gene system, the human TGF-β3 gene promotor was shown to be activated by the ER in the presence of ER antagonists (YANG et al. 1996). TGF-β3 is involved in bone metabolism; the effect of estrogens on TGF-β protein levels and of modulating isoform production (ROBINSON et al. 1996) is assumed to be one mechanism that contributes to the bone-preserving activity of estrogens (HUGHES et al. 1996).

Activation of a TGF-β3 promotor fragment that is placed in front of the CAT-reporter gene by an antiestrogen-loaded ER is mediated by a polypurine sequence close to the transcription start site of the TGF-β3 promoter and does

Fig. 6A–F. Study of phyto- and myco-estrogens in a transactivation assay performed in yeast. Yeast cells transformed with estrogen receptor (ER) expression vector and a reporter gene under the control of an estrogen response element (ERE)-containing promoter were teated with different concentrations of environmental estrogens or 17β-oestradiol (E2) in a concentration range of 0.01 nmol/l–10 μmol/l. The maximal induction by E2 was arbitrarily set at 100%. The myco-estrogens tested (**A**), coumestrol (**B**) and genistein (**C**) exhibit high estrogenic activity in this assay, the EC_{50} being 10- to 100-fold (myco-estrogens and coumestrol) or 1000-fold (genistein) higher than that of E2. Apigenin (**D**), naringenin (**E**) and phloretin (**F**) exhibit hardly detectable agonistic activity in this assay (HUSEMANN et al., unpublished observations) not require binding of the ER to DNA (YANG et al. 1996a). In a reporter gene assay performed in MG63 human ostesarcoma cells transfected with an ER expression vector together with a TGF-β3-promotor [nucleotides –301 to +110] CAT reporter plasmid, the non-steroidal antiestrogen raloxifen (Ely Lilly) induced the TGF-β3 gene. The stimulatory effect of raloxifen is seen as one possible mechanism contributing to the bone-protective activity of this "selective ER modulator" (YANG et al. 1996).

a) *Study of Raloxifen and ZM182,780 in the TGF-β3 Assay*

We established the TGF-β3 reporter gene assay to assess the effects of estrogens and antiestrogens on TGF-β3 gene expression. The study was per-formed with a reporter gene that consisted of a TGF-β3 promotor fragment (nucleotides –298 to 110) placed in front of the luciferase reporter gene (a 500-bp promotor fragment was kindly provided by Prof. M. SPORN) (LAFYATIS et al. 1990). MG63 cells were transfected by the lipofectin method (Life Technologies, Palo Alto, Calif., USA). Both antiestrogens tested, raloxifen and ZM182,780, caused about a threefold increase in reporter gene activity (Fig. 7). The estrogens oestradiol, oestrone (not shown) and oestriol (not shown) caused less than a 1.5-fold increase above the ethanol control.

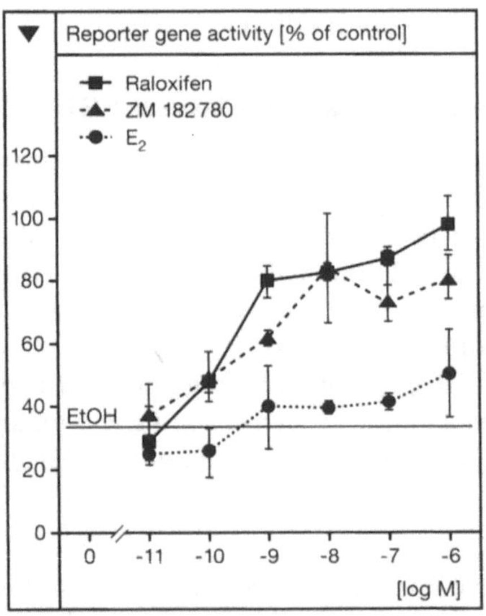

Fig. 7. Effect of ligand-occupied estrogen receptors (ER) on a transforming growth factor (TGF)-β3-luciferase reporter gene. Transactivation assay in MG63 human osteosarcoma cells transfected with HEGO and a luciferase reporter gene under the control of the human TGFβ-3 promotor (nucleotides –298 to 110). The effect of raloxifen, ZM182,780 and E2 on the reporter gene activity was measured in the concentration range 0.01–1000 nmol/l. The maximal effect of raloxifen was arbitrarily set at 100%

In contrast to raloxifen, a mixed agonist/antagonist with well-documented bone-sparing activity (BRYANT et al. 1996), ZM182,780, a "pure" antiestrogen, is not assumed to exhibit bone-protective activity (GALLAGHER et al. 1993). The two compounds behaved similarly in the TGF-β3 assay. It may be questioned whether the result of the TGF-β3 reporter gene assay is predictive of the bone-protective activity of ER-ligands. On the other hand, that the observed effect of raloxifen contributes to its specific pharmacological profile cannot be ruled out.

3. Repression of the IL-6 Gene Expression by the ER

Besides activating the expression of target genes, the ER controls transcriptional processes by inhibiting the expression of another subset of genes. Thus, estrogens were shown to suppress the cytokine interleukin-6 (IL-6) (POTTRATZ et al. 1993; STEIN and YANG 1995; GALIEN et al. 1996; RAY et al. 1997), a key mediator of immune and inflammatory responses and of osteoclastogenesis (SEHGAL 1992; JONES 1994; BRAKENHOFF 1995).

Estrogens suppress IL-6 production by osteoblasts and bone-marrow stromal cells. Given that IL-6 is a potent stimulator of osteoclast maturation, by inhibiting IL-6 synthesis, estrogens exert bone-sparing activity. IL-6 inhibition was shown to occur at the transcriptional level of the IL-6 gene (POTTRATZ et al. 1993; STEIN and YANG 1995; GALIEN et al. 1996; RAY et al. 1997). Many transcription factors are involved in IL-6 gene regulation. Binding sites for diverse transcription factors are present in the promotor. Among these is a binding site for the transcription factor NF-κB, but no ERE. The inhibitory effect of estrogens is mediated by NF-κB. This transcription factor is released from its inhibitor, IκB, in response to inflammatory signals (THANOS and MANIATIS 1995; DIDONATO et al. 1997; MANIATIS 1997). The ER directly interacts with activated NF-κB and prevents it from binding to the IL-6 promotor (STEIN and YANG 1995; GALIEN et al. 1996; RAY et al. 1997).

Similar to the effect on IL-6 expression, the ER is able to modulate other genes induced by NF-κB. We found rapid downmodulation of the COXII messenger ribonucleic acid (mRNA) levels in vessels of female ovex rats treated with E2 (Fig. 8). An IL-6 reporter gene assay was first described by POTTRATZ et al. (1993). An inhibitory effect of E2 on a reporter gene that consists of the 1.2-kb proximal region of the human IL-6 gene placed in front of the CAT reporter gene was demonstrated in HeLa and murine bone-marrow stromal cells (MBA 13.2) that were co-transfected with the reporter plasmid and an ER expression vector. Induction of the reporter gene occurred by treatment of the cells with PMA (phorbol 12-myristate 13-acetate) or a combination of TNF-α (tumor necrosis factor-α) and IL-1. The system was modified by STEIN and YANG (1995) and GALIEN et al. (1996).

GALIEN et al. (1996) found E2 inhibition of endogenously produced IL-6 secretion in MCF-7, Saos-2 and Hela cells expressing the ER. IL-6 production was induced by TNF-α. In Saos-2 cells, the mixed agonists/antagonists

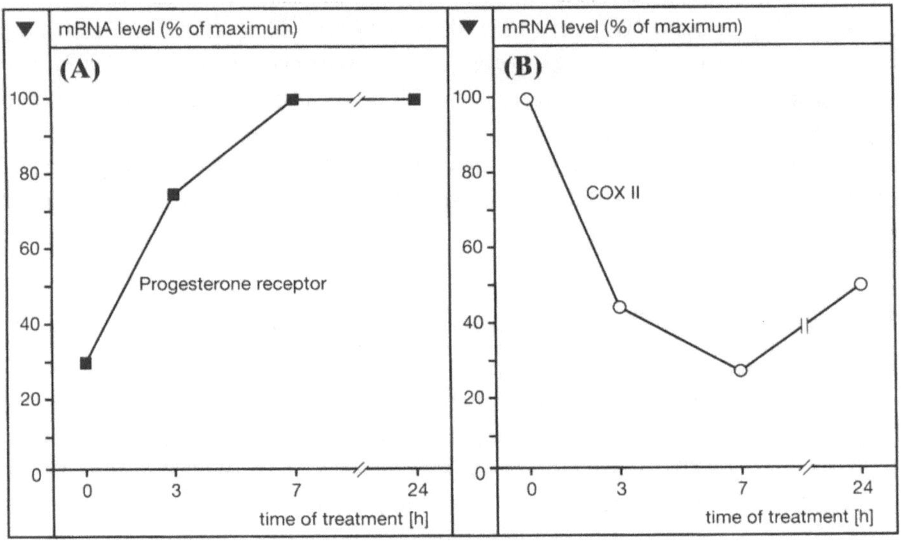

Fig. 8A,B. Effect of a single dose of E2 on the COXII (**B**) and progesterone receptor (PR, **A**) messenger ribonucleic acid (mRNA) level in female rat vena cava. Time-related effect of a single subcutaneous dose of E2 (1 μg/animal) on the COXII and PR mRNA level in the vena cava of OVEX female rats. Adult female rats were ovariectomised 14 days before treatment. Animals were sacrificed 0, 3, 7 and 24 h after administration of E2, RNA was isolated from the vena cava and the COXII and PR mRNA levels were determined by semiquantative polymerase chain reaction (PCR) as described (KNAUTHE et al. 1996). Maximal COXII and PR mRNA levels, respectively, were arbitrarily set at 100% (DIEL et al., unpubished observations)

OH-tamoxifen and raloxifen also caused inhibition of IL-6 production (about 50% of E2), whereas the pure antiestrogen RU58,668 had no effect. Experimental evidence from different groups, plus our own studies, indicate that the "IL-6 assay", either measuring the effect of ER ligands on endogenous IL-6 production or on an IL-6-promotor-containing reporter gene construct, is suitable for identifying and selecting ER ligands that exhibit bone-protective activity in vivo.

VI. Estrogen Effects on Expression of Endogenous Genes in Diverse Cell Lines

The expression of quite a number of genes was shown to be regulated by estrogens (see Table 3 for an overview of some of the best-characterised genes). The EREs in the promotor of several genes were analysed and used to construct estrogen-dependent reporter genes (see above). An alternative approach to establish in vitro models that are able to predict the pharmacological activity of estrogens and antiestrogens is to measure, in cell lines representing different organs ("organotypic" cell lines), the effect of ER ligands on the expression of ER target genes.

1. Alkaline Phosphatase and PR Measurement in Ishikawa Endometrial Carcinoma Cells

Human endometrial carcinomas have long been recognised to be estrogen responsive (SATYASWAROOP et al. 1997). However, human endometrial carcinoma cell lines in continuous culture have proved inconsistent in their response to steroid hormones, thus making it difficult to establish an in vitro model for estrogen regulation of these tumors (ALBERT et al. 1990).

NISHIDA et al. (1985) established the Ishikawa cell line. These cells were shown to respond to estrogen treatment with cell growth, induction of alkaline phosphatase activity (HOLINKA et al. 1986; ALBERT et al. 1990) and induction of the PR (JAMIL et al. 1990). Estrogen effects were antagonised by ZM164,384 in these cells. OH-tamoxifen exhibited agonistic activity on cell growth and the PR level (JAMIL et al. 1990). In contrast to ZM164,384, OH-tamoxifen was not able to inhibit the E2-stimulated alkaline phosphatase activity in Ishikawa cells.

2. The Endometrial Adenocarcinoma Cell Line RUCA-I as an In Vitro/In Vivo Tumour Model

Recently, an ER-positive endometrial adenocarcinoma cell line, RUCA-1 (VOLLMER and SCHNEIDER 1996), was tested for estrogen responsiveness in vitro and in vivo. In vitro estrogenic effects were observed on the expression of components of complement C3 and on fibronectin, an estrogen-repressed protein. ZM182,780 behaved as a complete antagonist in this model, whereas tamoxifen, like oestradiol, stimulated complement C3 production and repressed fibronectin.

In vivo, RUCA-I cells injected subcutaneously into rats caused the develpment of endometrial adenocarcinoma. Tumor size and the number of lung metastases were responsive to hormonal depletion (ovariectomy), E2 substitution and antiestrogen treatment, demonstrating estrogen dependency of growth and metastasis of the tumor cells.

The fact that, for RUCA-I cells, a good correlation exists between the in vitro and in vivo findings suggests that the cell line may be useful as an in vitro model for predicting the endometrial effects of new drugs.

3. Cathepsin D, PS2 and PR Induction in Mammary Cell Lines by Estrogens

The cathepsin D (CAVAILLES et al. 1993; AUGEREAU et al. 1994; WANG et al. 1997), pS2 (MORI et al. 1990) and PR (ECKERTS and KATZENELLENBOGEN 1982; NARDULLI et al. 1988; KATZENELLENBOGEN and NORMAN 1990) genes are induced by estrogen in breast carcinoma cell lines. Cathepsin D is a lysosomal asparty protease. Its expression is estrogen induced in breast cancer cell lines and it is produced in breast cancer cells in vivo. Its concentration in the primary tumor is correlated with increased risk of metastasis (CAVAILLES et al. 1993).

PS2 was identified as an estrogen-inducible protein in MCF-7 cells (Jakowlev et al. 1984). Although the pS2 gene is also expressed in other tissues, estrogen regulation of the gene was only observed in ER-positive breast carcinoma cells. In addition to responding to estrogen, the pS2 gene is regulated by a number of growth factors known to mediate estrogen's action (Nunez et al. 1989). The physiological role of the pS2 gene product is still unknown (Pilat et al. 1993).

Estrogen regulation of PR gene expression occurs in a tissue-specific manner. Estrogen effects on the PR level were found in many tissues, including uterus (Kraus et al. 1994), endometrial cancer cell lines (Jamil et al. 1990), brain (Shughrue et al. 1997b), blood vessels (Knauthe et al. 1996; Hegele-Hartung et al. 1997) and breast cancer cell lines (May et al. 1989).

a) Estrogen Effects on the PR Level in MCF-7 Cells

Eckerts and Katzenellenbogen (1982) studied the effect of E2, DES and several non-steroidal antiestrogens on the PR level in MCF-7 cells. The cells were incubated over five consecutive days with the compounds. The cells were then harvested and fractionated and the cytosol was assayed for PR, utilising 10 nmol/l [^{3}H]R5020 in the absence and presence of a 100-fold excess of radioinert R5020 as the ligand. A good correlation was found between ER affinity and the stimulation of the PR level.

b) Estrogen Effects on the Cathepsin D and pS2 mRNA Levels and on the PR Protein Level in MCF-7 Cells

Pilat et al. (1993) studied the effect of oestradiol derivatives on the cathepsin D and pS2 mRNA levels and on the PR protein level in MCF-7 breast carcinoma cells. Cathepsin D and pS2 mRNA levels were determined by "Northern-blot" analysis. The PR protein level was determined using a PR-EIA kit (Abbot Laboratories, Chicago, Ill., USA).

Cathepsin D and pS2 reacted differently to discrete changes in the structure of oestradiol-derived ligands, with ER-binding affinities that showed no correlation with their capacity to regulate pS2 and cathepsin D genes. In contrast, for most of the oestradiol derivatives tested, effects on the PR level correlated with the ER-binding affinity. In conclusion, the study provides evidence for gene-specific effects of oestradiol derivatives on the cathepsin D, pS2 and PR genes in MCF-7 cells.

4. Estrogen Regulation of Prolactin Expression in Pituitary Cells

In the rat, estrogens stimulate the expression of the prolactin gene in the anterior pituitary in vivo (Shull and Gorski 1984). In primary rat pituitary cells that express ERs a variety of ER ligands were examined for their effects on prolactin synthesis (Jordan and Lieberman 1984). The rat pituitary tumour cell line, GH_3 (ATCC), and the related cell line, GH_4C_1, express prolactin and

ERs. Estrogen regulation of prolactin synthesis was demonstrated in GH_3 cells grown in charcoal-stripped serum media. The addition of E2 (0.1–10 nmol/l) to the stripped medium caused up to a sixfold increase in prolactin mRNA levels (Rhode and Gorski 1991). Effects of tamoxifen and oestradiol on the prolactin and PR mRNA levels were studied in GH_4C_1 cells (Shull et al. 1992). Tamoxifen exhibited agonistic activity on the expression of the prolactin, but not the PR gene in these cells.

5. Estrogen-Regulated Genes in Liver Hepatoma Cells

To extend our understanding of direct estrogen effects on liver gene expression, we used the ER-expressing rat hepatoma cell line, Fe33. Fe33 cells were derived from FTO-2B cells by stable transfection with ER (Kaling et al. 1990). By applying the differential display reverse-transcription polymerase chain reaction (ddRT-PCR) method, we identified estrogen-regulated genes in this cell line. Insulin-like growth factor binding protein-1 (IGFBP-1), vitamin D-dependent calcium-binding protein (CaBP9k) and major acute phase glycoprotein (MAP) were found to be estrogen-regulated in these cells (Diel et al. 1995). As documented in Fig. 9, EE, RU39,411 and tamoxifen caused dose-related increases in the mRNA levels of IGFBP-1 and MAP. Similar dose–response curves were found for angiotensinogen and α-fibrinogen mRNA (Fig. 9). In contrast, CaBP9k was only stimulated by EE and a high concentration of RU39,411. In vivo, in the liver of E2-treated rats, estrogen stimulation was found for IGFBP-1, MAP and angiotensinogen (Diel et al. 1995). In contrast, the mRNA level of α-fibrinogen decreased by about 50% in the liver of E2-treated ovex rats (unpublished results) and CaBP9k was hardly detectable in the liver of female rats (Diel et al. 1995).

6. Induction of ERs and Vitellogenin Synthesis in Fish Liver Cells

There is increasing evidence that many chemicals, both natural and synthetic, exhibit estrogenic activity (Miksicek 1992; Miksicek 1994; White et al. 1994; Knight and Eden 1995; Dodge et al. 1996; Anderson 1997; Ramamoorthy et al. 1997; Routledge and Sumpter 1997). In view of the persistence of some of the synthetic chemicals, concerns have been raised about environmental effects that these compounds may have on wildlife. Relevant, sensitive, assay systems used to evaluate potentially estrogenic environmental chemicals have taken advantage of the specific response that vitellogenin synthesis in the liver of oviparous animals (specifically in fish) show to estrogenic stimuli. Vitellogenesis was induced in mature male Japanese quail following injection of estrogen analogues (Robinson and Gibbins 1984), in male carp after oestradiol administration (Hernandez et al. 1992), and ER and vitellogenin genes were upregulated after E2 injection in rainbow trout and other fish species (McKay et al. 1996; Flouriot et al. 1997).

Beside vitellogenin synthesis, the ER level can be used to monitor estrogen effects in fish hepatocytes (Pakdel et al. 1991; Smith and Thomas 1991).

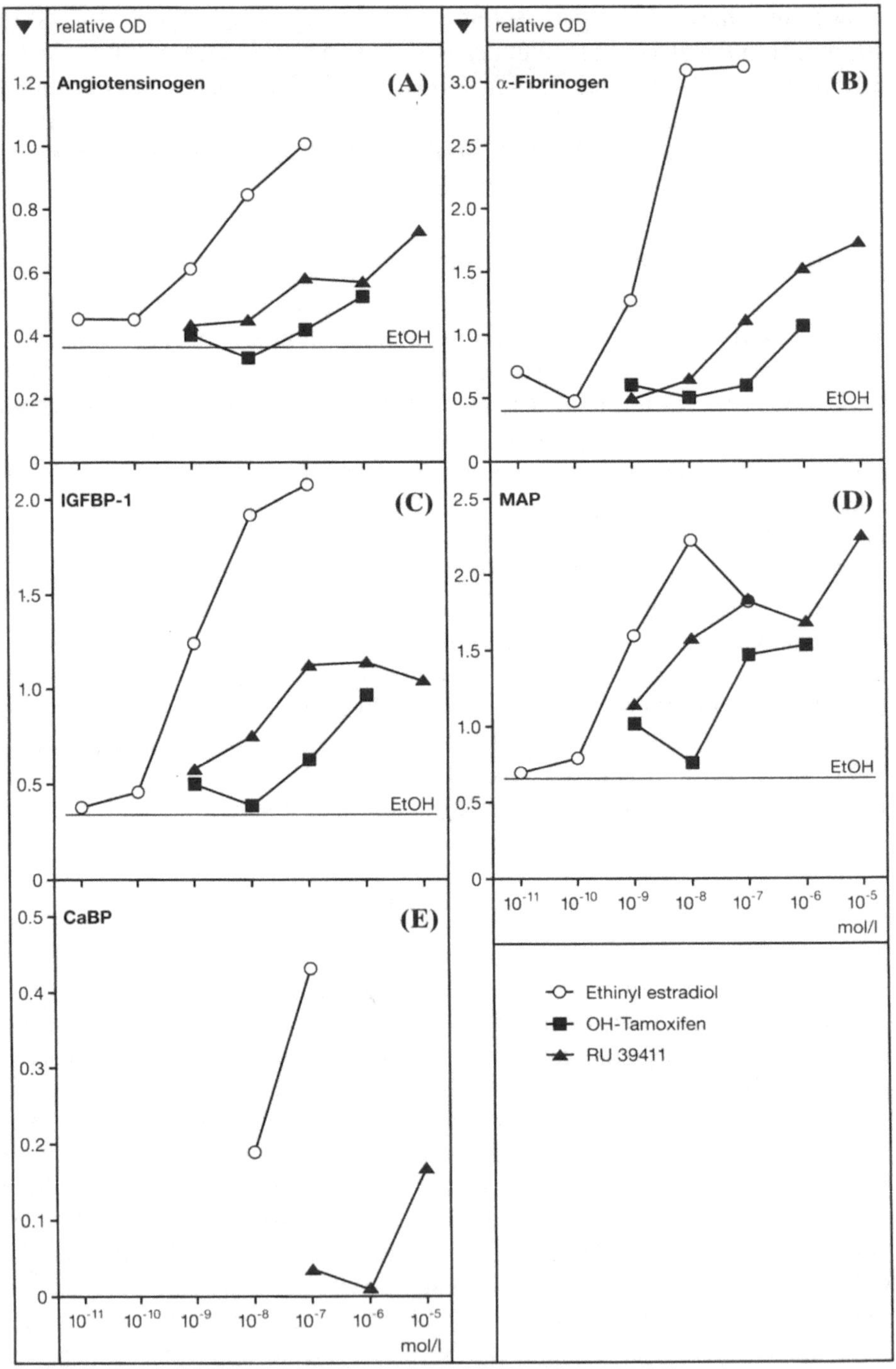

relative OD
Angiotensinogen
(A)
α-Fibrinogen
(B)
IGFBP-1
(C)
MAP
(D)
CaBP
(E)
EtOH
mol/l
Ethinyl estradiol
OH-Tamoxifen
RU 39411

An ERE was detected in the rainbow trout ER gene (LE DREAN et al. 1995). Reporter genes have been constructed that consist of trout ER promotor fragments placed in front of the CAT reporter gene. Dose-responsive effects of estrogen on these reporter constructs were found. Cell culture systems with rainbow trout hepatocytes were used to measure the estrogenic potency of classical estrogens (PELISSERO et al. 1993) and xenobiotics (FLOURIOT et al. 1995).

VII. Ligand-Induced ER Stabilisation or Destabilisation

Besides interfering with estrogen action at the transcriptional level (by inhibiting one or both of the activation functions of the ER), antiestrogens influence the ER level in target cells and tissues by affecting the turnover of ER protein. Compared with the ER bound by E2, the ER liganded by the pure antiestrogen ZM164,384 has a substantially shorter half-life (DAUVAIS et al. 1992). In contrast, ER stability tends to be enhanced by treatment with mixed agonists/antagonists (KIANG et al. 1989). Induction of ER degradation is a property shared by all known pure antiestrogens (PARCZYK and SCHNEIDER 1996). Reduction in ER protein levels upon treatment with pure antiestrogens is not accompanied by a decrease in ER mRNA levels (PARCZYK et al. 1995).

It was proposed by DAUVAIS et al. (1992) that the pure antiestrogen ZM164,384 inhibits transportation of the antagonist-occupied ER from the cytoplasm into the nucleus. Thus, ER molecules that diffuse out of the nucleus are prevented from being re-shuttled to the nucleus and are proteolytically degraded.

Evidence was provided for cross-talk between growth-factor- and ER-dependent pathways. Thus, it was demonstrated that the proliferative activity of EGF on mouse uterus is mediated by ER (IGNAR-TROWBRIDGE et al. 1992). Furthermore, dopamine activated an estrogen-responsive reporter gene. The effect could be blocked by the antiestrogen ZM164,384 (SMITH et al. 1993). The ability to induce ER disruption is assumed to contribute to the greater efficiency of pure antiestrogens versus mixed agonists/antagonists (tamoxifen), for inhibiting proliferation in experimental tumour models (THOMPSON et al. 1989; WAKELING 1989; PARCZYK and SCHNEIDER 1996).

Fig. 9A–E. Dose-related effect of ethinyl oestradiol (EE), OH-tamoxifen and RU39,411 on the messenger ribonucleic acid (mRNA) level of estrogen-sensitive endogenous genes in estrogen receptors (ERs) expressing Fe33 rat hepatoma cells. Fe33 cells were treated with increasing concentrations of EE and the mixed agonists/antagonists, respectively. The cells were harvested 24 h after treatment, RNA was extracted and was submitted to "Northern-blot" analysis. Specific mRNAs were detected by hybridisation with radioactive complementary deoxyribonucleic acid (cDNA) probes. Quantitation of radioactive signals was done by phosphoimage analysis. To correct for differences in RNA amounts, the blots were hybridised with glyceraldehyde phosphate dehydrogenase (GAPDH) cDNA, as described (DIEL et al. 1995)

The two mammary cell lines, MCF-7 (ECKERTS and KATZENELLENBOGEN 1982) and T47-D (KIANG et al. 1989), can be used to investigate the effect of ER ligands on ER protein levels (PARCZYK and SCHNEIDER 1996). Both cell lines express the human ERα endogenously. ER quantification in protein extracts can be accomplished by ligand-binding experiments that use radioactive ER ligands, such as [^{3}H]-oestradiol, by 'Western-blot' analysis or by using a commercially available Enzyme Immuno Assay (EIA; Abbot Laboratories, Chicago, USA).

1. Comparative Study of the Effects of Different Antiestrogens on the ER Level in T47-D Cells

A comparative study was performed to determine the effect of different antiestrogens on ER protein levels in T47-D cells (PARCZYK et al. 1995). Figure 10 shows the effects of different antiestrogens on the ER protein level. ZM182,780, ZK164,015 (VON ANGERER 1995b) and RU58,668 (VAN DE VELDE et al. 1994) caused a decrease in the ER protein levels to 10%, 20% and 10%, respectively, compared with the ethanol control. In contrast, 4-OH-tamoxifen and RU39,411 and ZK119,010 caused an increase in the ER protein level by 2- to 3-fold compared with the ethanol reference. E2 induces a slight decrease in ER protein.

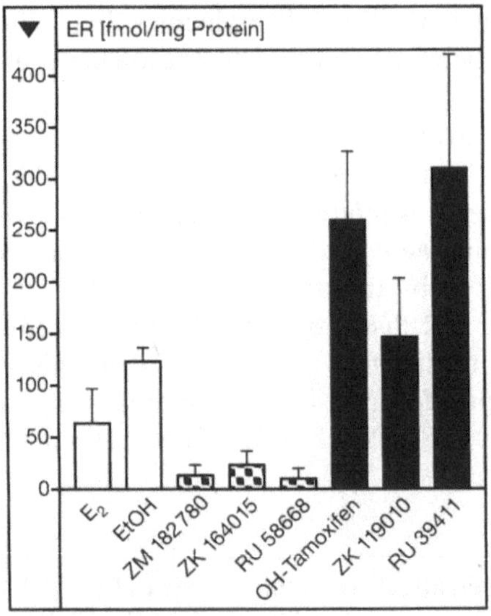

Fig. 10. Effect of antiestrogens on the estrogen receptor (ER) protein level in T47-D human breast carcinoma cells. T47-D cells were treated with antiestrogen (0.1 μmol/l), oestradiol (0.1 μmol/l) or ethanol. Twenty-four hours after addition of the compounds, the cells were harvested and protein extracts were prepared. After centrifugation, an aliquot of the supernatant was diluted to a protein concentration of 1.5 mg/ml and was used to determine the ER protein level by EIA (Abbot Laboratories, Chicago, USA) according to the manufacturer's advice

The samples were also investigated using "Western-blot" analysis. A commercially available ER antibody (B10a; Euromedex, Strasbourg, France) was used for the detection of the ER protein on blots of sodium dodecyl sulfate (SDS) polyacrylamide gels (10%). The mixed agonists/antagonists caused an increase in the ER protein band. In contrast, no ER was detectable in cells treated with "pure" antiestrogens (data not shown).

VIII. Effects of ER Ligands on the Proliferation of ER-Expressing Breast Cancer and Endometrial Cell Lines

Estrogens stimulate growth of ER-expressing tumour cell lines (ECKERTS and KATZENELLENBOGEN 1982; KATZENELLENBOGEN et al. 1987; THOMPSON et al. 1989; WAKELING 1989). This proliferative effect is mediated by ERs and can be inhibited by antiestrogens (ROCHEFORT 1987). This was shown in several mammary carcinoma cell lines (WESTLEY and MAY 1987; KIANG et al. 1989; WAKELING 1989) and in endometrial carcinoma cell lines (JAMIL et al. 1990; VOLLMER and SCHNEIDER 1996). The effect of estrogens and antiestrogens on cell proliferation can be used to characterise estrogenic and antiestrogenic compounds.

1. Inhibition of Estrogen-Stimulated Growth of MCF-7 Mammary Carcinoma Cells by Antiestrogens

We use low passage MCF-7 human mammary carcinoma cells to characterise antiestrogens. The cells are treated daily for 7 days with an estrogen or a combination of estrogen and antiestrogen. The proliferation of cells is determined by measuring the incorporation of [^{3}H]-thymidine, which is added to the cells 24 h before the end of the experiment.

Experimental Design

Cells are spread to a density of 16,000 cells/cm^2 and are maintained in medium containing 2% charcoal-stripped fetal calf serum (FCS) in RPMI 1640 (without neutral red) and 5 μg insulin/ml. Before the beginning of the experiment, the cells are transferred to 48-well plates at a density of 5000 cells/well. Four hours later, the compounds are added at different concentrations, ranging from 0.1 nmol/l to 1 μmol/l in 0.5 ml medium. Medium exchange is performed after 3 days and 6 days of culture. On day 6, [^{3}H]-thymidine (0.25 μCi) is added. On day 7, the medium is removed and the cells are washed with phosphate-buffered saline (PBS). They are lysed by the addition of 150 μl 0.1% SDS solution and 100 μl of the radioactive solution is transferred to scintillation vials for liquid scintillation counting.

Figure 11 shows the result of an experiment in which EE (0.1 nmol/l) was used to stimulate the growth of MCF-7 cells. Co-administration of OH-tamoxifen and ZM182,780 caused inhibition of EE-stimulated cell growth (Thierauch, Berlin, unpublished results).

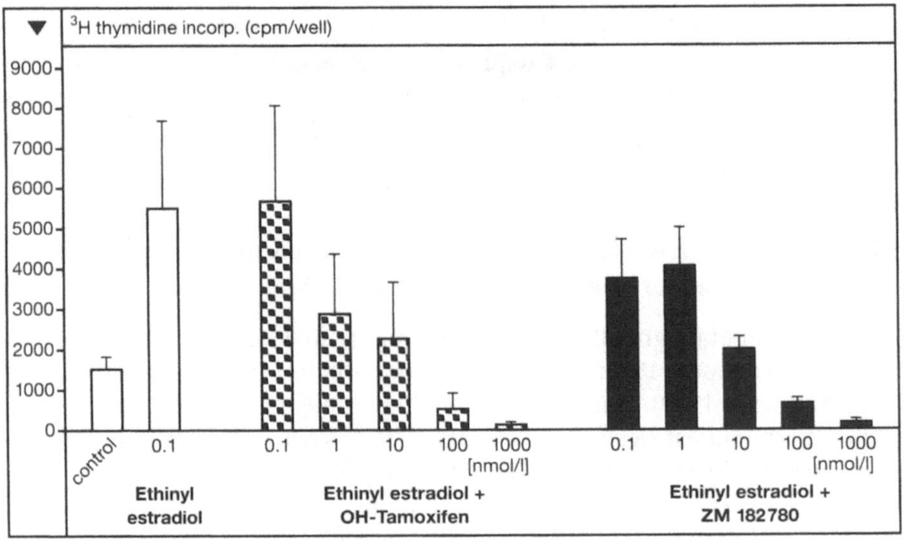

Fig. 11. Effect of antiestrogens on ethinyl oestradiol (EE)-stimulated growth of MCF-7 human mammary epithelial cells. Cells were spread to a density of 16,000 cells/cm^2 and were maintained in medium containing 2% charcoal stripped foetal calf serum (FCS) in RPMI 1640 without neutral red and 5 μg insulin/ml. Before the beginning of the experiment, the cells were transfered to 48-well plates at a density of 5000 cells/well. The compounds were added at different concentrations in the range 0.1 nmol/l–1 μmol/l, 4 h later, in 0.5 ml medium. Medium exchange was performed after 3 days and 6 days of culture. On day 6, [^{3}H]-thymidine (0.25 μCi) was added. On day 7, the medium was removed and the cells were washed with phosphate-buffered saline. They were lysed by the addition of 150 μl 0.1% SDS solution. 100 μl of the radioactive solution was transfered to scintillation vials for liquid scintillation counting. The two antiestrogens OH-tamoxifen and ZM182,780 both efficiently inhibit EE-stimulated growth of MCF-7 cells (THIERAUCH et al., unpublished results)

IX. Antioxidative Activity

Reactive oxygen intermediates (ROIs) play a role in numerous physiological and pathophysiological processes (HALLIWEL and GUTTERIDGE 1989). ROIs comprise radicals, such as the superoxide anion (O_2^-), the hydroxyl radical (OH) and the less reactive molecule hydrogen peroxide (H_2O_2). ROIs can covalently modify most biomolecules, leading for example to lipid peroxidation of membranes, oxidation of cysteine residues in proteins and to strand breaks and oxidation of guanosine residues in DNA (HALLIWEL and GUTTERIDGE 1989). All organisms have developed multicomponent defence mechanisms to protect cells from the toxic effects of ROIs. Many pathological conditions, such as viral and bacterial infections, injury and inflammatory processes, are accompanied by the production of ROIs. Excessive ROI production is supposed to occur and to be responsible for damage and cell death in various pathologies (KALTSCHMIDT et al. 1993; YOUDIM et al. 1994; YAN et al.

1996). It is assumed that antioxidative radical scavenging drugs could be effective in the treatment of such diseases (BEHL et al. 1994).

Due to their phenolic structure, estrogens exhibit "antioxidative activities" in vitro (SUBBIAH et al. 1993). This property is suggested to contribute to the vasculo- and neuroprotective activity of estrogens (WASHBURN et al. 1996; BEHL et al. 1997; WILCOX et al. 1997). In contrast to the classical estrogen effects, the antioxidative activity of estrogens is independent of their binding to the ER.

A variety of in vitro assays have been applied to measure the antioxidative activity of estrogen-like molecules (BEHL et al. 1997; RÖMER et al. 1997). The methodologies used do not define specific features of estrogens but, rather, identify the antioxidative and radical-scavenging effects that have been described for diverse compounds from different classes. We therefore confine our discussion to some of the available assays and otherwise refer the reader to applicable references.

1. Protection from Amyloid-Induced Neurotoxicity and Prevention of Glutamate-Induced Cell Death in Nerve Cells

Oxidative damage and lipid peroxidation can be caused by the neurotoxic amyloid β (Aβ) protein (BEHL et al. 1994). Thus, exogenously applied Aβ causes cell death in PC 12 neuronal cells that can be blocked by the antioxidants vitamin E and propylgallate (BEHL et al. 1992).

Glutamate is the most abundant excitatory neurotransmitter in the brain. However, under certain conditions, it may become a potent excitotoxin and contribute to neurogeneration (CHOI 1992; LIPTON and ROSENBERG 1994). The destructive effect of glutamate is mediated by glutamate receptors, specifically *N*-methyl-D-aspartate (NMDA)-type receptors. Activated NMDA receptors allow an influx of Ca^{2+} which, in excess, can activate a variety of potentially destructive processes.

Different estrogens were studied for their protective effect on rat primary hippocampal neurons and mouse hippocampal cells (HT22) that were challenged for 24 h with 2 μmol/l Aβ or 1 mM glutamate, respectively. Survival of neuronal cells was determined as a measure for the neuroprotective (antioxidative) activity of the estrogens (BEHL et al. 1997).

2. Attenuation of Lipid Peroxidation in Synaptosomal Membranes

Aβ-induced cell toxicity involves the formation of excess H_2O_2 (BEHL et al. 1994). H_2O_2 causes lipid peroxidation of cell membranes. The peroxidation of membranes can be monitored using *cis*-parinaric acid (CPA), a polyunsaturated fatty acid that, if incubated with membranes, is incorporated. The incorporated fatty acid is fluorescent and, upon peroxidation, the double-bond structure is lost, with a concomitant loss of fluorescence. Upon incubation of B12 suspension cells with Aβ, a decay in CPA fluorescence due to enhanced lipid peroxidation was observed. The decrease in CPA fluorescence was partly inhibited by vitamin E and the lipophilic spin-trapping agent *n*-*t*-butyl-phenyl-carbonate (BEHL et al. 1994).

An assay for (iron-induced) lipid peroxidation was used to measure the antioxidative activity of estrogens (RÖMER et al. 1997). Rat synaptosomal membranes were incubated with Fe(II) sulfate; H_2O_2 and lipid peroxidation was monitored by measuring the formation of 2-thiobarbituric acid (TBA)-reactive substances (BUEGE and AUST 1978; BRAUGHLER et al. 1989).

3. Fe(II) Autoxidation and Fe(III) Reduction Assays

If synaptosomal membranes are incubated with Fe-ions, measuring auto-oxidation and reduction of the ferric ions can be used to monitor the formation of free radicals. The formation of Fe(II) from Fe(III), and Fe(II) oxidation, respectively, are followed by the detection of Fe(II) as a Fe(II)-1,10-phenanthroline complex. Different estrogens were shown to inhibit Fe(II) auto-oxidation and to stimulate Fe(III) reduction reactions (RÖMER et al. 1997).

4. Oxidation of Low-Density Lipoprotein Cholesterol

In vivo oxidation of low-density lipoprotein (LDL) leads to a more rapid LDL uptake by arterial-wall macrophages and to the formation of foam cells. These processes are assumed to be involved in the pathogenesis of atherosclerosis (WEISER 1992). Based on the observation that estrogens exhibit antioxidative activity in vitro, it is assumed that the vasculoprotective activity of estrogens could be due, in part, to inhibition of LDL cholesterol oxidation in vivo (GUETTA and CANNON 1996).

The in vitro assay is based on the initiation of LDL oxidation by Cu^{2+} (ESTERBAUER 1989). The course of the oxidation process LDLox–LDL is continuously monitored by measuring the production of conjugated dienes. The effect of estrogens on the lag phase of cupric sulfate-induced LDL oxidation is assessed as a measure of antioxidative activity.

C. In Vivo Test Systems to Characterise Estrogens and Antiestrogens

In rodents and primates, a number of sensitive test systems are available by which the activity of estrogens and antiestrogens can be tested in reproductive tract organs, e.g. the vagina, uterus and hypothalamic–pituitary–ovarian axis. An overview is given in Tables 5–7.

I. Allen-Doisy Assay

1. Principle

In ovariectomised rodents, estrogens exhibit profound actions on the vagina and the uterus. Estrogenic compounds are effective in inducing proliferation and keratinisation of the vaginal epithelium of ovariectomised female rodents.

Table 5. Vagina: Assays to test estrogenicity in the vagina of rodents

Method	Principle	Endpoint determined
Allen-Doisy assay	Cornification and keratinisation of vaginal epithelium of ovariectomised rodents	Vaginal cytology
Vaginal mitosis and epithelial thickness	Increase in vaginal mitosis and epithelial thickness in ovariectomised rodents	Alteration in vaginal epithelial mitotic index
Vaginal tetrazolium reduction	Increase in vaginal reduction of tetrazolium in ovariectomised rodents measured colorimetrically	Tetrazolium reduction
Vaginal opening	The opening of the vagina in the immature rodent	Precocious or delayed vaginal opening
Vaginotrophic response	Increase in vaginal wet weight in the immature female rodent	Vaginal tissue growth response
Measurement of sialic acid	Estrogen-induced reduction of vaginal sialic acid production in ovariectomised female rodents	Reduction in vaginal sialic acid production

The epithelium of the vagina grows to a considerable thickness and to a cornified layer similar to that of the epidermis. Thus, the microscopic examination of vaginal smears is a reliable indicator of estrogenic activity. The Allen-Doisy test for vaginal cornification in rodents (Allen and Doisy 1923) is based on the observations of Stockard and Papanicolaou (1917), who first reported the cyclic vaginal cornification of guinea pigs. Their observations resulted in the Allen-Doisy bioassay for which cornification is accepted as the response in the rat. Using this method, the response is scored quantally, i.e. the animals either respond or they do not, and the results are, in effect, scored 1 or 0 (Stockard and Papanicolaou 1917).

2. Study Design

Adult rats or mice are ovariectomised and vaginal smears are examined 10–14 days thereafter to ensure complete ovariectomy. Only animals exhibiting dioestrus smears (see below) are used for the bioassay. The animals receive a single application of the estrogenic compound (s.c. or p.o.) dissolved in ethanol/arachis oil (sesame, olive or other oils can also be used) [1 + 9 v/v] or benzylbenzoat/castor oil [1 + 4 v/v] or other suitable vehicle. Alternatively, the test compound can be given in two to three portions within one day. Estrogens can also be applied locally when given per vaginam (Emmens 1940; Robson and Adler 1940). The application day is defined as d1.

Vaginal smears are taken at 48 h (d3), 54 h and 72 h (d4). On d4, after the last smear, the animals undergo autopsy to provide information about the vaginotrophic and uterotrophic response of the test compound. In this case,

Table 6. Uterus: Assays to test estrogenicity in the uterus of rodents and primates

Method	Principle	Endpoint determined
Uterine fluid imbibition (Astwood bioassay)	Specialised uterine growth test measuring early water uptake	Dose response increase in uterine wet weight
Uterine growth test	Increase in the weight of the uterus (wet and/or dry) of ovariectomised or immature rodents	Change in uterine weight
Regulation of steroid hormone receptors, proliferation, differentiation and protein expression in the rodent uterus	Measurement of cell cycle control parameters, histomorphometry and biochemical events are associated with estrogen-induced uterine growth in ovariectomised or immature rodents	PR-regulation, thymidin-incorporation into DNA; PCNA immunohistochemistry, protooncogenes, luminal epithelial cell height, glycogen content in response to estrogens, peroxidase activity in response to estrogens, enzyme activity (compl. C3) in response to estrogens
Withdrawal bleeding	In non-human primates, sudden deprivation or diminution of estrogen or progestin results in uterine bleeding	Induced uterine bleeding
Imaging of the primate uterus	With MRI and ultrasound the increase in endometrial and uterine volume induced by estrogens can be non-invasively visualised	Size of uterus, increase in endometrial and myometrial area
Endometrium: morphology and biochemistry (primates)	In non-human primates, estrogens lead to characteristic morphological and biochemical events in the endometrium	Ki 67/cell-cycle control parameters, morphology – endometrial secretion products

Table 7. Pituitary blockade: Assays to test the influence of estrogens on hypothalamic–pituitary–ovarian feedback

Method	Principle	Endpoint determined
Gonadal growth method	The administration of gonadal steroids to laboratory rodents leads to failure of growth of of both – the ovaries and testes	Change in gonadal weight
Gonadotropin and ovulation method	The preovulatory gonadotropin – peak and ovulation in female rodents can be inhibited by the administration of steroids (estrogens/progestins)	Change in follicular stimulating hormone/ luteinising hormone, change in ovulation

uteri and vagina are removed and the relative wet and dry weights are recorded (expressed as mg/100g rat or in mg/10g mouse).

3. Scoring Criteria

The vaginal smears (Fig. 12) are judged under the microscope and placed into the following categories:

0. (Dioestrus): leucocytes
1. (Dioestrus/Prooestrus): leucocytes, many nucleated cells (1:1)
2. (Prooestrus): nucleated cells only
3. (Oestrus): cornified cells only
4. (Metoestrus): a few cornified cells, many leucocytes, nucleated cells, mucin

4. Evaluation and Discussion

The appearance of prooestrus, oestrus or metoestrus smears are defined as a positive treatment result, whereas dioestrus and dioestrus/prooestrus smears are recorded as a negative result. KAHNT and DOISY (1928) adopted the view that a vaginal smear with a predominance of cornified cells (with some nucleated epithelial cells and very few leucocytes) should be scored as a positive response. BIGGERS and CLARINGBOLD (1954) further demonstrated that prooestrus smears with a predominance of nucleated cells should be classified as positive. Vaginal cornification usually occurs 48–72h after administration of the estrogenic compound. The onset and duration of the estrogen response are directly related to the dosing schedule and the pharmacokinetics of the administered estrogen. Some examples are given in Fig. 13. It can be seen that 17β-oestradiol induces keratinisation at 0.1μg in the rat, whereas oestriol is 30 times less potent. The synthetic estrogens 17α-ethinyl-oestradiol and cyclodiol (see Sect. I) are of comparable potency to 17β-oestradiol, when administered subcutaneously (Fig. 13). Alternatively, keratinisation can be evaluated by excising and fixing the vagina in formaldehyde, embedding it in paraffin, and then sectioning, staining and examining the vaginal epithelium under the microscope.

At d4, an increase in vaginal and uterine wet and dry weight also indicate the estrogenic potential of the estrogenic compound. However, the uterine and vaginal weights seem to be less sensitive to estrogenic stimulation than vaginal cornification. As emphasised by JONES and EDGREN (1973), vaginal keratinisation and cornification are among the most specific in vivo endpoints available for determining the estrogenic character of a compound. Other sex-steroid hormones, such as progestins and androgens, stimulate vaginal mucification but do not induce epithelial keratinisation or cornification (see also EDGREN 1994).

If vaginal and uterine weights are not determined at the end of the experiment, it is also possible that ovariectomised rats and mice can be maintained and used repeatedly for vaginal smear assays – provided that an adequate 'wash-out' period is allowed between assays. For example, ovariectomised

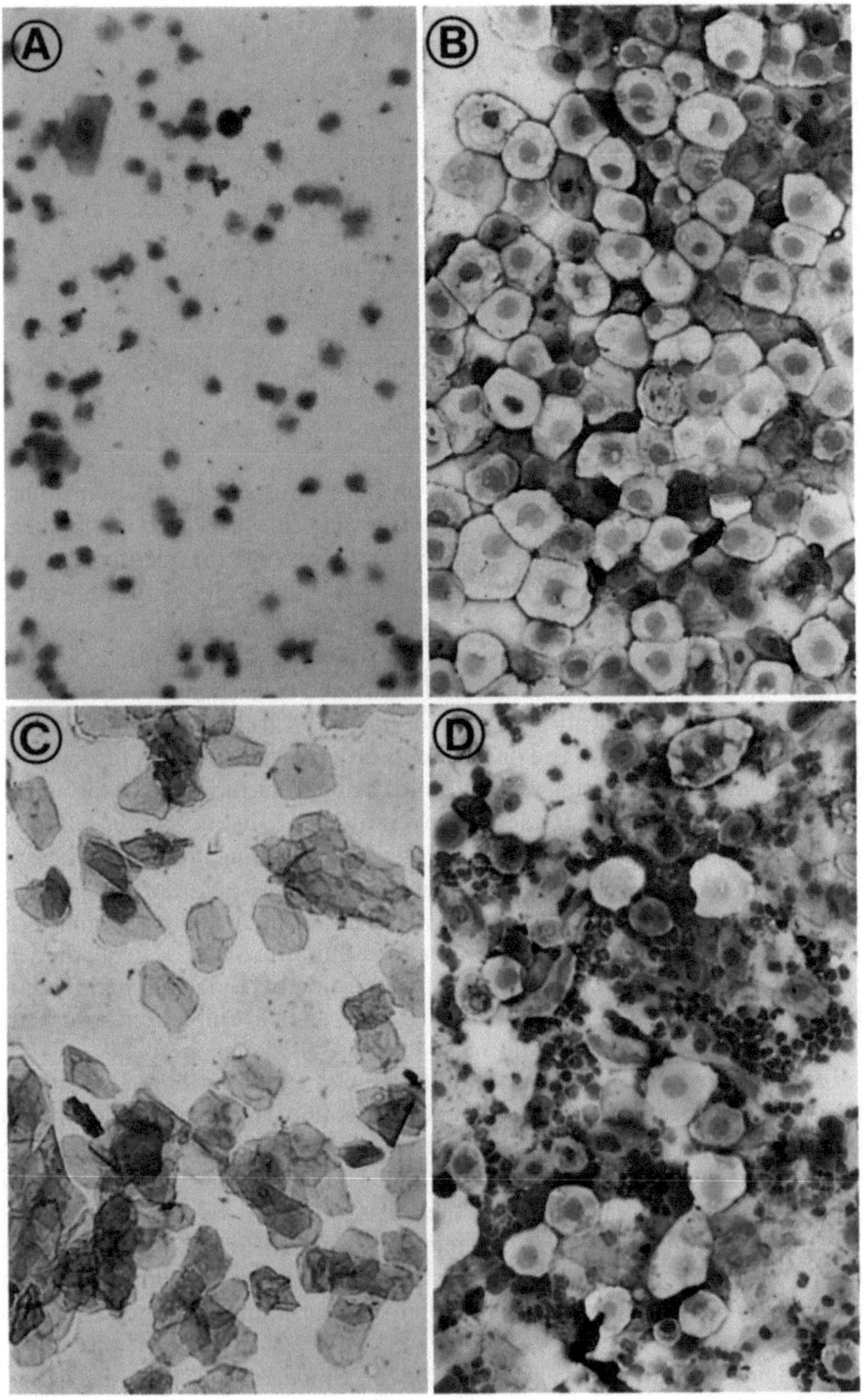
A
B
C
D

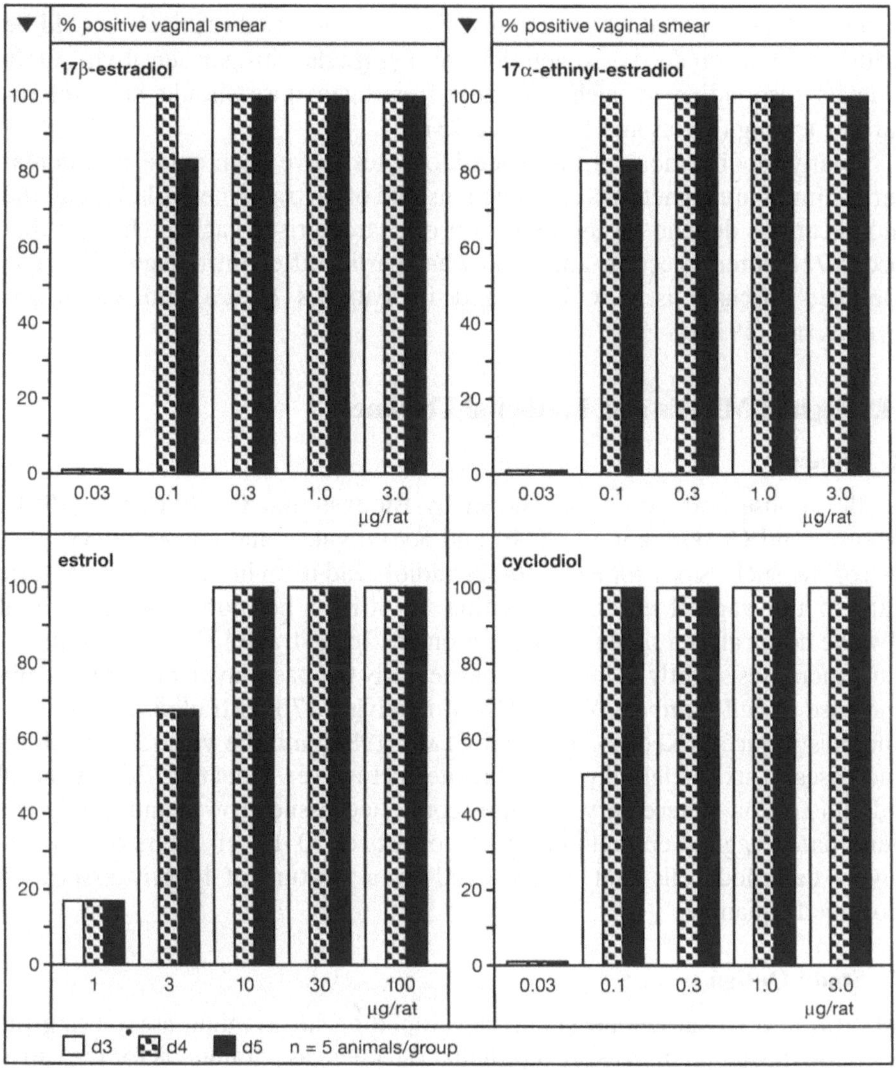

Fig. 13. Two synthetic and two natural estrogenic compounds are tested in the Allen-Doisy assay. Animals not responding are scored 0 (negative); animals that are in prooestrus, oestrus and metoestrus are scored 1 (positive). A 100% positive vaginal response indicates that all animals used showed a positive score

◄

Fig. 12A–D. Allen-Doisy assay: Vaginal smears with the typical changes during the oestrus cycle in the rat. **A** Dioestrus with leucocytes exclusively. **B** Prooestrus with nucleated squamous epithelial cells from the wall of the vagina. **C** Oestrus with cornified large squamous epithelial cells. **D** Metoestrus. A few cornified squamous epithelial cells are seen between leucocytes, mucin and nucleated epithelial cells animals may be employed every 10 days, depending on the duration of action of the estrogens tested. Ovariectomised animals not used routinely should be primed with a standard estrogen dose to determine estrogen sensitivity. Only animals responding to such a standard estrogen dose should be used for further testing (JONES and EDGREN 1973).

Many modifications of the Allen-Doisy test have been made, and immature animals are sometimes employed instead of castrates; nevertheless, all the modifications depend on the induction of characteristic vaginal changes that occur 72 h after injection and somewhat earlier after intra-vaginal dosage. Detailed discussions may be found in EMMENS (1962) and PEDERSEN-BJERGAARD (1939).

II. Vaginal Mitosis and Epithelial Thickness

1. Principle

In the mouse and rat, it was shown by BIGGERS and CLARINGBOLD (1954), MARTIN and CLARINGBOLD (1960b) and KRONENBERG and CLARK (1985), that estrogens, such as oestrone or 17β-oestradiol, lead to an increase in the mitotic rate of the vaginal epithelium within 16–36 h of hormone administration. Twelve hours after a single (s.c.) injection of 17β-oestradiol, the vaginal epithelium increases slightly in thickness, particularly the basal layer, but without any increase in cell number. Within 24 h of receiving 17β-oestradiol, keratin synthesis is induced (KRONENBERG and CLARK 1985) and the vaginal epithelium increases to six to eight layers in thickness. However, repeated 17β-oestradiol administration is necessary to cause continued tissue growth and a fully differentiated vaginal epithelium, which consists of 10–12 cell layers of progressively flattened cells that are covered by a stratum of loosely associated cornified squamae.

2. Study Design

MARTIN and CLARINGBOLD (1958) developed an intravaginal assay using the increase in vaginal mitosis and epithelial thickness as parameters for the estrogenic response. In this assay, ovariectomised albino mice are primed with a single (s.c.) injection of oestrone in peanut oil (or other oily vehicle) 2 weeks before use. Estrogens are administered in one intravaginal injection of 0.01 ml distilled water. Alternatively, estrogenic compounds may also be administered once subcutaneously or orally. Seven hours before the animals are killed, 0.1 mg colchicine in 0.05 ml distilled water are injected to arrest mitosis. Twenty-four hours after estrogen administration, the vagina is removed, fixed and processed for histological examination.

3. Evaluation and Discussion

Under the microscope, the total number of arrested mitoses are counted in five fields for each animal; the final score is the sum of five counts. For oestrone, the maximum response is elicited by 8×10^{-5} mg, and the response is linearly related to log dose in the range 0.5×10^{-5} to 8×10^{-5} mg oestrone (Fig. 14).

Martin and Claringbold (1960b) tested three natural and three synthetic estrogens in this assay (see Table 8). The relative potency of 17β-oestradiol in this assay was 1.86. Biggers (1951) found that intravaginal application of 17β-oestradiol had a potency of 1.75 times that of oestrone using cornification (Allen-Doisy assay) as the criterion of response. The fact that the relative potencies were similar in both bioassays is not surprising because vaginal

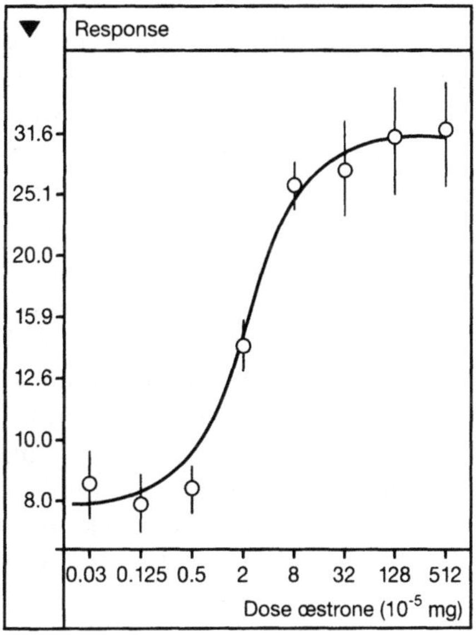

Fig. 14. The relationship of response (mean number of arrested mitoses per field) to log dose oestrone (10^{-5} mg), 24 h after administration of the hormone. The *vertical lines* indicate the SEM according to Martin and Claringbold (1960b)

Table 8. Relative activity of various estrogens, given once intravaginally on the mitotic count and epithelial thickness using oestrone as a standard (after Martin and Claringbold 1960b)

Estrogen	Test	MED ratio	Limits of error ($P = 0.95$)
17β-Oestradiol	Mitosis	1.86	1.17–2.96
Oestriol	Mitosis	9.99	0.72–1.36
Dienoestrol	Mitosis	0.89	0.64–1.24
Hexostrol	Mitosis	1.02	0.74–1.41
Diethylstilboestrol	Mitosis	1.08	0.78–1.49
17β-Oestradiol	Epithelial thickness	0.99	0.68–1.45
Oestriol	Epithelial thickness	0.69	0.45–1.05
Dienestrol	Epithelial thickness	0.57	0.38–0.86
Hexestrol	Epithelial thickness	0.68	0.47–0.98
Diethylstilboestrol	Epithelial thickness	1.12	0.75–1.38

mitosis and cornification represent interrelated cellular responses. It was claimed, however, that the mitotic index endpoint was ten times as sensitive and four times as precise as the cornification endpoint (MARTIN and CLARINGBOLD 1958). In addition, the mitotic index assay was of shorter duration than the cornification assay.

The measurement of proliferation requires excision of the vagina, and histological preparation and examination, whereas the cornification assay (the Allen-Doisy assay) requires only microscope readings of vaginal smears. The mitotic index assay also requires killing the test animals and more labor and technical skill than the cornification assay, for which animals can be used repeatedly.

In the Allen-Doisy assay, inhibition of responses to estrogens is seen with high doses of androgens or progesterone, with cortisone and hydrocortisone (SZEGO and ROBERTS 1953) and with various steroids related to 19-nortestosterone. Various tests were, therefore, performed to analyse possible inhibitory effects of these compounds in the present assay (EMMENS et al 1960). None except the synthetic 19-nortestosterone derivates had any effect. Up to 1 mg of progesterone, testosterone or hydrocortisone were tested subcutaneously and intravaginally as were up to 1 μg of 19-nortestosterone-derivatives. No inhibition was seen in any test; the higher doses of most 19-norsteroids caused mitotic increases instead. This is in line with the finding that some of the 19-nortestosterone derivatives are estrogenic in moderate doses, although they will inhibit at lower doses in the vaginal smear test. It may be concluded that, in view of the high sensitivity and specificity and the lack of interference with other steroids, the vaginal mitosis and epithelial thickness assays possess certain advantages over the conventional Allen-Doisy test.

III. Vaginal Tetrazolium Reduction

1. Principle

The reduction of 2,3,5-triphenyltetrazolium chloride to formazan in the vaginal epithelium of the mouse and rat was shown to increase linearly after a single dose of intravaginal oestrone and therefore provides a suitable basis for a highly sensitive assay to characterise estrogens (MARTIN 1960a; MARTIN 1964; SCHULTZE 1964). 2,3,5-Triphenyltetrazolium chloride is a pale yellow, non-toxic, water-soluble compound that is reduced by reduced nicotinamide adenine dinucleotide (NADH) or reduced nicotinamide adenine dinucleotide phosphate (NADPH) in vaginal epithelial cells to form a stable, water-insoluble, deep red pigment that is precipitated at the site of reaction. The reduction product, formazan, is soluble in various organic solvents and can be extracted from tissues and estimated colorimetrically.

2. Study Design

Adult albino mice are ovariectomised and treated 10–14 days later with a single intravaginal injection of estrogen. Single s.c. or oral applications can also be used. Twenty-four hours later, the animals are killed. Thirty minutes before being killed, mice are given 0.5 mg tetrazolium in 0.02 ml distilled water intravaginally. The vaginae are dissected out, cut open, washed in distilled water to remove any excess tetrazolium, dried on filter paper and placed in 1 ml of 3:1 ethanol-tetrachlorethylene (JARDETZKY and GLICK 1956). The amount of formazan is then estimated colorimetrically.

3. Evaluation and Discussion

A quite intensive survey of the tetrazolium technique has been published by MARTIN (1960a, 1964). 17β-oestradiol and DES are 2.8 times more potent than oestrone following intravaginal administration. Testosterone, progesterone and cortisol do not increase tetrazolium reduction at doses 1000 times that of oestrone, indicating specificity of this endpoint. In most respects, there is agreement between cell growth and the enhanced reduction of tetrazolium by the vaginal epithelium 24 h after administration. The tetrazolium reduction assay is comparable in sensitivity and accuracy to assays examining mitosis and epithelial thickness, but has the added advantage of simplicity. Nevertheless, this assay has the disadvantage that animals must be killed and the vagina excised.

IV. Vaginal Opening

1. Principle

The vagina of the immature rat is completely closed, the external third being a solid cord of cells. The opening of the vaginal orifice and the appearance of the first signs of oestrus may therefore be used to define sexual maturity. Precocious vaginal opening may be induced by the administration of various estrogens, but this endpoint has not been employed frequently for quantitative assays (EMMENS 1962). From the results of LAUSON et al. (1939) and EDGREN et al. (1966) it can nevertheless be seen that vaginal opening may be a sensitive and accurate index of estrogenic activity.

2. Study Design

EDGREN et al. (1966) noted that 50% of the vaginae of Charles River rats are open by the 40th day of life. These scientists further noted that the dose of estrogen required to produce 100% vaginal opening by d30 is a satisfactory criterion for assessing estrogenic activity. Therefore, subcutaneous injections of different doses of estrogens start on d25 and continued for 5 days. Alternatively, compounds can be administered by a local injection around the region of the future vaginal opening. Animals are checked for vaginal opening daily up to d30.

3. Evaluation and Discussion

EDGREN et al. (1966) compared the effects of 17β-oestradiol and 18-homooestradiol on vaginal opening. He found that daily doses of 0.03 μg

17β-oestradiol/rat were effective in causing vaginal opening on d30, whereas 0.3 μg 18-homooestradiol/rat were required to achieve the same effect. Thus, the 18-homo analogue of E2 exhibited only 10% of the potency displayed by 17β-oestradiol in this assay. This test has similar or slightly better accuracy and sensitivity than the Allen-Doisy assay. A further advantage of the method is that animals do not need to be killed at the end of the test.

V. Vaginotrophic Response

1. Principle

In addition to cornification and stimulation of vaginal epidermal growth by estrogens, an increase in vaginal weight can also be used as a parameter for measuring estrogenicity of a given compound in mice and rats. The increase in vaginal wet and dry weight in response to estrogens, however, is not frequently used as a quantitative endpoint. This may be due to the fact that it is usually more convenient to excise and weigh the uterus or the entire reproductive tract than it is to excise only the vagina, e.g. in connection with the Allen-Doisy test.

2. Study Design

Immature female mice and rats, 3–4 weeks old, or adult ovariectomised mice and rats, at 10–14 days following ovariectomy, are used in this assay. Subcutaneous injections of estrogenic compounds, dissolved in ethanol/arachis oil (alternatively, sesame, olive or other oils can also be used) [1 + 9 v/v] or benzylbenzoate/castor oil [1 + 4 v/v] or other vehicles, are given daily on three consecutive days (BULBRING and BURN 1935; FOLMAN and POPE 1966). The animals are killed and the vagina and uteri excised 24–48 h following the last injection. The relative wet and dry weight of the vagina and uterus in mg is calculated per 100 g body weight (rat) or per 10 g body weight (mouse) at death.

3. Evaluation and Discussion

FOLMAN and POPE (1966) investigated the action of potent estrogens, such as oestrone, oestriol, 17β-oestradiol and DES in the immature mouse. Whereas the vaginal weight of untreated control mice was, on average, 43% greater than the uterine weight, the maximum values obtained for either organ with estrogenic stimulation was 45 mg. For total oestrone doses (3-day dose) of 0.025, 0.05 and 0.1 μg, vaginal growth relative to controls was 78%, 158% and 262%. DES and 17β-oestradiol produced similar effects. In contrast, the short-acting oestriol produced 70% and 161% growth of the vagina at doses of 2 μg and 40 μg, indicating a weak potency relative to the other reference estrogens.

Our own data in the immature rat with different estrogenic partial agonists/antagonists are presented in Fig. 15. The sensitivity of the present assay is comparable with that of the Allen-Doisy assay. Because the animals are

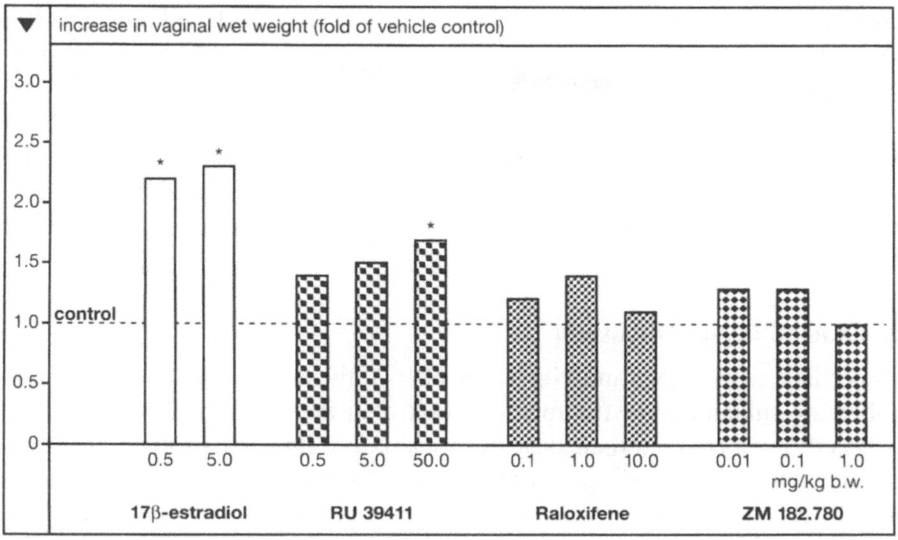

Fig. 15. Vaginotrophic response of pure and mixed antiestrogens in ovariectomised rats 1 day after a 3-day (s.c.) treatment phase. *$P < 0.05$ (ANOVA). Each column represents the mean of seven animals

killed at the end of the experiment, this assay should always be combined with the determination of uterine weights (see Sect. C. VIII).

VI. Measurement of Sialic Acid

1. Principle

The sialic acid content of the vagina is known to be correlated with mucus gland development of the vaginal epithelium (CARLBURG 1966). Its existence has been histochemically demonstrated in the mouse vagina using neuraminidase (WARREN and SPICER 1961). In intact animals, the sialic acid concentration is significantly higher in dioestrus than oestrus. In ovariectomised animals, sex steroids have characteristic effects on the sialic acid content. In general, estrogens decrease vaginal sialic acid concentrations in a dose-dependent fashion, whereas progesterone, given alone, has no effect. An increase in sialic acid concentration is induced only by combined estrogen-progesterone treatment. A detailed overview is given by NISHINO and NEUMANN (1974). The measurement of sialic acid content in the mouse vagina can be used as a quantitative parameter for testing the estrogenic potential of a given compound.

2. Study Design

Female NMRI-strain mice weighing about 30 g are generally used for the experiments. Animals are ovariectomised prior to steroid administration.

10–14 days following ovariectomy, subcutaneous or oral treatment with estrogens is initiated and continued for 3 days. Animals are sacrificed one day after the last treatment. Control animals are treated with vehicle (oil control) only. The vagina is immediately excised and vaginal wet weight determined. Afterwards, the vagina is put into a small tube in an appropriate volume of 0.1 N H_2SO_4 solution and hydrolysed at 80°C for 1 h, according to Svennerholm (1958). The sialic acid content in the hydrolysate is determined by the resorcinol method of Svennerholm (1957) and its concentration is expressed in terms of the wet weight.

3. Evaluation and Discussion

According to Nishino and Neumann (1974), the vaginal sialic acid content following subcutaneous treatment of castrated mice with 17β-oestradiol for 3 days shows a characteristic dose–response curve (Fig. 16). In summary, treatment with a very small dose of 0.003 μg 17β-oestradiol/animal produces a significant increase in the sialic acid concentration. With elevation of the dose, the sialic acid concentration reduces to a minimal level. At a 0.3 μg dose, no further decrease is observed.

Similar dose–response curves are obtained with other natural and synthetic estrogens, such as oestrone, oestriol, 17α-EE, mestranol, DES and conjugated estrogens (Nishino and Neumann 1974). Based on such studies, potency of these compounds, with most potent first, is as follows (Table 9): 17α-ethinyl-oestradiol, 17β-oestradiol, oestriol, oestrone, diethylstilboestrol

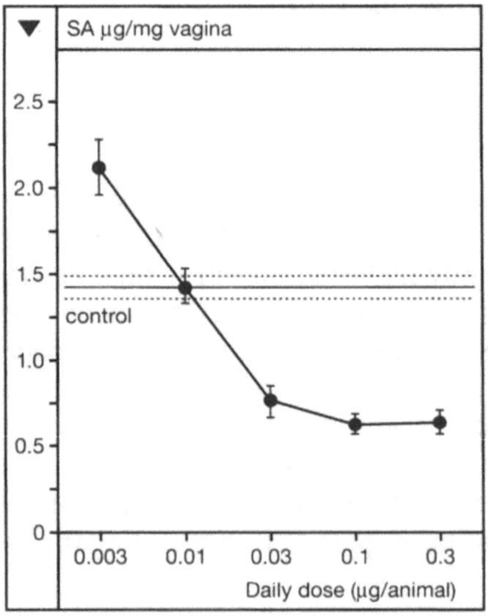

Fig. 16. The change in the vaginal sialic acid (*SA*) concentration after the subcutaneous treatment of castrated mice with 17β-oestradiol according to Nishino and Neumann (1974)

Table 9. Relative potencies of estrogenic compounds regarding the effect on the mouse vaginal and uterine SA concentration and organ weight (from NISHINO and NEUMANN 1974). Dose-response curves were not parallel

Estrogenic compounds	Relative potency		$(E_2=1.0)$	
	Vagina SA	Weight	Uterus SA	Weight
Estrone	0.16	0.09	0.14	0.09
17β-Oestradiol	1.0	1.0	1.0	1.0
Estriol	0.19	0.12	0.05	0.03
Ethinylestradiol	1.53	2.00	3.65	2.85
Mestranol	0.07	*	0.03	0.02
DES	0.08	0.12	*	0.26
AY-11483	0.04	0.03	0.03	0.03
Conjugated estrogens	0.38×10^{-3}	0.43×10^{-3}	0.12×10^{-3}	0.13×10^{-3}

(DES), mestranol, conjugated estrogens. Estimation of the mouse vaginal sialic acid content is of similar sensitivity as the Allen-Doisy assay and may be useful for identifying the estrogenic potential of compounds. The assay has the disadvantage that the animals have to be sacrificed and the vagina excised.

Progesterone and several synthetic progestogens oppose the action of estrogen on sialic acid content. They cause a dose-dependent increase in the vaginal sialic acid concentration in castrated animals, when injected simultaneously with 17β-oestradiol (NISHINO and NEUMANN 1974). Because the increase in vaginal sialic acid concentration is always proportional to the progestogen dose, it can also be used as a quantitative parameter for evaluating the biological activity of progestogens.

VII. Uterine Fluid Inhibition (Astwood Bioassay)

1. Principle

During the first few hours following a systemic injection of estrogen, the uterus of the immature rodent or ovariectomised adult rodent undergoes a rapid increase in wet weight. This increase in uterine weight is almost entirely due to an accumulation of water. Based on the quantitative aspects of this response, ASTWOOD (1938) devised a simple and accurate estrogen bioassay in the immature female rat.

2. Study Design

Immature female rats, 21–23 days old, weighing between 25 g and 50 g are used for the experiments. They are given a single subcutaneous injection of an estrogenic compound (17β-oestradiol) in 0.1 ml vehicle (sesame oil). The animals are killed at intervals thereafter, e.g. 0 h, 2 h, 4 h, and 6 h after the start of the treatment. Uteri are excised, blotted on absorbent paper and wet weight is quickly determined. Water determinations are made by dessicating the uteri at 60°C for 4h. Afterwards, uterine dry weights are determined. All uterine weights (wet and dry) are corrected for a body weight of 100g and defined as relative wet and dry weights.

3. Evaluation and Discussion

Characteristic changes in percentage of water, total (i.e. wet) weight and dry weight of immature rat uteri within 48h of a single 0.1 μg subcutaneous injection of 17β-oestradiol is shown in Fig. 17. During the first hour no effect is seen, but by the second hour a slight increase in uterine wet weight is observed. Thereafter, uterine wet weight increases rapidly, reaching a maximum at 6h. By 12h, one-fourth of the weight gain is lost, but during the next 18h, there is a second increase in weight that is maximal by 30h. The uterus then returns to a basal state.

During the early stages of the 17β-oestradiol response, the uteri are swollen, oedematous, and translucent, the vascularisation having somewhat increased. Water determinations confirm that the 6-h increase in weight is almost entirely due to increased water content, and that the fall at 12h is due to water loss. Moreover, the second weight increase is not accompanied by water uptake but, rather, represents a true growth response (ASTWOOD 1938). ASTWOOD (1938) also showed that a direct proportionality exists between uterine wet weight at 6h and the dose of 17β-oestradiol. Therefore, this bio-

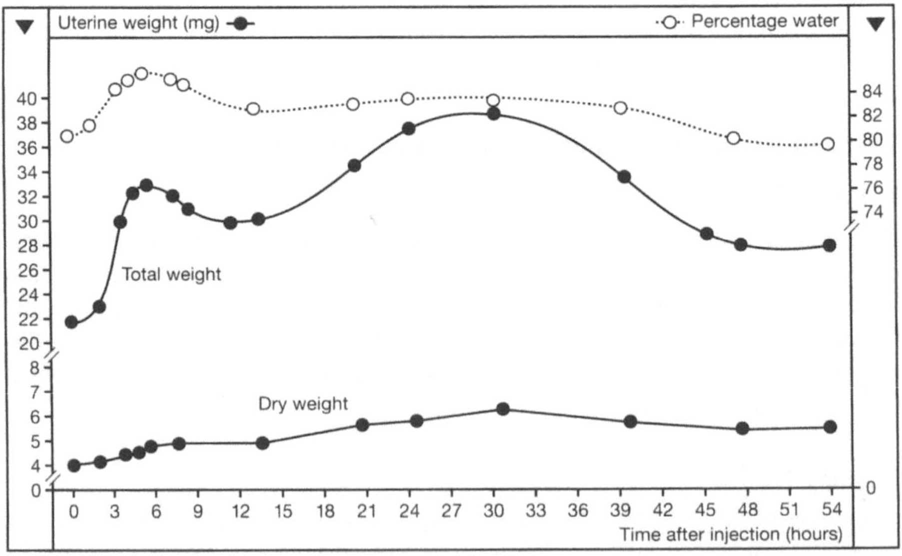

Fig. 17. Changes in amount of water (%), total weight, and dry weight of immature rat uteri over 48h after a single (s.c.) injection of 0.1 μg 17β-oestradiol (from ASTWOOD 1938)

assay is conducted by determining the minimal quantity of estrogen that gives a maximal uterine weight increase at 6h. In the hands of its author, the Astwood test included 0.006–0.1 μg doses of 17β-oestradiol, and 0.07–1 μg doses of oestrone (which exhibited 1/12 of the potency of 17β-oestradiol).

Hisaw (1959) compared the effectiveness of five natural estrogens (17β-oestradiol, oestrone, oestriol, equilin, and equilenin) and diethylstilboestrol (DES) for promoting fluid uptake and growth of the immature rat uterus. Some estrogens were more potent with respect to fluid inbibition and others more potent with respect to growth stimulation. The minimal dose in micrograms required to produce a 33% increase in uterine wet weight at 6h (Astwood unit) was for: 17β-oestradiol, 0.025; oestriol, 0.029; DES, 0.078; equilin, 0.312; oestrone, 0.45; and equilenin, 0.546. Conversely, the microgram dose required to produce a 70% gain in uterine wet weight at 6h were for: oestriol, 0.078; 17β-oestradiol, 0.1; DES, 0.156; equilenin, 0.625; oestrone, 1.25; and equilin, 1.25.

It is further noted by Hisaw (1959) that uterine wet weights obtained at 4h for high doses were greater than those obtained at 6h for the same dose. Thus, the time of maximal fluid inbibition occurred earlier for larger doses of estrogen than for smaller doses. These observations by Hisaw (1959) indicate that, to adequately evaluate the fluid response, it is necessary to compare dose–response curves at both 4h and 6h.

The 6-h fluid inbibition assay has the disadvantage that the animals have to be killed at the end of the experiment. In addition, it is impossible to distinguish between full- or long-acting estrogens, partial- or short-acting agonists. Moreover, the assay suffers from differences in the release of test compounds from the injection site, which affect the time course of the response.

VIII. Uterine Growth Test

1. Principle

The increase in uterine weight of immature or young ovariectomised rats or mice has been used by many workers to assess estrogenic activity. Short-acting estrogens, such as oestriol, dimethyl-stilboestrol (DMS) or 17α-oestradiol, stimulate early uterotrophic responses, while having little effect on true uterine growth. This pattern of activity was first observed by Hisaw (1959). Early uterotrophic responses are usually made by simply measuring the 4–6h gain in wet weight of the uteri following a single treatment. This increase in wet weight is mainly induced by water.

Most assays utilise 3–4 days of daily hormone administration to stimulate true uterine growth and to help minimise temporal differences in absorption of test compounds. Classical long-acting estrogens, such as oestrone, 17β-oestradiol and diethylstilboestrol (DES), lead to a clear increase in uterine weight after 3 days of treatment. Short-acting estrogens also respond after a 3-day treatment interval but will not produce similar dose–response curves.

Conversely, if short-acting estrogens are administered at more frequent intervals or given as a subcutaneous implant, full uterine growth is stimulated (CLARK and MANI 1991). Historically, BULBRING and BURN (1935) employed ovariectomised immature rats, DORFMAN et al. (1936); LAUSON et al. (1939) intact immature rats, DORFMAN and KINCL (1966) intact immature mice and EDGREN et al. (1966) and JONES and EDGREN (1973) ovariectomised pubertal and adult rats and immature mice to conduct the uterine growth assay.

2. Study Design

Immature female rats, 19–21 days old, weighing 35–50 g, are treated subcutaneously or orally once per day for three consecutive days with estrogens or vehicle alone. The estrogenic compounds are dissolved in a vehicle consisting of 10% ethanol in arachis oil (or other oils), a mixture of benzylbenzoate/castor oil (1:4 v/v), or other suitable vehicles. One day later (the fourth day), the animals are weighed and euthanised by carbon dioxide asphyxiation. The uteri are excised to remove intrauterine fluid and weighed. By this day, mean body weights range from 50 g to 60 g and mean uterine wet weights range from16 mg to 26 mg for the vehicle control group. Depending on the dose and potency of the compound administered, uterine wet weights may increase to a maximum of 110 mg. After determination of wet weight, uteri are dried for 4 h at 60°C and dry weights are determined. Afterwards, relative wet and dry weights of the uteri (in milligrams) are calculated per 100 g body weight.

Ovariectomised adult animals can also be used for the uterine growth assay. Before the start of the experiment, the ovariectomised animals should be given a rest of 10–14 days to decrease their endogenous sex steroid level.

3. Evaluation and Discussion

Characteristic results obtained with 17β-oestradiol, 17α-oestradiol, oestriol, equilin and cyclofenil are shown in Fig. 18. The effect of antiestrogens or mixed agonists/antagonists is shown in Fig. 22B. 17β-oestradiol exhibits potent estrogenic activity following subcutaneous application. 17α-oestradiol is substantially less potent (by a factor of 100) than 17β-oestradiol. The short-acting estrogen, oestriol, shows a shallow dose-response curve and is 10 times less potent than the reference estrogen, 17β-oestradiol. From the estrogenic partial agonists/antagonists tested (RU 39,411 and raloxifen), only RU 39,411 shows a clear estrogenic response; raloxifen is nearly inactive.

Although the uterotrophic response in the rat and mouse is commonly accepted as a sensitive measure of estrogenic activity, it suffers from a lack of specificity. For example, uterine weight can be increased by androgens, progesterone and various synthetic progestagens (JONES and EDGREN 1973).

In conclusion, the uterine growth bioassay, the vaginotrophic response and the Allen-Doisy assay remain the three "golden standards" for assessing estrogenic activity. The differential sensitivity of the vagina and uterus to estrogens

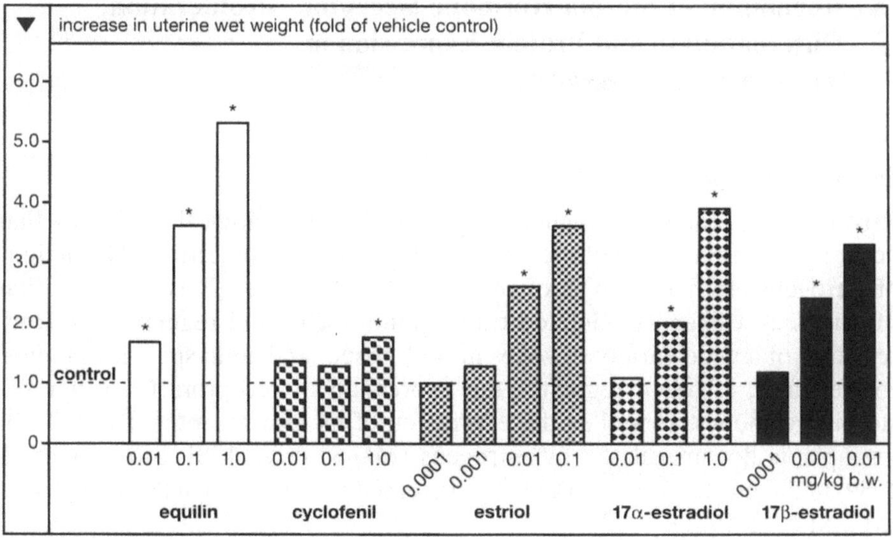

Fig. 18. Different estrogenic compounds tested in the uterine growth assay of the immature rat 1 day after a 3-day (s,c.) treatment phase. $*P < 0.05$ (analysis of variance, ANOVA). Each *column* represents the mean of seven animals

Table 10. The ratio of the vaginotrophic and uterotrophic potencies of estrogens as a dissociation index (from NISHINO and NEUMANN 1974)

Estrogenic compound	Dissociation index SA vagina:SA uterus	($E_2 = 1.0$) Vagina weight:uterus weight
E_1	1.13	0.94
E_2	1.0	1.0
E_3	3.34	3.61
Ethinyloestradiol	0.42	0.70
Mestranol	2.59	–
DES	–	0.48
AY-11483	1.22	0.94
Conjugated estrogens	3.20	3.20

can be used as a model for evaluating the dissociated effects of estrogens (NISHINO and NEUMANN 1974). NISHINO and NEUMANN (1974) compared the vaginotrophic and uterotrophic potencies of different estrogenic compounds (oestrone, 17β-oestradiol, oestriol, 17α-ethinyl oestradiol, diethylstilboestrol and conjugated estrogens). The ratio of vaginotrophic and uterotrophic potencies clearly shows (Table 10) that oestriol and conjugated estrogens have a relatively high value, meaning they are "dissociated" compounds. On the contrary, 17α-ethinyl oestradiol and diethylstilboestrol have low values, indicating that they are more effective on the uterus than on the vagina.

IX. Regulation of Steroid Hormone Receptors, Proliferation, Differentiation and Protein Expression in the Rodent Uterus

1. Principle

Steroid Hormone Receptors

Estrogens are known to regulate many of the physiological processes that occur at molecular and cellular levels in the rodent endometrium. The actions of estrogens are mediated through the nuclear ER. The ER is a protein that functions as a ligand-modulated transcription factor and regulates the transcription of a wide variety of genes in both tissue- and gene-specific manners. Interestingly, estrogen regulates the expression of receptors for other sex steroid hormones, as well as the expression of its own receptor. The ER, for example, is downregulated by estrogens (MANNI et al. 1981; SHUPNIK et al. 1989) in the uterus, whereas the PR is increased by estrogens in the same tissue (MANNI et al. 1981).

Proliferation

Immature or ovariectomised mice and rats have been used extensively as an experimental model for investigating the effects of estrogens on cellular proliferation in the endometrium. In this model, a single dose of 17β-oestradiol stimulates cell proliferation in all three tissue compartments of the uterus: the endometrial epithelium, stroma and myometrium (MARTIN et al. 1976; KIRKLAND et al. 1979). Oestradiol treatment of ovex mice causes a synchronised wave of luminal epithelial DNA synthesis, as determined by [3H]thymidine incorporation. This elevated rate begins 6h after administration and peaks after 12–15h. The [3H]thymidine incorporation wave is associated with an increased mitotic index in both the luminal and glandular epithelium at 24h, revealing that the cells had been stimulated to proliferate (CULLING and POLLARD 1988).

Protooncogenes

Recent reports indicate that estrogenic stimulation and proliferation of the rodent uterus involves regulation of expression of genes whose products control the cell cycle, such as jun and fos protooncogenes. It has been demonstrated (WEISZ and BRESCIANI 1988; WEBB et al. 1991) that 17β-oestradiol causes a rapid and pronounced increase in the uterine expression of these immediate early response genes, indicating that hormonal control of cell proliferation is linked to c-fos and c-jun expression. Both genes contain functional response elements capable of binding ER. The protein products of c-fos and c-jun form dimers, Jun-Jun or Jun-Fos, which act as the transcriptional regulator, AP-1, on a variety of genes (ANGEL and KARIN 1991).

Luminal Epithelial Cell Height

One of the most sensitive cell types in the rat uterus seems to be the luminal epithelial cells of the endometrium. 17β-oestradiol causes a nucleolar enlargement within 6h and, within 12h, the cytoplasmic polyribosomes and the rough endoplasmic reticulum are increased as a sign of a drastic increase in protein synthesis. After 24h, the microvilli of the apical surface appear taller and an increased number of golgi complexes, lyposomes and rough endoplasmic reticulum are found. Such a hypertrophic picture, with supranuclear granules and vacuoles together with an increase in the number of cells (i.e. hyperplasia) is characteristic for estrogen action (LJUNGKVIST 1971; HOSIE and MURPHY 1995).

Uterine Glycogen Deposition

Among the earliest and most pronounced responses of the immature rat uterus to estrogen are increased uptake, phosphorylation and utilisation of glucose, and enhanced glycogen synthesis (BITMAN et al. 1965; SMITH 1967). Glycogen accumulation is significantly increased at 6h, reaches a maximum at 12–24h, and declines to near control levels by 48h (GALAND et al. 1987).

Peroxidase Activity

Peroxidase is an enzyme that catalyses the metabolism and binding of [^{14}C]oestradiol to protein and other high-molecular-weight substances in the presence of H_2O_2. It was shown that the enzyme is absent from the uteri of immature rats and is induced by physiological doses of estrogen (LYTTLE and DESOMBRE 1977a,b). Peroxidase activity is primarily associated with increased uterine eosinophilia in the stromal and myometrial cells (BROCKELMANN 1969). In addition, it was shown that nafoxidine, an antiestrogen with partial estrogenic activity, blocks the 17β-oestradiol-stimulated induction of uterine peroxidase (MCNABB and JELLINCK 1976).

Complement Component C3

It was shown previously that complement component C3 is regulated by 17β-oestradiol in the rat uterus (SUNDSTROM et al. 1989). C3, an important component of the immune response, is a 180,000-Da-MW protein composed of two non-identical subunits (LAMBRIS 1988). In the uterus, C3 appears to be regulated at transcription and is produced and secreted in response to estrogen only in the luminal and glandular epithelial cells (SUNDSTROM et al. 1989). C3 synthesis is regulated throughout the oestrous cycle of a mature rat, beginning at prooestrus, reaching a maximum at oestrus, decreasing at metoestrus and being undetectable at dioestrus (LYTTLE et al. 1987). Moreover, the synthesis of C3 is inhibited by the co-administration of progesterone and oestradiol in the immature rat uterus.

2. Study Design

Immature mice and rats or ovariectomised adult animals are used for the experiments. The animals are treated with 17β-oestradiol or other estrogenic test compounds for up to 1 week, beginning at 20 days of age for immature animals or at 10 days following ovariectomy in adult animals. In all cases, uteri are removed at the end of treatment. 17β-oestradiol or another estrogenic compound can be administered as a single or daily subcutaneous injection in a mixture of castor oil (or other oils) and ethanol (9:1 v/v).

Steroid Hormone Receptors

Total RNA and nuclear extracts for the determination of ER and PR mRNA and proteins are prepared (Kirkland et al. 1993).

Proliferation

For the analysis of proliferation, immunohistochemical detection of e.g. BrdU (bromdesoxyuridine) or PCNA (proliferating cell nuclear antigen) can be carried out on uterine sections after fixation and paraffin-embedding, according to Rumpel et al. (1995). The number of e.g. PCNA-positive nuclei (i.e. the proliferation index) can be calculated per mm luminal and glandular epithelium. Similarly, the mitotic index (i.e. the number of mitoses per mm of epithelium) can be determined for the luminal and glandular epithelium in uterine histological sections stained with hematoxylin and eosin.

Protooncogenes

The determination of protooncogenes can be performed at the protein level, e.g. immunohistochemically on uterine sections with a specific antibody against c-fos and c-jun (Boettger-Tong et al. 1995), or on the mRNA level by "northern blot" analysis that follows the isolation of total RNA from whole uteri after homogenisation (Bigsby and Li 1994) or by the in situ hybridisation technique (Nephew et al. 1996).

Luminal Epithelial Cell Height

For histological evaluation, the uteri are removed at the end of the hormone treatment and placed in neutral buffered 3.7% formaldehyde solution for a minimum of 24 h. The uteri are then embedded in paraffin, cut into 4-μm transverse sections, and stained with hematoxylin and eosin. The uterine sections are then evaluated for luminal epithelium cell height, e.g. according to Branham and Sheehan (1995). In each experiment, a negative (vehicle) and positive (0.3 μg 17β-oestradiol/animal) control group should be included. The difference in epithelial cell height of positive and negative control groups is calculated and expressed as a percentage. The stimulation factor (x) of the estrogenic test-compound of interest is calculated according to the following formula:

$$x(\% \text{ of } 17\beta\text{-oestradiol}) = [\text{height (test compound)}-\text{height (vehicle)}] \times 100/[\text{height } (17\beta\text{-oestradiol})-\text{height (vehicle)}]$$

Uterine Glycogen Deposition

The uteri are quickly excised and blotted. Uteri from vehicle-treated animals serve as controls. Samples for glycogen determination (approximately 75 mg) are quickly weighed on a torsion balance, dropped into boiling 30% KOH and analysed by the anthrone method of SEIFTER et al. (1950).

Peroxidase Activity

At the appropriate time, rat uteri are dissected free of fat and connective tissue, blotted on filter paper and homogenised. In the homogenate, peroxidase activity is determined as described earlier (LYTTLE and DESOMBRE 1977a, b). Briefly, the assay mixture contained 13 mM guaiacol and 0.3 mM H_2O_2 in the extraction buffer. The reaction was started by the addition of 1.0 ml of the uterus extract. The initial rate (60 sec) of guaiacol oxidation was monitored on a Beckman spectrophotometer at 470 nm. An enzyme unit was defined as the amount of enzyme required to produce an increase of 1 absorbance unit/min under the assay conditions described. Enzyme activity is expressed per gram of tissue.

Complement Component C3

C3 can be detected in the rat uterus on the protein or mRNA levels. For detection of C3 protein, the uterus is split longitudinally and placed in a culture medium (e.g. Ham's F12 nutrient mixture) for metabolic labelling of secreted proteins. After 20 h incubation, the medium from each specimen is harvested, centrifuged to remove tissue debris and stored at –70°C until analysis by a specific immunoprecipitation technique (BIGSBY 1993). For detection of C3 mRNA, RNA is extracted from uteri, polyA is prepared and subjected to 'northern blot' analysis as described by SUNDSTROM et al. (1990).

3. Evaluation and Discussion

Steroid Hormone Receptors

It was recently found by KRAUS and KATZENELLENBOGEN (1993) that 17β-oestradiol causes rapid time- and dose-dependent increases in the levels of PR mRNA and PR protein in the rat uterus during a 6-day treatment interval on the one hand, and time- and dose-dependent decreases in the levels of ER mRNA and ER protein (Fig. 19A) on the other. This effect correlated with a steady increase in uterine weight over the treatment time. Interestingly, as little as a single 0.04 μg injection of 17β-oestradiol caused a 2.5-fold increase in PR mRNA after 1 day (Fig. 19B).

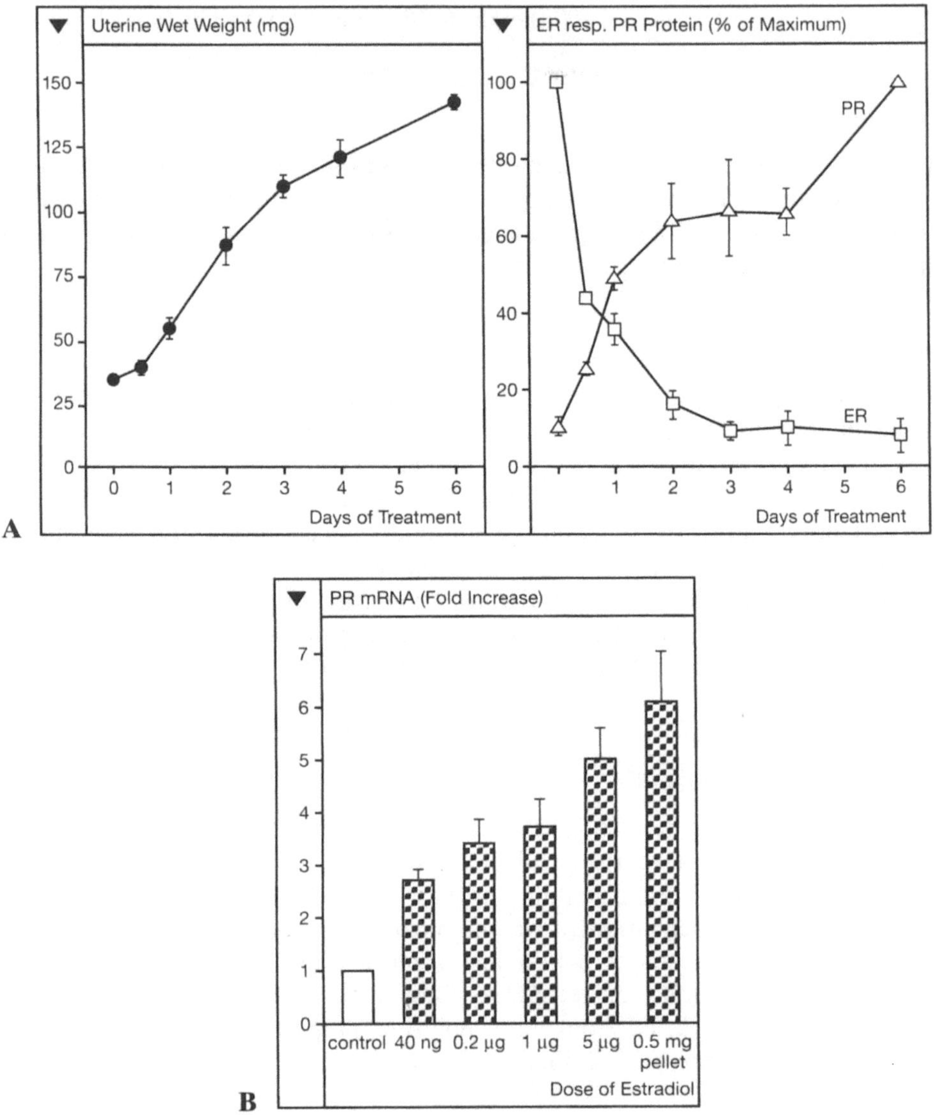

Fig. 19. A Time-dependent effect of 17β-oestradiol on uterine growth and the level of progesterone receptor (*PR*) and estrogen receptor (*ER*) protein in the immature rat uterus (from KRAUS and KATZENELLENBOGEN 1993). **B** Dose-dependent effect of 17β-oestradiol on PR mRNA expression in the immature rat uterus (from KRAUS and KATZENELLENBOGEN 1993)

Treatment with progesterone reversed stimulatory effects of 17*β*-oestradiol on both PR mRNA and PR protein (KRAUS and KATZENELLENBOGEN 1993). The antiprogestin, RU 486, acted as a potent antagonist of progesterone action, completely restoring the levels of 17*β*-oestradiol-stimulated PR mRNA and PR protein to the levels observed in the absence of progesterone (KRAUS and KATZENELLENBOGEN 1993). The partial estrogen agonist/antagonist LY 110,718 acted as both an estrogen antagonist and a partial estrogen agonist (KEEPING and LYTTLE 1982). It elicited a modest increase in PR mRNA and uterine growth when given alone, but was highly effective in preventing any further stimulation when administered in conjunction with 17*β*-oestradiol.

Proliferation

As shown in Fig. 20, cell proliferation, measured by quantitative immunohistochemistry, is very sensitive to 17*β*-oestradiol and to the estrogen agonist/antagonists RU 39,411 and raloxifene in the ovariectomised rat. After a 3-day treatment phase, the pure estrogen antagonist ZM 182,780 has no effect on the proliferation rate or mitotic index (data not shown) for the endometrial epithelium. 17*β*-oestradiol, however, significantly increases proliferation in the luminal and glandular epithelium. Interestingly, a maximal response of the proliferation rate can already be observed upon treatment with 0.1 *μ*g

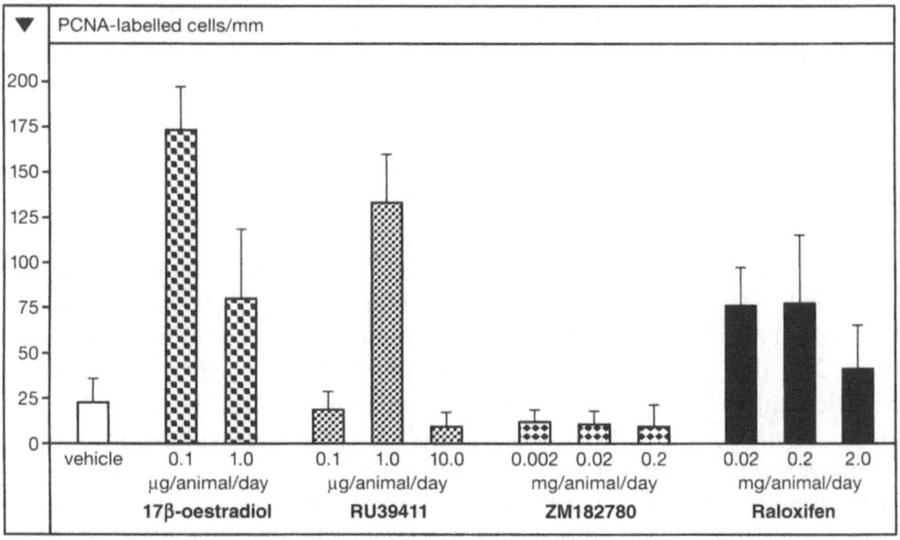

Fig. 20. Dose-dependent regulation of proliferation of luminal epithelial cells by different mixed agonists/antagonists and pure antiestrogens in the immature rat uterus 1 day after a 3-day (s.c.) treatment interval. Proliferation is measured by proliferating cell nuclear antigen (*PCNA*)-immunohistochemistry (ZK 5018: estradiol; ZK 139947: RU 39411. ZK 156901: ZM 182780. ZK 168367: Raloxifen)

17β-oestradiol, whereas the increase in uterine weights is maximal at 1 μg 17β-oestradiol. RU 39,411 stimulates a maximal uterine weight at 10 μg/animal, but a maximal proliferation response in the luminal epithelium at concentration that is 10 times lower (0.1 μg/animal). At 10 μg/animal, the proliferating potential of RU 39,411 has disappeared, probably due to the increasing antiestrogenic activity of the compound. Signs of significant estrogenic proliferation were observed for raloxifene at 0.02 mg/animal and 0.2 mg/animal doses in the luminal epithelium. This was not accompanied by any clear increase in uterine weights. Taken together, these examples clearly indicate that the luminal epithelium of the endometrium is much more sensitive to stimulation by estrogens than other cell populations within the uterus.

Protooncogenes

Studies of the immature rat uterus have shown that induction of c-fos mRNA transcripts and protein levels are increased 3 h after 17β-oestradiol treatment (LOOSE-MITCHELL et al. 1988; ETTGER-TONG et al. 1995). 17β-oestradiol induces c-fos expression in all major cell types of the immature rat uterus (Fig. 21). In addition, the results further indicate that the inductin of c-fos protein by 17β-oestradiol is most prominent in the endometrial epithelium and less in the stroma and myometrium. This pattern correlates with the degree of proliferative activity, which is highest in epithelial cells and lower in the stroma and myometrium after 17β-oestradiol treatment of immature animals. Protooncogenes suchs as c-fos and c-jun can also be induced by partial estrogen agonist/antagonists (KIRKLAND et al. 1993; NEPHEW et al. 1996). In addition, progesterone diminishes the induction of c-fos transcript levels by oestradiol (KIRKLAND et al. 1992), particularly in the endometrial epithelium. It may therefore be concluded that the luminal endometrial epithelium shows a high sensitivity towards estrogens, not only with respect to proliferation but also with respect to the expression of immediate early genes.

Luminal Epithelial Cell Height

Uteri from vehicle-treated ovariectomised rats are lined by cuboidal luminal epithelial cells. In contrast, the endometrium from 0.3 μg 17β-oestradiol/animal treated rats is characterised by a high columnar, pseudostratified epithelium (Fig. 22A) that has ~300% increase in epithelial cell height compared to a vehicle control. Raloxifene, an estrogen antagonist with a weak partial agonistic activity, only modifies luminal morphology from low columnar to focally pseudostratified. Morphometric measurements revealed an increase in epithelial cell height of ~50% (of 17β-oestradiol) at a dose of 0.1 mg/kg raloxifen. This stimulation could not be further increased by increasing the raloxifene concentration (Fig. 22A). For comparison, RU 39,411, an estrogen antagonist with pronounced agonistic activity, caused a dose-dependent increase in luminal epithelial cell height, with values of 130% and 140% (of 17β-oestradiol), respectively (Fig. 22A).

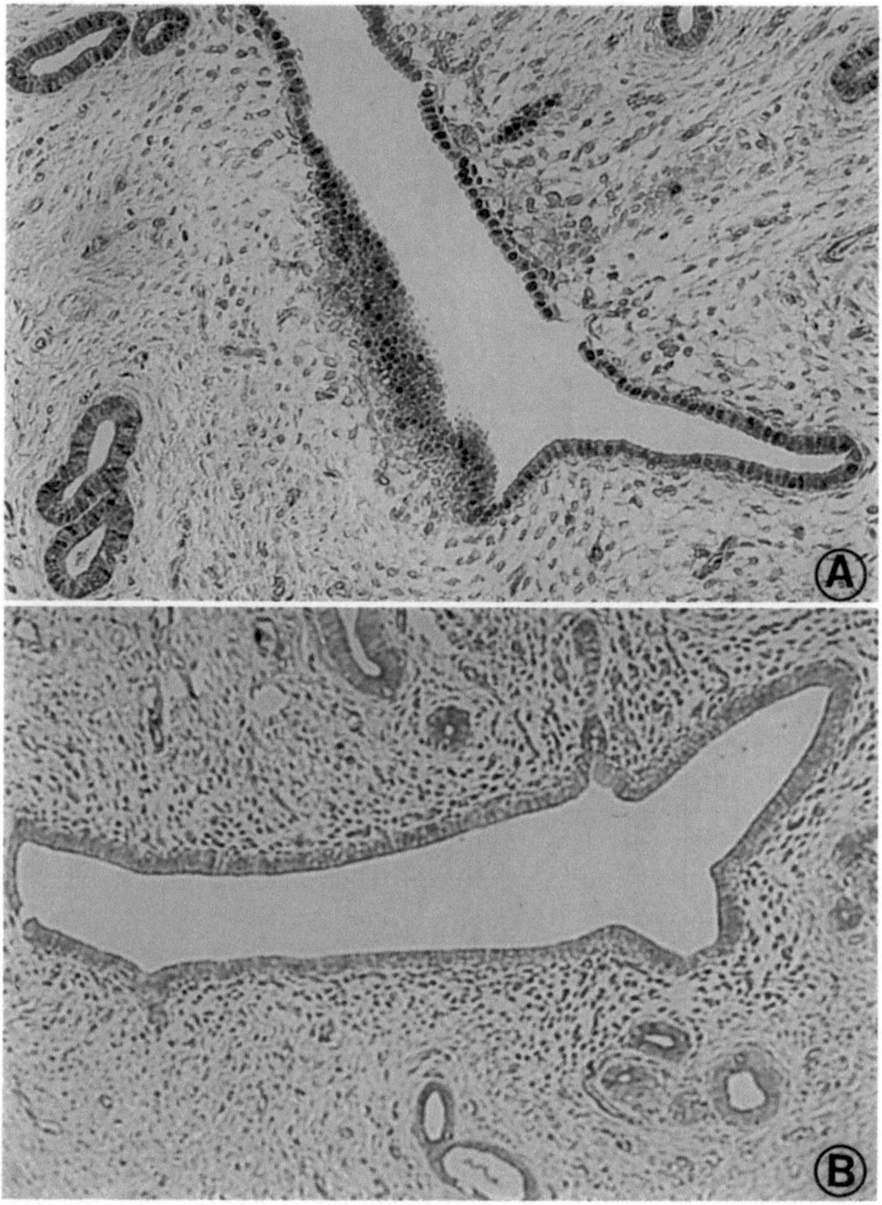

Fig. 21A,B. Immunohistochemical picture showing the protein-expression of c-fos in the uterus of ovariectomised rats, substituted with 1 μg/animal 17β-oestradiol (A) compared with vehicle-treated controls (B). Immunopositive cells are mainly confined to the nuclei of luminal and glandular epithelial cells. ×210

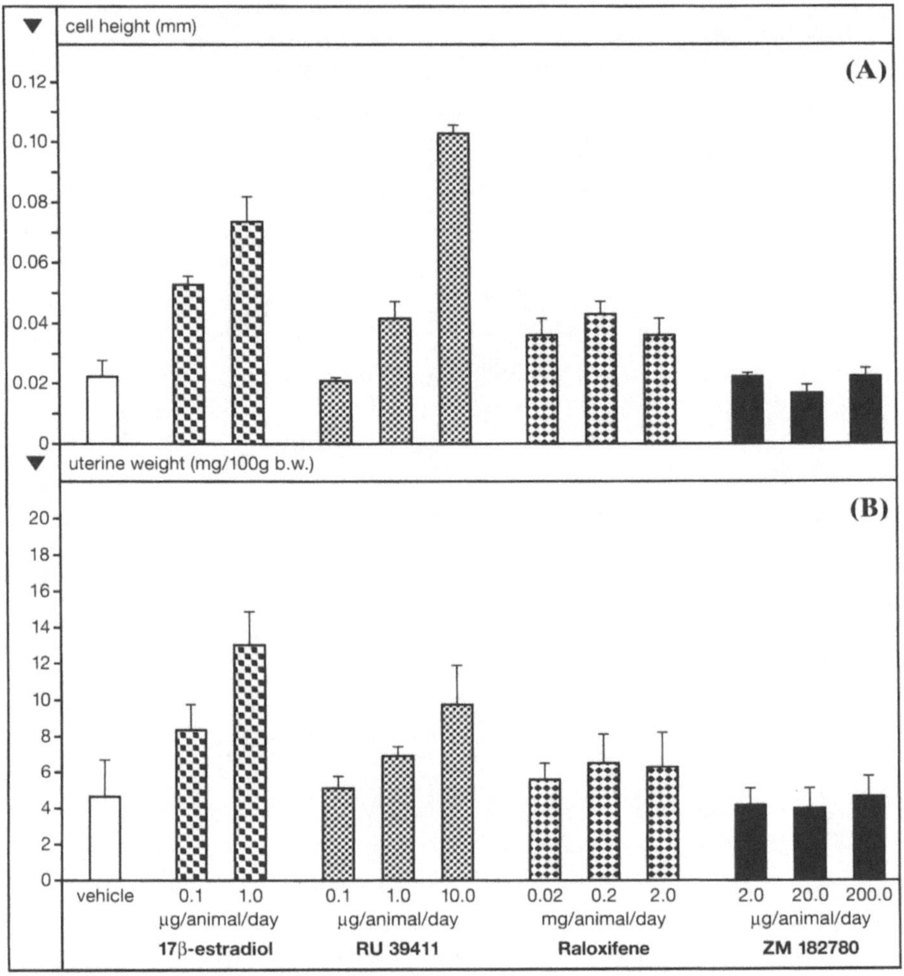

Fig. 22A,B. Epithelial cell height (**A**) and uterine dry weights (**B**) in ovariectomised rats 1 day after a 3-day (s.c.) treatment with different hormones/antihormones

Compared to uterine wet weights, for which 17β-oestradiol is known to dose-dependently increase uterine weights, weights were not useful to for distinguishing between raloxifen and RU 39,411 at 0.1–10 mg/kg (Fig. 22B). Therefore, histological analysis of the endometrial luminal epithelium is a more sensitive parameter than uterine weight for determining estrogenicity of a given compound. Similar findings were observed by others (Sato et al. 1996).

Uterine Glycogen Deposition

Bitman and Cecil (1970) used the 18-h glycogen deposition response in the immature rat uterus to evaluate the estrogenic activity of a series of diphenyl-

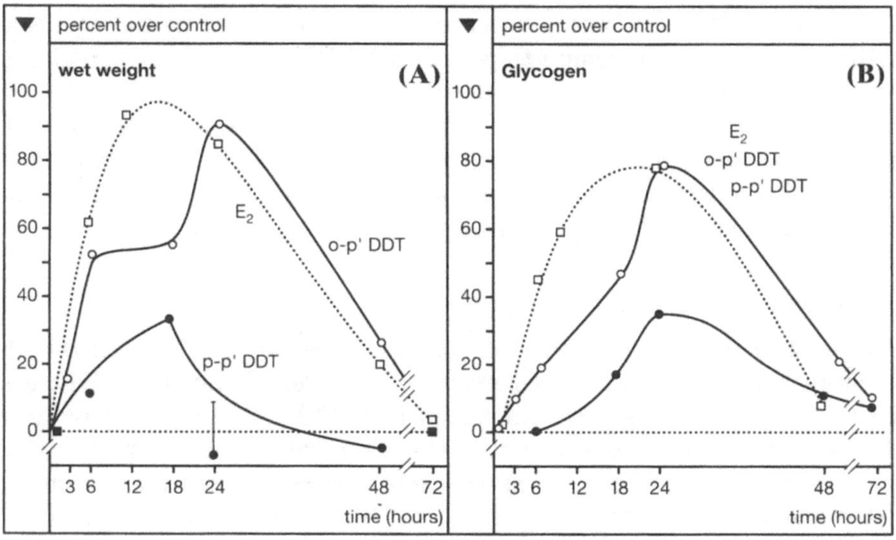

Fig. 23A,B. Time course of the effects of *o-p'*-dichlorodiphenyl-trichloroethane 2',3'-dideoxythymidine (DDT) (●) and *p-p'*-DDT (○) (12.5 mg/animal each) on uterine wet weight (*A*) and glycogen (*B*) content. The effect of 0.05 μg 17β-oestradiol/animal is shown for comparison (from GALAND et al. 1987)

methane, diphenylethane, and triphenylmethane compounds. Compounds in this series exhibited detectable estrogenic activity in this assay when the p- or p'-positions were unoccupied or occupied by a hydroxy or methoxy group. Halogen or alkyl groups in the p,p'-positions rendered the compounds inactive. Polychlorinated biphenyls and polychlorinated triphenyls were weakly estrogenic.

GALAND et al. (1987) confirmed that *o,p'*-DDT and *p,p'*-DDT at 12.5 mg/rat, when given once intraperitoneally, were active in the glycogen deposition assay, with *o,p'*-DDT being the more potent of the two (Fig. 23). Furthermore, the glycogen deposition response to a 12.5 mg dose of *o,p'*-DDT was quantitatively similar to that observed with a 0.05 μg/animal dose of 17β-oestradiol (Fig. 23). Hence, *o,p'*-DDT was about 4×10^{-6} as potent as 17β-oestradiol in the glycogen deposition assay. It can be concluded that the glycogen assay seems to be as sensitive to estrogenic stimulation as the uterine growth assay.

Peroxidase Activity

Within 24 h, 17β-oestradiol increases peroxidase activity in the immature rat; this activity lasts up to about one week (LYTTLE and DESOMBRE 1977a, b). Progesterone, when given alone, does not induce uterine peroxidase activity (DESOMBRE and LYTTLE 1980). However, co-treatment of rats with 17β-oestradiol plus progesterone or related progestins (e.g. R 5020 or norethindrone) markedly inhibits estrogen-induced uterine peroxidase activity (Desombre and Lyttle 1980). Tamoxifen, an antiestrogen with a clear partial agonistic activity, is a clear estrogenic compound that induces significant peroxidase activity, whereas LY 117,018, an antiestrogen with only a marginal estrogenic potency, induces only slight uterine peroxidase activity (Keeping and Lyttle 1982). Taken together, these data clearly indicate that uterine peroxidase activity is a suitable test for predicting estrogenicity of a given compound. Its sensitivity it is comparable to the uterine growth assay.

Complement Component C3

As already mentioned, the antiestrogens tamoxifen and LY 117,018 exert both agonistic and antagonist estrogen effects on the immature rat uterus. Whereas LY 117,018 is a weak agonist, tamoxifen exhibits strong estrogenic activity in the rat endometrium. With respect to C3, it was shown that both tamoxifen and LY 117,018 increase the synthesis and release of C3 in the rat uterus (Sundstrom et al. 1990). 0.1 μg of LY 117,018/immature rat (single injection) was sufficient to stimulate C3 synthesis and 1 μg/animal was at least as effective as an equivalent dose of 17β-oestradiol. The synthesis of C3 in response to oestradiol appears to saturate between 0.1 μg and 1 μg, with little or no increase at higher doses. Tamoxifen demonstrated similar potencies (Sundstrom et al. 1990).

To examine whether the concomitant administration of 17β-oestradiol and LY 117,018 is synergistic or whether LY 117,018 can antagonise 17β-oestradiol-stimulated C3, rats were given 1 μg E2 in combination with various doses of LY 117,018 (Sundstrom et al. 1990). The results demonstrate that LY 117,018 and 17β-oestradiol do not act synergistically. However, LY 117,018, at the doses tested, was unable to antagonise the 17β-oestradiol stimulation of C3.

Taken together, these data with ER mixed agonists/antagonists show that complement component C3 seems to be a highly sensitive estrogenic marker, much more sensitive than many other estrogenic uterine parameters, e.g. uterine growth or peroxidase activity. Immunohistochemical analysis revealed that C3 is produced exlusively by uterine luminal epithelial cells (Sundstrom et al. 1989). Given that tamoxifen increased both C3 synthesis and uterine epithelial hypertrophy, however, with no effect in stroma and myometrial cells (Sundstrom et al. 1990), it may be concluded that, within the uterus, luminal epithelial cell height and complement component C3 are the most sensitive estrogenic markers.

X. Withdrawal Bleeding (Primates)

1. Principle

In women or non-human primates, removal of both ovaries is usually followed a few days later by a brief period of vaginal bleeding from the uterus. Experiments have shown that sudden deprivation or diminution of either estrogen or progestin leads to uterine bleeding (EMMENS and MARTIN 1964; BURROWS 1994). It is also known that menstruation, which occurs only in primates, can be prevented by the continual administration in a sufficient dose of either estrogen or progestin. These observations have led to the development of an estrogen bioassay in primates that is used most frequently with the ovariectomised rhesus monkey (ECKSTEIN et al. 1952; SCHANE et al. 1972; HISAW and HISAW 1973).

2. Study Design

Mature female rhesus monkeys, weighing 2.5–4.5 kg, are bilaterally ovariectomised. After 4–6 weeks, 17β-oestradiol or other estrogenic compounds are administered subcutaneously in 0.2–0.5 ml of a 10% ethanol in arachis oil (v/v) vehicle. It is also possible to apply compounds suspended in Myrj or other vehicles, by gavage (i.g.), in 1.0–3.0 ml of vehicle. Each estrogen is administered daily for 10 days on a dose per kilogram body weight or on a per monkey per day basis using 1–4 dose levels and 3–4 monkeys per dose level. Vaginal smears are taken daily (for 6–12 days) to detect the start of bleeding at the end of treatment. In addition, the color of the perianal sex skin is rated subjectively using a quantal rating system (i.e. 0 = pale to 4 = intense red) on the first and last day after medication.

3. Evaluation and Discussion

SCHANE et al. (1972) studied the menstrual-like bleeding pattern in 18 ovariectomised monkeys following subcutaneous 17β-oestradiol treatment (10 μg/monkey/day for 10 days). The data are presented in Fig. 24. Bleeding began an average of 9.1 days after the last 17β-oestradiol injection. Bleeding never began before the sixth day following the last injection and, in all but two instances, began on or before day 15. In addition, SCHANE et al. (1972) have defined the dose-response relationship for subcutaneously injected 17β-oestradiol. In four monkeys injected subcutaneously with 2 μg 17β-oestradiol/day for 10 days, two showed a normal withdrawal bleeding pattern, one began bleeding during treatment, and one began bleeding earlier than normal (4 days after treatment stopped). At daily doses of 4 μg, 6 μg or 8 μg, 17β-oestradiol per monkey for 10 days (four monkeys/group), all monkeys had normal withdrawal bleeding.

The dose-response relationship for several orally active estrogens was also established. In particular, the oral doses that were completely ineffective in producing withdrawal bleeding were pinpointed. Ethinyl oestradiol (EE), mestranol, and oestrone produced withdrawal bleeding in all four of the monkeys when given 100 μg/day for 10 days. Premarin caused withdrawal bleeding in all four monkeys at 40 μg/day, whereas diethylstilboestrol (DES) was fully effective at a daily dose of 400 μg/day. The completely ineffective daily doses were: EE, 6.25 μg; mestranol, 25 μg; oestrone, 6.25 μg; premarin,

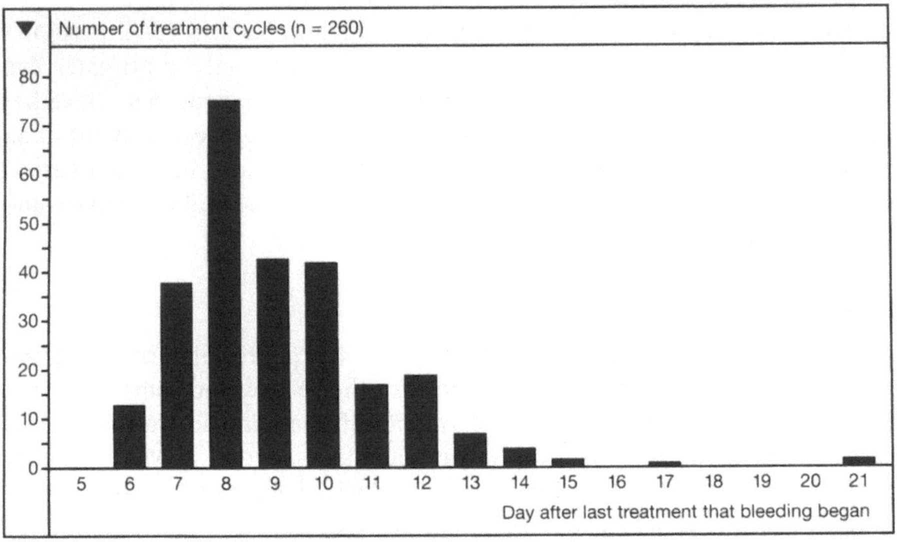

Fig. 24. Occurrence of withdrawal bleeding in rhesus monkeys following a constant regimen of 17β-oestradiol (10 μg/monkey/day × 10 days s.c.) (from SCHANE et al. 1972)

10 μg; and DES, 25 μg. Furthermore, EE, oestrone, premarin, and DES did not stimulate the development of perianal sexual skin coloring at daily doses below those required to produce withdrawal bleeding. In contrast, mestranol produced a change in color of the sexual skin at a daily dose below that which caused withdrawal bleeding. Taken together, the data indicated that EE, mestranol, oestrone, and premarin were approximately equipotent, whereas DES was of lesser potency in the withdrawal bleeding assay.

ECKSTEIN et al. (1952) estimated the minimal amounts of 11 estrogens required to induce uterine withdrawal bleeding when administered as a single intramuscular injection in 25 ovariectomised rhesus monkeys. Vaginal lavages were performed daily to detect the onset and duration of uterine bleeding. The latency to onset of withdrawal bleeding and the duration of bleeding in days at the threshold dose were used as the endpoints. The threshold doses then were used to estimate relative potencies. Esterification of estrogens increased potency (smaller threshold dose) and duration of bleeding, but diminished the latent interval between intramuscular injection and the onset of withdrawal bleeding. In summary, the more potent the estrogen, the shorter the latency period and the longer the duration of bleeding.

Generally, the estrogen withdrawal bleeding bioassay is used to confirm rodent bioassay data and to establish estrogenic activity and potency in primates. The assay has the advantage that data can be obtained from a primate species and that animals can be dosed orally or parenterally and can be used repeatedly. It has the disadvantage that non-human primates are difficult to handle and expensive to maintain. In addition, the uterine withdrawal bleeding endpoint is not specific to estrogens; progestins also can induce this response (KROHN 1956; HISAW and HISAW 1973).

XI. Imaging of the Primate Uterus

1. Principle

The evaluation of drugs that exhibit an estrogenic profile must be carried out in relevant test systems. Information from rodent species is valuable, but the uterine physiology of monkeys more closely resembles that of humans. For this reason, non-invasive imaging systems were developed (WATERTON et al. 1991), such as ultrasound imaging and quantitative MRI (magnetic resonance imaging), that can be used to measure the response of the uterus to estrogen stimulation. Similar to the rodent situation, estrogens lead to an increase in uterine volume as well as endometrial volume.

2. Experimental Design

Cynomolgus monkeys (*Macaca fascicularis*) or pig-tailed monkeys are ovariectomised at least 2–3 months prior to the study. They receive no treatment (vehicle only) or 17β-oestradiol or other estrogenic compounds dissolved, e.g. in 10% ethanol arachis oil (v/v) vehicle, daily for 14–21 days (or longer). This is done either on a dose per kilogram body weight or per monkey per day basis, using 1–4 dose levels and 3–4 monkeys per dose level. MRI or ultrasound examinations take place on days 0, 7, 14, and 21.

MRI Evaluation

The examination procedure is described in detail by WATERTON et al. (1991). It takes place before the animals receive their morning feed, to minimise artifactual noise associated with gut motion. Anaesthesia is induced with ketamine and maintained with halothane. In summary, high field (2.35 T), fat-suppressed, T2-weighted (TE 50) oblique methods are used to measure endometrial and myometrial volume. Slice thickness, location and angle are varied on each examination to obtain six contiguous slices between the cervix and fundus, regardless of size or orientation of the uterus.

Ultrasound Evaluation

Monkeys are scanned under anesthesia with ketamine. The ultrasound machine used can be a real-time instrument with a 7.5 MHz scanhead of high frequency and short focus. This scanhead allows detailed study of structures close to the transducer and is suitable for the study of monkeys. The transducer is placed on the lower midline of the animal to locate the uterus in a transverse and longitudinal plane. The innermost border of the endometrium, the interface between the endometrium and myometrium and the outermost border of the myometrium appeared as a prominent echogenic area (Fig. 25).

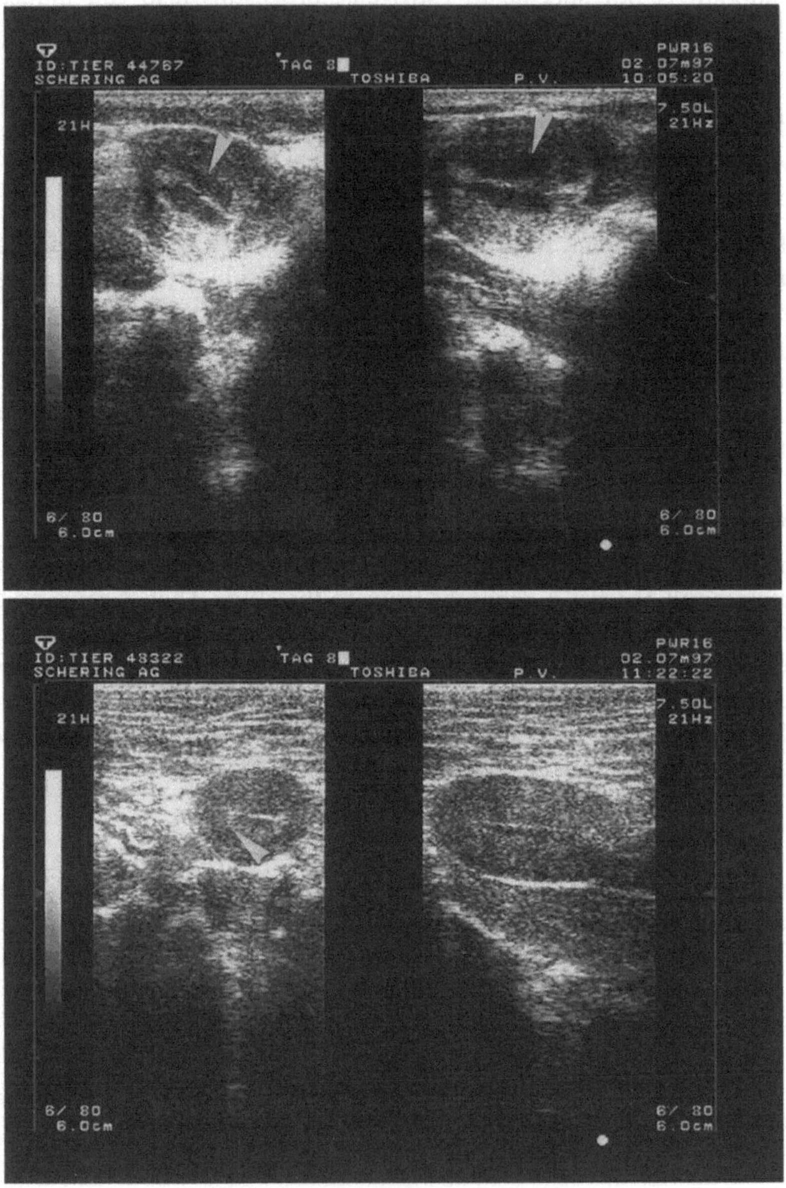

Fig. 25. Ultrasonograms showing the uterus from two ovariectomised Cynomolgus monkeys treated with 17β-oestradiol (10 μg/kg bw/day for 7 days). *Arrowheads* indicate the borderline between the endometrium and the myometrium

Antero–posterior diameter, transverse and longitudinal diameter of endometrium and myometrium can be measured; endometrial and myometrial area as well as whole uterus area can be calculated. In addition, an estimated endometrial and uterine volume are obtained by the product of the transverse, longitudinal, and antero-posterior diameters. A detailed description is given by MORGAN et al. (1987).

3. Evaluation and Discussion

According to WATERTON et al. (1991), the uterus is small in chronically ovariectomised monkeys. The endometrium occupies an extremely small volume of 0.051 ± 0.015cm^3, 2–7% of the total uterus volume. The myometrium occupies 1.31 ± 0.23cm^3. The dark junction zone between endometrium and myometrium is nearly absent. During seven days of treatment with 5μg 17β-oestradiol-benzoate/kg the endometrium grows 15.7 ± 3.6-fold to occupy a volume of 0.64 ± 0.14cm^3, or 8–23% of the total uterus volume. The myometrium also grows 3.1 ± 0.4-fold to occupy a volume of 3.9 ± 0.8cm^3. Endometrial and myometrial proliferation continued at a decreasing rate during the second week of estrogen administration (1.5-fold and 1.4-fold, respectively).

ZM182,780 is a potent antiestrogen. To characterise the potency and efficacy of ZM182,780, its effect on the uterus of 17β-oestradiol-treated pig-tailed monkeys have been measured using quantitative MRI (DUKES et al. 1992). It could be shown that repeated injections of 4mg ZM182,780/kg at intervals of 4 weeks provided an increasingly effective blockade of uterine proliferation. These studies demonstrated that ZM182,780 is a fully effective pure antiestrogen in a primate.

Stimulation of uterine growth by different doses of 17β-oestradiol in the ovariectomised, cynomolgus monkey is shown in Fig. 26. Whereas 0.1μg 17β-oestradiol/kg only produces a faint increase in uterine transverse area, 1.0μg/kg and 10μg/kg 17β-oestradiol lead to a clear increase in uterine transverse area. The most prominent increases are found within the first two treatment weeks. Thereafter, no further uterine stimulation can be observed.

Both MRI and ultrasound are ideal methods for studying estrogenicity in monkeys, given that they are quantitative and non-invasive; the consequences of interanimal variability are minimised by using each animal as its own control. This greatly reduces, therefore, the number of animals required to give a meaningful result. After a rest phase of about four weeks, the animals can be recruited for further experiments.

XII. Pituitary Blockade

Secretion of gonadotrophins by the anterior pituitary are normally counterbalanced by negative feedback from the gonads; secretion of steroid hormones, such as estrogens, tends to inhibit follicular stimulating hormone

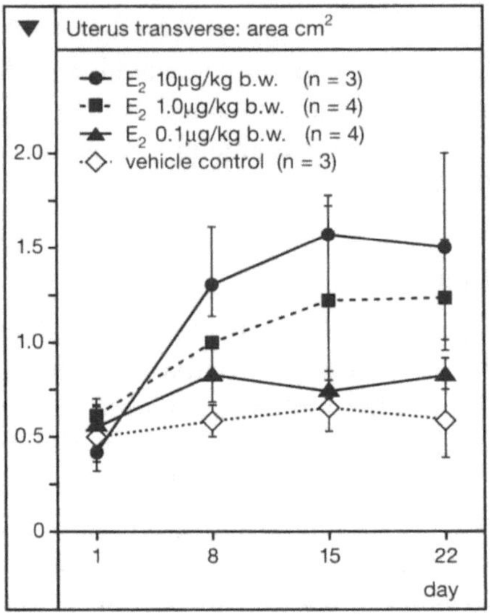

Fig. 26. Dose–response curve for 17β-oestradiol (treatment phase: 21 days) in the ovariectomised Cynomolgus monkey. Stimulation of uterine growth is measured by the ultrasound technique

(FSH) secretion and thus to hold down the blood-level of estrogens. A range of bioassay procedures have been employed to estimate the inhibiting effects of estrogens (and progestins) on the hypothalamic-pituitary system, in the belief that this is the primary mode-of-action for these drugs. Such assays have involved suppression of gonadal growth, measurement of circulating gonadotropins and evaluation of ovulation by oocyte counts. Based on the fact that these assays are not highly specific for estrogens, only a short overview of the test-principle is given.

Gonadal Growth Methods

The administration of estrogens and progestins to laboratory rodents leads to failure of growth for both the ovary and testes. Early controlled studies were based on parabiotic unions in which two animals, usually rats, were surgically joined at their side (KALLAS 1929; DAVSON and SEGAL 1980), i.e. made into "Siamese twins", to create intercommunication of their vascular supplies. In such pairs, the gonadotropic hormones passed readily across the union, whereas the gonadal steroids did not.

As a result, suppressed secretion of the gonadotropins can be measured by administrating the steroid to one member of the pair and evaluating gonadal size in the other individual. This technique is labor intensive and unions can only be created successfully using immature animals (i.e. prior to development of immunological capacity). These studies supported the contention that estrogens (and progestins) suppressed gonadotropin secretion.

In an effort to overcome the difficulties of a parabiotic system, an assay based upon compensatory ovarian hypertrophy in hemicastrated rats was developed by PETERSON et al. (1964). Using this technique, the compensatory hypertrophy caused by unilateral ovariectomy is evaluated by dual controls – one intact and one hemicastrated. This effect is suppressed by the administration of estrogens and progestins. The progestins produce simple linear curves that project below the baseline level of the intact controls, whereas estrogen suppression proceeds linearly to approximately 100% suppression level and then reverses to a growth phase that is associated with the formation of very large corpora lutea.

Gonadotropin Methods

The extensive work examining control of the oestrous cycle in rats has demonstrated that in females with regular four-day cycles, gonadotropins, especially luteal hormone (LH), rise to a peak on the afternoon of prooestrus. This increase, which induces ovulation early the next morning, can be inhibited by the adminstation of steroids.

Ovariectomy and orchidectomy of rodents – and also of primates – lead to an increase of gonadotropins, mainly LH and FSH, in peripheral blood. This is due to the negative feedback of estrogens, androgens and progestins on the production and secretion of LH and FSH. By the addition of sex steroids to ovariectomised animals, it is possible to suppress the elevated gonadotropin level (Fig. 27).

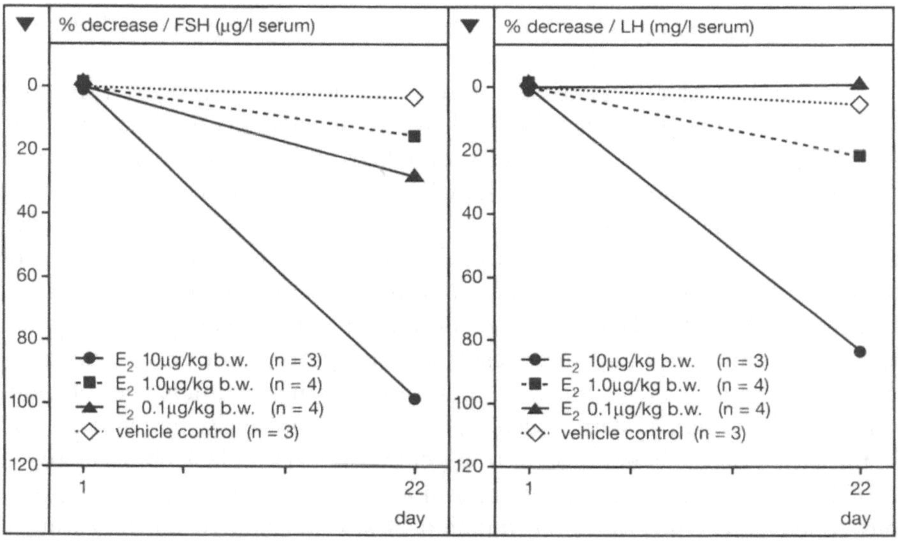

Fig. 27. Follicular stimulating hormone (FSH)/luteinising hormone (LH) inhibition after a 21-day subcutaneous treatment of ovariectomised Cynomolgus monkeys with different 17β-oestradiol doses

Ovulatory Methods

Sex steroids can be tested for their antiovulatory activity in rats. The antiovulatory potential of sex steroids is mediated by their negative feedback to the hypothalamus-pituitary axis; this leads to inhibition, especially of LH, thereby triggering ovulation. Female rats are treated for one cycle length (4 days) with estrogens or progestins. Around the time of expected ovulation, oviducts are removed. Crush preparations are prepared and examined by light microscopy for the presence of oocytes. At the end of the experiment, the percentage of animals in which ovulation is inhibited can be determined.

D. Transgenic Animals

Transgenic mice have become important model systems for studying molecular, cellular, organ, and whole animal physiology. Within the last 15 years we have seen rapid changes and increases in the use of transgenic mouse technology for studying mammalian development (for review see: PALMITER and BRINSTER 1986; BRADLEY et al. 1992; MATZUK et al. 1996; NISHIMORI and MATZUK 1996). Transgenic mice can be created to express wild-type, mutant, marker or cell lethal genes in a tissue-specific manner. In addition, homologous recombination strategies in embryonic stem cells permit more sophisticated manipulation of the mammalian genome, including functional deletion of specific genes in whole mice or in a specific tissue. Because all transgenic mice must be bred to study the consequences of transgene or mutant allel expression, a number of effects of those genome manipulations on the reproductive development and function of these mice have been uncovered recently (for review: NISHIMORI and MATZUK 1996).

The biological effects of sex steroid hormones, e.g. progesterone and 17β-oestradiol, are believed to be mediated by specific receptor proteins (PR and ER) that have great specificity and high affinity for their respective ligand. A target tissue for a given female sex steroid has been defined as an organ possessing functional levels of ER and PR protein. Several studies have demonstrated the presence of ER and PR in various tissues of the reproductive tract, as well as in several non-reproductive tissues (reviewed in CUNHA et al. 1991). Decades have described many roles fulfilled by the sex steroid hormones and their respective receptors – not only in development and reproductive function, but also in the maintenance of non-reproductive organ systems, such as bone, the cardiovascular system and brain. Based on the generation of transgenic animals that lack either estrogen or PR it has become possible to confirm as well as to contrast several previous conclusions concerning the function of sex steroids. In the following, a summary is given of the ERα-knockout (ERKO) mouse and the PR mutant mouse.

I. ERKO Mouse

The molecular design to disrupt the mouse ER gene has been described in detail by LUBAHN et al. (1993). Surprisingly, it was possible to generate mice of both sexes that are homozygous for ER gene disruption. No bias towards either male or female sex was observed, indicating that no effect on sexual differentiation can be found in this mouse. The mice survived to adulthood and developed gross-morphologically normal external genitalia. However, both male and female mice were infertile (LUBAHN et al. 1993).

Uterus

The adult uterus is known to undergo synchronised phases of proliferation and differentiation under the influence of estrogen and progesterone. Therefore, it was suspected that uteri of ERKO mice are unresponsive to estrogens. The uteri of ERKO mice were small and hypoplastic; however, all cell layers (myometrium, endometrium) were present. The uterotrophic response to 17β-oestradiol, with an initial increase in water uptake, vascular permeability, hyperemia, prostaglandin release, protein synthesis, followed by cell proliferation and hypertrophy within 24 h of exposure, could never be observed in adult female ERKO mice. Not only oestradiol, but also the partial estrogen agonist, tamoxifen, was unable to stimulate uterine growth (KORACH 1994). The dependency on the presence of ER for proper uterine cell proliferation following estrogen exposure was confirmed by a [3H]-thymidine incorporation assay in which no significant increase in DNA synthesis over control animals was observed in the ERKO mouse. This contrasts with a 10-fold increase in the wild-type mouse following a single estrogen treatment (COUSE et al. 1995). In addition to a lack of mitogenic action of oestradiol in the uteri of ERKO mice, further studies have shown resistance of known estrogen-induced genes, e.g. PR and lactoferrin (COUSE et al. 1995), after treatment with 17β-oestradiol.

Ovary

The adult ERKO ovary is characterised by the presence of primordial, primary and secondary follicles. However, preantral and ovulatory follicles and corpora lutea are missing. Rather than proceeding through stages of folliculogenesis to ovulation, the follicles in the ovaries of ERKO mice appear to ultimately become arrested in a preantral stage and to eventually become atretic, leading to large, hemorrhagic and cystic structures. The sexually mature ERKO female produces 10-fold higher levels of circulating oestradiol than normal (COUSE et al. 1995). Circulating levels of LH are elevated, whereas FSH appears to be within the normal range (SCULLY et al. 1997).

It has to be taken into acount that a relatively large amount of a second ER, termed ERβ, is found in the mouse (TREMBLAY et al. 1997), rat (KUIPER et al. 1996) and human (MOSSELMAN et al. 1996) ovary. ERβ is known to exhibit significant homology in the DNA and ligand-binding domains compared to the classic ER, termed ERα. In the rodent, granulosa cells of the ovary exhibits the highest levels of ERβ mRNA expression (Kuiper et al. 1996) and this seems to be regulated by LH (Tremblay et al. 1997). In the ERKO mouse, normal levels of ERβ mRNA are found (Couse et al. 1997). Obviously, ERβ seems to have no significant role in follicular development. However, further studies are warranted to determine the role of ERβ in the ovary. In the meantime, it can be concluded that ERα seems to be responsible for early processes during folliculogenesis.

Testis

Sexually mature ERKO males exhibit a significant, age-dependent reduction in testicular weight compared to wild-type animals. Changes in epididymis and accessory sex organs were not evident (Donaldson et al. 1996; Eddy et al. 1996). Histological evaluation of adult ERKO testes revealed the presence of severely atrophied and degenerated seminiferous tubules, with a dilated lumen and only a thin lining of sertoli cells. Only a few spermatogenic cells were present (Eddy et al. 1996). Testicular steroidogenesis in the ERKO mouse was not greatly altered, nor were the circulating levels of LH and FSH (Eddy et al. 1996).

Mammary Gland

The mammary gland of an adult wild-type female mouse consists of a network of epithelial ducts originating from the nipple. The branches terminate in the form of alveolar buds. ERKO females demonstrate a simple rudimentary branch structure confined to the nipple region, although oestradiol concentrations are 10 times the norm, as are physiological levels of progesterone. The mammary glands of ERKO mice possess low, but detectable, levels of PR mRNA but show no upregulation when treated with oestradiol (Korach, personal communication). Based on these data, it can be concluded that development and growth during puberty is heavily dependent on the actions of oestradiol via ERα.

Behaviour

Intact ERKO females tend to behave more like males with respect to maternal, aggressive, and sexual behavior. They showed a significant decrease in lordosis (Ogawa et al. 1996). In the ERKO male, little effect on mounting behavior and motivation toward wild-type females is found (Ogawa et al. 1997). However, an almost complete lack of intromission and ejaculation of the ERKO males was observed (Ogawa et al. 1997).

II. PR Mutant Mouse

O'MALLEY and colleagues (1995) have utilised homologous recombination in mouse embryonic stem cells to generate a line of transgenic mice that are homozygous for a disruption of the PR gene (LYDON et al. 1995). Both sexes were equally represented in the mutant mice, indicating no effect on viability or sexual differentiation (LYDON et al. 1995). Females homozygous for the PR disruption were completely infertile, whereas no effects on male fertility have been observed.

Uterus

Uteri of PR mutant mice developed normally and possessed an intact uterine histology. When ovariectomised adult wild-type and PR mutant animals were treated with daily injections of oestradiol (1 μg/kg) and progesterone (1 mg/kg) for 3 weeks, the wild-type responded with the characteristic enlarged, fluid-filled and highly developed uterus, whereas the uteri of the PR mutants were abnormally large, indicating of a hyper-estrogenic response (LYDON et al. 1995). Histological analysis of the uteri from treated PR mutant mice indicated extracellular edema, resulting in a thickened uterine wall, extensive proliferation of the glandular epithelia, an acute inflammatory response in the mucosa, submucosa, and stromal components of the endometrium, and an overall hyperplastic and disorganised epithelium compared to treated wild-type controls (LYDON et al. 1995). Therefore, it can be suggested that progesterone, by acting via the PR, is necessary for suppressing local inflammatory responses in the uterus.

Ovary

The ovaries of the PR mutant females were reported to appear normal upon gross and histological evaluation (LYDON et al. 1995). Nevertheless, PR mutant females were completely infertile when repeatedly mated with wild-type males. An in-depth study by LYDON et al. (1995) demonstrated that a superovulation treatment regimen with the exogenous gonadotropins, pregnant mare serum gonadotropin (PMSG) and human chorionic gonadotropin (hCG), resulted in no oocytes in the oviduct or uterine horn of PR mutant females; this contrasted with an average of more than ten oocytes/animal in similarly treated wild-type mice, indicating a lack of ovulation in the PR mutants (LYDON et al. 1995). The ovaries from superovulated PR mutant females are characterised by the presence of several unruptured preovulatory follicles, and the absence of corpora lutea (LYDON et al. 1995). These studies have now confirmed a role for progesterone in the process of ovulation, a hypothesis that was suggested by previous studies with progesterone-antagonists (LOUTRADIS et al. 1991). Based on the data from the PR mutant mice, it can be concluded that PR-signaling events are required in the final

Table 11. Structures of estrogens and antiestrogens most frequently mentioned in this chapter

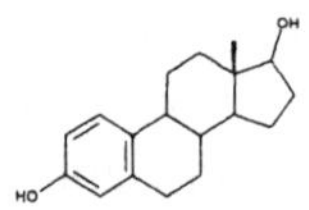
estradiol

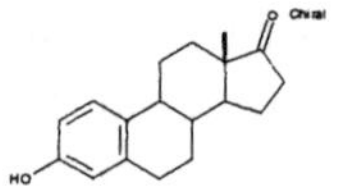
estrone

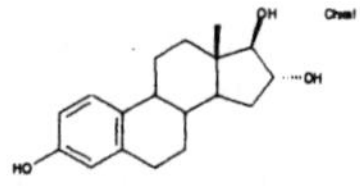
estriol

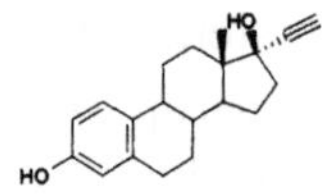
ethinyl estradiol

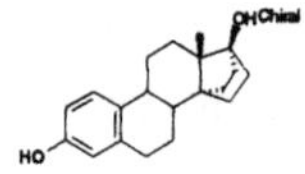
cyclodiol

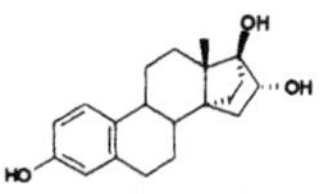
cyclotriol

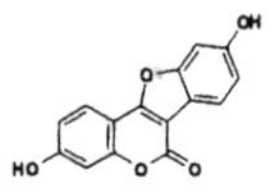
coumestrol

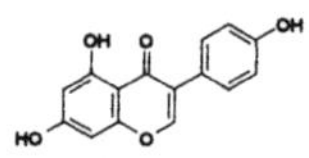
genistein

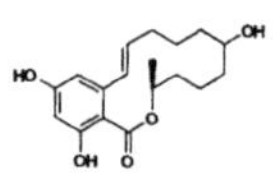
ß-zearalenol

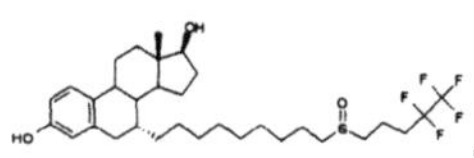
ZM 182,780

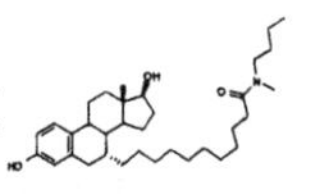
ZM 164,384

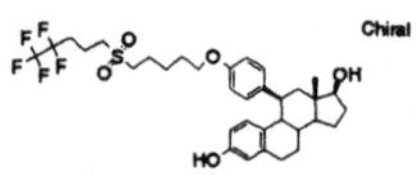
RU 58,688

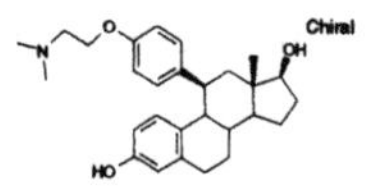
RU 39,411

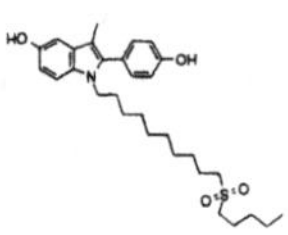
ZK 164,015

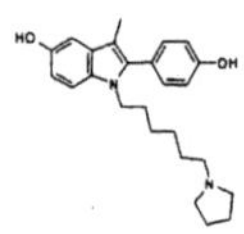
ZK 119,010

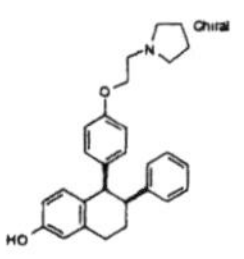
CP 336,156

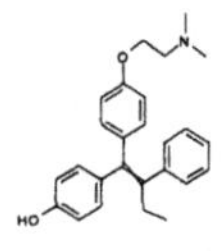
OH-tamoxifen

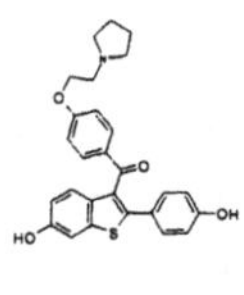
LY 117,018

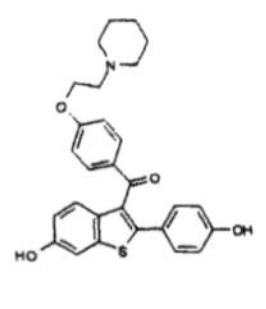
raloxifene

stages of the ovarian cycle to produce ovulation and the development of the corpora lutea.

Mammary Gland

The role of progesterone in mammary tissue has been thought to be mainly limited to differentiation of the gland. However, several reports have indicated mitogenic actions of progesterone in mammary tissue, such as an ability to augment the mitogenic response to estrogen when co-administered (IMAGAWA et al. 1994). The phenotype of the adult PR mutant female has provided evidence to support a dual role for progesterone (LYDON et al. 1995). The mammary gland of an adult virgin PR mutant female shows no gross difference in overall epithelial ductal development compared to the wild-type (LYDON et al. 1995). However, unlike in wild-type mice, when treated with both estrogen and progesterone over a period of 3 weeks, the mammary glands of mutant PR mice revealed a ductal structure characterised by less branching and a complete absence of lobuloalveolar structures at the termini of the branches (LYDON et al. 1995). These experiments have allowed LYDON et al. (1995) to conclude that progesterone does, indeed, play a role in the proliferation of mammary ductal branching and is critical to the development of the structures required for lactation (LYDON et al. 1995).

Behaviour

Studies by LYDON et al. (1995) with the PR mutant female have clearly shown that PR is essential for the lordosis response in mice.

References

Adan RA, Cox JJ, Beischlag TV, Burbach JP (1993) A composite hormone response element mediates the transactivation of the rat oxytocin gene by different classes of nuclear hormone receptors. Mol Endocrinol 7:47–57

Albert JL, Sundstrom SA, Lyttle CR (1990) Estrogen regulation of placental alkaline phosphatase gene expression in a human endometrial adenocarcinoma cell line. Cancer Res 50:3306–3310

Allen E, Doisy EA (1923) An ovarian hormone: preliminary report on its localization, extraction and partial purification, and action in test animals JAMA 81:819–821

Anderson JW (1997) Phytoestrogen effects in humans relative to risk for cardiovascular disease, breast cancer, osteoporosis, and menopausal symptoms. In: Pavlik EJ (ed) Estrogens, progestins and their antagonists, vol 1. Birkhäuser, Boston, pp 51–71

Angel P, Karin M (1991) The role of jun, fos and the AP-1 complex in cell-proliferation and transformation. Biochem Biophys Acta 1072:129–157

Arkbuckle ND, Dauvais S, Parker MG (1992) Effects of antiestrogens on the DNA binding activity of estrogen receptors in vitro. Nucleic Acids Res 20:3839–3844

Aronica SM, Katzenellenbogen BS (1993) Stimulation of estrogen receptor-mediated transcription and alteration in the phosphorylation state of the rat uterine estrogen receptor by estrogen, cyclic adenosine monophosphate, and insulin-like growth factor-I. Mol Endocrinol 7:743–752

Astwood EB (1938) A six-hour assay for the quantitative determination of estrogen. Endocrinology 23:25–31

Augereau P, Miralles F, Cavailles V, Gaudelet C, Parker M, Rochefort H (1994) Characterization of the proximal estrogen-responsive element of human cathepsin D gene. Mol Endocrinol 8:693–703

Azzi A, Galeotti T, Bartoli GM (eds) (1993) Cell proliferation and differentiation: modulation by free radicals. Mol Aspects Med 14:167–271

Barrack E, Coffey DS (1980) The specific binding of estrogens and androgens to the nuclear matrix of sex hormone responsive tissues. J Biol Chem 255:7265–7275

Batra S, Iosif CS (1989) Tissue specific effects of progesterone on progesterone and estrogen receptors in the female urogenital tract. J Steroid Biochem 32:35–39

Beato M, Herrlich P, Schutz G (1995) Steroid hormone receptors: many actors in search of a plot. Cell 83:851–857
Beekman JM, Allan GF, Tsai SY, Tsai MJ, O'Malley BW (1993) Transcriptional activation by the estrogen receptor requires a conformational change in the ligand binding domain. Mol Endocrinol 7:1266–1274
Behbakht K, Boyd J (1996) Estrogen and progesterone receptors in human endometrial cancer. In: Pavlik EJ (ed) Estrogens and progestins and their antagonists. Birkhäuser, Boston, pp 170–188
Behl C, Davis JB, Cole GM, Schubert D (1992) Vitamin E protects nerve cells from amyloid β-protein toxicity. Biochem Biophys Res Commun 186:944–952
Behl C, Davis JB, Lesley R, Schubert D (1994) Hydrogen peroxide mediates amyloid β-protein toxicity. Cell 77:817–827
Behl C, Skutella T, Lezoualc'h F, Post A, Widmann M, Newton CJ, Holsboer F (1997) Neuroprotection against oxidative stress by estrogens: structure activity relationship. Mol Pharmacol 51:535–541
Bergink EW, Kloosterboer HJ, van der Velden WHM, van der Vies J, de Winter MS (1983) Specificity of an estrogen binding protein in the human vagina compared with that of estrogen receptors in different tissues from different species. In: Jasonni VM et al. (eds) Steroids and endometrial cancer. Raven Press, New York, pp 77–84
Bergink EW, Kloosterboer HJ, van der Vies J (1984) Estrogen binding proteins in the female genital tract. J Steroid Biochem 20:1057–1060
Berry M, Nunez A-M, Chambon P (1989) Estrogen-responsive element of the human pS2 gene is an imperfectly palindromic sequence. Proc Natl Acad Sci USA 86:1218–1222
Berry M, Metzger D, Chambon P (1990) Role of the two activating domains of the estrogen receptor in the cell-type and promotor-context dependent agonistic activity of the anti-estrogen 4-hydroxytamoxifen. EMBO J 9:2811–2818
Bhavnani BR (1988) The saga of the B-ring unsaturated equine estrogens. Endocr Rev 9:396–416
Biggers JD (1951) Observations on the intravaginal assay of natural estrogens using aqueous egg albumin as the vehicle of administration. J Endocrinol 7:163–171
Biggers JD, Claringbold PJ (1954) Criteria of vaginal response to estrogens. J Endocrinol 7:277–284
Bigsby RM (1993) Progesterone and dexamethasone inhibition of estrogen-induced synthesis of DNA and complement in rat uterine epithelium: effects of antiprogesterone compounds. J Steroid Biochem Mol Biol 45:295–301
Bigsby RM, Li A (1994) Differentially regulated immediate early genes in the rat uterus. Endocrinology 134:1820–1826
Birge SJ (1997) The role of estrogen in the treatment of Alzheimer's disease. Neurology 48[Suppl 7]:S36–S41
Bitman J, Cecil HC (1970) Estrogenic activity of DDT analogs and polychlorinated biphenyls. J Agricultural Food Chem 18:1108–1112
Bitman J, Cecil HC, Mench ML, Wrenn TR (1965) Kinetics of in vivo glycogen synthesis in the estrogen-stimulated rat uterus. Endocrinology 76:63–69
Bocquel MT, Kumar V, Stricker C, Chambon P, Gronemeyer H (1989) The contribution of the N- and C-terminal regions of steroid receptors to activation of transcription is both receptor and cell-specific. Nucleic Acids Res 17:2581–2595
Boettger-Tong HL, Murthy L, Stancel GM (1995) Cellular pattern of c-fos induction by estradiol in the immature rat uterus. Biol Reprod 53:1398–1406
Bradley A, Hasty P, Davis A, Ramirez-Solis R (1992) Modifying the mouse: design and desire. Biotechnology 10:534–539
Brakenhoff JP (1995) Interleukin-6 receptor antagonists. DN & P 8:397–403
Branham WS, Sheehan DM (1995) Ovarian and adrenal contributions to postnatal growth and differentiation of the rat uterus. Biol Reprod 53:863–872
Brasier AR, Tate JE, Habener JF (1989) Optimized use of the firefly luciferase assay as a reporter gene in mammalian cell lines. Biotechniques 7:1116–1122

Braughler JM, Hall ED, Jacobson EJ, McCall JM, Means ED (1989) The 21-aminosteroids: potent inhibitors of lipid peroxidation for the treatment of central nervous system trauma and ischemia. Drug Future 14:141–152
Brockelmann J, Fawcett DW (1969) The localization of endogenous peroxidase in the rat uterus and its induction by estradiol. Biol Reprod 1:59–71
Bryant HU, Glasebrook AL, Yang NN, Sato M (1996) A pharmacological review of raloxifene. J Bone Miner Metab 14:1–9
Buege A, Aust SD (1978) Microsomal lipid peroxidation. Methods Enzymol 52:302–310
Bülbring E, Burn JH (1935) The estimation of oestrin and of male hormone in oily solution. J Physiol 85:320–333
Burch JB, Evans MI, Friedman TM, O'Malley PJ (1988) Two functional estrogen response elements are located upstream of the major chicken vitellogenin gene. Mol Cell Biol 8:1123–1131
Burrows H (1949) Biological actions of sex hormones, 2nd edn. Cambridge University Press, London
Butenandt A (1933) Zur Biologie und Chemie der Sexualhormone. Naturwissenschaften 21:49–54
Byers M, Kuiper GG, Gustafsson J-A, Park-Sarge OK (1997) Estrogen receptor-β mRNA expression in rat ovary: down-regulation by gonadotropins. Mol Endocrinol 11:172–182
Carlburg L (1966) Quantitative determination of sialic acids in the mouse vagina. Endocrinology 78:1093–1099
Carlson RA, Gorski J (1980) Characterization of a unique population of unfilled estrogen-binding sites associated with the nuclear fraction of immature rat uteri. Endocrinology 106:1776–1785
Carson-Jurica MA, Schrader WT, O'Malley BW (1990) Steroid receptor family: structure and functions. Endocr Rev 11:201–220
Cavailles V, Augereau P, Rochefort H (1993) Cathepsin D gene is controlled by a mixed promoter, and estrogens stimulate only TATA-dependent transcription in breast cancer cells. Proc Natl Acad Sci USA 90:203–207
Chen JD, Evans RM (1996) A transcriptional corepressor that interacts with nuclear hormone receptors. Nature 377:454–457
Cheskis BJ, Karathanasis S, Lyttle CR (1997) Estrogen receptor ligands modulate its interaction with DNA. J Biol Chem 272:11384–11391
Choi DW (1992) Excitotoxic cell death. J Neurobiol 23:1261–1276
Christman JK, Nehls S, Polin L, Brooks SC (1995) Relationship between estrogen structure and conformational changes in estrogen receptor/DNA complexes. J Steroid Biochem Mol Biol 54:201–210
Clark JH, Mani SK (1991) Actions of ovarian steroid hormones. In: Knobil E, Neill JD (eds) The physiology of reproduction, 2nd edn. Raven Press, New York, pp 1011–1059
Clark CR, MacLusky NJ, Naftolin F (1982) Unfilled nuclear estrogen receptors in the rat brain and pituitary gland. Endocrinology 93:327–338
Clarkson TB, Anthony MS (1997) Effects on the cardiovascular system: basic aspects. In: Lindsey R, Dempster DW, Jordan VC (eds) Estrogens and antiestrogens. Lippincott-Raven, Philadelphia, pp 89–118
Clemens MJ (1974) The regulation of egg yolk protein synthesis by steroidal hormones. Progr Biophys Mol Biol 28:71–107
Clinton GM, Hua W (1997) Estrogen action in human ovarian cancer. Crit Rev Oncol Hematol 25:1–9
Cooke PS, Young P, Hess RA, Cunha GR (1991) Estrogen receptor expression in developing epididymis, efferent ductules, and other male reproductive organs. Endocrinology 128:2874–2879
Couse JF, Curtis SW, Washburn TF, Lindzey J, Golding TS, Lubahn DB, Smithies O, Korach KS (1995) Analysis of transcription and estrogen insensitivity in the female mouse after targeted disruption of the estrogen receptor gene. Mol Endocrinol 9:1441–1454

Couse JF, Lindzey J, Grandien K, Kuiper GGJM, Gustafsson J-A, Korach KS (1997) Tissue distribution and quantitative analysis of estrogen receptor-α (ERα) and estrogen receptor-β (ERβ) mRNA in the wild-type and ERα-knockout mouse. Endocrinology 138:4613–4621

Cowley SM, Hoare S, Mosselman S, Parker MG (1997) Estrogen receptors α and β form heterodimers on DNA. J Biol Chem 272:19858–19862

Cullingford TE, Pollard JW (1988) RU 486 completely inhibits the action of progesterone on cell proliferation in the mouse uterus. J Reprod Fertil 83:909–914

Cunha GR, Cooke PS, Bigsby R, Brody JR (1991) Ontogeny of sex steroid receptors in mammals. In: Parker MG (ed) Nuclear hormone receptors: molecular mechanisms, cellular functions, clinical abnormalities. Academic Press, London, pp 235–268

Curtis SW, Korach KS (1991) Uterine estrogen receptor-DNA complexes: effects of different ERE sequences, ligands, and receptor forms. Mol Endocrinol 5:959–966

Dana SL, Hoener Pa, Wheeler DA, Lawrence CB, McDonnell DP (1994) Novel estrogen response elements identified by genetic selection in yeast are differentially responsive to estrogens and antiestrogens in mammalian cells. Mol Endocrinol 8:1193–1207

Darwish H, Krisinger J, Furlow D, Smith C, Murdoch FE, De Luca HF (1991) An estrogen-responsive element mediates the transcriptional regulation of calbindin D-9K gene in rat uterus. J Biol Chem 266:551–558

Dauvais S, Danielian PS, White R, Parker MG (1992) Antiestrogen ICI 164,384 reduces cellular estrogen receptor content by increasing its turnover. Proc Natl Acad Sci USA 89:4037–4041

Davson H, Segal MB (1980) Control of reproduction. Puberty and old age. In: Introduction to physiology, vol 5. Academic Press, London, pp 432–476

De Cupis A, Favoni RE (1997) Estrogen/growth factor cross-talk in breast carcinoma: a specific target for novel antiestrogens. Trends Pharmacol Sci 18:245–251

De Wet JR, Wood KV, De Luca M, Helinski DR, Subramani S (1987) Firefly luciferase gene: structure and expression in mammalian cell lines. Mol Cell Biol 7:725–737

Demirpence E, Duchesne M-J, Badia E, Gagne D, Pons M (1993) MVLN cells: a bioluminescent MCF-7-derived cell line to study the modulation of estrogenic activity. J Steroid Biochem Mol Biol 46:355–364

DeSombre ER, Lyttle CR (1980) Specific uterine protein synthesis as a guide to understanding the biologic significance of the estrogen-receptor interaction. In: Bresciani F (ed) Perspectives in steroid receptor research. Raven Press, New York, pp 167–182

Dickson RB, Lippman ME (1987) Estrogenic regulation of growth and polypetide growth factor secretion in human breast carcinoma. Endocr Rev 8:29–43

DiDonato JA, Hayakawa M, Rothwarf DM, Zandi E, Karin M (1997) A cytokine-responsive IkB kinase that activates the transcription factor NF-κB. Nature 388:548–554

Diel P, Walter A, Fritzemeier KH, Hegele-Hartung C, Knauthe R (1995) Identification of estrogen regulated genes in Fe33 rat hepatoma cells by differential display polymerase chain reaction and their hormonal regulation in rat liver and uterus. J Steroid Biochem Mol Biol 55:363–373

Dodds EC, Lawson W, Noble RL (1938) Biological effects of the synthetic estrogenic substance 4:4'-dihydroxy-α: β-diethylstilbene. Lancet 1:1389–1391

Dodds EC, Goldberg L, Lawson L (1939) Synthetic estrogenic compounds related to stilbene and diphenylethane. Proc R Soc Lond B Biol Sci 127:140–167

Dodge JA, Glasebrook AL, Magee DE, Phillips DL, Sato M, Short LL, Bryant HU (1996) Environmental estrogens: effects on cholesterol lowering and bone in the ovariectomized rat. J Steroid Mol Biol 59:155–161

Donaldson KM, Tong SYC, Washburn T, Lubahn DB, Eddy EM, Hutson JM, Korach KS (1996) Morphometric study of the gubernaculum in male estrogen receptor mutant mice. J Androl 17:91–95

Dorfman RI, Kincl FA (1966) Uterotrophic activity of various phenolic steroids. Acta Endocrinologica 52:619–626
Dorfman RI, Gallagher TF, Koch FC (1936) The nature of the estrogenic substance in human male urine and bull testis. Endocrinology 19:33–41
Dukes M, Miller D, Wakeling AE, Waterton JC (1992) Antiuterotrophic effects of a pure antiestrogen, ICI 182.780: magnetic resonance imaging of the uterus in ovariectomized monkeys. J Endocrinol 135:239–247
Eckerts RL, Katzenellenbogen BS (1982) Effects of estrogens and antiestrogens on estrogen receptor dynamics and the induction of progesterone receptor in MCF-7 human breast cancer cells. Cancer Res 42:139–144
Eckstein P, Krohn PL, Zuckerman S (1952) The comparative potency of natural and synthetic estrogens in the rhesus monkey. J Endocrinol 8:91–307
Eddy EM, Washburn TF, Bunch DO, Goulding EH, Gladen BC, Lubahn DB, Korach KS (1996) Targeted disruption of the estrogen receptor gene in male mice causes alteraton of spermatogenesis and infertility. Endocrinology 137:4796–4805
Edgren RA (1994) Issues in animal pharmacology. In: Goldzieher JW (ed) Pharmacology of the contraceptive steroids. Raven Press, New York, pp 81–97
Edgren RA, Peterson DL, Jones RC, Nagra CL, Smith H, Hughes GA (1966) Biological effects of synthetic gonanes. Recent Prog Horm Res 22:305–349
Ekman P, Barrack ER, Greene GL, Jensen EV, Walsh PC (1983) Estrogen receptors in human prostate: evidence for multiple binding sites. J Clin Endocrinol Metab 57:166–176
Emmens CW (1940) The differentiation of estrogens from pro-estrogens by the use of spayed mice possessing two separate vaginal sacs. J Endocrinol 3:174–177
Emmens CW (1962) Estrogens. In: Dorfman RI (ed) Methods in hormone research, vol II. Bioassay Academic Press, New York, pp 59–111
Emmens CW, Martin L (1964) Estrogens. In: Dorfman RI (eds) Methods in hormone research, vol III, steroidal activity in experimental animals and man, part A. Academic Press, New York, pp 1–80
Emmens CW, Cox RI, Martin L (1960) Antiestrogens. Recent Progr Horm Res 18:415–466
Esterbauer H, Striegel G, Puhl H, Rothender M (1989) Continuous monitoring of in vitro oxidation of human low density lipoprotein. Free Radic Res Commun 6:67–75
Fan JD, Wagner BL, McDonnell DP (1996) Identification of the sequences within the human complement 3 promotor required for estrogen responsiveness provides insight into the mechanism of tamoxifen mixed agonist activity. Mol Endocrinol 10:1605–1616
Feldmer M, Kaling M, Takahashi S, Mullins JJ, Ganten D (1991) Glucocorticoid- and estrogen-responsive elements in the 5′-flanking region of the rat angiotensinogen gene. J Hypertension 9:1005–1012
Flouriot G, Pakdel F, Ducouret B, Valotaire Y (1995) Influence of xenobiotics on rainbow trout liver estrogen receptor and vitellogenin gene expression. J Mol Endocrinol 15:143–151
Flouriot G, Pakdel F, Ducouret B, Ledrean Y, Valotaire V (1997) Differential regulation of two genes implicated in fish reproduction: vitellogenin and estrogen receptor genes. Mol Reprod Dev 48:317–323
Foegh ML (1992) Estradiol and myointimal proliferation. In: Ramwell P, Rubanyi GM, Schillinger E (eds) Sex steroids and the cardiovascular system. Springer, Berlin Heidelberg New York, pp 129–138
Folman Y, Pope GS (1966) The interaction in the immature mouse of potent estrogens with coumestrol, genistein and other utero-vaginotrophic compounds of low potency. J Endocrinol 34:215–225
Fuchs-Young R, Glasebrook AL, Short LL, Draper MW, Rippy MK, Cole HW, Magee DE, Termine JD, Bryant HU (1995) Raloxifene is a tissue-selective agonist antagonist that functions through the estrogen receptor. Ann NY Acad Sci 761:355–360

Furlow JD, Murdoch FE, Gorski J (1993) High affinity binding of the estrogen receptor to a DNA response element does not require homodimer formation or estrogen. J Biol Chem 268:12519–12525
Gaido KW, McDonnell DP, Korach KS, Safe SH (1997a) Estrogenic activity of chemical mixtures: is there synergism? CIIT Activities, Chemical Industry Institute of Toxicology, vol. 17:1–7
Gaido KW, Leonard LS, Lovell S, Gould JC, Babai D, Portier CJ, McDonnell DP (1997b) Evaluation of chemicals with endocrine modulating activity in a yeast-based steroid hormone receptor gene transcription assay. Toxicol Appl Pharmacol 143:205–212
Galand P, Mairesse N, Degraef C, Rooryck J (1987) o,p'DDT (1,1,1-trichloro-2(p-chlorophenyl)2-(o-chlorophenyl)ethane is a purely estrogenic agonist in the rat uterus in vivo and in vitro. Biochem Pharmacol 36:397–400
Galien R, Evans HF, Garcia T (1996) Involvement of CCAAT/enhancer-binding protein and nuclear factor-κB binding sites in interleukin-6 promotor inhibition by estrogens. Mol Endocrinol 10:713–722
Gallagher A, Chambers TJ, Tobias JH (1993) The estrogen antagonist ICI 182,780 reduces cancellous bone volume in female rats. Endocrinology 133:2787–2791
Geier A, Ginzburg R, Stauber M, Lunenfeld B (1979) Unoccuppied binding sites for oestradiol in nuclei from human breast carcinomatous tissue. J Endocrinol 80:281–288
Gill RK, Christakos S (1995) Regulation by estrogen through the 5′-flanking region of the mouse calbindin-D28k gene. Mol Endocrinol 9:319–326
Ginsburg M, Maclusky NJ, Morris ID, Thomas PJ (1977) The specificity of estrogen receptor in brain, pituitary and uterus. Br J Pharmacol 59:397–402
Girard A, Sandulesco G, Friedenson A, Rutgers JJ (1932a) Sur une nouvelle hormone sexuelle cristallisee. C R Acad Sci 195:982
Girard A, Sandulesco G, Friedenson A, Rutgers JJ (1932b) Sur une nouvelle hormone sexuelle cristallisee retiree de l'urine de juments gravides. C R Acad Sci 194:909
Gorill MJ, Marshall JR (1986) Pharmacology of estrogens and estrogen-induced effects on nonreproductive organs and systems. J Reprod Med 31:842–847
Gorman CM, Moffat LF, Howard BH (1982) Recombinant genomes which express chloramphenicol acetyltransferase in mammalian cells. Mol Cell Biol 2:1044–1051
Gorski J, Gannon F (1976) Current models of steroid hormone action: a critique. Annu Rev Biochem 38:425–450
Gottardis MM, Jiang SY, Jeng MH, Jordan VC (1989) Inhibition of tamoxifen-stimulated growth of an MCF-7 tumour variant in athymic mice by a novel steroidal antiestrogen. Cancer Res 49:4090–4093
Greco T, Duello T, Gorski J (1993) Estrogen receptors; estradiol; and diethylstilbestrol in early development: the mouse as a model for the study of estrogen receptors and estrogen sensitivity in embryonic development of male and female reeproductive tracts. Endocr Rev 14:59–71
Green S, Kumar V, Theulaz I, Wahli W, Chambon P (1988) The N-terminal DNA-binding "zinc finger" of the estrogen and glucocorticoid receptors determines target gene specificity. EMBO J 7:3037–3044
Guetta V, Cannon III RO (1996) Cardiovascular effects of estrogen and lipid-lowering therapies in post-menopausal women. Circulation 93:1928–1937
Habenicht UF, Schwarz K, Neumann F, el Etreby MF (1987) Induction of estrogen-related hyperplastic chages in the prostate of the cynomolgus monkey (*Macaca fascicularis*) by androstenedione and its antagonization by the aromatase inhibitor 1-methyl-androsta-1,4-diene-3,17-dione. Prostate 11:313–326
Halbreich U (1997) Role of estrogen in postmenopausal depression. Neurology 48[Suppl 7]:S16–S19
Halliwel B, Gutteridge JMC (1989) Comments on review of free radicals in biology and medicine. Free Radic Biol Med 12:93–95
Harper MJ, Walpole AL (1967) A new derivative of triphenylethylene: effect on implantation and mode of action in rats. J Reprod Fertil 13:101–119

Heery DM, Kalkhoven E, Hoare S, Parker MG (1997) A signature motif in transcriptional co-activators mediates binding to nuclear receptors. Nature 387:733–736
Hegele-Hartung C, Fritzemeier KH, Diel P (1997) Effects of a pure antiestrogen and progesterone on estrogen-mediated alterations of blood flow and progesterone receptor expression in the aorta of ovariectomized rabbits. J Steroid Biochem Mol Biol 63:237–249
Henderson D (1997) The epidemiology of estrogen replacement therapy and Alzheimer's disease. Neurology 48[Suppl 7]:S27–S35
Hernandez I, Pobletee A, Amthauer R, Pessot R, Krauskopf M (1992) Effect of seasonal acclimation on estrogen induced vitellogenesis and on the hepatic estrogen receptors in the male carp. Biochem Int 28:559–567
Hess RA, Bunick D, Lee Ki-Ho, Bahr J, Taylor JA, Korach KS, Lubhahn DB (1997) A role for estrogens in the male reproductive system. Nature 390:509–512
Hisaw FL Jr (1959) Comparative effectiveness of estrogens on fluid inhibition and growth of the rat's uterus. Endocrinology 64:276–289
Hisaw FL, Hisaw FL, Jr. (1973) Action of estrogen and progesterone on the reproductive tract of lower pimates. In: Young WC (ed) Sex and internal secretions, vol 1. Robert E. Krieger Publishing Company, Huntington New York, pp 556–589
Holinka CF, Hata H, Kuramoto H, Gurpide E (1986) Steroids and cancer: responses to estradiol in a human endometrial adenocarcinoma cell line (Ishikawa). J Steroid Biochem 24:85–89
Horigome T, Golding TS, Quarmby VE, Lubahn DB, McCarty KSr, Korach KS (1987) Purification and characterization of mouse uterine estrogen receptor under conditions of varying hormonal status. Endocrinology 121:2099–2111
Horlein AJ, Naar AM, Heinzel T, Torchia J, Gloss B, Kurokawa R, Ryan A, Kamei Y, Soderstrom M, Glass CK (1995) Ligand dependent repression by the thyroid hormone receptor mediated by a nuclear receptor co-repressor. Nature 377:397–404
Hosie MJ, Murphy CR (1995) A scanning and light microscope study comparing the effects of cloniphene citrate, estradiol 17-beta and progesterone on the structure of uterine luminal epithelial cells. Eur J Morphol 33:39–50
Hughes DE, Dai A, Tiffee JC, Li HH, Mundy GR, Boyce BF (1996) Estrogen promotes apoptosis of murine osteoclasts mediated by TGF-β. Nat Med 2:1132–1136
Iafrati MD, Karas RH, Aronovitz M, Kim S, Sullivan TR, Lubahn DB, O'Donnell TF, Korach KS, Mendelsohn ME (1997) Estrogen inhibits the vascular injury response in estrogen receptor alpha-deficient mice. Nat Med 3:545–548
Ignar-Trowbridge DM, Nelson KG, Bidwell MC, Curtis SW, Washburn TF, McLachlan JA, Korach KS (1992) Coupling of dual signaling pathways: epidermal growth factor action involves the estrogen receptor. Proc Natl Acad Sci USA 89:4658–4662
Imagawa W, Yang J, Guzman R, Nandi S (1994) Control of mammary gland development. In: Knobil E, Neill JD (eds) The physiology of reproduction, 2nd edn. Raven Press, New York, pp 1033–1063
Inhoffen HH, Logemann W, Hohlweg W, Serini A (1938) Untersuchungen in der Sexualhormon-Reihe. Berl Dtsch Chem Ges 71:1024–1032
Ivell R, Richter D (1984) Structure and comparison of the oxytocin and vasopressin genes from rat. Proc Natl Acad Sci USA 81:2006–2010
Jakowlev SB, Breathnach R, Jeltsch JM, Masiakowski P, Chambon P (1984) Sequence of the pS2 mRNA induced by estrogen in the human breast cancer cell line MCF-7. Nucleic Acids Res 12:2861–2878
Jamil A, Croxtall JD, White JO (1990) The effect of anti-estrogens on cell growth and progesterone receptor concentration in human endometrial cancer cells (Ishikawa). J Mol Endocrinol 6:215–221
Jardetzky CD, Glick D (1956) Studies in histochemistry XXXVIII. Determination of succinic dehydrogenase in microgram amounts of tissue and its distribution in rat adrenal. J Biol Chem 218:283–292

Jennings TS, Creasman WT (1997) Effects on the reproductive tract: clinical aspects. In: Lindsay R, Dempster DW, Jordan VC (eds) Estrogens and antiestrogen. Lippincott-Raven, Philadelphia, pp 223–242
Jensen EV, De Sombre ER (1973) Estrogen-receptor interaction. Science 182:126–134
Jensen EV, Jacobson HI (1960) Fate of steroid estrogens in target tissue. In: Pincus G, Vollmer EP (eds) Biological activities of steroids in relation to cancer. Academic Press, New York, pp 161–178
Jensen EV, Mohla S, Gorell TA, De Sombre ER (1974) The role of estrophilin in estrogen action. Vitam Horm 32:89–127
Jilka RL, Hangoc G, Girasole G, Passeri G, Williams DC, Abrams JS, Boyce B, Broxmeyer H, Manolagas SC (1992) Increased osteoclast development after estrogen loss: mediation by interleukin-6. Science 257:88–91
Jin L, Borras M, Lacroix M, Legros N, Leclerq G (1995) Antiestrogenic activity of two 11β-estradiol derivatives on MCF-7 breast cancer cells. Steroids 60:512–518
Jones TH (1994) Interleukin-6 an endocrine cytokine. Clin Endocrinol 40:703–713
Jones RC, Edgren RA (1973) The effects of various steroids on the vaginal histology in the rat. Fertil Steril 24:284–291
Jordan VC, Lieberman ME (1984) Estrogen-stimulated prolactin synthesis in vitro. Mol Pharmacol 26:279–285
Jordan VC, Murphy CS (1990) Endocrine pharmacology of antiestrogens as antitumour agents. Endocr Rev 11:578–610
Jordan VC, Mittal S, Gosden B, Koch R, Lieberman ME (1985) Structure-activity relationships of estrogens. Environ Health Perspect 61:97–110
Kahnt LC, Doisy EA (1928) The vaginal smear method of assay of the ovarian hormone. Endocrinology 12:760–768
Kaling M, Weimar-Ehl T, Kleinhaus M, Ryffel GU (1990) Transcription factors different from the estrogen receptor stimulate in vivo transcription from promotors containing estrogen response elements. J Steroid Biochem Mol Biol 37:733–739
Kallas H (1929) Puberté précoce par parabiose. Compt Rend Biol 100:974–980
Kaltschmidt B, Baeuerle PA, Kaltschmidt Ch (1993) Potential involvement of the transcription factor NF-κB in neurological disorders. Mol Aspects Med 14:171–190
Kamei Y, Xu L, Heinzel T, Torchia J, Kurokawa R, Gloss B, Lin SC, Heyman RA, Rose DW, Glass CK, Rosenfeld MG (1996) A CBP integrator complex mediates transcriptional activation and AP-1 inhibition by nuclear receptors. Cell 85:403–414
Karlsson R, Michaelsson A, Mattsson L (1991) Kinetic analysis of monoclonal antibody-antigen interactions with a new biosensor based analytical system. J Immunol Methods 145:229–240
Kasper P, Witzel H (1985) Steroid binding to the cytosolic estrogen receptor from rat uterus. Influence of the orientation of substituents in the 17-position of the 8β- and 8α-series. J Steroid Biochem:259–265
Kastner P, Krust A, Turcotte B, Stropp U, Tora L, Gronemeyer H, Chambon P (1990) Two distinct estrogen-regulated promotors generate transcripts encoding the two functionally different progesterone receptor forms A and B. EMBO J 9:1603–1614
Kato S, Tora L, Yamauchi, Masushige S, Rellard M, Chambon P (1992) A far upstream estrogen response element of the ovalbumin gene contains several half-palindromic 5′-TGACC-3′ motifs acting synergistically. Cell 68:731–742
Katzenellenbogen BS, Norman JJ (1990) Multihormonal regulation of the progesterone receptor in MCF-7 human breast cancer cells: interrelationships among insulin/insulin-like growth factor-I, serum, and estrogen. Endocrinology 126:891–898
Katzenellenbogen JA, Johnson HJJr, Myers NN (1973) Photoaffinity labels for estrogen binding proteins of rat uterus. Biochemistry 12:4085–4092
Katzenellenbogen BS, Kendra KL, Norman MJ, Bertois Y (1987) Proliferation, hormonal responsiveness and estrogen receptor content of MCF-7 human breast cancer cells grown in the short and long-term absence of estrogens. Cancer Res 47:4355–4360

Kauffman RK, Bryant HU (1995) Selective estrogen receptor modulators DN & P 8:531–539
Keeping HS, Lyttle CR (1982) Modulation of rat uterine progesterone receptor levels and peroxidase activity by tamoxifen citrate, LY117018 and estradiol. Endocrinology 111:2046–2054
Kelleher CJ, Cardozo L (1997) Estrogens, antiestrogens and the urogenital tract. In: Lindsay R, Dempster DW, Jordan VC (eds) Estrogens and antiestrogens. Lippincott-Raven, Philadelphia, pp 243–257
Kiang DT, Kollander RE, Thomas T, Kennedy BJ (1989). Up-regulation of estrogen receptors by nonsteroidal antiestrogens in human breast cancer. Cancer Res 49:5312–5316
Kirkland JL, LaPointe ML, Justin E, Stancel GM (1979) Effects of estrogen on mitosis in individual cell types of the immature rat uterus. Biol Reprod 21:269–272
Kirkland JL, Murthy L, Stancel GM (1992) Progesterone inhibits the estrogen-induced expression of c-fos messenger ribonucleic acid in the uterus. Endocrinology 130:3223–3230
Kirkland JL, Murthy L, Stancel GM (1993) Tamoxifen stimulates expression of the c-fos protooncogene in rodent uterus. Mol Pharmacol 43:709–714
Klein T, Potgieter HC, Spies JH, van der Watt JJ, Savage N (1985) Reversibility of the stabilization effect of sodium molybdate on uterine estrogen and progesterone receptors of the vervet monkey. J Recept Res 5:267–295
Klein-Hitpass L, Schorpp M, Wagner U, Ryffel GU (1986). An estrogen-responsive element derived from the 5′-flanking region of the Xenopus vitellogenin A2 gene functions in transfected human cells. Cell 46:1053–1061
Klein-Hitpass L, Ryffel GU, Heitlinger E, Cato ACB (1988) A palindrome is a functional estrogen responsive element and interacts specifically with estrogen receptor. Nucleic Acids Res 16:647–663
Knauthe R, Diel P, Hegele-Hartung C, Engelhaupt A, Fritzemeier KH (1996) Sexual dimorphism of steroid hormone receptor messenger ribonucleic acid expression and hormonal regulation in rat vascular tissue. Endocrinology 137:3220–3227
Knight DC, Eden LA (1995) Phytoestrogens – a short review. Maturitas 22:167–175
Korach KS (1994) Insights from the study of animals lacking functional estrogen receptor. Science 266:1524–1527
Korach KS, Couse JF, Curtis SW, Washburn TF, Lindzey J, Kimbro KS, Eddy EM, Migliaccio S, Snedeker SM, Lubahn DB, Schomberg DW, Smith EP (1996) Estrogen receptor gene disruption: molecular characterization and experimental and clinical phenotypes. Recent Prog Horm Res 51:159–186
Krattenmacher R, Knauthe R, Parczyk K, Walter A, Hilgenfeld U, Fritzemeier KH (1994) Estrogen action on hepatic synthesis of angiotensinogen and IGF-I: direct and indirect estrogen effects. J Steroid Biochem Mol Biol 48:207–214
Kraus WL, Katzenellenbogen BS (1993) Regulation of progesterone receptor gene expression and growth in the rat uterus: modulation of estrogen actions by progesterone and sex steroid hormone antagonists. Endocrinolgy 132:2371–2379
Kraus WL, Montano MM, Katzenellenbogen BS (1994) Identification of multiple, widely spaced estrogen-responsive regions in the rat progesterone receptor gene. Mol Endocrinol 8:952–969
Krohn PL (1956) The administration of estrogens and progesterone to monkeys. J Endocrinol 14:12–15
Kronenberg MS, Clark JH (1985) Changes in keratin expression during the estrogen-mediated differentiation of rat vaginal epithelium. Endocrinology 117:1480–1489
Kudolo GB, Elder MG, Myatt L (1984) A novel estrogen-binding species in rat granulosa cells. J Endocrinol:83–91
Kuiper GG, Carlsson B, Grandien K, Enmark E, Häggblad J, Nilsson S, Gustafsson J-A (1996a) Comparison of ligand binding specificity and transcript tissue distribution of estrogen receptors α and β. Endocrinology 138:863–870

Kuiper GG, Enmark E, Pelto-Huikko M, Nilsson S, Gustafsson JA (1996b) Cloning of a novel estrogen receptor expressed in rat prostate and ovary. Proc Natl Acad Sci USA 93:5925–5930
Kumar V, Chambon P (1988) The estrogen receptor binds tightly to its responsive element as a ligand-induced homodimer. Cell 55:145–156
Lafyatis R, Lechleider R, Kim S-J, Jakowlew S, Roberts A, Sporn M (1990) Structural and functional characterizationof the transforming growth factor $\beta 3$ promotor. J Biol Chem 265:19128–19136
Lambris JD (1988) The multifunctional role of C3, the third component of complement. Immunol Today 9:387–393
Lapiere CL (1978) Estrogene hormone. Pharmazeutische Zeitung 123:2227–2231
Lauson HD, Heller CG, Golden JB, Severinghaus EL (1939) The immature rat uterus in the assay of estrogenic substances, and a comparison of estradiol, estrone and estriol. Endocrinology 24:35–44
Le Drean Y, Lazennec G, Kern L, Saligaut D, Pakdel F, Valotaire V (1995) Characterization of an estrogen-responsive element implicated in regulation of the rainbow trout estrogen receptor gene. J Mol Endocrinol 15:37–47
Lee MO, Liu Y, Zhang XK (1995) A retinoic acid response element that overlaps an estrogen response element mediates multihormonal sensitivity in transcriptional activation of the lactoferrin gene. Mol Cell Biol 15:4194–4207
Li X, Schwartz PE, Rissman EF (1997) Distribution of estrogen receptor-β-like activity in rat forebrain. Neuroendocrinology 66:63–67
Lippman ME, Dickson RB (1989) Mechanisms of growth control in normal and malignant breast epithelium. Recent Prog Horm Res 45:383–440
Lipton SA, Rosenberg PA (1994) Excitatory amino acids as a final common pathway for neurological disorders. New Engl J Med 330:613–622
Liu D, Xiong F, Hew CL (1995) Functional analysis of estrogen-responsive elements in chinook salmon (*Oncorhynchus tschawytscha*) gonadotropin II beta subunit gene. Endocrinology 136:3486–3493
Liu Y, Teng CT (1992) Estrogen response module of the mouse lactoferrin gene contains overlapping chicken ovalbumin upstream promoter transcription factor and estrogen receptor-binding elements. Mol Endocrinol 6:355–364
Ljungkvist I (1971) Attachment reaction of rat uterine luminal epithelium. Acta Societatis Medicorum Upsaliensis 76:139–157
Lobo RA (1990) Estrogen and cardiovascular disease. Ann N Y Acad Sci 592:286–294
Loose-Mitchell DS, Chiappetta C, Stancel GM (1988) Estrogen regulation of c-fos messenger ribonucleic acid. Mol Endocrinol 2:946–951
Loutradis D, Bletsa R Aravantinos L, Kallianidis K, Michalas S, Psychoyos A (1991) Preovulatory effects of the progesterone antagonist mifepristone (RU486) in mice. Hum Reprod 6:1238–1240
Lubahn DB, Moyer JS, Golding TS, Couse JF, Korach KS, Smithies O (1993) Alteration of reproductive function but not prenatal sexual development after insertional disruption of the mouse estrogen receptor gene. Proc Natl Acad Sci USA 90:11162–11166
Lubke K, Schillinger E, Topert M (1976) Hormonrezeptoren. Angew Chem 88:790–798
Lukola A, Punnonen R (1983) Estrogen and progesterone receptors in human uterus and oviduct. J Endocrinol Invest 6:179–183
Lydon JP, DeMayo FJ, Funk CR, Mani D, Hughes AR, Montgomery CA Jr, Shyamala G, Conneely OM, O'Malley BW (1995) Mice lacking progesterone receptor exhibit pleiotropic reproductive abnormalities. Genes Dev 9:2266–2278
Lyttle CR, DeSombre ER (1977a) Generality of estrogen stimulation of peroxidase acitivity in growth response tissue. Nature 268:337–339
Lyttle CR, DeSombre ER (1977b) Uterine peroxidase as a marker for estrogen action. Proc Natl Acad Sci USA 74:3162–3166
Lyttle CR, Wheeler C, Komm BS (1987) Hormonal regulation of rat uterine secretory protein synthesis. In: Leavitt WW (ed) Cell and molecular biology of the uterus. Plenum, New York, pp 119–136

Maniatis T (1997) Catalysis by a IkB kinase complex. Science 278:818–819
Manni A, Baker R, Arafah BM, Pearson OH (1981) Uterine estrogen and progesterone receptors in the ovariectomized rat. J Endocrinol 91:281–287
Martin L (1960) The use of 2–3-5 triphenyltetrazolium chloride in the biological assay of estrogens. J Endocrinol 20:187–197
Martin L (1964) Further studies on the use of tetrazolium in the biological assay of estrogens. J Endocrinol 30:21–31
Martin L, Claringbold PJ (1958) Highly sensitive assay for estrogens. Nature 181:620–621
Martin L, Claringbold PJ (1960) The mitogenic action of estrogens in the vaginal epithelium of the ovariectomized mouse. J Endocrinol 20:173–186
Martin L, Pollard JW, Fagg B (1976) Oestriol, oestradiol-17β and the proliferation and death of uterine cells. J Endocrinol 69:103–115
Matzuk MM, Kumar TR, Shou W, Coerver KA, Lau A, Behringer RR, Finegold MJ (1996) Transgenic models to study the role of inhibins and activins in reproduction, oncogenesis, and development. Recent Prog Horm Res 51:123–157
May FE, Johnson MD, Wiseman LR, Wakeling AE, Kastner P, Westley BR (1989) Regulation of progesterone receptor mRNA by oestradiol and antiestrogens in breast cancer cell lines. J Steroid Biochem 33:1035–1041
Mayr U, Butsch A, Schneider S (1992) Validation of two in vitro test systems for estrogenic activities with Zearalenone, Phytoestrogens, and Cerea Extracts. Toxicology 74:35–149
McClelland RA, Gee JMW, Francis AB, Robertson JFR, Blamey RW, Wakeling AE, Nicholson RI (1996) Short-term effects of pure antiestrogen ICI182780 treatment on estrogen receptor, epidermal growth factor receptor and transforming growth factor-alpha protein expression in human breast cancer. Eur J Cancer 32A:413–416
McDonnell DP, Nawaz Z, Densmore C, Weigel NL, Pham TA, Clark JH, O'Malley BW (1991) High level expression of biologically active estrogen receptors in Saccharomyces Cerevisiae. J Steroid Mol Biol 39:291–297
McEwen BS, Alves SE, Bulloch K, Weiland NG (1997) Ovarian steroids and the brain: implications for cognition and aging. Neurology 48[Suppl 7]:S8–S15
McKay ME, Raelson J, Lazier CB (1996) Upregulation of estrogen receptor mRNA and estrogen receptor activity by estradiol in liver of rainbow trout and other teleostean fish. Comp Biochem Physiol C Pharmacol Endcrinol 115:201–209
McNabb T, Jellinck PH (1976) Effect of nafoxidine (U-11,100 A) on the induction of uterine peroxidase. Steroids 27:681–689
Metzger D, White JH, Chambon P (1988) The human estrogen receptor functions in yeast. Nature 334:31–36
Metzger D, Berry M, Ali S, Chambon P (1995) Effect of antagonists on DNA binding properties of the human estrogen receptor in vitro and in vivo. Mol Endocrinol 9:579–591
Miksicek RJ (1992) Commonly occurring plant flavonoids have estrogenic activity. Mol Pharmacol 44:37–43
Miksicek RJ (1994) Interaction of naturally occuring nonstroidal estrogens with expressed recombinant human estrogen receptor. J Steroid Biochem Mol Biol 49:153–160
Miller JH (ed) (1972) Experiments in molecular genetics. Cold Spring Harbor, New York, pp 352–355
Morgan PM, Hutz RJ,Kraus EM, Bavister B (1987) Ultrasonographic assessment of the endometrium in rhesus monkeys during the normal menstrual cycle. Biol Reprod 36:463–469
Mori K, Fujii R, Kida N, Takahashi H, Ohkubo S, Fujino M, Ohta M, Hayashi K (1990) Complete primary structure of the human estrogen-responsive gene (pS2) product. J Biochem 107:73–76
Morishima A, Grumbach MM, Simpson ER, Fisher C, Qin K (1995) Aromatase deficiency in male and female siblings caused by a novel mutation and the physiological role of estrogens. J Clin Endocrinol Metab 80:3689–3698

Mosselman S, Polman J, Dijkema R (1996) ERβ: identification and characterization of a novel human estrogen receptor. FEBS Lett 392:49–53

Nardulli AM, Greene GL, O'Malley BW, Katzenellenbogen BS (1988) Regulation of progesterone receptor messenger ribonuleic acid and protein levels in MCF-7 cells by estradiol: analysis of estrogen's effect on progesterone receptor synthesis and degradation. Endocrinology 122:935–944

Nathan L, Chaudhuri G (1997) Estrogens and atherosclerosis. Annu Rev Pharmacol Toxicol 37:477–515

Nativ O, Umehara T, Colvard DS, Therneau TM, Farrow GM, Spelsberg TC, Lieber MM (1997) Relationship between DNA ploidy and functional estrogen receptors in operable prostate cancer. Eur Urol 32:96–99

Nephew KP, Polek TC, Khan SA (1996) Tamoxifen-induced proto-oncogene expression persists in uterine endometrial epithelium. Endocrinology 137:219–224

Nishida M, Kashahara K, Kaneko M, Iwasaki H, Hayashi K (1985) Establishment of a new human endometrial carcinoma cell line containing estrogen and progesterone receptors. Acta Obstet Gynecol Jpn 37:1103–1111

Nishimori K, Matzuk M (1996) Transgenic mice in the analysis of reproductive development and function. Rev Reprod 1:203–212

Nishino Y, Neumann F (1974) The sialic acid content in mouse female reproductive organs as a quantitative parameter for testing the estrogenic and antiestrogenic effect, antiestrogenic depot effect, and dissociated effect of estrogens on the uterus and vagina. Acta Endocr 76[Suppl 187]:1–62

Noguchi S, Motomura K, Inaji H, Imaoka S, Koyama, H (1993) Up-regulation of estrogen receptor by tamoxifen in human breast cancer. Cancer 71:1266–1276

Norman AW, Litwack G (1987) Estrogens and progestins. In: Norman AW, Litwack G (eds) Hormones. Academic Press, New York, pp 516–564

Norris Daju Fan JD, Wagner BL, McDonnell DP (1996) Identification of the sequences within the human complement 3 promotor required for estrogen responsiveness provides insight into the mechanism of tamoxifen mixed agonist activity. Mol Endocrinol 10:1605–1616

Nunez AM, Berry M, Imler JL, Chambon P (1989) The 5'-flanking region of the pS2 gene contains a complex enhancer region responsive to estrogens, epidermal growth factor, a tumour promoter (TPA), the c-HA-ras oncoprotein and the c-jun protein. EMBO J 8:823–829

Ogawa S, Taylor JA, Lubahn DB, Korach KS, Pfaff DW (1996) Reversal of sex roles in genetic female mice by disruption of estrogen receptor gene. Neuroendocrinology 64:467–470

Ogawa S, Lubahn DB, Korach KS, Pfaff DW (1997) Behavioral effects of estrogen receptor gene disruption in male mice. Proc Natl Acad Sci USA 94:1476–1481

O'Malley BW, Schrader WT, Mani S, Smith C, Weigel NL, Conneely OM, Clark JH (1995) An alternative ligand-independent pathway for activation of steroid receptors. Recent Prog Horm Res 50:333–347

Onate SA, Tsai SY, Tsai MJ, O'Malley BW (1995) Sequence and characterization of a coactivator for the steroid receptor superfamily. Science 270:1354–1357

Onoe Y, Miyaura C, Ohta H, Nozawa S, Suda T (1997) Expression of estrogen receptor β in rat bone. Endocrinology 138:4509–4512

Pakdel F, Feon S, Le Gac F, Le Menn F, Valotaire Y (1991) In vivo estrogen induction of hepatic estrogen receptor mRNA and correlation with vitellogenin mRNA in rainbow trout. Mol Cell Endocrinol 75:205–212

Palmiter RD, Brinster RL (1986) Germ-line transformation of mice. Annu Rev Genet 20:465–499

Parczyk K, Schneider MR (1996) The future of antihormone therapy: innovations based on established principles, J Cancer Res Clin Oncol 122:383–396

Parczyk K, Madjno R, Angerer E von, Schneider MR (1995) Various pure antiestrogens downregulated the ER while partial agonists do not. In: Proceedings of 12th

International Symposium of the Journal of Steroid Biochemistry and Molecular Biology, Berlin

Peach K, Webb P, Kuiper GG JM, Nilsson S, Gustafsson J-A, Kushner PJ, Scanlan ThS (1997) Differential ligand activation of estrogen receptors ERα and ERβ at AP1-sites. Science 277:1508–1510

Pedersen-Bjergaard K (1939) Comparative studies concerning the strengths of estrogenic substances. Oxford University Press, London

Pelissero C, Flouriot G, Foucher JL, Bennetau, Dunogues J, Le Gac F, Sumpter JP (1993) Vitellogenin synthesis in cultured hepatocytes; an in vitro test for the estrogenic potency of chemicals. J Steroid Biochem Mol Biol 44:263–272

Peterson DL, Edgren RA, Jones RC (1964) Steroid-induced block of ovarian compensatory hypertrophy in hemicastrated female rats. J Endocrinol 29:255–264

Pham TA, Elliston JF, Nawaz Z, McDonnell DP, Tsai MJ, O'Malley BW (1991) Antiestrogen can establish nonproductive receptor complexes and alter chromatin structure at target enhancers. Proc Natl Acad Sci USA 88:3125–3129

Pike MC, Peters RK, Cozen W, Probst-Hensch NM, Felix JC, Wan PC, Mack TM (1997) Estrogen-progestin replacement therapy and endometrial cancer. J Natl Cancer Inst 89:1110–1116

Pilat MJ, Hafner MS, Kral LG, Brooks SC (1993) Differential induction of pS2 and cathepsin D mRNAs by structurally altered estrogens. Biochemistry 32:7009–7015

Poli V, Balena R, Fattori E, Markatos A, Yamamoto M, Tanaka H, Ciliberto G, Rodan GA, Costantini F (1994) Interleukin-6 deficient mice are protected from bone loss caused by estrogen depletion. EMBO J 13:1189–1196

Porter W, Wang F, Wang W, Duan R, Safe S (1996) Role of an Spl/ERE in E2-induced Hsp 27 gene expression. Mol Endocrinol 10:1371–1378

Potgieter HC, Spies JH, Klein T, Thierry M, Vanderwatt JJ (1985) Estrogen and progesterone receptors in the uterus of the vervet monkey. J Recent Res 5:193–218

Pottratz ST, Bellido T, Mocharla H, Crabb D, Manolagas SC (1993) 17β-Estradiol inhibits expression of human interleukin-6 promotor-reporter constructs by a receptor-dependent mechanism. J Clin Invest 93:944–950

Prichard KI (1997) Effects on breast cancer. In: Lindsay R, Dempster DW, Jordan VC (eds) Estrogens and antiestrogens. Lippincott-Raven, Philadelphia, pp 175–210

Punnonen R, Lukola A (1982) High-affinity binding of estrone, estradiol and estriol in human cervical myometrium and cervical and vaginal epithelium. J Endocrinol Invest 5:203–207

Quamby VE, Korach KS (1984) The influence of 17β-estradiol on patterns of cell division in the uterus. Endocrinology 114:694–702

Raisz LG (1996) Estrogen and bone: new pieces to the puzzle. Nat Med 2:1077–1136

Ramamoorthy K, Wang F, Chen Chen I, Norris JD, McDonnell DP, Leonard LS, Gaido KW, Bocchinfuso WP, Korach KS, Safe S (1997) Estrogenic activity of a dieldrin/toxaphene mixture in the mouse uterus, MCF-7 human breast cancer cells and yeast-based estrogen receptor assays: no apparent synergism. Endocrinology 138:1520–1527

Ramkumar T, Adler S (1995) Differential positive and negative transcriptional regulation by tamoxifen. Endocrinology 136:536–542

Ray P, Ghosh SK, Zhang D-H, Ray A (1997) Repression of interleukin-6 gene expression by 17β-estradiol: inhibition of the DNA-binding activity of the transcription factors NF-IL6 and NF-κB by the estrogen receptor. FEBS Lett 409:79–85

Revelli A, Massobrio M, Tesarik J (1998) Nongenomic actions of steroid hormones in reproductive tissues. Endocr Rev 19:3–17

Rhode PR, Gorski J (1991) Growth and cell cycle regulation of mRNA levels in GH3 cells. Mol Cell Endocrinol 82:11–22

Rishi AK, Shao ZM, Baumann RG, Li XS, Sheikh MS, Kimura S, Bashirelahi N, Fontana JA (1995) Estradiol regulation of the human retinoic acid receptor alpha gene in human breast carcinoma cells is mediated via an imperfect half-palindromic estrogen response element and SP1 motifs. Cancer Res 55:4999–5006

Robinson GA, Gibbins AM (1984) Induction of vitellogenesis in Japanese quail as a sensitive indicator of the estrogen-mimetic effect of a variety of environmental contaminants. Poult Sci 63:1529–1536

Robinson JA, Riggs BL, Spelsberg TC, Oursler MJ (1996) Osteoclasts and transforming growth factor-β: estrogen-mediated isoform-specific regulation of production. Endocrinology 137:615–621

Robson JM and Adler J (1940) Site of action of estrogens. Nature 146:160

Robzyk K, Kassir Y (1992) A simple and highly efficient procedure for rescuing autonomous plasmids from yeast. Nucleic Acids Res 20:3790

Rochefort H (1987) Nonsteroidal antiestrogens are estrogen-receptor targeted growth inhibitors that can act in the absence of estrogens. Hormone Res 28:196–201

Rodriguez C, Calle EE, Coates RJ, Miracle-McMahill HL, Thun MJ, Heath CW, Jr (1995) Estrogen replacement therapy and fatal ovarian cancer. Am J Epidemiol 141:828–835

Römer W, Oettel M, Menzenbach B, Droescher P, Schwarz S (1997) Novel estrogens and their radical scavenging effects, iron-chelating, and total antioxidative activities: 17α-substituted analogs of $\delta^{9(11)}$-dehydro-17β-estradiol. Steroids 62:688–694

Routledge EJ, Sumpter JP (1997) Structural features of alkylphenolic chemicals associated with estrogenic activity. J Biol Chem 272:3280–3288

Rumpel E, Michna H, Kühnel W (1995) PCNA-immunoreactivity in the uterus of rats after treatment with the antiestrogen tamoxifen. Ann Anat 177:133–138

Sabbah M, Gouilleux F, Sola B, Redeuilh G, Baulieu E-E (1991) Structural differences between the hormone and antihormone estrogen receptor complexes bound to the hormone response element. Proc Natl Acad Sci USA 88:390–394

Saiduddin S, Zassenhaus HP (1977) Estradiol-17beta receptors in the immature rat ovary. Steroids 29:197–213

Sarrel PM (1992) Vasoactive effects of estrogens. In: Crosignani PG, Paoletti R, Sarrel PM, Wenger NK (eds) Womens health in menopause. Endothelium 2:203–208

Sato M, Rippy MK, Bryant HU (1996) Raloxifene, tamoxifene, nafoxidine, or estrogen effects on reproductive and nonreproductive tissues in ovariectomized rats. FASEB J 10:905–912

Satyaswaroop PG (1997) Estrogenic and antiestrogenic actions of tamoxifen in the female reproductive tract. In: Lindsay R, Dempster DW, Jordan VC (eds) Estrogens and antiestrogens. Lippincott-Raven, Philadelphia, pp 211–221

Saunders PTK, Maguire SM, Gaughan J, Millar MR (1997) Expression of estrogen receptor beta (ERβ) in multiple rat tissues visualized by immunohistochemistry. J Endocrinol 154:R13–R16

Savouret JF, Bailly A, Misrahi M, Rauch C, Redeuilh G, Chauchereau A, Milgrom E (1991) Characterization of the hormone responsive element involved in the regulation of the progesterone receptor gene. EMBO J 10:1875–1883

Schane HP, Anzalone AJ, Potts GO (1972) A model for the evaluation of estrogens: withdrawal bleeding in ovariectomized rhesus monkeys. Fertil Steril 23:745–750

Schimke RT, McKnight GS, Shapiro DJ, Sullivan D, Palacios R (1975) Hormone regulation of ovalbumin synthesis in the chick oviduct. Recent Prog Horm Res 31:175–211

Schultze AB (1964) Triphenyltetrazolium reduction by uterine tissue of rats. Proc Soc Exp Biol Med 116:653–655

Scully KM, Gleiberman AS, Lindzey J, Lubahn DB, Korach KS (1997) Role of estrogen receptor α in the anterior pituitary gland. Mol Endocrinol 11:674–681

Sehgal PB (1992) Regulation of IL6 gene expression. Res Immunol:724–734

Seifter S, Dayton S, Novic B, Muntwyler E (1950) The estimation of glycogen with the anthrone reagent. Arch Biochem 25:191–200

Sherwin BB (1997a) Estrogen effects on cognition in menopausal women. Neurology 48[Suppl 7]:S21–S26

Sherwin BB (1997b) Estrogenic effects on the central nervous system: clinical aspects. In: Lindsey R, Dempster DW, Jordan VC (eds) Estrogens and antiestrogens. Lippincott-Raven, Philadelphia, pp 75–89

Shughrue PJ, Komm B, Merchenthaler I (1996) The tissue distribution of estrogen receptor-β mRNA in the rat hypothalamus. Steroids 61:678–681
Shughrue PJ, Lane MV, Merchenthaler I (1997a) Regulation of progesterone receptor messenger ribonucleic acid in the rat medial preoptic nucleus by estrogenic and antiestrogenic compounds: an in situ hybridization study. Endocrinology 138:5476–5484
Shughrue PJ, Lubahn DB, Negro-Vilar A, Korach K, Merchenthaler I (1997b) Responses in the brain of ERKO mice. Proc Natl Acad Sci USA 94:11008–11012
Shull JD, Gorski J (1984) Estrogen regulation of prolactin gene transcription in vivo: paradoxical effects of 17β-estradiol dose. Endocrinology 124:279–285
Shull JD, Beams FE, Baldwin TM, Gilchrist CA, Hrbek MJ (1992) The estrogenic and antiestrogenic properties of tamoxifen in GH_4C_1 pituitary tumour cells are gene specific. Mol Endocrinol 6:529–535
Shupnik MA, Gordon MS, Chin WW (1989) Tissue-specific regulation of rat estrogen receptor mRNAs. Mol Endocrinol 3:660–665
Smith DE (1967) Location of the estrogen effect on uterine glucose metabolism. Proc Soc Exp Biol Med 124:747–749
Smith JS, Thomas P (1991) Changes in hepatic estrogen receptor concentrations during the annual reproductive and ovarian cycles of a marine teleost, the spotted seatrout, Cynoscion nebulosus. Gen Comp Endocrinol 81:234–245
Smith CL, Conneely OM, O'Malley BW (1993) Modulation of the ligand-independent activation of the human estrogen receptor by hormone and antihormone. Proc Natl Acad Sci USA 90:6120–6124
Somasekhar MB, Gorski J (1988) An estrogen-responsive element from the 5′-flanking region of the rat prolactin gene functions in MCF-7 but not in HeLa cells. Gene 69:23–28
Stein B, Yang MX (1995) Repression of the interleukin-6 promotor by estrogen receptor is mediated by NF-κB and C/EBPβ. Mol Cell Biol 15:4971–4979
Steingold KA, Matt DW, deZiegler D, Sealy JE, Fratkin M, Reznikov S (1991) Comparison of transdermal to oral estradiol administration on hormonal and hepatic parameters in women with premature ovarian failure. J Clin Endocrinol Metab 73:275–280
Stockard CR, Papanicolaou GN (1917) The existence of a typical oestrous cycle in the guinea pig – with a study of its histological and physiological changes. Am J Anat 22:225–284
Subbiah MTR, Kessel B, Agrawal M, Rajan R, Abplanalp W, Rymaszewski Z (1993) Antioxidant potential of specific estrogens on lipid peroxidation. J Clin Endocrinol Metab 77:1095–1097
Summers MD, Smith GE (1987) A manual of methods for baculovirus vectors and insect cell procedures. Texas Agricultural Experiment Station Bulletin No. 1555
Sundstrom SA, Komm BS, Ponce-de-Leon H, Yi Z, Teuscher C, Lyttle CR (1989) Estrogen regulation of tissue specific expression of complement C3. J Biol Chem 264:16941–16947
Sundstrom SA, Komm BS, Xu Q, Boundy V, Lyttle R (1990) The stimulation of uterine complement component C3 gene expression by antiestrogens. Endocrinology 126:1449–1456
Sutherland RL, Watts CKW, Clarke CL (1988) Estrogen actions. In: Cooke BA, King RJB, Der Molen HJv (eds) Hormones and their actions. Part 1. Elsevier, Amsterdam, pp 197–215
Svennerholm L (1957) Quantitative estimation of sialic acids II. Colorimetric resorcinol hydrochloric acid method. Biochem Biophys Acta 24:604–611
Svennerholm L (1958) Quantitative estimation of sialic acids III. An anion exchange resin method. Acta Chem Scand 12:547–554
Szego CM, Roberts S (1953) Steroid action and interaction in uterine metabolism. Recent Progr Horm Res 8:419–469
Thanos D, Maniatis T (1995) NF-kappa B: a lesson in family values. Cell 80:529–532

Thompson EW, Katz D, Shima TB, Wakeling AE, Lippman ME, Dickson RB (1989) ICI164,384, a pure antagonist of estrogen-stimulated MCF-7 cell proliferation and invasiveness. Cancer Res 49:6929–6934

Tonetti DA, Jordan VC (1996) Targeted antiestrogens to treat and prevent diseases in women. Mol Med Today 380:218–223

Tong JH, Layne DS, Dostaler S, Williamson DG (1983) Comparative studies of the 17β-estradiol receptors in rabbit liver, kidney and uterus. J Steroid Biochem 18:273–279

Tora L, Gaub MP, Mader S, Dierich A, Bellard M, Chambon P (1988) Cell-specific activity of a GGTCA half-palindromic estrogen-responsive element in the chicken ovalbumin gene promotor. EMBO J 7:3771–3778

Tora L, Mullik A, Metzger D, Ponglikitmongkol M, Park I, Chambon P (1989) The cloned human estrogen receptor contains a mutation which alters its hormone binding properties. EMBO J 8:1981–1986

Torchia J, Rose DW, Inostroza J, Kamei Y, Westin S, Glass CK, Rosenfeld MB (1997) The transcriptional co-activator p/CIP binds CBP and mediates nuclear receptor function. Nature 387:677–684

Tremblay GB, Tremblay A, Copeland NG, Gilbert DJ, Jenkins NA, Labrie F, Giguere V (1997) Cloning, chromosomal localization and functional analysis of the murine estrogen receptor β. Mol Endocrinol 11:353–365

Tsai MJ, O'Malley BW (1994) Molecular mechanisms of action of steroid/thyroid receptor superfamily members. Annu Rev Biochem 63:451–486

Turner RT (1997) Effects on bone and mineral metabolism: basic aspects. In: Lindsey R, Dempster DW, Jordan VC (eds) Estrogens and antiestrogens. Lippcott-Raven, Philadelphia pp 129–150

Van de Velde P, Nique F, Bouchoux F, Bremaud J, Hameau MC, Lucas D, Moratille C, Viet S, Philibert D, Teusch G (1994) RU58,688, a new pure antiestrogen inducing a regression of human mammary carcinoma implanted in nude mice. J Steroid Biochem Mol Biol 48:187–196

Van der Walt LA, Sanfilippo JS, Siegel JE, Wittliff JL (1986) Estrogen and progestin receptors in human uterus: reference ranges of clinical conditions. Clin Physiol Biochem 4:217–228

Vander Kuur JA, Wiese T, Brooks SC (1993) Influence of estrogen structure on nuclear binding and progesterone receptor induction by the estrogen receptor complex. Biochemistry 32:7002–7008

Vollmer G, Schneider MR (1996) The rat endometrial adenocarcinoma cell line RUCA-I: a novel hormone-responsive in vivo/in vitro tumour model. J Steroid Biochem Mol Biol 58:103–115

von Angerer E (1995a) Compounds acting as mixed agonists/antagonists. In: von Angerer E (ed) The estrogen receptor as a target for rational drug design. Springer, Berlin Heidelberg New York, pp 50–95

von Angerer E (1995b) Development of pure antiestrogens. In: von Angerer E (ed) The estrogen receptor as a target for rational drug design. Springer, Berlin Heidelberg New York, pp 97–126

von Schoultz B, Carlstrom K, Collste L, Eriksson A, Henriksson P, Pousette A, Stege R (1989) Estrogen therapy and liver function – metabolic effects of oral and parenteral administration. Prostate 14:389–395

Wagner BL, McDonnell DP (1996) Saccharomyces Cerevisiae as a model system to study steroid hormone receptors. In: Pavlik EJ (ed) Estrogens, progestins and their antagonists, vol 2. pp 47–67

Wakeling AE (1985) Chemical structure and pharmacology of antiestrogens in oncology. Past, present and prospects. Excerpa Medica, Amsterdam Geneva Hongkong, pp 43–53

Wakeling AE (1989) Comparative studies on the effects of steroidal and nonsteroidal estrogen antagonists on the proliferation of human breast cancer cells. J Steroid Biochem 34:183–188

Wakeling AE, Bowler J (1987) Steroidal pure antiestrogens. J Endocrinol 112:R7–R10
Wakeling AE, Dukes M, Bowler J (1991) A potent specific pure antiestrogen with clinical potential. Cancer Res 51:3867–3873
Wang F, Porter W, Xing W, Archer TK, Safe S (1997) Identification of a functional imperfect estrogen -responsive element in the 5′-promotor region of the human cathepsin D gene. Biochemistry 24:7793–7801
Warren L, Spicer SS (1961) Biochemical and histochemical identification of sialic acid containing mucins of rodent vagina and salivary glands. J Histochem Cytochem 9:400–408
Washburn SA, Honore EK, Cline JM, Helman M, Wagner JD, Adams MR, Adelman SJ, Clarkson TB (1996) Effects of 17α-dihydroequilenin sulfate on atherosclerotic male and female rhesus monkeys. Am J Obstet Gynecol 175:341–351
Waterton JC, Miller D, Dukes M, Morrell JSW (1991) Oblique NMR imaging of the uterus in macaques: uterine response to estrogen stimulation. Magn Reson Med 20:228–239
Watts CKW, Parker MG, King RJB (1989) Stable transfection of the estrogen receptor gene into a human osteosarcoma cell line. J Steroid Biochem 34:483–490
Webb DK, Moulton BC, Khan SA (1991) Estrogen induced expression of the c-jun proto-oncogene in the mature and immature rat uterus. Biochem Biophys Res Commun 175:480–485
Webb P, Lopez GN, Uht RM, Kushner PJ (1995) Tamoxifen activation of the estrogen receptor/AP-1 pathway: potential origin for the cell-specific estrogen-like effects of antiestrogens. Mol Endocrinol 9:443–456
Webster NJGS, Green S, Jin JR, Chambon P (1988) The hormone binding domain of the estrogen and glucocorticoid receptors contain an inducible transcription activation function. Cell 54:199–207
Weigel NL, Zhang Y (1998) Ligand-independent activaiton of steroid hormone receptors. J Mol Med 76:469–479
Weiser B (1992) Oxidized low-density lipoproteins in atherogenesis: possible mechanisms of action. J Cardiovasc Pharmacol 19[Suppl]:S4–S7
Weisz A, Bresciani F (1988) Estrogen induces expression of c-fos and c-myc protooncogenes in rat uterus. Mol Endocrinol 2:816–824
Weisz A, Rosales R (1990) Identification of an estrogen response element upstream of the human c-fos gene that binds the estrogen receptor and the AP-1 transcription factor. Nucleic Acids Res 18:5097–5106
West N, Brenner R (1990) Estrogen receptor in the ductuli efferentes epididymis and testis of rhesus and cynomolgus macaques. Biol Reprod 42:533–538
Westley BR, May EB (1987) Estrogen regulates cathepsin D mRNA levels in estrogen responsive breast cancer cells. Nucleic Acids Res 15:3773–3786
White R, Jobling S, Hoare SA, Sumpter JP, Parker MG (1994) Environmentally persistent alkylphenolics are estrogenic. Endocrinology 135:175–182
Wilcox JG, Hwang J, Hodis HN, Sevanian A, Stanczyk FZ, Lobo RA (1997) Cardioprotective effects of individual conjugated equine estrogens through their possible modulation of insulin resistance and oxidation of low-density lipoprotein. Fertil Steril 67:57–62
Yamamoto KR (1985) Steroid recceptor regulated transcription of genes and gene networks. Annu Rev Genet 19:209–252
Yan SD, Chen X, Fu J, Chen M, Zhu H, Roher A, Slattery T, Zhao L, Nagashima M, Morser J, Migheli A, Nawroth P, Stern D, Schmidt AM (1996) Rage and amyloid-β peptide neurotoxicity in Alzheimer's disease. Nature 382:685–691
Yang NN, Bryant HU, Hardiger S, Sato M, Galvin RJS, Glasebrook AL, Termine JD (1996a) Estrogen and raloxifene stimulate transforming growth factor-β3 gene expression in rat bone: a potential mechanism for estrogen- or raloxifen mediated bone maintenance. Endocrinology 137:2075–2084

Yang NN, Venugopalan M, Hardikar S, Glasebrook A (1996b) Identification of an estrogen response element activated by metabolites of 17β-estradiol and raloxifene. Science 273:1222–1225
Youdim MB, Lavie L, Riederer P (1994) Oxygen free radicals and neurogeneration in Parkinson's disease: a role for nitric oxide. Ann NY Acad Sci 738:64–68
Ziel HK (1982) Estrogens role in endometrial cancer. Obstet Gynecol 60:509–515

CHAPTER 22

Estrogen Receptor β in the Pharmacology of Estrogens and Antiestrogens

G. KUIPER, M. WARNER, and J.-Å. GUSTAFSSON

A. Introduction

The discovery of the second type of estrogen receptor (ER), ERβ (KUIPER et al. 1996), changed the way we understand the physiological and pharmacological actions of estrogens and antiestrogens. The findings of multiple isoforms of ERβ (CHU and FULLER 1997; TREMBLAY et al. 1997; MARUYAMA et al. 1998; MOORE et al. 1998; OGAWA et al. 1998a,b; PETERSEN et al. 1998) and multiple sites on DNA where these receptors can act (WEBB et al. 1995; YANG et al. 1996; PAECH et al. 1997; PORTER et al. 1997; UHT et al. 1997; MONTANO et al. 1998) introduce levels of complexity to the mechanisms of estrogen action that were previously unimagined. Many aspects of the actions of estrogens and antiestrogens have mystified pharmacologists for a long time (KATZENELLENBOGEN et al. 1996). It was very difficult to understand how estradiol could increase the risk of breast cancer, whereas phytoestrogens would reduce it, if they were acting on only one type of ER (INGRAM et al. 1997; STEPHENS 1997; HILAKIVI-CLARKE et al. 1998). Equally puzzling were the differential tissue effects of the antiestrogen tamoxifen, which is an estrogen antagonist in the breast but an agonist in the uterus, cardiovascular system and bone (BRYANT and DERE 1998).

The role of ERs in the male is also not well understood. Although it is clear from the one human patient with no functional ERα that this receptor is crucial for normal bone structure in males (SMITH et al. 1994), the exact site of estrogen action in male bone (SLEMENDA et al. 1997) and the source of estrogen remains uncertain. With more than one type of ER, a better understanding of estrogen actions is achievable. The demonstration (WATANABE et al. 1997; BARKHEM et al. 1998) that tamoxifen, 4-OH-tamoxifen, raloxifene, and ICI 164,384 are partial agonists/antagonists on ERα but pure antagonists on ERβ, makes it now possible to talk about "good" estrogens/ligands and "bad" estrogens/ligands. These estrogens/ligands would be defined according to the receptor profile in a cell or tissue. In this chapter, we will summarize what is presently known about the ERβ receptor subfamily and consider the pharmacological usefulness of specific ERβ agonists or antagonists in certain tissues, i.e., brain, bone, breast, and urogenital and cardiovascular systems.

B. Basic Similarities Between ERα and ERβ

ERα and ERβ are products of two distinct genes located on different chromosomes, ERα on human chromosome 6 and ERβ on human chromosome 14 (ENMARK et al. 1997). As with all members of the nuclear receptor supergene family (ENMARK and GUSTAFSSON 1996), ERα and ERβ share a basic three-domain structure (Fig. 1); there is an N-terminal domain (A/B domain) which is highly immunogenic and contains sites for interaction with proteins involved in transcriptional activation, a central DNA-binding domain (C domain) and a C-terminal ligand-binding domain (E/F domain). The highest degree of homology between ERα and ERβ is in the DNA-binding domain, where there is 95% identity of amino acid sequence. In the A/B and E/F domains, the homologies are less than 20% and 60%, respectively (KUIPER et al. 1996). Both ERα and ERβ bind to estradiol with high affinity and specificity, and both activate transcription of reporter genes containing estrogen response elements (EREs) in a hormone-dependent manner (KUIPER et al. 1996, 1998; TREMBLAY et al. 1997; BARKHEM et al. 1998).

C. ERβ Isoforms

Although there is a single ERβ gene, differential splicing gives rise to several isoforms of ERβ, some of which have been detected only at the RNA level by means of reverse-transcription polymerase chain reaction (RT-PCR) and some of which have been detected as proteins. What is currently known about these isoforms is summarized in Table 1 and Figs. 1, 2. Although the tissue distribution of these ERβ isoforms has not yet been completely studied, the idea that expression of specific ERβ isoforms could be involved in tissue-specific responses to estrogen has not escaped the scrutiny of endocrinologists. Two of the ERβ isoforms, whose potential pharmacological significance is at once apparent, are the $ER\beta_{503}$ and $ER\beta_{cx}$ (OGAWA et al. 1998b; PETERSEN et al. 1998). The $ER\beta_{503}$ isoform (insertion of 18 amino acid residues in ligand-binding domain) binds estradiol with a 35-fold lower affinity than $ER\beta_{485}$ [the originally cloned ERβ cDNA (KUIPER et al. 1996)], and requires a 1000-fold higher concentration of estradiol to activate an ERE-driven reporter (PETERSEN et al.

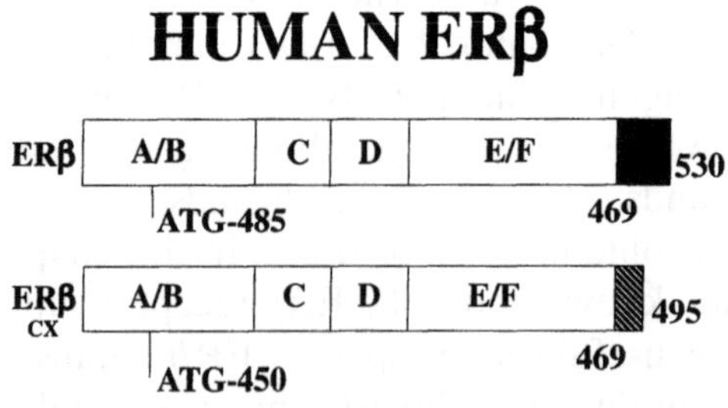

Fig. 1. Primary structures of estrogen receptor (ER)α and ERβ

Table 1. Isoforms of ERβ

No. of amino acids	Detection method	Predicted size of protein (kDa)	Species	Site of modification	Tissue
485 ERβ_{485}[a]	Western blot	55	Rat	Original ERβ cloned from rat prostate (Fig. 2)	Prostate
503 ERβ_{503}[b]	RT-PCR	57	Rat	18 amino acid insertion in the ligand-binding domain (Fig. 2)	Brain, lung, liver, kidney, fat, bone, uterus, prostate, ovary
485 ERβ_{485}/ 530 ERβ_{530}[c,d]	RT-PCR	57/61	Human	Predicted alternate initiation of translation from upstream ATG (45 amino acid extra) and downstream ATG (Fig. 1)	Spleen, testis, ovary, breast, uterus, fat
450/495 ERβ_{cx}[d,e]	RT-PCR/ Western blot	51/55	Human	C-terminal truncation of 61 amino acids and exchange with 26 novel amino acids (Fig. 1)	Testis, ovary, bone cells, thymus, spleen, Ishikawa cells, prostate tumor cells (PC3)
513[d]	RT-PCR	?	Human	C-terminal truncation of 61 amino acids compared with ERβ_{530} and exchange with novel 44 amino acids	Testis cDNA library (not in tissues?)
481[d]	RT-PCR	?	Human	Partial cDNA clone	Testis, ovary
472[d]	RT-PCR	?	Human	Partial cDNA clone	Testis, ovary
530/550[f,g]	RT-PCR	?	Rat and mouse	N-terminal extensions of rat ERβ_{485}	Prostate, ovary

[a] Kuiper et al. 1996.
[b] Maruyama et al. 1998; Petersen et al. 1998.
[c] Ogawa et al. 1998a.
[d] Moore et al. 1998.
[e] Ogawa et al. 1998b.
[f] Aldridge TC, GenBank AJ002602 (g) Leygue E GenBank AF067422.

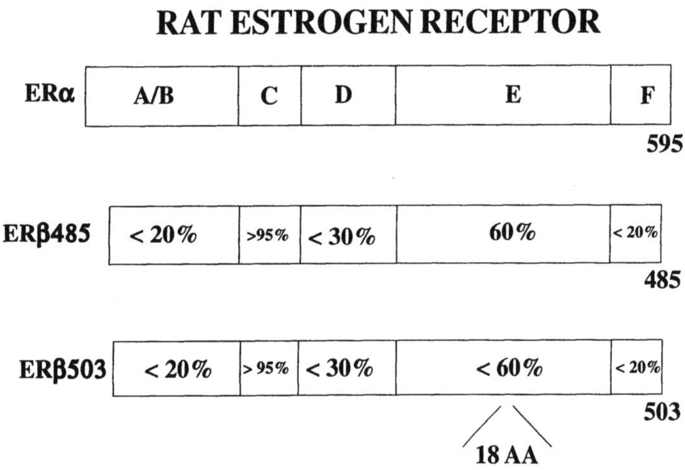

Fig. 2. Isoforms of estrogen receptor β

1998). The human $ER\beta_{cx}$ (C-terminal exchange) isoform does not bind estradiol at all and is not capable of activating estrogen-dependent transcription of an ERE-regulated reporter gene (OGAWA et al. 1998b). This is understandable since $ER\beta_{cx}$ lacks the C-terminal activation domain which is responsible for ligand-dependent transcriptional activation. It preferentially forms heterodimers with $ER\alpha$, and it has been suggested that $ER\beta_{cx}$ might be a repressor of $ER\alpha$ activity, assuming of course that both isoforms are present in the same cell types in vivo (OGAWA et al. 1998b).

One of the major problems now facing workers in this area is how to detect $ER\beta$ proteins in tissues and, even more problematic, how to identify the $ER\beta$ isoform profile of individual cells. At issue is how to prepare specific antibodies that distinguish between the various $ER\beta$ isoforms and which can be used for immunohistochemistry and Western blotting.

D. Role of Ligands in the Tissue-Specific Action of ERβ

Very intriguing to pharmacologists is the possibility that there are other so-far undiscovered ligands for these receptors, and that there is a whole new aspect of estrogen (natural and synthetic) pharmacology that remains to be explored. The first hint that this might be the case came from the surprising high affinity of $ER\beta$ for some of the plant estrogens, such as genistein (KUIPER et al. 1997, 1998). The described beneficial role of plant estrogens in human health (reduction of prostate and breast cancer risk) has long been questioned because, although they are estrogenic, their affinity for $ER\alpha$ is low, and it was questioned whether effective levels of these compounds could be achieved with a normal human diet (KURZER and XU 1997). In fact, for a long time, the beneficial effects of plant estrogens were thought to be due to their weak antagonist action on the ERα, i.e., they were thought of as estrogen-receptor blockers. The demonstrated high affinity of phytoestrogens for ERβ has changed our perception of the value of these dietary components to human health. It is now possible to imagine phytoestrogens binding to tissue-specific isoforms of ERβ and affecting transcription of a battery of genes that are not affected by ERα.

Further support for the notion that there might be tissue-specific effects of certain estrogens, depending on the ERβ isoform profile in a tissue, comes from the binding specificity of ERβ_{503}. Despite its low affinity for estradiol and genistein, ERβ_{503} retains a very high affinity for 4-OH-tamoxifen (PETERSEN et al. 1998). Specific ERβ isoforms in tissues such as the immune system and bone may prove to be extremely useful in specifically targeting these systems with isoform-selective ligands, either agonists or antagonists. It is well known that female bones benefit from estrogens, but the female immune system is at a disadvantage from high exposure to estrogens. This is evident from the higher incidence of autoimmune diseases in females (WILDER 1998) and the high incidence of osteoporosis in females after the menopause (CASTELO-BRANCO 1998). More recently, there has been a heightened interest in the role of estrogen in preventing or delaying the onset of Alzheimer's disease in females (SIMPKINS et al. 1997). If the correlation holds true, it may well be possible that estrogens can also be beneficial in the male brain (SEEMAN 1997). In that case, estrogens that specifically target the brain may be very useful.

Of course, if there is specificity for estrogen agonists, one obvious goal of pharmacologists would be to find ERβ-isoform-specific antagonists. Perhaps nowhere would such drugs be more important than in the treatment of breast cancer. In order to benefit from the effects of estrogens, the aim would be to target the breast and leave the bone, brain, skin, cardiovascular system and urogenital system unaffected.

E. Tissue Distribution of ERβ mRNA and Protein

Several peptide antibodies are commercially available and have been used with different degrees of success in different labs. SAUNDERS et al. (1997) used a peptide antibody raised against amino acids 196–213 of rat ERβ and detected nuclear staining in the ovary, uterus, lung, seminal vesicles, bladder, heart, adrenal and thymus. These are the tissues in which the presence of ERβ was found by means of in situ hybridization and RT-PCR (KUIPER et al. 1996, 1997). Interestingly, in male sex accessory tissue (prostate, testis and epididymis), ERβ appears to parallel androgen receptors in its localization (SAUNDERS et al. 1998; VAN PELT et al. 1999). This observation raises many questions about the interaction of these two hormonal systems in the male urogenital tract (SHARPE 1997).

In osteoblastic cells and osteosarcoma cell lines, ERs have been detected, although since these studies were conducted prior to the discovery of ERβ it is uncertain which subtype these cells contain (TURNER et al. 1994). A recent study using a human fetal osteoblastic cell line (SV-HFO) detected both ERα and ERβ mRNA (ARTS et al. 1997). In primary osteoblastic cells isolated from the bone of neonatal rats, both ERα and ERβ mRNA were detected (ONOE et al. 1997), and in situ hybridization with ERβ-specific riboprobes of neonatal rat bone showed expression in osteoblasts (Windahl et al., unpublished observations). It is clear that investigations into the role of ERβ in bone are essential to fully understand the mechanism of action of estrogen and mixed agonists/antagonists such as tamoxifen and raloxifene.

Some very convincing data regarding the expression of ERβ in rat brain have been published both by means of in situ hybridization (SHUGHRUE et al. 1997; HRABOVSZKY et al. 1998; LAFLAMME et al. 1998; ÖSTERLUND et al. 1998) and immunohistochemistry (SIMONIAN and HERBISON 1997; ALVES et al. 1998). Detection of specific isoforms of ERβ was not the subject of these studies. In general, all studies have revealed that ERβ is present in the brain and that ERα and ERβ are located in different populations of neurons, and may have distinct roles. An important brain region known to be regulated by estrogen, although ERα and its mRNA are absent, is the hypothalamic magnocellular neurosecretory system. The abundance of ERβ in the paraventricular nucleus (SHUGHRUE et al. 1997) suggests that estrogen may directly regulate genes present in the magnocellular subdivision, such as those for oxytocin and vasopressin, and thereby regulate their release from the posterior lobe of the pituitary. In support of this hypothesis, it was shown with combined immunocytochemistry and in situ hybridization experiments that ERβ mRNA is present in oxytocin and vasopressin neurons of the rat supraoptic and paraventricular nuclei (SHUGHRUE et al. 1997; SIMONIAN and HERBISON 1997; ALVES et al. 1998; HRABOVSKY et al. 1998; LAFLAMME et al. 1998; OSTERLUND et al. 1998). The observation that estrogen replacement therapy improves cognitive function in postmenopausal women was difficult to understand until the discovery that ERβ is highly expressed in regions associated with learning and memory, such as the neocortex, hippocampus and nuclei of the basal forebrain (SHUGHRUE et al. 1997).

F. Multiple Mechanisms of Transcriptional Activation and Repression by ERs

In transient transfection experiments with various cell lines, it was shown that both ERα and ERβ can activate transcription of reporter genes controlled by classical EREs; however, ERβ is less effective than ERα (BARKHEM et al. 1998; KUIPER et al. 1998; MONTANO et al. 1998). More recently it was found that ERs can regulate transcription of genes that lack EREs in their promoter region. Such novel response elements have been described in the collagenase promoter (PAECH et al. 1997), quinone reductase promoter (MONTANO et al. 1998), and transforming growth factor promoter (YANG et al. 1996). In the case of the collagenase promoter, it was shown that ERα and ERβ can mediate transcription from AP1 enhancer elements, which require ligand and the AP1 transcription factors Fos and Jun for transcriptional activation (WEBB et al. 1995; PAECH et al. 1997).

Interestingly, in transient transfection systems, ERα and ERβ, when complexed with estrogen, were shown to signal in opposite ways from an AP1 site, with estrogen activating transcription in the presence of ERα and inhibiting transcription in the presence of ERβ. The ER ligands tamoxifen, raloxifene and ICI-164384 were activators with ERβ and ERα, although the degree of agonism differed among cell types (PAECH et al. 1997). These results clearly suggest the possibility that in cells of the body, in which ERβ is expressed, transcription of certain genes may be repressed by estrogen but that the same genes would be activated by estrogens in cells containing ERα. Such a situation may well be important in tissues such as the breast, prostate and uterus, in which the two ERs co-exist, but where the cells that harbor these receptors have not yet been clearly defined.

G. What is the Phenotype of ERβ Knockout Mice

Mice lacking functional ERβ protein (BERKO) have been generated and their phenotypes are being characterized (unpublished results). As could be predicted from the high expression of ERβ in the ovary (BYERS et al. 1997), one of the most obvious defects in BERKOs can be found in this organ. There are no corpora lutea in the ovaries of BERKO females. The exact reason for the lack of ovulation is not clear, but the females are not fertile. As BERKOs get older (above 14 weeks of age), they seem to fail to accumulate abdominal and breast fat and lose their perirenal fat pad. This is true of both males and females. The males maintain their epididymal fat pad. The reason for the loss of body fat is not yet clear, but BERKOs appear to eat less than their wild-type litter mates. This is evident from the almost complete lack of evidence of food in the intestines. We speculate that the lack of ERβ in the brain causes defects in the secretion of certain neuropeptides that may affect appetite. Older BERKOs have highly distended bladders and appear to have a problem voiding. The previously mentioned studies describing co-localization of ERβ with vasopressin neurons suggest that part of the bladder phenotype could be due to low blood levels of antidiuretic hormone.

Acknowledgements. Several of the studies described in this review were from our laboratory and that research was supported by grants from the Swedish Cancer Fund and KaroBio AB.

References

Alves SE, Lopez V, McEwen BS, Weiland NG (1998) Differential colocalization of estrogen receptor-β with oxytocin and vasopressin in the paraventricular and

supraoptic nuclei of the female rat brain – an immunocytochemical study. Proc Natl Acad Sci USA 95:3281–3286
Arts J, Kuiper GGJM, Janssen JMMF, Gustafsson J-Å, Löwik C, Pols H, van Leeuwen JPTM (1997) Differential expression of estrogen receptors α and β mRNA during differentiation of human osteoblast SV-HFO cells. Endocrinology 138:5067–5070
Barkhem T, Carlsson B, Nilsson Y, Enmark E, Gustafsson J-Å, Nilsson S (1998) Differential response of estrogen receptor-α and estrogen receptor-β to partial estrogen agonists/antagonists. Mol Pharmacol 54:105–112
Bryant HU, Dere WH (1998) Selective estrogen receptor modulators: an alternative to hormone replacement therapy. Proc Soc Exp Biol Med 217:45–52
Byers M, Kuiper GG, Gustafsson J-Å, Park-Sarge OK (1997) Estrogen receptor-β mRNA expression in rat ovary: down-regulation by gonadotropins. Mol Endocrinol 11:172–182
Castelo-Branco C (1998) Management of osteoporosis. An overview. Drugs Aging 12[Suppl 1]:25–32
Chu S, Fuller PJ (1997) Identification of a splice variant of the rat estrogen receptor β gene. Mol Cell Endocrinol 132:195–199
Enmark E, Gustafsson J-Å (1996) Orphan nuclear receptors – the first eight years. Mol Endocrinol 10:1293–1307
Enmark E, Pelto-Huikko M, Grandien K, Lagercrantz S, Lagercrantz J, Fried G, Nordenskjöld M, Gustafsson J-Å (1997) Human estrogen receptor-β gene structure, chromosomal localization, and expression pattern. J Clin Endocrinol Metab 82:4258–4265
Hilakivi-Clarke L, Cho E, Clarke R (1998) Maternal genistein exposure mimics the effects of estrogen on mammary gland development in female mouse offspring. Oncol Rep 5:609–616
Hrabovszky E, Kallo I, Hajszan T, Shughrue PJ, Merchenthaler I (1998) Expression of estrogen receptor-β messenger ribonucleic acid in oxytocin and vasopressin neurons of the rat supraoptic and paraventricular nuclei. Endocrinology 139:2600–2604
Ingram D, Sanders K, Kolybaba M, Lopez D (1997) Case-control study of phytoestrogens and breast cancer Lancet 350:990–994
Katzenellenbogen J, O'Malley BW, Katzenellenbogen BS (1996) Tripartite steroid hormone receptor pharmacology: interaction with multiple effector sites as a basis for the cell- and promoter-specific action of these hormones. Mol Endocrinol 10:119–131
Kuiper GG, Enmark E, Pelto-Huikko M, Nilsson S, Gustafsson J-Å (1996) Cloning of a novel estrogen receptor expressed in rat prostate and ovary. Proc Natl Acad Sci USA 93:5925–5930
Kuiper GG, Carlsson B, Grandien K, Enmark E, Häggblad J, Nilsson S, Gustafsson J-Å (1997) Comparison of the ligand binding specificity and transcript tissue distribution of estrogen receptors alpha and beta. Endocrinology 138:863–870
Kuiper GGJM, Lemmen JG, Carlsson B, Corton JC, Safe SH, van der Saag P, van der Burg B, Gustafsson J-Å (1998) Interaction of estrogenic chemicals and phytoestrogens with estrogen receptor β. Endocrinology 139:4252–4263
Kurzer MS, Xu X (1997) Dietary phytoestrogens. Annu Rev Nutr 17:353–381
Laflamme N, Nappi RE, Drolet G, Labrie C, Rivest S (1998) Expression and neuropeptidergic characterization of estrogen receptors (ER-alpha and ER-beta) throughout the rat brain – anatomical evidence of distinct roles of each subtype. J Neurobiol 36:357–378
Maruyama K, Endoh H, Sasaki-Iwaoka H, Kanou H, Shimaya E, Hashimoto S, Kato S, Kawashima H (1998) A novel isoform of rat estrogen receptor β with 18 amino acid insertion in the ligand binding domain as a putative dominant negative regular of estrogen action. Biochem Biophys Res Commun 246:142–147
Montano MM, Jaiswal AK, Katzenellenbogen BS (1998) Transcriptional regulation of the human quinone reductase gene by antiestrogen-liganded estrogen receptor-α and estrogen receptor-β. J Biol Chem 273:25443–25449

Moore JT, McKee DD, Slentz-Kesler K, Moore LB, Jones SA, Horne EL, Su JL, Kliewer SA, Lehmann JM, Willson TM (1998) Cloning and characterization of human estrogen receptor β isoforms. Biochem Biophys Res Commun 247:75–78

Ogawa S, Inoue S, Watanabe T, Hiroi H, Orimo A, Hosoi T, Ouchi Y, Muramatsu M (1998a) The complete primary structure of human estrogen receptor β and its heterodimerization with ERα in vivo and in vitro. Biochem Biophys Res Commun 243:122–126

Ogawa S, Inoue S, Watanabe T, Orimo A, Hosoi T, Ouchi Y. Muramatsu (1998b) Molecular cloning and characterization of human estrogen receptor beta-cx – a potential inhibitor of estrogen action in human. Nucleic Acids Res 26:3505–3512

Onoe Y, Miyaura C, Ohta H, Nozawa S, Suda T (1997) Expression of estrogen receptor β in rat bone. Endocrinology 138:4509–4512

Österlund M, Kuiper GGJM, Gustafsson J-Å, Hurd Y (1998) Differential distribution and regulation of estrogen receptor ERa and b mRNA within the female rat brain. Mol Brain Res 54:175–180

Paech K, Webb P, Kuiper GG, Nilsson S, Gustafsson J.-Å, Kushner PJ, Scanlan TS (1997) Differential ligand activation of estrogen receptors ERalpha and ERbeta at AP1 sites. Science 277:1508–1510

van Pelt AMM, de Rooij DG, van der Burg B, van der Saag P, Gustafsson J-Å, Kuiper GGJM (1999) Ontogeny of estrogen receptor β expression in rat testis. Endocrinology 140:478–483

Petersen DN, Tkalcevic GT, Koza-Taylor PH, Turi TG, Brown TA (1998) Identification of estrogen receptor beta2, a functional variant of estrogen receptor beta expressed in normal rat tissues. Endocrinology 139:1082–1092

Porter W, Saville B, Hoivik D, Safe S (1997) Functional synergy between the transcription factor Sp1 and the estrogen receptor. Mol Endocrinol 11:1569–1580

Saunders PT, Maguire SM, Gaughan J, Millar MR (1997) Expression of estrogen receptor beta (ER beta) in multiple rat tissues visualised by immunohistochemistry. J Endocrinol 154:R13–R16

Saunders PTK, Fisher JS, Sharpe RM, Millar MR (1998) Expression of estrogen receptor β occurs in multiple cell types, including some germ cells, in the rat testis. J Endocrinol 156:R13–R17

Seeman MV (1997) Psychopathology in women and men: focus on female hormones. Am J Psychiatry 154:1641–1647

Sharpe RM (1997) Do males rely on female hormones? Nature 390:447–448

Shughrue PJ, Lane MV, Merchenthaler I (1997) Comparative distribution of estrogen receptor-alpha and -beta mRNA in the rat central nervous system. J Comp Neurol 388:507–525

Simonian SX, Herbison AE (1997) Differential expression of estrogen receptor alpha and beta immunoreactivity by oxytocin neurons of rat paraventricular nucleus. J Neuroendocrinol 9:803–806

Simpkins JW, Green PS, Gridley KE, Singh M, de Fiebre NC, Rajakumar. G (1997) Role of estrogen replacement therapy in memory enhancement and the prevention of neuronal loss associated with Alzheimer's disease. Am J Med 103:19S–25S

Slemenda CW, Longcope C, Zhou L, Hui SL, Peacock M, Johnston CC (1997) Sex steroids and bone mass in older men. J Clin Invest 100:1755–1759

Smith EP, Boyd J, Frank GR, Takahashi H, Cohen RM, Specker B, Williams TC, Lubahn DB, Korach KS (1994) Estrogen resistance caused by a mutation in the estrogen-receptor gene in a man. New Engl J Med 331:1056–1061

Stephens FO (1997) Breast cancer: aetiological factors and associations (a possible protective role of phytoestrogens). Austr N Z J Surg 67:755–760

Tremblay GB, Tremblay A, Copeland NG, Gilbert DJ, Jenkins NA, Labrie F, Giguere V (1997) Cloning, chromosomal localization, and functional analysis of the murine estrogen receptor beta. Mol Endocrinol 11:353–365

Turner RT, Riggs BL, Spelsberg TC (1994) Skeletal effects of estrogen. Endocrine Rev 15:275–296

Uht RM, Anderson CM, Webb P, Kushner PJ (1997) Transcriptional activities of estrogen and glucocorticoid receptors are functionally integrated at the AP-1 response element. Endocrinology. 138:2900–2908

Watanabe T, Inoue S, Ogawa S, Ishii Y, Hiroi H, Ikeda K, Orimo A, Muramatsu M (1997) Agonistic effect of tamoxifen is dependent on cell type, ERE promoter context, and estrogen receptor subtype. Biochem Biophys Res Commun 236:140–145

Webb P, Lopez GN, Uht RM, Kushner PJ (1995) Tamoxifen activation of the estrogen receptor/AP-1 pathway: potential origin for the cell-specific estrogen-like effects of antiestrogens. Mol Endocrinol 9:443–456

Wilder RL (1998) Hormones, pregnancy, and autoimmune diseases. Ann NY Acad Sci 840:45–50

Yang NN, Venugopalan M, Hardikar S, Glasebrook A (1996) Identification of an estrogen response element activated by matabolites of estradiol and raloxifene. Science 273:1222–1225 (erratum Science 275:1249)

CHAPTER 23

Interrelationship of Estrogens with Other Hormones or Endocrine Systems

N. NAVIZADEH and F.Z. STANCZYK

A. Introduction

Although there is much knowledge about hormones within a given endocrine system, there is a paucity of data pertaining to the relationship of a given hormone to other hormones or the endocrine system. The objective of the present chapter is to show what effect estrogens have on hormones from other endocrine systems and how these systems may affect estrogens. Specific systems that will be discussed include the thyroid, gastrointestine, pancreas, and parathyroid.

B. Thyroid Hormone Effects on Estrogen Levels

The incidence of thyroid disease is more common in women than in men (JACKSON and COBB 1986). It is not clear whether this is a direct effect of the different hormonal states found in women. However, thyroid hormone has several effects on the human female reproductive system. In addition, some steroid hormones have been shown to have an effect on thyroid function. Uterine cells contain thyroid-hormone receptors, and it has been demonstrated that thyroxine (T_4) decreases estradiol uptake and retention by the rat uterus (RUH et al. 1970). There has also been a report of reduced uterine response to estrogen in thyrotoxic rats (SCHULTZE and NOONAN 1970), which could be attributed to the presence of thyroid-hormone receptors in the uterus.

In human studies, women with thyrotoxicosis have elevated sex-hormone-binding globulin (SHBG) levels (AIN et al. 1987; COLON et al. 1988) and a decrease in the metabolic clearance rates of testosterone and estradiol (GORDON and SOUTHREN 1977). Also, there appears to be an increase in peripheral aromatization of androgens to estrogens in some thyrotoxic females (RIDGWAY et al. 1982). However, thyroxine seems to have little or no effect on aromatase activity (LONGCOPE 1982, 1987). Thyrotoxicosis also affects estrogen metabolism in females. Studies have shown an increase in excretion of 2-hydroxyestrone and a decrease in the excretion of 4-hydroxyestrone (RIDGWAY et al. 1975).

Thyroid hormone is also essential for proper function of the placental trophoblast in first-trimester pregnancy (MARUO et al. 1991). In early pregnancy, there is an increase in thyroid size, free T_4, and total tri-iodothyronine (T_3), as well as a decrease in thyroid-stimulating hormone (TSH) levels that, in part, are secondary to an increase in circulating serum thyroid-binding globulin (TBG) and human chorionic gonadotropin (hCG) levels (Burrow et al. 1994).

C. Estrogen Effects on Thyroid Function

Studies show that thyroid volume varies during the menstrual cycle, being largest in mid-cycle (de Remigis et al. 1990). However, circulating thyroid-hormone levels change only slightly during the menstrual cycle. These fluctuations are secondary to changes in blood estrogen levels during the cycle.

Administration of thyrotropin-releasing hormone (TRH) causes a greater increase in both TSH and prolactin in women than in men (Noel et al. 1974). Increases are greater in the pre-ovulatory phase than the luteal phase of the menstrual cycle (Sanchez-Franco et al. 1973). However, TSH and prolactin responses to TRH are only increased with respect to a given dosage of combined (estrogen plus progestin) oral contraceptive. In addition, TBG and reverse T_3 increase after administration of a combined oral contraceptive. Endogenous estrogen, however, was shown to increase only circulating levels of T_4, and this increase was secondary to a decreased clearance of TBG (Ramey et al. 1975; Pasini et al. 1987).

In normal pregnancy, there is an increase in iodine uptake and in the size of the thyroid gland due to hyperplasia and increased vascularity. For the most part, these transient changes are caused by an increase in circulating estrogen and an associated increase in serum TBG levels, as well as by elevated hCG. TSH levels decrease briefly in early pregnancy, but increase subsequently. Free T_4 and total T_3 levels increase in early pregnancy. Due to an increase in TBG, free T_4 and T_3 levels actually decrease during the same period (Glinoer et al. 1990; Hall et al. 1993; Burrow et al. 1994).

D. Estrogen Effects on the Gastrointestinal System

Many changes associated with pregnancy appear to be mediated by both hormonal and anatomical factors. The complex hormonal changes during this period are relevant to gastrointestinal (GI) physiology. hCG, progesterone, and estrogens are the major hormones that alter GI function. Serum concentrations of estradiol, estrone, and estriol are increased during pregnancy and have a great potential to increase uteroplacental blood flow. This effect is mediated by prostaglandins (Wald et al. 1982). Estrogens also play a role in relaxing smooth muscles, although they are less potent than progesterone. Furthermore, motility changes throughout the GI tract, including a reduction in lower esophageal sphincter pressure (Dodds et al. 1978) and its physiologic function, and a decrease in the rate of small-bowl and colonic transit, are mediated by progesterone, with estrogen acting as a primer (Wald et al. 1981).

There are several pieces of evidence that associate steroid hormones with nausea and vomiting during pregnancy. First, studies have shown a strong correlation between nausea and vomiting and intolerance to estrogen-based oral contraceptives. Other studies have demonstrated that women with a previous history of intolerance to oral contraceptives endure more episodes and a longer duration of nausea and vomiting during pregnancy than women with prior tolerance (DEPNE et al. 1987). Second, studies show a higher incidence of hyperemesis gravidum in nulliparous, non-smoking, and obese females who have higher serum and urinary estradiol levels. Estrogen and progesterone could contribute to this condition by having a relaxing effect on GI smooth muscle.

Cholestasis has been seen in both rats and humans after treatment with estrogen. The mechanism is not clear. A decrease in bile flow and Na/K-ATPase activity have also been reported during pregnancy (MULLER and KAPPAS 1964; O'MAILLE et al. 1984). Estrogen could induce cholestasis by altering the sodium pump.

E. Estrogen Effects on the Pancreas

The presence of estrogen seems to be necessary for the synthesis of pancreatic digestive enzymes. Studies have demonstrated that there is a marked depletion of the zymogen granules in the acinar cells of the pancreas of adrenalectomized and ovariectomized female rats. A complete restoration of these vesicles is seen after treatment. Estrogen also changes the pancreatic enzyme output. Administration of estrogen decreases the concentration of protein, increases the levels of zinc and bicarbonate, and increases activity of amylase lipase. The pancreas also retains radioactively labeled estradiol more than any other non-reproductive organ. Dietary estrogen can also influence the pancreas. Acute hemorrhagic pancreatic necrosis can be caused in female mice on a diet deficient in choline but rich in ethionine (GROSSMAN 1984).

Estrogen has been shown to stimulate insulin release via direct effects on the pancreatic B cells (Fig. 1). Estradiol enters the islet cells and binds to cytosolic receptors (POUSETTE et al. 1987). The estradiol receptor in the pancreas is different from that in uterus in that it has a lower molecular weight, a higher apparent dissociation constant, does not translocate estradiol to the nucleus, and requires an accessory factor as a co-ligand. The binding of estradiol to the cytoplasmic receptor influences enzyme induction (BOCTOR et al. 1983) and enhances the growth of pancreatic B cells.

The β cytotropic effect of estradiol observed in diabetic animals is also observed in non-diabetic animals. A large pancreatic islet hypertrophy and B-cell hyperplasia was noted following estradiol administration to ovariectomized mice (LEWIS et al. 1950).

In addition to affecting insulin secretion, estrogen seems to influence insulin action. In ovariectomized mice and rabbits, insulin sensitivity decreased

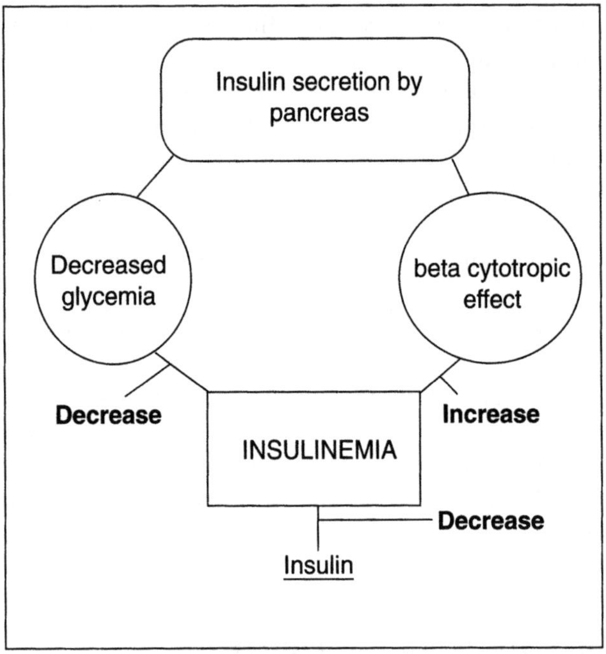

Fig. 1. Effects of estrogen on the regulation of insulin

and hypoglycemic effects of insulin increased after administration of estradiol (BAILEY and AHMED-SOROUR 1980). In a different study, glucose uptake by isolated skeletal muscle was accentuated by estradiol treatment. Therefore, when administered alone, natural estrogen is known to be anti-diabetogenic. This action is carried out via an increase in insulin sensitivity and an increase in circulating insulin concentration (Fig. 1).

The synthetic estrogens used as contraceptive agents have shown a different effect on insulin homeostasis. Some early studies reported a deterioration of glucose tolerance under the influence of oral contraceptive regimens. Synthetic estrogens raise the concentration of insulin in the circulation. Thus, the difference between natural and synthetic estrogen can be attributed to the different effects on insulin sensitivity. It appears that this effect is due to a decrease in insulin receptor concentrations and an increase in circulating insulin antagonists, such as growth hormone and glucocorticoids (ADAMS and OAKLEY 1972).

Pregnancy causes immense changes in both morphology and physiology of the pancreas. In pregnancy, circulating insulin concentrations are increased at all times. This increase is accompanied by pancreatic islet hypertrophy, B-cell hyperplasia and hypergranulation, increased insulin biosynthesis, and increased glucose-induced insulin release from isolated islets (AERTS and VAN ASCHE 1975). These pancreatic changes are considered to be a protective mechanism, due to a decrease in glucose tolerance under the influence of glucocorticoids and placental lactogen.

F. Estrogen Effects on the Parathyroid

There are a few recent studies of the effects of estrogen on circulating parathyroid hormone (PTH) levels; the results are contradictory. Studies have found that estrogen replacement therapy (ERT) results in an increase, decrease, or no change in PTH. In a study of elderly women (mean age 74 years), those taking estrogen had markedly lower serum PTH levels than untreated women, whereas in an earlier study of relatively younger women (mean age 51 years), ERT was associated with an increase in serum PTH levels. It has been hypothesized that, within the first 20 years after menopause, the increase in bone resorption is primarily due to a direct effect of estrogen on bone. In these women, ERT diminishes bone resorption directly, and this results in a compensatory increase in serum PTH levels. In the older postmenopausal women, ERT reverses the age-related increase in serum PTH (STEVENSON 1981; PRINCE et al. 1990). The latter effect could be due to the extraskeletal effects of estrogen on intestinal calcium absorption (GENNARI et al. 1990) and renal calcium resorption (MCKANE 1995), as well as direct effects on PTH secretion. This direct effect of estrogen on PTH could be dose related (GREENBERG et al. 1987).

A marked sex difference in circulating calcitonin levels normally exists, with a relative deficiency in women compared with men. It has been found that estrogens regulate calcitonin secretion and it appears likely that the loss of ovarian function at menopause accelerates the natural decline in calcitonin secretion that occurs with age.

Profound changes in calcium metabolism occur during pregnancy. In a cross-sectional study, plasma concentrations of the major calcium-regulating hormones, namely, calcitonin, PTH, 25-hydroxyvitamin D, and 1,25-dihydroxyvitamin D, were measured to establish their interrelationships during normal pregnancy. The major changes observed were increases in the circulating concentrations of 1,25-dihydroxyvitamin D and calcitonin. Concentrations of PTH and 25-hydroxyvitamin D remained within the normal range (HILLMAN et al. 1981; PITKIN 1983).

G. Conclusion

It is obvious from the above discussion that a great deal more needs to be learned about interrelationships of estrogens and different endocrine systems. Such knowledge will help us to gain a better understanding of how estrogens modulate hormones in different endocrine systems and also how these systems and hormones affect estrogens. Studies of interactions among endocrine systems in appropriate animal models should help gain insight into corresponding interactions in humans.

References

Adams PW, Oakley NW (1972) Oral contraceptives and carbohydrate metabolism. Clin Endocrinol Metabol 1:697–720

Aerts L, Van Asche FA (1975) Ultra structural changes of the endocrine pancreas in pregnant rats. Diabetologia 11:285

Ain KB, Mori Y, Refetoff S (1987) Reduced clearance rate of thyroxine-binding globulin (TBG) with increased sialylation: a mechanism for estrogen-induced elevation of serum TBG concentration. J Clin Endocrinol Metab 65:689

Bailey CJ, Ahmed-Sorour H (1980) Role of ovarian hormones in the long-term control of glucose homeostasis. Effects on insulin secretion. Diabetologia 19:475

Boctor AM, Band P, Grossman A (1983) Analysis of binding of estradiol to the cytosol fraction of rat pancreas: comparison with sites in the cytosol of uterus. Endocrinology 13:4553–4564

Burrow GN, Fisher DA, Larsen PR (1994) Maternal and fetal thyroid function. N Engl J Med 331:1072

Colon JM, Lessing JB, Yavetz C (1988) The effect of thyrotropin-releasing hormone stimulation on serum levels of gonadotropins in women during the follicular and luteal phases of the menstrual cycle. Fertil Steril 49:809

De Remigis P, Raggiunti B, Nepa A (1990) Thyroid volume variation during the menstrual cycle in healthy subjects. In: Hayes DK, Pauly JE, Reiter RJ (eds) Chronobiology: its role in clinical medicine, general biology, and agriculture. Wiley-Liss, New York, p 169

Depne RH, Bernstein L, Ross RK (1987) Hyperemesis gravidarum in relation to estradiol levels: pregnancy outcome, and other maternal factors. Am J Obstet Gynecol 156:137

Dodds W, Dent J, Hogan WJ (1990) Pregnancy and the LES. Gastroenterology 74:1334–1336

Gennari C, Agnusdei D, Nardi P, Civitelli R (1990) Estrogen preserves a normal intestinal responsiveness to vitamin D in oophorectomized women. J Clin Endocrinol Metab 71:1288–1293

Glinoer D, De Nayer P, Bourdoux P (1990) Regulation of maternal thyroid during pregnancy. J Clin Endocrinol Metab 71:276

Gordon GG, Southren AL (1977) Thyroid-hormone effects on steroid-hormone metabolism. Bull N Y Acad Med 53:241

Greenberg C, Kukreja SC, Bowser EN, Hargis GK, Henderson WJ, Williams GA (1987) Parathyroid hormone secretion: effect of estradiol and progesterone. Metabolism 36:151–154

Grossman A (1984) Review: an overview of pancreatic exocrine secretion. Comp Biochem Physiol B 78:1–13

Hall R, Richards CJ, Lazarus JH (1993) The thyroid and pregnancy. Br J Obstet Gynaecol 100:512

Hillman L, Sateesha S, Haussler M, Wiest W, Slatopolsky E, Haddad J (1981) Control of mineral homeostasis during lactation: interrelationships of 25-hydroxyvitamin D, 24,25-dihydroxyvitamin D, 1,25-dihydroxyvitamin D, parathyroid hormone, calcitonin, prolactin, and estradiol. Am J Obstet Gynecol 139:471–476

Jackson IMD, Cobb WE (1986) Disorders of the thyroid, In: Kohler PO (ed) Clinical endocrinology. John Wiley & Sons, New York, pp 67–74

Lewis JT, Foglia VG, Rodriguez RR (1950) The effect of steroids on the incidence of diabetes in rats after subtotal pancreatectomy. Endocrinology 46:111

Longcope C (1982) Methods and results of aromatization studies in vivo. Cancer Res 42[Suppl]:3307a–3311a

Longcope C (1987) Peripheral aromatization: studies on controlling factors. Steroids 50:253

Maruo T, Matsuo H, Mochizuki M (1991) Thyroid hormone as a biological amplifier of differentiated trophoblast function in early pregnancy. Acta Endocrinol 125:56

McKane WR, Khosla S, Burritt MF (1995) Mechanism of renal calcium conservation with estrogen replacement therapy in women in early postmenopause-a clinical research center study. J Clin Endocrinol Metab 80:3458–3464

Muller MN, Kappas A (1964) Estrogen pharmacology: the influence of estradiol and estriol on hepatic disposal of sulfobromophthalein (BSP) in man. J Clin Invest 43:1905–1914

Noel GL, Dimond RC, Wartofsky L (1974) Studies of prolactin and TSH secretion by continuous infusion of small amounts of thyrotropin-releasing hormone. J Clin Endocrinol Metab 39:6–17

O'Maille EDL, Kozmmy SV, Hofmannn AF, Guvantz D (1984) Differing effects of norcholate and cholate on bile flow and billary lipid secretion in the rat. Am J Physiol 246:G67–G71

Pasini F, Bassi P, Cavallini AR (1987) Effect of the hormonal contraception on serum reverse triiodothyronine levels. Gynecol Obstet Invest 23:133

Pitkin RM (1983) Endocrine regulation of calcium homeostasis during pregnancy. Clin Perinatol 10:575–592

Pousette A, Carlstrom K, Skoldefors H, Wilking N, Theve NO (1987) Purification and partial characterization of a 17B-esrradiol-binding macromolecule in the human pancreas. Cancer Res 42:633–637

Prince RL, Schiff I, Neer RM (1990) Effect of transdermal estrogen replacement on parathyroid hormone secretion. J Clin Endocrinol Metab 71:1284–1291

Ramey JN, Burrow GN, Polackwich RJ, Donabedian RK (1975) The effect of oral contraceptive steroids on the response of thyroid stimulating hormone to thyrotropin-releasing hormone. J Clin Endocrinol Metab 40:482

Ridgway EC, Longcope C, Maloof F (1975) Metabolic clearance and blood production rates of estradiol in hyperthyroidism. J Clin Endocrinol Metab 41:491

Ridgway EC, Maloof F, Longcope C (1982) Androgen and estrogen dynamics in hyperthyroidism. J Endocrinol 95:105

Ruh MF, Ruh TS, Klitgaard HM (1970) Uptake and retention of estrogens by uteri from rats in various thyroid states. Proc Soc Biol Med 134:558–561

Sanchez-Franco F, Garcia MD, Cacicedo L (1973) Influence of sex phase of the menstrual cycle on thyrotropin (TSH) response to thyrotropin-releasing hormone (TRH). J Clin Endocrinol Metab 37:736

Schultze AB, Noonan J (1970) Thyroxine administration and reproduction in rats. J Anim Sci 30:774

Stevenson JC (1981) Regulation of calcitonin and parathyroid hormones secretion by estrogens. Maturitas 4:1–7

Wald A, Vanb Thial DH, Hoechsteller L (1981) Gastrointestinal transit, the effect of menstrual cycle. Gastroenterology 80:1497–1500

Wald A, Vanb Thial DH, Hoechsteller L (1982) Effects of pregnancy on gastrointestinal transit. Dig Dis Sci 26:1013–1016

CHAPTER 24

Mammary Gland

G. Söderqvist and B. von Schoultz

A. Pharmacological and Physiological Effects of Estrogens in the Mammary Gland

I. Introduction

Estrogen is a growth hormone in target tissues such as breast, endometrium, vagina and bone, and has major effects on liver function. It has long been known from experimental studies that estrogen is important for the development and growth of the breast during fetal life, puberty and pregnancy. However, estrogen alone cannot exert a full effect on breast cell growth. In different animal models and in in vitro breast cell cultures, estrogen stimulates ductal development, while progesterone is necessary for the development of acini cells in the alveoli. Neither estrogen nor progesterone alone or in combination is sufficient for normal proliferation and differentiation of the mammary gland. This requires a number of other hormones, such as insulin, cortisol, thyroxin, prolactin and growth hormone (GH) (Russo et al. 1990; Dickson and Lippman 1995). However, basic knowledge of the effects of estrogen and progestogen on normal breast epithelium and on the regulation of proliferation is still lacking. Estrogen is generally accepted as a promoter of breast epithelial cells both in vitro and in vivo and is also involved in the development and growth of breast cancer (Colditz et al. 1993; Pike et al. 1993). However, there is not yet full consensus regarding the effects of progesterone/progestogens alone or in combination with estrogen on the breast. All over the world, women have made highly justified demands for information about exogenous hormonal treatment and individual risk assessments.

II. Hormone Action on the Development of the Mammary Gland

An increase in estrogen and other endogenous hormones stimulates the rudimentary breast epithelium, even during fetal life. This effect is reduced by fetal testosterone. There is an interaction between this epithelial rudiment at the nipple and the fatty stroma which stimulates the initial development of the breast prior to puberty. Local inductive factors seem to regulate this process, which is still poorly understood. During the developmental stage of puberty, GH and ovarian estrogen induce elongation and branching of mammary ducts. Progesterone is required for the initial development of the alveoli. Locally produced insulin-like growth factor-I (IGF-I) probably mediates the effect of GH. The estrogen effect is believed to be mediated by members of the transforming growth factor alpha (TGF-α) and epidermal growth factor (EGF) families. Ductal elongation and branching continues after puberty, influenced by hormones such as estradiol (E_2), progesterone, insulin, GH and thyroid hormones, prolactin and glucocorticoids (Russo et al. 1990; Dickson and Lippman 1995; Russo and Russo 1995; Harris et al. 1996).

The mammary gland consists of lobes which are divided into lobules by dense interlobular connective tissue. The interlobular connective tissue contains many blood vessels and groups of adipose cells. In the lobules, major ducts branch into terminal ducts which terminate in the secretory units, the alveoli. The ducts are surrounded by intralobular loose connective tissue with fine collagenous fibers and numerous fibroblasts.

Russo and Russo have proposed four different types of ducts; every type has a specific level of cell differentiation and proliferation. The immature duct which dominates during the years before and after puberty and has the highest level of proliferation is called type I. The more differentiated duct, which has a lower proliferation rate than type I and starts to develop at puberty, is the type-II duct. Type-III ducts develop in an increasing number during pregnancy and have a higher degree of differentiation and a lower proliferation rate than types II and I. The type-IV duct requires a completed pregnancy to differentiate and has the lowest proliferation rate. After lactation, at mammary involution, type-IV ducts disappear, leaving type III as the most differentiated and least proliferating duct in parous women. The higher the type number of the duct, according to this classification, the lower the susceptibility to carcinogenic stimuli. Russo and Russo also theorize that type-I ducts are the origin of ductal cancers and type-II ducts of lobular cancers, while type-III ducts are the origin of fibroadenomas and cysts (Russo et al. 1990; Russo and Russo 1995).

1. Pregnancy

E_2 and progesterone, GH, IGF-1, insulin, glucocorticoids and prolactin are the hormones which induce the marked lobuloalveolar development of the breast during pregnancy. The first half of pregnancy is characterized by a rapid proliferation of the breast epithelium, especially the alveolar structures. A reduction of the intralobular connective tissue and fat, leaving a much larger proportion of the gland to be occupied by epithelium, accompanies proliferation. Mainly during the second half of pregnancy, the differentiation of breast epithelium into functioning secretory units takes place. E_2, progesterone, placental lactogen, placental GH, prolactin and oxytocin, EGF and colony stimulating factor I (CSF-I) stimulate differentiation. Many hormones decline after parturition, especially E_2 and progesterone. As a result, prolactin increases and induces the synthesis of milk from the alveoli. When breast feeding ceases, there is a subsequent decline in prolactin which is followed by apoptosis and glandular involution (Russo et al. 1990; Dickson and Lippman 1995; Russo and Russo 1995; Harris et al. 1996).

III. Proliferation

There is still considerable uncertainty about the hormonal regulation of proliferation in the normal mature breast and the hormonal risk factors for the development of breast cancer. The basis of the risk associated with hormonal therapies may lie in regulation of cell proliferation. High rates of cellular proliferation increase the risk of transformation to the neoplastic phenotype within populations of cells both in vitro and in vivo (Butterworth and Goldsworthy 1991; Cohen et al. 1991). This discovery led to the development of the initiation–promotion model of carcinogenesis, in which cells genetically altered by some initiating event are promoted by a second, non-carcinogenic, proliferative stimulus (Pitot 1986). Moolgavkar et al. (1980) applied this model to the risk of developing breast cancer, which is supported by clinical evidence for the progression of benign proliferative breast lesions into overt breast carcinoma (London et al. 1992; Page et al. 1988; DuPont and Page 1985).

A proposed mechanism for an increased cancer risk is the enhancement of the rate of cell division by sex steroids, which would increase the risk that genetically damaged cells might divide and lose growth control. However, we still lack much basic knowledge concerning the effects of estrogen and progestogens on normal breast epithelium. In comparison with the uterus and endometrium, which have been extensively studied (Vihko and Isomaa 1989; Mäentausta et al. 1990; Pike 1990), there are remarkably few reports on physiologic hormone effects on the normal breast and on the regulation of proliferation.

It is believed that an interplay between hormones such as E_2, progesterone, corticosteroids, prolactin and insulin, hormone receptors, growth factors and their receptors, and interactions between stroma and breast epithelial cells regulates the in vivo proliferation of the normal breast epithelium (Dickson and Lippman 1995). However, in vitro and in vivo data are conflicting. The vast majority of in vitro studies of both breast cancer cell lines and normal breast cells in culture have shown that estrogens enhance breast cell proliferation and that the addition of a progestogen reduces this effect (Gompel et al. 1986; Mauvais-Jarvis et al. 1986). In contrast, most in vivo studies of the normal breast have shown the proliferation of breast epithelial cells to be highest during the luteal phase of the menstrual cycle (Vogel et al. 1981; Anderson et al. 1982, 1989; Longacre and Bartow 1986; Potten et al. 1988). Fig. 1 illustrates this paradox. In a French study, Gompel et al. (1986) found the growth of normal breast epithelial cells in culture to be stimulated by E_2 and the addition of the progestogen R-5020 to reduce breast cell growth.

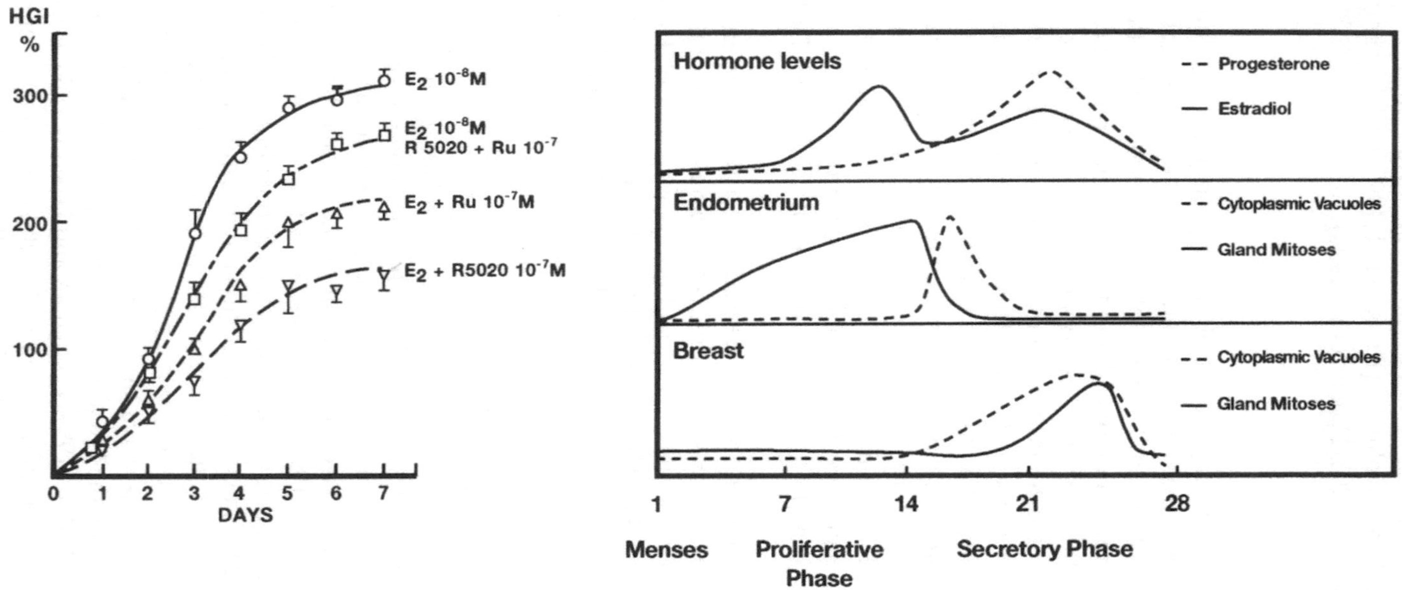

Fig. 1. Proliferation of breast epithelial cells in culture is stimulated by estradiol and reduced when the progestogen R-5020 is added (*left*). Breast cell proliferation was highest in the luteal phase in necropsy material from premenopausal women killed in accidents (*right*). *Left*, adapted from GOMPEL et al. 1986. *Right*, adapted from LONGACRE et al. 1986

In contrast, in the study by LONGACRE et al. (1986), where breast cell proliferation was assessed in necropsy material from premenopausal women killed in accidents, breast cell proliferation was highest in women who were in the luteal phase of the menstrual cycle at the time of death. A direct proliferative effect of progesterone is suggested by the increased mitotic activity during the luteal phase. This indicates an entirely different in vivo regulation of proliferation in the breast than in the endometrium, where progestogens have a clearly antiproliferative effect.

1. Progestogen Effects are Complex

Estrogens are clearly mitogenic for breast epithelial cells, whether benign or malignant, in vivo or in vitro The effects of progestogens are much more complex. In the endometrium, cell-cycle progression is inhibited early in the G1 phase by progestogens, while in the breast they may both stimulate and inhibit proliferation. The picture is further complicated by reports of different isoforms of both the estrogen and progesterone receptor (ER and PR) and that several growth factors, such as IGF-I, EGF and basic fibroblast growth factor (bFGF) also may initiate cell-cycle progression in the absence of sex steroids. In breast cancer cell lines, progestogens have been shown to increase only transiently the rate of cell-cycle progression; prolonged stimulation has been shown to turn the cell cycle off (CLARKE and SUTHERLAND 1990; SUTHERLAND et al. 1995). However, our group performed a study on macaques, which seems to refute this hypothesis in vivo. Macaque mammary glands bear many similarities to those of humans – in anatomic features, hormonal regulation and cytokeratin immunophenotype – which are not shared by the commonly used laboratory rodents. In our macaque study, two and a half years of prolonged continuous combined therapy with conjugated equine estrogens (CEE) and medroxyprogesterone acetate (MPA) significantly enhanced breast cell proliferation. The doses were equivalent, on a caloric basis, to a human dose of 0.625 mg CEE and 2.5 mg MPA daily. An interesting finding of this study was that long-term continuous combined estrogen/progestogen treatment induced marked proliferation, despite downregulation of both ERs and PRs. This may be explained either by the fact that receptor-mediated proliferation occurs even at very low receptor levels, or by a direct stimulating effect of sex steroids on growth factors and/or their receptors, or on intracellular tyrosine-kinase-receptor signal transduction.

Estrogens and progestogens can mediate transcription of the genes coding for EGF, IGF-II and TGF-α, and inhibition of TGF-β production (DICKSON et al. 1986; MURPHY et al. 1986, 1988; KNABBE et al. 1987, 1991; LIU et al. 1987; BATES et al. 1988; MURPHY and DOTSLAW 1989; KING et al. 1989; STEWART et al. 1990; MUSGROVE et al. 1991). Insulin-receptor protein and mRNA and IGF-II-receptor protein have been shown to be increased by progestogens, while the IGF-I receptor in the breast cancer cell line T_{47}-D was downregulated (GOLDFINE et al. 1992). In rat uterus, cyclic adenosine monophosphate (cAMP) and IGF-I can activate sex-steroid receptors by phosphorylation, without ligand binding of the steroid (Aronica and Katzenellenbogen 1993). Very little is known about these actions and interactions in normal human breast tissue in vivo.

IV. Apoptosis

To achieve total knowledge of mammary cell turnover, it is necessary to assess not only proliferation but also cell death. Deletion of cells in normal tissues is controlled by apoptosis, an actively and strictly regulated mode of cell immortalization. Apoptosis is distinct from necrosis and occurs in specific pathologic situations. Condensation of the cytoplasm, margination of nuclear chromatin and production of apoptotic bodies are the morphologic characteristics. The apoptotic bodies are phagocytosed by neighboring cells and degraded within lysosomes. C-myc is a protooncogene which induces mitosis in serum-rich media and apoptosis in low-serum or growth-factor-deprived media. E_2 can inhibit apoptosis by increasing the antiapoptotic protooncogene product Bcl-2 in vitro (Wang and Phang 1995). Maximum apoptosis was seen 2 days after estrogen ablation in MCF-7 cells in culture, and the level of apoptosis still remained higher than during estrogen treatment 14 days after withdrawal. However, the cells' ability to go into apoptosis after estrogen ablation was lost in some variants of MCF-7 cells; however, they retained their ability to go into cell cycle arrest (Kyprianu et al. 1995). Platelet-derived growth factor (PDGF) and IGFs suppressed c-myc-induced apoptosis independently of growth state, position within the cell cycle or whether apoptosis was triggered by low growth factors or by DNA damage. C-myc-induced apoptosis appears to require the presence of P_{53} (Evan 1995).

Clinical studies have suggested the luteal phase as the optimal time for surgical resection of breast cancers. The ability of estrogen ablation to induce apoptosis or cell cycle arrest may be highly relevant in this context. Three large studies (Badwe et al. 1991; Senie et al. 1991; Veronesi et al. 1994) have all shown follicular phase surgery (at approximately days 7–14 of the menstrual cycle) to have the worst and luteal phase surgery (at approximately days 18–33 of the menstrual cycle) to have the best prognosis. However, in none of these studies was the actual hormonal situation of the patients monitored. Apoptosis has been found to display its minimum in the follicular and maximum in the mid- and late luteal phases (days 25–27 of the menstrual cycle) (Anderson et al. 1982). The E_2/progesterone ratio is high during the late follicular phase and low in the mid- and late luteal phase. Very recently we found a positive correlation between serum E_2/progesterone ratio and IGF-I mRNA in breast tissue in women with normal menstrual cycles (Söderqvist et al., unpublished data). IGF-I is known to hamper apoptosis in transgenic mice during involution after lactation and in breast cancer cell lines (Streuli et al. 1997).

V. Estrogen Action During the Normal Menstrual Cycle

1. Estrogen and Progesterone Receptors

It is difficult to study cyclic variations in the breasts of healthy women. Reduction mammoplasty materials only provide the possibility to assess important parameters once in each woman. The possibility of repeated measurements in the same woman is provided by the fine-needle aspiration (FNA) biopsy technique, which allows cytologic cell monitoring in women with breast cancer as well as in healthy individuals. Proliferation, steroid receptors and other parameters can then be assessed by immunocytochemistry. In a study by MARKOPOULOS et al. (1988), in which FNA was performed, 21 of 35 women were found to have detectable ER levels during the first half of the cycle. In another group of 33 women, ER levels were not detectable during the second half of the cycle. A decline in ER detectability during the luteal phase was observed in our own study of healthy women undergoing FNA biopsies (SÖDERQVIST et al. 1993). In a majority of the women, but not in all, ERs were not detectable during this phase. ERs were found in about 65% of samples aspirated in the follicular compared with about 35% in the luteal phase. The proportion of ER-positive cells was significantly higher in the follicular than in the luteal phase. For PR-positive cells, the corresponding values were some 20–30% in both phases, with no significant difference. The percentage of women with detectable PRs was approximately 80% in both the follicular and luteal phase. Women with two evaluable aspirates during the same menstrual cycle and hormonally confirmed ovulation had a significant decline in ER levels, but the PR levels remained unchanged.

In both the breast and the endometrium, the ER level declines during the luteal phase. However, there is a striking difference with respect to PRs. Endometrial epithelial cells display a decline of detectable PRs during the luteal phase (MÄENTAUSTA et al. 1990). This does not occur in breast epithelium, in which the PR level is maintained at a high level throughout the cycle; this may be a mechanism for regulation of proliferation.

The difference in regulation of PRs makes it clear that the breast has a specific hormonal regulation different from the endometrium. A retained PR level under the influence of progesterone during the luteal phase was found previously by BATTERSBY et al. (1992). A downregulation of ER levels by progesterone in the breast, as in the uterus, was also found by this group.

2. Proliferation

In the endometrium, virtually all (80–100%) glandular endometrial cells proliferate at rather low estradiol levels ($\geq$ 200 pmol/l) during the follicular phase, and several progestogen effects combine during the luteal phase to inhibit proliferation and to differentiate into a secretory endometrium (KEY and PIKE 1988). In the breast, only a fraction of approximately 1–10% of breast epithelial cells proliferate. An increased proliferation in the luteal phase was observed in 25 women aspirated twice by means of FNA biopsy in the same menstrual cycle, in a study performed by our group (SÖDERQVIST et al. 1997). There was a significant increase in the mean percentage of the proliferation marker Ki-67 MIB-1-positive cells from the follicular to the luteal phase in these women. This increase remained significant in 18 of the women who had two adequate aspirates as well as a hormonally confirmed ovulation (Fig. 2). A significant positive correlation between proliferation and serum progesterone levels was also found in this study. This clearly indicates an additive or synergistic effect of estrogen and progesterone in vivo in healthy menstruating women. These data are also consistent with most of the earlier in vivo studies in this field.

ANDERSON et al. (1982, 1988), POTTEN et al. (1988) and LONGACRE et al. (1986) all found the highest proliferation rate of the breast epithelium in the luteal phase. However, there are also a few diverging findings during the menstrual cycle in vivo. VOGEL et. al. (1981) found the highest proliferation rate during the follicular phase. In our above-mentioned study, 4 of 18 women with confirmed ovulation and two aspirates during the same menstrual cycle had a decline of proliferating cells from the follicular to the luteal phase. Neither breast pathology nor hormonal aberrations were found in this subgroup of women with a regulation of proliferation apparently diverging from the majority (SÖDERQVIST et al. 1997).

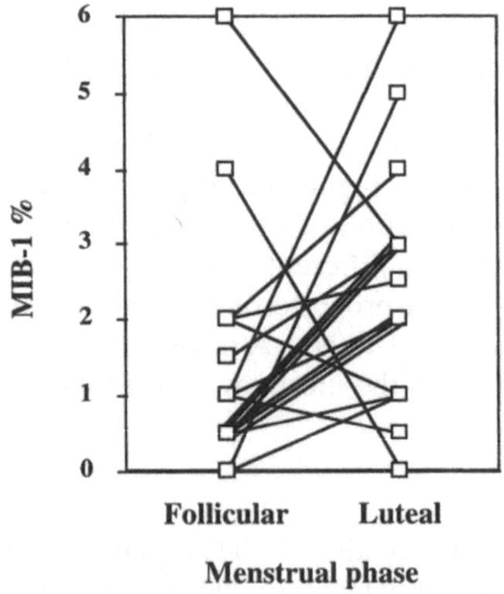

Fig. 2. Change in percentage of MIB-1-positive cells in the follicular and luteal phases in women ($n = 18$) with two consecutive fine-needle aspiration biopsies in same menstrual cycle and with hormonally confirmed ovulation (adapted from SÖDERQVIST et al. 1997)

B. Pharmacological Effects of Exogenous Estrogens and Progestogens in the Mammary Gland

I. Effects on Proliferation and Sex-Steroid Receptors

Estrogen is a growth hormone in mammary epithelial cells (normal and malignant), in vitro and in vivo, and it is also implicated in the development and growth of breast cancer (VOGEL et al. 1981; ANDERSON et al. 1982, 1989; GOMPEL et al. 1986; LONGACRE and BARTOW 1986; MAUVAIS-JARVIS et al. 1986; POTTEN et al. 1988; RUSSO et al. 1990; GOLDFINE et al. 1992; PIKE et al. 1993; DICKSON and LIPPMAN 1995; KYPRIANU et al. 1995; WANG and PHANG 1995; HARRIS et al. 1996).

Estrogens alone or in combination with progestogen are used by numerous women all over the world for contraception and hormone replacement therapy (HRT). For more than 30 years, HRT has been given to women for alleviation of postmenopausal hormone-deficiency symptoms. A lower risk of ischemic heart disease, prevention of bone loss, a reduced risk of fractures and a positive influence on mood, sexuality, vaginal dryness, incontinence and joints are the major beneficial health effects of HRT (HUTCHINSON et al. 1979; ERIKSEN 1986; KIEL et al. 1987; PAGANINI-HILL et al. 1991; COLDITZ et al. 1993; NATHORST-BÖÖS 1993). However, both E_2 and CEE given alone have been shown to increase the risk of endometrial cancer. This risk is lowered to a level even below that in untreated women by the addition of a progestogen (GAMBRELL 1988).

It is still controversial whether or not the breast is affected in the same way as the endometrium. Results from major epidemiological studies, such as the large Nurses Health cohort study and a recent meta analysis in the Lancet (COLDITZ et al. 1995; COLLABORATIVE GROUP ON HORMONAL FACTORS IN BREAST CANCER 1997), now evidently show that the long-term use of HRT increases the risk of breast cancer up to a relative risk level of 1.3–1.7, and that the addition of a progestogen gives no protection.

Elderly women may have a further increased risk level. Biological knowledge of the regulation of the normal breast during physiologic conditions and during treatment with exogenous hormones is crucial for the interpretation of epidemiological data and for individual assessment of the risk of breast cancer. Recently, when we studied surgically postmenopausal cynomolgus macaques, combined estrogen/progestogen treatment caused more proliferation and hyperplasia in the mammary epithelium than treatment with estrogen alone (CLINE et al. 1996). Treatment with CEE and MPA caused significantly higher proliferation than untreated controls in alveoli and major ducts. As much as 86% of animals given CEE + MPA presented with mammary hyperplasia, compared with lobular atrophy or hyperplasia in equal frequencies in animals given CEE alone ($P = 0.0065$). The group of monkeys receiving CEE + MPA also showed a significantly higher percentage of the breast to be occupied by epithelium (rather than

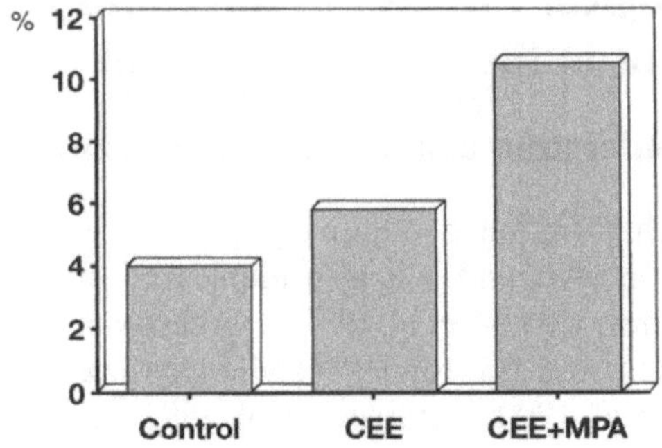

Fig. 3. Point-counting measurements of percentage of epithelium relative to stroma in the mammary gland of macaques. Measurements indicate glandular hyperplasia in the conjugated equine estrogens (*CEE*) + medroxyprogesterone acetate (*MPA*) group. The CEE + MPA group differs from the CEE group, ($P < 0.05$) and from controls ($P < 0.0001$)

connective tissue and fat) than the group receiving CEE alone, as illustrated in Fig. 3.

In another study of reduction mammoplasty material from 27 women, the women taking hormonal contraception showed a significantly higher percentage of the proliferation marker Ki-67 MIB-1-positive cells than those with normal cycles (Isaksson et al. unpublished data). This is in line with the study by Anderson et al. (1989), in which both combined and progestogen-only oral contraceptives caused a higher level of proliferation than that found during the normal menstrual cycle. In contrast, reduced breast epithelial cell proliferation was found after pretreatment with percutaneous progesterone plus E_2, in comparison with pretreatment with E_2 alone in women undergoing breast surgery (Chang et al. 1995). However, in this study, tissue concentrations of E_2 and progesterone considerably exceeded the physiological levels (de Boever et al. 1983). In a study of a model with human breast xenografts in athymic nude mice, E_2 – but not progesterone – increased proliferation (Laidlaw et al. 1995). Here, ER content was not affected by progesterone treatment and was remarkably low. The discrepant finding of proliferation could also be explained by differences in paracrine and hormonal influence in mice relative to women, presumably from adjacent fat and connective tissue. These are important for the in vivo regulation of proliferation in the mammary epithelium (Reed et al. 1991; Poutanen et al. 1995).

A contrast to the data received from women during their menstrual cycles (during which the PR level remained constant) was found in the study of cynomolgus macaques by Cline et al. (1996). Both ER and PR content in breast epithelium was downregulated during combined treatment with CEE + MPA. Endogenous cyclic progesterone and exogenous continuous MPA apparently differ in their effects on the PR in normal breast epithelium.

A rather low number of cells with detectable receptors is found in many different studies. Most likely, this finding is explained by the fact that the immunohistochemical techniques used do not detect very low levels of the receptors, although they are very sensitive. Recently, quantitative analysis of normal breast tissue revealed relatively low concentrations of ER and PR compared with the endometrium (ISAKSSON et al., unpublished data). It is possible that very low receptor levels are needed to mediate proliferation. A direct stimulating effect on growth factors and their receptors by sex steroids is another tentative mechanism. Hormonally treated women, with low receptor levels, undergoing reduction mammoplasty showed a positive correlation between IGF-I mRNA levels and proliferation (ISAKSSON et al., unpublished data).

II. Effects on Enzymatic Conversion of Estrogens in Mammary Tissue

Hormone metabolism and biosynthesis of estrogens in peripheral target tissues can strongly affect the intratissue hormone concentration, which thus may differ considerably from the plasma concentration (SÖDERQVIST et al. 1994; POUTANEN et al. 1995). In adipose tissue, estrone (E_1) is formed from androstenedione by aromatization (CARLSTRÖM 1984; SANTNER et al. 1984). By far, the most abundant estrogen in the peripheral circulation is estrone sulfate (E_1S), which is predominantly formed from E_1 in the liver by sulfotransferase. E_1S is not bound to the ER but may be converted to E_1 by sulfatase in target organs such as the breast. E_1, in turn, can be converted to the terminal biologically active hormone E_2 by another key enzyme: 17β-hydroxysteroid-dehydrogenase (17HSD) type 1. This conversion has been well documented in breast cancer tissue (REED et al. 1991), for which it has also been shown that the route from E_1S to E_2 is far more important than intratumoral aromatization (CARLSTRÖM 1984; SANTNER et al. 1984). In humans, four isoenzymes of 17HSD (1–4) have been characterized. Types 1 and 2 are associated with estrogen metabolism (POUTANEN et al. 1995). The type-1 enzyme is highly specific for estrogen, while 17HSD type 2 catalyzes reactions of both estrogens and androgens (WU et al. 1993). The two enzymes principally catalyze opposite reactions: type 1 forms E_2 from E_1 and type 2 forms E_1 from E_2 (WU et al. 1993; POUTANEN et al. 1995). There are few data on estrogen metabolism in normal breast tissue.

The conversion of E_1S to E_1 and E_2 was studied by our group during use of oral contraception (OC), during the menstrual cycle and postmenopausally. E_1 and E_2 were the main unconjugated compounds formed from E_1S by breast tissue. Conversion to [^{3}H]E_2 was significantly higher in the premenopausal women without OC than in postmenopausal women. Formation of [^{3}H]E_1 (sulfatase activity) was significantly higher in premenopausal women without OC than in OC users. In women without OC, a significant positive correlation was found between sulfatase activity and serum progesterone values

(SÖDERQVIST et al. 1994). The implication of this finding is that normal breast tissue has a significant ability to convert E_1S into potent E_2 and that this mechanism may be important for the development and growth of an early breast cancer. Progesterone and norethisterone acetate (NETA) in combined oral contraceptives seem to differ in their action on sulfatase, progesterone-enhancing and NETA-reducing sulfatase activity.

17HSD type-1 protein was assessed in normal breast tissue during the menstrual cycle, postmenopausally and during hormonal contraception in another study (SÖDERQVIST et al. 1998). 17HSD type-1 protein was detected during both the follicular and luteal phases of the menstrual cycle, and also in postmenopausal women. The maximum enzyme expression was in the early- and mid-luteal phase. There was no difference in enzyme protein expression between alveoli and ducts. In addition, a significantly negative correlation between serum E_2 levels and 17 HSD type 1 was found. Furthermore, there was an insignificant tendency for a negative correlation between serum progesterone values and 17HSD type 1 in premenopausal women.

The enzyme was absent from endometrial epithelial cells during the follicular phase when studied earlier, and there was a positive correlation between serum progesterone levels and 17HSD type 1 (MÄENTAUSTA et al. 1990). This indicates that the regulation of 17HSD type 1 in the normal breast is different from that previously shown in the endometrium.

The effect of exogenous sex steroids was also studied. Women on hormonal contraception had a significantly higher 17HSD type-1 expression in mammary epithelial cells than untreated women. Enhanced 17HSD type-1 protein expression might augment the conversion from E_1 to E_2 in normal mammary tissue during hormonal contraception. This could tentatively be related to the increased breast epithelial cell proliferation found previously by ANDERSON et al. (1989). However, it must be recalled that before any conclusions regarding intratissue E_2 levels can be drawn, the expression and regulation of the oxidative 17HSD type 2 must be assessed. This was not done in the current study. The negative correlation between the serum E_2 concentrations and 17HSD type-1 protein expression indicates a potential for greater tissue formation of E_2 from E_1 in subjects with low endogenous E_2 serum levels. This could be one of the regulatory mechanisms of intratissue E_2 concentration in normal mammary tissue and may thus be of importance for the estrogenic stimulation of the breasts.

References

Anderson TJ, Ferguson DJ, Raab GM (1982) Cell turnover in the "resting" human breast: influence of parity, contraceptive pill, age and laterality. Br J Cancer 46:376–382

Anderson TJ, Battersby S, King RJB, McPherson K, Going JJ (1989) Oral contraceptive use influences breast cell proliferation. Hum Pathol 20:1139–1141

Aronica SM, Katzenellenbogen BS (1993) Stimulation of estrogen receptor mediated transcription and alteration in the phosphorylation state of the rat uterine estro-

gen receptor by estrogen, cyclic adenosine monophosphate and insulin-like growth factor-1. Mol Endocrinol 7:743–752
Badwe RA, Gregory WM, Chaudary MA, Richards MA, Bentley AE, Rubens RD, Fentiman IS (1991) Timing of surgery during menstrual cycle and survival of premenopausal women with operable breast cancer. Lancet 337:1261–1264
Bates SE, Davidson NE, Valverius EM, Freter CE, Dickson RB, Tam JP, Kudlow JE, Lippman ME, Salomon DS (1988) Expression of transforming growth factor alpha and its mRNA in human breast cancer: its regulation by estrogen and its possible functional significance. Mol Endocrinol 2:543–555
Battersby S, Robertson BJ, Anderson TJ, King RJB, McPherson K (1992) Influence of menstrual cycle, parity and oral contraceptive use on steroid hormone receptors in normal breast. Br J Cancer 65:601–607
Butterworth BE, Goldsworthy TL (1991) The role of cell proliferation in multistage carcinogenesis. Proc Soc Exp Biol Med 198:683–687
Carlström K (1984) Influence of intratumoural estradiol biosynthesis on estrogen receptors. Recent Results Cancer Res 91:145–149
Chang KJ, Lee TTY, Linares-Cruz G, Fournier S, de Lignierès B (1995) Influences of percutaneous administration of estradiol and progesterone on human breast epithelial cell cycle in vivo. Fertil Steril 63:785–791
Clarke CL, Sutherland RL (1990) Progestin regulation of cellular proliferation. Endocr Rev 11:266–301
Cline JM, Söderqvist G, von Schoultz E, Skoog L, von Schoultz B (1996) Effects of hormone replacement therapy on the mammary gland of surgically postmenopausal macaques. Am J Obstet Gynecol 174:93–100
Cohen SM, Purtilo DT, Ellwein LB (1991) Pivotal role of increased cell proliferation in human carcinogenesis. Mod Pathol 4:371–382
Colditz GA, Egan KM, Stampfer MJ (1993) Hormone replacement therapy and risk of breast cancer: results from epidemiologic studies. Am J Obstet Gynecol 168:1473–1480
Colditz GA, Hankinson SE, Hunter DJ, Willett WC, Manson JE, Stampfer MJ, Hennekens C, Rosner B, Speizer FE (1995) The use of estrogens and progestins and the risk of breast cancer in postmenopausal women. N Engl J Med 332:1589–1593
Collaborative Group on Hormonal Factors in Breast Cancer (1997) Breast cancer and hormone replacement therapy: collaborative reanalysis of data from 51 epidemiological studies of 52705 women with breast cancer and 108411 women without breast cancer. Lancet 350:1047–1059
de Boever J, Verheugen C, Van Maele G, Vandekerckhove D (1983) Steroid concentrations in serum, glandular breast tissue, and breast cyst fluid of control and progesterone treated patients. In: Angeli A (ed) Endocrinology of cystic breast disease. Raven Press, New York, pp 93–99
Dickson RB, Lippman ME (1995) Growth factors in breast cancer. Endocr Rev 16:559–589
Dickson RB, Huff KK, Spencer EM, Lippman ME (1986) Induction of epidermal growth factor-related polypeptides by 17β-estradiol in MCF-7 human breast cancer cells. Endocrinology 118:138–142
DuPont WD, Page DL (1985) Risk factors for breast cancer in women with proliferative breast disease. N Engl J Med 312:146–151
Eriksen EF (1986) Normal and pathological remodeling of human trabecular bone: three dimensional reconstruction of the remodeling sequence in normals and in metabolic bone disease. Endocr Rev 7:379–408
Evan G (1995) The integrated control of cell proliferation and programmed cell death (apoptosis) by oncogenes (abstract). International symposium on the molecular biology of breast cancer, Lillehammer, Norway
Gambrell RD (1988) Studies of endometrial and breast disease with hormone replacement therapy. In: Studd JWW, Whitehead MI (eds) The menopause. Blackwell, Oxford, England, pp 247–261

Goldfine ID, Papa V, Vigneri R, Siiteri P, Rosenthal S (1992) Progestin regulation of insulin and insulin-like growth factor 1 receptors in cultured human breast cancer cells. Breast Cancer Res Treat 22:69–79

Gompel A, Malet C, Spritzer P, Lalardrie JP, Kuttenn F, Mauvais-Jarvis P (1986) Progestin effect on cell proliferation and 17β-hydroxysteroid dehydrogenase activity in normal human breast cells in culture. J Clin Endocrinol Metab 63:1174–1180

Harris JR, Lippman ME, Morrow M, Hellman S (1996) Diseases of the breast. Lippincott-Raven, Philadelphia

Hutchinson TA, Polansky SM, Feinstein AR (1979) Post-menopausal estrogens protect against fractures of hip and distal radius. A case-control study. Lancet 2:705–709

Key TJ, Pike MC (1988) The dose-effect relationship between "unopposed" estrogens and endometrial mitotic rate: its central role in explaining and predicting endometrial cancer risk. Br J Cancer 57:205–212

Kiel DP, Felson DT, Anderson JJ, Wilson PW, Moskowitz MA (1987) Hip fractures and the use of estrogens in postmenopausal women. N Engl J Med 317:1169–1174

King RJB, Wang DY, Daley RJ, Darbre PD (1989) Approaches to studying the role of growth factors in the progression of breast tumors from the steroid sensitive to the insensitive state. J Steroid Biochem 34:133–138

Knabbe C, Lippman ME, Wakefield LM, Flanders KC, Kasid A, Derynck R, Dickson RB (1987) Evidence that TGF β is a hormonally regulated negative growth factor in human breast cancer. Cell 48:417–428

Knabbe C, Zugmaier G, Schmahl M, Dietel M, Lippman ME, Dickson RB (1991) Induction of transforming growth factor beta by the antiestrogens droloxifen, tamoxifen, and toremifen in MCF-7 cells. Am J Clin Oncol 14:S15–S20

Kyprianu N, English H, Davidson NE, Isaacs JT (1995) Programmed cell death during regression of the MCF-7 human breast cancer following estrogen ablation. Cancer Res 51:162–166

Laidlaw IJ, Clarke RB, Howell A, Owen AWMC, Potten CS, Anderson E (1995) The proliferation of normal human breast tissue implanted into athymic nude mice is stimulated by estrogen but not progesterone. Endocrinology 136:164–171

Liu SC, Sanfilippo B, Perroteau I, Derynck R, Salomon DS, Kidwell WR (1987) Expression of transforming growth factor α (TGFα) in differentiated rat mammary tumors: estrogen induction of TGFα production. Mol Endocrinol 1:683–692

London SJ, Connolly JL, Schnitt SJ, Colditz GA (1992) A prospective study of benign breast disease and the risk of breast cancer. JAMA 267:941–944

Longacre TA, Bartow SA (1987) A correlative morphologic study of human breast and endometrium in the menstrual cycle. Am J Surg Pathol 10:382–393

Mäentausta O, Sormunen R, Isomaa V, Lehto VP, Jouppila P, Vihko R (1991) Immunohistochemical localization of 17β-hydroxysteroid dehydrogenase in the human endometrium during the menstrual cycle. Lab Invest 65:582–587

Markopoulos C, Berger U, Wilson P, Gazet JC, Coombes RC (1988) Estrogen receptor content of normal breast cells and breast carcinomas throughout the menstrual cycle. BMJ 296:1349–1351

Mauvais-Jarvis P, Kuttenn F, Gompel A (1986) Antiestrogen action of progesterone in breast tissue. Breast Cancer Res Treat 8:179–188

Moolgavkar SH, Day NE, Stevens RG (1980) Two-stage model of carcinogenesis: epidemiology of breast cancer in females. J Natl Cancer Inst 65:559–569

Murphy LC, Dotslaw H (1989) Regulation of transforming growth factor α messenger ribonuleic acid abundace in T47 D human breast cancer cells. Mol Endocrinol 3:611–617

Murphy LJ, Sutherland RL, Steed B, Murphy LC, Lazarus L (1986) Progestin regulation of epidermal growth factor receptor in human mammary carcinoma cells. Mol Endocrinol 46:728–734

Murphy LC, Murphy LJ, Dubik D, Bell GI, Shiu RPC (1988) Epidermal growth factor gene expression in human breast cancer cells: regulation of expression by progestins. Cancer Res 48:4555–4560

Musgrove EA, Lee CSL, Sutherland RL (1991) Progestins both stimulate and inhibit breast cancer cell cycle progression while increasing expression of transforming growth factor α, epidermal growth factor receptor, c-fos, and c-myc genes. Mol Cell Biol 11:5032–5043

Nathorst Böös J (1993) Hysterectomy, oophorectomy and estrogen replacement therapy. Thesis, Karolinska Institute, Stockholm, Sweden

Paganini-Hill A, Chao A, Ross RK, Henderson BE (1991) Exercise and other factors in prevention of hip fracture: the Lesure World study. Epidemiology 2:16–25

Page DL, DuPont WD, Rogers LW (1988) Ductal involvement by cells of atypical lobular hyperplasia in the breast: a long-term follow-up study of cancer risk. Hum Pathol 19:201–207

Pike MC (1990) Hormonal contraception with LHRH agonists and the prevention of breast and ovarian cancer. In: Mann RD (ed) Oral contraceptives and breast cancer. Parthenon, Lancaster, England, pp323–348

Pike MC, Spicer DV, Dahmoush L, Press MF (1993) Estrogens, progestogens, normal breast cell proliferation, and breast cancer risk. Epidemiol Rev 15:17–35

Pitot HC (1986) Fundamentals of oncology, 3rd edn. Marcel Dekker, New York

Potten CS, Watson RJ, Williams GT, et al. (1988) The effect of age and menstrual cycle upon proliferative activity of the normal human breast. Br J Cancer 58:163–170

Poutanen M, Isomaa V, Peltoketo H, Vihko R (1995) Regulation of estrogen action: role of 17β-hydroxysteroid dehydrogenases. Ann Med 27:675–682

Reed MJ, Singh A, Ghilchik MW, Coldham NG, Purohit A (1991) Regulation of estradiol 17β-hydroxysteroid dehydrogenase in breast tissues: the role of growth factors. J Steroid Biochem Mol Biol 39:791–798

Russo J, Russo IH (1995) Hormonally induced differentiation: a novel approach to breast cancer prevention. J Cell Biochem Suppl 22:58–64

Russo J, Gusterson BA, Rogers AE, Russo IH, Wellings SR, van Zwieten MJ (1990) Comparative study of human and rat mammary tumorigenesis. Lab Invest 62:244–278

Santner SJ, Feil PD, Santen RJ (1985) In situ estrogen production via the estrone sulfatase pathway in breast tumors: relative importance versus the aromatase pathway. J Clin Endocrinol Metab 53:29–33

Senie RT, Rosen PP, Rhodes P, Lesser ML (1995) Timing of breast cancer excision during the menstrual cycle influences duration of disease free survival. Ann Intern Med 115:337–342

Söderqvist G, von Schoultz B, Skoog L, Tani E (1993) Estrogen and progesterone receptor content in breast epithelial cells from healthy women during the menstrual cycle. Am J Obstet Gynecol 168:874–879

Söderqvist G, Olsson H, Wilking N, von Schoultz B, Carlström K (1994) Metabolism of estrone sulfate by normal breast tissue: influence of menopausal status and oral contraceptives. J Steroid Biochem Mol Biol 48:221–224

Söderqvist G, Isaksson E, von Schoultz B, Carlström K, Tani E, Skoog L (1997) Proliferation of breast epithelial cells in healthy women during the menstrual cycle. Am J Obstet Gynecol 176:123–128

Söderqvist G, Poutanen M, Wickman M, von Schoultz B, Skoog L, Vihko R (1998) 17β-hydroxysteroid dehydrogenase type 1 in normal breast tissue during the menstrual cycle and hormonal contraception. J Clin Endocrinol Metab 83:1190–1193

Stewart AJ, Johnson MD, May FEB, Westley BR (1990) Role of insulin-like growth factor and the type 1 insulin-like growth factor receptor in the estrogen-stimulated proliferation of human breast cancer cells. J Biol Chem 265:21172–21178

Streuli CH, Dive C, Hickman JA, Farrely N, Metcalfe A (1997) Control of apoptosis in breast epithelium. Endocrinc-Rclated Cancer 4:45–53

Sutherland RL, Hamilton JA, Sweeney KJE, Watts CKW, Musgrove EA (1997) Expression and regulation of cyclin genes in breast cancer. Acta Oncol 34:651–656

Veronesi U, Luini A, Mariani L, Del Vecchio M, Alvez D, Andreoli C, Giacobone A, Merson M, Pacetti G, Raselli R, et al. (1994) Effect of menstrual phase on surgical treatment of breast cancer. Lancet 343:1545–1547

Vihko R, Isomaa V (1989) Endocrine aspects of endometrial cancer. In: Voigt KD, Knabbe C (eds) Endocrine dependent tumors. Raven, New York, pp 197–214
Vogel PM, Georgiade NG, Fetter BF, Vogel FS, McCarty KS Jr (1981) The correlation of histologic changes in the human breast with the menstrual cycle. Am J Pathol 104:23–34
Wang TTY, Phang JM (1995) Effects of estrogen on apoptotic pathways in human breast cancer cell line MCF-7. Cancer Res 55:2487–2489
Wu L, Einstein M, Geissler WM, Chan HK, Elliston KO, Andersson S (1993) Expression cloning and characterization of human 17β-hydroxysteroid dehydrogenase type 2, a microsomal enzyme possessing 20α-hydroxysteroid dehydrogenase activity. J Biol Chem 268:12964–12969

CHAPTER 25

Cardiovascular System

G. SAMSIOE

A. Introduction

A bulk of evidence supports the notion that estrogens influence the cardiovascular system in several ways. Most of these effects point towards a reduction in risk factors associated with cardiovascular disease. Indeed, the conformity and consistency of close to 40 observational studies, using various designs and observational times, suggest that estrogens are cardioprotective. The anti-estrogens are estrogen antagonists in some organ systems, but work mainly as estrogen agonists in systems important for cardiovascular integrity. For a variety of reasons, data on classical anti-estrogens, such as tamoxifen and clomiphene, are limited and novel modifications of selective estrogen-receptor modulators (SERMS) have only been used in clinical development trials and in experimental animals. It is not until conclusive evidence is obtained from long-term observational studies (with definite endpoints) that the true effects of a given hormonal regimen, including SERMS, can be outlined. The study of surrogate endpoints may be helpful, but medical history tells us that we have a tendency to select those surrogate endpoints that vary with perceived outcome. In this review, the effects of different estrogens, estrogen dose, and mode of administration will be examined and the limited data demonstrating effects of anti-estrogens on the cardiovascular system will be discussed briefly.

B. Female Cardiovascular Morbidity

Until recently, cardiovascular disease was regarded predominantly as a male problem. In recent years, it has become increasingly clear that women are afflicted even more than men, although a decade later. In several studies, women also carried a poorer prognosis than men – a situation that may be due to the lower use of diagnostic tools with women, an embarrassing situation that is now about to be corrected. Of course, it has also been shown that women with significant coronary heart disease are less likely than men to undergo revascularisation (STEINGART et al. 1991).

Data from Framingham (EAKER et al. 1993) clearly suggested that menopause is a risk factor for coronary disease in women. Several large observational studies, both with case-control and cohort designs, consistently generated data indicative of coronary protection by estrogen monotherapy. The relative risk has been calculated to be between 0.5 and 0.6 (STAMPFER and COLDITZ 1991). However, these observational data are not without bias and confounding; a detailed analysis of one cohort revealed that even prior to estrogen therapy, women had fewer risk factors and were otherwise healthier than the corresponding controls (BROMBERGER et al. 1997). It has also been shown that compliance with a given therapy is a protective factor, even if the therapy is placebo (THE CORONARY DRUG PROJECT RESEARCH GROUP 1980). Nevertheless, the consistency of so many large well-designed and -conducted observational studies strongly support the view that estrogens are cardioprotective. Confounds and biases could probably influence the magnitude of the effects but not the general conclusions.

To further delineate this issue, randomised clinical trials are now underway. The HERS study aims at secondary prevention by conjugated equine estrogens (CEE), and the Women's Health Initiative aims at primary prevention and has been launched to discern whether cardioprotection occurs by means of hormone-replacement therapy (HRT). However, these studies commenced in the 1990s, when it was already publicly known that estrogens could be cardioprotective. Hence, it could be anticipated that women at somewhat higher risk of cardiovascular disease are unlikely to enter a randomised protocol in which they might be given placebo tablets for several years. The Women's Health Initiative should therefore be expected to reveal a somewhat smaller reduction in cardiovascular risk than has been shown in several observational studies.

Recently, several angiographic endpoint trials have been initiated to examine this relationship in more detail. The Estrogen Replacement and Atherosclerosis (ERA) trial is a three-arm randomised, double-blind angiographic endpoint trial that is examining the effects of conjugated equine estrogen (with or without continuous low-dose progestin) on the progression of coronary atherosclerosis in 300 postmenopausal women. The Women's Estrogen Progestin Lipid Lowering Hormone Atherosclerosis Regression Trial (WELL-HART) is of similar design, testing the efficacy of oestradiol [with or without cyclic medroxyprogesterone acetate (MPA)] in 214 women already on lipid-lowering therapy. The Angiographic Trials in Women (ATW) and the Estrogen and Graft Atherosclerosis Research (EAGAR) trials are two even newer angiographic endpoint trials examining questions regarding the joint and independent effects of estrogen and antioxidant therapy and of estrogen on the progression of graft atherosclerosis. These trials will complement the primary and secondary prevention trials already underway, providing a comprehensive assessment of the role of estrogen in the prevention of coronary heart disease.

Coronary angiography studies suggest that women have better coronary artery flow when on estrogen therapy (GRUCHOW et al. 1988; SULLIVAN et al. 1988; HONG et al. 1992). This finding was particularly prudent for women with severe, i.e. more than 70%, stenosis (Table 1). These observations are consis-

Table 1. Observational studies of estrogen use in women with angiographic defined coronary disease

First author	Year	Study design	Study size	End points	Risk estimates
HONG	1993	Cross sectional	18 Users, 72 non-users	Coronary stenosis	0.13
SULLIVAN	1988	Case control	Cases = 1.444	£70% Stenosis	0.44*
GRUCHOW<?1>	1988	Cross sectional	Users = 154	Severe occlusion	0.37*
			Non-users = 779	Moderate occlusion	0.59*
MCFARLAND	1989	Case control	Cases = 137	£70% Stenosis	0.50*

*$P < 0.05$

tent with two major mechanisms by which estrogens are believed to induce cardioprotection, i.e. by inhibiting the progress of arteriosclerosis or even reducing existing disease and by promoting blood flow through a direct vasodilatation of the coronary artery. A common denominator for these effects may exist, but the two major mechanisms will be discussed separately.

C. Anti-Atherosclerotic Effects

Arteriosclerosis is a multifactorial disease. Risk factors include (male) gender, age, smoking, obesity, perturbation of the haemostatic system, lipid metabolism, carbohydrate metabolism and blood-pressure changes. Some of these factors cannot be influenced; others are potentially changeable.

D. Lipid Metabolism

The bulk of data on cardiovascular risk factors deal with changes induced in serum lipid and lipoprotein composition. It is now well established that estrogens increase high-density lipoprotein (HDL) cholesterol and apolipoprotein A1 in a dose-dependent fashion. In particular, HDL2 formation is stimulated, which is believed to occur as a consequence of inhibition of hepatic lipase by estrogens (NABULSI et al. 1993). In doses used for clinical treatment of postmenopausal women, total cholesterol is commonly reduced, with the major reduction in low-density lipoprotein (LDL) cholesterol. There is a simultaneous reduction in apo B100 concentrations.

At higher estrogenic potencies, e.g. following ethinyloestradiol or during pregnancy, there is a marked increase in very-low-density lipoprotein (VLDL) and LDL cholesterol as well as in triglycerides (SAMSIOE 1991). Hence, a dual dose-dependent effect of estrogens exists. It is believed that the effect of higher doses on triglycerides is of hepatic origin, given that oral oestradiol augments triglycerides whereas transdermal oestradiol commonly decreases triglycerides and VLDL (CROOK et al. 1992). 17α ethinyloestradiol induces more triglyceride changes than do conjugated equine estrogens which, in turn, induce greater triglyceride elevation than oestradiol given in clinical doses (Crook et al. 1992). The reasons for differences in triglyceride concentrations are not fully understood. Hepatic lipase is inhibited by estrogens. This not only promotes HDL2 formation, but also impedes triglyceride and VLDL breakdown in the liver. However, the major part of triglyceride and VLDL catabolism takes place in peripheral cells under the influence of the peripheral lipase. The activity of this enzyme can be measured after heparin injection, as it is associated with heparin molecules in the capillary bed of most tissues.

It has been difficult to demonstrate any direct effect of estrogens or antiestrogens on post-heparin lipoprotein lipase activity (PHPLA) because estrogens also influence insulin sensitivity and resistance which, in turn, are linked to PHPLA activity. The increase in triglycerides is also due to an increase in VLDL synthesis and secretion rate, which occurs particularly with high estrogen concentrations and during pregnancy. Recent studies in the rat have shown that chylomicrons with arachidonic-acid-containing esters in the outer layers are relatively resistant to lipoprotein lipase (Melin et al. 1991). Given that estrogens stimulate the enzyme elongase, relatively more arachidonic acid than linoleic acid is available during estrogen treatment. For further details of lipid effects, the reader is referred to specialised literature on lipid metabolism (Crook 1996) (Table 2).

The reduction of total and LDL cholesterol, encountered especially after oral administration, is considered to be due to an increased cholesterol output in the bile. This, in turn, is thought to be a consequence of an estrogen-induced increase in hepatic concentrations of apolipoprotein B/E. Greater cholesterol lowering by oral preparations than oestradiol patches could be explained by differences in hepatic concentrations of estrogens following oral and transdermal administration (Crook et al. 1992).

Whether VLDL remnants play a part in trigging apolipoprotein B/E receptors in the liver during estrogen treatment is still under debate

Table 2. Summary of principal effects by various estrogens and selective estrogen-receptor modulators (SERMS) on serum lipids, lipoproteins and apo lipoproteins

	Ethinyl oestradiol	Oral oestradiol	Transdermal oestradiol	Conjugated equine estrogens	Tamoxifen/ raloxifene
HDL-C	++	+	+	++	0
HDL-2C	+++	++		++	0
LDL-C	–	–	–	–	–
TOT-C	–	–	–	–	–
TG	+++	+	–	++	+
LP(a)	–	–	0	–	–
APO Al	++	+	0	++	0
APO B	–	–	–		–

+ Increase; – decrease; 0 no effect

(BEISIEGEL 1995). In hypercholesterolaemic women, both oral estrogens and transdermal oestradiol have been shown to be efficacious. However, only the transdermal oestradiol lowers triglycerides. Oral estrogens should not be given in types IIB, III or IV without progestogen comedication, even if the patient is hysterectomised (SAMSIOE et al. 1996). Data on estrogen therapy, both oral (TIKKANEN et al. 1978; TONSTAD et al. 1995; DARLING et al. 1997) and transdermal (SAMSIOE et al. 1996), suggests a reduction of total and LDL cholesterol levels by about 20%; this approximately doubles the effect shown in normocholesterolaemic women. It has also been shown that estrogens reduce the concentrations of lipoprotein(a), which may be an independent risk factor and a link to the haemostatic system (SIRTORI and SOMA 1994).

E. Haemostatic System

Clinically overt intravasal clotting in the form of emboli or thrombi occur when procoagulatory mechanisms dominate over fibrinolysis. The coagulation cascade per se is very complex and only partly understood. In addition, several local factors play a vital role for the formation of a clot (WINKLER 1996). The rate of flow is also of utmost importance for the clotting process – at least on the venous side. There is also a profound difference between arterial coagulation and clotting occurring on the venous side. Arterial thrombus formation is mainly platelet driven, occurs at a higher blood pressure, at much higher flow rates and with different quantitative and qualitative functioning of the vascular wall. For these reasons, arterial and venous thrombosis will be discussed separately.

Pregnancy and oral contraceptives are linked to an increased risk of venous thromboembolism. Indeed, an estrogen dependency has been demonstrated at least for contraceptive pills (SAMSIOE 1994). The effect of estrogens in HRT has been much more debatable, due mainly to the lack of larger well-designed observational studies. This lack is due, in turn, to the low incidence of clinically manifest thromboembolic diseases in women, which may be as low as 1–2 cases per 10,000 women per year (GRODSTEIN et al. 1996). Four recent studies have demonstrated that there is a two- to fourfold increase in the occurrence of venous thrombosis (including pulmonary embolism) (DALY et al. 1996; GRODSTEIN et al. 1996; JICK et al. 1996; PEREZ GUTTHAM et al. 1997). Whether a difference in risk is associated with transdermal versus oral preparations given as HRT is yet to be discerned. There were small numbers of subjects involved in the published studies of transdermal therapy; the relative risks for transdermal therapy did not reach statistical significance (Table 3).

Haemostatic balance is of utmost importance for the survival of the individual. Hence, several mechanisms regulate this important phenomenon. When discussing effects induced in variables that are of importance for haemostatis, it should be remembered that a venous blood sample drawn at one site only takes into account general factors belonging to the coagulation

Table 3. Risk of venous thromboembolic events by estrogens according to four recent studies

Author	Type of study	No. of cases	Type of ERT	Odd ratio
JICK et al. 1996	Case control	42	CEE	3
DALY et al. 1996	Case control	103	CEE E2	3.5
GRODSTEIN et al. 1996	Cohort	67	CEE	2
PEREZ GUTTHAN et al. 1997	Case control	292	Various	2

CEE, conjugated equine estrogens; ERT, estrogen replacement therapy

Table 4. Principle effects of various estrogens and pregnancy on serum insulin and blood glucose

	Glucose tolerance	Insulin response
17β-oestradiol	+	–
Alkylated estrogens	+	–
Conjugated equine estrogens	0	0
Pregnancy	+	+

+, Increase; –, decrease; 0, no effect

cascade; it does not account for flow or local factors, known to be of great importance for the formation of a clot at a given site. The importance of local effects and the difference in function between arterial and venous walls follows from the fact that, although venous thrombotic events are increased, arterial thrombotic disease seems to be decreased.

Arterial thromboembolic events are less common in women with a past history of myocardial infarction (MI) treated with estrogen than in women with sustained MI who are not given HRT. The changes induced by estrogens may provoke thrombus formation at the venous side and impede such occurrences on the arterial side. Indeed, the changes induced by estrogens show a mixture of both procoagulant and anticoagulant activity. Most authors would agree that estrogens decrease factor VII, increase protein S and C activity and decrease anti-thrombin III. On the fibrinolytic side, fibrinogen may be increased but PAI-1 activity is decreased (Table 4) (NABULSI et al. 1993; GILABERT et al. 1995; LINDOFF et al. 1996). As with other surrogate endpoints, the clinical significance of changes in markers of the haemostatic system is disputable.

F. Carbohydrate Metabolism

Compared with haemostasis and lipid metabolism, much less data are available on carbohydrate metabolism. It would seem that ethinyl-oestradiol (LARSSON-COHN and WALLENTIN 1977) and conjugated equine estrogens

Table 5. Effects of age, menopause and hormone-replacement therapy (HRT) on some markers of the haemostatic system

Marker	Age	Menopause	HRT
Fibrinogen	+	+	–
Factor VIIc	+	++	–
Factor VIII	+	0	0
PAI-1	–	+	–
t-PA	+	–	+
AT III	+	0	–
Protein C	+	+	+
Protein S			
Plasmin activity	+	0	0
Thrombin activity	+	0	0

+, Increase; –, decrease; 0, no effect; t-PA, tissue plasminogen activator

(SPELLACY et al. 1978) in doses used for HRT commonly impair glucose tolerance and induce changes in C peptide and insulin resistance (LARSSON-COHN and WALLENTIN 1977). Oestradiol, especially when given transdermally, seems to increase hepatic insulin clearance, thus preventing hyperinsulinaemia associated with insulin resistance (GODSLAND et al. 1993). Oral preparations result in several-fold higher steroid concentrations in the liver than parenteral oestradiol administration. Given that the liver is a major site of insulin production, it is possible that the transdermal route is more favourable because of a lesser impact on the liver. Whether this is due to the first hepatic effect or to the differences in pharmacokinetic or pharmacodynamic profiles between the two forms of administration is not known. Nevertheless, at higher doses and during pregnancy, estrogens seem to deteriorate carbohydrate metabolism in a way that may increase the risk of cardiovascular disease (Table 5).

G. Blood Pressure

The overall clinical finding with estrogens is that blood pressure is unchanged. However, in large studies, it has been suggested that estrogens may lower blood pressure by 2–3 mmHg (LOBO 1987). This is particularly true for the diastolic blood pressure. Some data imply a more marked effect by transdermal oestradiol (AKKAD et al. 1997), but additional data are urgently needed (LOBO 1987). A shift of 2–3 mmHg may not seem like much and is hardly detectable in an individual, but for a large cohort, that kind of reduction may well decrease the risk of stroke. However, data from observational studies on incidence and prevalence of stroke and use of estrogens are conflicting.

Vasodilatory properties of estrogens have been suggested. Vasodilatation could also be due to calcium antagonism and the antioxidative effect, which is of utmost importance for the regulation of nitrous oxide synthesis. It has also been shown that estrogens induce nitric oxide (NO)-dependent increases in uterine blood flow. The mechanism for this is unknown, but the general vasodilatory properties of estrogens have been suggested. In addition, there are some data to link estrogens to an angiotensin-converting enzyme (ACE) inhibitory effect. Data also imply that estrogens could be regarded as calcium-channel blockers (COLLINS et al. 1993). Long-term (2 years) estrogen replacement therapy in postmenopausal female monkeys protects against impaired responses of atherosclerotic coronary arteries to acetylcholine. It is of interest to note that 17α-oestradiol, when injected into the coronary arteries, produces vasodilation in women but not in men (PROUDLER et al. 1995).

It has also been shown that estrogen may inhibit the activity of tyrosine kinase, which synthesizes catecholamines through feedback inhibition after being transferred to the catechol estrogens. Release of noradrenaline during sympathetic activation is modified by noradrenaline stimulation of α_2 receptors. Estrogen increases the density and enhances the function of the α_2 adrenoreceptors (COLLINS et al. 1995). There is also the possibility that gender differences in cardiovascular sympathetic activity may exist at a central level.

The activity of the cholinergic–muscarinic system in the central nervous system is potentiated by estrogens. The concentration of acetylcholine and the activity of choline acetyl transferase are increased in females when compared with males and are reduced following ovariectomy and increased by subsequent estrogen replacement therapy. Choline uptake is increased by estrogen and decreased following ovariectomy (PASQUALINI et al. 1991).

H. Direct Effects on the Arterial Wall

Several experiments, in both humans and experimental animals, have concluded that estrogens dilate vessels and promote blood flow through a given tissue. It has also been a long-standing observation in gynaecology that estrogens promote blood flow through the vaginal and uterine tissues. With the introduction of newer techniques, such as ultrasound and colour Doppler imaging, it has been possible to demonstrate an increased blood flow in almost every organ system, including the aorta, coronary arteries, carotid arteries, and the brain. Several mechanisms seem to contribute to this vasodilatation. Estrogen has been demonstrated to reduce the serum concentration of endothelin(s) (HONJO et al. 1992). Furthermore, estrogens have also been shown to induce changes in the content of serum lecithin fatty acids, which are precursors to the prostanoids. Urinary excretion of stable prostanoid metabolites indicate that the prostacyclin–thromboxane balance reflects increased amounts of prostacyclin and decreased amounts of thromboxane A2 (SHAY et al. 1993).

I. Antioxidative Effect

In recent years, oxidised LDL has gained particular interest because it rapidly promotes atherogenicity. In addition, oxidised LDL may be a link between the formation of arteriosclerosis and vascular function: the oxidised molecule inhibits the formation of NO but promotes endothelin activity (DIAZ et al. 1997) (Table 2). A growing body of evidence now suggests that all estrogens, including ethinyloestradiol, are antioxidants (WEIGRATZ et al. 1996) and that the antioxidant effects are not counteracted by the addition of a progestogen co-medication (SCHRÖDER et al. 1996).

Oxidatively modified LDL particles increase atherogenicity, but other properties of oxidised LDL may be at least as important as the increased arteriosclerosis. The cytotoxicity of oxidised LDL induces several changes in vascular wall metabolism. It has also been suggested that the oxidised LDL promotes the formation of unstable plaques which, in the presence of oxidised LDL, attract platelets and thereby form thrombotic lesions (DIAZ et al. 1997). Because estrogens are antioxidants, they could impair formation of arterial thrombi on an arteriosclerosis basis (SCHRÖDER et al. 1996; WEIGRATZ et al. 1996). Such effects have been offered as one explanation as to why estrogens do not uniformly reduce stroke; stroke consists both of haemorrhagic and thrombotic lesions.

A peculiar condition is often called cardiac syndrome X, which is commonly defined as angina pectoris, ST segment depression on electrocardiogram (ECG), and angiographically normal coronary arteries. This syndrome is more common in women than in men. It has been shown that estrogens could be of beneficial value in the treatment of angina pectoris in women with syndrome X and, in women with myocardial ischaemia, the time to reach 1 mm ST depression is much longer in those on oestradiol than on placebo; the same is true for the total exercise time that patients can endure (ROSANO et al. 1995).

J. Antiestrogens

To date, there is little published data on cardiovascular endpoints in humans. Tamoxifen, which has been used for a long time, is confined by its indication, i.e. in patients with breast cancer, which itself carries a higher risk of cardiovascular complication, especially thrombotic episodes. Hence, these data are difficult to evaluate. However, raloxifene (LOVE et al. 1991) and tamoxifen (ADACHI et al. 1997) seem to induce a lipid profile which could be in line with cardioprotection, i.e. a decrease in total and LDL cholesterol. In animal models, newer antiestrogens, such as raloxifene and levormeloxifen, were shown to increase blood flow. It is possible that these newer SERMS may be cardioprotective.

In summary, there is substantial evidence to suggest that estrogen monotherapy is cardioprotective, especially in women carrying risk factors for coronary heart disease, such as sustained MI and hypercholesterolaemia. We have no data definite endpoint data for anti-estrogens. A cardioprotective effect of both tamoxifen and raloxifene are suggested by the serum lipid profile. A key factor may well be the oxidative process which has links to both atheriogenesis and to vessel wall function.

References

Adachi J, Delmas P, Mitlak B, et al. (1997) Raloxifene has sustained, beneficial effects on serum lipid concentrations in healthy postmenopausal women. Menopause Rev 2:68

Akkad AA, Halligan AWF, Abrams K, Al-Azzawi F (1997) Differing responses in blood pressure over 24 hours in normotensive women receiving oral or transdermal estrogen replacement therapy. Obstet Gynecol 89:97–103

Beisiegel U (1995) Receptors for triglyceride-rich lipoproteins and their role in lipoprotein metabolism. Curr Opin Lipidol 6:117–122

Bromberger JT, Matthews KA, Euller LH, et al. (1997) Prospective study of the determinants of age at menopause. Am J Epidemiol 145:124–133

Collins P, Rosano GM, Jiang C, et al. (1993) Cardiovascular protection by estrogen – a calcium antagonist effect? Lancet 341:1264–1265

Collins P, Rosano GMC, Sarrel PM, et al. (1995) 17α-Estradiol attenuates acetylcholine-induced coronary arterial constriction in women but not men with coronary heart disease. Circulation 92:24–30

Crook D (1996) Post menopausal hormone replacement therapy, lipoprotein metabolism, and coronary heart disease. J Cardiovasc Pharmacol 28[Suppl]:46–50

Crook D, Cust MP, Gangar KF, et al. (1992) Comparison of transdermal and oral estrogen/progestin replacement therapy: effects on serum lipids and lipoproteins. Am J Obstet Gynecol 166:950–954

Daly E., Vessey MP, Hawkins MM, et al. (1996) Risk of venous thromboembolism in users of hormone replacement therapy. Lancet 348:977–980

Darling GM, Johns JA, McCloud PI, Davis SR (1997) Estrogen and progestin compared with simvastatin for hypercholesterolemia in postmenopausal women. N Engl J Med 337:595–601

Diaz MN, Frei B, Vita JA, Keaney JF Jr (1997) Antioxidants and atherosclerotic heart disease. N Engl J Med 337:408–416

Eaker C, Chesebro JH, Sacks FM, et al. (1993) Cardiovascular disease in women. Circulation 88:1999–2009

Gilabert J, Estells A, Cano A, et al. (1995) The effect of estrogen replacement therapy with or without progestogen on the fibrinolytic system and coagulation inhibitors in postmenopausal status. Am J Obstet Gynecol 173:1849–1854

Godsland I, Gangar K, Walton C, et al. (1993) Insulin resistance, secretion and elimination in postmenopausal women receiving oral or transdermal hormone replacement therapy. Metabolism 42:846–853

Grodstein F, Stampfer MJ, Goldhaber SZ, et al. (1996) Prospective study of exogenous hormones and risk of pulmonary embolism in women. Lancet 348:983–987

Gruchow HV, Anderson AJ, Barboriak JJ, Sobocinski KA (1988) Postmenopausal use of estrogen and occlusion of coronary arteries. Am Heart J 115:954–963

Hong MK, Romm PA, Reagan K, et al. (1992) Effects of estrogen replacement therapy on serum lipid values and angiographically defined coronary artery disease in postmenopausal women. Am J Cardiol 69:176–178

Honjo H, Tamura T, Matsumoto Y, et al. (1992) Estrogen as a growth factor to central nervous cells. Estrogen treatment promotes development of acetylcholinesterase-positive basal forebrain neurons transplanted in the anterior eye chamber. J Steroid Biochem Mol Biol 41:633–635

Jick H, Derby LE, Myers MW, et al. (1996) Risk of hospital admission for idiopathic venous thromboembolism among users of postmenopausal estrogens. Lancet 348:981–983

Larsson-Cohn U, Wallentin L (1977) Metabolic and hormonal effects of postmenopausal estrogen replacement treatment. Acta Endocrinol 86:583–596

Lindoff C, Petersson JF, Lecander I, et al.(1996) Transdermal estrogen replacement therapy: beneficial effects on hemostatic risk factors for cardiovascular disease. Maturitas 24:43–50

Lobo RA (1987) Estrogen replacement therapy and hypertension. Postgrad Med Sept 14:48–54

Love RR, Wiebe D, Newcomb P, et al. (1991) Effects of tamoxifen on cardiovascular risk factors in postmenopausal women. Ann Intern Med 115:860–864

Melin T, Qi C, Bengtsson-Olivecrona G, Kesson B, Nilsson (1991) Hydrolysis of chylomicron polyenoic fatty acid esters with lipoprotein lipase and hepatic lipase. Biochim Biophys Acta 1075:259–266

Nabulsi AA, Folsom AR, White A, et al. (1993) Association of hormone-replacement therapy with various cardiovascular risk factors in postmenopausal women. N Engl J Med 328:1069–1075

Nabulsi AA, Folsom AR, White A, et al. (1993) Association of hormone-replacement therapy with various cardiovascular risk factors in postmenopausal women. The atherosclerosis risk in communities study investigators. N Engl J Med 328:1069–1075

Pasqualini C, Leviel V, Guibert B, et al. (1991) Inhibitory actions of acute estradiol treatment on the activity and quantity of tyrosine hydroxylase in the median eminence of varioectomized rats. J Neuroendocrinol 3:575–580

Perez Guttham S, Garcia Rodriguez LA, Castellsague J, et al. (1997) Hormone replacement therapy and risk of venous thromboembolism: population based case-control study. BMJ 314:796–800

Proudler AJ, Hasib Ahmed AI, Crook D, et al. (1995) Hormone replacement therapy and serum angiotensin-converting-enzyme activity in post-menopausal women. Lancet 346:89–90

Rosano GMC, Collins P, Kaski JC, et al. (1995) Syndrome X in women is associated with estrogen deficiency. Eur Heart J 16:610–614

Samsioe G (1991) Lipid profiles in estrogen users. In: Sitruk-Ware R, Utian W (eds) The menopause and hormonal replacement therapy. Marcel Deckker, New York, pp 181–200

Samsioe G (1994) Coagulation and anticoagulation effects of contra-ceptive steroids. Am J Obstet Gynecol 170:1523–1527

Samsioe G, Balsell G, Berg A, Sandin K (1996) Transdermal oestradiol plus medroxyprogesteron acetates lower cholesterol in moderately hyper-cholesterolemic women (abstract). Maturitas 27[Suppl]:65

Schröder J, Dören M, Schneider B, Oettel M (1996) Are the antioxidative effects on 17α-oestradiol modified by concomitant administration of a progestogen? Maturitas 25:133–140

Shay J, Badrov N, Attele A, et al. (1993) Estrogen antagonises endethelin-1 vasoconstriction in rabbit basilar artery. Anesth Analg A561

Sirtori CR, Soma MR (1994) HRT and correction of lipoprotein disorders. In: Crosignani PG, et al. (eds) Women's health in menopause. Kluwer, Dordrecht, pp 159–169

Spellacy W, Buhi W, Birk S (1978) Effect of estrogen treatment for one year on carbohydrate and lipid metabolism. Am J Obstet Gynecol 131:87–90
Stampfer MJ, Colditz GA (1991) Estrogen replacement therapy and coronary heart disease: a quantitative assessment of the epidemiologic evidence. Prev Med 20:47–63
Steingart RM, Packer M, Hamm P, et al. (1991) Sex differences in the management of coronary artery disease. The survival and ventricular enlargement investigators. N Engl J Med 325:226–230
Sullivan JM, van der Zwaag RV, Lemp GF, et al. (1988) Postmenopausal estrogen use and coronary atherosclerosis. Ann Intern Med 108:358–366
The coronary drug project research group (1980) Influence of adherence to treatment and response of cholesterol on mortality in the coronary drug project. N Engl J Med 303:1038–1041
Tikkanen MJ, Nikkilö EA, Variainen E (1978) Natural estrogen as an effective treatment for type-II hyperlipoproteinaemia in post-menopausal women. Lancet II:490–492
Tonstad S, Ose L, Görbitz C, et al. (1995) Efficacy of sequential hormone replacement therapy in the treatment of hypercholesterolaemia among postmenopausal women. J Int Med 238:39–47
Weigratz B, Hertwig B, Jung-Hoffmann C, Kuhl H (1996) Inhibition of low-density lipoprotein oxidation in vitro and ex vivo by several estrogens and oral contraceptives. Gynecol Endocrinol 10[Suppl 2]:149–152
Winkler UH (1996) Hormone replacement therapy and haemostasis: principles of a complex interaction. Maturitas 24:131–145

CHAPTER 26
Bone

B. WINDING, H. JØRGENSEN, and C. CHRISTIANSEN

A. Bone

The skeleton, composed of bone and cartilage, protects vital organs including hematopoietic bone marrow and serves as a reservoir for fundamental ions, especially calcium, magnesium and phosphorus. Furthermore, the skeleton supports and possesses sites for muscle attachment.

The basic constituents of bone, as in all connective tissue, are the cells and the extracellular matrix. The latter is composed of type-I collagen fibers (90% of total protein weight), non-collagenous proteins, glycoproteins, proteoglycans and crystals of hydroxyapatite $[Ca_{10}(PO_4)_6(OH)_2]$ (BARON 1996).

Our knowledge of bone metabolism has increased tremendously over the last 10 years due to a still-growing body of research in the field of bones and minerals. The research has been prompted by the awareness of the social/economic effects of an increasing number of individuals developing osteoporosis in their senile life (see Chap. 39, JØRGENSEN et al.).

I. Macroscopic Anatomy

Bones are divided into two groups based on their morphological appearance: (1) flat bones, such as skull, scapula, and mandible; and (2) long bones, such as humerus, tibia, and femur. Long bones have an architectural structure specially designed to accomplish both the weight bearing, and metabolic and protective functions of bone. The growing long bone can be divided into three compartments (JEE 1983) (Fig. 1). The two wider extremities, epiphyses, are separated from the cylindrical diaphysis by a developmental zone, the metaphysis. The epiphysis and the metaphysis are again separated by a cartilaginous growth plate, responsible for the longitudinal growth of bones. The cortex of the diaphysis is composed of a thick and compact layer of calcified matrix, called the cortical or compact bone. As cortical bone is rather metabolically inert, the main function is mechanical and protective. Toward the epiphysis, the cortical bone becomes gradually thinner and the internal space is filled with a network of thin, partially calcified trabeculae, called the trabecular or cancellous bone. The trabecular bone is highly metabolic; in addition, a special orientation of the trabeculae results in a high mechanical strength. Both at the external (periosteum) and at the internal (endosteum) surfaces of cortical

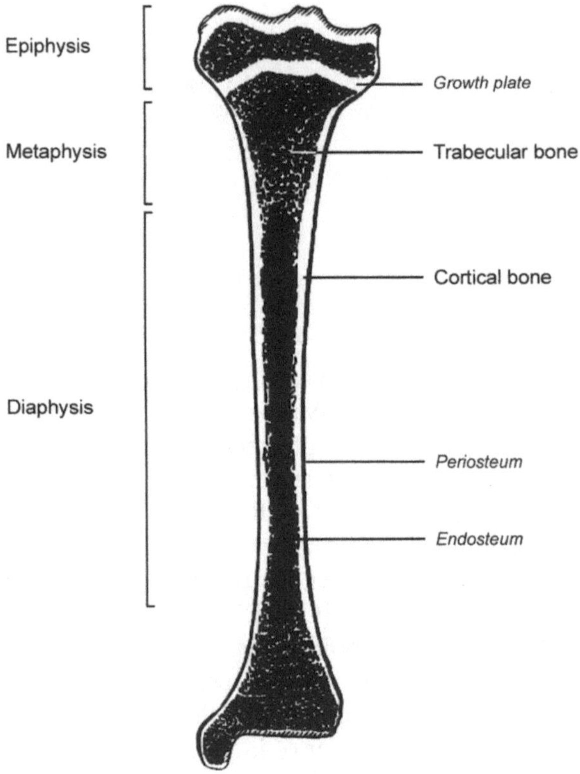

Fig. 1. Anatomy of long bone (adapted from JEE 1983)

bones and at the surfaces (endosteum) of trabecular bones, layers of osteogenic cells separate the bones from soft tissues and bone marrow, respectively.

II. Microscopic Anatomy

1. Matrix and Minerals

The major constituent of bone-matrix proteins is type-I collagen, which is a triple-helical coiled coil, containing two identical $\alpha 1$(I) chains and a structurally similar, albeit genetically different, $\alpha 2$(I) chain. The lamellar bone is tightly packed with well-aligned collagen fibrils. Proteoglycans and osteonectin are associated with the collagen fibrils, and this mixture of matrix proteins is mineralized with small crystallites of basic, carbonate-containing calcium phosphate, hydroxyapatite (TERMINE and ROBEY 1996).

Osteocalcin, a single chain of 46–50 amino acids, is the most abundant non-collagenous protein of bone synthesized by osteoblasts. Recently, it was reported that mice with osteocalcin deficiency have increased bone formation without impaired bone resorption (DUCY et al. 1996). Although the precise mechanism remains to be elucidated, it may suggest that osteocalcin is an autocrine regulator of osteoblastic bone formation in vivo.

Two of the smaller leucine-rich proteoglycans, decorin and biglycan, are also found in bone. Both proteins are reported to bind transforming growth factor beta (TGF-β) and other growth factors, and might be involved in the regulation of the activity of bound growth factors (TERMINE and ROBEY 1996). Mice with biglycan deficiency have a decreased bone density and a compensatory upregulation of decorin levels in bone (XU et al. 1997). Other non-collagenous matrix proteins, such as osteopontin, fibronectin, bone sialoprotein and thrombospondin, are involved in cellular attachment and cellular proliferation.

Excessive amounts of intact bone-matrix proteins are released to the blood during bone formation or in degraded forms after digestion by proteases during osteoclastic resorption. These bone-matrix constituents can be detected in serum and/or urine by means of biochemical assays. The commercially available assays are especially suited for monitoring bone turnover and are divided into two groups of assays based on their ability to detect changes in either bone formation or bone resorption (recently reviewed by CALVO et al. 1996; see also Chap. 39, JØRGENSEN et al.).

2. Bone Cells

Osteoblasts originate from local mesenchymal stem cells that, upon appropriate stimulation, proliferate and differentiate into mature osteoblasts. The mature osteoblasts are responsible for the production of most of the bone matrix, which is composed of collagen, non-collagenous proteins including latent growth factors, and ground substance. The bone-forming osteoblast, with one basal nucleus, has a cytoplasm rich in rough endoplasmic reticulum and Golgi complexes, and a cellular membrane rich in alkaline phosphatases. Osteoblasts are found in clusters of cuboidal cells, which line a layer of unmineralized (osteoid) bone matrix that the osteoblasts have formed.

Osteocytes originate from bone-forming osteoblasts that have been trapped in, or were designated to be embedded into, their own bone-matrix products, which later become calcified. They are connected to each other and to lining cells at the endosteal bone surfaces by small cellular extensions called canaliculi. Osteocytes have been hypothesized to serve as sensors of mechanical loading.

Mature osteoclasts, responsible for the active resorption of mineralized bone, are formed by fusion of mononuclear preosteoclasts of the granulocyte-macrophage lineage (for review see ROODMAN 1996). The mature multinucleated osteoclasts can easily be distinguished from other cells by a number of unique features: (1) the size can be up to 100 μm in diameter; (2) normally 5–20 nuclei are seen per cell; (3) formation of a ruffled border and resorption lacunae; (4) expression of calcitonin receptors and retraction in response to calcitonin; and (5) high expression of tartrate-resistant acid phosphatase (TRAP).

The shapes and activities of osteoclasts are controlled by systemic hormones and locally released cytokines. When the correct stimulatory signal is given, osteoclasts are recruited and adhere to the bone surface by a cell-integrin-bone surface interaction forming a tight seal between the cells and bone. The apical membrane adjacent to the bone surface will fold up, creating the unique ruffled border, thereby increasing the area of the apical membrane manifold and allowing fast exchange of large amounts of materials between the cell and the extracelluar space (HOLTROP et al. 1979).

The space between the cell and the bone is called the resorption lacunae. H^+ and bicarbonate are formed in the cytoplasm by hydration of CO_2, a process accelerated by carbonic anhydrase type II. Protons are actively pumped across the apical membrane into the resorption lacunae by H^+-adenosine triphosphatases, and bicarbonate is exchanged for chloride ions across the basolateral membrane. The acidification of the resorption lacunae (BARON et al. 1985), essential for the dissolution of hydroxyapatite crystals, is accompanied by the secretion of a range of proteinases that concert the degradation of demineralized matrix (DELAISSE et al. 1987) (Fig. 2b). Several proteinases have been identified in purified osteoclast including the cysteine proteinase cathepsin K (TEZUKA et al. 1994; INAOKA et al. 1995; BOSSARD et al. 1996), and the matrix metalloproteinases (MMP) MMP-9, MMP-12, and membrane type-1-MMP (REPONEN et al. 1994; HOU et al. 1997; SATO et al. 1997). Inhibitors to both cysteine- and matrix metalloproteinases have been found to decrease bone resorption in vivo and in vitro (DELAISSE et al. 1984; EVERTS et al. 1992; VOTTA et al. 1997), and may eventually become therapeutic agents used in the treatment of diseases characterized by increased bone resorption, e.g., Paget's disease, osteoporosis, and tumor-induced osteolysis.

3. Bone Remodeling

In the normal adult skeleton, bone is continuously renewed by osteoclastic resorption of old bone and osteoblastic formation of new bone (PARFITT 1979). The cells in the localized process of bone renewal constitute the bone-remodeling units (BMUs), in which resorption always precedes bone formation (Fig. 2a). The first event in bone remodeling involves recruitment of mature osteoclasts and generation of new osteoclasts from stem cells near the BMU. Osteoclasts will resorb a confined volume of old bone in the BMU. The resorption phase is finished in a yet undefined manner with a cessation of osteoclast activity followed by the reversal phase, in which mononuclear cells might be involved in forming a "cement line" on the cleaned bone surface. Bone-forming osteoblasts will follow the cement line, forming demineralized (osteoid) matrix. The BMU is completed by the mineralization of the osteoid mainly by fixation of hydroxyapatite crystals to the collagen fibers. In the normal skeleton, the volume of bone formed by osteoblasts equals the volume of bone resorbed by the osteoclasts (balanced remodeling).

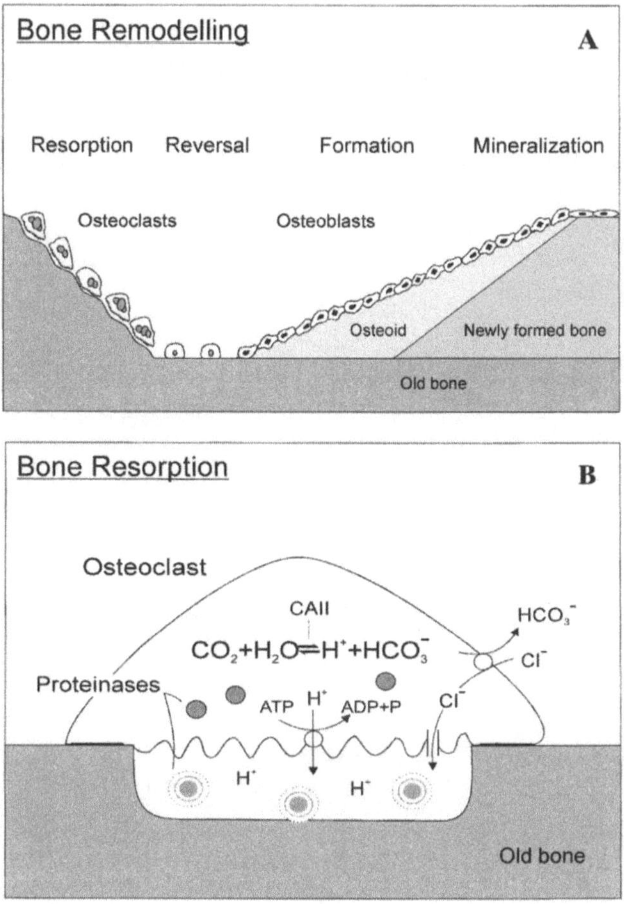

Fig. 2. A Schematic drawing showing the steps in normal bone remodeling at bone surfaces. The resorption of bone is initiated by activation and formation of osteoclasts, which adhere to the bone surfaces and form erosion cavities. After resorption has been completed, mononuclear bone-lining cells clean the bone surface (reversal). Subsequently, osteoblasts are activated, forming a seam of collagen, non-collagenous proteins, and ground substance (osteoid). The remodeling is completed by the mineralization of the osteoid, primarily by the deposition of hydroxyapatite crystals. **B** Schematic drawings of osteoclastic bone resorption. The degradation of old bone is accomplished by osteoclastic secretion of proteinases and protons to the resorption lacunae

III. Estrogen and Bone

Estrogen has a great impact on bone cell physiology. In postmenopausal women the cessation in ovarian estrogen secretion results in an accelerated, imbalanced bone remodeling (Fig. 3) with a subsequent decrease in bone mineral density (BMD). The postmenopausal-associated bone loss can be halted fully by administration of exogenous estrogen or estrogen-like substances (Christiansen et al. 1981; Delmas 1997).

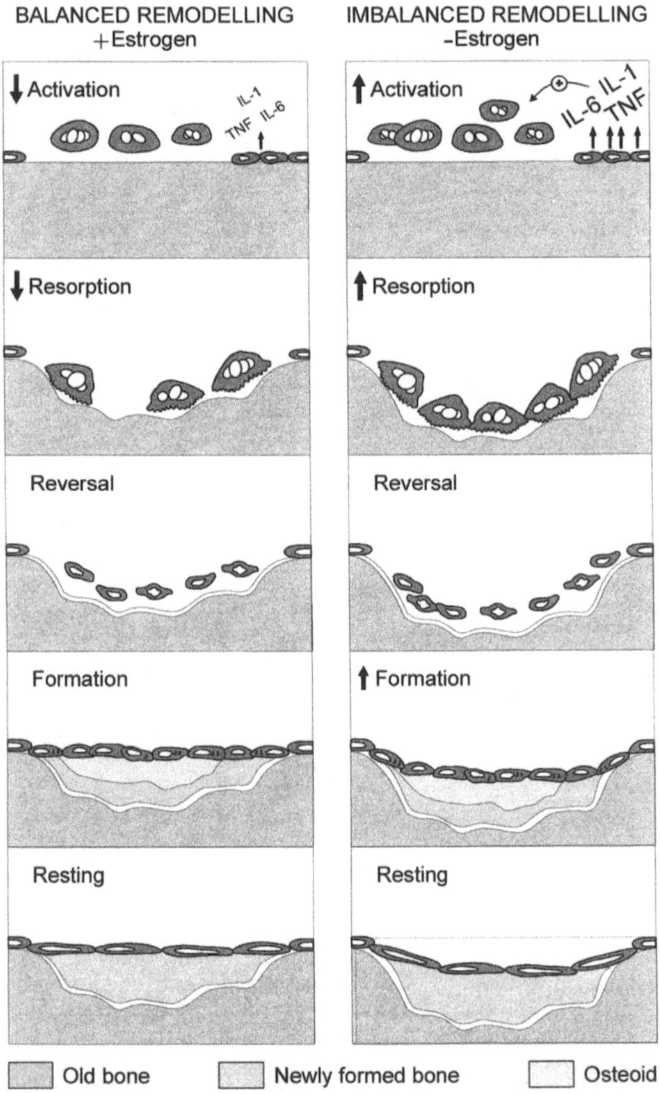

Fig. 3. Balanced versus imbalanced bone remodeling: several lines of evidence suggest that bone-cell-derived cytokines play a pivotal role in the imbalance in bone remodeling following postmenopausally or surgically induced estrogen deficiency. In postmenopausal women, the observed increase in bone resorption is due both to an increase in the number of bone remodeling units (BMUs) and an accelerated osteoclast activity within the single BMU. Although the bone formation will be partially increased, the volume of bone mass resorbed by osteoclasts will be greater than the bone mass formed by osteoblasts. This results in an imbalanced bone remodeling with a net loss in bone mass every time a BMU has been completed. The process seems to be reversible, as estrogen therapy can re-establish balanced bone remodeling in estrogen-deficient women

Several lines of evidence suggest a direct effect of estrogen on bone cells. First, estrogen receptors have been found in both osteoblasts and osteoclasts (Eriksen et al. 1988; Komm et al. 1988; Oursler et al. 1991). Second, estrogen regulates the activity of osteoblasts and osteoclasts in vitro (Gray et al. 1987; Benz et al. 1991; Oursler et al. 1993; Oursler et al. 1994; Pederson et al. 1997). Third, the in vivo expression of several cytokines in bone has been found to be up- or downregulated in response to surgically induced estrogen deficiencies (caused by ovariectomy) in animal models (Fig. 3) (Jilka et al. 1992; Kimble et al. 1994; Kitazawa et al. 1994).

References

Baron RE (1996) Anatomy and ultrastructure of bone. In: Favus, MJ (ed) Primer on the metabolic bone diseases and disorders of mineral metabolism, 3rd edn. Lippencott–Raven, Philadelphia, pp 3–10

Baron R, Neff L, Louvard D, Courtoy PJ (1985) Cell-mediated extracellular acidification and bone resorption: evidence for a low pH in the resorbing lacunae and localization of a 100-kD lysosomal membrane protein at the osteoclast ruffled border. J Cell Biol 101:2210–2222

Benz DJ, Haussler MR, Komm BS (1991) Estrogen binding and estrogenic responses in normal human osteoblast-like cells. J Bone Miner Res 6:531–541

Bossard MJ, Tomaszek TA, Thompson SK, Amegadzie BY, Hanning CR, Jones C, Kurdyla JT, McNulty DE, Drake FH, Gowen M, Levy MA (1996) Proteolytic activity of human osteoclast cathepsin K: expression, purification, activation, and substrate identification. J Biol Chem 271:12517–12524

Calvo MS, Eyre DR, Gundberg CM (1996) Molecular basis and clinical application of biological markers of bone turnover. Endocr Rev 17:333–368

Christiansen C, Christensen MS, Transbol I (1981) Bone mass in postmenopausal women after withdrawal of estrogen/gestagen replacement therapy. Lancet 1:459–461

Delaisse JM, Eeckhout Y, Vaes G (1984) In vivo and in vitro evidence for the involvement of cysteine-proteinase in bone resorption. Biochem Biophys Res Commun 125:441–447

Delaisse JM, Boyde A, Maconnachie E, Ali NN, Sear CHJ, Eeckhout Y, Vaes G, Jones SJ (1987) The effects of inhibitors of cysteine-proteases and collagenase on the resorptive activity of isolated osteoclasts. Bone 8:305–313

Delmas PD, Bjarnason NH, Mitlak BH, Ravoux AC, Shah AS, Huster WJ, Draper M, Christiansen C (1997) Effects of raloxifene on bone mineral density, serum cholesterol concentrations, and uterine endometrium in postmenopausal women. N Engl J Med 337:1641–1647

Ducy P, Desbois C, Boyce B, Pinero G, Story B, Dunstan C, Smith E, Bonadio J, Goldstein S, Gundberg C, Bradley A, Karsenty G (1996) Increased bone formation in osteocalcin-deficient mice. Nature 382:448–452

Eriksen EF, Colvard DS, Berg NJ, Graham ML, Mann KG, Spelsberg TC, Riggs BL (1988) Evidence of estrogen receptors in normal human osteoblast-like cells. Science 241:84–86

Everts V, Delaisse JM, Korper W, Niehof A, Vaes G, Beertsen W (1992) Degradation of collagen in the bone-resorbing compartment underlying the osteoclast involves both cysteine-proteinases and matrix metalloproteinases. J Cell Physiol 150:221–231

Gray TK, Flynn TC, Gray KM, Nabell LM (1987) 17β-Estradiol acts directly on the clonal osteoblastic cell line UMR106. Proc Natl Acad Sci USA 84:6267–6271

Holtrop ME, Raisz LG (1979) Comparison of the effects of 1,25-dihyroxy-cholecalciferol, prostaglandin E_2, and osteoclast-activating factor with parathyroid hormone on the ultrastructure of osteoclasts in cultured long bones of fetal rats. Calcif Tissue Int 29:201–205
Hou P, Ovejero MC, Sato T, Kumegawa M, Foged NT, Delaisse JM (1997) MMP-12, a proteinase that is indispensable for macrophage invasion, is highly expressed in osteoclasts (abstract). J Bone Miner Res 12:S417
Inaoka T, Bilbe G, Ishibashi O, Tezuka K, Kumegawa M, Kokubo T (1995) Molecular cloning of human cDNA for cathepsin K: novel cysteine proteinase predominantly expressed in bone. Biochem Biophys Res Commun 206:89–96
Jee WSS (1983) The skeletal tissues. In: Weiss L (ed) Histology, cell and tissue biology. Elsevier, New York, pp 200–255
Jilka RL, Hangoc G, Girasole G, Passeri G, Williams DC, Abrams JS, Boyce B, Broxmeyer H, Manolagas SC (1992) Increased osteoclast development after estrogen loss: mediation by interleukin-6. Science 257:88–91
Kimble RB, Vannice JL, Bloedow DC, Thompson RC, Hopfer W, Kung VT, Brownfield C, Pacifici R (1994) Interleukin-1 receptor antagonist decreases bone loss and bone resorption in ovariectomized rats. J Clin Invest 93:1959–1967
Kitazawa R, Kimble RB, Vannice JL, Kung VT, Pacifici R (1994) Interleukin-1 receptor antagonist and tumor necrosis factor binding protein decrease osteoclast formation and bone resorption in ovariectomized mice. J Clin Invest 94:2397–2406
Komm BS, Terpening CM, Benz DJ, Graeme KA, Gallegos A, Korc M, Greene GL, O'Malley BW, Haussler MR (1988) Estrogen binding, receptor mRNA, and biologic response in osteoblast-like osteosarcoma cells. Science 241:81–84
Oursler MJ, Osdoby P, Pyfferoen J, Riggs BL, Spelsberg TC (1991) Avian osteoclasts as estrogen target cells. Proc Natl Acad Sci USA 88:6613–6617
Oursler MJ, Pederson L, Pyfferoen J, Osdoby P, Fitzpatrick LA, Spelsberg TC (1993) Estrogen modulation of avian osteoclast lysosomal gene expression. Endocrinology 132:1373–1380
Oursler MJ, Pederson L, Fitzpatrick LA, Riggs BL, Spelsberg TC (1994) Human giant cell tumors of the bone (osteoclastomas) are estrogen target cells. Proc Natl Acad Sci USA 88:6613–6617
Parfitt AM (1979) Quantum concept of bone remodeling and turnover: implications for the pathogenesis of osteoporosis. Calcif Tissue Int 28:1–5
Pederson L, Kremer M, Foged NT, Winding B, Fitzpatrick LA, Oursler MJ (1997) Evidence of a correlation of estrogen receptor level and avian osteoclast estrogen responsiveness. J Bone Miner Res 12:742–752
Reponen P, Sahlberg C, Munaut C, Thesleff I, Tryggvason K (1994) High expression of 92-kD type IV collagenase (gelatinase B) in the osteoclast lineage during mouse development. J Cell Biol 124:1091–1102
Roodman GD (1996) Advances in bone biology: the osteoclast. Endocr Rev 17:308–332
Sato T, Ovejero M, Hou P, Heegard AM, Kumegawa M, Foged N, Delaisse JM (1997) Identification of the membrane-type matrix metalloproteinase MT1-MMP in osteoclasts. J Cell Sci 110:589–596
Termine JD, Robey PG (1996) Anatomy and ultrastructure of bone. In: Favus, MJ (ed) Primer on the metabolic bone diseases and disorders of mineral metabolism, 3rd edn. Lippencott–Raven, Philadelphia, pp 24–28
Tezuka K-I, Tezuka Y, Maejima A, Sato T, Nemoto K, Kamioka H, Hakeda Y, Kumegawa M (1994) Molecular cloning of a possible cysteine proteinase predominantly expressed in osteoclasts. J Biol Chem 269:1106–1109
Votta BJ, Levy MA, Badger A, Bradbeer J, Dodds RA, James IE, Thompson S, Bossard MJ, Carr T, Connor JR, Tomaszek TA, Szewczuk L, Drake FH, Veber DF, Gowen M (1997) Peptide aldehyde inhibitors of cathepsin K inhibit bone resorption both in vitro and in vivo. J Bone Miner Res 12:1396–1406

Xu T, Longneck G, Fisher L, Heegaard AM, Satomura K, Bianco G, Sommer B, Kulkarni A, Robey PG, Young MF (1997) Mice with targeted disruption of the biglycan gene exhibit decreased bone density and increased expression of decorin (abstract). J Bone Miner Res 12:S125

CHAPTER 27

Central Nervous System

R. GALLO, M. STOMATI, A. SPINETTI, F. PETRAGLIA, and A.R. GENAZZANI

A. Introduction

Gonadal hormones exert several effects on the central nervous system (CNS) throughout the lifespan and go beyond the traditional control of the reproductive function and the modulation of sexual behavior. The identification of estrogen, progestin and androgen receptors outside the classical CNS regions justifies their role in controlling different brain functions. In particular, specific receptors for gonadal steroids have been localized in the amygdala, hippocampus, cortex, basal forebrain, cerebellum, locus coeruleus, midbrain, rafe nuclei, glial cells and central gray matter, confirming an involvement of sex hormones in the control of psychophysiological well-being, cognitive functions and memory processes.

Estrogens can act by two mechanisms on the CNS target areas: genomic and non-genomic. The classical genomic activation is mediated via intracellular receptors. After the binding of the steroid to soluble cytoplasmic receptor proteins, which are normally concentrated near the membrane, the hormone–receptor complex moves to the nucleus, where cellular function is ultimately regulated through modifications in gene transcription and protein synthesis (GENAZZANI et al. 1996; SMITH 1993; ALONSO-SOLEIS et al. 1996; SHERWIN 1996). In this way, gonadal steroids modulate the synthesis, release and metabolism of the neuroactive transmitters and the expression of their receptors. In particular, norepinephrine, dopamine, γ-aminobutyric acid (GABA), acetylcholine and serotonin (5-HT) turnover is modulated by estrogens. Neuropeptides directly influenced by gonadal hormones include the opioid peptides, corticotropin-releasing factor (CRF), neuropeptide Y (NPY) and galanin (SPEROFF et al. 1995) (Table 1, Fig. 1).

Non-genomic effects of estrogen have been demonstrated in experimental studies on female rats, involving cell-membrane receptors. These activational effects of estrogens mediate the regulation of neuronal plasticity. Both genomic and non-genomic mechanisms are involved in the modulation of neuronal comunications and the neuroendocrine system by sex steroids, with a rapid short-term mechanism for non-genomic membrane effects and a slower long-term effect for the genomic (MCEWEN 1994a) (Fig. 2). Beginning during fetal life, estrogens influence growth, maturation, differentiation and functioning of brain cells. It seems that sex-steroid hormones' exposure during fetal

Table 1. Principal neurotransmitters and neuropeptides modulated by sex-steroid hormones

Neurotransmitters	Norepinephrine Dopamine Acetylcholine Serotonin γ-aminobutyric acid (GABA) Melatonin
Neuropeptides	Opioid peptides Neuropeptide Y (NPY) Galanin Corticotropin-releasing factor (CRF)

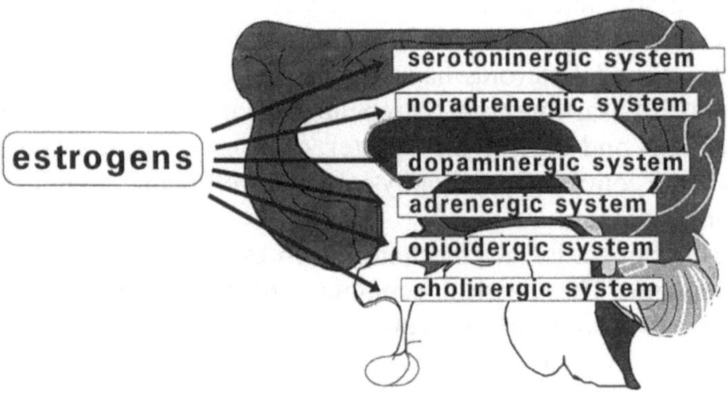

Fig. 1. Genomic and non-genomic mechanisms of action of estrogens in the central nervous system

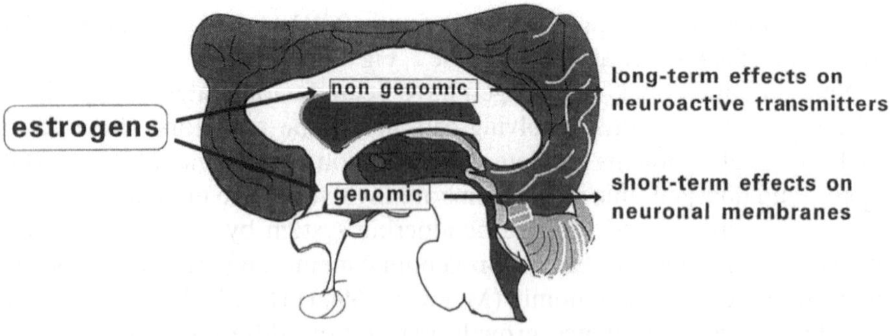

Fig. 2. Principal neuroendocrine systems modulated by estrogens

life could influence brain organization. In particular, estrogens are believed to be particularly responsible for the sexual differentiation of nervous tissues in specific area of the brain (TORAND-ALLERAND 1980; MCEWEN 1994b).

During reproductive life, in both sexes, gonadal hormones modulate the reproductive function, electively, by acting in the brain at the hypothalamic level, on the synthesis and release of gonadotropin-releasing hormone (GnRH), and also via the modulation of neurotransmitters and neuropeptides (SPEROFF et al. 1995). However, estrogens are of critical importance in women, not only in controlling the reproductive system, but also in many other biological functions. In fact, by modulating neurotransmitters and neuropeptide's synthesis and release, estrogens regulate the activity of the thermoregulatory, satiety, appetite and blood-pressure centers in the hypothalamus, while they are involved in the regulation of mood, behavior and psychological well-being in the limbic system (Table 2).

During the climacteric period, the decline in ovarian sex steroids is associated with an impaired turnover of classic neurotransmitters and a reduced activity of those neuropeptides involving CNS function. The greater part of classical knowledge and the recent research regarding gonadal hormones' effects on CNS have been carried out by means of in vitro studies and in vivo animals experiments. The attention is focused on menopause and post-menopausal periods, the physiological withdrawal of sex-steroid hormones and the hormonal replacement therapy. This represents a unique opportunity to investigate the actions of gonadal hormones on their specific receptors in reproductive organs, the cardiovascular system, bone and the CNS in humans. The major liabilities of chronic ERT in post-menopausal women are predominantly related to the adverse effects on reproductive tissues, breast and uterus, with an increased risk of various tumors.

Recently, there has been a great interest in identifying an ideal compound with the beneficial effects of estrogens but no stimulatory actions on

Table 2. Principal brain areas with functions modulated by sex steroids, and clinical effects of the neuronedocrine deregulation of the climacteric period and post-menopause. *CNS* central nervous system

Brain areas	Functions	Clinical effects at CNS level
Hypothalamus	Thermoregulation	Hot flushes and sweat
	Satiety	Obesity
	Hungry	Hypertension
	Blood pressure	
Limbic system		Changes of mood and behavior
	Mood and behavior	Anxiety
	Cognitive function	Depression
		Insomnia
		Headache/migraine
		Modification of cognitive function

reproductive tissues. Extensive research efforts have demonstrated that certain compounds, originally developed as estrogen antagonists for the prevention and treatment of breast cancer, produce estrogen agonist-like effects. Further developments in the pharmacology and molecular biology of antiestrogens have identified a class of compounds named selective ER modulators (SERMs).

B. Estrogens and CNS

I. Estrogens, Neurotransmitters and Neuropeptides

Sex-steroid hormones influence brain noradrenergic and dopaminergic systems. Studies conduced on rats showed an increased norepinephrine and dopamine turnover rate induced by estrogens on proestrus. In castrated female rats, an impairment of cathecholaminergic neurons with an increase in norepinephrine release and a decrease in dopamine has been demonstrated. Estrogen administration decreases hypothalamic norepinephrine release and increases dopamine release in the medio-basal hypothalamus (ETGEN and KARKANIAS 1994) (Fig. 1). The modification of the norepinephrine activity may be due to the decreased norepinephrine re-uptake, the inhibition of monoamine oxidase (MAO) or catechol-*O*-methyl-transferase activity.

Few data are available on the effects of progestagens: in ovariectomized female rats, the contemporary administration of progesterone suppresses estrogen's action on the noradrenergic neurons of the pineal glands (ALONSO-SOLEIS et al. 1996). Estrogen can also modify the concentration and the availability of 5-HT by increasing the rate of degradation of MAO, the enzyme that catabolizes 5-HT (LINE and MCEWEN 1997). Experimental data have demonstrated that estrogens displace tryptophan from its binding sites, thus, increasing its availability for the metabolization into 5-HT (PANAY et al. 1996). Moreover, gender differences in 5-HT concentration have been described in animal studies: higher concentrations of 5-HT have been demonstrated in female rats' forebrains, rafe, frontal cortex hypothalamus than in male rats' brain.

Brain activity is modified during periods of physiological ovarian-hormone fluctuation. In ovariectomized female rats, estrogen treatment positively affects the serotoninergic system by increasing postsynaptic responsiveness and receptor activity (DI PAOLO et al. 1983; JOHNSON and CROWLEY 1983). Estrogen affects basal forebrain and cholinergic neurons that project to the cerebral cortex and hippocampus in rat brain. Studies on steroid effects on the expression of cholinergic enzymes demonstrated that estrogens act as a cholinergic agonist, by inducing synthesis and activation of choline acetyltransferase (ChAT), the rate-limiting enzyme for acetylcholine formation. Estradiol also induces acetylcholinesterase, therefore, suggesting a general trophic effect on the cholinergic neurons (MCEWEN 1997).

Neuropeptides that are directly influenced by gonadal hormones include opioid peptides, CRF, NPY and galanin (SPEROFF et al. 1995). Endogenous opioid peptides are grouped into three main classes of neuropeptides, named endorphins, enkephalins and dynorphins which are involved in the regulation of several functions as well as the control of the (HPG) axis. Hypothalamic neurons producing β-endorphin (β-EP) have a key function in inhibiting GnRH activity. The β-EP pathway represents a target for the feedback action of gonadal steroids on GnRH secretion. The correlation between hypothalamic β-EP content and gonadal steroids is further supported by the evidence that the circadian changes in β-EP concentrations in the medial basal hypothalamus of female rats are abolished by ovariectomy and are restored by estradiol treatment (PETRLAIA et al. 1994).

Among those neuropeptides modulated by gonadal steroids, NPY influences several brain functions, including the control of food intake, sexual behavior and neuroendocrine functions. NPY participates in the neuroendocrine axis by stimulating the release of pulsatile GnRH and gonadotropins. Experimental studies have shown that estrogen is able to stimulate NPY synthesis and release in the hypothalamus. In castrated female rats, the deficiency of gonadal steroids reduces NPY production and secretion. Estrogen increases NPY content in the median eminence and the synthesis of NPY in the arcuate nucleus by inducing NPY gene expression (KARLA et al. 1996).

Recent findings have demonstrated several interactions between NPY and β-endorphin neurons at the hypothalamic level, suggesting that both estrogens and progestagens could exert indirectly modulatory effects on NPY inducing β-endorphin release. In addition, NPY is a potent stimulator of carbohydrate consumption, inducing an increase of food intake (KAPLAN 1988). Galanin is a neuropeptide isolated from the anterior pituitary gland of rat and man, the synthesis of which is under the control of sex steroids. Galanin's action has not been completely clarified. There is experimental evidence that demonstrates a role in the modulation of prolactin release and ACTH secretion (PRIEST and PFAFF 1995).

II. Estrogen Effects on Neuronal Plasticity

Experimental evidence indicates that estrogens and progestagens promote interactive communications among neurons (PANAY et al. 1996; PRIEST and PFAFF 1995). The effects on the neuronal plasticity occurs via stimulation of the dendritic growth and the number of dendritic spines. In fact, morphological studies in female rats under estrogen treatment showed an increase in dendritic spines and new synapses in the ventromedial hypothalamus, as well as in the density of dendritic spines of the CA1 pyramidal hippocampal neurons (MCEWEN et al. 1994c). Dendritic-spine density in female rats cyclically changes during the estrous cycle, indicating that synapses are formed rapidly and broken during the rat cycle in relation to estradiol production. Consequently, even though some brain regions, such as the hippocampus, express few ERs, with respect to the hypothalamus or pituitary, these regions must be

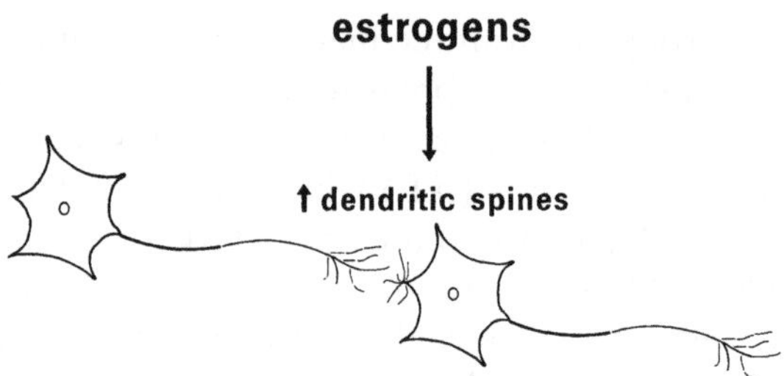

Fig. 3. Estrogens modify communication among neurons

considered when discussing the greatest sensitivity to estrogen treatment or during the natural estrus cycle.

Recent data support an involvement of *N*-methyl-D-aspartate (NMDA) receptors in the induction of new synaptic spines on neurons treated with estrogen. In fact, NMDA-receptor antagonists block the spine formation on the CA1 pyramidal neurons (WOOLEY and MCEWEN 1994). In this region, NMDA receptors are placed on the excitatory spine of the synapses and estradiol is able to increase the NMDA receptor binding sites, suggesting that the estrogen induction of NMDA receptors may be an important event in the synapses' formation (WOOLEY and MCEWEN 1994). NMDA receptors gate calcium ions, and this factor may be fundamental in the modification of synaptic spines. Moreover, estrogens induce the glutamic-acid decarboxylase enzyme mRNA in the inhibitory interneurons of the hippocampal neurons (WEILAND 1992) and some authors suggest that estrogens' inhibitory effect disinhibits the pyramidal neurons (WOOLEY and MCEWEN 1994). However, many questions remain unanswered, for example, the role of GABAa receptors or the effect of estrogens on other excitatory synapses, and the mechanisms that lead to the choice of the pathway are not clear. Recent in vitro studies showed that in those brain areas missing classical estrogen receptors, the synaptogenesis could be explained by a non-genomic mechanism; in fact, estrogens inhibit calcium-ion currents in isolated neurons and potentiate kainic acid-induced currents in hippocampal neurons (NAKAJIMA et al. 1995; MERMELSTEIN et al. 1996; GU et al. 1996) (Fig. 3).

III. Estrogens and Cognitive Functions

Clinical studies have reported a frequent decrease in attention and memory performances in climacteric and post-menopausal women. At present, the few studies available on this interesting argument demonstrate discrepant results of methodological problems (HACKMAN and GAILBRAITH 1977; FEDOR-FREYBERGH 1977). In fact, in various studies, cognitive functions have been evaluated using different tests that provide different and incomparable information. Recent studies in surgically post-menopausal women (SHERWIN 1988; PHILLIPS and SHERWIN 1992) have demonstrated specific memory impairment, independently from the presence of affective disorders or other symptoms related to menopause. In the last 10 years, epidemiological and clinical studies have shown an important interaction between aging and estrogen withdrawal for the development of cognitive disturbances and the dementia of Alzheimer's Type. In fact, aging women are more affected by chronic and degenerative diseases, such as reduction of short-term memory and increased incidence of Alzheimer's disease (AD), than age-matched men (PAGANINI-HILL and HENDERSON 1994; ARRET-CONNOR and KRITZ-SILVERSTEIN 1993). The long period of life with estrogen deprivation exerts a fundamental role in the physiopathological mechanism of these degenerative diseases (AMADUCCI 1996). Recent studies have shown a significantly decreased AD risk (about 50%) and an improvement of cognitive functions among estrogen users.

There are several mechanisms by which estrogen may affect cognitive function: estrogen may increase cerebral blood flow and glucose utilization, influence neurotransmitter systems and promote neuronal growth. On the basis of experimental data, the fall of plasma estrogen levels has been related to modifications involving hippocampal structures and neurotransmission of adrenergic and cholinergic systems. In particular, acetylcholine is the most important neurotransmitter involved in the neuroendocrine modulation of cognitive function and memory (LUINE 1985). In post-menopausal women, a reduction of chAT activity has been described (FILLIT et al. 1986). Actually, this is considered one of the most relevant elements responsible for the short-term memory impairment with parallel derangement of cognitive function.

Experimental studies in female rats have shown a significant reduction in dendritic-spine density on hippocampal cells of the CA1 region, in relation to the decrease in plasma estrogen levels during the estrus phase of the cycle. In contrast, the dendritic-spine density increases when circulating sex-steroid hormone levels rise (GOULD et al. 1990; WOOLLEY et al. 1990). These structural modifications occur over a short period of time: hours or days. It is possible that longer hormonal modifications, such as those occurring during menopause, provoke more remarkable morphological modifications. The behavioral implications of these phenomenons have not been completely explained: since the dendritic spines are a post-synaptic site, it is possible to hypothesize that morphological variations could have been translated in neuronal plasticity variations and then influence hippocampal functions, such as memory and learning processes.

C. Selective Estrogen Receptor Modulators (SERMs)

I. Introduction

Recent developments in the pharmacology and molecular biology of anti-estrogens have identified a class of compounds, named SERMs, that interact with ERs to exert genomic effects in subtly different ways from classical estrogens (SATO et al. 1995). The major classes of synthetic anti-estrogens include: triphenylethylene derivatives (clomiphene, tamoxifen, droloxifene, toremifene), benzothiophene derivatives (raloxifene), and pure anti-estrogens (the 7α-alkyl derivatives ICI 164, 384 and ICI 182,780 and 11β-amidoalkyl estradiol derivatives such as RU 51625). Triphenylethylene derivatives are non-steroidal estrogens consisting of an ethylene core (C=C) in which three hydrogen atoms are replaced by phenyl rings. Benzothiophene derivatives are sulfur-containing non-steroidal compounds. The 7α-alkyl and 11β-amidoalkyl estradiol derivatives are synthetic steroids that are structurally identical to estradiol except for the addition of a substituent at the 7- or 11- position (CLAUSSNER et al. 1992). Most synthetic estrogen antagonists are actually mixed agonist–antagonist compounds.

The ER resides in a transcriptionally inactive state within the nuclei of target cells when it is not bound to the hormone. After binding of the ligand, the receptor undergoes conformational changes that initiate a cascade of events, leading to the activation of a specific palindromic sequence of DNA termed an estrogen response element (ERE), which regulates gene transcription (KUNAR et al. 1987). The ER contains two activation domains, known as transcriptional activation function-1 (AF-1), located towards the amino terminus, and AF-2 contained within the ligand-binding domain of the receptor. These mediate the genomic activation in a different manner from cell to cell.

Regarding the effects of SERMs' binding on the ER, several studies have been carried out and only few data are available for each compound. Tamoxifen blocks estrogen actions by competing with endogenous estrogen by binding to its receptors. The tamoxifen binding to the ER inhibits AF-2 function and gene transcription; conversely, tamoxifen acts as an agonist when AF-1 is alone (BERRY et al. 1990; TZUKERMAN et al. 1994). Like tamoxifen, the other triphenylethylene derivatives, which display agonist activity in the uterus and antagonistic activity in the breast, all function as cell-specific AF-1 agonists in vitro. However, recent findings have shown that the genomic transduction signal is more complex, involving other transcriptional sites (TZUKERMAN et al. 1990). In addition, tamoxifen may exert extra-genomic actions such as the inhibition of proteine kinase-C and calmodulin-dependent cAMP phosphodiesterase (COLETTA et al. 1994). Similarly to tamoxifen, raloxifene, has a high affinity for the ER. This compound exerts an antagonist action in reproductive tissue, thus preventing the transcriptional activation of ERE-containing genes. However, the mechanism of action of raloxifene as an estrogen agonist may be explained by identifying novel genomic response elements, termed raloxifene inducible element (RIE). Raloxifene binding to ERs recruits RIE-specific binding to activate DNA transcription (YANG et al. 1990).

Another class of compounds, the steroidal estrogen antagonists, ICI-164384 and ICI-182780, exert a role of pure anti-estrogens, because they do not have any estrogenic properties. Their mechanism of action in vitro involves the destruction of newly synthesized ERs. Studies on female rats have demonstrated that ICI compounds acts as estrogen antagonists in the skeletal tissues, resulting in significant bone loss (GALLAGHER et al. 1993).

II. SERMs and CNS

At present, few data are available regarding the effects and the mechanism of action of SERMs in the brain. Since, at the neuroendocrine level, estrogens play a fundamental role in the modulation of the hormonal pituitary products, the role of anti-estrogen in regulating pituitary gonadotropin and prolactin secretion has been investigated. Raloxifene has been demonstrated to have a potent antagonist effect on both short-term suppressive and long-term stimulatory effects of estrogen on LH release in vitro (OORTMANN et al. 1988; ORTMANN et al. 1990; AWATA et al. 1992). Raloxifene in non-estrogen-treated cells had no effect on spontaneous LH release and, for this reason, it was termed a "pure anti-estrogen" in pituitary gonadotrophs (CLEMENS et al. 1983). In vivo, raloxifene treatment of ovariectomized rats, pre-treated with estrogen, reduces serum LH levels (PETRERSEN et al. 1989). However, there are no similar data regarding the inhibition of the proestrus afternoon peak of LH (PETERSEN et al. 1989; SIMARD et al. 1990).

Just as for estrogen, the regulatory role of raloxifene involves the hypothalamic preoptic area via specific ERs. In contrast, raloxifene seems to have no effect on the neuroactive transmitter modulation of LH release. In fact, experimental data do not support an involvement of norepinephrine-mediated LH stimuli by raloxifene (MEISL et al. 1987). Raloxifene antagonizes estrogen-induced prolactin release, both in vitro and partially in vivo (CLEMENS et al. 1983). The first preliminary studies on behavior have shown that, raloxifene implanted into the hypothalamic ventromedial nucleus of ovariectomized rats has an antagonistic action on the estrogen-induced lordotic behavior (MEISL et al. 1987). Clinical studies have demonstrated that among the tamoxifen analogs, toremifene or chlorotamoxifen have weak estrogen-like properties, thus inducing a slight reduction in serum LH and FSH levels and an increase of sex hormone binding globulin (HAMM et al. 1991; KIVINEN and MAENPAA 1990; SZAMEL et al. 1994).

Recent findings demonstrated, in cell cultures of embryonic hippocampal neurons, an estradiol-induced increase of the number of dendritic spines. This effect is antagonized by the anti-estrogen tamoxifen and by the NMDA receptor antagonist (APV), but not by the AMPA/kainate receptor antagonist DNQX. These data suggest an influence of the anti-estrogen compounds on the neuronal plasticity, dendritic spine formation and synaptic connections. Several studies indicated modulatory effects of gonadal hormones on the β-EP synthesis and release from the hypothalamus of female rats. The castration of female rats induces a reduction in β-EP, while estrogen treatment increases the content and the release of β-EP in these rats. Our previous studies regarding the effects of clomiphene and ciclophenile, two estrogen antagonists, have shown that a trend in intact female rats decreases hypothalamic β-EP content.

In contrast, in ovariectomized rats both anti-estrogens significantly increase β-EP release (GENAZZANI et al.(1990). The evidence that the treatment with anti-estrogen drugs induces an estrogen-like effect, suggests that an anti-estrogen exerts an agonistic effect in the absence of endogenous ligands.

D. Conclusions

The fundamental role of estrogen, progestin and androgen, in controlling several brain functions, has been largely demonstrated during the last three decades. In particular, recent developments in basic research have contributed to clarification of the mechanisms of action of sex steroids in different specific brain areas. Observational studies in post-menopausal women support a role for hormonal replacement therapy in the prevention of mood, cognitive and behavioral disturbances. The identification of selective ER modulators stimulated several clinical and experimental researches in all the target tissues. At present, few data are available on brain SERMs' effects and some studies are trying to clarify the role in this important target for women's well-being.

References

Alonso-Soleis R et al. (1996) Gonadal steroid modulation of neuroendocrine transduction: a transynaptic view. Cell Mol Neurobiol:357–382

Amaducci L (1996) Demenza ed estrogeni. Atti I Congresso Nazionale SNOG, Pisa 11–12 Marzo, p 50

Awata S (1992) Effects of RU486 and keoxifene on the dispersed pituitary cells of pregnant rats. Jpn J Fertil Steril 37:22–28

Barret-Connor E, Kritz-Silverstein D (1993) Estrogen replacement therapy and cognitive function in older women. JAMA 269:2637–2641

Berry M et al. (1990) Role of the two activating domains of the estrogen receptor in the cell-type and promoter-context dependent agonistic activity of the antiestrogen 4-hydroxytamoxifen. EMBO J 9:2811–2818

Claussner A et al. (1992)11β-amidoalkyl estradiols, a new series of pure antiestrogens. J Seroid Biochem Molec Biol 41:609–614

Clemens JA et al. (1983) Effects of a new antiestrogen, keoxifene LY156758, on growth of carcinogen induced mammary tumors and on carcinogen induced mammary tumors and on LH and prolactin levels. Life Sci:2869–2875

Colletta AA et al. (1994) Alternative mechanism of action of antiestrogens. Breast Cancer Res Treat 31:5–9

Di Paolo T et al. (1983) Effect of acute and chronic 17β-estradiol treatment on serotonin and 5- hydroxiindole acetic acid content of discrete brain nuclei of ovariectomized rats. Exp Brain Res 51:73–76

Etgen AM, Karkanias GB (1994) Estrogen regulation of noradrenergic signaling in the hypothalamus. Psychoneuroendocrinology 19:603–610

Fedor-Freybergh P (1988) The influence of estrogen on well being and mental performance in climateric and postmenopausal women. Acta Obstet Gynaecol Scand 1977 64 (Suppl):5–69

Fillit H et al. (1986) Observation in a preliminary open trial of estradiol therapy for senile dementia-Alzheimer type. Psychoneuroendocrinology 34:521–525

Gallagher A et al.(1993) The estrogen antagonist ICI 182,780 reduces cancellous bone volume in female rats. Endocrinology 133:2787–2791

Genazzani AR et al. (1990) Effect of steroid hormones and antihormones on hypothalamic beta- endorphin concentration in intact and castrated female rats. J Endocrinol Invest 13:91–96
Genazzani AR et al. (1996) The brain as source and target for sex steroid hormones. The Parthenon Publishing Group, Pearl Rivier NY
Gould E et al. (1990) Gonadal steroids regulate dendritic spine density in hippocampal pyramidal cells in adulthood. J Neurosci 10:1286–1291
Gu Q, Moss RL (1996). 17β-estradiol potentiates kainate-induced currents via activation of the cAMP cascade. J Neurosci 16:3620–3629
Hackman BW, Galbraith D (1977) Six month study of estrogen therapy with piperazine oestrone sulphate and its effect on memory. Curr Med Res Opin 4 (Suppl):21–27.
Hamm JT et al. (1991) Phase I study of toremifene in patients with advanced cancer. J Clin Oncol 9:2036–2041
Johnson MD, Crowley WR (1983) Acute effects of estradiol on circulating luteinizing hormone and prolactin concentrations and on serotonin turnover in individual brain nuclei. Endocrinology 113:1935–1941
Kaplan LM et al. (1988) Galanin is an estrogen-inducible, secretory product of the rat anterior pituitary. Proc Natl Acad Sci USA 85:7408–7412
Karla SP (1996) Gonadal steroid hormones promote interactive comunication. In: Genazzani AR, Petraglia F and Purdy RH (eds) The brain: source and target for sex steroid hormones. The Parthenon Publishing Group, Pearl Rivier NY, pp 257–276
Kivinen S, Maenpaa J (1990) Effect of toremifene on clinical hematological and hormonal parameters at different dose levels in healthy postmenopausal volunteers: phase I study. J Steroid Biochem 36:217–220
Kumar V et al. (1987) Functional domains of the human estrogen receptor. Cell 51:941–951
Levine AS et al. (1990) The effect of centrally administered naloxone on deprivation and drug induced feeding. Pharmacol Biochem Behav 36:409–412
Line VN, McEwen BS (1997) Effect of estradiol on turnover of type A monoamine oxidase in the brain. J Neurochem 28:1221–1227
Luine VN (1985) Estradiol increases choline acetyltransferase activity in specific basal forebrain nuclei and projection areas of female rats. Exper Neurol 89:484–490
McEwen BS (1994a) Steroid hormone action on the brain: when is genome involved. Horm Behav 4:396–405
McEwen BS (1994b) Ovarian steroids have diverse effects on brain structure and function. In: Berg G, Hammar M (eds) The modern management of the menopause. Parthenon Publishing Group, Pearl Rivier NY, pp 269–278
McEwen BS, Wooley CS (1994c) Estradiol and progesterone regulate neuronal structure and synaptic connectivity in adult as well as in developing brain. Exp Gerontol 29:431–436
McEwen BS et al. (1997) Ovarian steroids and the brain: implication for cognition and aging. Neurology 48 (Suppl 7):8–15
Meisl RL et al. (1987) Antagonism of sexual behaviour in female rats by ventromedial hypotzhalamic implants of antiestrogen. Neuroendocrinology 45:201–207
Mermelstein PG, Becker JB, Surmeier DJ (1996) Estradiol reduces calcium currents in rat neostriatal neurons via a membrane receptor. J Neurosci 16:595–604
Nakajima T (1995) 17β-estradiol inhibits the voltage dependent L-type Ca^{2+} currents in aortic smooth muscle cells. Eur J Pharmacol 294:625–635
Ortmann O et al. (1988) Inhibitory actions of keoxifene on luteinizing hormone secretion in pituitary gonadotrophs. Endocrinology 123:962–968
Ortmann O et al. (1990). Weak estrogenic activity of phenol red in the pituitary gonadotroph: reevaluation of estrogen and antiestrogen effects. J Steroid Biochem 35:17–22

Paganini-Hill A, Henderson V (1994) Estrogen deficiency and risk of Alzheimer's Disease in women. American Journal of Epidemiolog, 140:256–261
Panay N et al. (1996) Estrogen and behaviour. In: Genazzani AR, Petraglia F, Purdy RH (eds) The brain as source and target for sex steroid hormones. The Parthenon Publishing Group, Pearl Rivier NY, pp 257–276
Petersen SL et al. (1989) Medial preoptic microimplants of the antiestrogen, keoxifene, affect luteinizing hormone-releasing hormone mRNA levels, median eminence luteinizing hormone concentrations and luteinizing release in ovariectomized estrogen-treated rats. J Neuroendocrinol 1:279–283
Petersen SL, Barraclough CA (1989) Suppression of spontaneous LH surges in estrogen-treated ovariectomized rats by microimplants of antiestrogen into the preoptic brain. Brain Res 484:279–289
Petraglia F et al (1994) Neuroendocrine aspects in reproductive medicine. In: Fertility and sterility: research and practice.
Phillips SM, Sherwin BB (1992) Effects of estrogen on memory function in surgically menopausal women. Psychoneuroendocrinology 17:485–495
Priest CA, Pfaff DW (1995) Actions of sex steroid on behaviours beyond reproductive reflexes. Ciba Found Symp 191:74–84
Sato M et al. (1995). Raloxifene: a selective estrogen receptor modulator. J Bone Miner Metab 12 (Suppl 2):S9–S20
Sherwin BB (1988). Estrogen and/or androgen replacement therapy and cognitive functioning in surgically menopausal women. Psychoneuroendocrinology 13:345–357
Sherwin BB (1996. Hormones, mood and cognitive functioning in postmenopausal women. Obstet Gynaecol 87:20–26
Simard J, Labrie F (1985) Keoxifene shows pure antiestrogen activity in pituitary gonadotrophs. Mol Cell Endocrinol 34:141–144
Simard J et al. (1990) Pure antagonistic effect of a new steroidal antiestrogen in rat anterior pituitary cells in culture and in mouse uterus. Ann NY Acad Sci 595: 425–427
Smith SS (1993) Hormones, mood and neurobiology: a summary. In: Berg G and Hammar M (eds) The modern management of the menopause. The Parthenon Publishing Group, Pearl Rivier NY, pp 277–93
Speroff L et al. (1995) Clinical gynecological endocrinology and infertility, 5th edn. Williams and Wilkins, Baltimore, Maryland USA
Szamel I et al. (1994) Influence of toremifene on the endocrine regulation in breast cancer patients. Eur J Cancer 30 A:154–158
Torand-Allerand CD (1980) Sex steroids and the development of the newborn mouse hypothalamus and preoptic area in vitro, morphological correlates and hormonal specificity. Brain Res 189:413–427
Tzukerman M et al. (1990) Estrogen regulation of the insulin-like growth factor I gene transcription involves an AP-1 enhancer. J Biol Chem 269:16433–16442
Tzukerman M et al. (1994) Human estrogen receptor transactional capacity is determined by both cellular and promoter context and mediated by two functionally distinct intramolecular regions. Mol Endocrinol 81:21–30
Weiland NG (1992). Glutamic acid decarboxylase messenger ribonucleic acid is regulated by estradiol and progesterone in the hippocampus. Endocrinology 131:2697–2702
Wooley C, McEwen BS (1994). Estradiol regulates hippocampal dendritic spine density via an N-methyl-D-aspartate receptor-dependent mechanism. J Neurosci 14:7680–7687
Woolley CS et al. (1990) Naturally occurring fluctuation in dendritic spine density on adult hippocampal pyramidal neurons. J Neurosc 10:4035–403
Yang NN et al. Raloxifene, an antiestrogen simulates the effects of estrogen on inhibiting bone resorption through regulating TGFβ-3 expression in bone (abstract). J Bone Miner Res (Suppl I):S118

CHAPTER 28

Liver Inclusive Protein, Lipid and Carbohydrate Metabolism

L. SAHLIN and B. VON SCHOULTZ

A. Liver – A Non-Reproductive Target Organ for Estrogens

The mammalian liver contains specific estrogen receptors (ERs) (DUFFY and DUFFY 1976; ATEN et al. 1978; ERIKSSON 1982a; ERIKSSON 1982b; FREYSCHUSS et al. 1991), but is still quite different from other target organs for estrogens. The number of binding sites in the liver is only about one-eighth to one-tenth that of the normal receptor concentration in classical target organs such as the uterus (ERIKSSON 1982a; ERIKSSON 1983). Liver receptors, in significant concentration, can only be detected after the onset of puberty and, in rats, hypophysectomy results in a major decrease of hepatic estrogen receptors (ERIKSSON 1983). Activation of the receptor is not associated with the induction of a progesterone receptor in the liver and, compared with the estradiol doses required to elicit response in the endometrium, excessively high doses are necessary to achieve similar effects in the liver (MARR et al. 1980; LAX et al. 1983). EAGON et al. (1986) have found a circadian rhythm in the hepatic ER, although we have not been able to repeat these findings (SAHLIN and FREYSCHUSS, unpublished observations).

ER dimerization enables the complex to interact with estrogen responsive elements (EREs) to activate transcription of target genes. Although phosphorylated in the resting state, a receptor bound to DNA is further phosphorylated at several serine and possibly also tyrosine residues (WASHBURN et al. 1991). Changes in phosphorylation may function to regulate ER binding to ligands, DNA or other proteins (AUCHUS and FUQUA 1994). After binding to the ERE, the genes are transcribed and mRNAs from the estrogen responsive genes are produced. This finally results in an increased synthesis of specific proteins that mediate the biological effects of true hormone stimulation (CLARK and PECK 1979). The specificity of response is controlled by the amount of receptors within a cell, as well as co-regulators, and the accessibility to target genes in the chromatin (BANIAHMAN et al. 1994).

A direct relationship between receptor concentration and the degree of biological response has been described (EVANS et al. 1987; WEBB et al. 1992), although other groups have not found any correlation using their experimental systems (GAUBERT et al. 1986; DARBRE and KING 1987). It has also been shown that the threshold level of the receptor amount for one response is quite different from the concentration required for another response to occur (RABINDRAN et al. 1987; COOK et al. 1988; DONG et al. 1990). These results suggest that other factors, in addition to the receptor concentration, determine the biological effect of steroid hormones.

I. Regulation of the Hepatic Estrogen Receptor

Hepatic ER levels are very low, and the liver is unresponsive to estrogens in the absence of growth hormone (GH) (STEINBERG et al. 1967; NORSTEDT et al. 1981; THOMPSON et al. 1983). Therefore, the lack of hepatic response to estrogens in hypophysectomized rats may be due to loss of ERs, as well as loss of GH. This makes it difficult to discriminate between direct effects and indirect effects due to alterations in GH secretion, which are known to occur during estrogen treatment of rats (MODE and NORSTEDT 1982; PAINSON et al. 1992) and humans (DE LEO et al. 1993).

In ovariectomized rats treated with estradiol, the levels of ER and ER mRNA were significantly increased (SAHLIN et al. 1994). This effect was attenuated by dexamethasone pretreatment (SAHLIN 1995). In hypophysectomized rats GH increased both ER and ER mRNA (FREYSCHUSS et al. 1994). This effect of GH was enhanced when given together with dexamethasone. The synergy between GH and dexamethasone in the formation of the ER protein has also been shown in cultured hepatocytes (FREYSCHUSS et al. 1993). Thus, in ovariectomized rats, in which estrogen is the main regulator of ERs and ER mRNA, the effect of estradiol is attenuated by dexamethasone, whereas, in hypophysectomized rats, in which GH is the main regulator of ERs and ER mRNA, dexamethasone acts in synergy with GH.

Hepatic ERs and ER mRNA levels have been shown to decrease after thyroidectomy, compared with the levels in normal rats. GH treatment increased the ER level to half of the normal value, although ER mRNA was not affected (ERIKSSON and FREYSCHUSS 1988; FREYSCHUSS et al. 1994).

II. Exogenous Estrogens

Treatment with exogenous estrogens is known to affect several aspects of liver metabolism (ANDERSSON and KAPPAS 1982). Alterations in protein synthesis (Table 1), coagulation factors (Table 2), lipid metabolism (Table 3) and carbohydrate metabolism are of particular interest and may have important clinical implications. Changes in the synthesis of liver-derived proteins, such as angiotensinogen, high-density lipoprotein (HDL) and low-density lipoprotein (LDL), various coagulation factors and antithrombin III, may influence the risk of hypertension, hyperlipidemia, and hypercoagulability during estrogen treatment.

The effect of sex steroid hormones may be mediated principally in two ways: directly via receptor binding in target cells or indirectly by modifying the secretion of other hormones (VON SCHOULTZ and CARLSTRÖM 1989). In the liver, the given hormones are rapidly metabolized, processed and excreted

Table 1. Proteins affected by estrogen treatment

Protein	Effect
α1-Antitrypsin	+
Albumin	–
Alkaline phosphatase	+
Angiotensinogen	+
Bilirubin	+
Ceruloplasmin	+
Corticosteroid-binding globulin	+
χ-glutamyl transpeptidase	+
Growth hormone	+
Growth hormone binding protein	+
Insulin-like growth factor-I	–
Haptoglobin	–
Leucin aminopeptidase	+
α2-Macroglobulin	+
Orosomucoid	–
Pregnancy zone protein	+
Retinol binding protein	+
SHBG	+
Thyroxin-binding globulin	+
Transcortin	+
Transferrin	+

Table 2. Coagulation factors affected by estrogen treatment

Coagulation factor	Effect
Antithrombin III	–
Coagulation factor II	+
Coagulation factor VII-X	+
Coagulation factor XII	+
Complement reactive protein	+
Fibrinogen	+
Plasminogen	+
Protein C	+
Prothrombin time	–

Table 3. Lipids affected by estrogen treatment

Lipids	Effect
Apolipoprotein-A	+
HDL	+
LDL	–
Lecithin	+
Total lipids	+
Triglycerides	+

HDL, high-density lipoprotein; *LDL*, low-density lipoprotein mostly as sulfo- or glucuroconjugates. The interaction between exogenous sex steroids and the endogenous substances that are normally metabolized in the liver contributes a further stress to hepatic cells. It seems that only synthetic estrogens or high doses of natural estrogens, given orally, will saturate the hepatic metabolic capacity and cause a pharmacological response (STEINGOLD et al. 1986; MARR et al. 1980).

The impact of exogenous estrogens on liver metabolism is mainly dependent on two factors: the route of administration and the type and dose of estrogen (Fig. 1).

Oral treatment with estrogen pills is simple and convenient and has a well-documented therapeutic efficacy, but is also, in many ways, non-physiological (HOLST 1983; JUDD 1987). In the intestinal wall, about 70% of the ingested estradiol is metabolized to estrone, which has an approximate biological activity of about one-third that of estradiol (LYRENÄS et al. 1981). The intestinal absorption is rapid and yields high concentrations of hormone in the portal circulation. Very high concentrations are needed to saturate the hepatodigestive defense mechanisms before a general therapeutic effect can be achieved. Thus, it is necessary to use roughly a 20-fold higher dose via the oral route than parenterally. In fact, transdermal administration of 50–100 μg of estradiol provides the same therapeutic effect in postmenopausal women as 2 mg of estradiol given orally.

Available data imply that the multitude of effects following oral administration of high doses of synthetic estrogens to a major extent reflect a phar-

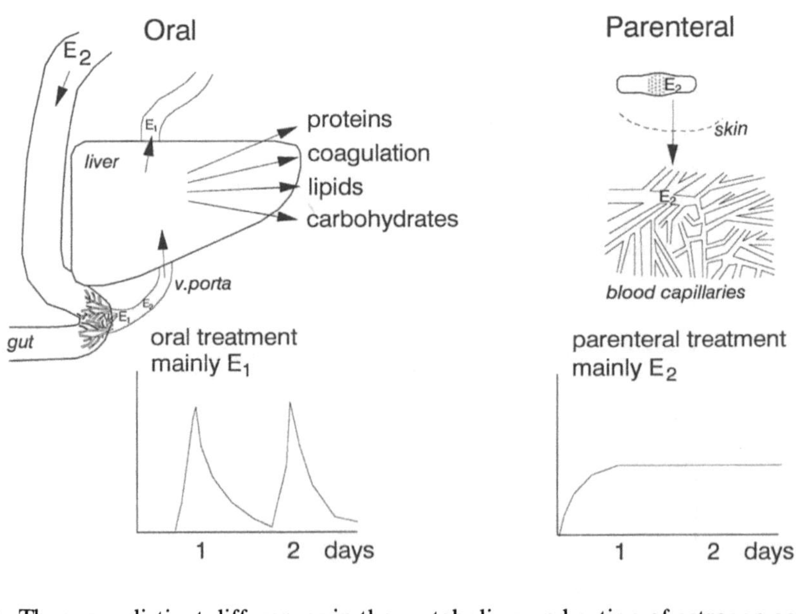

Fig. 1. There are distinct differences in the metabolism and action of estrogen according to the route of administration

macological rather than a physiological influence on liver metabolism. Specific differences in hepatic action between the native hormone estradiol-17β and synthetic estrogens have been clearly demonstrated (STEINGOLD et al. 1986). This concept has stimulated work on alternative non-oral delivery systems (HOLST 1983; JUDD 1987), and in developing derivatives of natural estrogens lacking the hepatic side effects (ELGER et al. 1995). Numerous clinical and animal studies have demonstrated that the hepatic impact of estrogen treatment can be reduced dramatically by modification of the type of estrogen and the route of administration (ELKIK et al. 1982; HOLST 1983; FÅHREUS and WALLENTIN 1983; JUDD 1987; DE LIGUIERES and BASDEVANT 1987; STEGE et al. 1987; ELGER et al. 1995).

Indeed, when the native estrogens are given parenterally, the effects on liver-derived plasma proteins, coagulation factors, lipoproteins and triglycerides are very weak or completely abolished (JUDD 1987; DE LIGUIERES and BASDEVANT 1987; STEGE et al. 1987; ELKIK et al. 1982; FÅHREUS and WALLENTIN 1983; HOLST et al. 1983). In a rat model, the estradiol-17β-sulfamate has proven to have a 90-fold elevated systemic effect, compared with estradiol-17β, when given orally, combined with a reduction of hepatic estrogenicity (ELGER et al. 1995). Systemic estrogenicity was quantitated by assessment of uterine weight, vaginal cornification and measurement of gonadotropins (ELGER et al. 1995). This substance may allow the use of native estrogens orally without affecting hepatic metabolism.

Estriol, given in low, daily doses, appears to be quite inert in stimulating the hepatic "estrogen-inducible" proteins, plasma renin activity and coagulation factors, as well as in modifying total cholesterol, HDL-cholesterol and triglycerides (ERRKOLA et al. 1978; BERGINK et al. 1981; CAMPAGNOLI et al. 1981).

B. Plasma Proteins

Estrogen treatment affects the synthesis of several proteins in the liver (Table 1), some of which are potentially important in development of disease or protection against disease (LAURELL and RANNEVIK 1979; VON SCHOULTZ et al. 1989; L'HERMITE 1990). Oral therapy increases the serum concentrations of several proteins in a dose-dependent way (VON SCHOULTZ 1988). When treatment is started, the serum concentration of the respective protein increases. After about 1–3 months, they have reached new values which remain stable for the rest of the treatment period (VON SCHOULTZ 1988). This serum protein induction is reversed by androgens and progestogens, particularly the 19-nor steroids. The net change in protein level may be used as an index of "estrogenicity" for different estrogen/progestogen combinations in clinical practice. Whether synthesis of these proteins is directly regulated by estrogens, or indirectly through alterations in secretion of other hormones, especially GH, has not yet been resolved (VON SCHOULTZ and CARLSTRÖM 1989). With regard to some of the estrogen-induced proteins, such as angiotensinogen and hepatic LDL receptor, the effects of estrogens are probably direct, at least partly, while other proteins may be regulated solely indirectly through other factors affected by estrogens, for instance GH secretion (Carlsson-Boustedt et al. 1987).

The regulation of protein metabolism can be achieved both at a transcriptional and translational level. In Xenopus liver cells, estrogen increases the half-life of vitellogenin mRNA from 16h to 3weeks (Brock and Shapiro 1983; Neilson and Shapiro 1990), resulting in a massive increase of vitellogenin mRNA. In contrast, estrogens destabilize the albumin mRNA in the same cells (Kazmaier et al. 1985; Wolffe et al. 1985). An estrogen-inducible ribonuclease activity has been identified in Xenopus liver (Pastori et al. 1991) which may mediate hormone regulating changes in mRNA stability in this tissue.

An increase in blood pressure is known to occur in some women during hormonal contraception and, also, but less frequently, during hormonal replacement therapy. While the physiological effect of estrogen is to promote blood flow and vasodilatation and rather to reduce blood pressure, the pharmacological action of exogenous estrogen on liver protein synthesis may cause an opposite effect. The influence on angiotensinogen is of particular interest in this respect. Estrogen treatment increases the synthesis/secretion of angiotensinogen in adult rat and human livers (Krakoff and Eisenfeld 1977; von Schoultz et al. 1989). A dose-dependent response in the range 50–300μg ethinylestradiol/day/rat has been shown (Hong-Brown and Deschepper 1993). Most data indicate a direct stimulatory effect of estrogens on the synthesis of angiotensinogen in the liver. Oral estrogens are stronger stimulators than parenterally administered ones (Dupont et al. 1991), and angiotensinogen secretion is increased by estrogen in isolated perfused livers (Nasjletti and Masso 1972). There is no response to estrogens with respect to angiotensinogen in livers of hypophysectomized or immature rats where hepatic ER levels are low (Krakoff and Eisenfeld 1977; Eisenfeld and Aten 1980). An ER-negative hepatoma cell line, unresponsive to estrogens, was stably transfected with functional ERs. Thereafter, estrogen treatment resulted in increased angiotensinogen secretion (Klett et al. 1992).

Elements conferring an estrogenic response have been identified in the promoter region of the angiotensinogen gene (Feldmer et al. 1991). Although even the relatively low doses of orally administered estrogens used in postmenopausal treatment today may increase circulating angiotensinogen levels, this does not seem to increase blood pressure (Dupont et al. 1991). Thus, the physiological/pathological importance of this phenomenon is not known, and the importance of estrogens in regulation of blood pressure is unresolved (L'Hermite 1990). In a recent paper Wang et al. (1995) have shown that physiological (endogenous) levels of estrogen do not affect or control angiotensinogen synthesis.

Circulating insulin-like growth factor-I (IGF-I) levels are reduced by oral estrogen therapy and this decrease has been proposed to result in the increased GH secretion due to reduced feedback inhibition by IGF-I (KELLY et al. 1993).

The antiestrogen tamoxifen, when given to subjects with estrogen deficiency, has mild estrogenic properties on protein metabolism (HELGASON et al. 1982). When added during estrogen therapy, it strongly counteracts the therapeutic effects of the estrogen. The estrogen-sensitive liver-derived pregnancy zone protein (PZP) showed the most pronounced reduction after three cycles of tamoxifen addition, whereas sex hormone-binding globulin (SHBG) levels were only moderately reduced (OTTOSSON 1984). Raloxifene lowered alkaline phosphatase in a short-term study in postmenopausal women (DRAPER et al. 1996). Progestogens are known to elicit antiestrogenic effects on protein metabolism (VON SCHOULTZ 1988).

I. Coagulation Factors

Estrogens influence liver-derived coagulation factors in a dose-dependent manner (Table 2). Estrogens used in the treatment of prostatic cancer and in oral contraceptives have been shown to increase the risk of thrombo-embolic diseases (MEADE 1981; HENRIKSSON et al. 1986). The risk has been related to high doses of orally administered synthetic estrogens, especially ethinylestradiol, but recently, the use of estradiol or conjugated equine estrogens for hormonal replacement therapy has also been associated with an increased incidence of thrombo-embolic disorders (WREN 1988; DALY et al. 1996). The exact mechanisms involved have not been fully elucidated, but include an increased synthesis of some coagulation factors and decreased synthesis of antithrombin III in the liver (HENRIKSSON et al. 1986). A possible direct activation of the coagulation system by estrogens has also been suggested, but is probably of less importance (INAUEN et al. 1991). The choice of the corresponding progestogen affects serum levels of some clotting factors, but the possible risk of thrombosis has not been shown convincingly (KUHL 1996). Parenteral administration of estrogens used for contraception or postmenopausal estrogen therapy seems to have less influence on the coagulation system, and it is unclear to what extent such therapy will increase the risk for thrombo-embolic disease (DEVOR et al. 1992). As for plasma proteins, the estrogenic effects on certain coagulation factors, in particular factors II, VII and protein C, are counteracted by androgens and 19-nor steroid progestogens (KUHL 1996).

Oral tamoxifen citrate treatment induced a modest decrease in anticoagulant proteins, but without biochemical signs of activation of coagulation and fibrinolysis (MANNUCCI et al. 1996). Tamoxifen has been shown to decrease fibrinogen levels, after both 2 years and 5 years of treatment (LOVE et al. 1994; GREY et al. 1995).

C. Lipids

Effects seen on lipid levels after oral estrogen therapy are summarized in Table 3. The LDL receptor is important in the regulation of serum cholesterol and is responsible for removal of most of the circulating cholesterol (RUDLING 1987). A 3-fold induction of hepatic LDL receptor levels has been reported in prostatic cancer patients receiving pharmacological doses of estrogens (ANGELIN et al. 1992). After treatment of rats with high doses of estradiol or ethinylestradiol a 5- to 10-fold induction of hepatic LDL receptor levels is observed, which is accompanied by drastically reduced serum cholesterol levels (KOVANEN et al. 1979; SRIVASTAVA et al. 1993). This effect of estrogens is observed neither in hypophysectomized rats (STEINBERG et al. 1967) nor in immature rats (PLONNÉ et al. 1993), in which hepatic ER levels are low. Substitution of hypophysectomized rats with GH, which increases ER levels, in combination with ethinylestradiol results in hepatic LDL receptor levels almost as high as those in intact rats treated with ethinylestradiol alone (RUDLING et al. 1992).

As in the case with synthesis of coagulation factors and angiotensinogen in the liver, the effects of oral estrogens on blood lipids in humans by far surpass the effects of injected or percutaneous estrogens (MOORJANI et al. 1991). The 5- to 10-fold induction of LDL receptors in the rat with ethinylestradiol treatment cannot be equaled by the use of other hormones, including GH in the absence of estrogens (BRINDLEY and SALTER 1991). These data suggest at least a partially direct action of estrogens via the ER on hepatic LDL receptor levels.

The indirect actions of estrogens through alterations in secretion of GH and perhaps other hormones are possibly of equal importance. For example GH injections to healthy individuals increase hepatic LDL receptors and decrease plasma cholesterol levels (RUDLING et al. 1992), and GH-deficient patients have an increased mortality in myocardial infarction (ROSÉN and BENGTSSON 1990). While the stimulatory effects of estrogens on hepatic LDL receptors in the rat require pharmacological doses of ethinylestradiol, variations in blood lipids in humans occur during the menstrual cycle (SCHIJF et al. 1993). On the other hand, variations in GH secretion have been reported during the menstrual cycle (FARIA et al. 1992). Therefore, the observed changes in blood lipids might also be due to alterations in serum concentrations of GH secondary to cyclic estradiol variations. Postmenopausal estrogen treatment has been reported to increase GH secretion (DE LEO et al. 1993), but the relevance of this finding with respect to effects on blood lipids in unclear, since orally administered estrogens have a greater impact on the lipid profile than transdermal estrogens (MOORJANI et al. 1991).

There is a clear difference in morbidity and mortality rates in cardiovascular disease between males and females (GANGAR et al. 1993). This difference almost disappears at the onset of menopause, after which the female mortality in cardiovascular disease rises abruptly (WITTEMAN et al. 1989).

Postmenopausal estrogen therapy reduces the risk for myocardial infarction by approximately 50% (SITRUK-WARE and DE PALACIOS 1989; NEWNHAM 1993; PSATY et al. 1993). Initially, it was believed that most, if not all anti-atherosclerotic effects of estrogens were mediated through the liver by induction of a "healthy" blood lipid profile (BRINDLY and SALTER 1991). Estrogen treatment decreases cholesterol and increases triglyceride levels in serum (BASDEVANT 1992), but most importantly, induces a "healthy lipid profile" by decreasing the LDL/HDL ratio (CROSIGNANY 1992). Decrease of serum LDL is accomplished by an increase in hepatic LDL receptor levels (RUDLING 1987). Serum HDL levels are believed to be increased by an increase in apolipoprotein A synthesis (MOORJANI et al. 1991) and a decrease in hepatic lipase activity (TIKKANEN et al. 1982). The second mechanism may actually decrease cholesterol transport from peripheral tissues to the liver and is not necessarily salubrious (VON SCHOULTZ et al. 1989).

Estrogens are often given in combination with progestogens, which diminish the positive lipid profile, and it has not yet been clarified whether the addition of progestogens affects the risk of developing cardiovascular disease during estrogen treatment (FALKEBORN et al. 1992; PSATY et al. 1993). The general belief, today, is that the positive lipid profile alone does not account for all effects of estrogens in cardiovascular disease, and that their effects on lipids may be of less importance than direct effects on blood vessels (L'HERMITE 1990). Estrogens have vasodilatory effects on arteries (MAGNESS and ROSENFELDT 1989; MÜGGE et al. 1993; RIEDEL et al. 1995), possibly mediated by stimulation of local prostacyclin or nitric oxide (NO) synthesis (GANGAR et al. 1993). ERs have been detected in blood vessels (CAMPISI et al. 1993) and in vascular smooth muscle (ORIMO et al. 1993). Another demonstrated effect of estrogens, possibly involved in slowing the artherosclerotic process, is inhibition of LDL oxidation (RIFICI and KHACHADURIAN 1992; MOORADIAN 1993).

Compared with the lipid effects of estrogens alone, a combined therapy with progestogens may induce atherosclerosis (PSATY et al. 1993). The addition of tamoxifen to estrogen-primed postmenopausal women induced significant effects on the lipoprotein pattern that should be considered as purely anti-estrogenic (OTTOSSON 1984). HDL cholesterol and apolipoprotein A levels were significantly reduced after 3 months tamoxifen treatment (OTTOSSON 1984). Total and LDL cholesterol levels fell significantly in women treated with tamoxifen only, improving the lipid profile (LOVE et al. 1994; GREY et al. 1995; MANNUCCI et al. 1996). Raloxifene, in a 8-week study in postmenopausal women, decreased LDL cholesterol in the same range as conjugated estrogens, and serum cholesterol was also decreased (DRAPER et al. 1996).

D. Carbohydrates

Several other hormones, such as insulin, glucagon, corticoids, growth hormone and catecholamines, are more important than estrogen in carbohydrate metabolism. The multitude of complex interactions makes it difficult to study the effect of estrogens only. Both "natural" and synthetic estrogens induce a decreased glucose tolerance and increased insulin levels. Data are contradictory, but the differences in effects are most likely dose dependent. The estrogens seem rather to have a transient reversible hyperglycemic rather than primarily diabetogenic effect (VAN KEEP et al. 1982). Diabetes is no absolute contraindication to estrogen treatment. The deterioration of carbohydrate tolerance that has been associated with oral contraceptives has been suggested to be caused mainly by the progestogen and not the estrogen (L'HERMITE 1990; GASPARD 1989). Hormone-replacement therapy with equine estrogens or estradiol valerate does not seem to induce impaired glucose tolerance (THOM et al. 1977; LARSSON-COHN and WALLENTIN 1977). It has even been suggested that the use of estrogen alone may improve glucose tolerance by enhancement of insulin receptor binding (SPELLACY et al. 1987). Improved glucose tolerance has been reported after 6 months of replacement therapy in women with reduced glucose tolerance (LUOTOLA et al. 1986).

An improvement of glucose tolerance following physiological estrogen treatment is also in agreement with the diminished insulin sensitivity, glucose intolerance and *acanthosis nigricans* found in a man with a disrupted ER and subsequent lack of estrogen responsiveness (SMITH et al. 1994). *Acanthosis nigricans* is a cutaneous marker of insulin resistance, especially when insulin resistance is associated with relative hyperandrogenism (BARBIERI and RYAN 1993). An increase in estrogens improves glucose tolerance by enhancing either target tissue responsiveness to insulin or insulin secretion (SHARP and DIAMOND 1993; LEITER et al. 1987; PROCHAZKA et al. 1986).

Insulin resistance associated with estrogen deficiency may in fact be reversed by physiological estrogen replacement. However, excessive estrogen action clearly deteriorates carbohydrate metabolism, possibly via increased glucocorticoid action. Also, progesterone and certain progestogens may induce insulin resistance and add to a pharmacological estrogenic effect (GODSLAND 1996).

References

Andersson KK, Kappas A (1982) Hormones and liver function. In: Schiff L, Schiff ER (eds) Diseases of the Liver. J.B. Lippincott Company, Philadelphia, pp 167–235

Angelin B, Olivercrona H, Reihner E, Rudling M, Stahlberg D, Eriksson M, Ewerth S, Henriksson P, Einarsson K (1992) Hepatic cholesterol metabolism in estrogen-treated men. Gastroenterology 103:1657–1663

Aten R, Weiberger MJ, Eisenfeld AR (1978) Estrogen receptor in rat liver: Translocation to the nucleus in vivo. Endocrinology 102:433–442

Auchus RJ, Fuqua SAW (1994) The estrogen receptor. Baillière's Clin. Endocrinol. and Metab. 8:433–449

Baniahman A, Tsai, MJ and Burris TP (1994) The nuclear hormone receptor superfamily In: Mechanism of steroid hormone regulation of gene transcription. Tsai M-J, O'Malley BW (eds) Molecular biology intelligence unit. CRC Press, USA

Barbieri RL, Ryan KJ (1993) Hyperandrogenism, insulin resistance, and *acanthosis nigricans* syndrome: a common endocrinopathy with distinct pathophysiologic features. Am J Obst Gynecol 147:90–101

Basdevant A (1992) Steroids and lipid metabolism: mechanisms of action. Int J Fertil 37:3–97

Bergink EW, Crona N, Dahlgren E, Samsioe G (1981) Effect of oestriol, oestradiol valerate and ethinylestradiol on serum proteins in estrogen-deficient women. Maturitas 3:241–247

Brindley DN, Salter AM (1991) Hormonal regulation of the hepatic low density lipoprotein receptor and the catabolism of low density lipoproteins: Relationship with the secretion of very low density lipoproteins. Progr Lipid Res 30:349–360

Brock ML, Shapiro DJ (1983) Estrogen stabilizes vitellogenin mRNA against cytoplasmic degradation. Cell 34:207–214

Campagnoli C, Prelato Tousijn L, Belforte P, Feruzzi L, Dolfin AM, Morra G (1981) Effects of conjugated equine estrogens and oestriol on blood clotting, plasma lipids and endometrial proliferation in post-menopausal women. Maturitas 3:241–247

Campisi D, Cutolo M, Carruba G, Locast M, Comito L, Granata OM, Valentino B, King RJB, Castagnetta L (1993) Evidence for soluble and nuclear site I binding of estrogens in human aorta. Atherosclerosis 103:267–277

Carlsson-Boustedt L, Fröhlander N, Eden S, Stigbrandt T, von Schoultz B (1987) Effects of estrogen and human growth hormone on pregnancy-associated plasma proteins in the rat. Acta Endocrinol 116:299–304

Clark JH, Peck EJ Jr (1979) Female Sex Steroids. Receptors and Function. Springer Verlag, Berlin

Cook PW, Swanson KT, Edwards CP, Firestone GL (1988) Glucocorticoid receptor-dependent inhibition of cellular proliferation in dexamethasone-resistant and hypersensitive rat hepatoma cell variants. Mol Cell Biol 8:1449–1459

Crosignani PG (1992) Effects of hormone replacement therapy. Int J Fertil 37:98–103

Daly E, Vessey MP, Hawkins MM, Carson JL, Gough P, Marsh S (1996) Risk of venous thromboembolism in users of hormone replacement therapy. Lancet 348:977–980

Darbre PD, and King RJB (1987) Progression to steroid insensitivity can occur irrespective of the presence of functional steroid receptors. Cell 51:521–528

DeLeo V, Lanzetta D, D'Antona D, Danero S (1993) Growth hormone secretion in premenopausal women before and after ovariectomy: effect of hormone replacement therapy. Fert Ster 60:268–271

de Liguieres B, Basdevant A (1987) Differential metabolic tolerance between oral and percutaneous administration of estradiol in postmenopausal women. In: Christiansen, Johansen JS, Riis BJ (eds) Osteoporosis vol.2. NØrhaven: Viborg, Denmark, pp 1120–1131

Devor M, Barrett CE, Renvall M, Feigal DJ, Ramsdell J (1992) Estrogen replacement therapy and the risk of venous thrombosis. Am J Med 92:275–282

Dong Y, Cairns W, Okret S, Gustafsson J-Å (1990) A glucocorticoid-resistant rat hepatoma cell variant contains a functional glucocorticoid receptor. J Biol Chem 265:7526–7531

Draper MW, Flowers DE, Huster WJ, Neild JA, Harper KD, Arnaud C (1996) A controlled trial of Raloxifene (LY139481) HCl: impact on bone turnover and serum lipid profile in healthy postmenopausal women. J Bone and Mineral Res 11: 835–842

Duffy MJ, Duffy GJ (1976) Estrogen receptors in human liver. J Steroid Biochem 9:122–235

Dupont A, Dupont P, Cusan L, Tremblay M, Riox J, Clotier D, Mailloux J, De Lignieres B, Gutkowska J, Boucher H, Bélanger A, Moyer DL, Moorjani S, Labrie F (1991) Comparative endocrinological and clinical effects of percutaneous estradiol and oral conjugated estrogens as replacement therapy in menopausal women. Maturitas 13:297–311

Eagon PK, DiLeo A, Polimeno L, Francavilla A, Van Thiel DH, Guglielmi F, Starzl TE (1986) Circadian rhythm of hepatic cytosolic and nuclear estrogen receptors. Chronobiol Internat 3:207–211
Eisenfeld AJ, Aten RF (1980) Estrogen receptor in the mammalian liver. In: Briggs MH, Corbin A (eds) Advances in steroid biochemistry and pharmacology. Academic Press, London, pp 91–117
Elger W, Schwarz S, Hedden A, Reddersen G, Schneider B (1995) Sulfamates of various estrogens are prodrugs with increased systemic and reduced hepatic estrogenicity at oral application. J Steroid Biochem Molec Biol 55:395–403
Elkik F, Gompel A, Mercier Bodard C, Kuttenn F, Guyenne PN, Corvol P, Mauvis-Jarvis P (1982) Effects of percutaneous estradiol and conjugated estrogens on the level of plasma proteins and triglycerides in postmenopausal women. Am J Obst Gyn 143:888–892
Eriksson HA (1982a) Different regulation of the concentration of estrogen receptors in the rat liver and uterus following ovariectomy. FEBS Letters 149:91–95
Eriksson HA (1982b) Estrogen-binding sites of mammalian liver: Endocrine regulation of estrogen receptor synthesis in the regenerating rat liver. J Steroid Biochem 17:471–477
Eriksson HA (1983) Regulation of estrogen receptor concentration in target organs of the rat. In: ERIKSSON H and GUSTAFSSON J-Å (eds) Steroid hormone receptors: Structure and function. Elsevier Publishers BV, pp 389–404
Eriksson H, Freyschuss B (1988) Effects of thyroid hormones on the receptor level in estrogen target organs. J Steroid Biochem 29:401–405
Errkola R, Lammintausta R, Punnonen R, Rauramo L (1978) The effect of estriol succinate therapy on plasma renin activity and urinary aldosterone in postmenopausal women. Maturitas 1:9–14
Evans MI, O'Malley PJ, Krust A, Burch JBE (1987) Developmental regulation of the estrogen responsiveness of five yolk protein genes in the avian liver. Proc Natl Acad Sci USA 84:8493–8497
Falkeborn M, Persson I, Adami H-O, Bergström R, Eaker H, Mohsen R, Naessen T (1992) The risk of acute myocardial infarction after estrogen and estrogen-progesterone replacement. Brit J Med 99:821–828
Faria ACS, Bekenstein Both RAJ, Vaccaro VR, Asplin CM, Veldhuis LD, Thorner MO, EVANS WS (1992) Pulsatile growth hormone release in normal women during the menstrual cycle. Clin Endocrinol 36:591–596
Feldmer M, Kaling M, Takahashi S, Mullins JJ, Ganten D (1991) Glucocorticoid- and estrogen-responsive elements in the 5′-flanking region of the rat angiotensinogen gene. J Hypertens 9:1005–1012
Freyschuss B, Sahlin L, Eriksson H (1991) Regulatory Effects of Growth Hormone, Glucocorticoids and Thyroid Hormone on the Estrogen Receptor Level in the Rat Liver. Steroids 56:367–374
Freyschuss B, Stavréus-Evers A, Sahlin L, Eriksson H (1993) Induction of Estrogen Receptor by Growth Hormone and Glucocorticoid Substitution in Primary Cultures of Rat Hepatocytes. Endocrinology 133:1548–1554
Freyschuss B, Sahlin L, Masironi B, Eriksson H (1994) The Hormonal Regulation of the Estrogen Receptor in Rat Liver: An Interplay Involving Growth Hormone, Thyroid Hormones and Glucocorticoids. J Endocrinol 142:285–298
Fåhreaus L, Wallentin L (1983) High density lipoprotein subfractions during oral and cutaneous administration of 17β-estradiol in menopausal women. J Clin Endocrinol Metab 56:797–801
Gangar KF, Reid BA, Crook D, Hillard TC, Whitehead MI (1993) Estrogens and atherosclerotic vascular disease-local vascular factors. Baillères Clin Endocrinol Metab 7:47–59
Gaspard UJ (1989) Carbohydrate metabolism, atherosclerosis and the selection of progestins in the treatment of menopause. In: Lobo RA, Whitehead MI (eds) Consensus development conference on progestogens. Int Proc J 1:223–229

Gaubert C-M, Carriero R, Shyamala G (1986) Relationships between mammary estrogen receptor and estrogenic sensitivity. Endocrinology 118:1504–1512

Godsland IF (1996) The influence of female sex steroids on glucose metabolism and insulin action. J Int Med 240 {Suppl 738}:1–60

Grey AB, Stapleton JP, Evans MC, Reid IR (1995) The effect of the anti-estrogen tamoxifen on cardiovascular risk factors in normal postmenopausal women. J Clin Endocrin Metab 80: 3191–3195

Helgason S, Wilking N, Carlström K, Damber MG, von Schoultz B (1982) A comparative study on the estrogenic effects of tamoxifen and 17β-estradiol in postmenopausal women. J Clin Endocrinol Metab 54:404–408

Henriksson P, Blombäck M, Bratt G, Edhag O, Eriksson A (1986) Activators and inhibitors of coagulation and fibrinolysis in patients with prostatic cancer treated with estrogen or orchidectomy. Thromb Res 44:783–791

Holst J (1983) Percutaneous estrogen therapy. Endometrial response and metabolic effects. Acta Obstet Gynecol Scand {Suppl 115}:7–30

Holst J, Cajander S, Carlström K, Damber MG, von Schoultz B (1983) A comparison of liver protein induction in postmenopausal women during oral and percutaneous estrogen replacement therapy. Br J Obstet Gynecol 90:355–360

Hong-Brown LQ, Deschepper CF (1993) Regulation of the angiotensinogen gene by estrogens in rat liver and different brain regions. Phys Soc Exp Biol Med 203:467–473

Inauen W, Stocker G, Haeberli A, Straub PW (1991) Effects of low and high dose oral contraceptives on blood coagulation and thrombogenesis induced by vascular subendothelium exposed to flowing human blood. Contraception 43:435–446

Judd, HL (1987) Efficacy of transdermal estradiol. Am J Obstet Gynecol 157:1326–1331

Kazmaier M, Brüning E, Ryffel GU (1985) Post-transcriptional regulation of albumin gene expression in Xenopus liver. EMBO J 4:1261–1266

Kelly JJ, Rajkovic IA, O'Sullivan AJ, Sernia C, Ho KKY (1993) Effects of different oral estrogen formulations on insulin-like growth factor-I, growth hormone and growth hormone binding protein in post-menopausal women. Clin Endocrinol 39:561–567

Klett C, Ganten D, Hellmann W, Kaling M, Ryffel GH, Weimar-Ehl T, Hackentahl E (1992) Regulation of hepatic angiotensinogen synthesis and secretion by steroid hormones. Endocrinology 130:3660–3668

Kovanen PT, Brown MS, Goldstein JL (1979) Increased binding of low-density lipoprotein to liver membranes from rats treated with 17α-ethinylestradiol. J Biol Chem 254:11367–11373

Krakoff LR, Eisenfeld AJ (1977) Hormonal control of plasma renin substrate; (angiotensinogen). Circulatory Res 41 {Suppl II}:43–46.

Kuhl H (1996) Effects of progestogens on haemostasis. Maturitas 24:1–19

Larsson-Cohn U, Wallentin L (1977) Metabolic and hormonal effects of postmenopausal estrogen replacement therapy. I. Glucose, insulin and human growth hormone levels during oral glucose tolerance tests. Acta Endocrinol 86:583–596

Laurell CB, Rannevik G (1979) A comparison of plasma protein changes induced by danazol, pregnancy and estrogens. J Clin Endocrinol Metab 49:719–725

Lax ER, Tamulevicius P, Müller A, Schriefers H (1983) Hepatic nuclear estrogen receptor concentrations in the rat -influence of age, sex, gestation, lactation and estrus cycle. J Steroid Biochem 19:1083–1088

Leiter EH, Beamer WG, Coleman DL, Longcope C (1987) Androgenic and estrogenic metabolites in serum of mice fed dehydroepiandrosterone: relationship to antihyperglycemic effects. Metabolism 36:863–869

L'Hermite M (1990) Risks of estrogens and progestogens. Maturitas 12:215–246

Love RR, Wiebe DA, Feyzi JM, Newcombe PA, Chappell RJ (1994) Effects of tamoxifen on cardiovascular risk factors in postmenopausal women after 5 years of treatment. J Nat Canc Inst 86:1534–1539

Luotola H, Pyörälä T, Liokkanen M (1986) Effects of natural estrogen/progesterone substitution therapy on carbohydrate and lipid metabolism in post-menopausal women. Maturitas 8:245–253

Lyrenäs S, Carlström K, Bäckström T, von Schoultz B (1981) A comparison of serum estrogen levels after percutaneous and oral administration of oestradiol-17β. Br J Obst. Gynaecol 88:181–187

Magness RR, Rosenfeldt CR (1989) Local and systemic estradiol 17β: effects on uterine and systemic vasodilatation. Am J Physiol 256:346–352

Mannucci PM, Bettega D, Chantarangkul V, Tripodi A, Sachini V, Veronesi U (1996) Effect of tamoxifen on measurements of hemostasis in healthy women. Arch Internal Med 156:1806–1810

Marr W, White JO, Elder MG, Lim L (1980) Nucleo-cytoplasmatic relationship of estrogen receptors in rat liver during the oestrus cycle and in response to administered and synthetic estrogen. Biochem J 188:17–25

Meade TW (1981) Oral contraceptives, clotting factors and thrombosis. Am J Obstet Gynecol 142:758–761

Mode A, Norstedt G (1982) Effect of gonadal steroids on the hypothalamo-pituitary-liver axis in the control of sex differences in hepatic steroid metabolism in the rat. J Endocrinol 95:181–187

Mooradian AD (1993) Antioxidant properties of steroids. J Steroid Biochem Mol Biol 45:509–511

Moorjani S, Dupont A, Labrie F, De Lignieres B, Cusan L, Dupont P, Mailloux J, Lupien P-J (1991) Changes in plasma lipoprotein and apolipoprotein composition in relation to oral versus percutaneous administration of estrogen alone or in cyclic association with utrogestan in menopausal women. J Clin Endocrinol Metab 73:373–379

Mügge A, Riedel M, Barton M, Kuhn M, Lichtlen PR (1993) Endothelium independent relaxation of human coronary arteries by 17β-estradiol in vitro. Cardiovasc Res 27:1939–1942

Nasjletti A, Masson GM (1972) Studies on angiotensinogen formation in a liver perfusion system. Circulatory Res 30:187–202

Neilson DA, Shapiro DJ (1990) Estradiol and estrogen receptor-dependent stabilization of mini-vitellogenin mRNA lacking 5,100 nucleotides of coding sequence. Mol Cell Biol 10:371–376

Newnham HH (1993) Estrogens and atherosclerotic vascular disease-lipid factors. Baillères Clin Endocrinol Metab 7:61–93

Norstedt G, Wrange Ö, Gustafsson J-Å (1981) Multihormonal regulation of the estrogen receptor in rat liver. Endocrinology 108:1190–1196

Orimo A, Satoshi I, Ikegami A, Hosoi T, Akishita M, Ouchi Y, Muramatsu M, Orimo H (1993) Vascular smooth muscle cells as targets for estrogens. Biochem Biophys Res Commun 195:730–736

Ottosson U-B (1984) Oral progesterone and estrogen/progesterone therapy. Acta Obst Gyn Scand {Suppl 127}:

Painson J-C, Thorner MO, Krieg RJ, Tannenbaum GS (1992) Short term adult exposure to estradiol feminizes the male pattern of spontaneous and growth hormone-releasing factor-stimulated growth hormone secretion in the rat. Endocrinology 130:511–519

Pastori RL, Moskaitis JE, Schoenberg DR (1991) Estrogen induced ribonuclease activity in Xenopus liver. Biochem 30:10490–10498

Plonnée D, Winkler L, Schröter A, Dargel R (1993) Low-density lipoprotein catabolism does not respond to estrogen in the fetal and newborn rat. Biol Neonate 63:230–235

Prochazka M, Premdas FH, Leiter EH, Lipson LG (1986) Estrone treatment dissociates primary versus secondary consequences of "diabetes" (db) gene expression in mice. Diabetes 35:725–728

Psaty BM, Heckbert SR, Atkins D, Sicovick DS, Koepsell TD, Wahl PW, Longstreth WT, Weiss NS, Wagner EH, Prentice R et al. (1993) A review of the association of

estrogens and progestins with cardiovascular disease in postmenopausal women. Arch Intern Med 153:1421–1427
Rabindran SK, Danielsen M and Stallcup MR (1987). Glucocorticoid-resistant lymphoma cell variants that contain the functional glucocorticoid receptors. Mol Cell Biol 7:4211–4217
Riedel M, Oeltermann A, Mügge A, Creutzig A, Rafflenbeul W, Lichtlen P (1995) Vascular responses to 17β-estradiol in postmenopausal women. Eur J Clin Invest 25:44–47
Rifici VA, Khachadurian AK (1992) The inhibition of low-density lipoprotein oxidation by 17β-estradiol. Metabolism 41:1110–1114
Rosén T, Bengtsson B-Å (1990) Premature mortality due to cardiovascular disease in hypopituitarism. Lancet 336:285–288
Rudling MJ (1987) Role of the liver for receptor-mediated catabolism of low-density lipoprotein in the 17α-ethinylestradiol treated rat. Biochem Biophys Acta 919:175–180
Rudling M, Norstedt G, Olivercrona H, Reihner E, Gustafsson J-Å, Angelin B (1992) Importance of growth hormone for the induction of hepatic low density lipoprotein receptors. Proc Natl Acad Sci USA 89:6983–6987
Sahlin L, Norstedt G, Eriksson H (1994) Estrogen Regulation of the Estrogen Receptor and Insulin-Like Growth factor-I in the Rat Uterus: A Potential Coupling Between Effects of Estrogen and IGF-I. Steroids 59:421–430
Sahlin L (1995) Dexamethasone Attenuates the Estradiol-Induced Increase of IGF-I mRNA in the Rat Uterus. J Steroid Biochem Molec Biol 55:9–15
Schijf CPT, van der Moren MJ, Doesburg WH, Thomas CMG, Rolland R (1993) Differences in serum lipids, lipoproteins, sex hormone binding globulin and testosterone between the follicular and the luteal phase of the menstrual cycle. Acta Endocrinol. 129:130–133
Sitruk-Ware R, Palacios P (1989) Estrogen replacement therapy and cardiovascular disease in post-menopausal women after menopause. A review. Maturitas 11:259–274
Sharp SC, Diamond MP (1993) Sex steroids and diabetes. Diabetes Rev 1:318–342
Smith EP, Boyd J, Frank GR, Takahashim H, Cohen RM, Specker B, Williams TC, Lubahn DB, Korach KS (1994) Estrogen resistance caused by a mutation in the estrogen receptor gene in a man. N Engl J Med 331:1056–1061
Spellacy WN (1987) Menopause, estrogen treatment and carbohydrate metabolism. In: Mishell DR Jr (ed) Menopause: physiology and pharmacology. Chicago: Year book, 1987, pp 253–260
Srivastava RA, Baumann D, Schonfeld G (1993) In vivo regulation of low-density lipoprotein receptors by estrogen differs at the post-transcriptional level in rat and mouse. Eur J Biochem 216:527–538
Stege R, Fröhlander N, Carlström K, Pousette Å, von Schoultz B (1987) Steroid sensitive proteins, growth hormone and somatomedin C in prostatic cancer: effects of parenteral and oral estrogen therapy. The Prostate 10:333–338
Steinberg M, Tolksdorf S, Gordon AS (1967) Relation of the adrenal and pituitary to the hypocholesterolemic effect of estrogens in rats. Endocrinology 81:340–344
Steingold KA, Cefalu W, Pardridge W, Judd HL, Chandhuri G (1986) Enhanced hepatic extraction of estrogens used for replacement therapy. J Clin Endocrin Metab 62:761–766
Thom M, Chakravarti S, Oram DH, Studd JWW (1977) Effect of hormone replacement therapy on glucose tolerance in postmenopausal women. Br J Obstet Gynaecol 84:776–784
Thompson C, Pearlie M, Hudson P, Lucier W (1983) Correlation of estrogen receptor concentrations and estrogen-mediated elevation of very low density lipoproteins. Endocrinology 112:1389–1397
Tikkanen MJ, Nikkilä EA, Kuusi T, Sipinen S (1982) High density lipoprotein 2 and hepatic lipase: Reciprocal changes produced by estrogen and norgestrel. J Clin Endocrinol Metab 54:1113–1117

van Keep PA, Utian WH, Vermeulen A (eds) The controversal climacteric. MTP Press Ltd, 1982

Wang E, Takano M, Okamoto T, Yayama K, Okamoto H (1995) Angiotensinogen synthesis in the liver is independent of physiological estrogen levels. Biol Pharm Bull 18:122–125

Washburn T, Hocutt A, Brautigan DL, Korach KA (1991) Uterine estrogen receptor in vivo: phosphorylation of nuclear-specific forms of serine residues. Mol Endocrinol 5:235–242.

Webb P, Lopez GN, Greene GL, Baxter JD, Kushner PJ (1992) The limits of the cellular capacity to mediate an estrogen response. Mol Endocrinol 6:157–167

Witteman JCM, Grobbee DE, Kok FJ, Hofman A, Falkenburg HA (1989) Increased risk for atherosclerosis in women after menopause. Br Med J 11:642–644

Wolffe AP, Glover, JF, Martin SC, Tenniswood PR, Williams JL, Tata JR (1985) Deinduction of transcription of *Xenopus* 74-kDa albumin genes and destabilization of mRNA by estrogen in vivo and in hepatocyte cultures. Eur J Biochem 146:489–496

von Schoultz B (1988) Potency of different estrogen preparations. In: Studd JW, Whitehead MI (eds) The menopause. Blackwell Scientific Publications, London, pp 130–137

von Schoultz B, Carlström K, Collste L, Eriksson A, Henriksson P, Pousette Å, Stege R (1989) Estrogen therapy and liver function. Metabolic effects of oral and parenteral administration. The Prostate 14:389–395

von Schoultz B, Carlström K (1989) On the regulation of sex-hormone binding globulin. A challenge of an old dogma and outlines of an alternative mechanism. J Steroid Biochem 32:327–334

Wren GB (1988) Hypertension and thrombosis with postmenopausal estrogen therapy. In: Studd JW, Whitehead MI (eds) The menopause. Blackwell Scientific Publications, London, pp 181–189

CHAPTER 29

Pharmacology of Antiestrogens

A.E. WAKELING

A. Introduction

The term antiestrogen defines the broad class of agents that antagonise the physiological effects of the female hormone 17β-estradiol (E2), for example, by reducing the trophic action of E2 on the uterus or by blocking estrogen-induced vaginal cornification. The antiestrogens discussed here are further defined as agents that act in direct competition with E2 for binding to the specific estrogen receptor (ER). Therefore, other effectors that may indirectly produce similar actions in animals by altering E2 synthesis or secretion are excluded.

In the four decades since the description of the first antiestrogen, MER 25 (LERNER et al. 1958), the chemical diversity and range of pharmacological actions described for antiestrogens has expanded the therapeutic horizon beyond the initial interest in their potential uses in fertility control (LERNER 1981; JORDAN 1988). Surprisingly, in view of the large number of antiestrogens described in the literature over the past four decades, only one compound, tamoxifen, is widely used clinically. Since it's first application in the therapy of advanced breast cancer (COLE et al. 1971), tamoxifen has been established as the treatment of choice for the endocrine therapy of breast cancer. Furthermore, its combination of easy administration, safety and efficacy has led to large-scale trials to determine whether it can prevent breast cancer (JORDAN 1997). The primary therapeutic target for antiestrogens is breast cancer and, for this indication, numerous analogues of tamoxifen have been described. One of these, toremiphene (KANGAS 1990), has been marketed whilst several others are in various stages of clinical evaluation (idoxifene, COOMBES et al. 1995; droloxifene, BRUNING 1992). Each of these molecules share chemical and pharmacological similarities; they are all non-steroidal, triphenylethylene-derived structures and are partial agonists (mixed agonist/antagonist activities). The term "partial agonist", whilst accurately conveying the idea that all tamoxifen-like antiestrogens have some estrogenic activity, does not adequately describe the complex pharmacology of these agents. The effect(s) of tamoxifen differ among species, target organs, cells and genes, depending on which effect of estrogen is measured (FURR and JORDAN 1984; WAKELING 1995). Differences in response between the rat and mouse were first described by HARPER and WALPOLE (1967a) in their original publication describing the properties of tamoxifen. This issue will be discussed further, but it should first be considered whether antiestrogens structurally distinct from tamoxifen also differ significantly in pharmacological properties.

The majority of chemical entities reported as antiestrogens are nonsteroidal in structure and encompass substituted tetrahydronaphthalenes, e.g. nafoxidene (DUNCAN et al. 1963) and trioxifene (JONES et al. 1979); indole derivatives, e.g. zindoxifene (STEIN et al. 1990) and ZK 119010 (VON ANGERER 1990); benzothiophenes, e.g. LY117018 (BLACK and GOOD 1980) and LY156758 (keoxifene) (CLEMENS et al. 1983); and benzopyrans (SHARMA et al. 1990; GRESE et al. 1996). One of the important considerations that has sustained a high level of interest in the synthesis of these and other new antiestrogens is the potential to reduce their partial-agonist (estrogenic) activity. This objective has been fulfilled, in part, with trioxifene, LY117018, LY156758 and ZK 119010 which show an ordered reduction in estrogenic activity compared with tamoxifen, to the point that ZK 119010 has been described as a "pure"' antiestrogen, i.e. completely lacking uterotrophic activity (VON ANGERER et al. 1990; NISHINO et al. 1991).

Whether or not complete elimination of estrogen-like actions is desirable will be discussed later. However, it should be noted that there is a resurgence of interest in molecules with a mixed spectrum of pharmacological effects (GRESE et al. 1996), particularly those that retain the beneficial estrogenic activities of tamoxifen on the bone and cardiovascular system, whilst eliminating uterine stimulatory activity. Such molecules have been termed selective estrogen receptor modulators (SERMs) (KAUFFMAN and BRYANT 1995) and include raloxifene, previously studied under the name keoxifene (BLACK et al. 1994); droloxifene (KE et al. 1995); and novel acrylic acid triphenylethylenes (WILLSON et al. 1994). The potential utility of raloxifene and droloxifene to prevent osteoporosis is being evaluated in clinical trials because such agents should reduce or eliminate the concerns of increased risk of endometrium- and breast cancer associated with conventional hormone-replacement therapy.

The continuing search for novel antiestrogens with reduced agonist activity led to the synthesis of steroid-derived compounds (BOWLER et al. 1989), in which addition of an alkylamide side-chain at the 7α-position in estradiol produced the first agent, ICI 164384, which completely blocked the trophic action of estradiol on the rat uterus. In addition ICI 164384 was correspondingly free of estrogen agonist activity and was, therefore, described as a pure (nonagonist) antiestrogen (WAKELING and BOWLER 1987). An important distinction between ICI 164384 and partial agonists, including other so-called pure antiestrogens, such as ZK 119010, was the demonstration that ICI 164384 also blocked the uterotrophic action of tamoxifen (WAKELING and BOWLER 1987). Subsequently, a number of other steroidal pure antiestrogens have been described including 11β-amidoalkoxyphenyl estradiol derivatives (NIQUE et al. 1994) and further modifications of the 7α-substituted estrogens (LEVESQUE et al. 1991), including one compound, ICI 182780 (WAKELING et al. 1991), which has progressed to clinical trial (HOWELL et al. 1995).

Continuing synthetic chemistry has produced non-steroidal molecules completely free of agonist activity; for example, ZM 189154, in which the alkyl-sulphinyl side-chain of ICI 182780 is substituted on a 2-methyl tetrahydro-naphthalene (DUKES et al. 1994), EM-800, a benzopyran derivative (LABRIE et al. 1996) and dichlorotriarylcyclopropanes (DAY et al. 1991). The pure antiestrogens are clearly differentiated from tamoxifen and SERMs both pharmacologically and in terms of their molecular mode of action. The pure antiestrogens block both transcription activation functions (AFs) of the ER, whereas partial agonists or SERMs appear to block AF2, whilst AF1 remains active (MCDONNELL et al. 1995). The remainder of this chapter will compare and contrast the pharmacological properties of partial-agonist (including SERMs) and pure-antagonist antiestrogens on a range of target organs, with particular emphasis on those properties germane to actual or potential clinical utility.

B. Reproductive Tract

The trophic response of the uterus and the stratification and cornification of the vaginal epithelium in ovariectomized or immature rats or mice treated with estradiol provide biological assays that allow accurate measurement of the relative estrogenic potency of newly synthesised molecules. Similarly, the antiestrogenic potency of novel compounds can be determined by concurrent treatment with a maximally stimulating dose of estrogen, together with test compound. The uterotrophic response is preferred over vaginal cornification for screening because of its simplicity, the size and reproducibility of the response and the precision conferred by the simple quantitative (tissue wet weight) output. For example, in immature rats, estradiol treatment for 3 days increases the (wet) weight of the uterus about fivefold, a response which is highly reproducible and readily detects the differences in trophic (estrogenic) action of a series of partial-agonist antiestrogens (WAKELING and SLATER 1981; WAKELING and VALCACCIA 1983). Correspondingly, when the same series of compounds were administered together with estradiol, it was readily apparent that their maximum antagonist activity (inhibition of the effect of unopposed estradiol) was determined by their intrinsic agonism. The net effect of each compound on the weight of the uterus accurately reflected the balance between agonist and antagonist activity.

The first pharmacological characterisation of tamoxifen by HARPER and WALPOLE (1967a) demonstrated clear differentials among species and tissues in thresholds of sensitivity to either agonist or antagonist effects in estrogen target organs. These original observations have been repeatedly validated in multiple tests with a variety of different non-steroidal antiestrogens (JORDAN and MURPHY 1990). Thirty years later, no satisfactory molecular explanation of these pharmacological differences has been achieved (WAKELING 1995; KATZENELLENBOGEN et al. 1996), although the availability of pure antiestrogens and modern molecular-biology methods offer much potential for achieving such understanding and, thus, a fully targeted approach to response-specific therapeutics.

The partial agonist effect of tamoxifen on the weight of the immature rat uterus reflects a differential tissue response among the epithelial, stromal and myometrial components of the organ. Estradiol stimulates hypertrophy and DNA synthesis to differing degrees in each tissue compartment, whereas tamoxifen induces substantial hypertrophy only of the luminal and glandular epithelium, and a small increase in DNA synthesis compared with estradiol (DIX and JORDAN 1980; HOLINKA et al. 1980; MARTIN 1981; FURR and JORDAN 1984). Tamoxifen-induced hypertrophy of the luminal epithelium was long lasting, whereas stimulation of the glands was transient, and tamoxifen appeared to be selectively toxic to gland cells since their number decreased (MARTIN 1981). These tissue-selective actions of tamoxifen in the immature rat are qualitatively and quantitatively different from those of estradiol. Differentials in tamoxifen action are further emphasised by comparisons between the mouse and rat uterus. Tamoxifen appears as a full agonist in the mouse. For example, it stimulated proliferation of luminal epithelium and, like estradiol, induced cystic hyperplasia of endometrial glands (MARTIN and MIDDLETON 1978. MARTIN 1981).

The synthesis of new steroid-derived antiestrogens (BOWLER et al. 1989) provided the first examples of compounds that have no uterotrophic activity in the immature rat uterus test. ICI 164384, a 7α-alkylamide derivative of estradiol, demonstrated a capacity to block estrogen-stimulated uterine growth in a dose-dependent and complete manner in immature and ovariectomized rats and mice (WAKELING and BOWLER 1987). The availability of antiestrogens devoid of uterotrophic activity permitted a test of the assumption that the stimulation of the uterus by tamoxifen is, like that of estrogen, mediated through the ER. Co-administration of tamoxifen with ICI 164384 showed a dose-dependent and complete blockade of tamoxifen-induced weight gain of the immature rat uterus (WAKELING and BOWLER 1987).

The term "pure" antiestrogen was coined to describe the pharmacological difference between a complete antagonist/antiestrogen, such as ICI 164384, and the partial agonists exemplified by tamoxifen, graphically demonstrated in the rat uterus assay (WAKELING and BOWLER 1987, 1988). The complete blockade of estrogen- or tamoxifen-mediated uterine growth by ICI 164384, by the more potent 7α-alkylsulphinyl steroidal pure antiestrogen, ICI 182780 (WAKELING et al. 1991), and by the non-steroidal pure antiestrogen ZM 189154 (DUKES et al. 1994), suggested that all of these molecules act through a common ER-mediated mechanism. Further evidence for this was provided by studies in transgenic mice, in which the ER was deleted. No uterine stimulation was seen following tamoxifen treatment of these animals (KORACH 1994). The latter observation also seems to preclude a role for the recently described

ER-β subtype in mediating the uterotrophic action of tamoxifen, since both ERα and ERβ are expressed in the uterus and tamoxifen binds to both receptors (KUIPER et al. 1997).

Interestingly, none of the other compounds that have been described as "pure" antiestrogens have yet been shown to block the uterotrophic action of tamoxifen. A final potentially important observation in the immature rat uterus assay was that both ICI 164384 and ICI 182780 given alone reduced the weight of the uterus to a level significantly less than that in vehicle-treated controls. This suggested that even in immature animals, the uterus is influenced by endogenous (or dietary) estrogens. This conclusion is supported by studies with the potent aromatase inhibitor, anastrozole, which also reduced the weight of the immature uterus below that in controls (DUKES et al. 1996). Since the immature rat uterus assay is currently being used as a bioassay in the investigation of the potential-endocrine toxicology of environmental estrogens (ODUM et al. 1997), it is important that investigators recognise that the uteri from control animals in such studies are not entirely unstimulated; thus, the assay may be not be sensitive enough to detect very weak estrogenicity. Furthermore, positive effects in such assays should be validated by testing whether tropism is blocked by co-administration of a pure antiestrogen.

The estrogen agonist activity of tamoxifen and other partial-agonist antiestrogens causes developmental abnormalities in neonatal rat, for example, premature vaginal opening (CHAMNESS et al. 1979; CILARK and GUTHRIE 1983). As might be anticipated, the pure antiestrogens, ICI 164384 and ICI 182780, did not cause developmental toxicity in neonatal rats and blocked such effects of tamoxifen (WAKELING 1988; BRANHAM et al. 1996). Thus, the pure antiestrogens may have toxicological advantages over partial agonists – a potentially important consideration in benign gynaecological applications. In other species, for example the guinea pig, rabbit and hamster, tamoxifen has either partial- or full-agonist activity, depending on the test endpoint (uterus, vagina, pregnancy), dose and duration of treatment (FURR and JORDAN 1984). In the primate, few studies have been reported but, using perineal swelling in ovariectomized monkeys as an endpoint, no agonist activity of tamoxifen was detected and co-administration with estradiol reduced estrogen-induced swelling (FURR and JORDAN 1984). More precise studies of antiestrogen effects in monkeys have been conducted using magnetic resonance imaging of the uterus in ovariectomized *Macaca nemestrina* treated with ICI 182780 (DUKES et al. 1992). In those studies, ICI 182780 completely blocked the trophic action of exogenous estradiol on the endometrium and myometrium. ICI 182780 was fully effective at a daily parenteral dose of 0.1/0.2 mg/kg and was equally effective given as a monthly injection of 4 mg/kg in a long-acting, oil-based formulation.

Differences between pure antiestrogens and tamoxifen are evident in adult (ovary-intact) animals. ICI 164384 treatment produced an ovariectomy-like involution of the uterus in rats, whereas the maximum involution in tamoxifen-treated rats was only 50–60% of that following ovariectomy

(Wakeling and Bowler 1988). This clearly illustrates the potential use of pure antiestrogens in the treatment of endometriosis and fibroids; conditions for which partial agonists such as tamoxifen are unsuitable. This potential was exemplified further in menstruating primates, in which ICI 182780 treatment completely blocked cyclical endometrial proliferation (Dukes et al. 1993), and in a monkey with adenomyosis successfully treated with ICI 182780 (Waterton et al. 1993). In a small clinical pharmacology study in premenopausal women, ICI 182780 blocked endometrial proliferation (Thomas et al. 1994). Further clinical studies to assess the potential value of ICI 182780 in the treatment of benign gynaecological conditions are in progress. In rats, tamoxifen disrupts the normal cyclical pattern of vaginal cornification, with most animals displaying a state of constant pro-oestrus; ICI 164384 completely blocked vaginal cornification (Wakeling and Bowler 1988). The thresholds of sensitivity of different reproductive tract end-points (uterus, vagina, ovulation) to complete blockade by pure antiestrogens differed (Wakeling et al. 1991; Dukes et al. 1994), consistent with the well-established differential-organ thresholds of estrogen action.

An important question raised by the trophic effect of tamoxifen on the uterus of rodents is whether or not such effects occur in women and, if so, what are the consequences? It is well-established that unopposed estrogen action in postmenopausal women receiving hormone replacement therapy can cause endometrial cancer. Model studies with a human-endometrial carcinoma grown in nude mice showed that tamoxifen, as well as estrogen, stimulated tumour growth (Gottardis et al. 1988a) and that pure antiestrogens inhibited both estrogen- and tamoxifen-stimulated growth (Gottardis et al. 1990). Paradoxically, tamoxifen has some efficacy in the treatment of advanced-endometrial carcinoma (White et al. 1993). Thus, reports of an increased incidence of endometrial tumours in breast-cancer patients on adjuvant tamoxifen therapy suggested that a causal connection might exist (Fisher et al. 1994).

Although there is an approximate twofold increase in the incidence of endometrial cancer in tamoxifen-treated breast-cancer patients, the balance of risk and benefit for the patient remains strongly in favour of continuing tamoxifen treatment (Jordan and Morrow 1994; Daniel et al. 1996). There is insufficient evidence to draw any conclusion about whether tamoxifen directly causes endometrial cancer, but a direct genotoxic effect in the human endometrium has been excluded (Carmichael et al. 1996). The latter observation is important because tamoxifen-induced liver tumours in rats (Greaves et al. 1993) have been attributed to a genotoxic mechanism associated with the formation of DNA adducts (Osborne et al. 1996). In women, comparison of the number of DNA adducts in the liver between control and tamoxifen-treated patients revealed no difference; numbers of adducts were very low compared with those in rats. Therefore, it appears that women are less susceptible than rats to tamoxifen-induced liver damage (Martin et al. 1995). The absence of any increase in the number of secondary cancers, except for those of the uterus, after adjuvant tamoxifen therapy is consistent with the a lack of genotoxicity of tamoxifen in women (CURTIS et al. 1996).

C. Breast and Breast Cancer

In the rat, mammary-gland development occurs post-pubertally; an extensively branched ductal tree grows to fill the mammary fat pad(s) over a period of approximately 3 weeks. This normal developmental process can be reproduced by estrogen treatment following ovariectomy in 30-day-old animals. In this model system, tamoxifen and other partial agonists supported full-ductal elongation, whereas ICI 164384 had no effect alone, although when co-administered with estrogen or tamoxifen, it completely blocked ductile growth (NICHOLSON et al. 1988). Mammary-gland growth in the mouse is also estrogen dependent (KORACH 1994). Local implantation of ICI 164384 or ICI 182780 into the mammary gland of adolescent animals caused regression of the ductile branches and blocked DNA synthesis in the terminal end buds (SILBERSTEIN et al. 1994). Partial-agonist antiestrogens have not been tested in this model. Thus, ICI 164384 and ICI 182780 act as pure antiestrogens in the rodent mammary gland.

In primates, there is some evidence that tamoxifen promotes a weak estrogen-like stimulation of the normal mammary gland (LEE and DUKELOW 1981), but in women with breast cancer, no evidence was found for a proliferative effect of tamoxifen on normal breast tissue (WALKER et al. 1991). In normal human breast tissue implanted into nude mice, epithelial proliferation was stimulated by oestradiol (LAIDLAW et al. 1995), consistent with the well-established growth dependence of malignant breast tissue on estrogens. The effect of tamoxifen in this model has not been reported.

To what extent, if any, the partial-agonist activity of tamoxifen influences its therapeutic efficacy in breast cancer is not known. The availability of (pure) antiestrogens, with no agonist activity, provides an opportunity to address this important question. Experimental studies with human breast cancer cells in vitro and in vivo predicted potential advantages for pure compared with partial-agonist antiestrogens, that might appear as more complete or longer-lasting responses (see WAKELING 1993 for a review). Both tamoxifen and the pure antiestrogens inhibit the growth of human breast cancer cells grown as xenografts in estrogen-treated athymic nude mice (OSBORNE et al. 1985; WAKELING et al. 1991; GOTTARDIS et al. 1988b; NIQUE et al. 1994; OSBORNE et al. 1995). There was some evidence of an efficacy advantage of the pure antiestrogens over tamoxifen, and ICI 182780 blocked tumour growth for twice as long as tamoxifen (OSBORNE et al. 1995). Also, resistance to treatment developed more slowly in ICI 182780-treated mice than in tamoxifen-treated mice (OSBORNE et al. 1995). Tamoxifen-resistant breast tumours, which grew out in nude mice after long-term tamoxifen treatment, remained sensitive to growth inhibition by pure antiestrogens (GOTTARDIS et al. 1989; OSBORNE et al. 1995; van de Velde et al. 1996), strongly suggesting that the development of resistance to tamoxifen in this model is due to an estrogenic action of the drug. Finally, in the nude-mouse tumour model system, the remarkable cell-specific pharmacological action of tamoxifen was evident in animals implanted on opposite flanks with a breast tumour and an endometrial tumour; tamoxifen inhibited breast tumour growth whilst stimulating the growth of the endometrial cancer (Gottardis et al. 1988a).

Clinical pharmacology studies with ICI 182780 have demonstrated anti-estrogenic effects in breast tumours (de Friend et al. 1994a) and provided some evidence of a greater efficacy than tamoxifen (de Friend et al. 1994b; McLelland et al. 1996). The therapeutic efficacy of ICI 182780 in breast cancer patients with tamoxifen-resistant tumours has been demonstrated in a small clinical trial (Howell et al. 1995, 1996), entirely consistent with the nude-mouse model studies. Further clinical trials will be required to discover whether the pure antagonists offer significant advantages over partial agonists in primary breast-cancer therapy.

D. Bone

The effects of antiestrogens on bone metabolism are of intense interest, because experimental studies with these agents have revealed the potential to design new treatments for osteoporosis that avoid the adverse effects of conventional estrogen-based hormone replacement (Kauffman and Bryant 1995; McDonnell and Norris 1997). Such agents, referred to as SERMs, prevent bone loss (estrogen agonist activity), without stimulating the uterus, and potentially prevent the development of breast cancer (estrogen-antagonist activity) (Black et al. 1994; Wilson et al. 1994; Grese et al. 1996; Keet et al. 1997). As yet, it is unclear whether the reduced agonist activity on the rat uterus of SERMs, such as raloxifene, compared with tamoxifen, will translate into an improved therapeutic profile (Draper et al. 1996).

The mechanisms by which SERMs act specifically as agonists on bone whilst having low or no agonist action on the reproductive tract and the breast remain unclear, but may involve mechanisms other than direct activation of estrogen-responsive genes (McDonell and Norris 1997; Willson et al. 1997). Studies with tamoxifen in rats, which, like women, experience a significant loss of bone following ovariectomy, showed that tamoxifen has mixed agonist/antagonist effects on bone density (Sibonga et al. 1996) determined by their endogenous estrogen status. Thus, in ovariectomized rats, tamoxifen treatment reduced bone loss, whereas in intact rats, bone density was reduced. These effects parallel those in women; bone loss is reduced by tamoxifen treatment in postmenopausal women, but induced in premenopausal women (Powles et al. 1996). Thus, the prediction from studies of SERMs in rats is that, like tamoxifen, these agents will prevent bone loss in postmenopausal women but cause bone loss in premenopausal women.

The effects of pure antiestrogens on bone density in women are still unknown, but their effects have been studied in the rat. ICI 182780 was reported to have no effect on gross bone density in ovary-intact rats at doses that produced an ovariectomy-like regression of the uterus (WAKELING 1993). Similarly, ICI 164384 did not affect bone turnover or trabecular bone volume (BAER et al. 1994), and EM-800 did not reduce bone density at an anti-uterotrophic dose (LUO et al. 1998b). These observations suggest either a difference in sensitivity to estrogen action between the bone and uterus (DUKES et al. 1994), or that estrogen's action in the bone is not mediated through the ERs. The former explanation is probably correct, since ER-knockout mice have reduced bone density (KORACH 1994), and other investigators have reported bone loss following pure antiestrogen treatment (GALLAGHER et al. 1993). Whether or not differential sensitivity between the uterus and bone to the action of pure antiestrogens can be demonstrated in women will clearly have a significant impact on the utility of these agents for the treatment of benign gynaecological conditions, such as fibroids and endometriosis, for which current therapy with gonadotrophin releasing hormone (GnRH) agonists is limited in duration by the concurrent loss of bone.

E. Brain

The well-established neuroendocrine effects of estrogens to control menstrual cyclicity through positive and negative feedback actions on the hypothalamus and pituitary gland contrast with the much less well understood role of estrogens in the control of mood and cognitive function (SHERWIN 1997). The immense clinical significance of the latter actions is emphasised by the recent reports that estrogens can reduce the onset or progression of Alzheimer's disease (SHERWIN 1997). Unfortunately, there are no animal models for such effects and, although it is possible to classify the pharmacological actions of antiestrogens by recording their effects on the hypothalamic-pituitary axis, this may not predict the balance of estrogenic and antiestrogenic effects on other estrogen target cells in the brain. Partial-agonist antiestrogens, such as tamoxifen, have mixed agonist and antagonist actions on the rat brain, for example, reducing food intake and luteinising hormone (LH) secretion in ovariectomized animals (agonist effects) whilst blocking estrogen-induced oestrous behaviour and ovulation in intact rats (WAKELING 1996).

In women, the neuroendocrine actions of tamoxifen are similarly mixed with effects, depending on menopausal status and endpoint measured; the most common side-effect of tamoxifen is an increased incidence of hot flashes (PRITCHARD 1997), presumed to be due to an antiestrogenic action in the brain. Little information is yet available on central effects of SERMs in women; one report indicated that a high dose of raloxifene increased vasodilation (hot flashes) in postmenopausal patients, consistent with an antiestrogenic effect (DRAPER et al. 1996). In clinical studies with the pure antiestrogen, ICI 182780, no central antiestrogenic effects [hot flashes, increased LH, follicle-stimulating hormone (FSH)] were reported (HOWELL et al. 1996; THOMAS et al. 1994). The clinical data are consistent with the failure of this compound to cross the blood–brain barrier in man, as demonstrated in rats indirectly, by the absence of increased LH secretion or weight gain (WAKELING et al. 1991) and, directly, by the failure of ICI 182780 (but not of tamoxifen) to block the uptake of radiolabelled E2 into the hypothalamus (WADE et al. 1993). This property of ICI 182780 could facilitate long-term treatment of both benign and malignant proliferative disease in premenopausal women, without initiating premature menopausal symptoms, such as mood changes, hot flashes, cognitive disability or stimulation of the hypothalamic–pituitary–ovarian axis through blockade of hypothalamic ERs. As yet, there are no clinical reports of other pure antiestrogens, but, in the case of EM-800, animal studies showed the anticipated stimulation of the hypothalamic–pituitary–ovarian axis (LUO et al. 1998a).

F. Cardiovascular System

It is generally accepted that hormone-replacement therapy reduces the risk of coronary heart disease (CHD) in postmenopausal women (COOPER and STEVENSON 1997); thus, antiestrogen treatment might be anticipated to increase the likelihood of CHD. Tamoxifen has clearly demonstrated that this is not the case. In two large clinical studies, tamoxifen reduced CHD in postmenopausal women (see CLARKSON and ANTHONY 1997 for review), consistent with an estrogen-like action on the cardiovascular system and with favourable changes in cardiovascular risk factors, for example, significant reduction of total and LDL-cholesterol and lipoprotein (a) (LOVE et al. 1994). Preclinical studies with the SERMs raloxifene and droloxifene showed that each agent has estrogen-like actions to lower cholesterol in rats (BLACK et al. 1994; KE et al. 1997), and a short-term clinical study with raloxifene demonstrated reduction of total and LDL-cholesterol (DRAPER et al. 1996). Interestingly, at a dose that provided effective palliation of breast cancer, ICI 182780 had no effect on total or HDL- or LDL-cholesterol (HOWELL et al. 1996), again suggesting that estrogen effects on the cardiovascular system are not mediated through ER or that the threshold for antiestrogen action on the cardiovascular system is different from that of the classical estrogen target organs.

G. Conclusions

Biological testing of antiestrogens belonging to different chemical and pharmacological classes, and developments in the understanding of mechanisms underlying the differential effects of antiestrogens on different organs and cells holds promise of providing more effective treatments for breast cancer and more selective hormone-replacement therapy for postmenopausal women. For breast cancer, a pure antiestrogen (ICI 182780) has proved efficacious in women whose disease has become resistant to tamoxifen, implying that tamoxifen resistance is caused by partial agonism. If that is the case, it can be predicted that pure antiestrogens might be superior to tamoxifen as the first-line endocrine treatment for breast cancer. Among the class of antiestrogens with partial agonism, the SERMs, several new compounds, for example raloxifene and droloxifene, may replace estrogens in the prevention of osteoporosis because of their selective agonism on bone and the cardiovascular system. Finally, clinical studies with these new agents will determine whether they have the potential to provide more effective treatment of benign gynaecological diseases, such as endometriosis and fibroids, by achieving involution of uterine tissues without concurrent bone loss.

References

Assikis VJ, Neven P, Jordan VC, Vergote I (1996) A realistic clinical perspective of tamoxifen and endometrial carcinogenesis. Eur J Cancer 32:1464–1476

Baer PG, Willson TM, Morris DC (1994) Lack of effect on bone of 28-days treatment of OVX and intact rats with a pure anti-estrogen (ICI 164384). Calcified Tissue Int 54:338 (Abstract)

Black LJ, Good RL (1980) Uterine bioassay of tamoxifen, trioxifene and a new estrogen antagonist (LY117018) in rats and mice. Life Sci. 26:1453–1458

Black LJ, Sato M, Rowley ER, Magee DE, Bekele A, Williams DC, Cullinan GJ, Bendele R, Kauffman RF, Bensch WR, Frolik CA, Termine JD, Bryant HU (1994) Raloxifene (LY139481 HCl) prevents bone loss and reduces serum cholesterol without causing uterine hypertrophy in ovariectomized rats. J Clin Invest 93:63–69

Bowler J, Lilley TJ, Pittam JD, Wakeling AE (1989) Novel steroidal pure antiestrogens. Steroids 54:71–99

Branham WS, Fisfman R, Streck RD, Medlock KL, DeGeorge JJ, Sheehan DM (1996) ICI 182,780 inhibits estrogen-dependent rat uterine growth and tamoxifen-induced developmental toxicity. Biol Reprod 54:160–167

Bruning PF (1992) Droloxifene, a new anti-estrogen in postmenopausal breast cancer: preliminary results of a double-blind dose-finding Phase II trial. Eur J Cancer 28 A:1404–1407

Carmichael PL, Ugwumadu AHN, Neven P, Hewer AJ, Poon GK, Phillips DH (1996) Lack of genotoxicity of tamoxifen in human endometrium. Cancer Res 56: 1475–1479

Chamness GC, Bannayan GA, Landry LA Jr, Sheridan PJ, McGuire WL (1979) Abnormal reproductive development in rats neonatally administered antiestrogen (tamoxifen). Biol Reprod 21:1087–1090

Clark JF, Guthrie SC (1983) The estrogenic effects of clomiphene during neonatal period in the rat. J Steroid Biochem 18:513–517

Clarkson TB, Anthony MS (1997) Effects on the cardiovascular system: basic aspects. In: Lindsay R, Dempster DW, Jordan VC (eds) Estrogens and Antiestrogens: Basic and Clinical Aspects. Lippincott-Raven, Philadelphia, pp 89–118

Clemens JA, Bennett DR, Black LJ, Jones CD (1983) Effects of a new antiestrogen, keoxifene (LY156758), on growth of carcinogen-induced mammary tumors and on LH and prolactin levels. Life Sci 32:2869–2875

Cole MP, Jones CTA, Todd IDH (1971) A new antiestrogenic agent in late breast cancer. An early clinical appraisal of ICI 46474. Br J Cancer 25:270–275

Coombes RC, Haynes BP, Dowsett M, Quigley M, English J, Judson IR, Griggs LJ, Potter GA, McCague R, Jarman M (1995) Idoxifene: report of a Phase I study in patients with metastatic breast cancer. Cancer Res 55:1070–1074
Cooper AJ, Stevenson JC (1997) Effects on the cardiovascular system: clinical aspects. In: Lindsay R, Dempster DW, Jordan VC (eds) Estrogens and Antiestrogens: Basic and Clinical Aspects. Lippincott-Raven, Philadelphia, pp 119–128
Cross SS, Ismail SM (1990) Endometrial hyperplasia in an oophorectomized woman receiving tamoxifen therapy. Case report. Br J Obstet Gynaecol 97:190–192
Curtis RE, Boice JD, Shriner DA, Hankey BF, Fraumeni JF (1996) Second cancers after adjuvant tamoxifen therapy for breast cancer. J Natl Cancer Inst 88:832–834
Daniel Y, Inbar M, Bar-Am A, Perser MR, Lessing JB (1996) The effects of tamoxifen treatment on the endometrium. Fert Steril 65:1083–1089
Day BW, Magarian RA, Jain PT, Pento JT, Mousissian GK, Meyer KL (1991) Synthesis and biological evaluation of a series of 1,1-dichloro-2,2,3-triarylcyclopropanes as pure antiestrogens. J Med Chem 34:842–851
DeFriend DJ, Howell A, Nicholson RI, Anderson E, Dowsett M, Mansel RE, Blamey RW, Bundred N, Robertson JF, Saunders C, Baum M, Walton P, Sutcliffe F, Wakeling AE (1995a) Investigation of a new pure antiestrogen (ICI 182780) in women with primary breast cancer. Cancer Res 54:408–414
DeFriend DJ, Anderson E, Bell J, Wilks DP, West CML, Mansel RE, Howell A (1994b) Effects of 4-hydroxytamoxifen and a novel pure antioestroge (ICI 182780) on the clonogenic growth of human breast cancer cells in vitro. Br J Cancer 70:204–211
Dix CJ, Jordan VC (1980) Subcellular effects of monohydroxytamoxifen in the rat uterus: steroid receptors and mitosis. J Endocr 85:393–404
Draper MW, Flowers DE, Huster WJ, Nield JA, Harper KD, Arnaud C (1996) A controlled trial of raloxifene (LY139481)HCl: impact on bone turnover and serum lipid profile in healthy postmenopausal women. J Bone Min Res 11:835–842
Dukes M, Miller D, Wakeling AE, Waterton JC (1992) Antiuterotrophic effects of a pure antiestrogen, ICI 182,780: magnetic resonance imaging of the uterus in ovariectomized monkeys. J Endocr 135:239–247
Dukes M, Waterton JC, Wakeling AE (1993) Antiuterotrophic effects of the pure antiestrogen ICI 182,780 in adult female monkeys *(Macaca nemestrina)*: quantitative magnetic resonance imaging. J Endocr 138:203–209
Dukes M, Chester R, Yarwood L, Wakeling AE (1994) Effects of a non-steroidal pure antiestrogen, ZM 189,154, on estrogen target organs of the rat including bones. J Endocr 141:335–341
Dukes M, Edwards PN, Large M, Smith IK, Boyle T (1996) The preclinical pharmacology of "Arimidex" (Anastrazole; ZD1033) – a potent, selective aromatase inhibitor. J Steroid Biochem Mol Biol 58:439–445
Duncan GW, Lyster SC, Clark JJ, Lednicer D (1963) Antifertility activities of two diphenyl-dihydronaphthalene derivatives. Proc Soc Exp Biol Med 112:439–442
Fisher B, Costantino JP, Redmond CK, Fisher ER, Wickerham DL, Cronin WM, Other NSABP contributors (1994) Endometrial cancer in tamoxifen-treated breast cancer patients: findings from the National Surgical Adjuvant Breast and Bowel project (NSABP) B-14. J Natl Cancer Inst 86:527–537
Furr BJA, Jordan VC (1984) The pharmacology and clinical uses of tamoxifen. Pharmac Ther 25:127–205
Gallagher A, Chambers TJ, Tobias JH (1993) The estrogen antagonist ICI 182,780 reduces cancellous bone volume in female rats. Endocrinol 133:2787–2791
Gauthier S, Caron B, Cloutier J, Dory YL, Favre A, Larouche D, Mailhot J, Ouellet C, Schwerdtfeger A, Lablanc G, Martel c, Simard J, Merand Y, Belanger A, Labrie C, Labrie F (1997) (S)-(+)-4-[7-(2,2-dimethyl-1-oxopropoxy)-4-methyl-2[4-[2-(1-piperidinyl)ethoxy]phenyl]-2H-1-benzopyran-3-yl]-phenyl 2,2-dimethylpropanoate (EM800): a highly potent, specific, and orally active nonsteroidal antiestrogen. J Med Chem 40:2117–2122

Gottardis MM, Robinson SP, Satyaswaroop PG, Jordan VC (1988a) Contrasting actions of tamoxifen on endometrial and breast tumor growth in the athymic mouse. Cancer Res 48:812–815
Gottardis MM, Robinson SP, Jordan VC (1988b) Estradiol-stimulated growth of MCF-7 tumors implanted in athymic mice: a model to study the tumoristatic action of tamoxifen, J Steroid Biochem 30:311–314
Gottardis MM, Jiang S-Y, Jeng M-H, Jordan VC (1989) Inhibition of tamoxifen-stimulated growth of an MCF-7 tumor variant in athymic mice by novel steroidal antiestrogens. Cancer Res 49:4090–4093
Gottardis MM, Ricchio ME, Satyaswaroop PG, Jordan VC (1990) Effect of steroidal and nonsteroidal antiestrogens on the growth of tamoxifen-stimulated human endometrial carcinoma (EnCa101) in athymic mice. Cancer Res 50:3189–3192
Greaves P, Goonetilleke R, Nunn G, Topham J, Orton T (1993) Two-year carcinogenicity study of tamoxifen in Alderley Park Wistar-derived rats. Cancer Res 53:3919–3924
Grese TA, Sluka JP, Bryant HU, Cole HW, Kim JR, Magee DE, Rowley ER, Sato M (1996) Benzpyran selective estrogen receptor modulators (SERMS): pharmacological effects and structural correlation with raloxifene. Biorg Med Chem Lett 6:903–908
Harper MJK, Walpole AL (1967a) A new derivative of triphenylethylene: effect on implantation and mode of action in rats. J Reprod Fert 13:101–119
Harper MJK, Walpole AL (1967b) Mode of action of ICI 46,474 in preventing implantation in rats. J Endocr 37:83–92
Holinka CF, Bressler RS, Zehr DR, Gurpide E (1980) Comparison of effects of estetrol and tamoxifen with those of estriol and estradiol on the immature rat uterus. Biol Reprod 22:913–926
Howell A, DeFriend D, Robertson J, Blamey R, Walton P (1995) Response to a specific antiestrogen (ICI 182780) in tamoxifen-resistant breast cancer. Lancet 345:29–30
Howell A, DeFriend DJ, Robertson JFR, Blamey RW, Anderson L, Anderson E, Sutcliffe FA, Walton P (1996) Pharmacokinetics, pharmacological and anti-tumour effects of the specific anti-estrogen ICI 182780 in women with advanced breast cancer. Br J Cancer 74:300–308
Jones CD, Suarez t, Massey EH, Black LJ, Tinsley J (1979) Synthesis and antiestrogenic activity of [3,4-dihydro-2-(4-methoxyphenyl)-1-naphthalenyl](4-(1-pyrolidinyl) ethoxyphenyl methanone, methanesulphonic acid salt. J Med Chem 22:962–966
Jordan VC (1978) Use of the DMBA-induced rat mammary carcinoma system for the evaluation of tamoxifen treatment as a potential adjuvant therapy. Rev Endocr Rel Cancer October Suppl pp 49–55
Jordan VC (1988) The development of tamoxifen for breast cancer therapy: a tribute to the late Arthur L Walpole. Breast Cancer Res Treat 11:197–209
Jordan VC, Murphy CS (1990) Endocrine pharmacology of antiestrogens as antitumor agents. Endocrine Rev 11:578–610
Jordan VC, Morrow M (1994) Should clinicians be concerned about the carcinogenic potential of tamoxifen? Eur J Cancer 30 A:1714–1721
Jordan VC (1997) Tamoxifen: the herald of a new era of preventive therapeutics. J Natl Cancer Inst 89:747–749
Kangas L (1990) Review of the pharmacological properties of toremiphene. J Steroid Biochem 36:191–195
Katzenellenbogen JA, O'Malley BW, Katzenellenbogen BS (1996) Tripartite steroid hormone receptor pharmacology: interaction with multiple effector sites as a basis for the cell- and promotor-specific action of these hormones. Mol Endoc 10:119
Kauffman RF, Bryant HU (1995) Selective estrogen receptor modulators. Drug News Perspect 8:531–539
Ke HZ, Chen HK, Simmons HA, Crawford DT, Pirie CM, Chidsey-Frink KL, Ma YF, Jee WSS, Thompson DD (1997) Comparative effects of droloxifene, tamoxifen,

and estrogen on bone serum cholesterol, and uterine histology in the ovariectomized rat model. Bone 20:31–39
Korach K (1994) Insights from the study of animals lacking functional estrogen receptor. Science 266:1524–1527
Kuiper GGJM, Carlsson B, Grandien K, Enmark E, Haggblad J, Nilsson S, Gustafsson J-A (1997) Comparison of the ligand binding specificity and transcript tissue distribution of estrogen receptors α and β. Endocrinol 138:863–870
Laidlaw IJ, Clarke RB, Howell A, Owen AWMC, Potten CS, Anderson E (1995) The proliferation of normal breast tissue implanted into athymic nude mice is stimulated by estrogen but not progesterone. Endocrinol 136:164–171
Lee AE, Dukelow WR (1981) Tamoxifen effects mammary gland morphology and ovarian activity in *Macaca fascicularis*. J Med Primatol 10:102–109
Lerner LJ (1981) The first nonsteroidal antiestrogen -MER25. In: RL Sutherland, VC Jordan (eds) Non-steroidal antiestrogens. Academic Press, Sydney, pp 1–16
Lerner LJ, Holthaus JF, Thompson (1958) A nonsteroidal estrogen antagonist (1-(p-2-diethylaminoethoxyphenyl)-1-phenyl-2-p-methoxyphenyl ethanol. Endocrinol 63: 295–318
Lerner LJ, Jordan VC (1990) Development of antiestrogens and their use in breast cancer: Eighth Cain Memorial Award Lecture. Cancer Res 50:4177–4189
Levesque C, Merand Y, Dufour JM, Labrie C, Labrie F (1991) Synthesis and biological activity of new halo-steroidal anti-estrogens. J Med Chem 34:1624–1630
Love RR, Wiebe DA, Feyzi JM, Newcomb PA, Chappell RJ (1994) Effects of tamoxifen on cardiovascular risk factors in postmenopausal women after 5 years of treatment. J Natl Cancer Inst 86:1534–1539
Luo S, Martel C, Sourla A, Gauthier S, Merand Y, Belanger A, Labrie C, Labrie F. Comparative effects of 28-day treatment with the new antiestrogen EM-800 and tamoxifen on estrogen-sensitive parameters in the intact mouse. Int J Cancer (in press)
Luo S, Sourla A, Labrie C, Belanger A, Labrie F. Combined effects of dehyrdoepiandrosterone and EM-800 on bone mass, serum lipids, and the development of dimethylbenz(a)anthracene(DMBA)-induced mammary carcinoma in the rat. (Endocrinology, in press)
Martin EA, Rich KJ, White INH, Wooda KL, Powles TJ, Smith LL (1995) ^{32}P-Postlabelled DNA adducts in liver obtained from women treated with tamoxifen. Carcinogenesis 16:1651–1654
Martin L (1981) Effects of antiestrogens on cell proliferation in the rodent reproductive tract. In: Sutherland RL, Jordan VC (eds) Non-steroidal antioestrogens. molecular pharmacology and antitumour activity. Academic Press, Sydney pp 143–163
Martin L, Middleton E (1978) Prolonged estrogenic and mitogenic activity of tamoxifen in the ovariectomized mouse. J Endocr 78:125–129
McClelland RA, Manning DL, Gee JMW, Anderson E, Clarke R, Howell A, Dowsett M, Robertson JFR, Blamey R, Wakeling AE, Nicholson RI (1996) Effects of short-term antiestrogen treatment of primary breast cancer on estrogen receptor mRNA and protein expression and on estrogen-regulated genes. Breast Cancer Res Treat 41:31–41
McDonnell DP, Clemm DL, Hermann T, Goldman ME, Pike JW (1995) Analysis of estrogen receptor function in vitro reveals three distinct classes of antiestrogens. Mol Endoc 9:659–669
McDonnell DP, Norris JD (1997) Analysis of the molecular pharmacology of estrogen receptor agonists and antagonists provides insights into the mechanism of action of estrogen on the bone. Osteoporosis Int Suppl. 1:S29–S34
Nicholson RI, Gotting KE, Gee J, Walker KJ (1988) Actions of estrogens and antiestrogens on rat mammary gland development: relevance to breast cancer prevention. J Steroid Biochem 30:95–103
Nishino Y, Schneider MR, Michna H, von Angerer E (1991) Pharmacological charcelerisation of novel estrogen antagonist, ZK 119010, in rats and mice. J Endocr 130:409–414

Nique F, Van de Velde P, Bremaud J, Hardy M, Philibert D, Teutsch G (1994) 11β-amidoalkoxyphenyl estradiols, a new series of pure antiestrogens. J Steroid Biochem Mol Biol 50:21–29

Odum J, Lefevre PA, Tittensor S, Paton D, Routledge EJ, Beresford NA, Sumpter JP, Ashby J (1997) The rodent uterotrophic assay: critical protocol features, studies with nonyl phenols, and comparison with a yeast estrogenecity assay. Reg Toxicol Pharmacol 25:226–231

Osborne CK, Hobbs K, Clark GM (1985) Effect of estrogens and antiestrogens on growth of human breast cancer cells in athymic nude mice. Cancer Res 45:584–590

Osborne CK, Coronado-Heinsohn EB, Hilsenbeck SG, McCue BL, Wakeling AE, McClelland RA, Manning DL, Nicholson RI (1995) Comparison of the effects of a pure steroidal antiestrogen with thos of tamoxifen in a model of human breast cancer. J Natl Cancer Inst 87:746–750

Osborne MR, Hewer A, Hardcastle IR, Carmichael PL, Phillips DH (1996) Identification of the major tamoxifen-deoxyguanosine adduct formed in the liver DNA of rats treated with tamoxifen. Cancer Res 56:66–71

Palkowitz AD, Glasebrook AL, Thrasher KJ, Hauser KL, Short LL, Phil;lips DL, Muehl BS, Sato M, Shetler PK, Cullinan GJ, Pell TR, Bryant HU (1997) Discovery and synthesis of [6-hydeoxy-3-[4-[2-(1-piperidinyl)ethoxy]phenoxy]-2-(4-hydroxyphenyl)]benzo[b]thiophene: a novel, highly potent, selective estrogen receptor modulator. J Med Chem 40:1407–1416

Powles TJ, Hickish T, Kanis JA, Tidy A, Ashley S (1996) Effect of tamoxifen on bone mineral density measured by dual-energy X-ray absorpsiometry in healthy premenopausal and postmenopausal women. J Clin Oncol 14:78–84

Pritchard K (1997) Effects on breast cancer. Clinical aspects. In: Lindsay R, Dempster DW, Jordan VC (eds) Estrogens and Antiestrogens: Basic and Clinical Aspects. Lippincott-Raven, Philadelphia, pp 175–210

Rutqvist LE, Johansson H, Signomklao T, Johansson U, Fornander T, Wilking N (1995) Adjuvant tamoxifen therapy for early stage breast cancer and second primary malignancies. J Natl Cancer Inst 87:645–651

Sharma AP, Saeed A, Durani S, Kapil RS (1990) Structure-activity relationship of antiestrogens. Phenolic analogues of 2,3-diaryl-2H-1-benzopyrans. J Med Chem 33:3222–3229

Sherwin BB (1997) Estrogenic effects on the central nervous system: clinical effects. In: Lindsay R, Dempster DW, Jordan VC (eds) Estrogens and Antiestrogens: Basic and Clinical Aspects. Lippincott-Raven, Philadelphia, pp 75–87

Sibonga JD, Evans GI, Hauck ER, Bell NH, Turner RT (1996) Ovarian status influences the skeletal effects of tamoxifen in adult rats. Breast Cancer Res Treat 41:71–79

Silberstein GB, Van Horn K, Shyamala G, Daniel CW (1994) Essential role of endogenous estrogen in directly stimulating mammary growth demonstrated by implants containing pure antiestrogens. Endocrinol 134:84–90

Stein RC, Dowsett M, Cunningham DC, Davenport J, Ford HT, Gazet J-C, von Angerer E, Coombes RC (1990) Phase I/II study of the anti-estrogen zindoxifene (D16726) in the treatment of advanced breast cancer. A Cancer Research Campaign PhaseI/II clinical trials committee study. Br J Cancer 61:451–453

Thomas EJ, Walton PL, Thomas NM, Dowsett M (1994) The effects of ICI 182,780, a pure anti-estrogen, on the hypothalamic-pituitary-ovarian axis and on endometrial proliferation in pre-menopausal women. Human Reprod 9:1991–1996

Van de Velde P, Nique F, Planchon P, Prevost G, Bremaud J, Hameau MC, Magnien V, Philibert D, Teutsch G (1996) RU 58668: further in vitro and in vivo pharmacological data related to its antitumoral activity. J Steroid Biochem Mol Biol 59:449–457

von Angerer E, Knebel N, Kager M, Ganss B (1990) 1-Aminoalkyl-2-phenylindoles as novel pure estrogen antagonists. J Med Chem 33:2635–2640

Wade GN, Blaustein JD, Gray JM, Meredith JM (1993) ICI 182,780: a pure antiestrogen that affects behaviours and energy balance in rats without acting in the brain. Am J Physiol 43:R1392–R1398

Wakeling AE (1993) The future of new pure antiestrogens in clinical breast cancer. Breast Cancer Res Treat 25:1–9
Wakeling AE (1995) Use of pure antiestrogens to elucidate the mode of action of estrogens. Biochem Pharmacol 49:1545–1549
Wakeling AE (1996) Physiological effects of pure antiestrogens. In: Pasqualini JR, Katzenellenbogen BS (eds) Hormone-Dependent Cancer, Dekker, New York, pp 107–118
Wakeling AE, Bowler J (1987) Steroidal pure antiestrogens. J Endocr 112:R7-R10
Wakeling AE, Bowler J (1988) Novel antiestrogens without partial agonist activity. J Steroid Biochem 31:645–653
Wakeling AE, Slater SR (1981) Biochemicel and biological aspects of anti-estrogen action. In: Lewis GP, Ginsburg M (eds) Mechanisms of steroid action. Macmillan Press, London, pp 159–171
Wakeling AE, Valcaccia B (1983) Antiestrogenic and antitumour activities of a series of non-steroidal antiestrogens. J Endocr 99:455–464
Wakeling AE, Dukes M, Bowler J (1991) A potent specific pure antiestrogen with clinical potential. Cancer Res 51:3867–3873
Walker KJ, Price-Thomas JM, Candlish W, Nicholson RI (1991) Influence of the antiestrogen tamoxifen on normal breast tissue. Br J Cancer 64:764–768
Waterton JC, Breen SA, Dukes M, Horrocks M, Wadsworth PF (1993) A case of adenomyosis in a pigtailed monkey diagnosed by magnetic resonance imaging and treated with the novel pure antiestrogen, ICI 182,780. Lab Animal Sci 43:247–251
Willson TM, Henke BR, Momtahen TM, Charifson PS, Batchelor KW, Lubahn DB, Moore LB, Oliver BB, Sauls HR, Triantafillou JA, Wolfe SG, Baer PG (1994) 3-[4-(1,2-diphenylbut-1-enyl)phenyl]acrylic acid: a non-steroidal estrogen with functional selectivity for bone over uterus in rats. J Med Chem 37:1550–1552
Willson TM, Norris JD, Wagner BL, Asplin I, Baer P, Brown HR, Jones SA, Henke B, Sauls H, Wolfe S, Morris DC, McDonnell DP (1997) Dissection of the molecular mechanism of action of GW5638, a novel estrogen receptor ligand, provides insights into the role of estrogen receptor in bone. Endocrinol 138:3901–3911
White JO, Owen GI, De Clerq NAM, Soutter WP (1993) Antiestrogens in the treatment of endometrial carcinoma. Rev Endocr Rel Cancer 44:47–57

CHAPTER 30

Oncology

E.V. JENSEN

A. Introduction

The principal association of estrogenic hormones with cancer is in the genesis and treatment of neoplasms of the female reproductive tract. These are discussed separately in the sections that follow. In addition to the specific literature references cited, more comprehensive information concerning the role of estrogenic hormones in oncology can be found in the following publications: LI et al. (1992, 1996); CASTAGNETTA et al. (1996); HOLLAND et al. (1997).

B. Estrogens and Cancer Etiology

I. General Considerations

Since the early finding that the administration of estrone leads to mammary cancer in male mice (LACASSAGNE 1932), much evidence has been obtained to indicate an association between estrogenic hormones and oncogenesis. In general, the estrogen-related tumors are in tissues of the reproductive tract, although, as discussed later, estrogens produce liver cancers in the hamster and are probably a factor in their occurrence in the human. Carcinogenesis is known to be a multi-stage process, involving both transformation (alteration of DNA) and promotion (proliferation of the altered cells). It is generally agreed that much of the effect of estrogenic hormones is on the promotion stage, especially in tissues in which growth and function is normally regulated by estrogen. Whether estrogens also cause genetic alteration in a manner similar to chemical carcinogens, e.g., dimethylbenzanthracene or nitrosourea, has been a subject of much experimentation and some controversy.

The human malignancy most studied in relation to estrogen is carcinoma of the breast, both because of its high incidence and because its involvement with estrogens is especially striking. Other tissues in which estrogens influence the development of neoplasms include uterus, especially the endometrium, ovary, vagina and liver.

II. Breast Cancer

A variety of evidence indicates that estrogenic hormones play an important role in the appearance of mammary cancer, not only in experimental animals, but also in the human (Persson 1996; Henderson et al. 1997). It has long been known that early menarche and/or late menopause increase the risk of breast cancer, suggesting a cumulatory effect of the number of ovulatory cycles on the incidence of the disease (Trichopoulos et al. 1972; Henderson et al. 1982; MacMahon et al. 1982). Artificial menopause, induced by oophorectomy or radiation, likewise reduces the risk. It is also recognized that a full-term pregnancy before the age of 20 years confers a significant protective effect, whereas nulliparous women have an increased susceptibility to breast cancer (MacMahon et al. 1970), but the precise basis of this phenomenon is not clearly understood.

The effect of hormone-replacement therapy on the risk of breast and other cancers has been the subject of considerable investigation (Persson 1996). Although there has not been complete consensus, from the most recent studies, it appears that, for breast cancer, the risk from unopposed estrogen is small and the addition of progestins to the regimen makes little difference.

The putative involvement of estrogens in the etiology of breast malignancies has afforded an approach to cancer prevention that is presently under study. After the introduction of tamoxifen (Harper and Walpole 1967), the first of the triphenylethylene-type antiestrogens to be tolerated on long-term administration, this agent was found to prevent the induction by dimethylbenzanthracene of mammary cancer in the rat (Jordan 1976), as well as appearance of cancer in the contralateral breast after mastectomy in humans (Jordan 1990). The proposal that antiestrogens, such as tamoxifen, might be used to prevent the initial tumor in women who have a high risk of developing breast cancer has recently been subjected to clinical trial with highly promising results (Fisher et al. 1998). Preliminary reports from an ongoing study with the related antiestrogen raloxifene (see Chap. 29) also appear to be favorable (Jordan et al. 1998).

III. Uterine and Cervical Cancer

It is well established that exposure to estrogen, unopposed by progestin, is a major factor in the occurrence of cancer of the uterine endometrium (Henderson et al. 1988). Use of unopposed estrogen-replacement therapy for more than 5 years results in an elevated risk of endometrial cancer (Ziel and Finkle 1975; Collins et al. 1980; Henderson et al. 1993), which persists for several years after medication has been discontinued (Paganini-Hill et al. 1989). The addition of progestin to the estrogen-replacement regimen substantially reduces the risk of endometrial cancer (Persson et al. 1989).

In comparison with cancers of the breast and uterine endometrium, involvement of estrogenic hormones in the etiology of cervical neoplasia is less clear. Most studies have focused on the effect of oral contraceptives, which generally have shown a correlation between the prolonged use of these agents and the incidence of cervical cancer (Ursin et al. 1994).

IV. Ovarian Cancer

As in the case of breast cancer, late menopause, giving a longer period of ovulatory activity, results in an increased risk of ovarian cancer. Similarly, pregnancy and the use of oral contraceptives, which decrease the number of ovulatory cycles, are protective factors (HENDERSON et al. 1993).

V. Vaginal Adenocarcinoma

During the period 1945 to 1955, hyperphysiological doses of estrogenic hormones were often administered to pregnant women with histories of miscarriage in the belief that this would protect against spontaneous abortion. Because orally active steroidal estrogens were not available at that time, the substance used was the synthetic estrogen, diethylstilbestrol (DES). In the early 1970s, a previously rare cancer, clear cell adenocarcinoma of the vagina, began to appear in daughters born to the DES-treated mothers (HERBST et al. 1971). This led to the impression that estrogens in general, and DES in particular, are carcinogens that should not be used for human medication. However, the amounts administered (0.5–1.0 g) were 5000–10,000 times the hormonally active dose, and the cancers produced were not in those persons receiving the estrogen but in their offspring, indicating that this was an in utero phenomenon; so, the action of the hormone is probably better described as teratogenic than carcinogenic. Longer follow-up has demonstrated a slightly increased incidence of breast cancer among the DES-treated mothers (HERBST et al. 1986), but not what one might expect from such a high dose of a true carcinogen. Genital abnormalities, such as hypospadia and cryptorchidism, were observed among the male offspring, in keeping with a teratogenic phenomenon.

VI. Liver Cancer

In certain animal species, such as the hamster, liver cancer can be induced by the simple administration of estrogens, 17a-ethynylestradiol being especially potent in this regard (LI and LI 1990). Although it is a major cause of death in Asia and South Africa, in the western world, human primary hepatocellular carcinoma is a relatively rare malignancy except for individuals with cirrhosis. However, the introduction of oral contraceptives has led to an increased incidence of liver tumors (BAUM et al. 1973; NISSEN and KENT 1975), especially after long-term use (NEUBERGER et al. 1986) and with preparations containing substantial amounts of estrogen (KEIFER and SCOTT 1977). It has been reported that hepatomas can arise from the use of DES therapy for prostatic cancer (BROOKS 1982).

C. Hormones and Cancer Therapy

I. Hormone-Dependent Neoplasms

When cancer occurs in tissues where growth and function depend on estrogenic hormones, in some cases, the malignant cells retain their hormone dependency while in others, they lose the need for continued stimulation. The precise basis for hormone dependency, and whether escape from regulation takes place during neoplastic transformation or during subsequent tumor progression, is not clearly established. For those cancers that retain hormone dependency, depriving them of estrogen can provide an effective palliative treatment, less traumatic than cytotoxic chemotherapy that is the only recourse for the majority of non-hormone-dependent metastatic cancers.

II. Breast Cancer

1. Endocrine Ablation

The first example of endocrine treatment for any type of cancer was reported more than a century ago (BEATSON 1896). On the basis of previous observations that the size of breast tumors varied with the menstrual cycle (COOPER 1836), and with remarkable insight for the time, the ovaries were removed from young women with advanced breast cancer with the result that some, but not all, showed striking remission of their disease. But the majority of breast cancers occur in the older patients, in whom the ovaries are no longer functional, and it was long suspected that, in this situation, the adrenal cortex was the source of supporting hormones. When cortisone became available, initially for the treatment of inflammatory diseases, it became feasible to remove the adrenal glands and maintain the patient with glucocorticoid replacement (HUGGINS and BERGENSTAL 1952). Soon thereafter, the use of hypophysectomy to eliminate the adrenal production of estrogen precursors by depriving them of adrenocorticotrophic hormone (ACTH) was reported (LUFT and OLIVECRONA 1953). Clinical experience has shown that about one-third of all patients have mammary tumors that undergo remission when deprived of supporting hormone by any of these procedures, and endocrine ablation became the first-line therapy for advanced breast cancer, especially after methods were developed to identify those cancers of the hormone-dependent type that were likely to respond to endocrine manipulation (see Sect. C.II.4.)

2. Antiestrogens

As discussed in more detail in Chap. 29, the discovery of antiestrogens provided a means of depriving tumor cells of hormonal stimulation at the molecular level, by preventing estrogen from binding to its receptor protein. After the advent of tamoxifen, the first of the antiestrogens to be tolerated on prolonged administration (JORDAN and MURPHY 1990), this reversible treatment has essentially replaced the irreversible endocrine ablation as first-line therapy in patients with estrogen-receptor-positive tumors. Although there are side effects from prolonged treatment, as well as a slightly increased risk of endometrial cancer (ASSIKIS and JORDAN 1995), the benefits appear to greatly outweigh the drawbacks. The most serious limitation of tamoxifen therapy is the development in many individuals of an "acquired tamoxifen resistance," in which the medication no longer inhibits but actually stimulates growth of the cancer. This appears to result from a mutation in the estrogen-receptor protein after prolonged tamoxifen administration (JORDAN et al. 1995).

3. Aromatase Inhibitors

Deprivation of an estrogen-dependent tumor of hormonal stimulation can be accomplished, not only by ablation of organs involved in estrogen production or by interfering with hormone action at the target level, but also by inhibiting enzymes involved in estrogen biosynthesis. This has the advantage over endocrine ablation in that it eliminates not only the estrogen arising from the ovary, or indirectly from the adrenal, but also that which, in some instances, appears to be produced in the tumor itself (LIPTON et al. 1987; O'NEILL and MILLER 1987). The first successful application of this approach (SANTEN et al. 1974) used aminoglutethimide, which, among other actions, inhibits the enzyme, aromatase, responsible for converting C_{19} androgenic precursors to C_{18} estrogenic steroids. However, the clinical application of aminoglutethimide was limited by the side effects induced in many patients. As discussed in more detail elsewhere (HARVEY and MANNI 1997), several improved agents have been developed recently, including fadrozole, letrozole, vorozole and arimidex, which show promise of increased activity with fewer side effects.

4. Estrogen Receptors

As discussed in detail in Chap. 6, vol. I, the stimulatory action of estrogens, like that of steroid hormones in general, is mediated by specific receptor proteins. These are latent transcription factors, which, on association with the hormone, are converted to an active form. Soon after the original demonstration of estrogen-binding components in female reproductive tissues, radioactive hexestrol, a synthetic estrogen, was administered to a small number of patients about to undergo adrenalectomy or oophorectomy for advanced breast cancer. It was found that tumors from those patients who responded favorably contained more radioactivity than those from patients who did not respond (FOLCA et al. 1961).

When procedures were developed for demonstrating hormone–receptor interaction in vitro, it became possible to determine estrogen receptors in excised tumor tissue or extracts thereof. It was found that cancers with low or negligible receptor content rarely respond to endocrine manipulation, whereas most, but not all, patients whose tumors have a substantial number of receptors show objective remission of their disease (JENSEN et al. 1971). It was later demonstrated that analysis of the primary tumor at the time of mastectomy can predict response to endocrine therapy if the cancer recurs in disseminated form (Block et al. 1978). As described in greater detail elsewhere (Jensen and DeSombre 1997), these findings were confirmed and extended in many laboratories, and shown to be applicable to antiestrogen therapy as well as to endocrine ablation. With the generation of specific antibodies to the estrogen-receptor protein (Greene et al. 1980), immunochemical and immunocytochemical procedures for receptor determination became commercially available; these have significant advantages over the original steroid-binding techniques. Receptor analysis is now carried out routinely to indicate whether a particular patient with advanced disease should be treated with endocrine therapy or with chemotherapy.

The finding (Walt et al. 1976; Knight et al. 1977) that mastectomy patients with estrogen-receptor-positive primary tumors show a lower incidence of cancer recurrence, and a longer disease-free interval if they do recur, provided an additional application of receptor analysis for the clinical management of breast-cancer patients. Taken together with nodal status, one has an indication of the risk of recurrence which is useful in deciding how aggressive an adjuvant therapy should be employed. The receptor status of the primary tumor also gives an indication of the most probable location of the first metastases if the cancer recurs (Walt et al. 1976; Koenders et al. 1991). Receptor-positive primary tumors show a greater tendency to spread to bone and receptor-negative to lung and liver. The basis for this selectivity is unclear.

III. Uterine and Cervical Cancer

Because the uterus is a tissue whose growth and development are stimulated by estrogen, attempts have been made to employ antiestrogen therapy for endometrial cancer in a manner analogous to that used for mammary cancer. However, the response rate to tamoxifen is reported to be low and variable (Moore et al. 1991). The most widely used hormonal therapy for this malignancy is treatment with progestagens. In the case of cervical cancer, endocrine therapy has found little application. Because this malignancy tends to be especially sensitive to radiation and does not metastasize aggressively, surgery and/or radiotherapy are the most commonly used therapeutic procedures.

D. Summary

Estrogenic hormones play a role in the development of cancer in the human and in experimental animals. For the most part, this occurs in tissues, such as uterus and breast, in which growth and function depend on estrogenic stimulation. The retention of hormone dependency in some, but not all, of these cancers permits an endocrinological approach to their treatment.

References

Assikis VJ, Jordan VC (1995) A realistic assessment of the association between tamoxifen and endometrial cancer. Endocrine Related Cancer 2:235–241

Baum JK, Holtz F, Bookstein JJ, Klein EW (1973) Possible association between benign hepatomas and oral contraceptives. Lancet 2:926–929

Beatson GT (1896) On the treatment of inoperable cases of carcinoma of the mamma: suggestions for a new method of treatment with illustrative cases. Lancet 2:104–107

Block GE, Ellis RS, DeSombre E, Jensen E (1978) Correlation of estrophilin content of primary mammary cancer to eventual endocrine treatment. Ann Surg 188:372–375

Brooks JJ (1982) Hepatoma associated with diethylstilbestrol therapy for prostatic carcinoma. J Urol 128:1044–1045

Castagnetta L, Nenci I, Bradlow HL (1996) Basis for cancer management, vol 784. Ann NY Acad Sci

Collins J, Donner A, Allen LH, Adams O (1980) Estrogen use and survival in endometrial cancer. Lancet 2:961–964

Cooper AP (1836) The principles and practice of surgery. Cox, London, pp 333–335

Fisher B, Costantino JP, Wickerham DL, Redmond CK, Kavanah M, Kronin WM, Vogel V, Robidoux A, Dimitrov N, Atkins J, Daly M, Wieand S, Tan-Chiu E, Ford L, Wolmark N (1998) Tamoxifen for the prevention of breast cancer: report of the National Surgical Adjuvant Breast and Bowel Project P-1 study. J Natl Cancer Inst 90:1371–1388

Folca PJ, Glascock RF, Irvine WT (1961) Studies with tritium-labelled hexoestrol in advanced breast cancer. Comparison of tissue accumulation with response to bilateral adrenalectomy and oophorectomy. Lancet 2:796–798

Greene GL, Nolan C, Engler JP, Jensen EV (1980) Monoclonal antibodies to human estrogen receptor. Proc Natl Acad Sci USA 77:5115–5119

Harper MJK, Walpole AL (1967) A new derivative of triphenylethylene: effect on implantation and mode of action in rats. J Reprod Fertil 13:101–119

Harvey HA, Manni A (1997) Clinical use of aromatase inhibitors in breast carcinoma. In: Holland JF, Frei E III, Bast R Jr, Kufe D, Morton D, Weichselbaum R (eds) Cancer medicine, 4th edn. Williams and Wilkins, Baltimore, pp 1113–1124

Henderson BE, Ross RK, Pike MC, Casagrande JT (1982) Endogenous hormones as a major factor in human cancer. Cancer Res 42:3232–3239

Henderson BE, Ross RK, Bernstein L (1988) Estrogens as a cause of human cancer. Cancer Res 48:246–253

Henderson BE, Ross RK, Pike MC (1993) Hormonal chemoprevention of cancer in women. Science 259:633–638

Henderson BE, Bernstein L, Ross RK (1997) Hormones and the etiology of cancer. In: Holland JF, Frei E III, Bast R Jr, Kufe D, Morton D, Weichselbaum R (eds) Cancer medicine, 4th edn. Williams and Wilkins, Baltimore, pp 277–292

Herbst AL, Ulfelder H, Pozkaner DC (1971) Adenocarcinoma of the vagina: association of maternal stilbestrol therapy with tumor appearance in young women. N Engl J Med 284:878–881

Herbst AL, Anderson S, Hubby M, Haenszel WM, Kaufman RH, Noller KL (1986) Risk factors for the development of diethylstilbestrol-associated clear cell adenocarcinoma: a case-control study. Am J Obstet Gynecol 154:814–822

Holland JF, Frei E III, Bast R, Jr, Kufe D, Morton D, Weichselbaum R (1997) Cancer medicine, 4th edn. Williams and Wilkins, Baltimore

Huggins C, Bergenstal DM (1952) Inhibition of human mammary and prostatic cancers by adrenalectomy. Cancer Res 12:134–141

Jensen EV, DeSombre ER (1997) Steroid hormone binding and hormone receptors. In: Holland JF, Frei E III, Bast R Jr, Kufe D, Morton D, Weichselbaum R (eds) Cancer medicine, 4th edn. Williams and Wilkins, Baltimore, pp 1049–1059

Jensen EV, Block GE, Smith S, Kyser K, DeSombre ER (1971) Estrogen receptors and breast cancer response to adrenalectomy. Natl Cancer Inst Monograph 34:55–70

Jordan VC (1976) Effect of tamoxifen (ICI 46474) on initiation and growth of DMBA-induced rat mammary carcinomata. Eur J Cancer 12:419–424

Jordan VC (1990) Long term adjuvant tamoxifen therapy for breast cancer. Breast Cancer Res Treat 15:125–136

Jordan VC, Murphy CS (1990) Endocrine pharmacology of antiestrogens as antitumor agents. Endocr Rev 11:578–610

Jordan VC, Catherino WH, Wolf DM (1995) A mutant receptor as a mechanism of drug resistance to tamoxifen response. Ann NY Acad Sci 761:138–147

Jordan VC, Glusman JE, Eckert S, Lippman WE, Powles T, Costa A, Morrow M, Norton L (1998) Incident primary breast cancers are reduced by raloxifene: integrated data from multicenter, double blind, randomized trials in 12,000 postmenopausal women (abstract). ASCO Proceedings 17:122, No.466

Keifer W, Scott J (1977) Liver neoplasms and oral contraceptives. Am J Obstet Gynecol 128:448–454

Knight WA III, Livingston RB, Gregory EJ, McGuire WL (1977) Estrogen receptor as an independent prognostic factor for early recurrence in breast cancer. Cancer Res 37:4669–4671

Koenders PG, Beex LVAM, Langans R, Kloppenborg PWC, Smals AGH, Benraad TJ (1991) Steroid hormone receptor activity of primary human breast cancer and pattern of the first metastasis. Breast Cancer Res Treat 18:27–32

Lacassagne A (1932) Apparition de cancers de la mamelle chez la souris mâle, soumise à des injections de folliculine. Compt Rend Acad Sci 195:630–632

Li JJ, Li SA (1990) Estrogen carcinogenesis in hamster tissues: a critical review. Endocr Rev 11:524–531

Li JJ, Nandi S, Li SA (1992) Hormonal carcinogenesis. Springer, Berlin Heidelberg New York

Li JJ, Li SA, Gustafsson J-Å, Nandi S, Sekely LI (1996) Hormonal carcinogenesis II. Springer, Berlin Heidelberg New York

Lipton A, Santner SJ, Santen RJ, Harvey HA, Feil PD, White-Hershey D, Bartholomew MJ, Antle CE (1987) Aromatase activity in primary and metastatic human breast cancer. Cancer 59:779–782

Luft R, Olivecrona H (1953) Experiences with hypophysectomy in man. J Neurosurg 10:301–316

MacMahon B, Cole P, Lin TM, Lowe CR, Mirra AP, Ravihar B, Salber EJ, Valaoras VG, Yuasa S (1970) Age at first birth and cancer of the breast: a summary of an international study. WHO Bull 43:209–221

MacMahon B, Trichopoulos D, Brown J, Andersen AP, Aoki K, Cole P, de Waard F, Kauraniemi T, Morgan RW, Purde M, Rahivar B, Stormby N, Westlund K, Woo N-C (1982) Age at menarche, probability of ovulation and breast cancer risk. Int J Cancer 29:13–16

Moore TD, Phillips PH, Nerenstone SR, Cheson BD (1991) Systemic treatment of advanced and recurrent endometrial carcinoma – current and future directions. J Clin Oncol 9:1071–1088

Nissen ED, Kent DR (1975) Liver tumors and oral contraceptives. Obstet Gynecol 46:460–467

Neuberger J, Forman D, Doll R, Willians R (1986) Oral contraceptives and hepatocellular carcinoma. BMJ 292:1355–1357

O'Neill JS, Miller WR (1987) Aromatase activity in breast adipose tissue from women with benign and malignant breast diseases. Br J Cancer 56:601–604

Paganini-Hill A, Ross RK, Henderson BE (1989) Endometrial cancer and patterns of use of estrogen replacement therapy: a cohort study. Br J Cancer 59:445–447

Persson I (1996) Breast cancer incidence in women exposed to estrogen and estrogen-progestin replacement therapy. In: Li JJ, Li SA, Gustafsson J-Å, Nandi S, Sekely

LI (eds) Hormonal carcinogenesis II. Springer, Berlin Heidelberg New York, pp 91–98
Persson I, Adami HO, Bergkvist K, Lundgren A, Petterson B, Hoover R, Schairer C (1989) Risk of endometrial cancer after treatment with estrogens alone or in conjunction with progesterone: results of a prospective study. BMJ 298:147–151
Santen RJ, Lipton A, Kendall J (1974) Successful medical adrenalectomy with aminoglutethimide. Role of altered drug metabolism. JAMA 230:1661–1665
Trichopoulos D, MacMahon B, Cole P (1972) The menopause and breast cancer. J Natl Cancer Inst 48:605–613
Ursin G, Peters RK, Henderson BE, d'Ablaing G III, Monroe KR, Pike MC (1994) Oral contraceptive use and adenocarcinoma of the cervix. Lancet 344:1390–1399
Walt AJ, Singhakowinta A. Brooks SC, Cortez A (1976) The surgical implications of estrophile protein estimations in carcinoma of the breast. Surgery 80:506–512
Ziel HK, Finkle WD (1975) Increased risk of endometrial carcinoma among users of conjugated estrogens. N Engl J Med 293:1167–1170

CHAPTER 31
Hormonal Resistance in Breast Cancer

S.R.D. JOHNSTON, E. ANDERSON, M. DOWSETT, and A. HOWELL

A. Introduction

Most current endocrine therapies for breast cancer produce inhibitory effects on tumour growth either by depriving the cells of estrogen, i.e. ovarian ablation or aromatase inhibition, sufficient to reduce estrogen receptor (ER) activity or by blocking receptor action, i.e. tamoxifen, ICI 182780. Resistance to ovarian ablation was first described over 100 years ago (BEATSON, 1896). Tamoxifen is the most commonly prescribed drug for breast cancer. For over two decades its role has expanded from primary treatment for advanced metastatic disease (MOURIDSEN et al. 1979) to established adjuvant therapy following surgery for early breast cancer which prolongs both disease-free and overall survival (Early Breast Cancer Trialists Collaborative Group 1992). Not all patients with breast cancer respond to tamoxifen, and those who do will probably acquire a resistance to the drug. Other endocrine therapies are now available, although until now these have been reserved for second-line therapy after tamoxifen fails. However, new aromatase inhibitors are now being studied both as first-line therapy in advanced breast cancer and in the adjuvant setting in early breast cancer. Therefore, it will become increasingly important to understand the basis for resistance to these endocrine agents as well.

Although resistance was demonstrated early in the development of endocrine therapies, any understanding of the mechanisms for resistance had to await the discovery of the ER in breast cancers (JENSEN et al. 1966) and the unravelling of the complex molecular biology of the receptor protein (GREEN et al. 1986; GREENE et al. 1986). The latter has also been linked closely to parallel developments in experimental biology, with the isolation of ER-positive human tumour cell lines in culture and their transplantation into nude mice. In recent years, this has allowed the biological basis for acquired tamoxifen resistance to be studied in depth (JOHNSTON 1997). Approximately 30–40% of primary breast carcinomas have very low or absent levels of ERs and, invariably, this is associated with primary (complete) resistance. The reason for the absence of ER expression is unknown, but appears to occur very early in tumour ontogeny. For example, large cell, in-situ (comedo) carcinomas are mostly ER negative and, when transplanted into athymic nude mice, fail to respond to appropriated serum concentrations of estrogen and antiestrogens administered by subcutaneous pellets (HOLLAND et al. 1997). After an initial response to endocrine therapy, some ER-positive tumours respond to second-line endocrine therapy at progression and, thus, have secondary (partial) resistance (Fig. 1). Equally, some tumours, although they express the ER protein, are primarily resistant to endocrine therapy from the outset. It is unlikely that a single mechanism can explain hormonal resistance in all breast cancer patients. Fundamental differences in tumour biology, therefore, probably do exist between those with partial resistance and those with complete

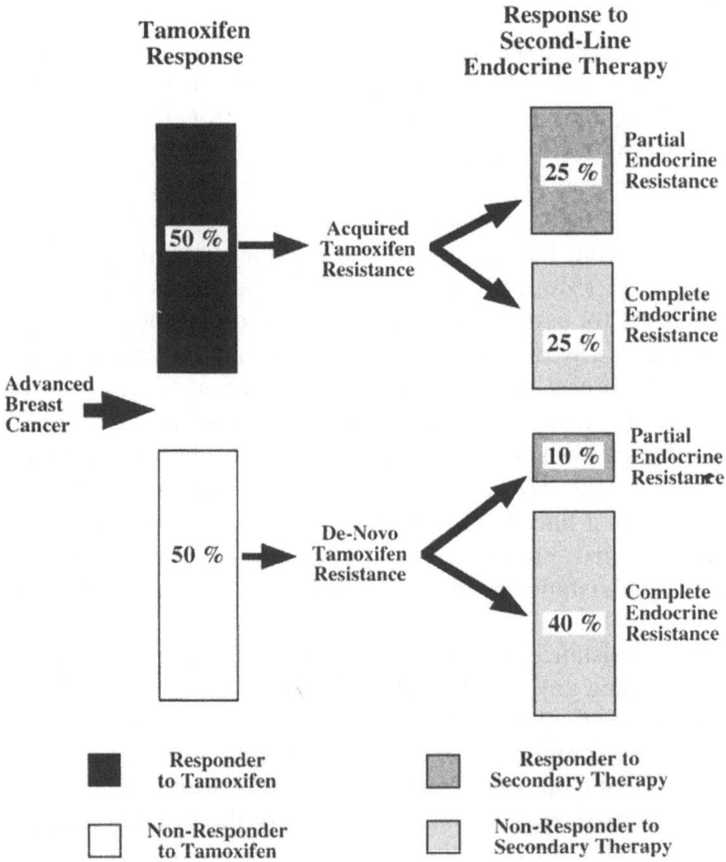

Fig. 1. Response to second-line endocrine therapy after tamoxifen failure. Approximately 50% of patients with advanced breast cancer will respond to tamoxifen. When these patients relapse with aquired tamoxifen resistance, half (25% of original) will respond to further endocrine therapy and, thus, by definition, have only partial endocrine resistance. In contrast, complete endocrine resistance may exist de novo in 40% patients who do not respond to tamoxifen or second-line therapy, and in a further 25% who develop endocrine resistance after an initial response to tamoxifen. A small subgroup (10%) may respond to second-line therapy despite never responding to tamoxifen

resistance. While a functional ER remains the key to hormone sensitivity, with complete endocrine resistance in many cases being due to a lack of ER and dependence on other mitogenic pathways, the basis for either complete or partial resistance in apparent ER-positive tumours has been the focus of recent studies of the ER pathway.

B. The Estrogen Receptor

The estrogen receptor (ER) is a ligand-dependent gene-transcription factor. The ligand (estradiol) binds to the hormone-binding domain of the receptor molecule situated in the nucleus of the target cell. This initiates dimerisation with another ER molecule and the dimer-hormone complex then binds to specific short DNA sequences called estrogen response elements (EREs) in the promoter regions of target genes (KUMAR and CHAMBON 1988). A series of steroid receptor-specific co-activators and co-repressors (the type and combination that is both target tissue and gene specific) bind to the ER and to each other to stimulate or repress transcription (HORWITZ et al. 1996; HEERY et al. 1997). Thus, whether an ER-inducible gene is transcribed depends on several variables, including the concentration and type of ligands, the ER structure including any post-translational modifications, the relative concentrations of co-activators and co-repressors, the structure and number of EREs in the promoter region of a given gene and, finally, the presence of ERE-independent pathways through which the gene may also be induced or repressed. Inappropriate activation of any of these molecular events that are involved in the estrogen/ER-regulated pathway may facilitate hormonal escape.

One consequence of the ER's binding to DNA is a mitogenic signal to the cell, possibly involving the release of local peptide growth factors, such as transforming growth factor alpha (TGFα) (DICKSON and LIPPMAN 1988; BATES et al. 1988). For example, structural alteration in the ER may cause an uncoupling of the protein's ligand activation from its DNA-binding function and growth-regulatory activity and result in a non-functional, transcriptionally inactive ER. Alternatively, unconstrained ERs may exist that are constitutively active or downstream processes that are initially under hormonal control may start to function independently. Understanding how this complex ER-dependent pathway may change in the context of hormonal resistance is only partially understood and is reviewed below.

I. The ER's Ligands

In ER-positive human mammary tumour cell lines, estradiol shows a bell-shaped dose-response curve, stimulating or inhibiting growth, depending on the dose administered. Tamoxifen has a similarly shaped dose-response curve (REDDEL and SUTHERLAND 1984), whereas the pure antiestrogen ICI 182,780 is inhibitory only (DEFRIEND et al. 1994). In vivo, the normal breast and breast tumours appear to respond to the circulating level of estrogens, although the levels of estradiol can also be regulated within normal and malignant breast cells. Tumour cells tend to metabolise estrogenic precursors to estradiol, whereas normal luminal epithelial cells tend to metabolise estradiol to inactive sulphates (ANDERSON and HOWELL 1995). Therefore, cells may be capable of changing the amount of ligand delivered to ER, providing a possible mechanism by which tumours could obtain a growth advantage over normal epithelial cells.

Tamoxifen undergoes conversion in the liver to a series of biologically active and quantitatively significant metabolites (RUENITZ et al. 1984; JACOLOT et al. 1991). Some of these, including *cis*-hydroxytamoxifen and metabolite E, are estrogenic or less antiestrogenic than tamoxifen (LYMAN and JORDAN 1985; LIEBERMAN et al. 1983). Although formation of estrogenic metabolites is a potential mechanism of tamoxifen resistance (OSBORNE et al. 1992; WIEBE et al. 1992), this was shown to be an unlikely route when WOLF et al. (1993) demonstrated resistance to fixed-ring tamoxifen-derivatives which cannot be metabolised to estrogenic compounds. Metabolic tolerance has been used to describe an alternative mechanism for resistance whereby reduced intracellular tamoxifen levels develop (OSBORNE et al. 1991; OSBORNE et al. 1994; WIEBE et al. 1995). Plasma levels of tamoxifen remain constant during prolonged therapy in vivo (LANGHAN FAHEY et al. 1990), but there is both experimental and clinical evidence that reduced intra-tumoral accumulation may occur following continuous tamoxifen exposure. A tenfold reduction in intra-tumoral tamoxifen was observed in MCF-7 xenografts which had become tamoxifen-resistant compared with responding tumours, in the absence of any change in the serum concentrations (OSBORNE et al. 1991). Likewise, lower intra-tumoral levels of tamoxifen were found in a proportion of resistant tumours from breast cancer patients (OSBORNE et al. 1992; JOHNSTON et al. 1993).

No mechanism by which these reduced intra-tumoral concentrations occur has been described. Alternative antiestrogen binding sites (AEBS) within the cytosol have been described that bind tamoxifen (but not estrogen) in a saturable fashion (SUTHERLAND et al. 1980). One study has suggested that enhanced AEBS activity could be one biochemical resistance mechanism to prevent tamoxifen's interaction with ER (PAVLIK et al. 1992). These remain potential mechanisms of endocrine resistance since they may reduce intracellular tamoxifen concentrations to a level where the drug is unable to antagonise estradiol competitively or, alternatively, where it acts as an estrogen agonist as a result of its bell-shaped dose-response curve.

II. ER Structure: Mutants, Variants and Post-Translational Changes

The structural organisation of the ER is outlined in Fig. 2. In theory, ER function could be altered either by mutation at the DNA level, disruption of ER gene transcription or post-translational modification of the protein. Unlike many oncogenes, there is no evidence of ER gene amplification or rearrange-

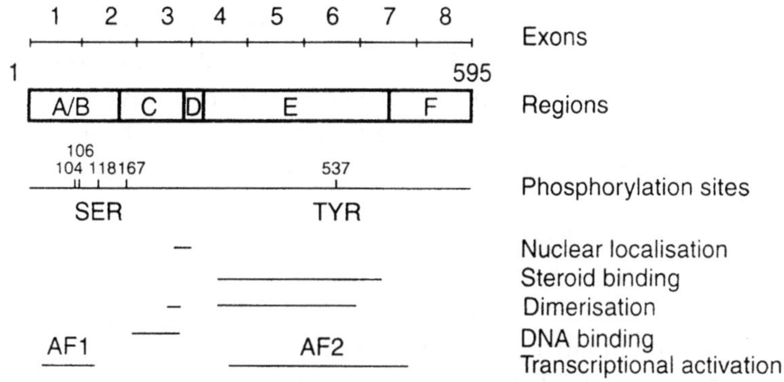

Fig. 2. Organisation of the human estrogen receptor gene. The gene comprises eight exons which produce six functional domains (A–F) on the ER protein. Region C is the DNA binding domain and region E the hormone binding domain. After binding to the receptor and induction of dimerisation, oestradiol activates both activating factors (AF1 and AF2) which then stimulate transcription of target genes. Tamoxifen inhibits the activity of AF2 but not AF1, whereas ICI 182780 inhibits both AF1 and AF2. The receptor may also be activated independently of ligand binding by, for example, phosphorylation on serine and tyrosine induced by growth-factor pathways

ment in breast cancer (Koh et al. 1989). Point mutations in specific regions of ER, generated by site-directed mutagenesis in vitro, can significantly alter the pharmacological response of ER to tamoxifen from that of an antagonist to a full agonist (Mahfoudi et al. 1995). A point mutation in the ER's ligand-binding domain has been found in a tamoxifen-stimulated MCF-7 xenograft, accounting for 80–90% of the total ER and representing the first identification of a mutant ER protein from a transplantable tamoxifen-stimulated human tumour line (Wolf and Jordan 1994). While these experimental data imply a possible role for ER mutants in rendering a tumour "resistant" to tamoxifen, there is little evidence of these critical mutations in human breast cancer in vivo (Kamick et al. 1994; Dowsett et al. 1997; Anderson et al. 1997). However, one recent report has described a missense mutation in human breast cancer at codon 537, which in vitro was associated with estradiol-independent transcriptional activity (Zhang et al. 1997).

During transcription of the ER gene and synthesis of its mRNA (messenger RNA), alternative splicing with deletions of whole exons can occur which generates shorter mRNA transcripts. A variety of different ER mRNA splice variants have been identified in human breast tumours (Fuqua et al. 1991; Fuqua et al. 1992; Peffer et al. 1995; Zhang et al. 1993; Wang and Mikisicek 1991), which may have a range of different properties within the cell (McGuire et al. 1991). The ER variant may be completely inactive, as occurs with the exon-2 deletion variant (ERΔE2), at which a frameshift results in premature termination of translation, giving a transcriptionally inactive protein (Wang and Mikisicek 1991). Other ER variants may behave as a "dominant–negative" receptors capable of inhibiting the transcriptional activity of the wild-type receptor, as occurs following in-frame deletion of exon 3 (Mikisicek et al. 1993). Alternatively, variant mRNA transcripts may code for a "dominant–positive" receptor that becomes constitutively active, independent of ligand, as exemplified by the ERΔE5 variant which has been shown in vitro to be transcriptionally active, independent of estradiol (Fuqua et al. 1991).

The significance of ER mRNA splice variants in relation to tamoxifen resistance has been the subject of several recent studies. One group found that transfection of the ERΔE5 variant into MCF-7 cells conferred a tamoxifen-resistant phenotype in vitro (Fuqua and Wolf 1995), but others reported that inducible expression of ERΔE5 in their MCF-7 cells failed to prevent the growth-inhibitory effects of tamoxifen or ICI 182,780 (Rea and Parker 1996). In human breast carcinomas, no difference was found in expression of ERΔE5 between tamoxifen-resistant and untreated tumours (Daffada et al. 1995). Taken together, these recent data imply that acquired tamoxifen resistance is only infrequently related to overexpression of this particular constitutively active ER variant. The ER protein may be phosphorylated at several sites, particularly in the AF1 and AF2 transcriptional activation domains (Fig. 3). Phosphorylation mediates transcriptional activation by estrogens and antiestrogens (LeGoff et al. 1994). Indeed, transcription via the ER can be activated in the absence of ligand by phosphorylation induced by the ras/raf mitogen-activated protein (MAP) kinase pathway, e.g. secondary to stimula-

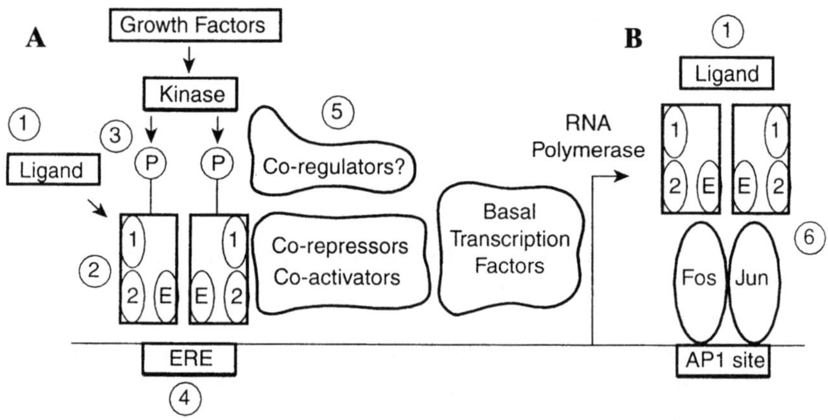

Fig. 3A,B. Multiple levels of control of gene activation by estrogens and antiestrogens. **A** ERE-mediated transactivation. **B** Non-ERE-mediated transactivation. *1*, Concentration and metabolism of ligand; *2*, receptor conformation change and activation of AF1 and AF2; *3*, post-translational modification of ER, e.g. phosphorylation by growth factor activity; *4*, binding of ER dimer to various EREs; *5*, translation of the activation state of ER to basal transcription factors by co-activators, co-repressors and, possibly, co-regulators; *6*, ER binding to other transcription factors that act through response elements other than ERE. In this example, we have shown the fos/jun family of transcription factors: another example is alteration of activity by ER of the NF-κB transcription factor for IL-6 tion with epidermal growth factor (EGF), insulin-like growth factor (IGF-I) (Kato et al. 1995; Pietras et al. 1995), protein kinases A or C and the src family of kinases (Arnold et al. 1995). Activation of ERs via these growth-factor pathways could be a major cause of complete endocrine resistance, since many of them now appear to be active via AF1. Tamoxifen also activates AF1, so it may synergise with growth factors causing phosphorylation at this site.

III. Transcriptional Activity: Co-Activators and Co-Repressors

Specific proteins associate with the ligand/ER/ERE complex to regulate transcriptional activation of estrogen-responsive genes (Fig. 3). These co-activators appear to bind via an LXXLL motif to a conserved helix (helix 12) of the ER, required for AF-2 function (Heery et al. 1997). Several proteins have been identified including receptor interacting protein (RIP140) (Cavailles et al. 1994) and steroid receptor co-activator (SRC-1)/ER-associated proteins (ERAP-160) (Onate et al. 1995). Likewise a series of co-repressors has been identified, including Ssn6 which acts as a repressor of ER-mediated gene transcription in yeast (McDonell et al. 1992). Co-activator function is induced by pure estrogen agonists and inhibited by the pure antiestrogen ICI 182,780. Deletion of Ssn6 strongly enhances the transcriptional effects of ER on AF1, and allows ICI 164,384 (a less potent pure antiestrogen related to ICI 182780) and nafoxidine to act as agonists. Co-activators, co-repressors and co-integrators (postulated to integrate the effects of peptides and steroids at the receptor) may well be the mechanism by which the signals from an activated ER are read in a gene- and cell-specific manner. For example, they may explain how tamoxifen can inhibit transcription of growth-regulatory genes, yet stimulate progesterone-receptor synthesis in a single cell. In addition, they might explain the tissue-specific effects of tamoxifen, for example, how it might inhibit mammary cell growth, whilst being an agonist for growth in the uterus. While the inhibitory effects of antiestrogens on ER may be mediated by co-repressors within mammalian cells (Horwitz et al. 1996), it is possible that absence of co-repressors or increased expression of co-activators could be a mechanism for hormonal resistance. There are no published data regarding whether levels of co-activator/co-repressor protein change with acquired resistance.

IV. EREs and Promoter Elements

The ER's transcriptional activity within individual cells may also be determined by the promoter context, which in turn may dictate whether tamoxifen acts predominantly as an antagonist, or as an agonist (Berry et al. 1990). The archetypal ERE sequence is a 5-bp palindrome, separated by a 3-bp spacer region (5′-GGTCA nnn TGACC-3′). Modifications of the ERE by sequence variation, their number or their orientation can affect the efficiency with which estrogen- or antiestrogen-responsive genes are transcribed

(PONGLIKITMONGOL et al. 1990). For example, when the *C* in the 5′ part of the palindromic sequence was replaced, 4-hydroxy-tamoxifen became fully estrogenic (DANA et al. 1994). Thus, changes in the promoter sequences of genes involved in growth could, conceivably, be responsible for antiestrogen resistance in breast cancer cells, although this remains to be proven. Likewise, a model has been proposed to explain the tissue-specific, partial-agonist effects of tamoxifen, based on the different activities of the ER's two transcriptional-activation domains (TAF1 and TAF2), following binding of either estrogen or tamoxifen (TZUKERMAN et al. 1994).

V. Non-estrogen Response Element-Dependent Pathways

Finally, evidence is emerging that ERs may activate gene transcription independent of classical ERE-regulated pathways through the AP-1 pathway. The ER has been shown to interact with Jun/Fos proteins at the AP-1 promoter site of certain genes, including the collagenase gene (WEBB et al. 1995). Furthermore, tamoxifen was shown to act as an agonist through AP-1 in a tissue-specific manner, thus paralleling tamoxifen agonism in vivo. MCF-7 cells that acquired resistance to tamoxifen in vitro have increased AP-1 DNA binding (DUMONT et al. 1996), and a recent report has confirmed elevated AP-1 DNA binding in human tamoxifen-resistant tumours (LU et al. 1997). Whether this relates to increased levels of Jun or Fos proteins, or enhanced kinase activity of the complex remains to be determined. However, this alternative pathway for ER-mediated gene transcription may be relevant to the stimulatory response observed with tamoxifen in acquired resistance, specifically, if such non-ERE-dependent pathways are predominant regulators of growth. Equally, ERs occupied by the non-steroidal antiestrogen raloxifene have been shown to interact with adapter proteins and bind to non-ERE DNA sequences (the so-called raloxifene-response element). This may account for the differential effects of the ligand on tissues such as the bone and endometrium (YANG et al. 1996). This mechanism awaits confirmation.

C. Secondary Resistance

I. Tumour Adaptation to Estrogen Levels

The clinical observation that many patients who develop acquired resistance to tamoxifen remain sensitive to further endocrine therapies, including pure antiestrogens (HOWELL et al. 1995) and aromatase inhibitors (SMITH et al. 1981), suggests that ERs may be expressed and remain functional in many tumours with partial endocrine resistance. In a series of 72 breast cancer patients treated with tamoxifen, we found that 61% of previous responders remained ER positive at relapse, whereas all non-responders were ER negative (JOHNSTON et al. 1995). One hypothesis that could account for repeated responses is adaptation of the tumour, possibly by clonal selection, to the prevailing estrogen/tamoxifen concentration.

Endocrine-dependent tumours in both pre- and postmenopausal women respond to either estrogen blockade or deprivation. However, in a recent experimental report, MCF-7 cells deprived of estradiol for 1–6 months subsequently showed enhanced sensitivity to estradiol, becoming maximally stimulated at 10^{-14} M in contrast to 10^{-10} M for wild-type cells (MASAMURA et al. 1995). Such a shift towards the left of the bell-shaped dose-response curve could explain some secondary endocrine responses in breast cancer. For example, in premenopausal women with advanced breast cancer who respond to oophorectomy (with estradiol levels falling from approximately 400 pmol/l to 30 pmol/l), a secondary response can be achieved at relapse by using further estrogen deprivation, achieved by aromatase inhibitors (where estradiol levels fall to <10 pmol/l). This sequential approach to endocrine therapy, in particular step-wise estrogen deprivation, has proved effective with 4-hydroxyandostenedione following secondary resistance to both aminoglutethimide in postmenopausal women (MURRAY et al. 1995) and GnRH agonists in premenopausal women (DOWSETT et al. 1992). The molecular mechanism for any enhanced sensitivity is not known and it is difficult to conceive how ERs with a Kd of around 10^{-10} M could be responsible without some additional concentration of amplification mechanism.

II. Tamoxifen Stimulation and Withdrawal Responses

Equally, within a heterogeneous endocrine-sensitive tumour, clones of cells may exist with differing sensitivity to tamoxifen (HOWELL et al. 1990), such that cells that are simulated by tamoxifen to grow may be selected during prolonged therapy. Clonal selection of cells with differential expression of progesterone receptor (PgR) has been demonstrated by flow cytometry to occur in breast cancer cell lines treated with tamoxifen in vitro (GRAHAM et al. 1992). MCF-7 cells growing as tumours on the flanks of athymic-oophorectomised nude mice that are initially inhibited by tamoxifen treatment, eventually regrow and are stimulated by tamoxifen in a dose-dependent manner. This regrowth can be inhibited by the pure antiestrogen ICI 164, 394, suggesting that tamoxifen is acting as an estrogen agonist in this situation (GOTTARDIS and JORDAN 1988). Remodelling of a heterogeneous tumour with selection of cells with an altered sensitivity to tamoxifen, such that they are stimulated by the drug, is a theoretical possibility that may explain not only tamoxifen-stimulated growth, but also withdrawal responses that have been observed following cessation of tamoxifen (HOWELL et al. 1992) (Fig. 4).

D. Clinical Implications and New Endocrine Agents

Non-steroidal antiestrogens, some of which have a purer or more advantageously selective antagonist profile than tamoxifen, are in clinical

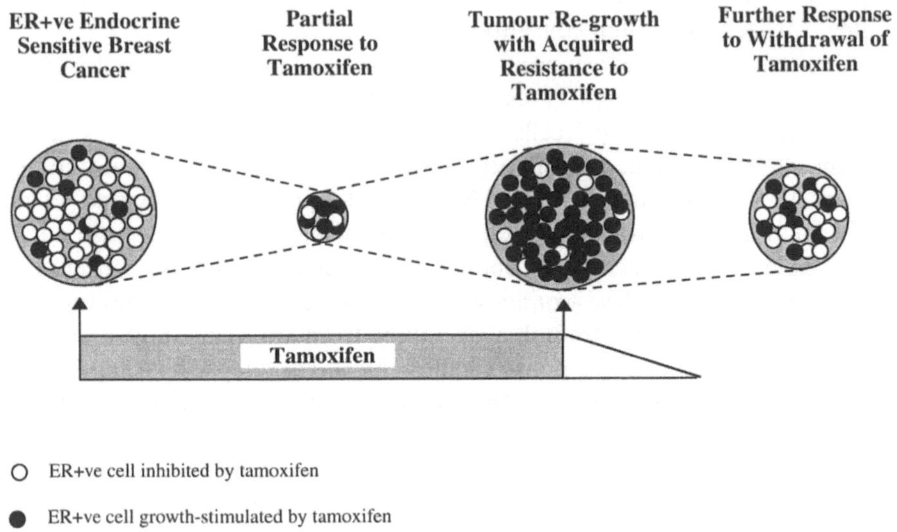

Fig. 4. Model of acquired tamoxifen resistance due to clonal selection of ER + ve cells, which are growth stimulated by tamoxifen. While the original tumour may be heterogeneous in terms of cells with different inherent sensitivity to tamoxifen, initial therapy with tamoxifen will eliminate hormone-sensitive cells and favour the selective outgrowth of cells that are growth stimulated by tamoxifen. This will allow the emergence of an ER + ve tumour that perceives tamoxifen as a growth stimulus. Further endocrine responses may be achieved either by withdrawal of the stimulatory tamoxifen or by use of a second-line endocrine agent with no agonist activity. Thus, the ER + ve growth-stimulated cells would remain sensitive to effective blockade or deprivation of hormonal signals

development and include raloxifene (Jones et al. 1984), idoxifene (COOMBES et al. 1995), droloxifene (LOSER et al. 1985) and toremifene (BLACK et al. 1983). The structure, relative binding affinity for the ER and inherent agonist activity of these compounds in comparison with tamoxifen is shown in Fig. 5. In addition, a pure steroidal antiestrogen ICI 182,780 has entered clinical development, having shown complete absence of agonist activity in preclinical studies (WAKELING et al. 1987).

Evidence suggests that the ligand/receptor conformation induced by some of these compounds imparts a different transcriptional response within the cell than the tamoxifen/ER complex. In particular, the agonist activity of ER, normally mediated by tamoxifen through TAF-1-dependent gene transcription, was prevented by both raloxifene and the pure steroidal antiestrogen ICI 164,384 (MCDONELL et al. 1995). Furthermore, experimental evidence exists that, while breast carcinoma cells established in clonogenic assay from patients who relapsed on tamoxifen can be stimulated in vitro by either estrogen or 4-hydroxytamoxifen, they remain completely inhibited by ICI 182,780 (DEFRIEND et al. 1994). The MCF-7 xenograft data imply that more effective estrogen antagonism with either ICI 182,780 or idoxifene may delay the

	Relative Binding Affinity (relative to E2, RBA=100)	% Agonism (immature rat uterus assay without E2)

Tamoxifen

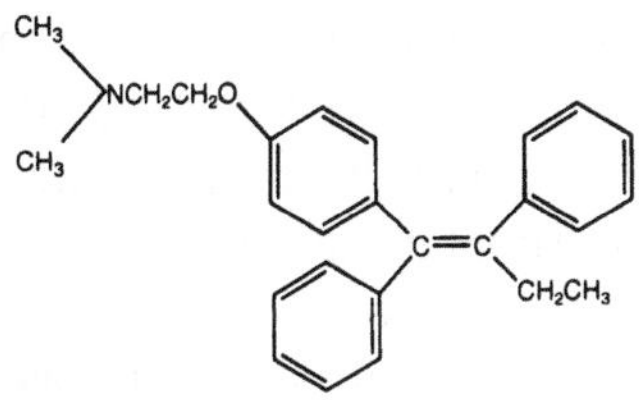

5 50

Toremifene

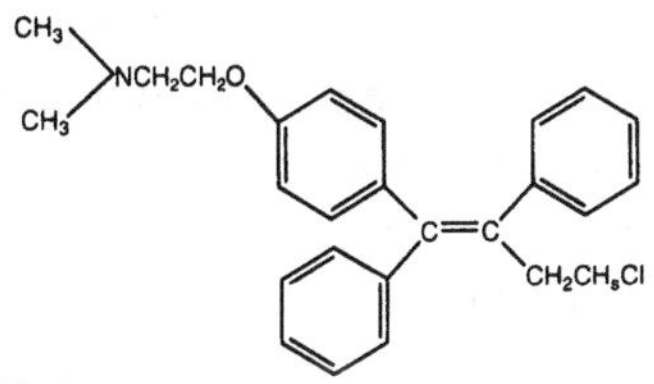

5 43

Droloxifene

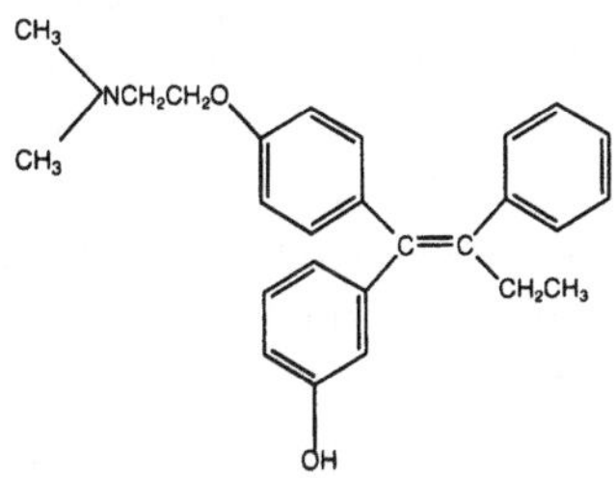

7.5 35

Raloxifene

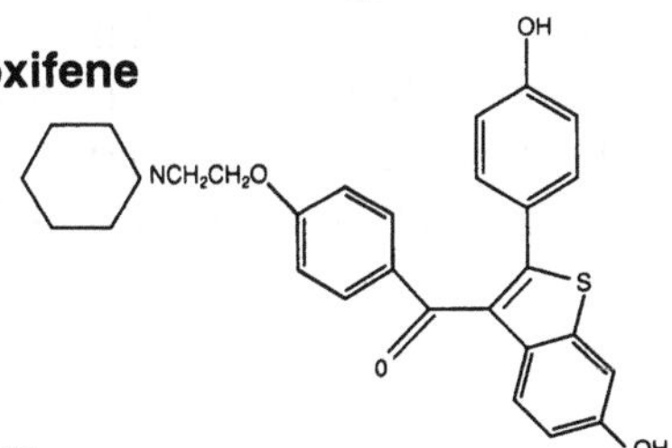

>100 5

Idoxifene

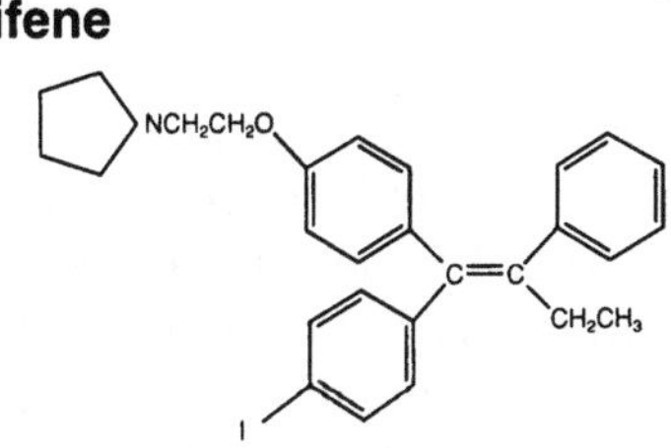

12.5 15

Fig. 5. Structures, relative binding affinities for ER, and percentage agonism (in the absence of oestradiol) of the five non-steroidal antiestrogens available clinically or in clinical development emergence of acquired resistance (OSBORNE et al. 1995; JOHNSTON et al. 1997). In clinical practice, this could translate into a significant benefit for patients with advanced breast cancer, for whom these new agents may prolong the time to disease progression.

The effectiveness of adjuvant tamoxifen in both node-positive and node-negative breast cancer was established in the world overview of 30,000 women from 133 randomised controlled clinical trials (Early Breast Cancer Trialists Collaborative Group 1992), although the optimal duration of adjuvant tamoxifen therapy in breast cancer has been the cause of much recent debate. A significant improvement in disease-free and overall survival was found in patients treated with tamoxifen for 5 years compared with 2 years (Swedish Breast Cancer Cooperative Group 1996), whereas tamoxifen for more than 5 years was of no further benefit and could be detrimental (FISHER et al. 1996). The acquisition of tamoxifen-stimulated breast cancer growth could account for these adjuvant data. These data are important when considering the potential role of novel endocrine therapies in delaying relapse in the adjuvant setting. The relative merits of pure steroidal antiestrogens or non-steroidal tamoxifen analogues, such as raloxifene or idoxifene, as opposed to aromatase inhibitors in terms of a beneficial effect on bone mineral density, endometrium and lipid profile, may then become important in the choice of an effective long-term adjuvant endocrine agent. Based on our evolving understanding of hormonal resistance, especially partial resistance, a new endocrine strategy could emerge comprising either alternating hormonal therapies or on/off periods of treatment with conventional agents such as tamoxifen or aromatase inhibitors. This would take advantage of these drugs' different modes of action and, if sequenced correctly, might prevent clonal selection and adaptive mechanisms by the tumour from inducing endocrine failure. Clinical trials will be required to see if this approach can provide more effective endocrine control and prolonged progression-free survival in patients with metastatic breast cancer.

E. Conclusions

There are likely to be several different mechanisms of hormonal resistance in human breast cancer in vivo and more than one might be operative in an individual breast carcinoma. The ER remains crucial in determining a tumour's primary endocrine response and lack of expression of a functionally active ER appears to be the major factor responsible for de-novo resistance. In contrast, acquired resistance in many tumours is often associated with maintained ER expression. Several potential sites for alteration in the ER-mediated signal-transduction pathway have been examined in vivo, and recent experimental data have postulated further downstream molecular processes that may be involved. While these studies have identified potential novel molecular targets in endocrine-resistant tumours, for the immediate future, several new endocrine therapies may allow clinical strategies to be developed to circumvent or delay hormonal resistance in the clinic.

References

Anderson E, Howell A (1995) Estrogen sulphotransferases in malignant and normal human breast tissue. Endocrine Related Cancer 2:227–23

Anderson I, Wooster R, Laake K, Collins N, Warren W, Skrede M, Eeeles R, Tveit KM, Johnston SRD, Dowsett M, Olsen AO, Moller P, Stratton MR, Borresen-Dale A-L (1997) Screening for estrogen receptor gene mutaions in breast and ovarian cancer patients. Hum Mutat 9:531–536

Arnold SF, Obourn JD, Jaffe H, Notides AC (1995) Phosphorylation of the human estrogen receptor on tyrosine 537 in vivo and by src family tyrosine kineses in-vitro. Mol Endocrinol 9:24–33

Bates SE, Davidson NE, Valverius EM, Dickson RB, Freter CE, Tam JP et al. (1988) Expression of transforming growth factor alpha and mRNA in human breast cancer: its regulation by estrogen and its possible functional significance. Mol Endocrinol 2:543–555

Beatson GT (1896) On the treatment of inoperable cases of carcinoma of the mamma: suggestions for a new method of treatment, with illustrative cases. Lancet II: 104–107

Berry M, Metzger D, Chambon P (1990) Role of the two activating domains of the estrogen receptor in the cell-type and promoter-context dependent agonisticacyivity of the anti-estrogen 4-hydroxytamoxifen. EMBO J 9:2811–2818

Black LJ, Jones CD, Falcone JF (1983) Antagonism of estrogen action with a new benzothiopene derived antiestrogen. Life Sci 32:1031–1036

Cavailles V, Dauvois S, Danielian PS, Parker MG (1994) Interaction of proteins with transcriptionally active estrogen receptors. Proc Natl Acad Sci 91:10009–10013

Coombes RC, Haynes BP, Dowsett M, Quigley M, English J, Judson I et al. (1995) Idoxifene: report of a phase I study in patients with metastatic breast cancer. Cancer Res. 55:1070–1074

Daffada AAI, Johnston SRD, Smith IE, Detre S, King N, Dowsett M (1995) Exon-5 Deletion Variant Estrogen Receptor mRNA Expression in Tamoxifen Resistant Breast Cancer and its Association with PgR/pS2 Status. Cancer Res 55:288–293

Dana SL, Hoener PA, Wheeler DA, Lawrence CB, McDonnell DP (1994) Novel estrogen response elements identified by genetic selection in yeast are differentially responsive to estrogens and anti-estrogens in mammalian cells. Mol Endocrinol 8:1193–1207

DeFriend DJ, Anderson E, Bell J, Wilks DP, West CML, Howell A (1994) Effects of 4-hydroxytamoxifen and a novel pure antiestrogen (ICI 182780) on the clonogenic growth of human breast cancer cells in vitro. Brit J Cancer 70:204–211

Dickson RB, Lippman ME (1988) Control of human breast cancer by estrogen, growth factors and oncogenes. In: Lippmann ME, Dickson RB (eds) Breast Cancer: Cellular and Molecular Biology. Kluwer Academic Publishers, Boston, pp 119–165

Dowsett M, Stein RC, Coombes RC (1992). Aromatase inhibition alone or in combination with GnRH agonists for the treatment of premenopausal breast cancer patients. J Steroid Biochem 43:155–159

Dowsett M, Daffada AAI, Chan CMW, Johnston SRD (1997) Estrogen receptor mutants and variants in breast cancer. Eur J Cancer 33:1177–1183

Dumont JA, Bitonti AJ, Wallace CD, Baumann RJ, Cashman EA, Cross-Doersen DE (1996) Progression of MCF-7 breast cancer cells to antiestrogen resistant phenotype is accompanied bty elvated levels of AP-i DNA-binding activity. Cell Growth Diff 7:351–359

Early Breast Cancer Trialists Collaborative Group (1992) Systemic treatment of early breast cancer by hormonal, cytotoxic or immune therapy. Lancet 39:1–15, 71–85
Fisher B, Dignam J, Wieand S, Wolmark N, Wickerham DL et al. (1996) Five versus more than five years of Tamoxifen therapy for Breast Cancer Patients with negative lymph nodes and estrogen-receptor positive tumours. J Natl Cancer Instit 88:1529–1542
Fuqua SAW, Fitzgerald SD, Allred DC, Elledge RM, Nawaz Z, McDonnell DP et al. (1992) Inhibition of estrogen receptor action by a naturally occurring variant in human breast tumors. Cancer Res 52:483–486
Fuqua SAW, Fitzgerald SD, Chamness GC, Tandon AK, McDonnell DP, Nawaz Z et al. (1991) Variant human breast tumour estrogen receptor with constitutive transcriptional activity. Cancer Res 51:105–109
Fuqua SAW, Wolf DM (1995) Molecular aspects of estrogen receptor variants in breast cancer. Breast Cancer Res Treat 35:233–241
Gottardis MM, Jordan VC (1988) Development of tamoxifen stimulated growth of an MCF-7 tumour in athymic mice after long term antiestrogen administration. Cancer Research 48:5183–5187
Graham II ML, Smith JA, Jewett PB, Horwitz KB (1992) Heterogeneity of progesterone receptor content and remodeling by tamoxifen characterise subpopulations of cultured human breast cancer cells: analysis by quantitative dual parameter flow cytometry. Cancer Res 52:593–602
Green S, Walter P, Kumar V, Krust A, Bornert JM, Argos P, Chambon P (1986) Human estrogen receptor cDNA: sequence, expression and homology to v-erb-A. Nature 320:134–139
Greene GL, Gilna P, Waterfield M, Baker A, Hort Y, Shine J (1986) Sequence and expression of human estrogen receptor complementary DNA. Science 231:1150–1154
Heery DM, Kalkhoven E, Hoare S, Parker MG (1997) A signature motif in transcriptional coactivators mediates binding to nuclear receptors. Nature 387:733–736
Holland PA, Knox WF, Potten CS, Howell A, Anderson E, Baildam AD, Bundred NJ (1997) Comedo DCIS is hormone independent and may not benefit from antiestrogen therapy. J Natl Cancer Inst 89:1059–1065
Horwitz KB, Jackson TA, Bain DL, Richer JK, Takimoto GS, Tung L (1996) Nuclear receptor coactivators and corepressors. Mol Endocrinol 10:1167–1175
Howell A, Dodwell DJ, Laidlaw I, Anderson H, Anderson E (1990) Tamoxifen as an agonist for metastatic breast cancer. In: Goldhirsch A (ed) Endocrine Therapy of Breast Cancer. Springer-Verlag, New-York, pp 49–58
Howell A, Dodwell DJ, Anderson H, Radford J (1992) Response after withdrawal of tamoxifen and progestins in advanced breast cancer. Ann Oncol 3:611–617
Howell A, De Friend D, Robertson J, Blamey R, Walton P (1995) Response to the pure antiestrogen ICI 182,780 in tamoxifen resistant breast cancer. Lancet 345:29–30
Jacolot F, Simon I, Dreano Y, Beaune P, Riche C, Berthou F (1991) Identification of the cytochrome p450 IIIa family as enzymes involved in the N-demethylation of tamoxifen in human liver micrososmes. Biochem Pharmacol 41:1911–1919
Jensen EV, Jacobson HI, Flesher JW (1966) Estrogen receptors in target tissues. In: Nakao T, Pincus G, Tait J (eds) Steroid Dynamics. Academic Press, New York, pp 133–157
Johnston SRD, Haynes BP, Smith IE, Jarman M, Sacks NPM, Ebbs SR, Dowsett M (1993) Acquired tamoxifen resistance in human breast cancer and reduced intratumoural drug concentration. Lancet 342:1521–1522
Johnston SRD, Saccani-Jotti G, Smith IE, Salter J, Newby J, Coppen M, Ebbs SR, Dowsett M (1995). Changes in ER, PgR, pS2 Expression in Tamoxifen Resistant Human Breast Cancer. Cancer Research 55:3331–3338
Johnston SRD, Smith IE, Haynes BP, Jarman M, Dowsett M (1997) The effects of a novel antiestrogen Idoxifene on the growth of tamoxifen-sensitive and resistant MCF-7 xenografts in athymic mice. Br J Cancer 75:804–809

Johnston SRD (1997) Acquired tamoxifen resistance in breast cancer – mechanisms and clinical implications. Anti-Cancer Drugs (to be published December 1997)

Jones CD, Jevnikar MG, Pike AJ (1984) Antiestrogens. 2. Structure-activity studies in a series of 3-aroxyl-2-arylbenzo[b]thiophene derivatives leading to {6-hydroxy-2-(4-hydroxyphenyl)benzo[b]thien-3-yl][2-(piperidinyl)ethoxy-phenyl-methanone hydrochloride} (LY156758), a remarkably effective estrogen antagonist with only minimal intrinsic estrogenicicty. J Med Chem 27:1057–1066

Karnick PS, Kulkarni S, Liu X-P, Budd GT, Bukowski RM (1994) Estrogen receptor mutations in tamoxifen-resistant breast cancer. Cancer Res 54:349–353

Kato S, Endoh H, Masuhiro Y, Kitamoto T, Uchlyama S, Sasaki H, Masushige S, Gotoh Y, Nishida E, Kawashima H, Metzger D, Chambon P (1995) Activation of the estrogen receptor through phosphorylation by mitogen-activated protein kinase. Science 270:1491–1494

Koh EH, Wildrick DM, Hortobagyi GN, Blick M (1989) Analysis of the estrogen receptor gene structure in human breast cancer. Anticancer Res 9:1841–1846

Kumar V, Chambon P (1988) The estrogen receptor binds tightly to its responsive element as a ligand-induced homodimer. Cell 55:145–156

Langhan-Fahey SM, Tormey DC, Jordan VC (1990) Tamoxifen metabolites in patients on long-term adjuvant therapy for breast cancer. Eur J Cancer 26:883–888

Le Goff P, Montano MM, Schodin DJ, Katzenellenbogen BS (1994) Phosphorylation of the human estrogen receptor: identification of hormone regulated sites and examination of their influence on transcriptional activity. J Biol Chem 269:4458–4466

Lieberman ME, Jordan VC, Fritsch M, Santos MA, Gorski J (1983) Direct and reversible inhibition of estradiol stimulated prolactin synthesis by antiestrogens in-vitro. J Biol Chem 258:4734–4740

Loser R, Seibel K, Roos W, Eppenberger U (1985) In vivo and in vitro antiesotrogenic action of 3-hydroxytamoxifen, tamoxifen and 4-hydroxytamoxifen. Eur J Cancer Clin Oncol 21:985–990

Lu B, Johnston SRD, Scott GK, Kushner PJ, Dowsett M, Benz CC (1997) Increased AP-1 DNA binding activity in tamoxifen-resistant human breast tumours. Proc Am Assoc Cancer Res 38:437 (A2920)

Lyman SD, Jordan VC (1985) Metabolism of tamoxifen and its uterotrophic activity. Biochem Pharmacol 34:2787–2794

Mahfoudi A, Roulet E, Dauvois S, Parker MG, Wahli W (1995) Specific mutations in the estrogen receptor change the properties of antiestrogens to full agonists. Proc Natl Acad Sci USA 92:4206–4210

Masamura S, Santner SJ, Heitjan DF, Santen RL (1995) Estrogen deprivation causes estradiol hypersensitivity in human breast cancer cells. J Clin Endocrinol Metab 80:2918–2925

McDonnell DP, Vegeto E, O'Malley BW (1992) Identification of a negative regulatory function for steroid receptors. Proc Natl Acad Sciences USA 89:10563–10567

McDonnell, DP, Clemm DL, Hermann T, Goldman ME, Pike JW (1995) Analysis of estrogen receptor function in vitro reveals three distinct classes of antiestrogen. Mol Endocrinol 9:659–669

McGuire WL, Chamness GC, Fuqua SAW (1991) Estrogen receptor variants in clinical breast cancer. Mol Endocrinol 5:1571–1577

Miksicek RJ, Lei Y, Wang Y (1993) Exon skipping gives rise to alternatively spliced forms of the estrogen receptor in breast tumor cells. Breast Cancer Res Treat 26:163–174

Mouridsen HT, Ellemann K, Mattsson W, Palshof T, Daehnfeldt JL, Rose C (1979) Therapeutic effect of tamoxifen versus tamoxifen combined with medroxyprogesterone acetate in advanced breast cancer in postmenopausal women. Cancer Treat Rep 63:171–175

Murray R, Pitt P (1995) Aromatase inhibition with 4-OH-androstenedione after prior aromatase inhibition with aminoglutethimide in women with advanced breast cancer. Breast Cancer Res Treat 35:249–253

Onate SA, Tsai SY, Tsai MJ, O'Malley BW (1995) Sequence and characterisation of a coactivator for the steroid hormone receptor superfamily. Science 270:1354–1357
Osborne CK, Coronado E, Allred DC, Wiebe V, DeGregorio M (1991) Acquired tamoxifen resistance: correlation with reduced breast tumor levels of tamoxifen and isomerisation of trans-4-hydroxytamoxifen. J Natl Cancer Inst 83:1477–1482
Osborne CK, Wiebe VJ, McGuire WL, Ciocca DR, DeGregorio MW (1992) Tamoxifen and the isomers of 4-hydroxytamoxifen in tamoxifen-resistant tumors from breast cancer patients. J Clin Oncol 10:304–310
Osborne CK, Jarman M, McCague R, Coronado EB, Hilsenbeck SG, Wakeling AE (1994) The importance of tamoxifen metabolism in tamoxifen-stimulated breast tumor growth. Cancer Chemother Pharmacol 34:89–95
Osborne CK, Coronado-Heinsohn EB, Hilsenbeck SG, McCue BL, Wakeling AE, McClelland RA et al. (1995) Comparison of the effects of a pure steroidal antiestrogen with those of tamoxifen in a model of human breast cancer. J Natl Cancer Inst 87:746–750
Pavlik EJ, Nelson K, Srinivasan S, Powell DE, Kenady DE, DePriest PD et al. (1992) Resistance to tamoxifen with persisting sensitivity to estrogen: possible mediation by excessive antiestrogen binding site activity. Cancer Res 52:4106–4112
Pfeffer U, Fecarotta E, Vidali G (1995) Coexpression of multiple estrogen receptor variant messenger RNAs in normal and neoplastic breast tissues and in MCF-7 cells. Cancer Res 55:2158–2165
Pietras RJ, Arboleda J, Reese DM, Wongvipat N, Pegram MD, Ramos L, Gorman CM, Parker MG, Sliwkowski MX, Slamon DJ (1995) HER-2 tyrosine kinase pathway targets estrogen receptor and promotes hormone-independent growth in human breast cancer cells. Oncogene 10:2435–2446
Ponglikitmongkol M, White JH, Chambon P (1990) Synergistic activation of transcription by the human estrogen receptor bound to tandem responsive elements. EMBO Journal 9:2221–2231
Rea D, Parker MG (1996) Effects of Exon 5 Variant of the Estrogen Receptor in MCF-7 Breast Cancer Cells. Cancer Res 56:1556–1563
Reddel RR, Sutherland RL (1984) Tamoxifen stimulation of human breast cancer cell proliferation in vitro: a possible model for tamoxifen tumour flare. Eur J Cancer Clin Oncol 11:1419–1424
Ruenitz PC, Bagley JR, Pape CW (1984) Some chemical and biochemical aspects of liver microsomal metabolism of tamoxifen. Drug Metab Dispos 12:478–483
Smith IE, Harris AL, Morgan M (1981) Tamoxifen versus aminoglutethimide in advanced breast carcinoma: a randomised cross-over trial. Br Med J 283:1432–1434
Sutherland RL, Murphy LC, Foo MS, Green MD, Whybourne AM, Krozowski ZS (1980) High affinity anti-estrogen binding site distinct from the estrogen receptor. Nature 288:273–275
Swedish Breast Cancer Cooperative Group (1996) Randomised trial of 2 versus 5 years of adjuvant tamoxifen in post-menopausal early-stage breast cancer. J Natl Cancer Instit 88:1543–1549
Tzukerman MT, Esty A, Santiso-Mere D, Danielian P, Parker MG, Stein RB, Pike JW, McDonnell DP (1994) Human estrogen receptor transactivational capacity is determined by both cellular and promoter context and mediated by two functionally distinct intramolecular regions. Mol Endocrinol 8:21–30
Wakeling AE, Bowler J (1987) Steroidal pure antiestrogens. J Endocrinol 112:R7-R10
Wang Y, Miksicek RJ (1991) Identification of a dominant negative form of the human estrogen receptor. Mol Endocrinol 5:1707–1715
Webb P, Lopez GN, Uht R, Kushner PJ (1995) Tamoxifen activation of the estrogen receptor/AP-1 pathway: potential origin for the cell-specific estrogen like effects of antiestrogens. Mol Endocrinol 9:443–456
Wiebe VJ, DeGregorio MW, Osborne CK (1995) Tamoxifen metabolism and resistance. In: Dickson RB and Lippmann ME (eds) Drug and Hormonal Resistance in Breast Cancer. Ellis Horwood, London, pp 115–131

Wiebe VJ, Osborne CK, McGuire WL, DeGregorio MW (1992) Identification of estrogenic tamoxifen metabolite(s) in tamoxifen-resistant human breast tumors. J Clin Oncol 10:990–994

Wolf DM, Jordan VC (1994) The estrogen receptor from a tamoxifen stimulated MCF-7 tumour variant contains a point mutation in the ligand binding domain. Breast Cancer Res Treat 31:129–138

Wolf DM, Langhan-Fahey SM, Parker CJ, McCaque R, Jordan VC (1993) Investigation of the mechanism of tamoxifen-stimulated breast tumor growth with nonisomerisable analogues of tamoxifen and metabolites. J Natl Cancer Inst 85:806–812

Yang NN, Venugopalan M, Hardikar S, Glasebrook A (1996) Identification of an estrogen response element activated by metabolies of 17beta-estradiol and raloxifene. Science 273:1222–1225

Zhang Q-X, Borg A, Fuqua SAW (1993) An exon 5 deletion variant of the estrogen receptor frequently coexpressed with wild-type estrogen receptor in human breast cancer. Cancer Res 53:5882–5884

Zhang Q-X, Borg A, Wolf DM, Oesterreich S, Fuqua SAW (1997) An estrogen receptor mutant with strong hormone-independent activity from a metastatic breast cancer. Cancer Res 57:1244–1249

CHAPTER 32

Pharmacology of Inhibitors of Estrogen Biosynthesis

A.S. BHATNAGAR and W.R. MILLER

A. Introduction

Since 1896, when Sir George Beatson demonstrated that ovariectomy caused regression of mammary tumours in women, the key aim of endocrine breast cancer therapy has been to deprive the body of estrogen (BEATSON 1896). Ovariectomy accomplishes this by removing the gland that is the predominant source of estrogens in pre-menopausal women. Thus, estrogen deprivation is an effective therapy for breast cancer. It was only over half a century later that the mechanism of this therapy began to be explained. The elegant studies of JENSEN and his collaborators (1982) demonstrated that estrogen action was mediated by the estrogen receptor (ER). Binding of estrogen to ERs stimulated growth, and the attenuation of activity in this signal pathway led to estrogen deprivation and growth inhibition, thus explaining the therapeutic effects that BEATSON reported following ovariectomy. The knowledge generated through the years following Jensen's first report on the ER can be visualised in Fig. 1.

Thus, to effect estrogen deprivation pharmacologically, one can either antagonise the binding of estrogen with ERs, which is the mechanism of action for anti-estrogens such as tamoxifen, or reduce the production of estrogen through the use of estrogen synthesis inhibitors, such as inhibitors of the enzyme aromatase.

B. Non-Steroidal and Steroidal Aromatase Inhibitors

Aromatase is a cytochrome P-450-dependent enzyme, with a haem–iron-binding site and a steroid-binding site. The substrate, androstenedione, occupies the steroid-binding site, and its aromatisation to estrogen is catalysed by the enzyme through oxidative hydroxylations at the haem–iron-binding site. Thus, there are two major ways of inhibiting aromatase:

1. By occupying the steroid-binding site of the enzyme with a compound that has a steroidal structure (steroidal mimics)
2. By binding the haem–iron with nitrogen-containing compounds that have a non-steroidal structure

The structures of aromatase inhibitors commercially available for the treatment of breast cancer are shown in Fig. 2.

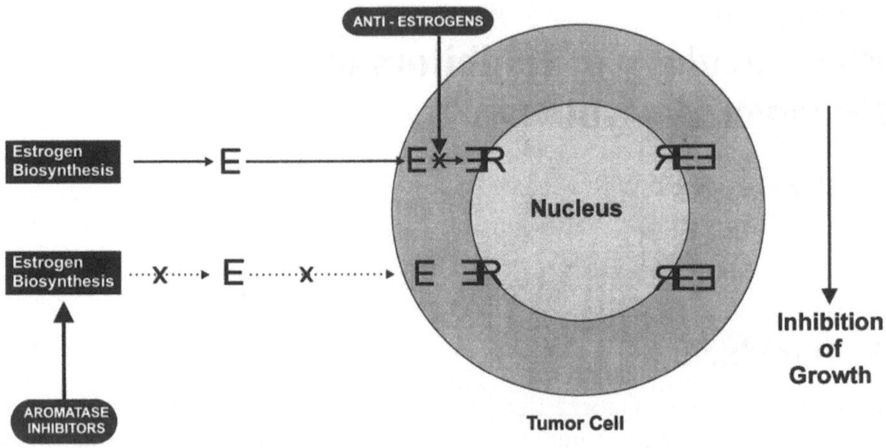

Fig. 1. Inhibition of estrogen (*E*)-dependent, estrogen receptor (*ER*)-mediated growth

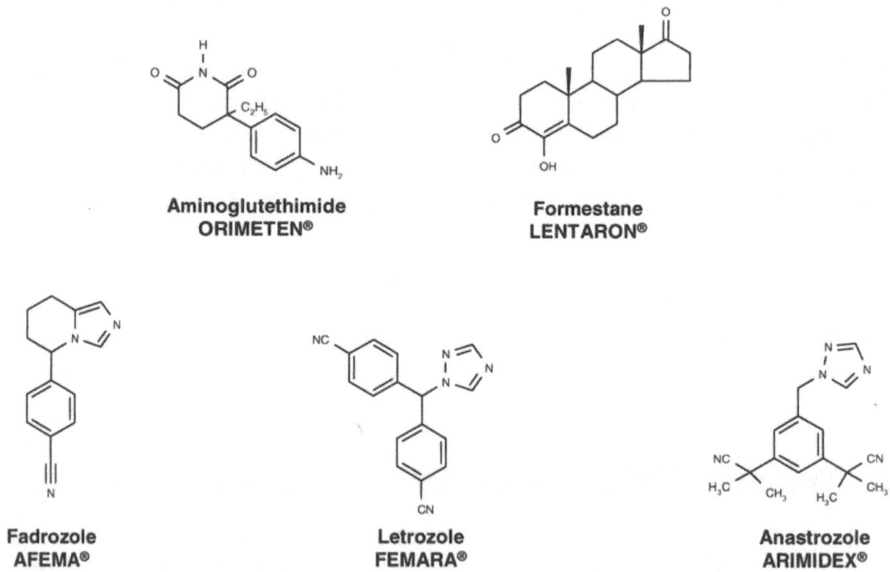

Fig. 2. Structures of commercially available aromatase inhibitors

Aminoglutethimide (AG) is a non-steroidal orally active aromatase inhibitor which has long been used to treat advanced breast cancer (STUART-HARRIS and SMITH 1984). However, its inhibition of aromatase is not selective. The AG molecule does not optimally fit the catalytic site of aromatase; thus, it also binds to, and thereby inhibits, several other cytochrome P-450-dependent enzymes in the pathways of steroidogenesis (Fig. 3), leading to an

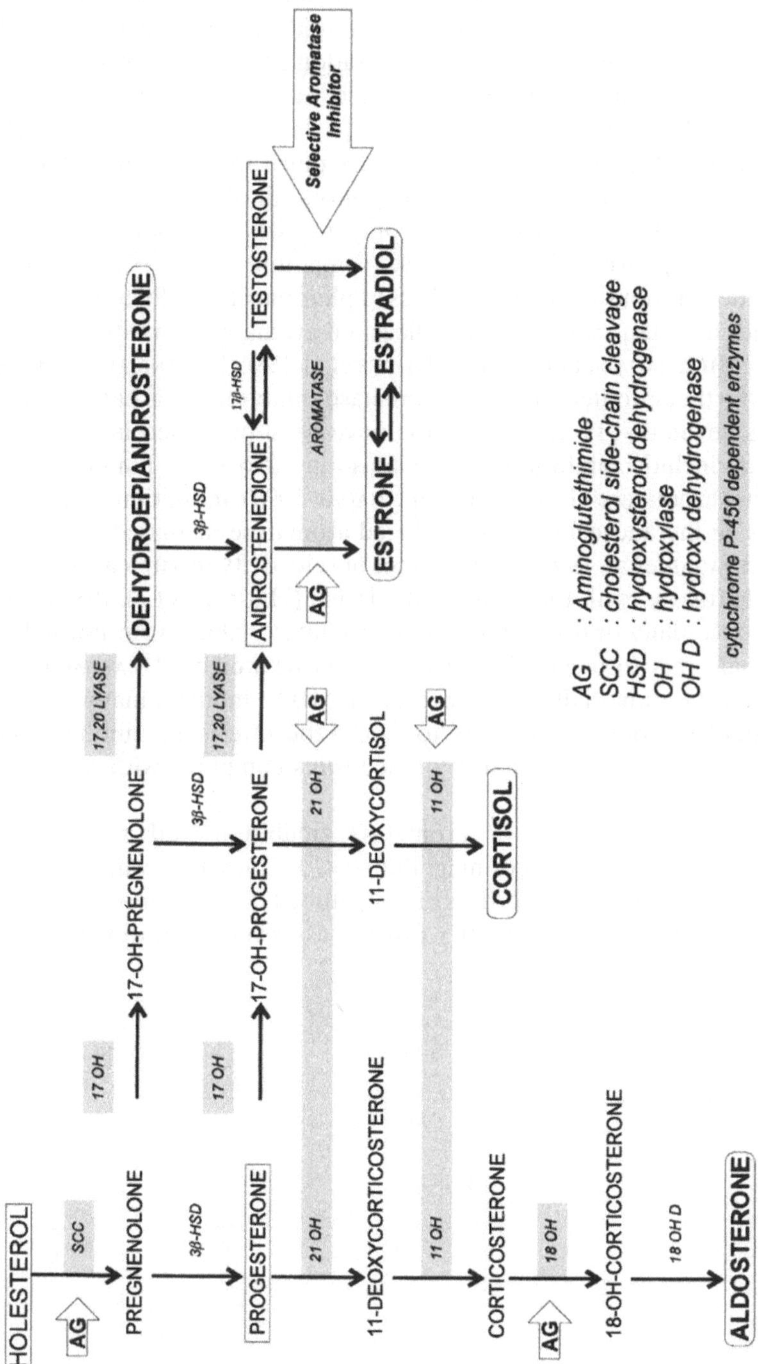

Fig. 3. Major pathways of steroidogenesis

attenuation in the production of important adrenal steroids. Formestane, however, is a highly effective and selective steroidal aromatase inhibitor which is administered parenterally; oral administration does not reproduce the high efficacy in serum estrogen suppression seen with parenteral administration (Dowsett et al. 1989).

Fadrozole, letrozole and anastrozole are much more recently developed orally active non-steroidal aromatase inhibitors, all of which show enhanced potency, selectivity and efficacy when compared with AG. Several reviews of the preclinical pharmacology of these compounds have been published, including descriptions of the methods used pharmacologically to profile these compounds as potent, selective and efficacious aromatase inhibitors (Plourde et al. 1994; Bhatnagar et al. 1996a,b; Dukes et al. 1996; Bhatnagar et al. 1997). We have further profiled all three aromatase inhibitors in a battery of standard tests developed in our labs. These have been published in detail previously. They include inhibition of aromatase in vitro using human placental microsomal aromatase (Purba and Bhatnagar 1990); inhibition of aromatase in vivo using the androstenedione-induced aromatase-mediated uterotrophic test (Bhatnagar et al. 1990a); assessment of selectivity in vitro using hamster ovarian (Häusler et al. 1989b) and rat adrenal (Häusler et al. 1989a) tissue; a 14-day once-daily oral treatment with aromatase inhibitor in normal adult cyclic female rats to assess the effect on uterine weight (Bhatnagar et al. 1990b); and the dimethylbenzanthracene (DMBA)-induced mammary carcinoma model in adult female rats to assess the effects on tumour volume (Schieweck et al. 1988). A summary of the results, compared with AG, is shown in Fig. 4.

In Fig. 4, comparisons of aromatase inhibition and selectivity are expressed in terms of orders of magnitude (×10). Thus, for example, all three aromatase inhibitors are two orders of magnitude (100-fold) more potent than AG in inhibiting aromatase in vitro. Efficacy assessments, however, are made

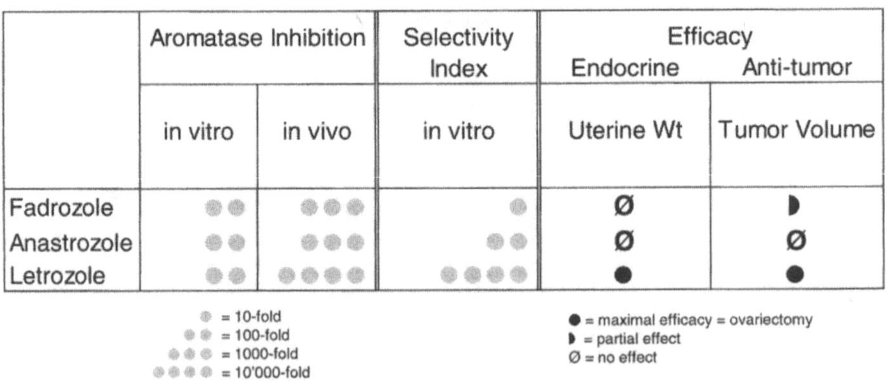

Fig. 4. Preclinical pharmacological profiles compared with aminoglutethimide and ovariectomy

with reference to ovariectomy. It is clear from Fig. 4 that all three compounds show comparable potency for inhibiting aromatase; however, there are major differences in in vivo efficacy in both non-tumour bearing and tumour-bearing adult female rats. Whereas letrozole mimics the endocrine sequelae of ovariectomy (BHATNAGAR et al. 1990b), neither fadrozole nor anastrozole have any significant effects on uterine weight at doses equal to or 10 times greater than the maximally effective dose of letrozole (BHATNAGAR et al. 1996b). In the DMBA model, letrozole also demonstrates a maximal efficacy in causing mammary tumour regression that is comparable to that observed for ovariectomy. Anastrozole, in contrast, attenuates tumour growth but does not result in a regression of tumour volume (BHATNAGAR et al. 1996b).

C. Inhibition of Intracellular Aromatase

One possible explanation for the observed differences in relative efficacy lies in differences in inhibitor metabolism/half-life in the rat. It is known that the half-life of fadrozole is 1–2 h, compared with 14 h for letrozole. Thus, fadrozole cannot express its full efficacy potential after a once-daily treatment. There are no published data available on the half-life of anastrozole in the rat.

Another potential explanation for differences in efficacy centres on findings that breast tumours possess aromatase activity and are, therefore, capable of producing their own estrogen (MILLER and FORREST 1974; MILLER 1986). Thus, in the post-menopausal woman, breast tumours can be exposed to estrogen derived not only from peripheral conversion of adrenal androgens in adipose tissue (then brought to the tumour cell by the circulation) (GRODIN et al. 1973), but from local synthesis by the intratumoural aromatase. This

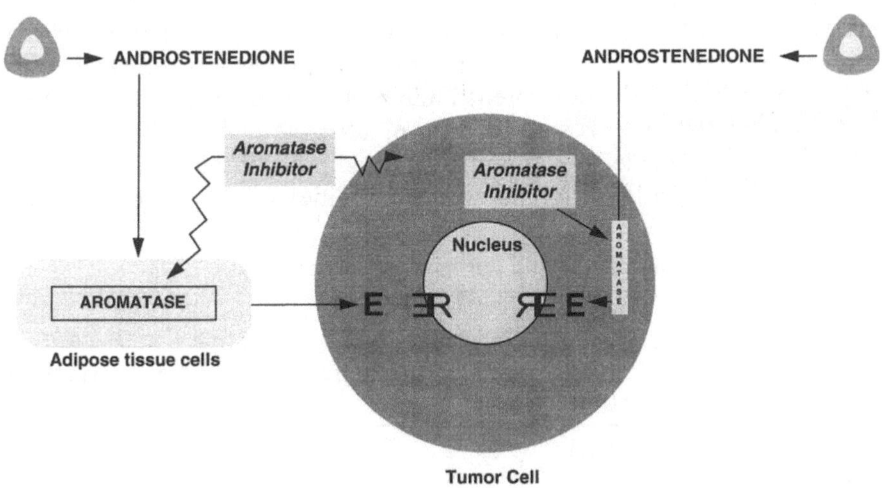

Fig. 5. Intratumoural production of estrogen

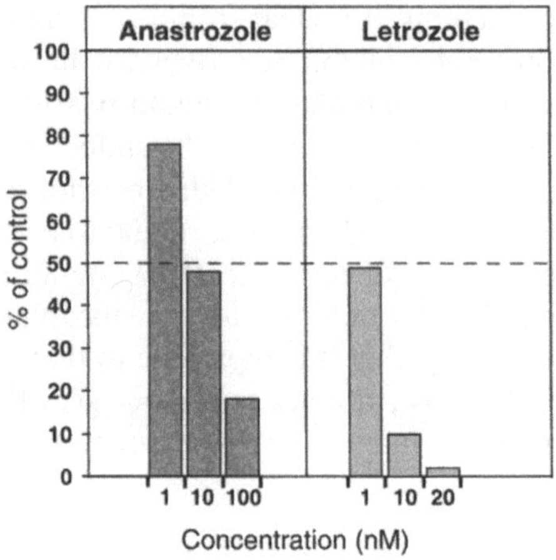

Fig. 6. Comparative effects of letrozole and anastrozole on aromatase in vitro in cultured adipose fibroblasts

concept is depicted in Fig. 5 which shows how an aromatase inhibitor maximally deprives the tumour cell of estrogen by effectively inhibiting both adipose tissue cell aromatase and intratumoural aromatase. Differences in the in vivo efficacy of aromatase inhibitors would more likely be related to the potency with which they inhibit intracellular aromatase in intact cells, rather than the potency with which they inhibit aromatase from isolated enzyme preparations (such as placental microsomal aromatase).

The in vitro results using non-tumour cells obtained from hamster ovary and human adipose fibroblasts are presented here; the methods have been published previously (HÄUSLER et al. 1989b; MILLER and MULLEN 1993). Table 1 compares aromatase inhibition observed in cell-free human placental microsomal aromatase with the inhibition of estrogen production in hamster ovarian cells. Whereas fadrozole, anastrozole and letrozole inhibit microsomal aromatase with a similar potency, in the hamster ovarian cells, letrozole and fadrozole are over 20 times more potent inhibitors of estrogen production than anastrozole. Similarly, as is shown in Fig. 6, letrozole is again 10 times more potent than anastrozole in cultured human adipose fibroblasts. These data demonstrate substantial differences in the abilities of the three aromatase

Table 1. Inhibition of aromatase and estrogen production in vitro

Compound	IC_{50} (n*M*)	
	Human placental aromatase (cell free)	Hamster ovarian tissue fragments
Aminoglutethimide	1900	13,000
Anastrozole	23	600
Fadrozole	5	30
Letrozole	11	20

inhibitors to inhibit estrogen synthesis in intact cells. Together with potential differences in half-lives, this may account for the significant differences seen in the in vivo efficacy of these compounds (in animals).

It remains to be determined whether the differences in potency of aromatase inhibitors in model systems are reflected in differences of anti-tumour efficacy in the clinic. The results of clinical trials now underway, which will directly compare the anti-tumour efficacy of letrozole and anastrozole in breast cancer patients, are awaited with interest.

References

Beatson GT (1896) On the treatment of inoperable cases of carcinoma of the mamma: suggestions for a new method of treatment with illustrative cases. Lancet 2:104–107

Bhatnagar AS, Häusler A, Schieweck K (1990a) Inhibition of aromatase in vitro and in vivo by aromatase inhibitors. J Enzyme Inhib 4:179–186

Bhatnagar AS, Häusler A, Schieweck K, Lang M, Bowman R (1990b) Highly selective inhibition of estrogen biosynthesis by CGS 20 267, a new non-steroidal aromatase inhibitor. J Steroid Biochem 37:1021–1027

Bhatnagar AS, Batzl C, Häusler A, Schieweck K, Lang M, Trunet PF (1996a) Pharmacology of nonsteroidal aromatase inhibitors. In: Pasqualini JR, Katzenellenbogen BS (eds) Hormone-dependent cancer. Marcel Dekker, New York, pp 155–168

Bhatnagar AS, Häusler A, Schieweck K, Batzl-Hartmann C, Lang M, Trunet P (1996b) Estrogen depletion in advanced breast cancer: why, how and where are we going? In: Rubens RD (ed) Advanced breast cancer: reassessing hormonal therapy. Parthenon, London, pp 21–32

Bhatnagar AS, Bowman RM, Schieweck K, Batzl-Hartmann C, Häusler A, Lang M, Trunet PF (1997) Letrozole: from test tube to patients. In: Mouridsen HT (ed) New options for the therapy of advanced breast cancer. Parthenon, London, pp 9–14

Dowsett M, Cunningham DC, Stein RC, Evans S, Dehennin L, Hedley A, Coombes RC (1989) Dose-related endocrine effects and pharmacokinetics of oral and intramuscular 4-hydroxyandrostenedione in postmenopausal breast cancer patients. Cancer Res 49:1306–1312

Dukes M, Edwards PN, Large M, Smith IK, Boyle T (1996) The preclinical pharmacology of "arimidex" (anastrozole; ZD1033) – a potent, selective aromatase inhibitor. J Steroid Biochem Mol Biol 58:439–445

Grodin JM, Siiteri PK, MacDonald PC (1973) Source of estrogen production in the postmenopausal woman. J Clin Endocrinol Metab 36:207–214

Häusler A, Monnet G, Borer C, Bhatnagar AS (1989a) Evidence that corticosterone is not an obligatory intermediate in aldosterone biosynthesis in the rat adrenal. J Steroid Biochem 34:567–570

Häusler A, Schenkel L, Krähenbühl C, Monnet G, Bhatnagar AS (1989b) An in vitro method to determine the selective inhibition of estrogen biosynthesis by aromatase inhibitors. J Steroid Biochem 33:125–131

Jensen EV, Greene GL, Closs LE, DeSombre ER, Nadji M (1982) Receptors reconsidered. A 20-year perspective. Recent Prog Horm Res 38:1–34

Miller WR (1986) Steroid metabolism in breast cancer. In: Stoll BA (ed) Breast cancer: treatment and prognosis. Blackwell, Oxford, pp 156–172

Miller WR, Forrest APM (1974) Oestradiol synthesis from C19 steroids by human breast cancer. Br J Cancer 33:16–18

Miller WR, Mullen P (1993) Factors influencing aromatase activity in the breast. J Steroid Biochem Mol Biol 44:597–604

Plourde PV, Dyroff M, Dukes M (1994) Arimidex®: a potent and selective fourth-generation aromatase inhibitor. Breast Cancer Res Treat 30:103–111

Purba HS, Bhatnagar AS (1990) A comparison of methods measuring aromatase activity in human placenta and rat ovary. J Enzyme Inhib 4:169–178

Schieweck K, Bhatnagar AS, Matter A (1988) CGS 16949 A, a new nonsteroidal aromatase inhibitor: effects on hormone-dependent and -independent tumors in vivo. Cancer Res 48:834–838

Stuart-Harris R, Smith IE (1984) Aminoglutethimide in the treatment of advanced breast cancer. Cancer Treat Rev 11:189–204

CHAPTER 33

Pharmacology of Inhibition of Estrogen-Metabolizing Enzymes

M.J. REED and A. PUROHIT

A. Rationale for Development of Steroid Sulphatase and Estradiol 17β-Hydroxysteroid Dehydrogenase Inhibitors

In addition to the aromatase enzyme complex, steroid sulphatase and estradiol 17β-hydroxysteroid dehydrogenase type 1 (17βHSD1) have important roles in regulating the tissue availability of biologically active steroids. The development of improved therapies for the treatment of endocrine-dependent tumours has been a major stimulus to research in this field. Growing awareness that steroid sulphatase is involved in the regulating part of the immune response, some aspects of reproduction and cognitive function has added impetus to the development of potent inhibitors of this enzyme.

The importance of estrogens in supporting the growth of breast tumours has led to the introduction of a number of drugs that aim to either block their ability to interact with the estrogen receptor (ER) in malignant tissues or to inhibit their biosynthesis. The antiestrogen tamoxifen is widely used in breast cancer as the first-line endocrine therapy. Unfortunately, most breast tumours eventually become resistant to its antiproliferative effects, necessitating the use of second- and third-line therapies. The development of new inhibitors for use in women with breast cancer who no longer respond to antiestrogens is therefore an important therapeutic target.

Three main enzymes are involved in estrogen synthesis in peripheral tissues, including normal and malignant breast tissues (Fig. 1) (JAMES et al. 1987). Breast cancer occurs most frequently in postmenopausal women with high tumour-estrogen concentrations, resulting from the actions of these enzymes (BONNEY et al. 1983). The aromatase complex is responsible for the formation of estrone from androstenedione and estradiol from testosterone. 17βHSD1 converts estrone, which is a weak estrogen, to its biologically active form, estradiol (LUU-THE et al. 1990; ANDERSSON 1995). Much of the estrone formed from androstenedione can be converted to estrone sulphate (E1S) by estrone sulphotransferase (HOBKIRK 1993). Blood and tissue concentrations of E1S are much higher than those of unconjugated estrogens (NOEL et al. 1981; PASQUALINI et al. 1989). Furthermore, the half-life of E1S in blood (10–12 h) is considerably longer than that of estrone or estradiol (20–30 min) (RUDER et al. 1972). The high concentrations of E1S in blood and tissues are thought

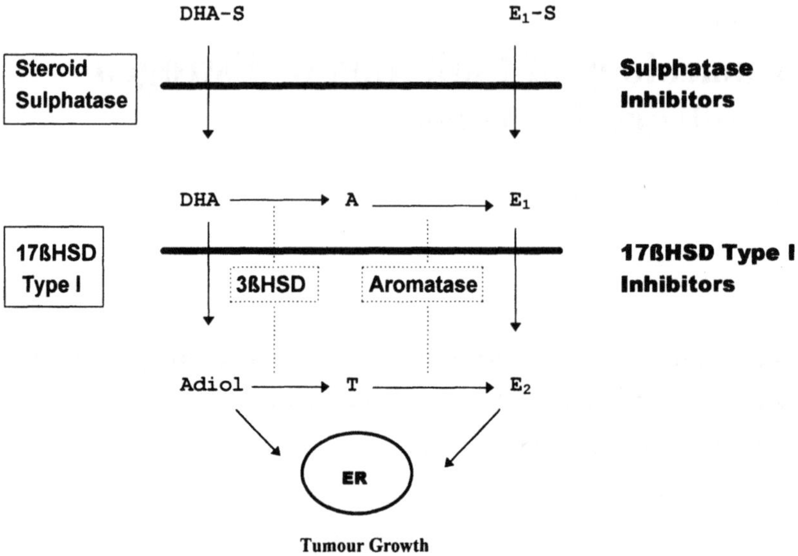

Fig. 1. Sites of action of steroid sulphatase and 17β-hydroxysteroid dehydrogenase type-1 (17βHSD type 1) inhibitors in blocking estrogen synthesis. Both 5-androstene-3β,17β-diol (*Adiol*) and estradiol (*E_2*) can bind to the estrogen receptor (*ER*) and stimulate breast tumour growth. *DHA*, dehydroepiandrosterone; *DHA-S*, DHA-sulphate; *E_1*, estrone; *E1S*, E1-sulphate; *A*, androstenedione; *T*, testosterone; *3βHSD*, 3β-hydroxysteroid dehydrogenase

to act as a reservoir for the formation of estrone by the action of estrone sulphatase (REED and PUROHIT 1993; REED and PUROHIT 1994).

So far, as reviewed elsewhere in this monograph (see Chapt. 29), most attention has focussed on the development of potent aromatase inhibitors. However, in a recent phase-III trial, in which a new potent aromatase inhibitor was evaluated, the response rate of 11% [complete response (CR) and partial response (PR)] in patients with ER-positive tumours was disappointing (JONAT et al. 1996; CASTIGLIONE-GERTSCH 1996). A possible reason for the low response rate in aromatase inhibitor-treated patients is that other sources of estrogenic steroids may be available to support tumour growth. As previously discussed, estrone can be formed from E1S, and there is evidence that estrone sulphatase activity in breast tumours is much higher than aromatase activity (JAMES et al. 1987). Furthermore, formation of estrone via the sulphatase pathway provides at least ten times as much tumour estrone as that synthesised by the aromatase route (SANTNER et al. 1984). In three recent studies, in which the effects of the new aromatase inhibitors CGS 169, 49A, fadrozole hydrochloride and vorozole were examined, plasma E1S concentrations remained between 400 pg/ml and 1000 pg/ml during therapy (SVENSTRUP et al. 1994; DOWSETT et al. 1994; JOHNSTON et al. 1994). The importance of the sulphatase route for estrogen synthesis in breast tumours has stimulated a search for a potent steroid sulphatase inhibitor.

A further impetus to the development of a steroid sulphatase inhibitor has been the growing realisation that if enzyme inhibitor therapy is to succeed in postmenopausal women, it will be necessary to block the synthesis of 5-androstene-3β,17β-diol (Adiol). This steroid, although an androgen, can bind to the ER and stimulate the growth of ZR-75–1 breast cancer cells and dimethyl benz (a) anthracene (DMBA)-induced mammary tumours in ovariectomised rats (POULIN and LABRIE 1986; DAUVOIS and LABRIE 1989). Over 90% of Adiol in postmenopausal women originates from dehydroepiandrosterone sulphate (DHA-S) after it has been hydrolysed to DHA. There is evidence that only one steroid sulphatase is responsible for the hydrolysis of E1S and DHA-S (PUROHIT et al. 1994). Therefore, the ability to inhibit the hydrolysis of E1S and DHA-S in vivo should reduce production not only of estrone, but also of Adiol.

In addition to the role that steroid sulphatase may have in breast cancer, this enzyme is also important in regulating part of the immune response (DAYNES et al. 1993; ROOK et al. 1994). It may also regulate some aspects of reproductive (LANGLAIS et al. 1981) and cognitive functions (LI et al. 1995).

Although the aromatase and steroid sulphatase complexes are required for the formation of estrone and DHA, it is the action of 17βHSD1 that converts these precursors to their more biologically active forms, i.e. estradiol and Adiol. While estrone can bind to the ER with low affinity, the efficacy of 17βHSD1 inhibitors in vivo remains to be established.

There is now convincing evidence that a further response can be achieved in women with breast cancer, subsequent to the failure of antiestrogen therapy, but the mechanism by which this occurs remains unclear. SANTEN (1996), however, has suggested that deprivation of breast cancer cells of estrogen leads to the development of adoptive mechanisms to increase their level of sensitivity to estrogens. The use of an inhibitor of estrogen biosynthesis for second-line therapy permits a further response to be achieved by reducing estrogen levels, leading to the inhibition of growth of tumour cells that have enhanced sensitivity to estrogens. While there is experimental evidence to support this hypothesis (MASAMURA et al. 1995), in order to confirm its validity in patients, it will be essential to develop steroid sulphatase and 17βHSD1 inhibitors for use in combination with the potent aromatase inhibitors that are already available.

B. Steroid Sulphatase Inhibitors

I. Substrate Analogue Inhibitors

A number of endocrine drugs, such as danazol (CARLSTROM et al. 1984) and some progestogens (PASQUALINI et al. 1994), can inhibit steroid sulphatase activity. The first steroid-based compound to be specifically synthesized and tested as a steroid sulphatase inhibitor was estrone-3-*O*-methylthiophosphonate (Fig. 2a, E1-MTP) (DUNCAN et al. 1993). This compound was selected as

a Estrone-3-*O*-methylthiophosphonate

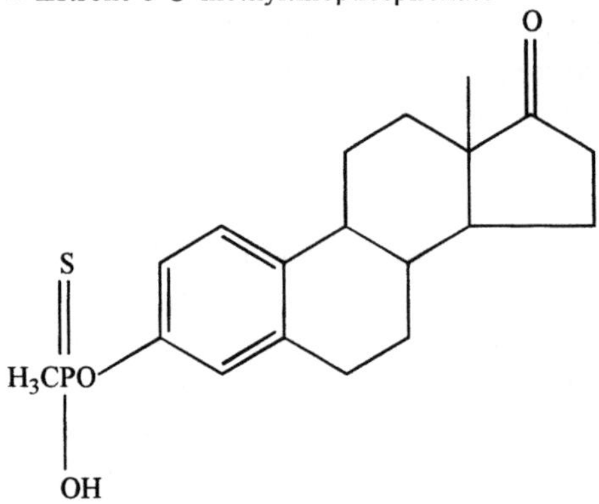

b Estrone-3-*O*-sulphonyl chloride

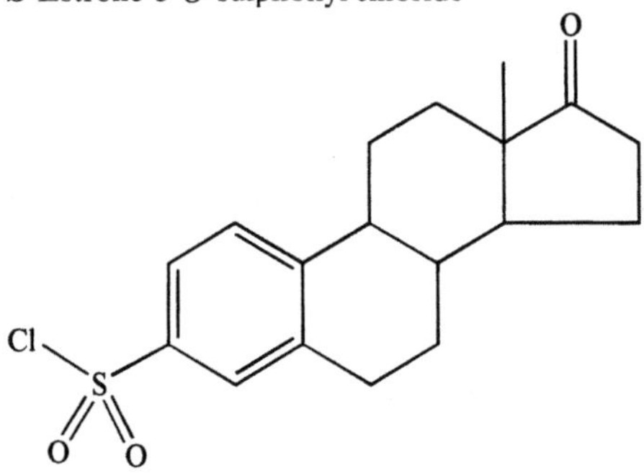

c Estrone-3-*O*-sulphamate (EMATE)

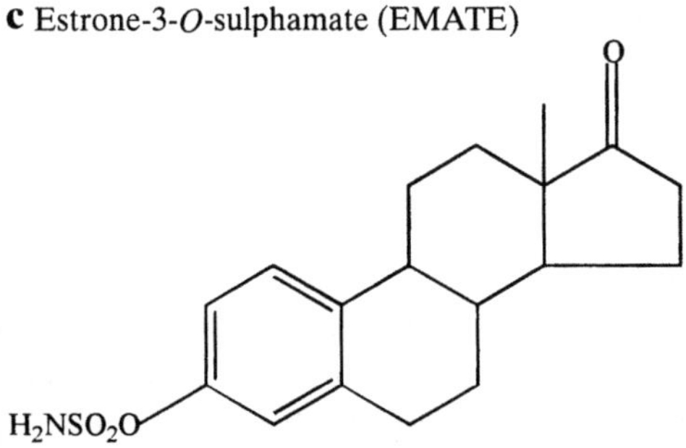

Fig. 2a–l. Structure of steroid sulphatase and 17β-hydroxysteroid dehydrogenase type-1 inhibitors

a potential steroid sulphatase inhibitor as it acted as a steroid sulphate structural mimic when used in the generation of antibodies that recognised E1S (Cox et al. 1979). E1-MTP was 14 times more potent than danazol in its ability to inhibit sulphatase activity in MCF-7 cells. In a preliminary in vivo study, E1-MTP (1 mg/kg per day for 7 days s.c.) resulted in a 50% lowering of plasma estradiol concentrations in rats (Reed et al. 1996).

d 4-Methylcoumarin-7-*O*-sulphamate

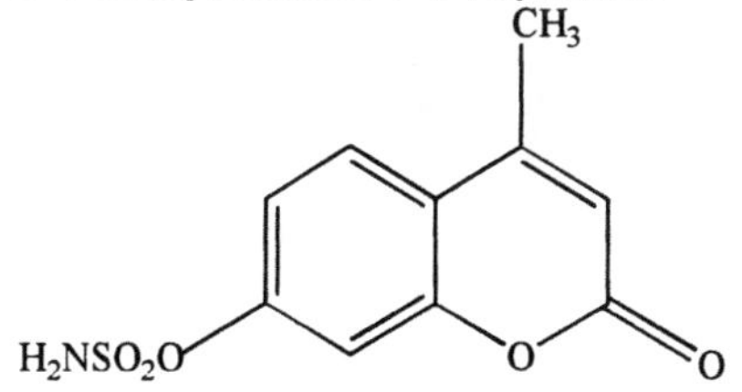

e (*p*-*O*-sulphamoyl)-*N*-tetradecanoyl tyramine

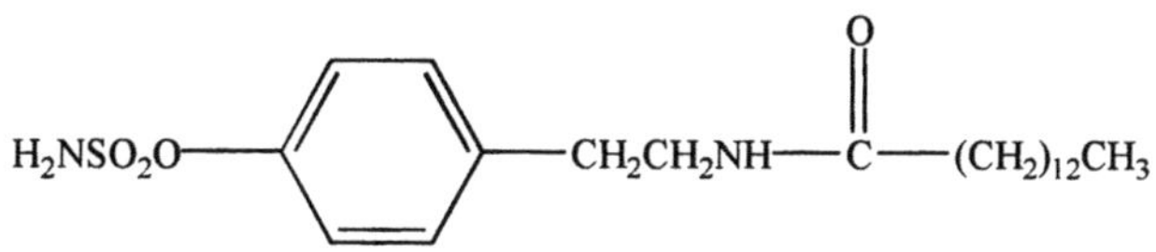

f 3-Hydroxy-1,3,5(10)-estratrien-[17,16c]pyrazole

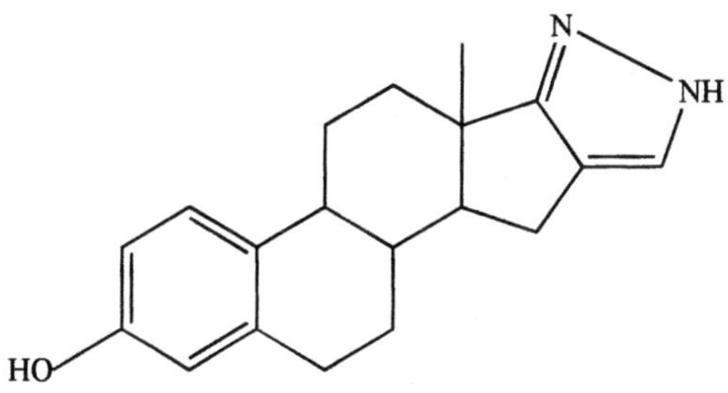

g 3-Hydroxy-1,3,5(10)-estratrien-[16,17-d]-isoxazole

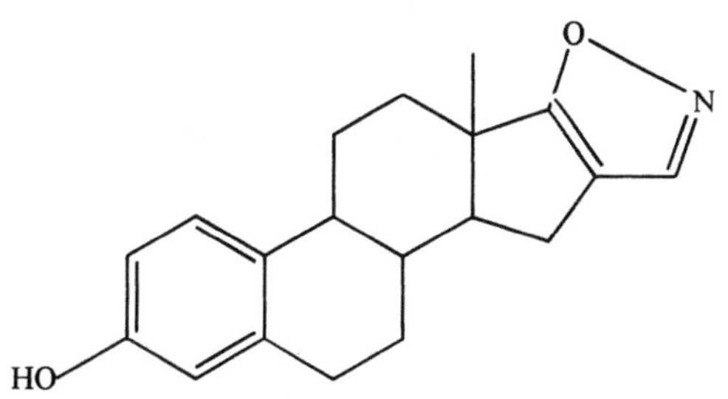

Fig. 2a–l. (*Continued*)

A series of steroid-sulphatase inhibitors, designed from modifications to the enzyme substrate, were also synthesized by LI and his colleagues (LI et al. 1993). A sulphonyl-chloride analogue, estrone-3-*O*-sulphonyl chloride (Fig. 2b) was the most potent compound in this series, causing 91.5% inhibition of placental-microsome sulphatase activity at 300 μM. E1-MTP and the other substrate-based inhibitors provided some indication as to the types of steroid derivatives that might be useful as steroid-sulphatase inhibitors.

h EM 139

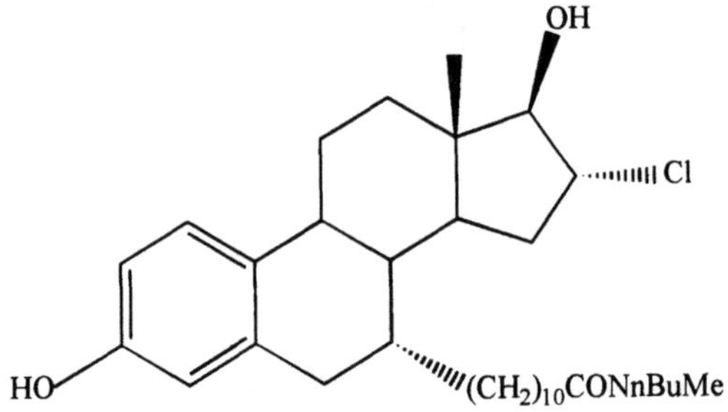

i EM 140

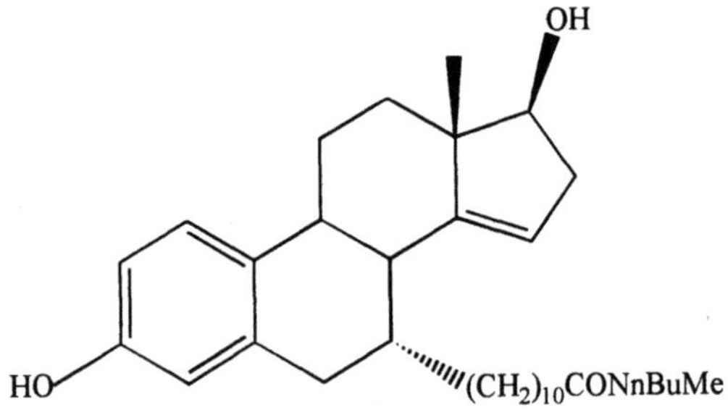

j 16α-(bromopropyl)-estradiol

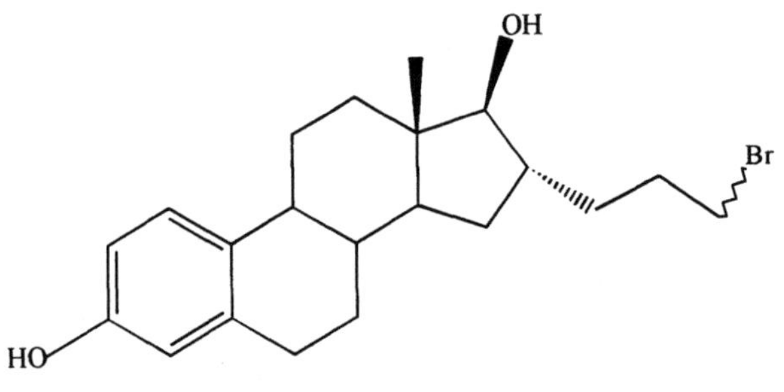

Fig. 2a–l. (*Continued*)

II. Identification of the Active Pharmacophore Required for Potent Steroid Sulphatase Inhibition

Subsequent to finding that E1-MTP was a moderately potent steroid sulphatase inhibitor, a series of related steroid sulphate surrogates were synthesized (REED et al. 1996). Most compounds in this series had a potency similar to that of E1-MTP, but one compound, estrone-3-*O*-sulphamate (Fig. 2c, EMATE) and its *N*-methylated derivatives emerged as being extremely

k *N*-Butyl,*N*-methyl, 9-[3',17'β-(dihydroxy)-1',3',5'(10')-estratriene-16'α-yl]-7-bromononamide

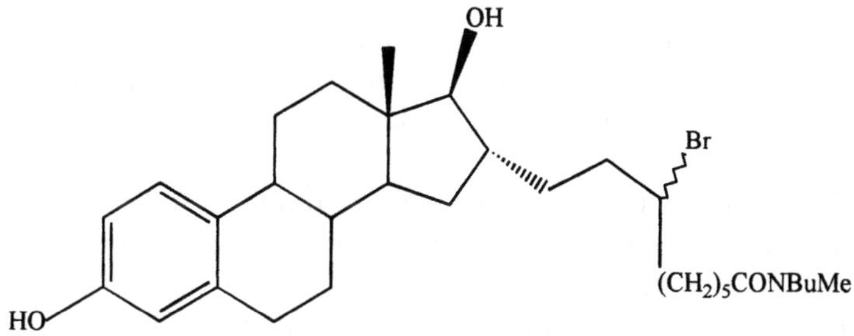

l *N*-Butyl,*N*-methyl, 9-[3',17'β-dihydroxy-1',3',5'(10')-estratriene-6'β-yl]-7- II thiaheptanamide

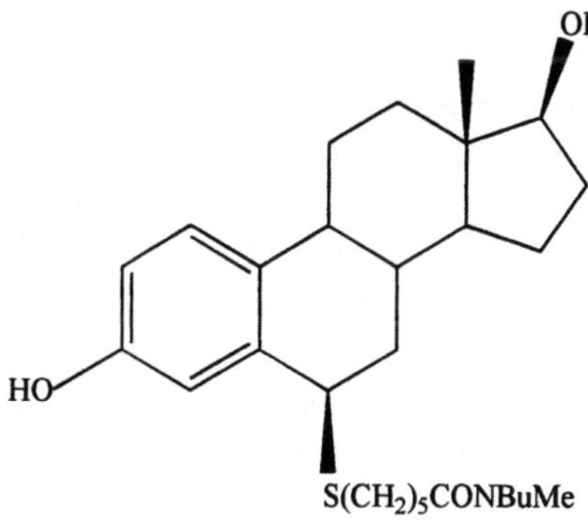

Fig. 2a–l. (*Continued*)

potent. Using intact MCF-7 cells and physiological concentrations of E1S or DHA-S, EMATE (1 μM) almost completely inhibited the hydrolysis of both steroid sulphates (Howarth et al. 1995; Purohit et al. 1995a). The IC_{50} for the inhibition of sulphatase activity in MCF-7 cells by EMATE was 65pM. Enzyme-kinetic analysis, using placental steroid sulphatase, revealed that EMATE inhibited steroid sulphatase activity in a time- and concentration-dependent manner (Purohit et al. 1995a).

EMATE was also active in vivo and at a dose of 1 mg/kg, when administered either by the oral or subcutaneous routes for 7 days, effectively abolished both estrone- and DHA-sulphatase activities in all tissues examined in the rat (Purohit et al. 1995b). Furthermore, EMATE has a prolonged duration of action. After a single dose (10 mg/kg), liver steroid sulphatase activity remained completely inactivated, with only a small recovery in activity (<10%) occurring 7 days after administration of the drug. Therefore, the pharmacophore identified for potent in vitro and in vivo inhibition of steroid sulphatase was a phenolic ring to which a sulphamoyl group is attached.

III. Development of Potent Non-steroidal, Non-estrogenic Steroid Sulphatase Inhibitors

Since a major goal of current steroid sulphatase inhibitor research is to develop a drug for breast cancer therapy, it may not be advisable to use an estrogen derivative as an inhibitor. It is also possible that, during the inactivation of sulphatase activity by EMATE, the release of estrone may occur (PUROHIT et al. 1995a).

Having identified the active pharmacophore required for potent steroid sulphatase inhibition, a search was made to find a non-steroid inhibitor. Further impetus to this research was provided by the finding that, unexpectedly, EMATE possesses potent estrogenic properties in vivo (ELGER et al. 1995). A series of sulphamate derivatives of a number of different ring systems, one of which was phenolic, were therefore synthesized. From this series, 4-methylcoumarin-7-*O*-sulphamate (Fig 2d, COUMATE) was identified as a potent non-steroidal inhibitor with therapeutic potential (Woo et al. 1996). In MCF-7 cells, the IC_{50} for sulphatase inhibition was 380 nM. COUMATE had no proliferative effect on MCF-7 cell growth, indicating that it was not estrogenic. In the ovariectomised rat, COUMATE did not stimulate uterine growth, confirming that it is devoid of estrogenic activity (PUROHIT et al. 1996).

Using the active pharmacophore required for potent steroid-sulphatase inhibition, LI and his colleagues (1996) have developed a series of non-steroidal (*p-O*-sulphamoyl)-*N*-alkanoyl-tyramine inhibitors. Of this series (*p-O*-sulphamoyl)-*N*-tetradecanoyl tyramine (Fig. 2e) was the most potent, with an IC_{50} of 56 nM when tested using a placental–microsomal sulphatase. This compound inhibited the E1S-stimulated growth of MCF-7 breast cancer cells in a dose-dependent manner (SELCER 1997).

IV. Pharmacological Effectiveness of Steroid Sulphatase Inhibition

EMATE and, to a lesser extent, its *N*-methylated derivatives were able to inhibit the growth of E1S stimulated nitrosomethylurea (NMU)-induced mammary tumours in ovariectomised rats (PUROHIT et al. 1995b). Estrone-sulphatase activity in tumours and livers of treated animals was almost completely suppressed. Although not yet tested in the NMU-tumour model, COUMATE blocked the ability of E1S to stimulate uterine growth in ovariectomised rats effectively, demonstrating the therapeutic potential of this inhibitor (PUROHIT et al. 1996).

Inhibition of steroid sulphatase activity may also have a beneficial effect on cognitive function. High levels of DHA-S are present in the human brain (MATHUR et al. 1993), in which it acts as a negative allosteric modulator of the γ-aminobutyric acid-A ($GABA_A$) receptor (MELLON 1994). DHA-S can reverse scopolamine-induced amnesia, and EMATE (LI et al. 1995) or the non-steroidal (*p-O*-sulphamoyl)-*N*-alkanoyl tyramine (LI et al. 1997) inhibitor markedly potentiated the reversal induced by DHA-S.

A new potential role for the steroid sulphatase inhibitors is their ability to modulate the immune response. T-helper (Th) lymphocytes can progress to either a Th1 or Th2 phenotype with each subset of cells secreting a characteristic profile of cytokines (ROMAGNANI 1991; ABBAS et al. 1996). There is good evidence that a number of autoimmune diseases, such as insulin-dependent diabetes or rheumatoid arthritis, arise from an imbalance in the production of Th lymphocytes (LIBLAU et al. 1995). Whether Th cells progress to a Th1 or Th2 phenotype is governed by the balance of DHA to glucocorticoid production (DAYNES et al. 1990; ROOK et al. 1994). Maturation of Th cells occurs in lymphoid tissue and DHA sulphatase in macrophages within such tissue regulates the formation of DHA from DHA-S. Thus inhibition of DHA sulphatase will reduce the balance of DHA to glucocorticoid and should shift the Th response towards the Th2 phenotype. Preliminary support for such a role for steroid sulphatase was obtained in Balb/c mice. Administration of EMATE inhibited the ear swelling seen in vivo after contact sensitisation and reversed the augmentation seen with DHA-S (SUITTERS et al. 1997).

C. Estradiol 17β-Hydroxysteroid Dehydrogenase Inhibitors

I. Substrate Analogue Inhibitors

The final step in the biosynthesis of the biologically active estrogen, estradiol, and also of Adiol which can bind to the ER, is regulated by 17βHSD1 (Fig. 1). Initial investigations into the regulation of E2DH activity employed a number of steroidal and non-steroidal compounds related to the substrate for this enzyme. 17-Deoxyestradiol was found to be a potent inhibitor of estradiol oxidation with an IC_{50} of 0.12 μM (JARABAK and SACK 1969). Interestingly, in view of the current concern regarding the potential estrogenic effects of pesticides, *o,p'*-DDD, was also identified as a potent inhibitor (IC_{50} = 0.16 μM) (JARABAK and SACK 1969). Other compounds including ethinylestradiol, diethylstilbestrol and danazol were also shown to possess 17βHSD inhibiting properties with Ki values of 0.3 μM, 0.4 μM and 0.6 μM respectively (BLOMQUIST et al. 1984).

II. Irreversible Inhibitors

Some of the first compounds identified as potent irreversible inhibitors of 17βHSD were detected after their synthesis as potential affinity-labelling reagents to identify amino acids at the active site of the enzyme. Two such compounds, 16α-bromoacetoxyestradiol-3-methyl ether and 12β-bromoacetoxy-4-estrene-3, 17-dione were found to inhibit 17βHSD in an irreversible manner (CHIN and WARREN 1975; WARREN et al. 1977). 16-Methylene estradiol was also synthesized to evaluate its potential as a 17βHSD inhibitor (THOMAS et al. 1983). The kinetics of inactivation of 17βHSD revealed that this compound acted as an active site-directed inhibitor.

III. Fused A- or D-Ring Pyrazole and Isoxazole Inhibitors

Steroids with A-ring fused pyrazol or isoxazole rings can have potent effects on the endocrine system, e.g. danazol, an isoxazole derivative of 17α-ethinyl testosterone. A number of 2,3- and 3,4-steroid-fused pyrazoles inhibited 3(17)β-HSD activity from *Pseudomonas testosteroni* (LEVY et al. 1987). The mechanisms by which these and other types of inhibitors act was recently reviewed by PENNING (1996). Hydrogen bonding between the steroid pyrazol and imidazole group of a histidine at the active site of the enzyme is thought to stabilize the complex.

A series of D-ring pyrazole and isoxazole derivatives of estrone were tested as potential inhibitors of 17βHSD (SWEET et al. 1991). 3-Hydroxy-1,3,5(10) estratriene–[16,17d] pyrazole (Fig. 2f, Ki 4.08 μM) was more active than the corresponding–[16,17c] isoxazole (Fig 2.g, Ki 69.4 μM).

Some of the 17βHSD inhibitors identified from this early phase of research did possess reasonable inhibitory properties. However, their specificity as type 1-, i.e. reductive, or type 2-, i.e. oxidative, inhibitors was not always investigated. Furthermore, the extent to which many of the compounds, as estrogen derivatives, retained their intrinsic estrogenicity was not fully explored.

IV. Inhibitors with Antiestrogenic Properties

The finding that some antiestrogens, such as tamoxifen and ICI 164,384 (SANTNER and SANTEN 1993), possess weak 17βHSD-inhibiting properties suggested that it may be possible to develop a potent 17βHSD1 inhibitor that is also antiestrogenic. Based on the observation that derivatives of estradiol possessing 7α substituents can bind to the ER, a series of 7α alkyl estradiol derivatives were tested in vivo for their ability to inhibit E2DH and for their antiestrogenic properties (LABRIE et al. 1992). Unsaturated D-ring derivatives of estradiol with a 7α undecamide side chain and 16α halogen atom (Fig. 2h) or similar unsaturated D-ring compounds lacking a 16α halogen atom (Fig. 2i) inhibited both the reductive and oxidative functions of this enzyme in vivo. Most compounds in this series possessed reasonable antiestrogenic properties, demonstrating the feasibility of developing an inhibitor with dual properties.

A further series of 16α/β-halogen alkyl derivatives also inhibited 17βHSD1 (PELLETIER et al. 1996). Of these, the iodo- and bromo-derivatives were more potent than the chloro- and fluoro-analogues. 16α-(Bromopropyl)-estradiol (Fig. 2j) was the most potent inhibitor, but retained some residual estrogenic activity. Further derivatives were therefore synthesized that possessed the alkylamide group required for antiestrogenic activity, and the bromoalkyl group required for 17βHSD1 inhibition, on the 16α side chain (PELLETIER and POIRIER 1996). Of this series *N*-butyl, *N*-methyl,9-[3′,17′β-(dihydroxy)-1,3,5(10)-estratrien-16′α-yl]-7-bromononamide (Fig. 2k) was the best compromise for a dual-action inhibitor. This compound moderately inhibited 17βHSD1 activity in placental cytosol (IC_{50} = 10.4 μM), possessed no estrogenic activity and exhibited antiestrogenic activity in ZR-75–1 cells. However, the inhibitory potency of this compound was lower than that of 16α-(bromopropyl)-estradiol (Fig. 2j IC_{50} = 0.49 μM). The addition of the bulky alkylamide side chain was thought to decrease the affinity of compounds with dual functional groups at the 16α position and to account for their relatively weak potency.

Recently, a series of 17βHSD1 inhibitors with a thia-alkanamide chain at C6 of the steroid nucleus have been synthesized (POIRIER et al. 1997). The analogue with an intermediate side-chain length ($CH_{2(5)}$) caused the strongest inhibition (Fig. 2l, IC_{50} = 0.26 μM). The orientation of the side chain had a marked effect on inhibitor potency with the 6β derivative (IC_{50} = 0.17 μM), being 70 times more active than the 6α analogue. This compound is a reversible, but selective, 17βHSD1 inhibitor and could have considerable therapeutic potential.

V. Flavone and Isoflavone Inhibitors

A number of flavones and isoflavones have been examined as potential 17βHSD inhibitors. Using a purified recombinant 17βHSD1, coumestrol was found to be a potent inhibitor (IC_{50} = 0.12 μM) with the isoflavanoid genistein being less potent (IC_{50} = 1.2 μM). (MAKELA et al. 1995). However, compounds that were able to inhibit the reduction of estrone in T47D cells were not able to prevent the estrone-stimulated proliferation of these cells. Similar studies, but employing a placental 17βHSD preparation, revealed the flavonoids apigenin and narigenin (IC_{50}s = 18.3 μM and 19.9 μM, respectively) to be more potent than genistein (IC_{50} = 195 μM) (LE LAIN et al. 1996).

D. Future Perspectives

In order to improve the response rate to enzyme-inhibitor therapy, it is now generally acknowledged that it will be necessary to achieve complete estrogen deprivation. This has been a major stimulus to the search for steroid sulphatase and 17βHSD1 inhibitors. A number of potent steroid sulphatase inhibitors have now been identified and should enter a phase-I trial for the treatment of women with breast cancer in the near future. In view of the important roles that steroid sulphatase has in regulating part of the immune response and some aspects of reproductive function, these inhibitors should prove to be a powerful tool to investigate the pathophysiology of this enzyme.

Although research to design and synthesize 17βHSD1 inhibitors has been in progress for many years, a potent inhibitor with clinical potential remains to be identified. It also remains to be established whether this type of inhibitor can in fact block stimulation of cells or tumour growth by estrone. However, a number of core structures and side-chain substituents required for 17βHSD1 inhibition have now been identified. Such information will be of considerable value in the design of a clinically useful inhibitor.

References

Abbas AK, Murphy KM, Sher A (1996) Functional diversity of helper T lymphocytes. Nature 383:787–793

Andersson S (1995) 17β-Hydroxysteroid dehydrogenase: isoenzymes and mutations. J Endocrinol 146:197–200

Blomquist CH, Lindemann NJ, Hakanson EY (1984) Inhibition of 17β-hydroxysteroid dehydrogenase activities of human placenta by steroids and non-steroidal hormone agonists and antagonists. Steroids 43:571–586

Bonney RC, Reed MJ, Davidson K, Beranek PA, James VHT (1983) The relationship between 17β-hydroxysteroid dehydrogenase activity and estrogen concentrations in human breast tumours and in normal breast tissue. Clin Endocrinol 19:727–739

Carlstrom K, Doberl A, Pousette A, Rannevik G, Wilking N (1984) Inhibition of steroid sulphatase by danazol. Acta Obstet Gynaec Scand 123 (Suppl):107–111

Castiglione-Gertsch M (1996) New aromatase inhibitors: more selectivity, less toxicity, unfortunately, the same activity. Eur J Cancer 32 A:393–395

Chin C-C, Warren JC (1975) Synthesis of 16α-bromoacetoxy estradiol 3-methyl ether and study of the steroid binding site of human placental estradiol 17β-dehydrogenase. J Biol Chem 250:7682–7686

Cox RI, Hoskinson RM, Wong MSF (1979) Antisera reactive directly with estrone sulfate. Steroids 33:549–562

Dauvois S, Labrie F (1989) Androstenedione and androst-5-ene-3β, 17β-diol stimulate DMBA-induced mammary tumours – role of aromatase. Breast Cancer Res Treat 13:61–69

Daynes RA, Araneo BA, Ershler WB, Maloney C, Li G-Z, Ryn S-Y (1993) Altered regulation of IL-6 production with normal aging. J Immunol 150:5219–5230

Daynes RA, Dudley DJ, Araneo BA (1990) Regulation of murine lymphokine production in vivo: II. Dehydroepiandrosterone is a natural enhancer of interleukin 2 synthesis by helper T cells. Eur J Immunol 20:793–803

Dowsett M, Smithers D, Moore J, Trunet PF, Coombes RC, Powles TJ, Rubens R, Smith IE (1994) Endocrine changes with the aromatase inhibitor fadrozole hydrochloride in breast cancer. Eur J Cancer 30 A:1453–1458

Duncan LJ, Purohit A, Howarth NM, Potter BVL, Reed MJ (1993) Inhibition of estrone sulfatase activity by estrone-3-methylthiophosphonate: a potential therapeutic agent in breast cancer. Cancer Res 53:298–303

Elger W, Schwarz S, Hedden A, Reddersen G, Schneider B (1995) Sulphamates of various estrogens are prodrugs with increased systemic and reduced hepatic estrogenicity at oral applications. J Steroid Biochem Mole Biol 55:395–403

Hobkirk R (1993) Steroid sulfation. Trends Endocrinol Metab 4:69–79

Howarth NM, Purohit A, Reed MJ, Potter BVL (1994) Estrone sulfamates: potent inhibitors of estrone sulfatase with therapeutic potential. J Med Chem 37:219–221

James VHT, McNeill JM, Lai LC, Newton CJ, Ghilchik MW, Reed MJ (1987) Aromatase activity in normal breast and breast tumour tissue: in vivo and in vitro studies. Steroids 50:269–279

Jarabak J, Sacks GH (1969) Soluble 17β-hydroxysteroid dehydrogenase from human placenta – the binding of pyridine nucleotides and steroids. Biochem 8:2203–2212

Johnston SRD, Smith IE, Doody D, Jacobs S, Robertshaw H, Dowsett M (1994) Clinical and endocrine effects with the oral aromatase inhibitor vorozole in postmenopausal women with advanced breast cancer. Cancer Res 54:5875–5881

Jonat W, Howell A, Blomquist C, Eiermann W, Winblad G, Tyrrel C, Mauriac L, Lundgren S, Hellmund R, Azab M (1996) A randomized trial comparing two doses of the new selective inhibitor Anastrozole (Arimidex) with megestrol acetate in postmenopausal patients with advanced breast cancer. Eur J Cancer 32 A:404–412

Labrie C, Martel C, Dufour J-M, Levesgne C, Merand Y, Labrie F (1992) Novel compounds inhibit estrogen formation and action. Cancer Res 52:610–615

Langlais J, Zollinger M, Plante L, Chapdelaine A, Bleau G, Roberts KD (1981) Localization of cholestrerol sulfate in human spermatozoa in support of a mechanism of capacitation. Proc Natl Acad Sci USA 78:7266–7270

Le Lain R, Ahmadi M, Smith HJ, Nicholls PJ, Whomsley R (1996) Some flavones and isoflavones as inhibitors of human placental 17β-hydroxysteroid dehydrogenase in vitro. Pharm Sciences 2:21–23

Levy MA, Holt DA, Brandt M, Metcalf BW (1987) Inhibition of 3(17)β-hydroxysteroid dehydrogenase from *Pseudomonas testosteroni* by steroidal A ring fused pyrazoles. Biochem 26:2270–2279

Li, PK, Milano S, Kluth L, Rhodes ME (1996) Synthesis and sulfatase inhibiting properties of non-steroidal estrone sulfatase inhibitors. J Steroid Biochem Molec Biol 59:41–48

Li PK, Pillai R, Young BL, Bender WH, Martino DM, Liu F-T (1993) Synthesis and biochemical studies of estrone sulfatase inhibitors. Steroids 58:106–111

Li PK, Rhodes ME, Jagannathan S, Johnson DA (1995) Reversal of scopolamine induced amnesia in rats by the steroid sulfatase inhibitor estrone-3-*O*-sulfamate. Cognit Brain Res 2:251–259

Li PK, Rhodes ME, Burke AM, Johnson DA (1997) Memory enhancement mediated by the steroid sulfatase inhibitor (*p*-*O*-sulfamoyl)-*N*-tetradecanoyl tyramine. Life Sci 60:PL45–51

Liblau RS, Singer SM, McDevitt HO (1995) Th1 and Th2 CD4 + T cells in the pathogenesis of organ specific autoimmune diseases. Immunol Today 16:34–38

Luu-The V, Labrie C, Simard J, Lachance Y, Zhao HF, Couet J, Leblanc G, Labrie F (1990) Structure of two in tandem human 17β-hydroxysteroid dehydrogenase genes. Molec Endocrinol 4:268–275

Makela S, Poutanen M, Lehtimaki J, Kostian M-L, Santti R, Vikko R (1995) Estrogen-specific 17β-hydroxysteroid oxidoreductase Type 1 (E.C. 1.1.1.62) as a possible target for the action of phytoestrogens. Proc Soc Expt Biol Med 208:51–59

Masamura S, Santner SJ, Heitjan DF, Santen RJ (1995) Estradiol deprivation causes estradiol hypersensitivity in human breast cancer cells. J Clin Endocrinol Metab 80:2918–2925

Mathur C, Prasad VVK, Raju VS, Welch M, Leiberman S (1993) Steroids and their conjugates in the mammalian brain. Proc Natl Acad Sci USA 90:85–88

Mellon SH (1994) Neurosteroids: biochemistry, modes of action and clinical relevance. J Clin Endocrinol Metab 78:1003–1008

Noel CT, Reed MJ, Jacobs HS, James VHT (1981) The plasma concentration of oestrone sulphate in postmenopausal women: lack of diurnal variation, effect of ovariectomy, age and weight. J Steroid Biochem 14:1101–1105

Pasqualini JR, Gelly C, Nguyen B-L, Vella C (1989) Importance of estrogen sulphates in breast cancer. J Steroid Biochem 34:155–163

Pasqualini JR, Maloche C, Maroni M, Chetrite G (1994) Effect of progestogen Promegestone (R-5020) on mRNA of the oestrone sulphatase in MCF-7 human mammary cancer cells. Anticancer Res 14:1589–1594

Pelletier J, Poirier D (1996) Synthesis and evaluation of estradiol derivatives with 16α(-bromoalkylamide), 16α-(bromoalkyl) or 16α-(bromoalkynyl) side chains as inhibitors of 17β-hydroxysteroid dehydrogenase Type 1 without estrogenic activity. Bioorg Med Chem 4:1617–1628

Poirier D, Dionne P, Auger S (1998) A 6β-(thiaheptanamide) derivative of estradiol as inhibitor of 17β-hydroxysteroid dehydrogenase Type 1. J Steroid Biochem Molec Biol 64:83–90

Poulin R, Labrie F (1986) Stimulation of cell proliferation and estrogenic response by adrenal C19-delta-5-steroids in the ZR-75–1 human breast cancer cell line. Cancer Res 46:4933–4937

Purohit A, Dauvois S, Parker MG, Potter BVL, Williams GJ, Reed MJ (1994) The hydrolysis of oestrone sulphate and dehydroepiandrosterone sulphate by human steroid sulphatase expressed in transfected COS-1 cells. J Steroid Biochem Molec Biol 50:101–104

Purohit A, Williams GJ, Howarth NM, Potter BVL, Reed MJ (1995a) Inactivation of steroid sulfatase by an active site-directed inhibitor, estrone-3-*O*-sulfamate. Biochem 34:11508–11514

Purohit A, Williams GJ, Roberts CJ, Potter BVL, Reed MJ (1995b) In vivo inhibition of oestrone sulphatase and dehydroepiandrosterone sulphatase by oestrone-3-*O*-sulphamate. Int J Cancer 62:106–111

Purohit A, Woo LWL, Singh A, Winterborn CJ, Potter BVL, Reed MJ (1996) In vivo activity of 4-methylcoumarin-7-*O*-sulfamate, a nonsteroidal, nonestrogenic steroid sulfatase inhibitor. Cancer Res 56:4950–4955

Reed MJ, Purohit A (1993) Sulphatase inhibitors: the rationale for the development of a new endocrine therapy. Rev Endocr-Rel Cancer 45:51–62

Reed MJ, Purohit A (1994) Inhibition of steroid suphatase. In: Sandler M and Smith J (eds) Design of Enzyme Inhibitors as Drugs, vol 2. Oxford University Press, Oxford, pp 481–494

Reed MJ, Purohit A, Woo LWL, Potter BVL (1996) The development of steroid sulphatase inhibitors. Endocr-Rel Cancer 3:9–23

Romagnani S (1991) Human Th1 and Th2 subsets: doubt no more. Immunol Today 12: 256–257

Rook GAW, Hernandez-Pando R, Lightman S (1994) Hormones, peripherally activated prohormones and regulation of Th1/Th2 balance. Immunol Today 15:301–303

Ruder HJ, Loriaux DL, Lipsett MB (1972) Estrone sulfate: production rate and metabolism in man. J Clin Invest 51:1020–1023

Santner SJ, Feil PD, Santen RJ (1984) In situ estrogen production via estrone sulfatase pathway in breast tumours: relative importance versus aromatase pathway. J Clin Endocrinol Metab 59:29–33

Santner SJ, Santen RJ (1993) Inhibition of estrone sulfatase and 17β-hydroxysteroid dehydrogenase by antiestrogens. J Steroid Biochem Molec Biol 45:383–390

Santen RJ (1996) Estrogen synthesis inhibitors for breast cancer: an introductory overview. Endocr-Rel Cancer 3:1–8

Selcer KW, Hegde PV, Li PK (1997) Inhibition of estrone sulfatase and proliferation of human breast cancer cells by nonsteroidal (*p*-*O*-sulfamoyl)-*N*-alkanoyl tyramines. Cancer Res 57:702–707

Suitters AJ, Shaw S, Wales MR, Porter JR, Leonard J, Woodger R, Brand H, Bodmer M, Foulkes R (1997) Immune enhancing effects of dehydroepiandrosterone and dehydroepiandrosterone sulphate and the role of steroid sulphatase. Immunol 91:314–321

Svenstrup B, Herrstedt J, Brünner N, Bennett P, Wachman H, Dombernowsky P (1994) Sex hormone levels in postmenopausal women with advanced metastatic breast cancer treated with CGS 169 49 A. Eur J Cancer 30 A:1254–1258

Sweet F, Boyd J, Medina O, Konderski L, Murdock GL (1991) Hydrogen bonding in steroidogenesis: studies on new heterocyclic analogs of estrone that inhibit human estradiol 17β-dehydrogenase. Biochem Biophys Res Comm 180:1057–1063

Thomas JL, La Rochelle MC, Covey DF, Strickler RC (1983) Inactivation of human placental 17β,20α-hydroxysteroid dehydrogenase by 16-methylene estrone, an affinity alkylator enzymatically generated from 16-methylene estradiol-17β. J Biol Chem 258:11500–11504

Warren JC, Mueller JR, Chin C-C (1977) Structure of the steroid-binding site of human placental estradiol-17β-dehydrogenase. Am J Obstet Gynec 129:788–794

Woo LWL, Purohit A, Reed MJ, Potter BVL (1996) Active site-directed inhibition of estrone sulfatase by non-steroidal coumarin sulfamates. J Med Chem 39:1349–1351

CHAPTER 34

Pharmacology of Different Administration Routes – Oral vs Transdermal

R. SITRUK-WARE

A. Rationale for Various Routes of Estrogen Delivery

For a drug to produce its effect, it must be present in an appropriate concentration at the site of action. This concentration will be influenced by the dose, the rate and the amount of absorption, how the drug is distributed in the body, how it is metabolized and how rapidly it is excreted.

Estrogens are lipid-soluble drugs and, as such, are easily absorbed from the gastrointestinal tract. However, the rate of gastric emptying and the intestinal transit time may influence the rate of absorption and the amount of steroid absorbed. Also, natural estrogens may be broken down by enzymes in the intestinal mucosa and the liver, and fail to reach the systemic circulation (JAMES 1996).

In order to avoid the variability of absorption that may occur after oral administration of an estrogen in a tablet form, other routes of administration were sought that would escape the initial hepatic breakdown, also called the first-pass effect.

Among other parenteral routes of administration of estrogens, subcutaneous, transdermal and vaginal delivery were tested, both for contraceptive use and for replacement therapy, in endocrinological situations of estrogen deficiency.

Plasma levels of estradiol, obtained after various modes of the steroid delivery, vary dramatically. Oral therapy leads to very unstable levels. A large bolus of estrogens, predominantly estrone in high concentrations, appears in the systemic circulation, with a peak reached after 1–4 h of ingestion (YEN et al. 1975; BAKER 1994). After these peak levels, the concentration rapidly declines. Therefore, peaks and troughs are observed; the former may induce side effects, and the latter could lead to treatment inefficiency.

Absorption through the skin is slower, and lipid-soluble drugs such as estrogens are favored. The plasma estradiol levels after transdermal application are more stable than after oral ingestion, but may still vary during the 3 day to 4-day lifetime of a transdermal patch application.

B. Estrogens Available for Therapy

Among the molecules available to the prescriber, the natural estrogens are usually preferred to synthetic steroids for substitutive estrogen therapy.

I. Artificial Estrogens

1. Ethinyl Estradiol

The principal synthetic estrogen that is in fact used for oral contraception, but no longer for hormone-replacement therapy (HRT), is ethinyl estradiol (EE). Since crystallized estradiol is rapidly destroyed in the liver after oral administration, the adjunct of an ethinyl radical in C17 allowed the molecule to resist liver metabolism and to have a half-life of about 48h (Henzl 1986).

2. Mestranol

Mestranol must be demethylated in C3 to be transformed in EE, and quinestrol is a cyclophenyl ester of EE with a very long half-life.

3. Others

Diethylstilbestrol (DES) is a stilbene derivative that is no longer used in clinical practice.

All the synthetic molecules are far more potent than the natural estrogens, and it has been shown that EE is about 1000 times more potent than conjugated equine estrogens (CEE) in stimulating the liver proteins as sex hormone-binding globulin (SHBG) and renin substrate (RS) (Mashchak et al. 1982; von Schoultz 1988), when administered orally. Avoiding the liver first-pass effect is not as beneficial with this synthetic steroid, as it has been shown that 50 μg EE administered vaginally is equal to 10 μg EE given orally in the stimulation of hepatic molecules such as SHBG (Goebelsman et al. 1985).

As mentioned, the artificial molecules are no longer used in menopausal therapy, given their metabolic effects, and natural steroids are preferred in HRT.

II. Natural Estrogens

1. Conjugated Equine Estrogens (CEE)

The most commonly used natural estrogens are the CEEs, which are mainly composed of estrone and estrone sulfate and extracted from pregnant-mare urine. These CEEs are commonly prescribed orally, but also as vaginal creams.

After oral ingestion of 0.625 mg CEE or 1.25 mg E1S or 1 mg micronized E2, similar levels are reached within 30–40 pg/ml of E2 and 150–250 pg/ml of E1 (Barnes and Lobo 1987). CEEs contain various compounds and, in addition to E1, equilin sulfate is present for 20–25%. The latter compound is metabolized in equilenin and 17-hydroxyequilenin. Equilin levels can reach 1.25 mcg/ml after ingestion of 1.25 mg CEE. It can be stored in adipose tissue and released for several weeks after treatment withdrawal. These metabolites account for a great part of the estrogenic effect of CEE and, although low levels of E2 are achieved, the total amount of estrogens is much higher.

2. Estradiol

Estradiol has been successfully micronized and was shown to be active over 24 h when given orally (YEN et al. 1975). Estradiol valerate is also being prescribed orally. All these estrogens, when given orally, result in higher serum levels of E1 and its conjugates than E2. Levels of E1 as high as 466 pg/ml have been published after oral ingestion of 2 mg micronized estradiol (YEN et al. 1975). The conversion takes place mainly in the intestinal mucosa (BUSTER 1985) and then in the liver, where glucuronidation occurs for about 30% of the initial ingested dose. A rapid urinary and biliary excretion then occurs (BUSTER 1985; BARNES and LOBO 1987).

To avoid this intensive first-pass metabolism, other routes of administration of 17β estradiol have been sought and delivery through injections, implants, vaginal rings, percutaneous gels and transdermal systems has been successfully realized. With these systems of parenteral E2 administration, premenopausal serum levels of E2 are achieved with lower levels of E1, resulting in a more physiologic E2/E1 ratio.

The principal estrogen acting at the cellular and nuclear levels appears to be E2 (GURPIDE 1978). Only free E2 unbound to circulating proteins and SHBG is biologically active. After penetration into the target cells, it binds to the receptor present in the nucleus, where the nuclear retention allows chromatin modifications and specific messages to be induced. The duration of the nuclear retention of the complex formed by the hormone and its receptor determines the potency of the steroid. It has been shown by CLARCK and PECK (1979) that the nuclear retention of E2 is far longer than that of E1, which itself is retained longer than E3. Accordingly, at the cellular level, estrogenic potency is highest for E2 > E1 > E3 (CLARCK and PECK 1979). However, continuous delivery of a weak estrogen through infusion or repeated administration results in the same metabolic effects as with more potent estrogens.

It has been shown that after oral intake of CEE, although E1 was the main circulating estrogen, the estrogen present in the cytosolic fraction of endometrial cells is E2 (GURPIDE 1978; KING et al. 1980).

3. Parenteral Estrogens

After parenteral administration of estrogens, the first-pass gastrointestinal and liver metabolism is avoided. Therefore, the plasma levels measured reflect the dose delivered and absorbed more accurately.

a) Subdermal implantation of estradiol pellets leads to more sustained serum E2 levels of 50–70 pg/ml (LOBO et al. 1980) after a peak effect at insertion.
b) The vaginal epithelium rapidly absorbs estrogens, and E2 can be delivered through silastic rings (MISHELL et al. 1978), giving relatively stable E2 levels of 100–150 pg/ml over several months. Vaginal creams of E2 (RIGG et al: 1978) can also lead to high plasma E2 levels and physiological E2/E1 ratios. The inconvenience of such systems is mainly a quite unstable level of E2

from one day to another, probably because of coincidental, but striking changes in vaginal secretion and vascularization.

c) Percutaneous application and transdermal administration of E2 via a patch provides controlled delivery of the steroid. Rate-controlled systems for administering E2 through intact skin have been designed to be applied either twice per week or, more recently, once a week. They are programmed to release various doses of estradiol in vivo, for relatively small surfaces of application.

Among the skin systems developed so far, the first type used for skin delivery of E2 was the percutaneous application of E2 dissolved into a water–alcohol solvent in the form of a gel containing 3 mg of E2 per 5 g of gel. The application on a fixed surface area of that dose led to E2 plasma levels of 110 ± 24 pg/ml and the usually recommended daily dose of 1.5 mg per day leads to plasma levels of about 50–60 pg/ml which allow the relief of symptoms (SITRUK-WARE et al. 1980; MACINTYRE et al. 1988). The absorption through the skin is proportional, however, to the surface of application (MARTY et al. 1980), and inadequate dosing may lead to inter- and intraindividual fluctuations (WHITEHEAD et al. 1980). This mode of skin delivery of E2 is described as percutaneous administration and should be differentiated from the transdermal delivery using transdermal delivery systems (TDS), also called transdermal therapeutic systems (TTS).

The first generation of rate-controlled systems for administration has been designed to be applied twice weekly (POWERS et al. 1985). They are programmed to release, in vivo, 0.025, 0.05 and 0.1 mg of E2 per day, according to patient needs. These TTS have surface areas of 5 cm^2, 10 cm^2 and 20 cm^2, and they contain 2 mg, 4 mg and 8 mg of E2, respectively, and 100 mg, 200 mg and 400 mg of ethanol, respectively.

Pharmacokinetic studies in postmenopausal women indicated that therapeutic E2 levels in serum were achieved less than 4 h after application and persisted until the TTS removal. In the initial study of POWERS et al. (1985), plasma E2 levels rose from 7 pg/ml to 25 pg/ml, 40 pg/ml or 75 pg/ml, respectively, during the use of these three systems. The E2/ E1 ratio was elevated and maintained close to 1 in the premenopausal range. Further studies have indicated levels in the range 80–90 pg/ml with the 0.05 mg daily dose (ERKKOLA et al. 1995).

More recently, second-generation systems have been introduced using the matrix dispersion-type systems (BARACAT et al. 1995; BORG et al. 1995; SETNIKAR et al. 1996; STUDD et al. 1995). Among these, one is available at a release rate of 0.375 mg, 0.05 mg, 0.75 mg and 0.1 mg estradiol per day and has been approved recently for worldwide use (STUDD et al. 1996). Other matrix dispersion-type systems have been introduced, including one designed for a once weekly application.

The main goal of the development of second-generation TTS was to avoid the ethanol enhancer of the reservoir system, potentially responsible for the skin irritation observed in 14.2% of the first-generation-system users that was severe enough to induce treatment discontinuation in about 5% of the cases (CHEANG et al. 1993). From the presently available literature, although a certain decrease in the rate of slight irritation is observed, there was no striking difference in patients' withdrawal for skin problems between the first- and the second-generation systems. It must be borne in mind that occlusion of the skin is not a physiological condition, especially for several consecutive days.

4. Potency of Various Estrogens According to the Route of Administration

Comparisons of variations in lipoproteins or other liver-produced proteins to assess estrogen potency prove true only when oral preparations are compared (HELGASON 1982). Indeed, the liver is a target organ for steroids and contains E2 receptors, and it has been shown that extraction of circulating estrogens was higher in the liver than in the uterus or the brain, indicating why the hepatic effect of estrogen replacement therapy (ERT) can be supraphysiological, but that a physiological effect is obtained in target organs like the uterus (STEINGOLD et al. 1986).

Parenteral administration of E2, which bypasses the liver, has been shown to achieve therapeutic levels of estrogens, without adversely affecting hepatic globulins (MISHELL et al. 1978; CHETKOWKI et al. 1986; ALKJAERSIG et al. 1988, DE LIGNIERES et al. 1986). However, these therapeutic systems are very effective and, when delivering E2, which is the most potent estrogen at the cellular level, they still avoid modification of liver proteins. Therefore, using change in liver proteins to assess the potency of estrogens is artificial and can no longer be used when non-oral routes of estrogen delivery are considered.

The relevance of the increased synthesis of liver proteins is not known. It is obvious that an increase in circulating levels of SHBG can decrease the availability of free E2 to target cells, which might be considered either undesirable or beneficial. Also, an increase in high-density lipoprotein (HDL) has been related to a decreased risk of cardiovascular disease (BUSH et al 1987) and may be considered beneficial. However, an increase in triglycerides (TG) and an increase in RS, possibly leading to increased blood pressure may be deleterious.

III. Pharmacodynamic Effects of Oral and Non-oral Estrogens

1. On the Endometrium

Unopposed estrogens, whatever the route of administration, would lead to endometrial-cell multiplication and endometrial hyperplasia within a few months of continuous administration. From the progestin/estrogen postmenopause intervention (PEPI) trial, conducted for over 3 years, it was shown that 0.625 mg of CEE administered daily would give up to 62% of endometrial hyperplasia after 3 years of continuous intake. (The Writing Group for the PEPI Trial 1996). As far as transdermal estradiol is concerned, unopposed daily treatment over 2 years would lead to hyperplasia in about 40% of the cases (CLISHAM et al 1992).

The addition of progestin therapy for 12–14 days per month has been shown to protect the endometrium, whatever the estrogen and its route of application. (WOODRUFF and PICKAR 1994; WISEMAN and MCTAVISH 1994).

2. On Bone

Long-term double-blind placebo-controlled trials have documented the bone-sparing effect of estrogens, whether given orally or parenterally (CHRISTIANSEN et al. 1980, ETTINGER et al. 1992; FIELD et al. 1992; REGINSTER et al. 1992). From these studies, it has been established that the minimal effective daily dose of estrogen therapy to prevent bone loss would be 0.625 mg of CEE, 1–2 mg of oral estradiol, 50 μg of transdermal estradiol and 1.5–3 mg of percutaneous estradiol.

HASSAGER et al. (1994) compared several long-term studies and indicated that oral estradiol or conjugated oral-equine estrogens or percutaneous estradiol would maintain the bone mineral density and prevent the acceleration of bone loss observed under placebo. Less than 1.2% of the patients would not respond to therapy and would still exhibit bone loss of 1% per year of their total bone mass.

A higher percentage of bone losers was observed by HILLARD et al. (1994) when comparing oral CEE and a transdermal combination of estradiol and norethisterone acetate. Both therapies gave similar results, and a total of 12% of the patients did not respond significantly to the treatment.

A difference of effect has been suggested between oral or non-oral estrogen therapy in women who smoke. It has been shown that oral estrogen users who smoke would loose the beneficial effect of estrogens (KIEL et al. 1992). However, JENSEN et al. (1988) did not find any deleterious effect of smoking on the bone-sparing effect of percutaneous estradiol. No direct comparison has been made between oral and non-oral estrogens in women who smoke.

3. On Cardiovascular Risk Factors

The cardioprotective effect of ERT in the postmenopausal female has been suggested from the results of long-term prospective cohort studies (COLDITZ et al. 1987; STAMPFER and COLDITZ 1991). However, no clear-cut demonstration has been made and a secondary prevention trial was negative (HULLEY et al. 1998).

Despite this, several surrogate markers of cardiovascular risk have been shown to be positively modified by estradiol (CHEANG et al. 1994). Direct vascular effects of estradiol have been identified, both receptor and non-receptor mediated. The latter effects seem to be similar to calcium-antagonist effects (COLLINS et al. 1993).

a) Acute Effects of Estradiol

Different effects are observed either after acute or chronic administration of estradiol: sublingual administration of estradiol has been shown to acutely increase the coronary flow. In postmenopausal women with documented coronary disease, the treadmill exercise was significantly improved after a single dose of sublingual estradiol (ROSANO et al.1993). Plasma levels observed after ingestion of 1 mg estradiol sublingually reached about 900 pg/ml 1 h after intake. Therefore, the observed effects of improved treadmill exercise in coronary-diseased patients were most likely of a pharmacological nature.

In contrast, HOLDRIGHT et al. (1995) evaluating the effect of trandermal estradiol at doses of 50 μg per day, in a double-blind, cross-over trial showed no effect of the therapy on treadmill exercise, 24 h post treatment. In those cases, the plasma levels were of about 70 pg/ml.

b) *Chronic Effects of Estradiol*

Under chronic administration, endothelium-dependent effects have been shown to be induced by estradiol. YLIKORKALA et al. (1995) showed that E2 administered either transdermally at the dose of 50 μg per day or orally at the dose of 2 mg per day, leading to plasma levels of about 60–80 pg/ml, induced a significant decrease in endothelin-1 (ET-1), a vasoconstrictive factor. Also, similar levels of estradiol, obtained after administration of 1.5 mg per day of percutaneous E2 were shown to increase the vasodilator factors prostacyclins, (FOIDART et al. 1991).

Estrogens also induce changes in lipoproteins and, according to the route of administration, those changes are of various magnitude (CROOK et al. 1992). Oral estrogens, either CEE or E2, induce an increase in HDL and in TG and a decrease in total cholesterol and LDL. These changes are observed after 3 weeks of intake, due to the direct first-pass effect of estrogens on liver metabolism. In contrast, parenteral administration of E2, either through implants or by transdermal delivery, lead to similar changes in the lipid profile, except for TGs that do not increase under parenteral administration. These changes are observed to a lesser extent than after oral intake and appear only after 12–24 weeks of chronic administration (JENSEN et al. 1987; STANCZYK et al. 1988; SITRUK-WARE and IBARRA DE PALACIOS 1989). Futhermore, oral and non-oral estrogens have no effect on blood pressure, although a slight decrease in systolic and diastolic blood pressure has been described under transdermal-estrogen therapy (PANG et al. 1993)

Glycemia and insulinemia have also been studied under estrogen therapy (CAGNACCI et al. 1992). An intravenous glucose-tolerance test was modified by oral therapy, but not by transdermal administration of steroids in the study of GODSLAND et al. (1993). However, the increase in insulin response, indicating a decrease in glucose tolerance, that was observed in oral steroid users was likely related to the progestin associated with estrogens rather than the estrogen or its route of administration (WYNN et al. 1979).

Finally, clotting factors evaluated under several routes of administration of estradiol were slightly modified by oral treatment, with a decrease in antithrombin-III activity (DE LIGNIERES et al. 1986); however, no change in fibrinogen activity was observed with long-term use of CEE (THE PEPI

TRIAL 1995). The relevance in terms of thromboembolic events is not clear (PEREZ-GUTTHAN et al. 1995).

C. Conclusion

Among the various ways of administering estradiol, the most popular used today are the transdermal systems or the oral tablets. Acceptability of long-term therapy will depend upon the ease of use of therapy and the absence of unwanted side effects. Other routes of administration, such as vaginal rings, nasal sprays or sublingual tablets, have not yet proven their acceptability, but are possible options for the future.

References

Alkjaersig N, Fletcher AP, De Ziegler D, Steingold KA, Meldrum DR, Judd HL (1988) Blood coagulation in postmenopausal women given estrogen treatment. Comparison of transdermal and oral administration. J Lab Clin Med 111:224–228

Baker V (1994) Alternatives to oral estrogen replacement. Obstet Gynecol Clin North Amer 21(2):271–297

Baracat E, Tufik S, Castelo A, Rodrigues de Lima G, Haidar M, Casoy J, Peloso U(1995) Comparative bioavailability of a matrix versus a reservoir transdermal estradiol patch in healthy postmenopausal women. Curr Ther Res 56(4):358–368

Barnes RD, Lobo RA (1987) Pharmacology of estrogens. In: Mishell DR Jr (ed) Menopause. Physiology and Pharmacology. Chicago, Year Book Medical Publishers, pp 301–315

Borg ML, Gimona A, Renoux A, Douin MJ, Djebbar F, Panis-Rouzier R (1995). Pharmacokinetics of estradiol and estrone following repeated administration of Menorest, a new estrogen transdermal delivery system, in menopausal women. Clin Drug Invest 10(2):96–103

Bush TL, Barrett-Connor E, Cowan LD, Criqui MH, Wallace RB, Suchindran CM et al (1987) Cardiovascular mortality and non contraceptive use of estrogen in women: results from the lipid research clinics program follow-up study. Circulation:75:1102–1109

Buster JE (1985) Estrogen metabolism. In: Sciarra JJ (ed) Gynecology and Obstetrics. Philadelphia, Harper and Row

Cagnacci A, Soldani R, Carriero PL, Pauletti AM, Fioretti P, Melis GB (1992) Effects of low doses of transdermal 17β-estradiol on carbohydrate metabolism in postmenopausal women. J Clin Endocrinol Metab 74:1396–1400

Cheang A, Sitruk-Ware R, Utian W (1993) A risk-benefit appraisal of transdermal estradiol therapy. Drug Safety 9(5):365–379

Cheang A, Sitruk-Ware R, Samsioe G (1994) Transdermal oestradiol and cardiovascular risk factors. 101:571–581

Chetkowski RJ, Meldrum DR, Steingold KA, Randle R, Lu JK, Eggena P, Hershman JM, Alkjaersig NK, Fletcher AP, Judd HL (1986) Biologic effects of transdermal estradiol. N Engl J Med 314:1615–1620

Christiansen C, Christensen MS, McNair P, Hagen C, Stocklund KE, Transbol I (1980) Prevention of early postmenopausal bone loss: a controlled 2-year study in 315 normal females. Eur J Clin Invest 10:273–279

Clarck JH, Peck EJ Jr (1979) Female Sex Steroids. Receptors and Function. Monographs on Endocrinology. New York, Springer

Clisham P, Cedars MI, Greendale G, Fu YS, Gambone J, Judd H (1992) Long-term transdermal estradiol therapy: effects on endometrial histology and bleeding patterns. Obstet Gynecol 79:196–201
Colditz GA, Willett WC, Stampfer MJ, Rosner B, Speizer FE, Hennekens CH (1987) Menopause and the risk of coronary heart disease in women. N Engl J Med 316:1105–1110
Collins P, Rosano GMC, Jiang C, Lindsay D, Sarrel PM, Poole-Wilson PA (1993) Cardiovascular protection by estrogen, a calcium antagonist effect. 341:1264–1265
Crook D, Cust MP, Gangar KF, Worthington M, Hillard TC, Stevenson JC, Whitehead MI, Wynn V (1992) Comparison of transdermal and oral estrogen-progestin replacement therapy: effects on serum lipids and lipoproteins. Am J Obstet Gynecol 166:950–955
De Lignieres B, Basdevant A, Thomas G, Thalabard JC, Mercier-Bodard C, Conard J, Guyenne TT, Mairon N, Corvol P, Guy-Grand B, Mauvais-Jarvis P, Sitruk-Ware R (1986) Biological effects of estradiol 17β in postmenopausal women: oral versus percutaneous administration. J Clin Endocrinol Metab 62:536–541
Erkkola RU, Hirvonen EJ, Juntunen K, Saastamoinen JA, Tuimala RJ, Rinne R, Paakkari I (1995) Comparison of transdermal to oral hormone replacement therapy. Eur Menopause J 2:21–27
Ettinger B, Genant HK, Steiger P, Madvig P (1992) Low dosage micronised 17β estradiol prevents bone loss in postmenopausal women. Am J Obstet Gynecol 166:479–488
Field CS, Ory SJ, Wahner HW, Hermann RR, Judd HL, Riggs BL (1992) Preventive effects of transdermal 17β-estradiol on osteoporotic changes after surgical menopause: a two year placebo-controlled trial. Am J Obstet Gynecol 168:114–121
Foidart JM, Dombrowicz N, de Lignières B (1991) Urinary excretion of prostacylin and thromboxane metabolites in postmenopausal women treated with percutaneous estradiol (Oestrogel) or conjugated estrogens (Premarin): In: Dusitsin N et al. (eds) Physiological hormone replacement therapy. Carnforth Parthenon, pp 99–107
Godsland IF, Gangar K, Walton C, Cust MP, Whitehead MI, Wynn V, Stevenson JC (1993) Insulin resistance, secretion and elimination in postmenopausal women receiving oral or transdermal hormone replacement therapy. Metabolism 42:846–853
Goebelsmann U, Mashchak A, Mishell DR Jr (1985) Comparison of hepatic impact of oral and vaginal administration of ethinyl estradiol. Am J Obstet Gynecol 151:868–877
Gurpide E (1978) Metabolic influences of the action of estrogens: therapeutic implications. Pediatrics (Suppl) 62:1114–1120
Hassager C, Jensen SB, Christiansen C (1994) Non-responders to hormone replacement therapy for the prevention of postmenopausal bone loss: Do they exist? Osteoporosis Int 4:36–41
Helgason S (1982) Estrogen replacement therapy after the menopause. Estrogenicity and metabolic effects. Acta Obstet Gynecol Scand (Suppl) 107:5–29
Henzl MR (1986) Contraceptive hormones and their clinical use. In: Yen SSC, Jaffe RB (eds) Reproductive Endocrinology 2nd edition. Saunders, Philadelphia, pp 643–682
Hillard TC, Whitcroft SJ, March MS, Ellerington MC, Lees B, Whitehead MJ, Stevenson JC (1994) Long-term effects of transdermal and oral hormone replacement therapy on postmenopausal bone loss. Osteoporosis Int 4:341–348
Holdright DR, Sullivan AK, Wright CA, Sparrow JL, Cunningham D, Fox KM (1995) Acute effect of estrogen replacement therapy on treadmill performance in postmenopausal women with coronary artery disease. Eur Heart J 16:1566–1570
Hulley S, Grady D, Bush T, Furberg C, Herrington O, Riggs B, Vittinghoff E (1998) Heart and estrogen/progestin replacement study (HERS) research group.

Randomized trial of estrogen plus progestin for secondary prevention of coronary heart disease in postmenopausal women. JAMA 208:605–613
James IM (1996) Pharmacological principles of drug action. In: Ginsburg J (ed) Drug therapy in Reproductive Endocrinology. London, Arnold, pp 1–20
Jensen J, Christiansen C (1988) Effects of smoking on serum lipoproteins and bone mineral content during postmenopausal hormone replacement therapy. Am J Obstet Gynecol 159:820–825
Jensen J, Riis BJ, Strom V, Nilas L, Christiansen C (1987) Long-term effects of percutaneous estrogens and oral progesterone on serum lipoproteins in postmenopausal women. Am J Obstet Gynecol 156:66–71
Kiel DP, Baron JA, Anderson JJ, Hannan MT, Felson DT (1992) Smoking eliminates the protective effect of oral estrogens on the risk for hip fracture among women. Ann Intern Med 116:716–721
King RJB, Dyer G, Collins WP (1980) Intracellular estradiol, estrone and estrogen receptor levels in endometria from postmenopausal women receiving estrogens and progestins. J Steroid Biochem 13:377–382
Lobo RA, March CM, Goebelsmann V, Krauss RM, Mishell DR Jr (1980) Subdermal estradiol pellets following hysterectomy and oophorectomy. Am J Obstet Gynecol 138:714–719
MacIntyre I, Stevenson JC, Whitehead MI, Wimalawansa SJ, Banks LM, Healy MJR (1988) Calcitonin for prevention of postmenopausal bone loss. Lancet 1:900–902
Marty JP, James M, Hajo N, Wepierre J (1980) Percutaneous absorption of oestradiol and progesterone: pharmacokinetic studies. In: Mauvais-Jarvis P, Vickers CFH, Wepierre J (eds) Percutaneous absorption of steroids. London, Academic Press, pp 205–218
Mashchak CA, Lobo RA, Dozono-Takano R (1982) Comparison of pharmacodynamic properties of various estrogen formulations. Am J Obstet Gynecol 144:511–518
Mishell DR Jr, Moore DE, Roy S, Brenner PF, Page MA (1978) Clinical performance and endocrine profiles with contraceptive vaginal rings containing a combination of estradiol and d-norgestrel. Am J Obstet Gynecol 130:55–62
Pang SC, Greendale GA, Cedars Ml, Cambone JC, Lozano K, Eggena P, Judd HL (1993) Long term effects of transdermal estradiol with and without medroxyprogesterone acetate. Fertil Steril 59:76–82
Perez-Gutthann S, Garcia-Rodriguez LA, Duque-Oliart A, Castellsague-Pique J (1995) HRT and the risk of venous thromboembolic event. Pharmacoepidemiol Drug Safety 4(Suppl 1 Abstract 118):S53
Powers MS, Schenkel L, Darley PE, Good WR, Balestra JC, Place VA (1985) Pharmacokinetics and pharmacodynamics of transdermal dosage forms of 17β-estradiol: comparison with conventional oral estrogen used for hormone replacement. Am J Obstet Gynecol 152:1099–1106
Reginster JY, Sarlet N, Deroisy R, Albert A, Gaspard U, Franchimont P (1992) Minimal levels of serum estradiol prevent postmenopausal bone loss. Calcif Tissue Int 51:340–343
Rigg LA, Hermann H, Yen SSC (1978) Absorption of estrogens from vaginal creams. N Engl J Med 298:195–197
Rosano GMC, Sarrel PM, Poole-Wilson PA, Collins P (1993) Beneficial effects of estrogen on exercise-induced myocardial ischemia in women with coronary artery disease. Lancet 342:133–136
Setnikar I, Rovati LC, Thebault JJ, Guillaume M, Mignot A, Renoux A, Gualano V (1997) Pharmacokinetics of estradiol and estrone during application of three strengths of an estradiol transdermal patch with active matrix. Arzneim Forsch/Drug Res 47(II)n7:859–865
Sitruk-Ware R, Ibarra de Palacios P (1989) Estrogen replacement therapy and cardiovascular disease in postmenopausal women. A review. Maturitas 11:259–274
Sitruk-Ware R, De Lignieres B, Basdevant A, Mauvais-Jarvis P(1980) Absorption of percutaneous oestradiol in postmenopausal women. Maturitas 2:207–211

Stampfer MJ, Colditz GA (1991)Estrogen replacement therapy and coronary heart disease: A quantitative assessment of the epidemiologic evidence. Prev Med 20:47–63

Stanczyk FZ, Shoupe D, Nunez V, Macias-Gonzales P, Vijod MA, Lobo A (1988) A randomized comparison of non-oral estradiol delivery in postmenopausal women. Am J Obstet Gynecol 159:1540–1546

Steingold KA, Cefalu W, Pardrige W, Judd HL, Chaudhuri G (1986) Enhanced hepatic extraction of estrogens used for replacement therapy. J Clin Endocrinol Metab 62:761–766

Studd JWW, McCarthy K, Zamblera D, Burger HG, Silberberg S, Wren B, Dain MP, Le Lann L, Vandepol C (1995) Efficacy and tolerance of Menorest compared to Premarin in the treatment of postmenopausal women. A randomised, multicentre, double-blind, double-dummy study. Maturitas 22:105–114

The writing group for the PEPI trial. (1995) Effects of estrogen or estrogen/progestin regimens on heart disease risk factors in postmenopausal women: the Postmenopausal Estrogen/Progestin Intervention (PEPI) Trial. JAMA 273:199–208

The writing group for the PEPI Trial (1996) Effects of hormone replacement therapy on endometrial histology in postmenopausal women. JAMA 275:370–375

Von Schoultz B (1988) Potency of different estrogen preparations. In: Studd JWW, Whitehead MI (eds) The Menopause. Oxford, Blackwell Scientific, pp 130–137

Whitehead MI, Townsend PT, Kitchin Y (1980) Plasma steroid and protein hormone profiles in postmenopausal women following topical administration of estradiol 17β. In: Mauvais-Jarvis P, Vickers CFH, Wepierre J (eds) Percutaneous Absorption of Steroids. London, Academic Press, pp 231–248

Wiseman LR, McTavish D (1994) Transdermal Estradiol/Norethisterone. Drugs and Aging 4:238–256

Woodruff JD, Pickar J(1994) Incidence of endometrial hyperplasia in postmenopausal women taking conjugated estrogens (Premarin) with medroxy progesterone acetate or conjugated estrogens alone. Am J Obstet Gynecol 70:1213–1223

Wynn V, Adams PW, Godsland I, Melrose J, Niththyananthan R, Oakley NW, Seed M (1979) Comparison of effects of different combined oral contraceptive formulations on carbohydrate and lipid metabolism. Lancet 1:1045–1049

Yen SSC, Martin PL, Burnier AM, Czekala NM, Greney MO Jr, Callantine MR (1975) Circulating estradiol, estrone and gonadotropin levels following the administration of orally active 17β-estradiol in postmenopausal women. J Clin Endocrinol Metab 40:518–520

Ylikorkala 0, Orpana A, Puolakka I, Pyöräla T, Viinikka L (1995) Postmenopausal hormonal replacement decreases plasma levels of endothelin I. J Clin Endocrinol Metab 80:3384–3387

Part 6
Kinetics and Toxicology of Estrogens and Antiestrogens

CHAPTER 35

Pharmacokinetics of Exogenous Natural and Synthetic Estrogens and Antiestrogens

W. KUHNZ, H. BLODE, and H. ZIMMERMANN

A. Natural Estrogens

In this chapter, the pharmacokinetics of exogenously administered natural estrogens will be described. Natural estrogens considered here include:

17β-estradiol as the main-acting estrogen in humans.

Esters of 17β-estradiol, such as estradiol valerate, estradiol benzoate and estradiol cypionate. Esterification aims at either better absorption after oral administration or a sustained release from the depot after intramuscular administration. During absorption, the esters are cleaved by endogenous esterases and the pharmacologically active 17β-estradiol is released; therefore, the esters are considered as natural estrogens.

Estrone.

Estriol.

Conjugated equine estrogens, which are not natural estrogens for humans, but, because of their therapeutic use in estrogen replacement therapy, will be treated as natural estrogens.

I. Analytical Methods

The most common methods to determine the natural estrogens 17β-estradiol, estrone and estriol are radioimmunoassays and enzyme immunoassays, and a number of them are commercially available (e.g., GHOSH 1988; HENDERSON et al. 1995; LONNING and EKSE 1995; PODESTA et al. 1996; TAMATE et al. 1997; ZIMMERMANN et al. 1998a). Other methods described include a chemiluminescent immunoassay (SATO et al. 1996); receptor-binding assays (IIDA et al. 1991); high-performance liquid chromatography with photometric detection (CASTAGNETTA et al. 1991; DESTA 1988; SLIKKER et al. 1981; TOWNSEND et al. 1988); chemiluminescence detection (NOZAKI et al. 1988); electrochemical detection (NOMA et al. 1991) or fluorimetric detection (KATAYAMA and TANIGUCHI 1993; CHANDRASEKARAN et al. 1996); adsorptive stripping voltammetry (HU et al. 1992); recombinant cell bioassay (KLEIN et al. 1994); and gas chromatography–mass spectrometry (GC–MS) (SPINK et al. 1990; VANLUCHENE et al. 1983; ZIMMERMANN et al. 1998b).

However, most of these assays have not been developed for pharmacokinetic studies and their associated requirements. In the last few years, many reports have dealt with the problems of accuracy, specificity and/or variability of different immunological methods especially for estradiol determination (CARLSTRÖM 1996; COOK and READ 1995; LICHTENBERG et al. 1992; MIKKELSEN et al. 1996; POTISCHMAN et al. 1994; THOMAS et al. 1993; ZIMMERMANN et al. 1998a). In summary, big differences were found in the estradiol concentrations when different kits were used for analysis of the same sample. Especially when women are under estrogen replacement therapy, most of the commercially available assays do not yield reliable results (LOBO et al. 1996). Even if extraction procedures are used, estradiol concentrations measured with immunological methods should be considered with caution. Thus, the extremely high estradiol concentrations measured by TEPPER et al. (1994) are probably confounded by cross-reacting compounds.

In addition to the other endogenous steroids that might interfere with the antibody, there may be cross-reacting metabolites of estradiol or of the co-administered progestin as well as non-specific interferences by matrix constituents, which may result in incorrect measurements. In particular, the high amount of conjugated estrone (up to nearly 1000-fold of the estradiol concentrations), which is produced after oral administration of estradiol or its esters, may cross-react with the antibody. Therefore, it is recommended that assay validation of estradiol should be performed in serum samples spiked with estradiol, a 5-fold concentration of estrone and a 100 to 1000-fold concentration of conjugated estrone. Another possibility to demonstrate the accuracy and specificity of a radioimmunoassay is a cross-validation against a specific and sensitive GC–MS method. In recent years, highly specific and sensitive GC–MS methods have been developed (BORG et al. 1995) and they are the methods of choice for pharmacokinetic studies with natural estrogens. Nevertheless, so far, most of the knowledge of estrogen pharmacokinetics has been acquired using immunoassays. The limit of quantitation (LLQ) of most of the immunoassays and GC–MS methods is in the range of 5–20 pg/ml and, in one case, even a LLQ of 2 pg/ml was reported for a recombinant cell bioassay (KLEIN et al. 1994).

II. Pharmacokinetics in Animals

The pharmacokinetics of exogenously administered estradiol and other natural estrogens has been investigated in several animal species, including dogs, monkeys, rats, pigs, guinea pigs, hamsters and rabbits.

1. Absorption

After oral administration, estradiol valerate and estradiol are completely absorbed, but the drug is metabolized extensively during both the absorption process in the gastrointestinal tract and the first liver passage and is, therefore, subject to a substantial first-pass effect (DÜSTERBERG and NISHINO 1982). In pigs, estradiol represented only 6% of the total estrogens detected in the portal vein after administration of estradiol into the stomach (RUOFF and DZIUK

1994). In rats, estradiol valerate was completely split into estradiol and valeric acid during passage through the duodenal wall (DÜSTERBERG and NISHINO 1982). Following intragastric administration of estradiol or estradiol valerate to rats, the bioavailability of estradiol was in the range of 0.5–6.6% (PATEL et al. 1995; KUHNZ and PUTZ 1989; LOKIND et al. 1991). In marmoset monkeys, an approximate 10% bioavailability was achieved after sublingual administration of estradiol using the intramuscular route as reference (KHOLKUTE et al. 1987).

When the prodrug *O*-(saccharinylmethyl)-17β-estradiol was administered to rats, the oral bioavailability of estradiol increased to approximately 16% (PATEL et al. 1995). With other prodrugs of estradiol administered to dogs, a 5-fold (estradiol-3-anthranilate) or 17-fold (estradiol-3-acetylsalicylate) increase in the bioavailability was observed compared with unmodified estradiol (HUSSAIN et al. 1988). When LOKIND et al. (1991) tested the above-mentioned and other prodrugs in the rat, they found no relevant increase in the bioavailability compared with estradiol valerate.

2. Distribution

Estradiol is rapidly distributed in the body. In plasma, estradiol is bound to albumin, and in primates and rabbits also to sex hormone-binding globulin (SHBG), although in the rabbit the affinity of estradiol for SHBG is very low (BOURGET et al. 1984). According to LITTLETON and ANDERSON (1972), the distribution of estradiol in rats is characterized by two disposition phases. Following intravenous administration of the drug, estradiol concentrations in the plasma decline in a first disposition phase with a half-life of about 5 min, indicating mixing of estradiol with the blood and, thereafter, in a second disposition phase with a half-life of about 30 min, representing the utilization and metabolism of estradiol by target tissues.

Distribution of estradiol in the tissues is influenced by receptor binding. Therefore, tissue distribution of estradiol was found to be different between ovariectomized and intact animals, adult and immature animals and animals of each gender. After intraperitoneal injection of estradiol to ovariectomized rats, total estradiol uptake in the tissues correlated well with the receptor content. It was highest in the uterus, followed by anterior pituitary and hypothalamus (GOMEZ-BENITEZ et al. 1984). After intravenous administration of estradiol to rats, HEFFNER (1976) found the highest concentrations in the anterior pituitary, followed by the uterus and the hypothalamus. When total radioactivity was measured after intravenous injection of ^{3}H-estradiol, BANERJEE et al. (1973) found the highest activity in the pituitary, followed by the uterus, liver, vagina, adrenals, hypothalamus, serum and cerebellum. Similarly, JEHAN et al. (1982) investigated the distribution of radiolabel after intravenous administration of ^{3}H-estradiol in rats, and they found the highest activities in the uterus and liver, followed by the spleen, kidney and plasma, and finally the muscle tissue.

The age-dependent tissue distribution of ^{3}H-estradiol, administered intraperitoneally to rats, was investigated by Woolley et al. (1969). In adult rats, they found the highest levels of radioactivity in the liver, followed by the uterus and anterior pituitary, and then other brain regions and the plasma. In immature rats they found the highest radioactivity in the plasma. The differential uptake of estradiol in various tissues was supposed to be due to the presence of tissue-specific receptor systems characterized by different capacities (Luttge and Whalen 1972). In mice, the concentration of estradiol in the pituitary and in the testes was also much higher than in plasma or fat (Bollingier et al. 1972).

In monkeys suffering from endometriosis, the radioactivity in endometriosis tissue was nearly as high as in the endometrium and was about ten times higher than in subcutaneous fat, skeletal muscle or plasma. All of the radioactivity in the endometrium and endometriosis tissue was represented by estradiol (Eisenfeld et al. 1971). Estradiol passes the blood–brain barrier in rhesus monkeys. The average concentration of ^{3}H-estradiol was about 3.5% of the total concentration in plasma, representing roughly the unbound fraction of estradiol (Marynick et al. 1976).

Balikian et al. (1968) investigated the distribution of radioactivity in dogs after intravenous administration of ^{3}H-estrone. They observed preferential concentration of estrone in omentum and no preferential concentration of estradiol in any tissue.

Estradiol undergoes enterohepatic recirculation in rats and mice (Bolt 1979; Fullerton et al. 1987). Especially glucuronide, but also sulfate conjugates of estradiol are excreted with the bile and reach the intestinal tract where the aglycones are liberated by enzymes of the gut microflora. Brewster et al. (1977) demonstrated a 50% reduction in the degree of enterohepatic circulation in rats following treatment with the antibiotic neomycin.

3. Metabolism

The metabolism of estradiol and estrone in animals and humans is very similar. The major metabolic pathway of estradiol metabolism is the conversion to estrone. Other important metabolic reactions are the 2- and 16-hydroxylation of estrone and estradiol. Further metabolites of the 2- and 16-hydroxylated estrogens include estriol, 16-epi-estriol, 2-hydroxy-estriol, 2-methoxy-estrone, 2-methoxy-estradiol and 2-methoxy-estriol (Bolt 1979). The main conjugation products are sulfates and glucuronides. Unlike humans, the interconversions of estrone and estrone sulfate are nearly the same in rhesus monkeys, whereas in humans the conversion of estrone to estrone sulfate is predominant (Longcope et al. 1994).

The main metabolites found in plasma of mice were estriol-16-glucuronide, estrone-3-glucuronide, estriol-17-sulfate, estrone-3-sulfate, estriol, estriol-3-sulfate and estradiol-17-glucuronide. The metabolite pattern was different after oral and intravenous administration. (Fullerton et al.

Table 1. Metabolic clearance rates (CL) of estradiol (E2) and estrone (E1) and conversion ratios of E2 to E1and vice versa

Reference	Species	E2 CL (ml/min/kg)	E1 CL (ml/min/kg)	E2 ⇒ E1 (%)	E1 ⇒ E2 (%)
DE HERTOGH et al. 1970	Rat	75 (p) 120 (b)	63 (p) 99 (b)	14–20	8–8.8
LARNER and HOCHBERG 1985	Rat	174 (s)	–	–	–
BOURGET et al. 1984	Rabbit	31 (b)	41 (b)	27	9
FRASER et al. 1976	Rabbit	61.5 (b)	–	27.6	–
ELSAESSER et al. 1982	Pig	48.5 (p)	–	15.9	–
KAZAMA and LONGCOPE 1972	Sheep	63 (b)	60 (b)	15.6	13.8

E2 ⇒ E1, conversion of E2 to E1; E1 ⇒ E2, conversion of E1 to E2; s, serum; b, blood; p, plasma

1987). The main urinary metabolites of estriol in baboons were estriol-16-glucuronide, estriol-3-glucuronide, estriol-3-sulfate and estriol 3-sulfate-16-glucuronide (MUSEY et al. 1973). The metabolic clearance rates (CL) of estradiol and estrone from several species are summarized in Table 1.

4. Excretion

In rodents, estradiol is mainly excreted in metabolized form with the feces (BOLT 1979; FULLERTON et al. 1987), whereas in monkeys, dogs and pigs, urinary excretion is predominant (ELSAESSER et al. 1982; YAMAMOTO et al. 1979; WASSER et al. 1994). Both, estradiol and estrone undergo rapid and extensive metabolism in the rat, and the metabolites are excreted mainly as glucuronide and sulfate conjugates via the bile. Within 90 min of intravenous administration of estrone to rats, about 40% of the dose was excreted with the bile (TAKIKAWA et al. 1996).

III. Pharmacokinetics in Humans

The pharmacokinetics of natural estrogens in the human has been investigated after intravenous, oral, transdermal, intramuscular, subcutaneous and intravaginal administration. Basic pharmacokinetic parameters of estradiol and estrone obtained after intravenous administration are shown in Tables 2 and 3.

The route of administration not only has an impact on the absorption and distribution of estradiol and other natural estrogens, but also strongly influences the extent of metabolite formation. This is of importance, since some of these metabolites, e.g., estrone, contribute to the pharmacodynamic response.

Because of the low oral bioavailability of estradiol, different approaches were taken to overcome this disadvantage. Micronization or esterification of

Table 2. Pharmacokinetic parameters of estradiol (mean values) obtained after intravenous administration

Parameter		Reference
CL ($ml\,min^{-1}\,kg^{-1}$)	14.8*, y	LONGCOPE et al. 1985
	29.9, y	KUHNZ et al. 1993
	15*, y	ANDERSON 1993
	10.5*, p	ANDERSON 1993
	11.9*, y	HEMBREE et al. 1969
	16.3*, m	HEMBREE et al. 1969
	14.9*, ep	HEMBREE et al. 1969
	9.3*, lp	HEMBREE et al. 1969
	11.2, p	DÜSTERBERG et al. 1985
	30*, p	GABRIELSSON et al. 1995
	13.5*, y	LONGCOPE and WILLIAMS 1974
V_z ($L\,kg^{-1}$)	1.17*, y	KUHNZ et al. 1993
	0.85*, p	GABRIELSSON et al. 1995
$t_{1/2}$ (h)	1.7, y	KUHNZ et al. 1993
	2.2, p	DÜSTERBERG et al. 1985
	0.33, p	GABRIELSSON et al. 1995
f (%)	5.5, y	KUHNZ et al. 1993
	3, p	DÜSTERBERG et al. 1985

*Recalculated from original data (if no body weight was given, 60 kg was used for calculation in females and 70 kg in males).
Absolute bioavailability (*f*) was determined from intravenous and oral administration of estradiol to the same subjects. *y* young women.
p, postmenopausal women; m, male subjects; ep, early menopausal; lp, late menopausal; $t_{1/2}$, terminal half-life; CL, total clearance from plasma/serum; V_z, volume of distribution during terminal disposition phase

Table 3. Clearance values of estrone (mean values) obtained after intravenous administration

CL ($ml\,min^{-1}\,kg^{-1}$)	Reference
25.5*, y	ANDERSON 1993
18.5*, p	ANDERSON 1993
21.6*, y	LONGCOPE and WILLIAMS 1974
18.9*, y	HEMBREE et al. 1969
23.9*, m	HEMBREE et al. 1969
19.8*, ep	HEMBREE et al. 1969
13.6*, lp	HEMBREE et al. 1969

*Recalculated from original data (if no body weight was given, 60 kg was used for calculation in females and 70 kg in males.
y, young women; p, postmenopausal women; m, male subjects; ep, early menopausal; lp, late menopausal

estradiol did not substantially improve oral bioavailability. Other approaches to improve the bioavailability were either to administer prodrugs of estradiol, such as long-chain fatty esters (SCHUBERT et al. 1993) or sulfamates (ELGER et al. 1995); to choose alternative routes of administration, such as sublingual (PRICE et al. 1997), vaginal (KALUND-JENSEN and MYREN 1984; NASH et al. 1997; SCHMIDT et al. 1994) or intramuscular administration (DÜSTERBERG and NISHINO 1982); or to use subcutaneous pellets (SUHONEN et al. 1993).

The transdermal route of estradiol administration has attracted increasing attention during the last few years, not only because the first-pass effect could be avoided, but also because, in contrast to the oral route, a more physiological ratio of estradiol to estrone concentrations was achieved in the plasma.

There is a considerable inter- and intraindividual variation in the pharmacokinetic parameters of estradiol. After transdermal administration of estradiol to women, BOYD et al. (1996) found an interindividual variation of 44% and an intraindividual variation of only 20% for the area under the serum concentration–time curve (AUC) for estradiol. The interindividual variation of the same parameter observed by other authors using different transdermal formulations was in the range of 20–40% (LEROUX et al. 1995; MÜLLLER et al. 1996; ROHR et al. 1995; SCOTT et al. 1991). After oral administration of estradiol, much higher interindividual variations were reported for the AUC of estradiol: 27.6% (SCHUBERT et al. 1995), 43.4% (SCOTT et al. 1991), 56.5% (HOON et al. 1993), 77.3% (PRICE et al. 1997), 127.3% (KUHNZ et al. 1993a), 54.7% (ZIMMERMANN et al. 1998a) and 54.2% (ZIMMERMANN et al. 1998b). The higher interindividual variability after oral administration is not surprising because of the high first-pass effect. The intraindividual variation after oral administration, however, seems to be slightly lower. In one study, intraindividual variation of the AUC for estradiol was found to be 26.1% in postmenopausal women (ZIMMERMANN et al. 1998a), whereas in another study, where estradiol was administered orally to young women, values of 56%, 33% and 16% were calculated for estradiol, unconjugated estrone and total estrone, respectively. These data suggest little intraindividual variation in the extent of absorption of estradiol; however, a larger variation in the systemic availability of the parent compound itself (KUHNZ et al. 1993a).

1. Absorption

Following oral administration of therapeutic doses, estradiol is rapidly and completely absorbed. (FISHMAN et al. 1969). Estradiol valerate is very quickly split into 17β-estradiol and the fatty acid during the absorption process and/or the first passage through the liver (DÜSTERBERG and NISHINO 1982). The influence of particle size on rate and extent of absorption of estradiol was demonstrated by KVORNING and CHRISTENSEN (1981). At a single dose of 2 mg estradiol, maximum estradiol levels in the plasma of about 30–50 pg/ml were observed within 6–10 h of administration to young women. Drug levels increased linearly up to a dose of 4 mg, whereas, after a dose of 8 mg, absorption was incomplete (KUHNZ et al. 1993a).

Estradiol undergoes an extensive first-pass effect, and a considerable part of the dose administered is metabolized in the gastrointestinal mucosa. Together with the presystemic metabolism in the liver, about 95% of the orally administered dose becomes metabolized before entering the systemic circulation. The main metabolites are estrone, estrone sulfate and estrone glucuronide. According to LONGCOPE et al. (1985); 15% of the administered dose of estradiol is absorbed as estrone, 25% as estrone sulfate, 25% as estradiol glucuronide and 25% as estrone glucuronide. In particular, glucuronidation seems to play an important role after oral administration, because the percentage of glucuronides in plasma is much higher after oral than after parenteral administration (LONGCOPE et al. 1985). Estrogens are excreted to a significant degree in bile and are partly reabsorbed from the intestine. The enterohepatic recycling of natural estrogens has a delaying effect on their final elimination from the body (ADLERCREUTZ et al. 1979; ADLERCREUTZ and MARTIN 1980).

2. Distribution

In serum, 38% of estradiol is bound to SHBG, 60% to albumin and only about 2–3% circulate in free form (SIITERI et al. 1982). Estrone, estriol and estrone sulfate all bind poorly to SHBG, but have greater affinity for albumin than estradiol (LEVRANT and BARNES 1994). The time course of estradiol plasma levels following intravenous administration of either estradiol or estradiol valerate can be described, in most cases, by a two-compartment model. A rapid distribution phase of about 6 min is followed by a terminal disposition phase of about 2 h (DÜSTERBERG et al. 1985; KUHNZ et al. 1993a).

After oral administration, the situation is completely different. Because of the large circulating pool of estrogen sulfates and glucuronides, on the one hand, and the enterohepatic recirculation, on the other, the terminal half-life of estradiol represents a composite parameter that is dependent on all of these processes and is in the range of about 13–20 h (KUHNZ et al. 1993a; PRICE et al. 1997; ZIMMERMANN et al. 1998a,b). Because of this, most authors do not calculate half-lives for estradiol and estrone after oral administration.

During repeated oral administration of estradiol- or estradiol valerate-containing drugs, an increase in the trough serum levels of estradiol was observed within the first few days, which is a result of normal accumulation which can be expected on the basis of terminal half-life and dosage interval. No further increase in estradiol levels was seen between the 16th day of administration and after 3 months of daily administration of a sequential combination of estradiol valerate and chlormadinone acetate, and no deviation of linear pharmacokinetics was observed for estradiol or unconjugated and conjugated estrone (ZIMMERMANN et al. 1998b).

3. Metabolism

Estrone sulfate is the principal circulating metabolite of estradiol in plasma, irrespective of the route of administration. About 15% of orally administered estradiol is converted to estrone and about 65% to estrone sulfate. On the one hand, estrone sulfate is a metabolite of estradiol and, on the other, it serves as a pool from which estradiol can be regenerated by the enzyme 17β-hydroxysteroid dehydrogenase. About 5% of estrone and 1.4% of estrone sulfate can be converted to estradiol. Another 21% of estrone sulfate can be broken down to estrone, whereas the conversion of estrone to estrone sulfate is about 54% (LOBO and CASSIDENTI 1992).

FISHMAN et al. (1969) compared the metabolism of estradiol after oral and intravenous administration. They concluded that a certain portion of the orally administered estradiol was promptly glucuronidated in the intestinal tract and in the liver during its first pass before and after conversion to estrone. The major part of the dose of estradiol was metabolized in the same way as after intravenous injection. The hydroxylation of estradiol and, in particular, estrone is under the control of cytochrome P450 monooxygenases and hydroxylations in positions C-2, C-4 and C-16 are of major importance.

The hydroxylation in position C-2 plays the most important role in humans (BALL and KNUPPEN 1990), followed by the C-16 hydroxylation. SCHNEIDER et al. (1982) looked at the estradiol metabolism in positions C-17 (oxidation) and subsequent hydroxylations at C-2 and C-16, and found oxidation rates of 73%, 32.7% and 14.9%, respectively. The 2-hydroxylation is catalyzed by cytochromes of the CYP3A family (KERLAN et al. 1992) and, to a lesser extent, by cytochromes of the CYP1A family (MARTUCCI and FISHMAN 1993). According to the in vitro studies of SHOU et al. (1997), these two cytochrome families are also the most important ones for estrone metabolism. But the authors found that, in case of estrone, CYP1A2 was the enzyme mainly responsible for the 2-hydroxylation, whereas CYP3A4 preferentially formed 16α-estrone and an unknown hydroxylation product. Further metabolites of the 2- and 16-hydroxylated estrogens include estriol, 16-epi-estriol, 2-hydroxy-estriol, 2-methoxy-estrone, 2-methoxy-estradiol and 2-methoxy-estriol. The catechol-estrogens (2-hydroxy- and 4-hydroxy-estrogens) can be further metabolized in addition to the methylation to ortho-semiquinones (BOLT 1979).

ADLERCREUTZ et al. (1994) reported ethnic differences in the metabolism of estradiol between Caucasian and Oriental women. The 16α-hydroxylation of estrone was significantly increased in the Oriental women, whereas 2-, 4- and 16β-hydroxylation were similar in both groups. TAIOLI et al. (1996) reported ethnic differences in the ratio of the urinary metabolites 2-hydroxy-estrone/16α-hydroxy-estrone as 2.25 for Caucasians and 1.42 for African-American women.

4. Excretion

Following parenteral administration of ^{14}C-estradiol and ^{14}C-estrone, about 50% of the dose is excreted via the bile, but only 7% is excreted via the feces, whereas more than 80% is found in the urine. (SANDBERG and SLAUNWHITE 1957). The same ratio between renal and fecal excretion was observed after the administration of an oral dose of 2 mg ^{3}H-estradiol valerate, where 54% of the dose was recovered in the urine and 6% in the feces within 24 h (DÜSTERBERG and NISHINO 1982). The most likely explanation is the efficient enterohepatic recirculation of biliary excreted metabolites which, following reabsorption, are metabolized further and eventually excreted in the urine.

LONGCOPE et al. (1985) compared the urinary excretion following oral and intravenous administration of radiolabeled estradiol. For all metabolites identified, the ratio between glucuronides and acid-hydrolyzable conjugates was around 0.6 (estriol, 16α-hydroxy-estrone, 2-hydroxy-estrone, 2-hydroxy-estradiol, 2-methoxy-estrone, 2-methoxy-estradiol), with the exception of estrone and estradiol where it was about 2. After oral administration of ^{3}H-estradiol, the main fractions (expressed as a percentage of the administered dose) found in urine were glucuronides of estrone (13–30%), 2-hydroxy-estrone (2.6–10.1%), estradiol (5.2–7.5%), estriol (2.0–5.9%) and 16α-hydroxy-estrone (1.0–2.9%). After intravenous administration, the ranking of the above-mentioned glucuronides was 2-hydroxy-estrone, estrone, estriol, estradiol and 16α-hydroxy-estrone. Following intravenous administration of ^{3}H-estradiol, ZUMOFF et al. (1980) found, in the urine of healthy male subjects, glucuronides and other conjugates (expressed as a percentage of the dose administered) of estrone (9.2%), 2-hydroxy-estrone (7.1%), estriol (5.4%), estradiol (2.5%) and 2-methoxy-estrone (1.3%), whereas the ranking in male subjects with prostate cancer was estriol (14%), estrone (9.2%), 2-hydroxy-estrone (7.4%), estradiol (3.4%) and 2-methoxy-estrone (2.7%). Further metabolites identified in urine include 2-hydroxy-estradiol, 2-methoxy-estrone, 2-methoxy-estradiol and 4-hydroxy-estrone (LONGCOPE et al. 1985).

In the feces of pregnant women, the main metabolites were identified as estriol, estradiol, 15α-hydroxy-estradiol, 16-epi-estriol and estrone. They occurred mainly in the unconjugated form (ADLERCREUTZ et al. 1979).

IV. Comparison of Different Routes of Administration

Because of the limited oral bioavailability of natural estrogens, other routes of administration have been used to avoid the high first-pass effect. The transdermal delivery has probably become the most important parenteral route for estradiol in clinical practice. At present, reservoir patches that have to be applied every 3–4 days, matrix patches that have to be applied every 3–4 days or once weekly, and a gel that has to be applied daily are available. Daily delivery rates of estradiol are typically in the range of 25–100 μg, which corresponds to an oral dose of 1–2 mg. However, in contrast to oral administration, a more physiological estrone to estradiol ratio close to 1 is achieved, which is similar to that observed in premenopausal women. SCOTT et al. (1991) calculated an estrone to estradiol ratio of about 1 after administration of an estradiol-containing gel and a patch, and a ratio of about 5 for orally administered estradiol. Although, the use of patches avoids the necessity of daily oral drug intake, it should be pointed out that, after transdermal administration, the fluctuation in estradiol serum levels is almost as large as after multiple oral administration of estradiol (ZIMMERMAN et al., unpublished results).

Intramuscularly administered depot preparations of estradiol esters are also available. Elevated estradiol plasma levels were observed for about 12 days after administration of estradiol valerate or estradiol benzoate (DÜSTERBERG et al. 1985). ORIOWO et al. (1980) compared estradiol and estrone plasma levels after intramuscular administration of 5 mg estradiol valerate, estradiol benzoate and estradiol cypionate. Peak levels of estradiol were reached after 2 days with the benzoate and valerate ester and after 4 days with estradiol cypionate. The estradiol levels returned to baseline values after 4–5 days with the benzoate, after 7–8 days with the valerate and after 11 days with the cypionate ester. WIEMEYER et al. (1986) measured elevated estradiol levels up to 31 days after an intramuscular dose of 10 mg estradiol enanthate.

Vaginal administration is another way of circumventing the oral first-pass effect. The absorption of estrogens is dependent on the condition of the vaginal mucosa and a significantly higher absorption was observed in women with an atrophic vaginal mucosa (PSCHERA et al. 1989). Vaginal absorption of estradiol after single dose administration was lower in premenopausal women than in postmenopausal women; however, after multiple doses, estradiol absorption in postmenopausal women was reduced to premenopausal levels (CARLSTRÖM et al. 1988). Following vaginal administration of 0.5 mg estradiol suspended in saline to postmenopausal women, 860 pg/ml estradiol and 210 pg/ml estrone were observed in plasma after 3 h, declining to levels of 250 pg/ml and 130 pg/ml 6 h postdose, respectively (SCHIFF et al. 1977). ENGLUND and JOHANSSON (1981) measured maximum estradiol concentrations of 675 pg/ml in postmenopausal women, 1 h after intravaginal administration of 0.25 mg estradiol in solution, and maximum estrone concentrations of 100 pg/ml were reached after 2 h. Thus, via the vaginal route, relatively high plasma levels of estradiol, which are by far exceeding those of estrone, can be achieved.

Somewhat divergent results have been reported for the intravaginal administration of estradiol via silastic rings. ENGLUND et al. (1981) used silastic rings that delivered 200 μg estradiol daily. Within the first 24 h, estradiol concentrations in plasma were higher than those of estrone, whereas during the subsequent period of 21 days, both estradiol and estrone concentrations were equal and in the range of 100–200 pg/ml. SCHMIDT et al. (1994) used a different vaginal ring with a daily delivery of about 7–8 μg estradiol. Following an initially high release of estradiol, relatively low estradiol levels of 3 pg/ml were measured from day 3 onwards, compared with estrone levels of 177 pg/ml and total estrone levels of 912 pg/ml. Using the same vaginal-ring system, similar low release rates of about 4 μg/day were reported by JOHNSTON (1996).

In one study of oophorectomized women, the nasal delivery of 0.34 mg estradiol resulted in initially high serum levels of estradiol of about 1360 pg/ml with estrone/estradiol ratios below unity (HERMENS et al. 1991).

A comparative study of sublingual versus oral administration of 1 mg estradiol was performed by PRICE et al. (1997). Following sublingual administration, they measured high initial estradiol concentrations of about 500 pg/ml which declined to about 60 pg/ml after 6 h. The AUC of estradiol was 2.5-fold higher than after oral administration, whereas the AUC of estrone was only 20% higher than after oral dosage. FRIDRIKSDOTTIR et al. (1996) observed maximum serum concentrations of 154 pg/ml, 15 min after sublingual administration of 100 μg estradiol.

Somewhat limited information is available on the delivery of estradiol from subdermal implants. SUHONEN et al. (1993) reported estradiol and estrone levels in the range of 30–70 pg/ml for a period of at least 12 weeks and STUDD and SMITH (1993) described estradiol levels in the range of 40–10 pg/ml during a period of 6 months, estrone levels being lower than those of estradiol.

V. Other Estrogens

1. Conjugated Equine Estrogens

The main ingredients of the equine estrogens are the sodium sulfates of estrone (50.6%), equilin (23.6%), 17α-dihydroequilenin (15.3%), 17α-estradiol (3.8%), equilenin (3.3%), 17β-dihydroequilenin (2.3%) and 17β-estradiol (1.1%); further ingredients are 17α-dihydroequilenin, 17β-dihydroequilenin and δ8,9-dehydroestrone (ANSBACHER 1993).

POWERS et al. (1988) compared the concentrations of estradiol and estrone after administration of 2 mg micronized estradiol and 1.25 mg conjugated equine estrogens, and found that average concentrations of estradiol and estrone were about twice as high after estradiol administration. LOBO et al. (1983), however, found similar serum estradiol and estrone levels after oral administration of 0.625 mg conjugated equine estrogens and 1 mg estradiol. According to O'CONNELL (1995), equal serum estradiol levels were achieved after oral administration of 1.25 mg conjugated equine estrogens, 2 mg estradiol and 2 mg estradiol valerate. The pharmacokinetics of conjugated equine estrogens is complicated because of the many ingredients administered (see above), which are also undergoing intensive first-pass metabolism.

BHAVNANI and CECUTTI (1993, 1994) and BHAVNANI et al. (1994) investigated the pharmacokinetics of equilin, equilenin, 17β-dihydroequilin and 17β-dihydroequilenin and its sulfates in postmenopausal women. The interconversion between 17β-dihydroequilin and equilin was catalyzed by 17β-hydroxysteroid dehydrogenase in analogy to the interconversion of estradiol and estrone. The major conversion products of equilin were free and sulfate-conjugated 17β-dihydroequilin. At the receptor level, 17β-dihydroequilin was found to be a much more potent estrogen than estradiol.

TROY et al. (1994) measured total estrone, estrone, total equilin and equilin after administration of 0.625 mg conjugated equine estrogens. Estradiol and 17β-dihydroequilin were not measured in this study. Maximum serum total estrone concentrations of 5.7 ng/ml occurred 7.7 h after administration; maximum serum estrone concentrations of 135 pg/ml were reached after 8.6 h; maximum serum total equilin concentrations of 4.1 g/ml were reached after 5.8 h; and maximum serum equilin concentrations of 70 pg/ml were reached after 7.3 h. The observed ratio between serum estrone and equilin levels was in disagreement with STUMPF (1990), who stated that serum levels of equilin were many times higher than those of estradiol or estrone after administration of the same preparation.

2. Estriol

The pharmacokinetics of estriol after oral and intravaginal administration was investigated by ENGLUND et al. (1982), HEIMER and ENGLUND (1984, 1986b), RAURAMO et al. (1978), SCHIFF et al. (1980), ENGLUND et al. (1984) and TOLINO et al. (1990), and reviewed by HEIMER (1987). Estriol has been administered orally at doses of 4–10 mg and a fairly large range of maximum concentrations of estriol in the range of 80–340 pg/ml was observed by different authors in the serum, 1–2 h after administration (ENGLUND et al. 1982, 1984; SCHIFF et al. 1980). Elevated estriol levels could be measured up to 24 h post administration. The occurrence of a second peak in the plasma concentration–time curve indicated an enterohepatic recirculation of estriol, which could be confirmed by HEIMER and ENGLUND (1986a). The bioavailability of orally administered estriol was compared with that after vaginal administration and was found to be about 10%. When an equal dose of 4 mg estriol was administered either orally or intravaginally, maximum estriol concentrations of 336 pg/ml were observed 6 h after oral administration, whereas after intravaginal administration, the absorption rate was more rapid, resulting in maximum drug levels of 687 pg/ml as soon as 2 h after administration (SCHIFF et al. 1980). A rapid absorption of estriol was also observed by MATTSON and CULLBERG (1983) after intravaginal administration of 0.5 mg estriol formulated as a cream. Maximum estriol concentrations in the plasma of about 160 pg/ml were observed 2 h after administration.

HEITHECKER et al. (1991) compared estriol levels in the plasma of women after intramuscular administration of estriol and its dipropionate and dihexanoate esters in an oily solution. After injection of 1 mg estriol, basal estriol levels in the plasma were reached between 8 h and 24 h, with a terminal half-life of 1.5–5.3 h, whereas elevated estriol levels were maintained up to 4 days and up to 20–50 days in those women who received the corresponding dipropionate and dihexanoate ester, respectively.

VI. Drug Interactions

1. Interaction with Progestins

Progestins are known to stimulate the enzyme 17β-hydroxysteroid dehydrogenase, especially in the endometrium (TSENG and GURPIDE 1975) and other target cells such as the breast (FOURNIER et al. 1982). MURUGESAN et al. (1989) demonstrated a severalfold increase in the interconversion of estradiol to estrone in rabbit uteri in vitro following treatment with norethisterone or norgestrel. The circulating levels of estradiol in women, however, seem not to be affected since AEDO et al. (1989) found that co-administration of estradiol and levonorgestrel had no influence on the circulating levels of unconjugated estradiol and estrone.

Since hydroxylation reactions of estradiol are catalyzed by enzymes of the CYP3A and CYP1A gene families, there may be a potential for an interaction with progestins, such as levonorgestrel and norethisterone, which are used in combination with estrogens in hormone replacement therapy. However, so far, there are no results available from clinical studies which would indicate such an interaction by changes in circulating plasma levels of estradiol.

2. Interaction of Estradiol with its Own Metabolism

Interaction of estradiol with its own metabolism can be excluded, because no deviation from linear pharmacokinetics was observed after single oral administration of estradiol valerate, after 16 days of administration of estradiol valerate alone and after three cycles of a sequential regimen with chlormadinone acetate (ZIMMERMANN et al. 1998b).

3. Interaction of Estradiol with Other Drugs

There are relatively few reports in the literature dealing with the interaction of exogenously administered natural estrogens with other drugs. Antibiotics that suppress the intestinal microflora can impair the enterohepatic recirculation of estrogens. BREWSTER et al. (1977) found that in rats treated with neomycin there was an impairment of the recirculation of estradiol by about 50%. VAN LOOK et al. (1981) investigated the influence of amoxicillin and ampicillin on endogenous estrogen levels in the plasma of pregnant women. They found an influence on neither unconjugated estriol nor unconjugated and conjugated estradiol, but a short-lasting decrease in conjugated estriol levels.

Rifampicin is a well known inducer of CYP3A4. BOLT et al. (1975) found that liver microsomes from patients treated with rifampicin showed an approximate 4-fold increase in their ability to hydroxylate estradiol. The anticonvulsant phenytoin, another inducer of CYP3A4, was reported to reduce plasma estrone and estradiol concentrations in subjects on oral estrogen replacement therapy (NOTELOVITZ et al. 1981).

Long-term treatment of breast cancer patients with aminoglutethimide, a non-specific aromatase inhibitor, increased estrone sulfate CL, whereas estradiol CL was not influenced (LONNING et al. 1986, 1987).

Cimetidine, a histamine H2-receptor antagonist, used in the treatment of peptic ulcer disease, inhibits the 2-hydroxylation of estradiol (GALBRAITH and MICHNOVICZ 1989). This inhibition has been suggested to explain the findings of gynecomastia (JENSEN et al. 1983) and sexual dysfunction PEDEN et al. (1979) reported in some men after treatment with cimetidine. The proton-pump inhibitor omeprazole, also used for the treatment of peptic ulcer disease, had no effect on estradiol metabolism in man (GALBRAITH and MICHNOVICZ 1993).

HARRIS et al. (1996) compared the pharmacokinetics of prednisolone and erythromycin in postmenopausal women under estrogen therapy and in non-substituted postmenopausal women. No influence of estrogen treatment on either drugs was observed. This contradicts GUSTAVSON et al. (1986) who found a decrease in the CL of prednisolone when postmenopausal women were taking conjugated estrogens.

EUGSTER et al. (1993) demonstrated that estradiol was a potent inhibitor of CYP1A1 in vitro, whereas the effect on CYP1A2 was much less pronounced. The CYP1A1-dependent metabolism of caffeine was inhibited by estradiol, whereas the CYP1A2-dependent metabolism was not. Whether these in vitro results are also relevant in vivo and whether they can adequately explain the observed increased elimination half-life of caffeine in pregnant women remains to be shown.

The effect of smoking on estradiol metabolism in vivo was investigated by CASSIDENTI et al. (1990). After oral administration of 1 or 2 mg estradiol, the serum concentration–time curves of estradiol and free estrone were similar in smokers and non-smokers, but estrone sulfate concentrations were higher in smokers. The estradiol binding capacity of SHBG was higher in smokers and, therefore, the concentrations of free estradiol were lower in smokers both prior to and after administration of estradiol. However, a dose adjustment of estradiol in smokers was not deemed necessary by the authors.

GINSBERG et al. (1995) reported a decrease in estradiol CL caused by ethanol ingestion in women who received estradiol transdermally. Also, after oral administration of 1 mg estradiol to postmenopausal women using estrogen replacement therapy, up to 3-fold higher levels of circulating estradiol were measured after alcohol ingestion than in the control group of women who received the same dose of estradiol without alcohol. However, alcohol did not change estradiol serum levels in the controls who did not receive estrogen replacement therapy. No significant increase in corresponding estrone levels was noted (GINSBERG et al. 1996). The underlying mechanisms and possible consequences have yet to be identified.

After co-infusion of dehydroepiandrosterone sulfate, 5-androstenediol or cortisol with ^{3}H-estradiol, REED et al. (1988) found an approximate 18%

decrease in the metabolic CL rate of estradiol, while the conversion ratio of estradiol to estrone increased by 45%. As possible reasons, they discussed changes in the distribution of estradiol, differences in the metabolism of estradiol bound to albumin or SHBG, and an effect of androgens or cortisol on the uptake of estradiol by the liver.

Unlike synthetic estrogens, conjugated equine estrogens did not decrease antipyrine CL up to a dose of 0.625 mg (SCAVONE et al. 1988), whereas during pregnancy, high estradiol to progesterone ratios in women corresponded to decreased plasma CL values of antipyrine (LOOCK et al. 1988).

A low-fat diet decreased the urinary elimination of estriol and 16α-hydroxyestrone and increased the urinary elimination of catechol estrogens without influencing the metabolic CL rates of estradiol or the interconversion of estradiol to estrone, estrone sulfate, estrone glucuronide or estradiol glucuronide (LONGCOPE et al. 1987).

SCHUBERT et al. (1994, 1995) tested the influence of grapefruit juice on estradiol metabolism in vitro and in vivo. The flavonoids naringenin, quercetin and kaempferol were added to human liver microsomes and showed a dose-dependent inhibition of estradiol metabolism. The formation of estrone was not inhibited, whereas the estriol formation was. In vivo, they found a significantly increased AUC for estrone, but not for estradiol after oral administration of a single dose of 2 mg estradiol when grapefruit juice was co-administered.

Dietary indoles, present in cruciferous vegetables can induce CYP450 enzymes. KALL et al. (1996) found an increase in the average urinary 2-hydroxyestrone/16-hydroxyestrone ratio after 12 days of daily intake of broccoli. MICHNOVICZ and BRADLOW (1990) were able to show that, in rats and in humans, indole-3-carbinol, which is an ingredient of cruciferous vegetables, induced the 2-hydroxylation of estradiol following repeated oral administration of this compound.

B. Synthetic Estrogens

Only a limited number of synthetic estrogens have gained therapeutic relevance. The most important non-steroidal estrogen is diethylstilbestrol. However, since diethylstilbestrol has been found to induce clear cell adenocarcinoma of the vagina in female offspring of women who received this drug during pregnancy, it is no longer in widespread use.

Steroidal estrogens are more important. A major step in increasing the poor oral bioavailability of 17β-estradiol was achieved by the introduction of the ethinyl group in 17α-position which prevents rapid metabolism during the first liver passage. Ethinylestradiol (EE) has become the most widely used estrogenic component of combination oral contraceptives. Historically, the 3-methyl ether of EE, mestranol, was present in the first combination oral contraceptives that appeared on the market. Later, however, it turned out that mestranol is rapidly demethylated in the liver and that metabolically derived EE is the pharmacologically active compound. Since the use of mestranol offered no advantage over EE, the former has markedly lost importance. Quinestrol, the 3-cyclopentyl ether of EE, however, is another potent oral estrogen that also only achieved limited distribution. Following oral administration, quinestrol becomes partly dealkylated to yield EE. Due to its pronounced lipophilicity, quinestrol is stored in and slowly released from body fat, which makes it a long-acting estrogen that can be administered at longer dosing intervals. EE sulfonate, another lipophilic derivative of EE, follows the same principle.

The main focus of this chapter will be the pharmacokinetics of EE and, where appropriate, also of mestranol.

I. Analytical Methods

With the advent of the newer low-dose oral contraceptive formulations containing 30 μg of EE or less per pill, the demands on the analytical methods for the determination of EE in body fluids have become more stringent in terms of sensitivity and specificity. A number of methods exist for the determination of EE in plasma or serum. These comprise of a high-performance liquid chromatography (HPLC) method with electrochemical detection (FERNANDEZ et al. 1993), competitive protein-binding assays (WARREN and FOTHERBY 1973), an enzyme-mediated immunoassay (TURKES et al. 1981) and several radioimmunoassays (BACK et al. 1979; HÜMPEL et al. 1979; DYAS et al. 1981; LEE et al. 1987; STANCZYK et al. 1980). The latter differ either in the antisera used or the sample preparation procedures employed. The lower limit of quantitation (LOQ) of these assays is in the range of 5–20 pg/ml. However, the underlying acceptance criteria for the claimed LOQs are not uniform and are sometimes difficult to recognize in the publications.

Although radioimmunoassays generally provide the necessary sensitivity, it was the lack of specificity of the antisera used, which often limited their use. In addition to the endogenous steroids which might interfere with the antibody, there may be cross-reacting metabolites of either EE itself or the co-administered progestogen, as well as non-specific interferences by matrix constituents which can give rise to false positive results. Because of these limitations, the implementation of additional chromatographic purification or extraction steps prior to sample analysis was advocated. In fact, several assays, including, e.g., Celite-column chromatography (STANCZYK et al. 1980), HPLC (LEE et al. 1987), immunoadsorption (DYAS et al. 1981) or extraction with organic solvents (BACK et al. 1979; HÜMPEL et al. 1979) were developed.

Valuable information on the specificity of the antiserum, in the presence of metabolites of EE and those of the co-administered progestogen, can be obtained from the analysis of ex vivo serum samples obtained after the administration of the EE-containing contraceptive formulation to animals or humans. A generally applicable procedure of how to demonstrate specificity has been suggested by KUHNZ et al. (1993b). Serum samples were extracted and the extracts were separated on a reversed-phase HPLC column. Fractions of the eluate were collected and submitted to radioimmunoassay. Thus, the radioimmunoassay was used as a detector, revealing all compounds that would contribute to the overall radioimmunological response, in addition to EE, if the same extracts were analyzed directly by radioimmunoassay. In the case of endogenous hormones or when particular metabolites of EE or the progestogens are available, the identity of a cross-reacting compound can be elucidated by a comparison of the retention times of the metabolite and the cross-reacting compound under the same chromatographic conditions. This approach allows for the detection of metabolites that only occur in vivo and that are the most likely candidates to interfere with the assay.

Another way of confirming the specificity of a radioimmunoassay is the cross-validation with another independent method of acknowledged specificity, such as GC–MS. Several mass spectrometric assays have been developed for EE (AKPOVIRORO and FOTHERBY 1980; SIEKMANN et al. 1980; TETSUO et al. 1980). However, the complete procedure has not been published in every case in detail (KUHNZ et al. 1993b). Due to the powerful separation efficiency of capillary GC, on the one hand, and the selective registration of drug-specific mass fragments on the other, GC–MS is a highly specific method for the analysis of EE in biological samples with a sensitivity comparable to radioimmunoassay. LOQs of 5–10 pg/ml have been reported. With the optional selection of different selective derivatization and ionization procedures, the choice of different diagnostic fragments and the use of stable isotope-labeled internal standards, GC–MS has not only become a valid alternative to the established radioimmunoassays, but even serves as the "gold standard". However, since GC–MS requires a considerable investment in equipment and skilled personnel, radioimmunoassay is still the more widely used method for EE analysis.

II. Pharmacokinetics in Animals

The pharmacokinetics of EE has been investigated in several experimental animal species, including rat, rabbit, guinea-pig, dog and monkey, and has been reviewed recently (KUHNZ and BLODE 1994). A summary of some basic pharmacokinetic parameters of EE is presented in Table 4.

1. Absorption

EE is completely and rapidly absorbed, but during the absorption process in the gut and the first liver passage, the drug is subject to an extensive presystemic elimination. Studies performed in rats indicated that after intraduodenal administration, about 40–50% of the drug is metabolized in the gut wall, mainly by glucuronidation (REED and FOTHERBY 1979; HIRAI et al. 1981; SCHWENK et al. 1982). The bioavailability is in the range of about 1–10% in

Table 4. Pharmacokinetic parameters (mean values) of EE in selected animal species

Species	f (%)	$t_{1/2}$ (h)	CL (ml/min/kg)	V_z (l/kg)	Reference
Rat	3	2.5	64	73	DÜSTERBERG et al. 1986
	9.9	10.4	73.9	66.5	KUHNZ, unpublished
Rabbit	4	1.4	52.5	7.5	BACK et al. 1980d
	–	1.7	28.3	4.8	BACK et al. 1982b
	0.3	3.0	37	31.7	DÜSTERBERG et al. 1986
	–	0.7–1.9#	80–116#	6.6–13.8#	FERNANDEZ et al. 1996
Dog	9	2.3	25	16.8	DÜSTERBERG et al. 1986
Monkey					
Rhesus	0.6	2.7	17	6	DÜSTERBERG et al. 1986
Rhesus	–	–	21	–	BACK et al. 1982c
Baboon	2	2.5	22	6.9	DÜSTERBERG et al. 1986
Baboon	61*	2.2*	47.5*	9*	NEWBURGER et al. 1983

#Dose range of 30–100 μg/kg; *Recalculated from the original data
f, absolute bioavailability; $t_{1/2}$, terminal half-life; CL, total clearance from plasma/serum; V_z, volume of distribution during terminal disposition phase

rat, rabbit, dog and monkey. Somewhat contradictory is the reported bioavailability of about 64% in the baboon in one study, whereas in two other studies with baboons and rhesus monkeys, oral bioavailability was found to be 2% and 0.6%, respectively (NEWBURGER et al. 1983; DÜSTERBERG et al. 1986).

2. Distribution

The distribution of EE is characterized by at least two disposition phases. Following intravenous administration of the drug, EE concentrations in the plasma decline in a first disposition phase with a half-life of about 0.2–0.5 h, indicating rapid distribution and, thereafter, in a second disposition phase with a half-life of about 1–3 h. Total plasma CL rates almost exclusively represent metabolic CL, since practically no unchanged drug is excreted in the urine. The volume of distribution (Vz) (expressed as l/kg) was highest in the rat, followed by dog, rabbit and monkey. The high values indicate a wide distribution of the drug into tissues and organs. This finding correlates with the plasma protein-binding characteristics of EE, which have been investigated for a number of animal species, including rats, rabbits, dogs, guinea-pigs and monkeys (AKPOVIRORO et al. 1981). In all species, more than 90% of EE was bound almost exclusively by albumin and in none of the species was there any binding to SHBG. CL values of 60–75 ml/min/kg in the rat and 37–116 ml/min/kg in the rabbit exceeded the hepatic plasma flow in both species, indicating an additional contribution of extrahepatic metabolism. In the beagle dog, a total CL of 25 ml/min/kg was observed which is close to the hepatic plasma flow.

EE undergoes enterohepatic recirculation in rats and rabbits. Glucuronide and sulfate conjugates of EE and its metabolites, mainly 2-hydroxy-EE, are excreted via the bile and reach the intestinal tract where the aglycones are liberated by β-glucuronidases and sulfatases of the gut microflora. The reabsorbed EE appears as a secondary peak in the plasma level–time curve between approximately 5 h and 8 h post administration. This enterohepatic circulation of EE can be significantly reduced by pretreatment with antibiotics, such as ampicillin or neomycin (BREWSTER et al. 1977; BACK et al. 1982b; MAGGS et al. 1983b).

3. Metabolism

The metabolism of EE and other synthetic estrogens in animals has been reviewed previously and comprises studies in baboons, rabbits, guinea-pigs and rats (FOTHERBY 1973). Similar to the human situation, hydroxylation and conjugation reactions are of major importance. EE is extensively metabolized by the rat and 2-hydroxylation is the predominant pathway. The metabolites are excreted mainly as glucuronide and sulfate conjugates via the bile. Following enzymatic hydrolysis, EE, 2-hydroxy-EE, 16-hydroxy-EE, 2-methoxy-EE and 2-hydroxy-mestranol could be identified, with 2-methoxy-EE being the principal metabolite. Only a small fraction of the biliary excreted radiolabel was freely extractable and the principle components were EE and 2-methoxy-EE (MAGGS et al. 1982). Similar metabolite patterns were also observed for EE and mestranol during incubation with rat liver slices in vitro (BALL et al. 1973).

In the baboon, the 3-sulfate and the 3,17-disulfate were the major circulating metabolites after intravenous administration of EE, whereas after intragastric administration, the 3-glucuronide and the 3,17-diglucuronide were also observed (NEWBURGER et al. 1983). Following intravenous administration of radiolabeled EE to pregnant rhesus monkeys, methoxy-EE, EE-3-sulfate and EE-3-glucuronide were observed in the plasma in addition to the parent compound (SLIKKER et al. 1982).

Besides EE, the homoannulated derivatives D-homoestrone and D-homoestradiol have been observed in the urine of rabbits after subcutaneous administration of both EE and mestranol (ABDEL-AZIZ and WILLIAMS 1969). The formation of the homoannulated derivative as well as the loss of the ethinyl moiety was also observed in incubations of mestranol with hepatic microsomes from female rhesus monkeys (SCHMID et al. 1983). In contrast to the situation in humans, de-ethinylation occurs to a larger extent in experimental animals. This has, for example, been shown for the baboon in vitro (RAITANO et al. 1981) and the rhesus monkey in vivo (HELTON et al. 1977).

When incubated with hamster liver microsomes in vitro, $[^{14}C]$-EE was metabolized to 2-hydroxy-EE, 4-hydroxy-EE, 7α-hydroxy-EE, D-homoestrone and two further catechol metabolites. The main metabolite, accounting for 47% of the total metabolites formed, was 2-hydroxy-EE (HAAF et al. 1988).

4. Excretion

EE undergoes rapid and extensive metabolism in the rat and the metabolites are excreted mainly as glucuronide and sulfate conjugates via the bile. Following intravenous administration of [^{3}H]-EE to bile duct-cannulated male rats, nearly 60% of the drug was eliminated via the bile (MAGGS et al. 1982). Within 3 days after intravenous administration of [^{14}C]-mestranol to female rats, about 45% and 3.3% of the dose were excreted in the feces and urine, respectively; after 3 weeks, 55% of the radiolabel was recovered from the feces (BOLT and REMMER 1972a, b).

When [^{3}H]-EE was administered orally to rabbits, about 44% of the radiolabel appeared in the feces and only 12.4% was eliminated in the urine after 7 days (BEACH et al. 1967). Following oral administration of radiolabeled EE to female guinea-pigs, about 52% of the radiolabel was excreted in the feces and about 27% was eliminated in the urine within 7 days (REED and FOTHERBY 1975).

In contrast to other species, a major part of the radiolabel (77–83%), mainly represented by glucuronide conjugates, was recovered from urine when ^{14}C-labeled EE was administered intravenously to female baboons (KULKARNI 1970). However, when ^{14}C-labeled mestranol was administered intravenously to the same species, only about 5–21% of the radiolabel was excreted in the urine, suggesting that EE is metabolized differently than mestranol by the baboon (KULKARNI 1976).

III. Pharmacokinetics in Humans

The pharmacokinetics of EE and mestranol in the human has been investigated after intravenous and after oral administration in numerous studies. The basic pharmacokinetic parameters of EE obtained after intravenous administration are summarized in Table 5. The CL of mestranol seems to be similar to or slightly higher than the one observed for EE (MILLS et al. 1974; BIRD and CLARK 1973; LONGCOPE and WILLIAMS 1977).

There is a considerable variation in the pharmacokinetics of EE. Both, inter- and intraindividual coefficients of variation for parameters, such as AUC (0–24 h), were found to be in the range 47–57% and 41–42%, respectively, which is consistent with the observations of others (BRODY et al. 1989; GOLDZIEHER 1989; FOTHERBY 1991). It was also shown in a clinical study that, in spite of adhering to the most stringent inclusion and exclusion criteria, no substantial reduction in the interindividual variability could be achieved (STADEL et al. 1980). Although a marked interindividual variability can be expected for a drug with such a high first-pass effect, the factors that account for the large intraindividual variability of EE are not completely understood.

Table 5. Pharmacokinetic parameters of EE (mean values) obtained after intravenous administration

Parameter		Reference
CL ($ml\,min^{-1}\,kg^{-1}$)	6.3*	Back et al. 1979
	4.5*	Hümpel et al. 1979
	2.3*	Timmer et al. 1990
	5.5*	Orme et al. 1991
	4.3	Baumann et al. 1996
	7.0	Kuhnz et al. 1996
V_z ($L\,kg^{-1}$)	3.8	Back et al. 1979
	3.4	Grimmer et al. 1986
	2.8*	Timmer et al. 1990
	4.9*	Orme et al. 1991
	8.6*	Baumann et al. 1996
	5.8	Kuhnz et al. 1996
$t_{1/2}$ (h)	6.8	Back et al. 1979
	25.5	Hümpel et al. 1979
	19.0	Timmer et al. 1990
	26.1	Baumann et al. 1996
	9.7	Kuhnz et al. 1996
f (%)	42	Back et al. 1979
	43	Hümpel et al. 1979
	45	Grimmer et al. 1986
	59	Orme et al. 1991
	36	Baumann et al. 1996

*Recalculated from original data.
Bioavailability (*f*) was determined from intravenous and oral administration of EE to the same subjects

1. Absorption

Following oral administration of therapeutic doses, EE is rapidly and completely absorbed. At a dose of 30 μg EE, maximum drug levels in the plasma of about 90–130 pg/ml are observed within 1–2 h postadministration. Drug levels increase linearly with a dose in the range of about 20–100 μg EE. EE undergoes a marked first-pass effect. Besides an extensive presystemic metabolism in the liver, direct sulfation of the drug in the gut wall seems to play an important role. The mean oral bioavailability of EE is about 45%, with a large interindividual variation of about 20–65% (Orme et al. 1989).

Following oral administration of mestranol, which itself is pharmacologically inactive, the drug is rapidly converted into EE by demethylation. This demethylation is not complete and it has been found that the plasma levels of EE observed after a dose of 50 μg mestranol are similar to those of 35 μg EE. Only a slight delay in the time to reach maximum plasma levels has been noted after treatment with mestranol (Goldzieher 1990; Goldzieher et al. 1990). It has also been pointed out that due to the large interindividual variation in the pharmacokinetics of EE, plasma levels generated by 35 μg EE in one person may be indistinguishable from those produced by 50 μg in another (BRODY et al. 1989).

2. Distribution

In serum, EE is mainly bound to albumin and only about 2–5% circulate in free form; EE does not bind to SHBG (AKPOVIRORO et al. 1980; KUHNZ et al. 1990b). The time course of EE plasma levels following oral administration can be described, in most cases, by a two-compartment model. A rapid distribution phase is followed by a terminal disposition phase that is characterized by a half-life in the range of about 5–30 h (ORME et al. 1989). Occasionally, a secondary peak can be observed at about 10–14 h after administration, which is attributed to the enterohepatic circulation of the drug.

The large range in the terminal half-life of EE is partly due to interindividual variation and partly because the half-life determination is compromised by several factors (NEWBURGER and GOLDZIEHER 1985). Modern combination oral contraceptives contain only low doses of less than 50 μg EE and, therefore, the lower limit of quantification of EE in serum is usually reached within 24 h postadministration. A terminal half-life longer than about 10 h can, therefore, not be determined with any accuracy, which is a limitation of most published half-life data of EE. In two studies, where higher doses of EE were administered, the concentrations could be measured up to 48 h and mean half-life values of 13 h and 23 h, respectively, were reported (HÜMPEL et al. 1979; BACK et al. 1980e). However, at doses of EE up to a few milligrams, it is questionable whether linear pharmacokinetics can still be assumed.

In a recent publication, it was shown that, following single oral doses of 120 μg and 240 μg EE to postmenopausal women, the calculated bioavailability of EE increased from 36% to 71% when compared with an intravenously administered dose of 60 μg EE, pointing to a saturation of metabolism at higher doses. At the same doses, the mean terminal half-lives increased from16.8 h to 28.9 h, which indicates non-linear pharmacokinetics (BAUMANN et al. 1996). Thus, there is considerable uncertainty with regard to the "true" terminal half-life of EE, but a range between 10 h and 20 h can serve as a reasonable estimate.

During repeated oral administration of EE-containing oral contraceptives, an increase in the trough serum levels of EE can be observed until about day 5–10, with only little further change until the end of the treatment cycle (KAUFMANN et al. 1981; HAMMERSTEIN et al. 1993). An increase in EE serum levels of between 30% and 60% within a treatment cycle has been described in numerous studies with different oral contraceptive preparations and can be explained by pharmacokinetic acccumulation, which is determined by the ratio of the terminal half-life and the dosing interval. With a dosing interval of 24 h and a terminal half-life in the range 10–20 h, accumulation factors ranging from 1.23 to 1.77 can be calculated for EE. In other words, compared with single-dose administration, an accumulation between 23% and 77% can be expected during repeated administration.

An alternative hypothesis to explain the increasing EE plasma levels assumed an interference of EE with its own metabolism, caused by an irreversible inhibition of cytochrome P-450 enzymes (JUNG-HOFFMANN and KUHL 1989). Such an inhibitory effect of EE has been observed during in vitro experiments with microsomes of rat and human liver (WHITE and MÜLLER-EBERHARD 1977; BÖCKER et al. 1991). However, it could be shown in a clinical pharmacokinetic study that this in vitro finding had no relevance to the in vivo situation, since total CL of EE before and after repeated oral administration of EE remained unchanged (KUHNZ et al. 1996). Interestingly, it has already been shown, using a similar study design, that prolonged oral administration of mestranol has no effect on its metabolic CL rate (MILLS et al. 1974).

3. Metabolism

Following oral administration, EE is subject to a considerable first-pass metabolism. EE-sulfate is the main metabolite formed during passage of the gut wall, and it could be demonstrated that the intestinal conjugation of EE accounts for about 65% of the observed first-pass effect in humans (BACK et al. 1982a). Also, EE-3-sulfate is the principal circulating metabolite in plasma, exceeding the EE levels by a factor of 6–22 (BIRD and CLARK 1973; BACK et al. 1980e).

The major pathway of metabolism of EE in humans is the cytochrome P450-dependent 2-hydroxylation and the formation of the catechol estrogen 2-hydroxy-EE (BOLT et al. 1973; BOLT 1979). The 2-hydroxylation of EE is catalyzed by cytochromes of the CYP2C, CYP2E and CYP3A gene families (GUENGERICH 1988; BALL et al. 1990). It has been estimated that an average of 29% of ingested EE is ortho-hydroxylated (2- and 4-hydroxylation); however, this figure may rise up to 64% in some individuals (BOLT 1979). Several other pathways of minor quantitative importance, involving the hydroxylation at positions 6α and 16β have also been reported from in vitro and in vivo studies in humans (BOLT et al. 1974; WILLIAMS and WILLIAMS 1975; WILLIAMS et al. 1975; PURBA et al. 1987). The formation of the homoannulated derivative (D-homo-estradiol) and the loss of the ethinyl moiety (formation of estradiol, estrone, estriol, 2-methoxy-estradiol) has also been observed (WILLIAMS et al. 1975). However, only 1–2% of the dose of EE administered to women was found to be de-ethinylated (WILLIAMS et al. 1975) and only traces of the homoannulated metabolite were detected (ABDEL-AZIZ and WILLIAMS 1970). There are substantial interindividual and interethnic variations in the metabolism of EE in humans, with regard to the preferred site of oxidation at the steroid molecule and the excretion of the different EE-glucuronides (WILLIAMS and GOLDZIEHER 1980).

Orally administered mestranol is rapidly metabolized by demethylation in humans and about 54% of the dose was converted to EE (BOLT and BOLT 1974). The metabolism of mestranol is closely related to that of EE and basically the same metabolites, free and conjugated, have been described for mestranol in vitro and in vivo (BALL and KNUPPEN 1974; WILLIAMS and WILLIAMS 1975). In vitro studies with human microsomal preparations have indicated that CYP2C9 is mainly responsible for the conversion of mestranol to EE, whereas CYP3A4 is mainly involved in the further hydroxylation reactions of EE (SCHMIDER et al. 1997).

4. Excretion

Following oral administration of radiolabeled ^{3}H-EE to women, about 62% of the totally excreted dose was eliminated with the feces and about 38% with the urine (SPECK et al. 1976). Other investigators arrived at similar figures for the extent of renal excretion of radiolabeled EE and mestranol (REED et al. 1972; KAMYAB and FOTHERBY 1969; ABDEL-AZIZ and WILLIAMS 1970; KULKARNI and GOLDZIEHER 1970; MAGGS et al. 1983a). Only about 6% (range 3.2–8.4%) of the dose appeared as unchanged EE in the urine and about 7% (range 2.4–16.5%) was excreted as EE-conjugates (REED et al. 1972). The major part of the radiolabeled compounds that was excreted in the urine was represented by conjugated and non-conjugated metabolites of EE. Approximately 80% of the conjugates were glucuronides (3-glucuronides, 17-glucuronides and 3,17-diglucuronides) and only 8–10% were sulfates (WILLIAMS and GOLDZIEHER 1980; KAMYAB and FOTHERBY 1969; CARGILL et al. 1969).

The following metabolites were identified after administration of EE or mestranol: 2-hydroxy-EE, 2-methoxy-EE, 2-hydroxy-3-methoxy-EE, 6α-hydroxy-EE, 16β-hydroxy-EE and, to a minor extent of up to 1–2% of the dose, de-ethinylated metabolites (WILLIAMS JG et al. 1975; WILLIAMS and WILLIAMS 1975). In the feces, about 9% of the dose administered (corresponding to about 30% of radiolabel excreted with feces) was excreted as unchanged EE. Following a cleavage of the conjugates, only marginal amounts of EE were liberated (REED et al. 1972). EE-glucuronide and EE-sulfate were found to be the predominant components in bile (CARGILL et al. 1969), but 2-methoxy-EE and 2-hydroxy-EE were also excreted to a significant extent as glucuronides and sulfates (MAGGS et al. 1983a; MAGGS and PARK 1985).

IV. Comparison of Different Routes of Administration

Only a few animal and human studies have investigated the influence of different routes of administration on the pharmacokinetics of EE, mainly to find ways of avoiding the high first-pass effect of the liver. A marked increase in the bioavailability of EE (80–85%) compared with the intraduodenal route (15–25%) was observed in rats following nasal administration of the drug (BAWARSHI-NASSAR et al. 1989). In a crossover study, the bioavailability of EE was investigated in five women, following oral, vaginal and intravenous administration of an EE-containing preparation. Although a somewhat slower rate of absorption was noted after vaginal than after oral administration, a slightly increased mean bioavailability of about 74% was observed after vaginal administration and about 62% after oral administration of the preparation (BACK et al. 1987).

The observed small reduction in the first-pass effect of EE after vaginal administration correlates well with the known substantial contribution of direct sulfation in the gut wall to the first-pass effect observed after oral administration of EE. A similar result was obtained in another clinical study, comparing the bioavailability of EE released from contraceptive vaginal rings with oral administration. A bioavailability about 1.2 times higher was observed after vaginal administration (TIMMER et al. 1990). In another study, the serum level–time course of EE was investigated following oral administration and vaginal administration of the drug to two groups of postmenopausal women. It was found that the AUC after vaginal administration amounted to only 36% of the corresponding value observed after oral administration. However, this discrepancy was thought to be due to the fact that vaginal absorption of EE was less efficient because the drug was contained in a polyethylene glycol suppository, whereas a tablet had been administered orally (GOEBELSMANN et al. 1985).

V. Drug Interactions

1. In Vitro

a) Interaction with Progestins

It has been shown that 2-hydroxylation of EE is catalyzed by cytochromes of the CYP2C, CYP2E and CYP3A gene families (GUENGERICH 1988; BALL et al. 1990). Of those, CYP3A4 seems to be of major importance. Since a number of different progestins, such as levonorgestrel, gestodene, norethisterone, desogestrel, norgestimate and dienogest, are used in combination with EE in oral contraceptive preparations, it is important to know whether there is an interaction of these progestins with the 2-hydroxylation of EE. Except for dienogest, all of these progestins carry an ethinyl group at position 17, and previous studies have indicated the potential of some of these progestins inactivating cytochrome P-450 by a mechanism-based inactivation (WHITE and MÜLLER-EBERHARD 1977; ORTIZ DE MONTELLANO et al. 1979). Subsequently, there were several in vitro studies that examined the influence of different progestins on the 2-hydroxylation of EE in human microsomal preparations (GUENGERICH 1990; BACK et al. 1991; BÖCKER et al. 1991; BÖCKER and LEPPER 1991).

The results of these studies indicated that all of the investigated progestins had the propensity to inhibit not only 2-hydroxylation of EE, but also other metabolic pathways in different marker compounds in vitro, although the extent of inhibition observed depended strongly on the experimental conditions used. Among the 19-nortestosterone-type progestins, gestodene seemed to have the highest inhibitory potency in vitro, whereas for dienogest, which carries a 17α-cyanomethyl function, almost no inhibition of P-450 catalyzed reactions was observed (Böcker and Lepper 1991). Thus, it was suggested that the acetylene moiety at position 17 of the steroids was responsible for the inhibition observed in vitro, possibly due to the formation of a reactive intermediate which irreversibly binds to cytochrome P-450 either at the heme or the apoprotein site of the enzyme.

b) Interaction of EE with Its Own Metabolism

EE itself has also been found to exert an inhibitory effect on P-450-dependent enzymes during in vitro experiments with microsomal preparations of rat and human liver (White and Müller-Eberhard 1977; Guengerich 1988; Böcker et al. 1991). Based on these in vitro findings, it has been suggested that EE may also inhibit its own metabolism in vivo, which might be largely responsible for the observed accumulation of EE in the serum of women during oral contraceptive therapy (Jung-Hoffmann and Kuhl 1989).

However, the crucial and general question is whether results obtained from in vitro experiments can be readily transferred to the clinical situation in vivo. The in vitro experiment should be regarded primarily as an indicator of potential interactions that may be of relevance in vivo, but which have to be examined further in appropriate clinical studies.

2. In Vivo

a) Interaction with Progestins

Until recently, there was no evidence available from clinical studies to suggest that the pharmacokinetics of EE could be altered due to interactions with a co-administered progestin. However, in one case during a comparative study, it was observed that women who received a combination of gestodene and EE showed higher serum levels of EE than women who received a combination of desogestrel and EE, despite the fact that the EE dose was the same in both formulations (Jung-Hoffmann and Kuhl 1989). This finding seemed to be supported by in vitro studies of the inhibition of the 2-hydroxylation of EE by progestins, where a higher inhibitory potency was observed for gestodene than for desogestrel. However, a number of subsequently performed clinical pharmacokinetic studies, dedicated to unravel this particular interaction, failed to confirm the finding of Jung-Hoffmann and Kuhl (Hümpel et al. 1990; Kuhnz et al. 1990a; Dibbelt et al. 1991; Kuhnz et al. 1991; Orme et al. 1991; Hammerstein et al. 1993). There was no interaction of either gestodene or desogestrel with the pharmacokinetics of EE following single-dose administration (Orme et al. 1991). Similarly, no differential effect of these two progestogens on EE serum levels could be detected in women following single- or multiple-dose administration of the corresponding combination oral contraceptives.

Although progestins do not seem to interfere with the pharmacokinetics of EE, there are indications that EE may interfere with the pharmacokinetics of progestins. A decrease in the CL of unbound levonorgestrel and gestodene has been noted during chronic administration of corresponding EE-containing combination preparations, whereas, when each of the two progestogens was administered repeatedly, alone, without EE, no change in the unbound CL values occurred (KUHNZ et al. 1992a, b, 1993c, 1994). Therefore, it seemed likely that EE had an influence on the metabolic disposition of both progestogens in vivo, although this hypothesis needs further experimental verification.

b) Interaction of EE with Its Own Metabolism

In vitro studies revealed an inhibitory potential of EE on cytochrome P-450-dependent enzymes. If those enzymes were also affected in vivo, this might cause an impairment of the hepatic metabolic CL of EE, and one would expect an increase in the AUC of EE during chronic treatment. This hypothesis was examined in a clinical pharmacokinetic study of young women given single intravenous administrations of ^{13}C-labeled EE prior to and at the end of a treatment period of 8 days with daily oral doses of 60 μg EE. Terminal half-life, Vz and CL were determined after both intravenous doses. No significant difference was observed in any of these parameters between the first and the second intravenous administration of ^{13}C-EE. Since the CL in particular remained unchanged, it could be excluded that EE inhibited its own metabolism in vivo (KUHNZ et al. 1996).

c) Drugs Interfering with the Metabolism of EE

There is a large body of literature dealing with the interaction of drugs with oral contraceptives and this has been presented in several comprehensive reviews (D'ARCY 1986; BACIEWICZ 1985; BACK and ORME 1990, 1994; SHENFIELD 1993). It is beyond the scope of the present monograph to cover the complete subject and, therefore, only those interactions that are of particular interest with regard to EE will be presented.

d) Interaction with Anticonvulsant Drugs

A number of anticonvulsants, such as phenobarbital, phenytoin and carbamazepine are known to be potent inducers of cytochrome P-450-dependent enzymes, including those which are involved in the metabolism of EE (CYP3A4, CYP2C). Concomitant use of these anticonvulsants and EE-containing oral contraceptives is, therefore, likely to result in an accelerated metabolism and a fall in the EE serum concentrations. This has been investigated in a few clinical pharmacokinetic studies and was confirmed for the three anticonvulsants (BACK et al. 1980b; CRAWFORD et al. 1990). It is, therefore, recommended that women receiving phenobarbital, phenytoin or carbamazepine should use oral contraceptive preparations with higher doses of both EE and the progestin in order to avoid contraceptive failure.

e) *Interaction with Rifampicin*

Rifampicin is known to be a potent inducer of liver enzymes (Bolt et al. 1975). There is evidence that rifampicin induces CYP3A4 which is one of the major enzymes involved in the 2-hydroxylation of EE (Guengerich 1988; Combalbert et al. 1989). During treatment of those women using combination oral contraceptives with rifampicin, the half-life of EE decreased by about 50% and a concomitant fall in EE serum levels by about 40% was noted (Bolt et al. 1977; Back et al. 1980c). However, in another study, almost no effect of rifampicin treatment was observed on the EE levels in the plasma of women using a low-dose combination oral contraceptive (Joshi et al. 1980).

f) *Interaction with Antibiotics*

There is an ongoing debate regarding the association between reported pregnancies in women using oral contraceptives and a concomitant therapy with antibiotics. Numerous reports point towards an impairment of the contraceptive efficacy by antibiotics. Except for those which are known to interfere with the metabolism of contraceptive steroids due to their enzyme-inducing activity (e.g., rifampicin), it remains unclear by which mechanisms broad-spectrum antibiotics, such as penicillins and tetracyclines, could interfere with the pharmacokinetics of the contraceptive steroids.

With regard to the interaction of antibiotics with EE, it is important to know that, in contrast to the progestins, EE undergoes enterohepatic recirculation. The sulfate and glucuronide conjugates of EE are excreted via the bile, reach the colon and can be hydrolyzed by the bacterial flora present. The liberated EE can then be reabsorbed and reaches the systemic circulation again. Antibiotics markedly reduce the gut microflora and, thus, interrupt the enterohepatic recirculation of EE. As a consequence, the polar conjugates of EE are now excreted with the feces. However, it is questionable whether this has any relevance in terms of contraceptive efficacy because:

The contribution of recirculated EE to the total serum levels of the drug seems to be less than 10% (Back et al. 1979) and it cannot even be observed consistently (Kuhnz et al. 1990a);

It is the progestogenic component that is mainly responsible for inhibition of ovulation.

In fact, a number of studies have failed to show an influence of antibiotics, such as ampicillin, cotrimoxazole, tetracycline temafloxacin or clarithromycin, on the pharmacokinetics of EE (Back and Orme 1994; Fernandez et al. 1997). Although the interaction with antibiotics does not seem to be of general clinical relevance, it may become important in individual subjects.

g) *Interaction with Paracetamol*

Paracetamol is metabolized primarily by sulfation in the gut wall and may therefore compete with EE for the available sulfation capacity. In an in vitro study with sections of human ileum, a significant decrease in the extent of EE sulfation was observed in the presence of paracetamol (ROGERS et al. 1987a). When a single dose of 1 g paracetamol was administered to healthy volunteers who received an oral contraceptive, a significant increase in the EE serum levels was produced within less than 3 h (ROGERS et al. 1987b). Parallel analysis of EE-sulfate concentrations revealed a significant decrease following treatment with paracetamol compared with the control, thus supporting the proposed mechanism.

h) Interaction with Grapefruit Juice

Meanwhile, it has been shown for a number of drugs and steroid hormones that there is an interaction of several flavonoids present in grapefruit juice with cytochrome P-450-dependent enzymes and, in particular, with CYP3A4 and CYP1A2 (BAILEY et al. 1994; SCHUBERT et al. 1995; LEE et al. 1996). Since CYP3A4 is also the major enzyme involved in the metabolism of EE, this interaction may well be of relevance. Recently, a study was published reporting an increased bioavailability of EE in young women who took an oral contraceptive together with grapefruit juice. An average increase of the EE serum levels of about 28% was observed compared with the control group (WEBER et al. 1996). However, since the interindividual variation in the EE serum levels of women under oral contraceptive therapy is at least in the same range, the clinical relevance of this finding remains to be established.

i) Interaction with Ascorbic Acid

Previous studies in few women have suggested that the administration of high doses of ascorbic acid interferes with the intestinal sulfation of EE, leading to higher blood levels of EE (BACK et al. 1980a; 1981). EE is extensively conjugated with sulfate and this occurs to a large extent in the gut mucosa. Ascorbic acid also undergoes conjugation in the gut wall and it has been suggested that the oral intake of gram quantities of vitamin C may cause depletion of sulfate or saturation of ATP-sulfurylase capacity (ROGERS et al. 1987a). The possibility of such an interaction has been investigated in a large collective of 37 women using a combination oral contraceptive for two consecutive cycles. Concomitant daily administration of 1 g ascorbic acid shortly before intake of the oral contraceptive was randomly assigned to the first or second cycle. EE was analyzed in the women's serum and the pharmacokinetic parameters maximum plasma concentration (C_{max}) and $AUC_{(0-12\,h)}$ of EE were determined. No significant effect of ascorbic acid on the serum levels of EE was observed. The serum concentrations of EE-sulfate exceeded those of EE by a mean factor of 12.8 and only on one day during the observation period a significantly lower AUC value for EE-sulfate was noted in the presence of ascorbic acid. Thus, the competition between ascorbic acid and EE for sulfation does not lead to an increased systemic availability of EE and is, therefore, unlikely to be of any clinical importance (ZAMAH et al. 1993).

j) Interference of EE with the Metabolism of Other Drugs

There are numerous reports on the altered disposition of drugs during concomitant use of oral contraceptives. This comprises drugs such as theophylline (TORNATORE et al. 1982), caffeine (MEYER et al. 1989; BALOGH et al. 1995), benzodiazepines (JOCHEMSEN et al. 1982; ROBERTS et al. 1979; ABERNETHY et al. 1982), antipyrine (PAZZUCCONI et al. 1991), aminopyrine (HERZ et al. 1978) and prednisolone (BOEKENOOGEN et al. 1983; FREY and FREY 1985; MEFFIN et al. 1984). In general, concomitant use of oral contraceptives led to a decrease in the total CL and to a prolongation in the terminal half-life of these drugs, although this was not observed consistently. The impairment of drug metabolism was found to be reversible shortly after discontinuation of oral contraceptive therapy. However, in none of these studies was the influence of the single components of the different combination oral contraceptives on the disposition of these drugs investigated.

Only in two studies (BALOGH et al. 1990; BALOGH et al. 1991) was it shown that the progestins levonorgestrel and dienogest, given alone, had no effect. When these were administered in combination with EE, there was an impairment of caffeine and metamizol elimination, suggesting that EE was responsible for the observed effect. The relative contribution of EE and the different progestins to the observed impairment of metabolism of other drugs remains to be established. Furthermore, it has to be examined whether these interactions are solely due to direct drug interactions (inhibition) with cytochrome P-450-dependent enzymes or if there is, for example, also an interference at the level of regulation of P-450 synthesis (MACKINNON et al. 1977).

C. Antiestrogens

The present overview of pharmacokinetics of estrogen antagonists comprises representatives of three compound classes with antiestrogenic properties, namely derivatives of triphenylethylenes, a benzothiophene and a chroman.

Among the triphenylethylene derivatives, tamoxifen is well known and is most extensively used in endocrine treatment of breast cancer. Other compounds of this class, such as toremifene and droloxifen, were developed subsequently for this indication and the pharmacokinetic properties of these substances were reviewed comprehensively with regard to their clinical implications (LONNING et al. 1992; LONNING and LIEN 1993; WISEMAN and GOA 1997). More recent publications providing somewhat limited information on the human pharmacokinetics of panomifene, a further triphenylethylene derivative, was also included in the present review where appropriate.

The benzothiophene derivative raloxifene is currently under evaluation in the treatment and prevention of osteoporosis. Although publications on the pharmacokinetics of this substance in humans are available only in the form of abstracts, they were also considered in the present review.

Another compound with antiestrogenic activity is the chroman derivative centchroman, which has been used for oral contraception and is presently under development for the treatment of advanced breast cancer (LAL et al. 1996).

I. Analytical Methods

Several analytical methods have been published regarding the determination of tamoxifen and its metabolites in plasma/serum. An overview was given by LONNING et al. (1992), and further methods were described by LIM et al. (1993), FRIED and WAINER (1994), MACCALLUM et al. (1996), EL YAZIGI and LEGAYADA (1997) and SANDERS et al. (1997). The majority of methods is based on HPLC separation and fluorescence detection of phenanthrene derivatives which were generated by either pre- or postcolumn ultraviolet (UV) irradiation of the analytes. Most of these methods allow for the simultaneous determination of metabolites and parent drug with detection limits between less than 0.1 ng/ml and 8 ng/ml. Other methods describe the determination of tamoxifen and different numbers of metabolites by means of GC–MS with detection limits ranging between 0.05 ng/ml and about 10 ng/ml, or thin layer chromatography (TLC) with a sensitivity of 2 ng/ml (LONNING et al. 1992) and capillary electrophoresis with an LOQ of 10 pg tamoxifen (SANDERS et al. 1997).

For the determination of toremifene and its metabolites, several HPLC methods were developed that use either direct UV detection (WEBSTER et al. 1991) with a detection limit of about 200 ng/ml or fluorescence detection after conversion of the triphenylethylene compounds to phenanthrenes by UV-irradiation with a detection limit of 5–10 ng/ml for the parent drug (HOLLERAN et al. 1987; ANTTILA et al. 1995; BERTHOU and DREANO 1993; LIM et al. 1994a). Additionally, a TLC method was described for the determination of the parent compound and two of its metabolites also using the UV light-mediated conversion to phenanthrene derivatives (ANTTILA et al. 1990). Recently, a liquid chromatography–mass spectrometry (LC–MS) method was published for the analysis of toremifene and five of its metabolites with a lower limit of quantification for toremifene of 10 ng/ml (MARTINSEN and GYNTHER 1996).

Droloxifene and its metabolites can be determined similarly to tamoxifen and toremifene, using HPLC separation and fluorescence detection after post-column UV-mediated conversion to phenanthrene compounds with an LLQ of 25 pg/ml for the parent drug (TESS et al. 1995). Other authors described quantitation limits of 0.3 ng/ml and 1 ng/ml using methods that allow for the simultaneous determination of several droloxifene metabolites (GRILL and POLLOW 1991; TANAKA et al. 1994; LIEN et al. 1995a).

Panomifene plasma concentrations were determined with a sensitivity of 1 ng/ml using a similar technique of HPLC separation and fluorescence detection (ERDELYI-TOTH et al. 1994). Concentrations of raloxifene were determined in native and hydrolyzed human plasma by means of a HPLC method with UV detection (LANTZ and KNADLER 1993) or by a LC-MS/MS method

(FORGUE et al. 1996). A different HPLC method with amperometric detection was described for the analysis of raloxifene in plasma samples of rats, dogs and monkeys (LINDSTROM et al. 1984). In neither case was the sensitivity of the respective assay given.

In the case of centchroman, direct fluorescence detection was used after HPLC separation for the determination of the parent compound and its metabolite 7-desmethyl centchroman. The sensitivity was 1–2 ng/ml in serum and 2.5 ng/ml in milk samples (LAL et al. 1994; PALIWAL and GUPTA 1996).

II. Pharmacokinetics in Animals

1. Absorption

Tamoxifen was almost completely absorbed after oral administration to laboratory animals. Following oral or intraperitoneal administration of radiolabeled tamoxifen to mice, rats, dogs and monkeys, maximum concentrations of radioactive compounds were in general achieved within 1–6 h, and a second maximum was observed 24–44 h after dosing (FROMSON et al. 1973a), which was attributed to enterohepatic circulation of the drug and/or its metabolites. However, no information regarding the absolute bioavailability of tamoxifen in laboratory animals was available. The same holds true for toremifene, which was also almost completely absorbed after oral administration to rats, as indicated by similar amounts of radiolabeled compounds excreted renally after intravenous and oral administration (SIPILÄ et al. 1990).

Droloxifene was rapidly and almost completely absorbed after oral administration to female mice, rats and monkeys. Maximum concentrations of the drug in serum/plasma were reached 1 h, 1.3 h and 1 h after oral dosing to rats, monkeys and mice, respectively. The absolute bioavailability was 18% in rats, 11% in monkeys and 8% in mice, indicating that the drug is subject to a marked first-pass effect (TANAKA et al. 1994). Similar results were obtained by NICKERSON et al. (1997) who reported an oral bioavailability of 27% in rats. They also showed that absorption of droloxifene was almost complete and that the drug was not glucuronidated in the intestine.

Following oral administration of raloxifene to rats, dogs and monkeys, maximum concentrations of the parent compound in plasma were reached after 0.5 h, 1 h and 2 h, respectively, thus indicating rapid absorption of the drug in these species. A considerable first-pass metabolism was observed and the bioavailability of the drug was determined to be 39% in rats, 17% in dogs and 5% in monkeys (LINDSTROM et al. 1984).

The absorption rate of centchroman was moderate after oral administration of the drug to female rats and maximum concentrations in plasma were reached about 12 h after drug administration (PALIWAL and GUPTA 1996). Data on the absolute bioavailability of the drug in animals have not been published.

2. Distribution

Information on tamoxifen pharmacokinetics in animals is limited to studies with the radiolabeled drug, therefore, no defined pharmacokinetic parameters of the parent drug are available. Following oral administration, postmaximum serum concentrations of radiolabel, representing tamoxifen and its metabolites, decreased biphasically with terminal half-lives of 170h, 62h and 25h in mice, rats and rhesus monkeys, respectively. Considerable interindividual differences were observed in dogs, where terminal half-lives of 29h, 35h and 130h were reported in one study. Reabsorption of biliary-excreted tamoxifen and/or its metabolites was demonstrated in rats and dogs, which explains the occurrence of the secondary peak after oral administration, on the one hand, and the apparent persistence of radiolabeled compounds indicated by the long elimination half-lives of 3–18 days in excreta, on the other (FROMSON et al. 1973a). In rats, concentrations of tamoxifen and its metabolites in most tissues were 8- to 70-fold higher than in serum. The highest levels were observed in lung and liver, and considerable amounts were also recovered from kidney and fat (LIEN et al. 1991a). Storage of the drug and its metabolites in fat may also contribute to the slow process of excretion.

Toremifene and its metabolites were also distributed to several tissues in rats, and elevated concentrations were observed in lung, brain and tumor tissues (LONNING et al. 1992). Investigations with transdermally administered toremifene in nude mice, baboons and a horse showed that the distribution of the drug to tissues surrounding the application site is much higher than the distribution to other organs via the circulation. Thus, transdermal application of toremifene may improve the therapeutic efficiency of the drug due to high drug concentrations in tumor tissue at the application site. On the other hand, there may be reduced side effects since systemic drug concentrations are expected to be low due to the distribution of the drug to fat tissue near the site of application from which it is released only slowly (SOE et al. 1997).

Droloxifene concentrations decreased with a terminal half-life of 2h in rats, 5h in monkeys and 1.4h in mice after intravenous administration, whereas corresponding values of 4.3h, 10.6h and 1.6h were determined, respectively, after oral administration (TANAKA et al. 1994). Further investigations in rats employing a more sensitive analytical method for droloxifene revealed a prolonged terminal half-life of the drug being 8.8h after intravenous and 11h after oral administration (NICKERSON et al. 1997). The mean total CL was 50.9ml/min/kg, 23.5ml/min/kg and 93.8ml/min/kg in rats, monkeys and mice, respectively, in one study (TANAKA et al. 1994), which was higher than the plasma flow to the liver in these species, thus indicating involvement of extrahepatic elimination. In another study, the mean total CL was determined to be 32.4ml/min/kg in rats (NICKERSON et al. 1997). The drug is widely distributed to tissues in animals as demonstrated by the large Vz of 8.8l/kg in rats, 10.2l/kg in monkeys and 11.4l/kg in mice. This was also demonstrated after oral administration of radiolabeled droloxifene to rats, where maximum values of radiolabel in most tissues and organs were reached within 1–4 h. High concentrations of the parent compound and its metabolites were detected in the liver and the gastrointestinal contents, as well as in lung, spleen and the adrenal gland (TANAKA et al. 1994).

Centchroman plasma levels increased gradually within 12 h following oral administration to female rats, without reaching a pronounced maximum, and decreased monoexponentially, thereafter, with a half-life of about 1 day. The highest centchroman concentrations were observed in the lung, where the drug was more than 200-fold enriched compared with plasma. Spleen, liver and adipose tissue also contained considerable amounts of the drug leading to more than 100-fold higher centchroman concentrations than in plasma, while drug concentrations in the uterus were only about 40 times higher than in plasma. The major metabolite 7-desmethyl-centchroman reached peak concentrations in plasma at 8 h and 24 h after oral administration to female rats. Plasma levels then decreased with a half-life of about 1.5 days. The tissue distribution pattern of 7-desmethyl-centchroman was similar to the parent compound; however, at any given time, the centchroman concentrations in plasma and tissues were higher than the corresponding concentrations of the metabolite (PALIWAL and GUPTA 1996). In plasma of monkeys, the drug was bound to albumin, but not to SHBG or corticoid binding globulin. However, the amount of drug bound was not determined (AGNIHOTRI et al. 1996).

3. Metabolism

Tamoxifen is subject to extensive metabolism in laboratory animals and only low amounts of unchanged drug were recovered in excreta. First investigations in rats, mice, dogs and monkeys showed that metabolic reactions involve hydroxylation of the aromatic systems, *N*-demethylation and deamination of the side chain, as well as oxidation of the ethenyl group (FROMSON et al. 1973a). An overview of the metabolic pathways was given by LONNING et al. 1992. Additionally, the formation of acidic metabolites arising from deaminohydroxy-tamoxifen by further oxidation was detected in rats (RUENITZ and BAI 1995).

In vitro, the major metabolites formed by rat liver microsomes were 4-hydroxy-tamoxifen, 4′-hydroxy-tamoxifen, *N*-desmethyl-tamoxifen and tamoxifen-*N*-oxide. In addition to these, 3,4-epoxy-tamoxifen and 3′,4′-epoxy-tamoxifen and their hydrolyzed derivatives 3,4-dihydrodihydroxy-tamoxifen and 3′,4′-dihydrodihydroxy-tamoxifen were detected in lower amounts. The same major metabolites were also formed in vitro by mouse liver microsomes, where formation of tamoxifen-*N*-oxide was predominant. The 3,4-epoxy metabolite, was also formed in mouse liver microsomes, but at lower levels than in rat liver microsomes (LIM et al. 1994b). The *N*-demethylation reaction is catalyzed by cytochromes P450 1A, 2C and 3A in rat liver microsomes (MANI et al. 1993a) while the formation of the *N*-oxide is catalyzed by the flavin containing monooxygenase (MANI et al. 1993b). Treatment of rats with tamoxifen increased the mRNA levels of several phase-I and phase-II drug metabolizing enzymes (NUWAYSIR et al. 1995; HELLRIEGEL et al. 1996).

Metabolic activation of tamoxifen to reactive compounds binding irreversibly to microsomal proteins was reported by MANI and KUPFER (1991). HAN and LIEHR (1992) showed formation of DNA adducts in tamoxifen-treated rats and hamsters. The involvement of different metabolic pathways (e.g., α-hydroxylation, formation of epoxides and arene oxide formation) in the process of tamoxifen activation has been discussed recently (WHITE et al. 1997; LIM et al. 1997; TANNENBAUM 1997), but the nature of the reactive metabolites of tamoxifen in vivo has yet to be elucidated.

The metabolism of toremifene in rats was similar to tamoxifen with regard to *N*-demethylation, 4-hydroxylation and side-chain modifications. It was shown that the chlorine atom was still present in all identified metabolites of toremifene, which indicates its metabolic stability. It was proposed that the presence of the chlorine atom may inhibit the formation of DNA-reactive metabolites (SIPILÄ et al. 1990); however, low levels of DNA adducts were detected in rat liver after toremifene treatment (WISEMAN and GOA 1997).

In the case of droloxifene, at least eight and six metabolites were observed in the feces of rats and monkeys, respectively. In rat feces, the major metabolites were 3-methoxy-4-hydroxydroloxifene and 2-hydroxydroloxifene, which were also detected in monkey feces, but in smaller amounts. *N*-desmethyldroloxifene, which has an antiestrogenic activity like droloxifene, was present in feces of monkeys, but not in rat feces. This compound was, however, detected in serum of rats, mice and monkeys (TANAKA et al. 1994).

Raloxifene was found to be mainly glucuronidated in rats and monkeys. About 50% of the drug was excreted as raloxifene-6-glucuronide with the bile, in rats, after oral administration. In monkeys, about 80% of the radioactivity in plasma after oral administration of the radiolabeled drug was present as a glucuronide conjugate; however, no information regarding the binding site of the glucuronyl moiety was given (LINDSTROM et al. 1984).

Centchroman metabolism was investigated in vitro using rat liver homogenates and seven metabolites have been characterized, including 7-desmethyl-centchroman, two didesmethyl-centchroman compounds and several dephenylated derivatives (RATNA et al. 1986).

4. Excretion

After oral or intraperitoneal administration of tamoxifen to mice, rats, dogs and rhesus monkeys, the major route of excretion was via the feces, and only minor amounts were excreted renally. In rats, dogs and mice, between 2% and 10% and, in rhesus monkeys, between 16% and 21% of the dose was excreted via the kidneys. The rates of excretion of radiolabeled compounds were characterized by half-lives of 3–10 days in rats, 18 days in mice, 12 days in monkeys and 11–17 days in dogs (FROMSON et al. 1973a).

In the case of toremifene, total excretion amounted to about 70% of the dose within 13 days after oral or intravenous administration to rats. About 90% of the excreted compounds were recovered in feces (SIPILÄ et al. 1990).

The main route of excretion of droloxifene and its metabolites in rats and monkeys was also via the feces, and only about 1% and 3% of the dose, respectively, were excreted with urine. As shown by whole body autoradiography in rats, concentrations of droloxifene and its metabolites in tissues and organs decreased with, and most of these were excreted within 120 h after oral dosing (TANAKA et al. 1994).

Excretion of raloxifene and its metabolites following oral administration of the drug to rats and dogs was primarily via the feces. Less than 1% of the dose was excreted in urine, and excretion was almost complete after 72 h in rats and after 48 h in dogs (LINDSTROM et al. 1984).

Centchroman and its metabolites were mainly excreted via the feces (82%) and only minor amounts were recovered in urine after oral administration to rats (RATNA et al. 1994).

III. Pharmacokinetics in Humans

1. Absorption

Early studies with orally administered radiolabeled tamoxifen showed that maximum concentrations of radioactive compounds in serum were reached within 4 h (FROMSON et al. 1973b). This was confirmed later using selective analytical methods, where maximum concentrations of tamoxifen in serum were reached about 5 h after oral administration (DE VOS et al. 1989; LONNING et al. 1992). The absolute bioavailability of the drug in humans is not known (LONNING et al. 1992).

Toremifene was rapidly absorbed after oral administration of doses between 10 mg and 680 mg. Maximum serum concentrations were reached 2–6 h after oral administration of toremifene, given either as single dose or as daily doses over a period of 5 days to postmenopausal women. Drug levels in terms of AUC and C_{max} values increased dose dependently after single and repeated daily administration (TOMINAGA et al. 1990; ANTTILA et al. 1990). At steady state, maximum concentrations of toremifene were reached between 1.5 h and 4.5 h after oral administration (WIEBE et al. 1990).

In the case of droloxifene, maximum concentrations were reached, on average, within 2–5 h after single and multiple oral administration (GRILL and POLLOW 1991, BREITBACH et al. 1994). A dose-linear increase in C_{max} and AUC values was observed within the dose range 20–100 mg (GRILL and POLLOW 1991).

Following oral administration of panomifene to healthy postmenopausal volunteers, maximum concentrations of the drug were reached, on average, after 3.6 h. The AUC and C_{max} values increased with increasing doses in the range 24–96 mg (ERDELYI-TOTH et al. 1997).

Raloxifene was well absorbed after oral administration and underwent extensive first-pass metabolism, mainly due to glucuronidation. Maximum raloxifene concentrations in plasma were reached, on average, after 6h; however, after acid hydrolysis of conjugates, total raloxifene concentrations reached a maximum as soon as 1h after administration (FORGUE et al. 1996; NI et al. 1996).

Centchroman was absorbed at a moderate rate and maximum serum concentrations were reached, on average, 4–5h after oral administration (PALIWAL et al. 1989; LAL et al. 1995, 1996; GUPTA et al. 1996).

Data on the absolute bioavailability of these drugs in humans were not available.

2. Distribution

Following cessation of chronic oral tamoxifen treatment, a monophasic elimination from serum was observed for tamoxifen and its pharmacologically active metabolites. The mean terminal half-life of tamoxifen was 9.6 ± 2.5days and was similar to that of *N*-desmethyl-tamoxifen (10.3 ± 2.9days), deaminohydroxy-tamoxifen (11.9 ± 4.6days) and 4′-hydroxy-tamoxifen (10.3 ± 3.3 days), but was shorter than that of *N*-didesmethyl-tamoxifen (17.5 ± 9.6days). The concentrations of 4-hydroxy-tamoxifen were low and, therefore, the half-life of this metabolite could not be evaluated (DE VOS et al. 1992). After single administration, a similar mean terminal half-life of 8.75days was determined for tamoxifen in healthy postmenopausal women (FUCHS et al. 1996).

Following repeated daily administration, tamoxifen serum concentrations reached steady-state levels after about 1month of treatment (LONNING et al. 1992; LIEN et al. 1995b). In patients undergoing chronic tamoxifen therapy, serum levels of *N*-desmethyl-tamoxifen were, in general, about twice as high as those of the parent compound, while the concentrations of *N*-didesmethyl-tamoxifen and deaminohydroxy-tamoxifen were only about one tenth that of the parent drug; concentrations of 4-hydroxy-tamoxifen were even lower (JORDAN 1982; KEMP et al. 1983; LONNING et al. 1992). Since the absolute bioavailability of tamoxifen is not known, CL cannot be estimated with any certainty after oral administration. Apparent CL/f values of about 8–11l/h, which were calculated after single (FUCHS et al. 1996) and multiple (LIEN et al. 1990) oral administration, should be regarded with some reservations. Tamoxifen and its metabolites were detected in human liver, lung, adipose tissue, pancreas, skin, bone, brain and tumor tissues. The highest concentrations were detected in liver and lung and concentrations were, in general, 10- to 60-fold higher in tissues than in serum (LIEN et al. 1991a, b; LONNING et al. 1992). This correlates well with the tissue distribution observed in animals. In plasma, tamoxifen is more than 98% bound to albumin (LONNING and LIEN 1993).

After single and multiple oral administration, postmaximum serum concentrations of toremifene decreased biphasically with mean half-lives of 4h and 4–6 days (ANTTILA et al. 1990, 1995; SOTANIEMI and ANTTILA 1997). The mean apparent Vz/f ranged between 4571 and 9581 in healthy subjects and increased with age. The mean oral CL/f was 2.7–5.1 l/h (ANTTILA et al. 1990; SOTANIEMI and ANTTILA 1997). These values, however, should be judged with caution, since the absolute bioavailability of toremifene after oral administration is not known.

Following daily oral administration of 10, 20, 40 and 60 mg toremifene to breast cancer patients for a period of 8 weeks, steady-state serum concentrations of toremifene were reached within 3–4 weeks of treatment, whereas it took only 1–2 weeks of treatment to achieve steady-state concentrations at doses of 200 mg/day and 400 mg/day. The steady-state concentrations of the main metabolite *N*-desmethyl-toremifene were reached within 2–4 weeks of treatment under these conditions and were about 3.5–4.5 times higher than those of the parent compound. The terminal half-life was estimated after cessation of treatment in a limited number of patients only. Following administration of 20, 60 and 400 mg toremifene/day, the half-life was 5–10 days ($n = 3$), 4–5 days ($n = 3$) and 4.7 days ($n = 1$), respectively, for the parent compound, and to 4.5–15 days ($n = 3$), 4.3–6.0 days ($n = 3$) and 6.3 days ($n = 1$), respectively, for *N*-desmethyl-toremifene. The metabolite 4-hydroxy-toremifene was detectable only after administration of the higher doses of 200 mg and 400 mg toremifene/day. Steady-state concentrations of this compound were reached within 2–3 weeks of treatment and reached about 25–30% of the toremifene steady-state concentrations (WIEBE et al. 1990; KOHLER et al. 1990; BISHOP et al. 1992).

Toremifene is almost completely (99.7%) bound to serum proteins. It was shown by agarose gel electrophoresis, that about 92% of the protein-bound drug is bound to albumin, about 6% is bound to the β_1 globulin fraction and about 2% is probably bound to α_1-acid glycoprotein (ANTTILA et al. 1990; SIPILÄ et al. 1988).

For droloxifene, a terminal half-life in the range 19–37 h was reported in different studies. During chronic treatment, droloxifene plasma concentrations reached a steady state within the first 3–5 days of treatment. Droloxifene and its major metabolite in plasma, droloxifene-glucuronide, accumulated about twofold during chronic treatment. The average concentrations of droloxifene-glucuronide at steady-state were about three times higher than those of the parent compound. Compared with tamoxifen, droloxifene was absorbed faster, showed only little accumulation, and was eliminated more rapidly from plasma (GRILL and POLLOW 1991, BREITBACH et al. 1994). In the absence of data on the absolute bioavailability, the apparent total CL/f of droloxifene after oral administration was determined to be about 42 l/h (TANAKA et al. 1994).

After oral administration, panomifene plasma concentrations followed either a mono- or a biphasic disposition pattern. A distribution phase characterized by a half-life of about 18 h was followed by a terminal phase with a half-life of about 70 h (ERDELYI-TOTH et al. 1997).

In the case of raloxifene, multiple peaks were observed in plasma for both the parent drug and the major plasma metabolites raloxifene-4'-glucuronide and raloxifene-6-glucuronide, thus indicating enterohepatic recirculation of the drug. The mean terminal half-life ranged from 11 h to 27 h, and the time courses in plasma were almost parallel for raloxifene and its metabolites (NI et al. 1996). In vitro, raloxifene was bound more than 95% to human plasma proteins (KNADLER et al. 1995).

Postmaximum centchroman serum levels declined biexponentially with a mean terminal half-life of about 7 days. The drug was widely distributed into the tissues, as indicated by the large Vz/f which, on average, was determined to be 1110–1328 l in different studies. The mean total CL/f after oral administration was about 6 l/h (PALIWAL et al. 1989; LAL et al. 1995, 1996; GUPTA et al. 1996). Again, these data should be regarded with caution, because no data on the absolute bioavailability of centchroman are available. The drug is secreted into the breast milk, where it reaches maximum concentrations about 8.5 h after oral administration. The mean milk/serum AUC ratio was 1.5 and the average dose reaching the infant via the breast milk was estimated to be 7.4% of the maternal dose (GUPTA et al. 1996). In human plasma, centchroman is protein bound in a non-saturable manner, most likely to albumin, and shows no affinity to SHBG and corticoid binding globulin. However, the degree of binding has not been described (SRIVASTAVA et al. 1984).

As far as data are available, the antiestrogens described are characterized by a wide distribution into the tissues, as indicated by their high apparent Vz.

3. Metabolism

The metabolism of tamoxifen has been intensively studied and several metabolites were identified. Tamoxifen undergoes extensive hepatic metabolism through hydroxylation and *N*-demethylation. The metabolites detected in the blood of tamoxifen-treated patients are *N*-desmethyl-tamoxifen, *N*-didesmethyl-tamoxifen, 4-hydroxy-tamoxifen, 4-hydroxy-*N*-desmethyl-tamoxifen and deamino-hydroxy-tamoxifen (LONNING et al. 1992). The highest metabolite concentrations were reached by the *N*-desmethyl and *N*-didesmethyl derivatives of tamoxifen in patients with long-term therapy. These metabolites have low affinities for the estrogen receptor and exhibited antiestrogenic activity (JORDAN 1982; KEMP et al. 1983). Formation of 4-hydroxy-tamoxifen is an important metabolic pathway since the antiestrogenic potency of this metabolite is higher than that of the parent compound. However, steady-state serum concentrations of this metabolite are very low (KEMP et al. 1983; LONNING et al. 1992). As shown by in vitro experiments with human liver microsomes and human P450 enzymes expressed in different cells, this reaction is catalyzed by CYP2D6, 2C9 and 3A4 (CREWE et al. 1997; DEHAL and KUPFER 1997).

Toremifene is extensively metabolized and the main metabolite in serum is the *N*-desmethyl derivative of the drug. In addition, *N*-didesmethyl-toremifene, deaminohydroxy-toremifene and 4-hydroxy-toremifene were determined in serum (ANTTILA et al. 1990; WIEBE et al. 1990; BISHOP et al. 1992). Based on in vitro studies with human liver microsomes, the formation of *N*-desmethyl-toremifene and 4-hydroxy-toremifene is mainly mediated by CYP3A4 (BERTHOU et al. 1994). Up to ten toremifene metabolites were described in human feces; however, these structures were not presented (ANTTILA et al. 1990). Four unconjugated and three glucuronide-conjugated metabolites were detected in human urine (WATANABE et al. 1989).

Droloxifene is metabolized in humans mainly by direct glucuronidation to droloxifene-glucuronide. Besides this major metabolite, *N*-desmethyl-droloxifene, 4-methoxy-droloxifene and deamino-hydroxy-droloxifene were also detected in human serum. Except for the glucuronide, all of these metabolites bind to the estrogen receptor (GRILL and POLLOW 1991; HASMAN et al. 1994).

The major pathways of raloxifene metabolism in humans are the glucuronide conjugations of the 4′- and 6-hydroxyl group of the parent drug, with the formation of raloxifene-4′-glucuronide predominating in serum. The concentrations of both metabolites in human serum are higher than those of the parent drug. Human urinary metabolites are raloxifene-4'-glucuronide, raloxifene-6,4′-diglucuronide and comparably low amounts of raloxifene-6-glucuronide. In human feces, raloxifene is the major compound; however, it was shown that glucuronide conjugates of the drug were hydrolyzed to raloxifene when added in vitro to feces samples, thus indicating that cleavage of glucuronides was the major source of raloxifene in feces instead of non-absorbed parent drug (KNADLER et al. 1995; NI et al. 1996; FORGUE et al. 1996).

No published data were available describing the metabolism of panomifene and centchroman in humans.

4. Excretion

There is only limited information available regarding the excretion of different antiestrogens. In the case of tamoxifen, data were obtained only from two women after oral administration of radiolabeled tamoxifen. Excretion was followed for 8 days in one of these women and for 13 days in the other. Although recovery was incomplete, within these time intervals, about threefold higher amounts were excreted with feces than with urine. The radiolabeled compounds formed were excreted slowly with half-lives of 4 days and 9 days (FROMSON et al. 1973b).

Toremifene and/or its metabolites are mainly (70%) excreted via the feces (ANTTILA et al. 1990, 1995).

Following oral administration of radiolabeled raloxifene, the majority of the radioactivity was excreted within 5 days and was found primarily in the feces. Only about 6% of the dose was excreted via the kidneys (KNADLER et al. 1995).

There are no reports describing the excretion of droloxifene, panomifene and centchroman in humans.

IV. Special Populations

1. Influence of Age

Age dependency of tamoxifen pharmacokinetics was suggested in a study showing increased levels of *N*-desmethyl-tamoxifen and *N*-didesmethyl-tamoxifen in postmenopausal breast cancer patients compared with premenopausal women with breast cancer (LIEN et al. 1995b). This was confirmed in a subsequent study comprising breast cancer patients between 29 years and 85 years of age, showing an increase of steady-state levels of tamoxifen and its major metabolites *N*-desmethyl-tamoxifen, deaminohydroxy-tamoxifen and *N*-didesmethyl-tamoxifen with age (PEYRADE et al. 1996). This is further supported by a comparison of the results of two independent studies. A prolonged terminal half-life of about 8.75 days was observed after single administration of tamoxifen to healthy postmenopausal women (average age 54.8 years), whereas in healthy male volunteers with a mean age of 30.8 years, mean half-lives between 4.33 days and 5.33 days were observed (FUCHS et al. 1996; LORKOWSKI et al. 1994).

The influence of age on toremifene pharmacokinetics was investigated following single oral administration of 120 mg toremifene to either ten healthy young men (24.4 ± 4.4 years) or ten elderly women (66.9 ± 3.1 years). While maximum toremifene concentrations in serum were similar in both groups and were reached at comparable times, the terminal half-life of the drug was prolonged in the elderly women compared with the young men (mean values 7.2 days versus 4.2 days). This change was attributed to an increase in the apparent Vz, since the AUC and the apparent oral CL were comparable in both groups (SOTANIEMI and ANTTILA 1997). These findings are in accordance with the known changes of body fat with age and the distribution of these lipophilic drugs into fat tissue, thus leading to a higher Vz in the elderly.

2. Influence of Altered Liver and Kidney Function

Only limited information is available on the pharmacokinetics of tamoxifen in patients with impaired liver and kidney function. In a large study with 316 breast cancer patients, comprising 112 women with altered hepatic function, no obvious effect on plasma levels of tamoxifen and its metabolites was observed except for a slight reduction in the relative proportion of tamoxifen (PEYRADE et al. 1996). Additionally, one single report describing higher plasma levels of tamoxifen, *N*-desmethyl-tamoxifen and 4-OH-tamoxifen in one patient with liver obstruction and another report showing normal tamoxifen blood concentrations in one patient with impaired renal function were published (LONNING et al. 1992).

In the case of toremifene, the pharmacokinetics of the parent compound and its major metabolites *N*-desmethyl-toremifene and deaminohydroxy-toremifene was investigated in a comparative study involving ten subjects with normal liver and kidney function, ten subjects with impaired liver function, ten subjects with enhanced metabolic activity due to anticonvulsant therapy (resulting in cytochrome P450 enzyme induction) and ten subjects with impaired kidney function, who received a single oral administration of 120mg toremifene. The pharmacokinetics of toremifene, *N*-desmethyl-toremifene and deaminohydroxy- toremifene was not altered in subjects with reduced kidney function compared with the control group. In contrast, subjects with impaired liver function showed a prolongation of the terminal half-life of toremifene, but not of the metabolites, while all other pharmacokinetic parameters of the parent compound and the metabolites were similar to those of the control group.

V. Drug Interactions

Tamoxifen itself is a potent inhibitor of some mixed function oxidases (e.g., ethylmorphine and aminopyrine *N*-demethylase) in vitro, inhibiting its own metabolism and the metabolism of other drugs (Meltzer et al. 1984; Lonning and Lien 1993).

The interaction of tamoxifen with aminoglutethimide, an aromatase inhibitor that is used for breast cancer treatment, was investigated in two clinical studies and showed a reduction in the serum levels of tamoxifen and its metabolites in the presence of aminoglutethimide. In patients receiving chronic tamoxifen treatment, the apparent oral CL/f of tamoxifen increased about threefold after co-administration of aminoglutethimide, which may be attributed to the induction of microsomal mixed-function oxidases by aminoglutethimide. However, tamoxifen did not influence the pharmacokinetics of aminoglutethimide (Lien et al. 1990). In a separate study with patients under tamoxifen therapy, shorter mean serum half-lives were determined for tamoxifen (3.0 ± 1.5 days) and *N*-desmethyl-tamoxifen (6.0 ± 1.2 days) in the presence of aminoglutethimide compared with treatment with tamoxifen alone. However, the corresponding values of *N*-didesmethyl-tamoxifen, deaminohydroxy-tamoxifen and 4′-hydroxy-tamoxifen were 16.1 ± 7.6 days, 8.4 ± 2.3 days and 7.2 ± 1.3 days, respectively, and thus remained nearly unchanged when aminoglutethimide was co-administered (de Vos et al. 1992). It was assumed, that this interaction may be the reason why the combined tamoxifen/aminoglutethimide therapy of breast cancer failed to increase the response rate significantly when compared with tamoxifen treatment alone (Lien et al. 1990; Lonning and Lien 1993).

Life-threatening interactions between tamoxifen and warfarin have been reported (Lodwick et al. 1987; Tenni et al. 1989; Ritchie and Grant 1989), but the underlying mechanisms are not clear. In these cases, co-administration of tamoxifen to patients who were adjusted to a stable prothrombin time with warfarin, led to a serious increase in prothrombin time. This effect could be reversed by reducing the warfarin dose.

An interaction between tamoxifen and digitoxin was assumed based on observations in two patients, who had elevated digitoxin levels while receiving concomitant tamoxifen therapy. Neither patient showed signs of digitoxin overdose or intoxications (MIDDEKE et al. 1986).

General concern was expressed with regard to antibiotics that influence the intestinal flora, since they may reduce serum concentrations of tamoxifen and/or metabolites by decreasing the intestinal reabsoption (LONNING et al. 1992). Concentrations of *N*-desmethyl-tamoxifen were found to be temporarily lowered after concomitant administration of medroxyprogesterone acetate and tamoxifen when compared with treatment with tamoxifen alone. However, similar *N*-desmethyl-tamoxifen levels were measured after 6 months of treatment with tamoxifen alone or with the combination, but the clinical relevance of this effect is not clear (REID et al. 1992). Subjects undergoing anticonvulsant therapy showed a significant reduction in the terminal half-lives and the AUC values of the parent compound and *N*-desmethyl-toremifene. The apparent oral CL of toremifene was significantly increased in these subjects. All other pharmacokinetic parameters were similar to the control group (ANTTILA et al. 1995).

With regard to toremifene, droloxifene, panomifene, raloxifene and centchroman there are no published data suggesting interactions with other drugs.

References

Abdel-Aziz MT, Williams KIH (1969) Metabolism of 17α-ethynylestradiol and its 3-methyl ether by the rabbit; an in vivo D-homoannulation. Steroids 13:809–820

Abdel-Aziz MT, Williams KIH (1970) Metabolism of radioactive 17α-ethynylestradiol by women. Steroids 15:695–710

Abernethy DR, Greenblatt DJ, Divoli M, Arendt R, Ochs HR, Shader RI (1982) Impairment of diazepam metabolism by low-dose estrogen-containing oral contraceptive steroids. New England J Med 306:791–792

Adlercreutz H, Gorbach SL, Goldin BR, Woods MN, Dwyer JT, Hamalainen E (1994) Estrogen metabolism and excretion in oriental and caucasian women. J Natl Cancer Inst 86:1076–1082

Adlercreutz H, Martin F (1980) Biliary excretion and intestinal metabolism of progesterone and estrogens in man. J Steroid Biochem 13:231–244

Adlercreutz H, Martin F, Jarvenpaa P, Fotsis T (1979) Steroid absorption and enterohepatic recycling. Contraception 20:201–223

Aedo AR, Sunden M, Landgren BM, Diczfalusy E (1989) Effect of orally administered estrogens on circulating profiles in postmenopausal women. Maturitas 11:159–168

Agnihotri A, Srivastava AK, Kamboj VP (1996) Characterization of centchroman binding protein in plasma of rhesus monkey (*macaca mulatta*), J Med Primatol 25:53–56

Akpoviroro JO, Fotherby K (1980) Assay of ethynyloestradiol in human serum and its binding to plasma proteins. J Steroid Biochem 13:773–779

Akpoviroro JO, Mangalam M, Jenkins N, Fotherby K (1981) Binding of the contraceptive steroids medroxyprogesterone acetate and ethynyloestradiol in blood of various species. J Steroid Biochem 14:493–498
Anderson F (1993) Kinetics and pharmacology of estrogens in pre- and postmenopausal women. Int J Fertil 38 (Suppl) 1:53–64
Ansbacher R (1993) Bioequivalence of Conjugated Estrogen Products. Clin Pharmacokinet 24:271–274
Anttila M, Valavaara R, Kivinen S, Mäenpää J (1990) Pharmacokinetics of toremifene. J. Steroid Biochem. 36:249–252
Anttila M, Laakso S, Nyländen P, Sotaniemi EA (1995) Pharmacokinetics of the novel antiestrogenic agent toremifene in subjects with altered liver and kidney function. Clin Pharmacol Ther 57:628–635
Back DJ, Breckenridge AM, Crawford FE, MacIver M, Orme MLE, Rowe PH, Watts MJ (1979) An investigation of the pharmacokinetics of ethynylestradiol in women using radioimmunoassay. Contraception 20:263–273
Back DJ, Bates M, Breckenridge AM, Crawford FE, Ellis A, Hall JM, McIver M, Orme MLE, Rowe PH (1980a) Drug metabolism by gastrointestinal mucosa: Clinical aspects. In: Prescot, Nimmo, (eds) Drug absorption. Sydney, Adis Press, pp 80–87
Back DJ, Bates M, Bowden A, Breckenridge AM, Hall MJ, Jones H, MacIver M, Orme MLE, Perucca A, Richens A, Rowe PH, Smith E (1980b) The interaction of phenobarbital and other anticonvulsants with oral contraceptive steroid therapy. Contraception 22:495–503
Back DJ, Breckenridge AM, Crawford FE, Hall JM, MacIver M, Orme MLE, Rowe PH, Smith E, Watts MJ (1980c) The effect of rifampicin on the pharmacokinetics of ethynylestradiol in women. Contraception 21:135–143
Back DJ, Breckenridge AM, Crawford FE, Orme MLE, Rowe PH (1980d) Phenobarbitone interaction with oral contraceptive steroids in the rabbit and rat. Br J Pharmacol 69:441–452
Back DJ, Bolt HM, Breckenridge AM, Crawford FE, Orme MLE, Rowe PH, Schindler AE (1980e) The pharmacokinetics of a large (3 mg) oral dose of ethinylestradiol in women. Contraception 21:145–153
Back DJ, Breckenridge AM, MacIver M, Orme MLE, Purba H, Rowe PH (1981) Interaction of ethinylestradiol with ascorbic acid in man. Br Med J 282:1516
Back DJ, Breckenridge AM, MacIver M, Orme MLE, Purba HS, Rowe PH, Taylor I (1982a) The gut wall metabolism of ethinyloestradiol and its contribution to the pre-systemic metabolism of ethinyloestradiol in humans. Br J Clin Pharmacol 13:325–330
Back DJ, Breckenridge AM, Cross KJ, Orme MLE, Thomas E (1982b) An antibiotic interaction with ethynyloestradiol in the rat and rabbit. J Steroid Biochem 16:407–413
Back DJ, Barkfeldt JO, Breckenridge AM, Odlind V, Orme MLE, Park BK, Purba H, Tjia J, Victor A (1982c) The enzyme inducing effect of rifampicin in the rhesus monkey and its lack of interaction with oral contraceptive steroids. Contraception 25:307–316
Back DJ, Grimmer SFM, Rogers S, Stevenson PJ, Orme MLE (1987) Comparative pharmacokinetics of levonorgestrel and ethinyloestradiol following intravenous, oral and vaginal administration. Contraception 36:471–479
Back DJ, Orme MLE (1990) Pharmacokinetic drug interactions with oral contraceptives. Clin Pharmacokinet 18:472–484
Back DJ, Houlgrave R, Tjia JF, Ward S, Orme MLE (1991) Effect of the progestogens, gestodene, 3-keto desogestrel, levonorgestrel, norethisterone and norgestimate on the oxidation of ethinyloestradiol and other substrates by human liver microsomes. J Steroid Biochem Molec Biol 38:219–225
Back DJ, Orme MLE (1994) Pharmacokinetic drug interactions with oral contraceptives. In: Snow R, Hall P (eds) Steroid Contraceptives and Women's Response. New York, Plenum Press, pp 103–107

Baciewicz AM (1985) Oral contraceptive drug interactions. Ther Drug Mon 7: 26–35

Bailey DG, Arnold JMO, Spence JD (1994) Grapefruit juice and drugs how significant is the interaction? Clin Pharmacokinet 26:91–98

Balikian H, Southerland J, Howard CM, Preedy RK (1968) Estrogen metabolism in the male dog. Uptake and disappearance of specific radioactive estrogens in tissues and plasma following estrone-6,7-^{3}H Administration. Identification of estriol-16α, 17αin tissues and urine. Endocrinology 82:500–510

Ball P, Gelbke HP, Haupt O, Knuppen R (1973) Metabolism of 17α-ethynyl-[4-^{14}C]oestradiol and [4-^{14}C]mestranol in rat liver slices and interaction between 17α-ethynyl-2-hydroxyoestradiol and adrenalin. Hoppe-Seyler's Z Physiol Chem 354:1567–1575

Ball P, Knuppen R (1974) Metabolism of ethynyloestradiol and mestranol in man and interaction between 2-hydroxy-ethynyloestradiol and epinephrine. Acta Endocr (Kbh) (Suppl) 184:32

Ball SE, Forrester LM, Wolf CR, Back DJ (1990) Differences in the cytochrome P-450 isoenzymes involved in the 2-hydroxylation of oestradiol and 17α-ethinylestradiol Biochem J 267:221–226

Ball P, Knuppen R (1990) Formation, metabolism, and physiologic importance of catecholestrogens. Am J Obstet Gynecol 163:2163–2170

Balogh A, Liewald T, Liewald S, Schröder S, Klinger G, Splinter FC, Hoffmann A (1990) Zum Einfluß eines neuen Gestagens – Dienogest – und seiner Kombination mit Ethinylestradiol auf die Aktivität von Biotransformationsreaktionen. Zbl Gyn 112:735–746

Balogh A, Irmisch E, Wolf P, Letrari F, Splinter K, Hempel G, Klinger G, Hoffmann A (1991) Zum Einfluß von Levonorgestrel und Ethinylestradiol sowie deren Kombination auf die Aktivität von Biotransformationsreaktionen. Zbl Gyn 113:1388–1396

Balogh A, Klinger G, Henschel L, Börner A, Vollanth R, Kuhnz W (1995) Influence of ethinylestradiol-containing combination oral contraceptives with gestodene or levonorgestrel on caffeine elimination. Eur J Clin Pharmacol 48:161–166

Banerjee RC, Brazeau P, Saucier R, Husain SM (1973) Effects of norethindrone (17-ethinyl-17β-hydroxy-4-estren-3-one) and norgestrel (dl-13β-ethyl-17-ethynyl-17β-hydroxy-4-gonen-3-one) on the tissue distribution of ^{3}H-estradiol-17β in ovariectomized rats. Steroids 21:133–145

Baumann A, Fuhrmeister A, Brudny-Klöppel M, Draeger C, Bunte T, Kuhnz W (1996) Comparative pharmacokinetics of two new steroidal estrogens and ethinylestradiol in postmenopausal women. Contraception 54:235–242

Bawarshi-Nassar RN, Hussain A, Crooks PA (1989) Nasal absorption of 17α-ethinylestradiol in the rat. J Pharmacol 41:214–215

Bhavnani BR, Cecutti A (1993) Metabolic Clearance Rate of Equilin Sulfate and Its Conversion to Plasma Equilin, Conjugated and Unconjugated Equilenin, 17β-Dihydroequilin, and 17β- Dihydroequilenin in Normal Postmenopausal Women and Men Under Steady State Conditions. J Clin Endocrinol Metab 77:1269–1274

Bhavnani BR, Cecutti A (1994) Pharmacokinetics of 17β-Dihydroequilin Sulfate and 17β-Dihydroequilin in Normal Postmenopausal Women. J Clin Endocrinol Metab 78:197–204

Bhavnani BR, Cecutti A, Wallace D (1994) Metabolism of [^{3}H] 17β-dihydroequilin and [^{3}H] 17β-dihydroequilin sulfate in normal postmenopausal women. Steroids 59:389–394

Beach VL, Steinetz BG, Giannina T, Meli A (1967) Fate of orally administered quinestrol and ethinyl estradiol in rabbits. Int J Fertil 12:148–154

Berthou F, Dréano Y (1993) High-performance liquid chromatographic analysis of tamoxifen, toremifene and their major human metabolites. J Chromatogr Biomed Appl 616:117–127

Berthou F, Dreano I, Belloc C, Kangas L, Gautier J-C, Beaune P (1994) Involvement of cytochrome P450 3A enzyme family in the major metabolic pathways of toremifene in human liver microsomes. Biochem Pharmacol 47:1883–1895

Bird CE, Clark AF (1973) Metabolic clearance rates and metabolism of mestranol and ethynylestradiol in normal young women. J Clin Endocr Metab 36:296–302

Böcker R, Kleingeist B, Eichhorn M, Lepper H (1991) In vitro interaction of contraceptive steroids with human liver cytochrome P-450 enzymes. Adv Contraception 7:140–148

Böcker R, Lepper H (1991) Mechanism-based inhibition of human liver cytochrome P-450 by several synthetic steroids. Naunyn-Schmiedeberg's Arch Pharmacol (Suppl) 343:R13

Boekenoogen SJ, Szefler SJ, Jusko WJ (1983) Prednisolone disposition and protein binding in oral contraceptive users. J Clin Endocrinol Metab 56:702–709

Bollengier WE, Eisenfeld AJ, Gardner WU (1972) Accumulation of ^{3}H-estradiol in testes and pituitary glands of mice of strains differing in susceptibility to testicular interstitial cell and pituitary tumors after prolonged estrogen treatment. J Natl Cancer Inst 49:847–852

Bolt HM, Remmer H (1972a) The accumulation of mestranol and ethynyloestradiol metabolites in the organism. Xenobiotica 2:489–498

Bolt HM, Remmer H (1972b) Retention, metabolism and elimination of 17α-ethynylestradiol-3-methyl ether (mestranol). Xenobiotica 2:77–88

Bolt HM, Kappus H, Remmer H (1973) Studies on the metabolism of ethynylestradiol in vitro and in vivo: the significance of 2-hydroxylation and the formation of polar products. Xenobiotica 3:773–785

Bolt HM, Kappus H and Bolt M (1975) Effect of rifampicin treatment on the metabolism of oestradiol and 17α-ethinyloestradiol by human liver microsomes. Eur J Clin Pharmacol 8:301–307

Bolt HM, Bolt WH (1974) Pharmacokinetics of mestranol in man in relation to its estrogenic activity. Eur J Clin Pharmacol 7:295–305

Bolt WH, Kappus H, Bolt HM (1974) Ring A oxidation of 17α-ethynylestradiol in man. Horm Metab Res 6:432

Bolt HM, Bolt M, Kappus H (1977) Interaction of rifampicin treatment with pharmacokinetics and metabolism of ethinyloestradiol in man. Acta Endocrinol 85:185–197

Bolt HM (1979) Metabolism of estrogens-natural and synthetic. Pharmacol Ther 4:155–181

Borg ML, Gimona A, Renoux A, Douin MJ, Djebbar F, Panisrouzier R (1995) Pharmacokinetics of estradiol and estrone following repeated administration of menorest®, a new estrogen transdermal delivery system, in menopausal women. Clin Drug Invest 10:96–103

Breitbach GP, Reister C, Droege H, Bastert G (1994) Pharmakokinetik von Droloxifen im Vergleich zu Tamoxifen. Onkologie 17(Suppl 1):49–53

Brewster D, Jones RS, Symons AM (1977) Effects of neomycin on the biliary excretion and enterohepatic circulation of mestranol and 17β-oestradiol. Biochem Pharmacol 26:943–946

Brody SA, Turkes A, Goldzieher JW (1989) Pharmacokinetics of three bioequivalent norethindrone/mestranol-50μg and three norethindrone/ethinyl estradiol-35μg oc formulations: are "low-dose" pills really lower? Contraception 40: 269–284

Bourget C, Flood C, Longcope C (1984) Steroid dynamics in the rabbit. Steroids 43:225–233

Cargill DI, Steinetz BG, Gosnell E, Beach VL, Mell A, Fujimoto GI, Reynolds BM (1969) Fate of ingested radiolabeled ethynylestradiol and ist 3-cyclopentyl ether in patients with bile fistulas. J Clin Endocr Metab 29:1051–1061

Boyd RA, Yang BB, Abel RB, Eldon MA, Sedman AJ, Forgue ST (1996) Pharmacokinetics of a 7-day 17β-estradiol transdermal delivery system: Effect of application site and repeated applications on serum concentrations of estradiol and estrone. J Clin Pharmacol 36:998–1005

Carlström K, Pschera H, Lunell N-O (1988) Serum levels of estrogens, progesterone, FSH and SHBG during simultaneous vaginal administration of 17β-oestradiol and progesterone in the pre- and postmenopause. Maturitas 10:307–316

Carlström K (1996) Low endogenous estrogen levels – analytical problems. Acta Obstet Gynecol Scand (Suppl) 163:11–15

Cassidenti DL, Vijod AG, Vijod MA, Stanczyk FZ, Lobo RA. (1990) Short-term effects of smoking on the pharmacokinetic profiles of micronized estradiol in postmenopausal women. Am J Obstet Gynecol 163:1953–1960

Castagnetta LA, Granata OM, Casto ML, Calabro M et al. (1991) Simple approach to measure metabolic pathways of steroids in living cells. J Chromatogr 572:25–39

Combalbert J, Fabre I, Fabre G, Dalet I, Derancourt J, Cano JP, Maurel P (1989) Metabolism of cyclosporin A. IV. Purification and identification of the rifampicin-inducible human liver cytochrome P450 (cyclosporin A oxidase) as a product of P450IIIA gene subfamily. Drug Metab Disp 17:197–207

Chandrasekaran, A, Osman, M, Adelman, SJ et al. (1996) Determination of 17α-dihydroequilenin in rat, rabbit and monkey plasma by high-performance liquid chromatography with fluorimetric detection. J Chromatogr B 676:69–75

Cook NJ, Read GF (1995) Oestradiol measurement in women on oral hormone replacement therapy: the validity of commercial test kits. Br J Biomed Sci 52:97–101

Crawford P, Chadwick DJ, Martin C, Tjia J, Back DJ, Orme M (1990) The interaction of phenytoin and carbamazepine with combined oral contraceptives. Br J Clin Pharmacol 30:892–896

Crewe KH, Ellis SW, Lennard MS, Tucker GT (1997) Variable contribution of cytochromes P450 2D6, 2C9 and 3A4 to the 4-hydroxylation of tamoxifen by human liver microsomes. Biochem Pharmacol 53:171–178

D'Arcy PF (1986) Drug interactions with oral contraceptives. Drug Intell Clin Pharmacy 20:353–362

Dehal SS, Kupfer D (1997) CYP2D6 catalyzes tamoxifen 4-hydroxylation in human liver. Cancer Re 57:3402–3406

deHertogh R, Ekka E, Vanderhayden I, Hoet JJ (1970) Metabolic clearance rates and the interconversion factors of estrone and estradiol-17βin the immature and adult female rat. Endocrinology 87:874–880

Desta B (1988) Complete separation of nine equine estrogens by high-performance liquid chromatography. J Chromatogr 435:385–390

Dibbelt L, Knuppen R, Jütting G, Heimann S, Klipping CO, Parikka-Olexik H (1991) Group comparison of serum ethinyl estradiol, SHBG and CBG levels in 83 women using two low-dose combinationoral contraceptives for three months. Contraception 43:1–21

DeVos D, Mould G, Stevenson D (1989) The bioavailability of tamoxifen:new findings and their clinical implications. Current Therapeutic Res 46:703–708

De Vos D, Slee PHTJ, Stevenson D, Briggs RJ (1992) Serum elimination half-life of tamoxifen and its metabolites in patients with advanced breast cancer, Cancer Chemother Pharmacol 31:76–78

Düsterberg B, Nishino Y (1982) Pharmacokinetic and pharmacological features of oestradiol valerate. Maturitas 4:315–324

Düsterberg B, Schmidt-Gollwitzer M, Hümpel M (1985) Pharmacokinetics and Biotransformation of estradiol valerate in ovarectomized women. Horm Res 21:145–154

Düsterberg B, Kühne G, Täuber U (1986) Half-lives in plasma and bioavailability of ethinylestradiol in laboratory animals. Drug Res 36:1187–1190

Dyas J, Turkes A, Read GF, Riad-Fahmy D (1981) A radioimmunoassay for ethinyl oestradiol in plasma incorporating an immunosorbent, pre-assay purification procedure. Ann Clin Biochem 18:37–41

Eisenfeld AJ, Gardner WU, vanWagenen G (1971) Radioactive estradiol accumulation in endometriosis of the rhesus monkey. Am J Obstet Gynecol 109:124–130
Elger W, Schwarz S, Hedden AM, Reddersen G and Schneider B (1995) Sulfamates of various estrogens are prodrugs with increased systemic and reduced hepatic estrogenicity at oral application. J Steroid Biochem Mol Biol 55:395–403
Elsaesser F, Stickney K, Foxcroft G (1982) A comparison of metabolic clearance rates of oestradiol-17β in immature and peripubertal female pigs and possible implications for the onset of puberty. Acta Endocrinol 100:506–512
El Yazigi A, Legayada E (1997) Direct liquid chromatographic micro-measurement of tamoxifen in plasma of cancer patients. J Chromatogr B 691:457–462
Englund D, Heimer G, Johansson EDB (1984) Influence of food on oestriol blood levels. Maturitas 6:71–75
Englund DE, Elamsson KB, Johansson EDB (1982) Bioavailability of oestriol. Acta Endocrinol 99:136–140
Englund DE, Johansson ED (1981) Oral versus vaginal absorption in oestradiol in postmenopausal women. Effects of different particles sizes. Ups J med Sci 86:297–307
Englund DE, Victor A, Johansson ED (1981) Pharmacokinetics and pharmacodynamic effects of vaginal oestradiol administration from silastic rings in postmenopausal women. Maturitas 3:125–133
Erdélyi-Tóth V, Pap E, Kralovánszky J, Bojti E, Klebovich I (1994) Determination of panomifene in human plasma by high-performance liquid chromatography. J Chromatogr A 668:419–425
Erdélyi-Tóth V, Gyergyay F, Számel I, Pap E, Kralovánszky J, Bojti E, Csörgö M, Drabant S, Klebovich I (1997) Pharmacokinetics of panomifene in healthy volunteers at phase I/a study. Anti-Cancer Drugs 8:603–609
Eugster HP, Probst M, Wurgler FE, Sengstag C (1993) Caffeine, estradiol, and progesterone interact with human CYP1A1 and CYP1A2. Evidence from cDNA-directed expression in *Saccharomyces cerevisiae*. Drug Metab Dispos 21:43–49
Fernandez N, Garcia JJ, Diez MJ, Teran MT, Sierra M (1993) Rapid high-performance liquid chromatographic assay of ethynyloestradiol in rabbit plasma. J Chromatogr Biomed Appl 619:143–147
Fernandez N, Sierra M, Diez MJ, Teran T, Sahagun AM, Garcia JJ (1996) Pharmacokinetics of ethinyloestradiol in rabbits after intravenous administration. Contraception 53:307–312
Fernandez N, Sierra M, Diez MJ, Teran T, Pereda P, Garcia JJ (1997) Study of the pharmacokinetic interaction between ethinylestradiol and amoxicillin in rabbits. Contraception 55:47–52
Fishman J, Goldberg S, Rosenfeld RS, Zumoff B, Hellman L, Gallagher TF (1969) Intermediates in the transformation of oral estradiol. J Clin Endocrinol Metab 29:41–46
Forgue ST, Rudy AC, Knadler MP, Basson RP, Nelson JE, Henry DP, Allerheiligen SR (1996) Raloxifene pharmacokinetics in healthy postmenopausal women. Pharm Res 13 No. 9 (Suppl):S429
Fotherby K (1973) Metabolism of synthetic steroids by animals and man. Acta endor (Copenh Suppl) 185:119–147
Fotherby K, Akpoviroro JO, Siekmann L, Breuer H (1981) Measurement of ethynylestradiol by radioimmunoassay and by isotope dilution-mass spectrometry. J Steroid Biochem 14:499–500
Fotherby K (1991) Intrasubject variability in the pharmacokinetics of ethynyloestradiol. J Steroid Biochem Molec Biol 38:733–736
Fournier S, Kuttenn F, de Cicco F, Baudot N, Malet C, Mauvais Jarvis P (1982) Estradiol 17β-hydroxysteroid dehydrogenase activity in human breast fibroadenomas. J Clin Endocrinol Metab 55:428–433
Fraser IS, Challis JRG, Thorburn GD (1976) Metabolic clearance rate and production rate of oestradiol in conscious rabbits. J Endocrinol 68:313–320

Frey BM, Frey FJ (1985) The effect of altered prednisolone kinetics in patients with the nephrotic syndrome and in women taking oral contraceptive steroids on human mixed lymphocyte cultures. J Clin Endocrinol Metab 60:361–369
Fried KM, Wainer IW (1994) Direct determination of tamoxifen and its four major metabolites in plasma using coupled column high-performance liquid chromatography. J Chromatogr B 655:261–268
Fridriksdottir H, Loftsson T, Gudmundsson JA, Bjarnason GJ, Kjeld M, Thorsteinsson T (1996) Design and in vivo testing of 17β-estradiol-HP beta CD sublingual tablets. Pharmazie 51:39–42
Fromson JM, Pearson S, Bramah S (1973a) The metabolism of tamoxifen (ICI 46474) Part I: In laboratory animals. Xenobiotica 3:693–709
Fromson JM, Pearson S, Bramah S (1973b) The metabolism of tamoxifen (ICI 46474) Part II: In female patients. Xenobiotica 3:711–714
Fuchs WS, Leary WP, van der Meer MJ, Gay S, Witschital K, von Nieciecki A (1996) Pharmacokinetics and bioavailability of tamoxifen in postmenopausal healthy women. Arzneim Forsch/Drug Res 46:418–422
Fullerton FR, Greenman DL, Young JF (1987) Influence of a purified diet and route of administration on the metabolism and disposition of estradiol in B6C3F1 mice. Drug Metab Dispos Biol Fate Chem 15:602–607
Gabrielsson J, Wallenbeck I, Larsson G, Birgerson L, Heimer G (1995) New kinetic data on estradiol in light of the vaginal ring concept. Maturitas 22:S35-S39
Galbraith RA, Michnovicz JJ (1989) The effects of cimetidine on the oxidative metabolism of estradiol. N Engl J Med 321:269–274
Galbraith RA, Michnovicz JJ (1993) Omeprazole Fails to Alter the Cytochrome-P450-Dependent 2-Hydroxylation of Estradiol in Male Volunteers. Pharmacology 47:8–12
Ginsburg ES, Mello NK, Mendelson JH et al. (1996) Effects of alcohol ingestion on estrogens in postmenopausal women. JAMA 276:1747–1751
Ginsburg ES, Walsh BW, Shea BF, Gao X, Gleason RE, Barbieri RL (1995) The effects of ethanol on the clearance of estradiol in postmenopausal women. Fertil Steril 63:1227–1230
Goebelsmann U, Mashchak A, Mishell DR (1985) Comparison of hepatic impact of oral and vaginal administration of ethinyl estradiol. Am J Obstet Gynecol 151:868–877
Goldzieher JW (1989) Pharmacology of contraceptive steroids: a brief review. Am J Obstet Gynecol 160:1260–1264
Goldzieher JW (1990) Selected aspects of the pharmacokinetics and metabolism of ethinyl estrogens and their clinical implications. Am J Obstet Gynecol 163:318–322
Gomez-Benitez J, Sosa-Gonzales A, Diaz-Chico BN (1984) Relations between ^{3}H-estradiol uptake and receptor content of estrogen responsive tissues of castrated female rats. Revista Espanola de Fisiologia 40:311–318
Goldzieher JW, Brody SA (1990) Pharmacokinetics of ethinyl estradiol and mestranol. Am J Obstet Gynecol 163:2114–2119
Ghosh SK (1988) Production of monoclonal antibodies to estriol and their application in the development of a sensitive nonisotopic immunoassay. Steroids 52:1–14
Grill HJ, Pollow K (1991) Pharmacokinetics of droloxifene and its metabolites in breast cancer patients. Am J Clin Oncol 14 (Suppl 2):S21-S29
Grimmer SFM, Back DJ, Orme MLE, Cowie A, Gilmore I, Tjia J (1986) The bioavailability of ethinyloestradiol and levonorgestrel in patients with an ileostomy. Contraception 33:51–59
Guengerich FP (1988) Oxidation of 17α-ethynylestradiol by human liver cytochrome P-450. Mol Pharmacol 33:500–508
Guengerich FP (1990) Mechanism-based inactivation of human liver microsomal cytochrome P-450 IIIA4 by gestodene. Chem Res Toxicol 3:363–371
Gupta RC, Nityanand S, Asthana OP, Lal J (1996) Pharmacokinetics of centchroman in nursing women and passage into breast milk. Clin Drug Invest 11:305–309

Gustavson LE, Legler UF, Benet LZ (1986) Impairment of prednisolone disposition in women taking oral contraceptives or conjugated estrogens. J Clin Endocrinol Metab 62:234–237
Haaf H, Metzler M, Li JJ (1988) Influence of α-naphthoflavone on the metabolism and binding of ethinylestradiol in male syrian hamster liver microsomes: possible role in hepatocarcinogenesis. Cancer Res 48:5460–5465
Hammerstein J, Daume E, Simon A, Winkler UH, Schindler AE, Back DJ, Ward S, Neiss A (1993) Influence of gestodene and desogestrel as components of low-dose oral contraceptives on the pharmacokinetics of ethinyl estradiol (EE2), on serum CBG and on urinary cortisol and 6β-hydroxycortisol. Contraception 47:263–281
Han X, Liehr JG (1992) Induction of covalent DNA adducts in rodents by tamoxifen. Cancer Res 52:1360–1363
Harris RZ, Tsunoda SM, Mroczkowski P, Wong H, Benet LZ (1996) The effects of menopause and hormone replacement therapies on prednisolone and erythromycin pharmacokinetics. Clin Parmacol Ther 59:429–435
Hasman M, Rattel B, Löser R (1994) Preclinical data for droloxifene. Cancer Lett 84:101–116
Heffner LJ (1976) ^{3}H-estradiol uptake and retention by target tissues of light-sterilized rats. Neuroendocrin 20:319–327
Heimer GM (1987) Estriol in the postmenopause. Acta Obstet Gynecol Scand (Suppl) 139:1–23
Heimer GM, Englund DE (1984) Enterohepatic recirculation of oestriol studied in cholecystectomized and non-cholecystectomized menopausal women. Upsala J Med Sci 89:107–115
Heimer GM, Englund DE (1986a) Enterohepatic recirculation of oestriol: inhibition by activated charcoal. Acta Endocrinol 113:93–95
Heimer GM, Englund DE (1986b) Plasma oestriol following vaginal administration: morning versus evening insertion and influence of food. Maturitas 8:239–243
Heithecker R, Aedo AR, Landgren BM, Cekan SZ (1991) Plasma estriol levels after intramuscular injection of estriol and two of its esters. Horm Res 35:234–238
Hellriegel ET, Matwyshyn GA, Fei P, Dragnev KH, Nims RW, Lubet RA, Kong A-NT (1996) Regulation of gene expression of various phase I and phase II drug-metabolizing enzymes by tamoxifen in rat liver. Biochem Pharmacol 52:1561–1568
Helton ED, Williams EC, Goldzieher JW (1977) Oxidative metabolism and de-ethynylation of 17α-ethynyestradiol by baboon liver microsomes. Steroids 30:71–83
Hembree WC, Bardin CW, Lipsett MB (1969) A study of estrogen metabolic clearance rates and transfer factors. J Clin Invest 48:1809–1819
Henderson KM, Camberis M, Hardie AH (1995) Evaluation of antibody- and antigen-coated enzymeimmunoassays for measuring oestrone-3-glucuronide concentratios in urine. Clin Chim Acta 243:191–203
Hermens WA, Belder CW, Merkus JM, Hooymans PM, Verhoef J, Merkus FW (1991) Intranasal estradiol administration to oophorectomized women. Eur J Obstet Gynecol Reprod Biol 40:35–41
Herz R, Koelz H, Haemmerli U, Benes I, Blum AL (1978) Inhibition of hepatic demethylation of aminopyrine by oral contraceptive steroids in humans. Eur J Clin Invest 8:27–30
Hirai S, Hussain A, Haddadin M, Smith RB (1981) First-pass metabolism of ethinyl estradiol in dogs and rats. J Pharmac Sci 70:403–406
Holleran WM, Gharbo SA, DeGregorio MW (1987) Quantitation of toremifene and its major metabolites in human plasma by high performance liquid chromatography following fluorescent activation. Anal Lett 20:871–879
Hoon TJ, Dawood MY, Khandawood FS, Ramos J, Batenhorst RL (1993) Bioequivalence of a β-Hydroxypropyl-β-Cyclodextrin Complex in Postmenopausal Women. J Clin Pharmacol 33:1116–1121
Hu S, He Q, Zhao Z (1992) Determination of trace amounts of estriol and estradiol by adsorptive cathodic stripping voltammetry. Analyst 117:181–184

Hümpel M, Nieuweboer B, Wendt H, Speck U (1979) Investigations of pharmacokinetics of ethinyloestradiol to specific consideration of a possible first-pass effect in women. Contraception 19:421–432

Hümpel M, Täuber U, Kuhnz W, Pfeffer M, Brill K, Heithecker R, Louton T, Steinberg B (1990) Comparison of serum ethinyl estradiol, sex hormone-binding globulin, corticoid-binding globulin and cortisol levels in women using two low-dose combined oral contraceptives. Horm Res 33:35–39

Hussain MA, Aungst BJ, Shefter E (1988) Prodrugs for improved oral β-estradiol bioavailability. Pharm Res 5:44

Iida K, Imai A, Tamaya T (1991) Estriol binding in uterine corpus cancer and in normal uterine tissues. Gen Pharmacol 22:491–493

Jehan Q, Srivasta S, Akhlaq M et al. (1982) Kinetics and distribution and retention of ^{3}H-oestradiol-17βin rat tissues: A comparative study with free oestradiol and after its incorporation into liposomes. Endokrinologie 80:8–12

Jensen RT, Collen MJ, Pandol SJ et al. (1983) Cimetidine-induced impotence and breast changes in patients with gastric hypersecretory states. N Engl J Med 308:883–887

Jochemsen R, Van der Graaff M, Boeijinga JK, Breimer DD (1982) Influence of sex, menstrual cycle and oral contraception on the disposition of nitrazepam. Br J Clin Pharmacol 13:319–324

Johnston A (1996) Estrogens – pharmacokinetics and pharmacodynamics with special reference to vaginal administration and the new estradiol formulation – Estring. Acta Obstet Gynecol Scand (Suppl)163:16–25

Joshi JV, Joshi UM, Sankolli GM, Gupta K, Rao AP, Hazari K, Sheth UK, Saxena BN (1980) A study of interaction of a low-dose combination oral contraceptive with anti-tubercular drugs. Contraception 22:643–652

Jordan CV (1982) Metabolites of tamoxifen in animals and man: identification, pharmacology, and significance. Breast Cancer Res Treat 2:123–138

Jung-Hoffmann C, Kuhl H (1989) Interaction with the pharmacokinetics of ethinylestradiol and progestogens contained in oral contraceptives. Contraception 40:299–312

Kall MA, Vang O, Clausen J (1996) Effects of dietary broccoli on human in vivo drug metabolizing enzymes: Evaluation of caffeine, oestrone and chlorzoxazone metabolism. Carcinogenesis 17:793–799

Kalund-Jensen, H, Myren, CJ (1984) Vaginal absorption of oestradiol and progesterone. Maturitas 6:359–367

Kamyab S, Fotherby K (1969) Metabolism of 4-^{14}C-ethynyl oestradiol in women. Nature 221:360–361

Katayama M, Taniguchi H (1993) Determination of Estrogens in Plasma by High-Performance Liquid Chromatography After Pre-Column Derivatization with 2-(4-carboxyphenyl)-5,6-dimethylbenzimidazole. J Chromatogr 616:317–322

Kaufmann JM, Thiery M, Vermeulen A (1981) Plasma levels of ethinylestradiol (EE) during cyclic treatment with combined oral contraceptives. Contraception 24:589–602

Kazama N, Longcope C (1972) Metabolism of estrone and estradiol-17β in sheep. Endocrinology 91:1450–1454

Kemp JV, Adam HK, Wakeling AE, Slater R (1983) Identification and biological activity of tamoxifen metabolites in human serum. Biochem Pharmacol 32:2045–2052

Kerlan V, Dreano Y, Bercovici JP, Beaune PH, Floch HH, Berthou F (1992) Nature of cytochromes P450 involved in the 2-/4-hydroxylations of estradiol in human liver microsomes. Biochem Pharmacol 44:1745–1756

Kholkute SD, Kumar TCA, Puri CP (1987) Pharmacokinetic and pharmacodynamic studies on estradiol-17β administered sublingually or intramuscularly to adult male marmosets. Adv Contra Delv Sys 3:195–204

Klein KO, Baron J, Colli MJ, Mcdonnell DP, Cutler GB (1994) Estrogen levels in childhood determined by an ultrasensitive recombinant cell bioassay. J Clin Invest 94:2475–2480

Knadler MP, Lantz RJ, Gillespie TA, Allerheiligen SRB, Henry DP (1995) The disposition and metabolism of 14C-labeled raloxifene in humans. Pharm-Res 12 No. 9 (Suppl):S372

Kohler PC, Hamm JT, Wiebe VJ, DeGregorio MW, Shemano I, Tormey DC (1990) Phase I study of the tolerance and pharmacokinetics of toremifene in patients with cancer. Breast Cancer Res Treat 16 (Suppl):S19–S26

Kuhnz W, Putz B (1989) Pharmacokinetic interpretation of toxicity tests in rats treated with oestradiol valerate in the diet. Pharmacol Toxicol 65:217–222

Kuhnz W, Hümpel M, Schütt B, Louton T, Steinberg B, Gansau C (1990a) Relative bioavailability of ethinyl estradiol from two different oral contraceptive formulations after single oral administration to 18 women in an intraindividual cross-over design. Horm Res 33:40–44

Kuhnz W, Pfeffer M, Al-Yacoub G (1990b) Protein binding of the contraceptive steroids gestodene, 3-ketodesogestrel and ethinylestradiol. J Steroid Biochem 35:313–318

Kuhnz W, Back DJ, Power J, Schütt B, Louton T (1991) Concentration of ethinyl estradiol in the serum of 31 young women following a treatment period of three months with two low-dose oral contraceptives in an intraindividual cross-over design. Horm Res 36:63–69

Kuhnz W, Gansau C, Fuhrmeister A (1992a) Pharmacokinetics of gestodene in 12 women who received a single oral dose of 0.075 mg gestodene and, after a washout phase, the same dose during a treatment cycle. Contraception 46:29–40

Kuhnz W, Al-Yacoub G, Fuhrmeister A (1992b) Pharmacokinetics of levonorgestrel in 12 women who received a single oral dose of 0.15 mg levonorgestrel and, after a wash-out phase, the same dose during a treatment cycle.Contraception 46:443–454

Kuhnz W, Gansau C, Mahler M (1993a) Pharmacokinetics of Estradiol, Free and Total Estrone, in Young Women Following Single Intravenous and Oral Administration of 17β-Estradiol. Arzneimittelforsch 43–2:966–973

Kuhnz W, Louton T, Back DJ, Michaelis K (1993b) Radioimmunological analysis of ethinylestradiol in human serum. Validation of the method and comparison with a gas chromatographic/mass spectrometric assay. Arzneim Forsch/Drug Res 43:16–21

Kuhnz W, Baumann A, Staks T, Dibbelt L, Knuppen R, Jütting G (1993c) Pharmacokinetics of gestodene and ethinylestradiol in 14 women during three months of treatment with a new tri-step combination oral contraceptive: serum protein binding of gestodene and influence of treatment on free and total testosterone levels in the serum. Contraception 48:303–322

Kuhnz W, Blode H (1994) Pharmacokinetics of selected contraceptive steroids in various animal species. In: Goldzieher J (ed) Pharmacology of the contraceptive steroids. Raven Press, New York, pp 41–51

Kuhnz W, Staks T, Jütting G (1994) Pharmacokinetics of levonorgestrel and ethinylestradiol in 14 women during three months of treatment with a tri-step combination oral contraceptive: serum protein binding of levonorgestrel and influence of treatment on free and total testosterone levels in the serum. Contraception 50:563–579

Kuhnz W, Hümpel M, Biere H, Gross D (1996) Influence of repeated oral doses of ethinyloestradiol on the metabolic disposition of [$^{13}C_2$]-ethinyloestradiol in young women. Eur J Clin Pharmacol 50:231–235

Kulkarni BD (1970) Metabolism of [^{14}C]ethynylestradiol in the baboon. J Endocrinol 48:91–98

Kulkarni BD, Goldzieher JW (1970) Urinary excretion pattern and fractionation of radioactivity after injection of 4-^{14}C-mestranol in women. Contraception 1:131–136

Kulkarni BD (1976) Steroid contraceptives in non-human primates (I). Metabolic fatew of synthetic estrogens in the baboon before exposure to oral contraceptives. Contraception 14:611–623

Kvorning I, Christensen MS (1981) Bioavailability of four oestradiol suspensions with different particle-sizes – in vivo/in vitro correlation. Drug Dev Ind Pharm 7:289–303

Lal J, Paliwal JK, Grover PK, Gupta RC (1994) Simultaneous liquid chromatographic determination of centchroman and its 7-demethylated metabolite in serum and milk. J Chromatogr B 658:193–197

Lal J, Asthana OP, Nityanand S, Gupta RC (1995) Pharmacokinetics of centchroman in healthy female subjects after oral administration. Contraception 52:297–300

Lal J, Nityanand S, Asthana OP, Gupta RC (1996) Comparative bioavailability of two commercial centchroman tablets in healthy female subjects. Indian J Pharmacol 28:32–34

Lantz RJ, Knadler MP (1993) HPLC Determination of raloxifene before and after hydrolysis of human plasma. Pharm Res 10 No. 10 (Suppl):S50

Larner JM, Hochberg RB (1985) The clearance and metabolism of estradiol and estradiol-17-esters in the rat. Endocrinology 117:1209–1214

Lee GJL, Oyang MH, Bautista J, Kushinsky S (1987) Determination of ethinylestradiol and norethindrone in a single specimen of plasma by automated high-performance liquid chromatography and subsequent radioimmunoassay. J Liquid Chromatogr 10:2305–2318

Lee YS, Lorenzo BJ, Koufis T, Reidenberg MM (1996) Grapefruit juice and its flavonoids inhibit 11β-hydroxysteroid dehydrogenase. Clin Pharmacol Ther 59:62–71

Leroux Y, Borg ML, Sibille M et al. (1995) Bioavailability study of Menorest®, a new estrogen transdermal delivery system, compared with a transdermal reservoir system. Clin Drug Invest 10:172–178

Levrant SG, Barnes RB (1994) Pharmacology of estrogens. In: Lobo RA (ed) Treatment of the postmenopausal woman. Basic and clinical aspects. Raven Press, New York, pp 57–68

Lichtenberg V, Schulte-Baukloh A, Lindner C and Braendle W (1992) Die Bestimmung von 17-β-Östradiol (E_2) im Serum von Frauen unter einer Östrogen-Substitutionstherapie liefert diskrepante Ergebnisse mit verschiedenen Immunoassay-Kits. Lab Med 16:412–418

Lien EA, Anker G, Lonning PE, Solheim E, Ueland PM (1990) Decreased serum concentrations of tamoxifen and ist metabolites induced by aminoglutethimide. Cancer Res 50:5851–5857

Lien EA, Solheim E, Ueland PM (1991a) Distribution of tamoxifen and ist metabolites in rat and human tissues during steady-state treatment. Cancer Res 51:4837–4844

Lien EA, Wester K, Lonning PE, Solheim E, Ueland PM (1991b) Distribution of tamoxifen and metabolites into brain tissue and brain metastases in breast cancer patients. Br J Cancer 63:641–645

Lien EA, Anker G, Lonning PE, Ueland PM (1995a) Determination of droloxifene and two metabolites in serum by high-pressure liquid chromatography. Ther Drug Monit 17:259–265

Lien EA, Anker G, Ueland PM (1995b) Pharmacokinetics of tamoxifen in premenopausal and postmenopausal women with breast cancer. J Steroid Biochem Mol Biol 55:229–231

Longcope C, Williams KIH (1977) Ethynylestradiol and mestranol: their pharmacodynamics and effects on natural estrogens. In: Garattini S, Berendes HW (eds) Pharmacology of steroid contraceptive drugs. Raven Press, New York, pp 89–98

Lim CK, Chow LCL, Yuan Z-X, Smith LL (1993) High performance liquid chromatography of tamoxifen and metabolites in plasma and tissues. Biomed Chromatogr 7:311–314

Lim CK, Yuan Z-X, Ying K-C, Smith LL (1994a) High performance liquid chromatography of toremifene and metabolites. J Liquid Chromatogr 17:1773–1783

Lim CK, Yuan Z-X, Lamb JH, White INH, DeMatteis F, Smith LL (1994b) A comparative study of tamoxifen metabolism in female rat, mouse and human liver microsomes. Carcinogenesis 15:589–593

Lim CK, Yuan Z-X, Jones RM, White INH, Smith LL (1997) Identification and mechanism of formation of potentially genotoxic metabolites of tamoxifen: study by LC-MS/MS. J Pharmaceut Biomed Anal 15:1335–1342

Lindstrom TD, Whitacker NG, Whitacker GW (1984) Disposition and metabolism of a new benzothiophene antiestrogen in rats, dogs and monkeys. Xenobiotica 14:841–847

Littleton GK, Anderson RR (1972) Characterization of 17β-estradiol-^{3}H single injection disappearance curves in rat plasma and red cells (36602). Proc Soc Exp Biol Med 140:1015–1020

Lobo RA, Brenner P, Mishell DR, Jr (1983) Metabolic parameters and steroid levels in postmenopausal women receiving lower doses of natural estrogen replacement. Obstet Gynecol 62:94–98

Lobo RA, Cassidenti DL (1992) Pharmacokinetics of oral 17β-estradiol. J Reprod Med 37:77–84

Lobo RA, Ettinger B, Hutchinson KA, Knopp RH, Lindsay R, Nachtigall LE, Santoro N, Studd J, Hutchinson KA et al. (1996) Estrogen replacement. The evolving role of transdermal delivery. J Reprod Med 41:781–796

Lodwick R, McConkey B, Brown AM (1987) Life threatening interaction between tamoxifen and warfarin. Br Med J 295:141

Lokind KB, Lorenzen FH, Bundgaard H (1991) Oral Bioavailability of 17β-Estradiol and Various Ester Prodrugs in the Rat. Int J Pharmaceut 76:177–182

Longcope C, Flood C, Tast J (1994) The metabolism of estrone sulfate in the female rhesus monkey. Steroids 59:270–273

Longcope C, Gorbach S, Goldin B et al. (1985) The metabolism of estradiol; oral compared to intravenous administration. J Steroid Biochem 23:1065–1070

Longcope C, Gorbach S, Goldin B et al. (1987) The effect of a low fat diet on estrogen metabolism. J Clin Endocrinol Metab 64:1246

Longcope C, Williams KIH (1974) The metabolism of estrogens in normal women after pulse injections of ^{3}H-estradiol and ^{3}H-estrone. J Clin Endocrinol Metab 38:602–607

Lonning PE, Ekse D (1995) A sensitive assay for measurement of plasma estrone sulphate in patients on treatment with aromatase inhibitors. J Steroid Biochem Mol Biol 55:409–412

Lonning PE, Kvinnsland S, Thorsen T, Ekse D (1986) Aminoglutethimide as an inducer of microsomal enzymes. Part 2: Endocrine aspects. Breast Cancer Res Treat 7 (Suppl):S77–S82

Lonning PE, Kvinnsland S, Thorsen T, Ueland PM (1987) Alterations in the metabolism of estrogens during treatment with aminoglutethimide in breast cancer patients. Clin Pharmacokinet 13:393–406

Lonning PE, Lien EA, Lundgren S, Kvinnsland S (1992) Clinical pharmacokinetics of endocrine agents used in advanced breast cancer. Clin Pharmacokinet 22:327–358

Lonning PE, Lien EA (1993) Pharmacokinetics of anti-endrocrine agents. In: Workman P, Graham MA (guest eds) Cancer Surveys, Volume 17, Pharmacokinetics and cancer chemotherapy. Cold Spring Harbour Laboratory Press, pp 343–370

Loock W, Nau H, Schmidt Gollwitzer M, Dvorchik BH (1988) Pregnancy-specific changes of antipyrine pharmacokinetics correlate inversely with changes of estradiol/progesterone plasma concentration ratios. J Clin Pharmacol 28:216–221

Lorkowski G, Lücker PW, Petersen G, Schnitzler M, Wetzelsberger N (1994) Bioequivalence of two commercially available tamoxifen tablet formulations in healthy male volunteers. Meth Find Exp Clin Pharmacol 16:443–447

Luttge WG, Whalen RE (1972) The accumulation, retention and interaction of oestradiol and oestrone in central neural and peripheral tissues of gonadectomized female rats. J Endocrinol 52:379–395

MacCallum J, Cummings J, Dixon JM, Miller WR (1996) Solid-phase extraction and high-performance liquid chromatographic determination of tamoxifen and its major metabolites in plasma. J Chromatogr B 678:317–323

Mackinnon M, Sutherland E, Simon FR (1977) Effects of ethinyl estradiol on hepatic microsomal proteins and the turnover of cytochrome P-450. J Lab Clin Med 90:1096–1106

Maggs JL, Grabowski PS, Rose ME, Park BK (1982) The biotransformation of 17α-ethynyl[3H]estradiol in the rat: irreversible binding and biliary metabolites. Xenobiotica 12:657–668

Maggs JL, Grimmer SFM, Orme MLE, Breckenridge AM, Park BK (1983a) The biliary and urinary metabolites of 3H-17α-ethynylestradiol in women. Xenobiotica 13:421–431

Maggs JL, Grabowski PS, Park BK (1983b) The enterohepatic circulation of the metabolites of 17α-ethynyl[^{3}H]estradiol in the rat. Xenobiotica 13:619–626

Maggs JL, Park BK (1985) A comparative study of biliary and urinary 2-hydroxylated metabolites of [6,7–3H] 17α-ethynylestradiol in women. Contraception 32:173–182

Mani C, Gelboin HV, Park SS, Pearce R, Parkinson A, Kupfer D (1993a) Metabolism of the antimammary cancer antiestrogenic agent tamoxifen I. Cytochrome P-450-catalyzed N-demethylation and 4-hydroxylation. Drug Metab Dispos 21:645–656

Mani C, Hodgson E, Kupfer D (1993b) Metabolism of the antimammary cancer antiestrogenic agent tamoxifen II. Flavin-containing monooxygenase-mediated N-oxidation. Drug Metab Dispos 21:657–661

Mani C, Kupfer D (1991) Cytochrome P450 mediated action and irreversible binding of the antiestrogen tamoxifen to proteins in rat and human liver: possible involvement of the flavin-containing monooxygenases in tamoxifen activation. Cancer Res 51:6052–6058

Martinsen A, Gynther J (1996) Liquid chromatography-thermospray mass spectrometry of toremifene and its derivatives. J Chromatogr A 724:358–363

Martucci CP, Fishman J (1993) P450 Enzymes of Estrogen Metabolism. Pharmacol Ther 57:237–257

Marynick SP, Havens WW, Ebert MH et al. (1976) Studies on the transfer of steroid hormones across the blood-cerebrospinal fluid barrier in the rhesus monkey. Endocrinology 99:400–405

MeffinPJ, Wing LMH, Sallustio BC, Brooks PM (1984) Alterations in prednisolone disposition as a result of oral contraceptive use and dose. Br J Clin Pharmac 17:655–664

Mattsson LA, Cullberg G (1983) Vaginal absorption of two estriol preparations. Acta Obstet Gynecol Scand 62:393–396

Meltzer NM, Stang P, Sternson LA (1984) Influence of tamoxifen and its N-desmethyl and 4-hydroxy metabolites on rat liver microsomal enzymes. Biochem Pharmacol 33:115–123

Meyer FP, Walther H, Canzler E, Giers H (1989) Einfluß der oralen Kontrazeptiva Minisiston/Trisiston auf die Pharmakokinetik von Coffein – ein intraindividueller Langzeitversuch. Z Klin Mrd 44:239–240

Middeke M, Remien C, Lohmüller G, Holzgreve H, Zöllner N (1986) Interaction between tamoxifen and digitoxin? Klin Wschr 64:1211

Mikkelsen AL, Borggaard B, Lebech PE (1996) Results of serial measurement of estradiol in serum with six different methods during ovarian stimulation. Gynecol Obstet Invest 41:35–40

Mills TM, Lin TJ, Hernandez-Ayup S, Greenblatt RB, Ellegood JO, Mahesh VB (1974) The metabolic clearance rate and urinary excretion of oral contraceptive drugs. Am J Obstet Gynecol 120:773–778

Müller P, Botta L, Ezzet F (1996) Bioavailability of estradiol from a new matrix and a conventional reservoir-type transdermal therapeutic system. Eur J Clin Pharmacol 51:327–330

Murugesan K, Vij U, Lal B, Farooq A (1989) Effect of progestins, estradiol, and coenzymes NAD and NADPH on the interconversion of estradiol and estrone in rabbit uterus in vitro. Steroids 53:695–712

Musey PI, Kirdani RY, Bhanalaph T, Sandberg AA (1973) Estriol metabolism in the baboon: Analysis of urinary and biliary metabolites. Steroids 22:795–817

Nash HA, Brache V, Alvarez-Sanchez F, Jackanicz TM, Harmon TM (1997) Estradiol delivery by vaginal rings: Potential for hormone replacement therapy. Maturitas 26:27–33

Newburger J, Castracane VD, Moore PH, Williams MC, Goldzieher JW (1983) The pharmacokinetics and metabolism of ethynyl estradiol and its three sulfates in the baboon. Am J Obstet Gynecol 146:80–87

Newburger J, Goldzieher JW (1985) Pharmacokinetics of ethynyl estradiol: a current view. Contraception 32:33–44

Ni L, Allerheiligen SRB, Basson R, Knadler MP, Latz J, Geiser J, Lantz R, Rash J, Henry DP (1996) Pharmacokinetics of raloxifene in men and postmenopausal women volunteers: Pharm Res 13 No. 9 (Suppl):S430

Nickerson DF, Tess DA, Toler SM (1997) First-pass metabolism and biliary recirculation of droloxifene in the female Sprague-Dawley rat: Xenobiotica 27:257–264

Noma J, Hayashi N; Sekiba K (1991) Automated direct high-performance liquid chromatographic assay for estetrol, estriol, cortisone and cortisole in serum and amniotic fluid. J Chromatogr A 568:35–44

Notelovitz M, Tjapkes J, Ware M (1981) Interaction between estrogen and dilantin in a menopausal women. N Engl J Med 304:788

Nozaki O, Ohba Y, Imai K (1988) Determination of serum estradiol by normal-phase high-performance liquid chromatography with peroxyoxalate chemiluminescence detection. Anal Chim Acta 205:255–260

Nuwaysir EF, Dragan YP, Jefcoate CR, Jordan VC, Pitot HC (1995) Effects of tamoxifen administration on the expression of xenobiotic metabolizing enzymes in rat liver. Cancer Res 55:1780–1786

O'Connell, MB (1995) Pharmacokinetic and pharmacologic variation between different estrogen products. J Clin Pharmacol 35:S18–S24

Orme MLE, Back DJ, Ball S (1989) Interindividual variation in the metabolism of ethynylestradiol. Pharmacol Ther 43:251–260

Oriowo MA, Landgren BM, Stenström B, Diczfalusy E (1980) A comparison of the pharmacokinetic properties of three estradiol esters. Contraception 21:415–424

Orme M, Back DJ, Ward S, Green S (1991) The pharmacokinetics of ethinylestradiol in the presence and absence of gestodene and desogestrel. Contraception 43:305–316

Ortiz de Montellano PR, Kunze K, Yost GS, Mico BA (1979) Self-catalized destruction of cytochrome P-450: covalent binding of ethynyl sterols to prosthetic heme. Proc Natl Acad Sci USA 76:746–749

Pazzucconi F, Malavasi B, Galli G, Franceschini G, Calabresi L, Sirtori CR (1991) Inhibition of antipyrine metabolism by low-dose contraceptives with gestodene and desogestrel. Clin Pharmacol Ther 49:278–284

Paliwal JK, Gupta RC, Grover PK, Asthana OP, Srivastava JS, Nityanand S (1989) High performance liquid chromatographic (HPLC) determination of centchroman in human serum and application to single-dose pharmacokinetics. Pharmaceut Res 6:1048–1051

Paliwal JK, Gupta RC (1996) Tissue distribution and pharmacokinetics of centchroman. A new nonsteroidal postcoital contraceptive agent and ist 7-desmethyl metabolite in female rats after single oral dose. Drug Metab and Dispos 24:148–155

Patel J, Katovich MJ, Sloan KB, Curry SH, Prankerd RJ (1995) A prodrug approach to increasing the oral potency of a phenolic drug. 2. pharmacodynamics and preliminary bioavailability of an orally administered o(imidomethyl) derivative of 17β-estradiol. J Pharm Sci 84:174–178

Peden NR, Cargill J, Browning MC et al. (1979) Male sexual dysfunction during treatment with cimetidine. Br Med J 1:659

Peyrade F, Frenay M, Etienne M-C, Ruch F, Guillemare C, Francois E, Namer M, Ferrero J-M, Milano G (1996) Age-related difference in tamoxifen disposition. Clin Pharmacol Ther 59:401–410

Podesta A, Smith CJ, Villani C, Montagnoli G (1996) Shared reaction in solid-phase immunoassay for estriol determination. Steroids 61:622–626

Potischman N, Falk RT, Laiming VA, Siiteri PK, Hoover RN (1994) Reproducibility of laboratory assays for steroid hormones and sex hormone-binding globulin. Cancer Res 54:5363–5367

Powers MS, Schenkel L, Darley PE, Good WR, Balestra JC, Place VA (1988) Pharmacokinetics and pharmacodynamics of transdermal formulations of estradiol. Münch Med Wochenschr 130:7–14

Price TM, Blauer KL, Hansen M, Stanczyk F, Lobo R, Bates GW (1997) Single-dose pharmacokinetics of sublingual versus oral administration of micronized 17β-estradiol. Obstet Gynecol 89:340–345

Pschera H, Hjerpe A, Carlstrom K (1989) Influence of the maturity of the vaginal epithelium upon the absorption of vaginally administered estradiol-17β and progesterone in postmenopausal women. Gynecol Obstet Invest 27:204–207

Purba HS, Maggs JL, Orme MLE, Back DJ, Park BK (1987) The metabolism of 17α-ethinyloestradiol by human liver microsomes: formation of catechol and chemically reactive metabolites. Br J Clin Pharmac 23:447–453

Raitano LA, Slikker W, Hill DE, Hadd HE, Cairns T, Helton ED (1981) Ethynyl cleavage of 17α-ethynylestradiol in the rhesus monkey. Drug Metab Disp 9:129–134

Ratna S, Roy SK, Ray S, Kole PL, Salman M, Madhusudan KP, Sircar, KP, Anand N (1986) Centchroman: In vitro metabolism by rat liver homogenate. Drug Develop Res 7:173–178

Ratna S, Mishra NC, Ray S, Roy SK (1994) Centchroman: tissue distribution and excretion profile in albino rats after oral and intravenous administration. J Basic Appl Biomed 2:31–36

Rauramo L, Punnonen R, Kaihola H-L, Grönroos M (1978) Serum oestriol, oestrone and and oestradiol concentrations during oral oestriol succinate treatment in ovariectomized women. Maturitas 1:71–78

Reed MJ, Fotherby K, Steele SJ (1972) Metabolism of ethynyloestradiol in man. J Endocr 55:351–361

Reed MJ, Fotherby K (1975) Metabolism of ethynyloestradiol and oestradiol in the guinea-pig. J Steroid Biochem 6:121–125

Reed MJ, Fotherby K (1979) Intestinal absorption of synthetic steroids. J Steroid Biochem 11:1107–1112

Reed MJ, Lai LC, Ghilchik MW, James VHT (1988) The effects of androgens and cortisol on the in vivo metabolism of oestradiol. J Steroid Biochem 30:489–492

Reid AD, Horobin JM, Newman EL, Preece PE (1992) Tamoxifen metabolism is altered by simultaneous administration of medroxyprogesterone acetate in breast cancer patients. Breast Cancer Res Treat 22:153–156

Ritchie LD, Grant SMT (1989) Tamoxifen-warfarin interaction: the Aberdeen hospitals drug file. Br Med J 298:1253

Roberts RK, Desmond PV, Wilkinson GR, Schenker S (1979) Disposition of chlordiazepoxide: sex differences and effects of oral contraceptives. Clin Pharmacol Ther 25:826–831

Rogers SM, Back DJ, Orme MLE (1987a) Intestinal metabolism of ethinyloestardiol and paracetamol in vitro: studies using Ussing Chambers. Br J Clin Pharmacol 23:727–734

Rogers SM, Back DJ, Stevenson PJ, Grimmer SFM, Orme MLE (1987b) Paracetamol interaction with oral contraceptive steroids: increased plasma concentrations of ethinyloestradiol. Br J Clin Pharmacol 23:721–725

Rohr UD, Nauert C, Ehrly AM (1995) Kinetics of a new patch for transdermal administration of 17β-estradiol. Zbl Gynäkol 117:531–539

Ruenitz PC, Bai X (1995) Acidic metabolites of tamoxifen. Aspects of formation and fate in the female rat. Drug Metab Dispos 23:993–998

Ruoff WL, Dziuk PJ (1994) Absorption and metabolism of estrogens from the stomach and duodenum of pigs. Dom Anim Endocrinol 11:197–208

Sandberg AA, Slaunwhite WR (1957) Studies on phenolic steroids in human subjects. II. The metabolic fate and hepatobiliary-enteric circulation of ^{14}C-estrone and ^{14}C-estradiol in women. J Clin Invest 36:1266–1278

Sanders JM, Burka LT, Shelby MD, Newbold RR, Cunningham ML (1997) Determination of tamoxifen and metabolites in serum by capillary electrophoresis using a nonaqueous buffer system J Chromatogr B 695:181–185

Sato H, Mochizuki H, Tomita Y, Kanamori T (1996) Enhancement of the sensitivity of a chemiluminescent immunoassay for estradiol based on hapten heterology. Clin Biochem 29:509–513

Scavone JM, Greenblatt DJ, Blyden GT (1988) Antipyrine pharmacokinetics in women receiving conjugated estrogens. J Clin Pharmacol 28:463–466

Schiff I, Tulchinsky D, Ryan KJ (1977) Vaginal absorption of estrone and 17β-estradiol. Fertil Steril 28:1063–1066

Schiff I, Tulchinsky D, Ryan KJ et al. (1980) Plasma estriol and its conjugates following oral and vaginal administration of estriol to postmenopausal women: correlations with gonadotropin levels. Am J Obstet Gynecol 138:1137–1141

Schmid SE, Au WYW, Hill DE, Kadlubar FF, Slikker W (1983) Cytochrome P-450-dependent oxidation of the 17α-ethynyl group of synthetic steroids. Drug Metab Dispos 11:531–536

Schmider J, Greenblatt DJ, von Moltke LL, Karsov D, Vena R, Friedman HL, Shader RI (1997) Biotransformation of mestranol to ethynyl estradiol in vitro: the role of cytochrome P-450 2C9 and metabolic inhibitors. J Clin Pharmacol 37:193–200

Schmidt G, Andersson SB, Nordle O, Johansson CJ, Gunnarsson PO (1994) Release of 17β-oestradiol from a vaginal ring in postmenopausal women: pharmacokinetic evaluation. Gynecol Obstet Invest 38:253–260

Schneider J, Kinne D, Fracchia A et al. (1982) Abnormal oxidative metabolism of estradiol in women with breast cancer. Proc Natl Acad Sci USA 79:3047–3051

Schubert W, Eriksson U, Edgar B, Cullberg G, Hedner T (1995) Flavonoids in grapefruit juice inhibit the in vitro hepatic metabolism of 17β-estradiol. Eur J Drug Metab Pharmacokinet 20:219–224

Schubert W, Cullberg G, Edgar B, Hedner T (1994) Inhibition of 17β-estradiol metabolism by grapefruit juice in ovariectomized women. Maturitas 20:155–163

Schubert W, Cullberg G, Hedner T (1993) Pharmacokinetic evaluation of oral 17β-oestradiol and two different fat soluble analogues in ovariectomized women. Eur J Clin Pharmacol 44:563–568

Schwenk M, Schiemenz C, Del Pino VL, Remmer H (1982) First pass biotransformation of ethinylestradiol in rat small intestine in situ. Naunyn-Schmiedeberg's Arch Pharmacol 321:223–225

Scott RT, Ross B, Anderson C, Archer DF (1991) Pharmacokinetics of percutaneous estradiol: A crossover study using a gel and a transdermal system in comparison with oral micronized estradiol. Obstet Gynecol 77:758–764

Shenfield GM (1993) Oral contraceptives are drug interactions of clinical significance? Drug Safety 9:21–37

Shou M, Korzekwa KR, Brooks EN, Krausz KW, Gonzales FJ, Gelboin HV (1997) Role of human hepatic cytochrome P450 1A2 and 3A4 in the metabolic activation of estrone. Carcinogenesis 18:207–214

Siekmann L, Siekmann A, Breuer H (1980) Measurement by isotope dilution mass spectrometry of 17α-ethinyl-estradiol-17β and norethisterone in serum of women taking oral contraceptives. Biomed Mass Spectrometry 7:511–514

Siiteri PK, Murai JT, Hammond GL, Nisker JA, Raymoure WJ, Kuhn RW (1982) The serum transport of steroid hormones. Recent Prog Horm Res 38:457–510

Sipilä H, Näntö V, Kangas L, Anttila M, Halme T (1988) Binding of toremifene to human serum proteins. Pharmacol Toxicol 63:62–64

Sipilä H, Kangas L, Vuorilehto L, Kalapudas A, Eloranta M, Södervall M, Toivola R, Anttila M (1990) Metabolism of toremifene in the rat. J Steroid Biochem 36:211–215
Slikker W, Bailey JR, Newport GD, Lipe GW, Hill DE (1982) Placental transfer and metabolism of 17α-ethynylestradiol-17β and estradiol-17β in the rhesus monkey. J Pharmacol Exp Ther 223:483–489
Slikker W, Lipe GW, Newport GD (1981) High-performance liquid chromatographic analysis of estradiol-17 and metabolites in biological media. J Chromatogr 224:205–219
Soe L, Wurz GT, Mäenpää JU, Hubbard GB, Cadman TB, Wiebe VJ, Theon AP, DeGregorio MW (1997) Tissue distribution of transdermal toremifene. Cancer Chemother Pharmacol 39:513–520
Sotaniemi EA, Anttila MI (1997) Influence of age on toremifene pharmacokinetics. Cancer Chemother Pharmacol 40:185–188
Speck U, Wendt H, Schulze PE, Jentsch D (1976) Bioavailability and pharmacokinetics of cyproterone acetate-14C and ethinyloestradiol-3H after oral administration as a coated tablet (SH B 209 AB). Contraception 14:151–163
Spink DC, Lincoln II DW, Dickermann HW, Gierthy JF (1990) 2,3,7,8-Tetrachlorodibenzo-p-dioxin causes an extensive alteration of 17β-estradiol metabolism in MCF-7 breast tumor cells. Proc Natl Acad Sci USA 87:6917–6921
Srivastava AK, Agnihotri A, Kamboj VP (1984) Binding of centchroman – a nonsteroidal antifertility agent to human plasma proteins. Experientia 40:465–466
Stadel BV, Sternthal PM, Schlesselman JJ, Douglas MB, Dallas Hall W, Kaul L, Ahluwalia B (1980) Variation of ethinylestradiol blood levels among healthy women using oral contraceptives. Fertil Steril 33:257–260
Stanczyk FZ, Gale JA, Goebelsmann U, Nerenberg C, Matin S (1980) Radioimmunoassay of plasma ethinylestradiol in the presence of circulating norethindrone. Contraception 22:457–470
Studd JWW Smith RNJ (1993) Oestradiol and testosterone implants. Bailliere'S Clin Endocrinol Metab 7:203–223
Stumpf PG (1990) Pharmacokinetics of estrogen. Obstet Gynecol 75:9S–14S
Suhonen SP, Allonen HO, Lahteenmaki P (1993) Sustained-release subdermal estradiol implants: A new alternative in estrogen replacement therapy. Am J Obstet Gynecol 169:1248–1254
Taioli E, Garte SJ, Trachman J et al. (1996) Ethnic differences in estrogen metabolism in healthy women. J Natl Cancer Inst 88:617
Takikawa H, Sano N, Tadokoro K, Yamanaka M (1996) Biliary excretion of estrone metabolites in the rat. Int Hepatol Commun 5:251–258
Tanaka Y, Sekiguchi M, Sawamoto T, Hata T, Esumi Y, Sugai S, Ninomiya S (1994) Pharmacokinetics of droloxifene in mice, rats, monkeys, premenopausal and postmenopausal patients. Eur J Drug Metab Pharmacokin 19:7–58
Tannenbaum SR (1997) Comparative metabolism of tamoxifen and DNA adduct formation and invitro studies on genotoxicity. Seminars in Oncology 24 (Suppl 1):S1–81–S1–86
Tamate K, Charleton M, Gosling JP et al. (1997) Direct colorimetric monoclonal antibody enzyme immunoassay for estradiol-17β in saliva. Clin Chem 43:1159–1164
Tenni P, Lalich DL, Byrne MJ (1989) Life threatening interaction between tamoxifen and warfarin. Br Med J 298:93
Tepper R, Goldberger S, Cohen I et al. (1994) Estrogen replacement in postmenopausal women: are we currently overdosing our patients? Gynecol Obstet Invest 38:113–116
Tess DA, Cole RO, Toler SM (1995) Sensitive method for the quantitation of droloxifene in plasma and serum by high-performance liquid chromatography employing fluorimetric detection. J Chromatogr B 674:253–260
Tetsuo M, Axelson M, Sjovall J (1980) Selective isolation procedures for GC/MS analysis of ethinylsteroids in biological material. J Steroid Biochem 13:847–860

Thomas CMG, van den Berg RJ, Segers MFG et al. (1993) Inaccurate measurement of 17β-Estradiol in serum of female volunteers after oral administration of milligram amounts of micronized 17β-Estradiol. Clin Chem 39:2341–2342
Timmer CJ, Apter D, Voortman G (1990) Pharmacokinetics of 3-keto-desogestrel and ethinyletradiol released from different types of contraceptive vaginal rings. Contraception 42:629–642
Tolino A, Ronsini S, Granata P, Gallo FP, Riccio S, Montemagno U (1990) Topical treatment using estriol of post-menopausal atrophic vaginitis. (Franz). Rev Fr Gynecol Obstet 85:692–697
Tominaga T, Abe O, Izuo M, Nomura Y (1990) A phase I study of toremifene. Breast Cancer Res Treat 16 (Suppl):S27-S29
Tornatore KM, Kanarkowski R, McCarthy TL, Gardner MJ, Yurchak AM, Jusko WJ (1982) Effect of chronic oral contraceptive steroids on theophylline disposition. Eur J Clin Pharmacol 23:129–134
Townsend RW, Keuth V, Embil K et al. (1988) High-performance liquid chromatographic determination of conjugated estrogens in tablets. J Chromatogr 450:414–419
Troy SM, Hicks DR, Parker VD, Jusko WJ, Rofsky HE, Porter RJ (1994) Differences in pharmacokinetics and comparative bioavailability between Premarin® and Estratab® in healthy postmenopausal women. Curr Ther Res 55:359–372
Tseng L, Gurpide E (1975) Induction of human endometrial estradiol dehydrogenase by progestins. Endocrinology 97:825–833
Turkes A, Dyas J, Read GF, Riad-Fahmy D (1981) Enzyme immunoassay for specific determination of the synthetic estrogen ethynyl estradiol in plasma. Clin Chem 27:901–905
Vanluchene E, Vandekerckhove D, Jonckheere J et al. (1983) Steroid profiles of body fluids other than urine, obtained by capillary gas chromatography. J Chromatogr 279:573–580
Van Look PF, Top Huisman M, Gnodde HP (1981) Effect of ampicillin or amoxycillin administration on plasma and urinary estrogen levels during normal pregnancy. Eur J Obstet Gynecol Reprod Biol 12:225–233
Warren RJ, Fotherby K (1973) Plasma levels of ethynylestradiol after administration of ethynylestradiol or mestranol to human subjects. J Endocrinol 59:369–370
Wasser SK, Monfort SL, Southers J et al. (1994) Excretion rates and metabolites of oestradiol and progesterone in baboon (Papio cynocephalus cynocephalus) faeces. J Reprod Fertil 101:213–220
Watanabe N, Irie T, Koyama M, Tominaga T (1989) Liquid chromatographic-atmospheric pressure ionization mass spectrometric analysis of toremifene metabolites in human urine. J Chromatogr 497:169–180
Weber A, Jäger R, Börner A, Klinger G, Vollanth R, Matthey K, Balogh A (1996) Can grapefruit juice influence ethinylestradiol bioavailability? Contraception 53:41–47
Webster LK, Crinis NA, Stokes KH, Bishop JF (1991) High-performance liquid chromatographic method for the determination of toremifene and its major human metabolites. J Chromatogr Biomed Appl 565:482–487
White I, Müller-Eberhard H (1977) Decreased liver cytochrome P-450 in rats caused by norethindrone or ethynylestradiol. Biochem J 166:57–64
White INH, Martin EA, Styles J, Lim C-K, Carthew P, Smith LL (1997) The metabolism and genotoxicity of tamoxifen. In: Aldaz CM, Gould MN, McLachlan J, Slaga TJ (eds) Etiology of breast and gynecological cancers. Wiley-Liss Inc., New York Chichester Weinheim Brisbane Singapore Toronto, pp 257–270
Wiebe V, Benz CC, Shemano I, Cadman TB, DeGregorio MW (1990) Pharmacokinetics of toremifene and its metabolites in patients with advanced breast cancer. Cancer Chemother Pharmacol 25:247–251
Wiemeyer JC, Fernandez M, Moguilevsky JA, Sagasta CL (1986) Pharmacokinetic studies of estradiol enanthate in menopausic women. Arzneimittelforsch 36:1674–1677

Williams JG, Williams KIH (1975) Metabolism of 2-^{3}H- and 4-^{14}C-17α-ethynylestradiol 3-methyl ether (mestranol) by women. Steroids 26:707–720

Williams JG, Longcope C, Williams KIH (1975) Metabolism of 4-^{3}H- and 4-^{14}C-17α-ethynylestradiol 3-methyl ether (mestranol) by women. Steroids 25:343–354

Williams MC, Helton ED, Goldzieher JW (1975) Chromatographic profiling and identification of ethynyl and non-ethynyl compounds. Steroids 25:229–246

Williams MC, Goldzieher JW (1980) Chromatographic patterns of urinary ethynyl estrogen metabolites in various populations. Steroids 36:255–282

Wiseman LR, Goa KL (1997) Toremifene A review of ist pharmacological properties and clinical efficiency in the management of advanced breast cancer. Drugs 54:141–160

Woolley DE, Holinka CF, Timiras PS (1969) Changes in ^{3}H-estradiol distribution with development in the rat. Endocrinology 84:157–161

Yamamoto T, Honjo H et al. (1979) The metabolic fate of estradiol benzoate in female dog. J Steroid Biochem 11:1287–1294

Zamah NM, Hümpel M, Kuhnz W, Louton T, Rafferty J, Back DJ (1993) Absence of an effect of high vitamin C dosage on the systemic availability of ethinyl estradiol in women using a combination oral contraceptive. Contraception 48:377–391

Zimmermann H, Koytchev R, Mayer O, Börner A, Mellinger U, Breitbarth H (1998a) Pharmacokinetics of orally administered estradiol valerate. Results of a single-dose cross-over bioequivalence study in postmenopausal women. Arzneimittelforsch 48(11):941–947

Zimmermann H et al. (1998b) Single and multiple dose pharmacokinetics of a sequential estradiol valerate / chlormadinone acetate preparation for hormone replacement therapy. (publication in preparation)

CHAPTER 36

Toxicology of Estrogens and Antiestrogens

H. ZIMMERMANN

A. General Aspects

The toxicological assessment of estrogens is complicated by the fact that the endocrine glands of the body interact by complex feedback mechanisms in order to maintain normal status, so that the administration of exogenous hormone can have repercussions that may not be attributable directly to the hormone itself (HEYWOOD and WADSWORTH 1980). EDGREN (1994) stated that predicting clinical effects from laboratory animal research was particularly difficult in the case of contraceptive steroids, and HEYWOOD (1986) concluded that it was unlikely that no-effect levels could be established with any steroid administered to give greater-than-physiological levels. Further problems in the toxicological assessment of estrogens arise from species differences in reproduction processes that make it difficult to extrapolate from one species to another, especially to humans. These species differences include:

- Different serum estradiol concentrations within normal reproductive cycles. In primates the physiological estradiol levels are one or two orders of magnitude higher than in rats, mice or dogs (GUNZEL et al. 1989).
- Different estrogen/progestogen ratio. Estrogens and progestogens act synergistically in the regulation of nearly all reproductive processes; they do this only in certain ratios to each other, and these ratios are markedly different between the species (BEIER et al. 1983). The optimal estradiol/progesterone ratio for the maintenance of physiological functions of target organs is about 1:20,000 for rats, 1:800 for rabbits and 1:50 for primates (NISHINO and NEUMANN 1980). From these data it becomes clear that, in rodents, estrogenic effects will markedly dominate when a combined oral contraceptive is tested in the estrogen/progestogen ratio as it is used in humans. Therefore, the testing of estrogen/progestogen combinations would not be useful in pharmacological and toxicological studies. LEHMANN et al. (1989) concluded that the results obtained from the testing of sex-steroid combinations did not provide further useful information beyond that already known from the administration of the individual components.
- Feedback mechanisms that are affected in different ways in the various animal species (BEIER et al. 1983).
- Different functions of some hormones that are released to an increased or decreased extent under the influence of sexual hormones. Whereas in dogs

only progestogen is necessary for the lobuloalveolar growth of the mammary gland, in other species estrogen and progestogen are necessary (BOLT 1976). Prolactin, to give another example, is the principal luteotropic hormone in rats and mice, but not in primates. Therefore, pseudopregnancy can be induced in the rat and mouse by estrogen administration via the positive feedback on prolactin, whereas this is not possible in primates (BEIER et al. 1983).
- Differences in receptor content of some target organs. The estrogen-binding capacity of rat liver is much higher than that of human liver (BOJAR et al. 1982).
- Different susceptibility of certain species, strains or individuals. Pituitary tumors are found in rats and mice after long-term estrogen treatment, whereas in dogs and monkeys no increase in pituitary tumors is found in spite of the fact that prolactin synthesis and excretion is increased in these species (EELETRBY et al. 1979).
- Differences in pharmacokinetics. The oral bioavailability of ethinyl estradiol, for example, ranges from less than 1% in Rhesus monkeys and rabbits to 40% in man, with the total clearance being lowest in man at 4.5 ml/min/kg and highest in the rat at 61 ml/min/kg (HUMPEL et al. 1986).

All these physiological differences must be considered in the risk assessment of estrogens and, therefore, no simple extrapolation of experimental results to humans is possible. GUNZEL et al. (1989) concluded in their review article on comparative aspects of steroid toxicology that results derived from experimental studies, at least quantitatively, were only relevant to the system in which the test was carried out. LEHMANN et al. (1989) questioned the relevance of toxicity studies of contraceptive steroids to human-risk assessment. Because the widest use of estrogens is that in combined oral contraceptives, together with a progestogen, their conclusions are summarized here. Epidemiological studies have suggested the following conditions in humans to be associated with the use of oral contraceptives:

- Increased incidence of venous thromboembolism
- Cardiovascular complications
- Influence on glucose tolerance
- Increased incidence of benign liver-cell tumors in rare individual cases

None of these effects has been predicted by the toxicity studies in experimental animals (LEHMANN et al. 1989). However, most of the beneficial effects of contraceptives have not been predicted by toxicity studies either. But, what can be derived from experimental models is the demonstration of whether a given sex steroid, such as an estrogen, reveals the anticipated species-specific response or whether it exerts additional, unexpected adverse effects (LEHMANN et al. 1989).

B. Toxicology of Estrogens

I. Acute Toxicity

The acute toxicity of all estrogens is very low after oral and parenteral administration. The acute toxicity of ethinyl estradiol in mice, rats and dogs was recently investigated by MAKINO et al. (1990) and OHNO et al. (1991). The acute toxic symptoms they found include decrease in spontaneous activity, hair loss, lacrimation, salivation, exophthalmus, corneal opacity, convulsion and dyspnea. The majority of animal deaths were found within 24 h. No remarkable changes were found at autopsy of animals that died during the study, but in surviving animals hypertrophy of liver, adrenals and uterus, as well as atrophy of thymus, testis and prostate were revealed. Table 1 shows the LD_{50} values of some estrogens.

II. Subchronic and Chronic Toxicity

Effects seen in subchronic- and chronic-toxicity studies can generally be attributed to physiological effects of estrogens or exaggerated estrogenic effects.

Table 1. Lethal dose-50% (LD_{50}) values of different estrogens in various species

Estrogen	Species (sex)	LD_{50} oral (mg/kg)	LD_{50} i.p. (mg/kg)	LD_{50} s.c. (mg/kg)	Reference
EE	Mouse	2500	700	>2600	LEHMANN et al. 1989
	Mouse (f)	970	462	>3000	MAKINO et al. 1990
	Mouse (m)	955	559	>3000	MAKINO et al. 1990
	Mouse	>5000			OHNO et al. 1991
	Rat	>5000			LEHMANN et al. 1989
	Rat (f)	960	500	>2000	MAKINO et al. 1990
	Rat (m)	1200	471	>2000	MAKINO et al. 1990
	Rat	>5000			OHNO et al. 1991
	Dog	>1000			LEHMANN et al. 1989
	Dog (f)	>2000			OHNO et al. 1991
EV	Mouse	>4000		>4000	LEHMANN et al. 1989
	Mouse (f)	>4000	ca. 4000	>2000	GÜNZEL 1971
	Rat	>5000	>4000		LEHMANN et al. 1989
	Dog	>1000		i.m.: >250	SEIBERT and GÜNZEL 1994
DES	Mouse	>3000	538		ENGEL et al. 1983
	Rat	>3000	750		ENGEL et al. 1983
Mestranol	Mouse (f)	>10000	>3200	2500–5000	MINESITA et al. 1970
	Mouse (m)	>10000	>3000	>5000	MINESITA et al. 1970
	Rat (f)	>10000	>2400	>5000	MINESITA et al. 1970
	Rat (m)	>10000	>3000	>5000	MINESITA et al. 1970

EE, ethinylestradiol; EV, estradiol valerate; DES, diethylstilbestrol; i.p., intraperitoneal; s.c., subcutaneous; i.m., intramuscular injection; f, female; m, male

The most salient clinical and laboratory findings observed in subchronic- and chronic-toxicity studies of estrogens in mice and rats are dose-related reductions in food consumption and body-weight gain (HEYWOOD and WADSWORTH 1980). Furthermore, hair loss (KRAMER and GUNZEL 1979) and depressed righting and placing reflexes were seen in rats (GIBSON et al. 1967). The dose-dependent development of anemia is a typical estrogen effect in rats (KRAMER and GUNZEL 1979). HEYWOOD and WADSWORTH (1980) found elevated alkaline phosphatase and decreased transaminases, whereas MINESITA et al. (1970) observed increased transaminases and a decrease in red blood cell counts, hematocrit and hemoglobin. In addition to the above-mentioned anemic changes, SEIBERT and GUNZEL (1994) observed reduced thrombocyte and leukocyte counts after the administration of estradiol valerate. NAGAE et al. (1992) treated ovariectomized rats s.c. with estradiol at doses of 0.1, 1 and 10 μg/animal/day, producing estradiol plasma levels of 10, 75 and 400 pg/ml. Compared with the vehicle-treated control, they found reduced red blood cell and white blood cell counts, shortened mean prothrombin time, prolonged partial thromboplastin time, increased transaminases and total cholesterol, and decreased alkaline phosphatase at 1 and 10 μg/animal/day. The majority of these changes were not observed when compared with nonovariectomized rats, and/or were also noted when the values in the non-ovariectomized rats were compared with those in the vehicle control.

Findings on total cholesterol are inconsistent; MINESITA et al. (1970) and HEYWOOD and WADSWORTH (1980) found no changes, whereas LEHMANN et al. (1989) found a decrease after synthetic estrogens that is confirmed by our own unpublished data, and NAGAE et al. (1992) found an increase, but did so after treating ovariectomized rats s.c. with the natural estrogen 17β-estradiol. GIBSON et al. (1967) found faint signs of toxicity in hematological and clinical biochemical parameters in rats after the administration of conjugated equine estrogens and diethylstilbestrol. A slight reduction in hemoglobin and hematocrit in the highest diethylstilbestrol dose group was the only remarkable finding. LEHMANN et al. (1989) observed no effects on glucose levels in rats. Our own data suggest no effect of estrogens on glucose levels in Wistar rats and an increase in glucose in Sprague Dawley rats. Because there seem to be several strain- and sex-dependent effects in rats, the main findings of our own studies using different strains are summarized in Table 2.

In rats, bile flow is reduced with estrogens, but jaundice has never been reported in rodents (HEYWOOD and WADSWORTH 1980).

The most significant macroscopic postmortem findings in rats and mice are pituitary enlargement in both sexes; uterine enlargement often associated with hydrometra and pyometra; inhibition of ovarian-follicle maturation, thymic atrophy and proliferation of the mammary glands in females; and reduced size of seminal vesicles, testes and prostate in males (MINESITA et al 1970; HEYWOOD and WADSWORTH 1980; HART 1988; LEHMANN et al. 1989; COOKSON 1994). Other reported findings in rats include adrenocortical hyperplasia (HEYWOOD and WADSWORTH 1980), atrophy of the fascicular zone of the

Table 2. Main effects of estrogens observed in 2- to 4-week-long rat toxicity studies (own unpublished data)

	Wistar		Sprague Dawley	
	Male	Female	Male	Female
Body weight	–	–	–	–
Food consumption	–	–	–	–
Hemoglobin		–		–
Hematocrit		–		–
Prothrombin time	+	+	+	+
Gamma glutamyl transpeptidase	+	+		+
Alkaline phosphatase	(+)	+		
Aspartate aminotransferase	+	+		
Total protein		(+)	(+)	+
Glucose			+	+
Total cholesterol		–		–
Testes weight	–		–	
Epididymis weight	–		–	
Seminal vesicle and prostate weight	–		–	
Uterus weight		+		+
Ovary weight		–		–
Adrenal weight	+		+	
Thymus weight		–		–
Alopecia		+	+	+

–, decrease; +, increase

adrenals and hypertrophy of the reticular zone (Seibert and Gunzel 1994) and increased liver weight (Nagae et al. 1992; Hart 1988). Furthermore, Cookson (1994) reported pyelitis, pyelonephritis and hydronephrosis in mice and stomach ulcers in mice and rats after 32 weeks of s.c. treatment with estradiol cypionate. Neoplastic changes will be discussed later.

In a 4-week study, even the lowest dose of 1.2 mg/kg estradiol valerate significantly reduced food consumption, body-weight gain and terminal body weight. With this dose, the systemic burden of exogenous estradiol was 1.6-fold that of the endogenous estradiol levels, thus demonstrating the extreme sensitivity of rats towards exogenous estradiol (Kuhnz and Putz 1989).

In a 30-day study of mestranol, even the lowest tested dose of 0.4 mg/kg caused significant changes in body-weight gain, food consumption, red blood cell count, adrenal and thymic weights, serum glutamic pyruvate transaminase (GPT) and total protein (Minesita et al. 1970).

The dog is especially susceptible to the bone-marrow toxicity of estrogens. In dogs, a general finding is a decrease in values of erythrocytic parameters (Lehmann et al. 1989; Kramer and Gunzel 1979). Hematological changes in dogs caused by ethinyl estradiol were investigated by Capel-Edwards et al. (1971), who found leukocytosis and thrombocytopenia resulting in hemorrhagic diathesis in the 5-mg/kg dose group after 14–17 days; thus, the animals had to be killed for humane reasons. Mortality was also increased in dogs treated with 12 mg/kg estradiol valerate (SEIBERT and GUNZEL 1994). Estrogens induce bilateral alopecia in dogs (HEYWOOD and WADSWORTH 1980; WEIKEL and NELSON 1977). Other clinical findings include vomiting, diarrhea, drop in blood pressure, hemorrhages of different tissues and blood in the urine and feces (SEIBERT and GUNZEL 1994). In male dogs, estrogens cause a reduction in testicular size, together with mammary-gland development and the induction of pendulous prepuce. In female animals, the mammary glands and vulva become swollen and a mucoid or blood-tinged vaginal discharge is evident. Macroscopic postmortem findings are enlargement of the uterus and prostate and reduction in testis and ovary sizes (HEYWOOD and WADSWORTH 1980; SEIBERT and GUNZEL 1994). No effects of estrogens on total cholesterol were found by HEYWOOD and WADSWORTH (1980), whereas LEHMANN et al. (1989) and SEIBERT and GUNZEL (1994) observed an increase.

ATTIA et al. (1996) compared estrogenic effects of 5 mg/kg estradiol and 0.5–2 mg/kg ethinyl estradiol in dogs and monkeys and concluded that the monkey was better suited as a nonrodent species for toxicological investigations of estrogens. Thrombocytopenia and alopecia were only observed in dogs and not in monkeys. Mortality in dogs was mainly due to thrombocytopenia, whereas no mortality was observed in monkeys. Decreased body weight, small testes, mammary-gland swelling, brownish/blackish pigments on the abdomen and anemia were observed in both species. In the Food and Drug Administration (FDA) guidelines for testing of contraceptive steroids (JORDAN 1992), there is also a recommendation for the monkey as a nonrodent species.

In monkeys, estrogens cause only minor changes in food consumption and body weight. Estrogen administered to male and female nonhuman primates caused the development of sex skin and of the mammary gland (HEYWOOD and WADSWORTH 1980). GEIL and LAMAR (1977) found mild anemia with mestranol in Rhesus monkeys, whereas GOLDZIEHER and KRAEMER (1972) found no effects of ethinyl estradiol on any of the hematological and coagulation parameters tested. Changes in recorded blood-biochemistry parameters include an increase in alanine aminotransferase and leucine aminopeptidase (HEYWOOD AND WADSWORTH 1978), a decrease in alkaline phosphatase and inorganic phosphorus and an increase in triglycerides (ATTIA et al. 1996). However, GOLDZIEHER and KRAEMER (1972) observed no influence of ethinyl estradiol on liver enzymes. Squirrel monkeys developed mesotheliomas after estrogen treatment, whereas no mesotheliomas were observed in rhesus monkeys (HEYWOOD and WADSWORTH 1980). In Cynomolgus monkeys, white blood cell counts, antithrombin III and fibrinogen were slightly increased after 4-week oral treatment with 2 mg/kg ethinyl estradiol. Increased triglyceride concentrations were the only blood chemistry finding (own unpublished data). Macroscopic postmortem findings include effects on the uterus, cervix, fallopian tubes, vagina and mammary glands (HEYWOOD and WADSWORTH 1980). Effects observed in the mammary glands are hyperplasia of the ducts followed by the formation of acini. However, nodular or neoplastic changes were never seen in primates after long-term estrogen treatment (GEIL and LAMAR 1977).

There were reductions in thymic weights in Cynomolgus monkeys (own unpublished data). After very high doses of estrogens, piloerection, alopecia, weight loss and decreased food consumption were also observed (own unpublished data).

III. Genotoxicity

For the assessment of a possible drug-related mutagenic potential, a test battery, consisting of a gene-mutation assay in bacterial and mammalian cells in vitro, an in vitro assay for chromosome mutations with and without metabolic activation by mammalian microsomes or liver S-9 fraction and an in vivo assay covering a cytogenetic endpoint, are generally proposed. The most widely used in vitro standard tests are the Ames Salmonella test, gene-mutation tests using Chinese hamster V79 cells or Chinese hamster ovary (CHO) cells, cytogenetic analysis of chromosomal aberrations in human lymphocytes or permanent mammalian cell lines. The mouse micronucleus test for the indirect detection of chromosome and genome mutations is usually performed on erythrocytes of the bone marrow, in vivo. All of these standard tests and various additional tests were performed with different estrogens.

Some results of genotoxicity tests are summarized in Tables 3–6. It is quite clear that estrogens never gave positive results in gene-mutation tests in bacteria. Most of the estrogens tested were also negative in mammalian gene-mutation tests. The only exceptions are the phytoestrogens coumestrol and genistein which were positive in V79 cells, whereas the phytoestrogen daidzein in the same test was negative. More confusing is the situation concerning cytogenetic endpoints: there is agreement that estrogens act like colcemide by

Table 3. Selection of reported gene-mutation tests with different estrogens

Species/ cells		Out- come	Substance	Reference
S.	In vitro	–	E2	SEIBERT and GÜNZEL 1994
S.	In vitro	–	Cyclodiol, Cyclotriol	LANG and REIMANN 1993
S.	In vitro	–	EE, Cyclodiol, Cyclotriol	HUNDAL et al. 1997
S.	In vitro	–	E2	DHILLON and DHILLON 1995
S.	In vitro	–	Mestranol	DHILLON et al. 1994
S.	In vitro	–	EE	ITOH et al. 1991
S., E. coli	In vitro	–	DES metabolites indenestrol A and B	ISHIKAWA et al. 1996
SHE	In vitro	–	E2	TSUTSUI et al. 1987
V79	In vitro	–	Cyclodiol, Cyclotriol	LANG and REIMANN 1993
V79	In vitro	+	Coumestrol, Genistein	KULLING and METZLER 1997
V79	In vitro	–	E2, EE, E1, DES	DREVON et al. 1981
V79	In vitro	–	Daidzein	KULLING and METZLER 1997

S., Salmonella; SHE, Syrian hamster embryo cells; V79, V79 cell line of Chinese hamster lung; E1, E2, estrone, estradiol; DES, diethylstilbestrol; EE, ethinylestradiol

Table 4. Selection of reported micronucleus tests with different estrogens

Species/cells		Out-come	Substance	Reference
V79	In vitro	+[1]	17β-E2	ECKERT and STOPPER 1996
V79	In vitro	+	Coumestrol, Genistein	KULLING and METZLER 1997
V79	In vitro	–	Daidzein	KULLING and METZLER 1997
SHE	In vitro	+[1]	DES, E2	SCHNITZLER et al. 1994
PC/NCE	In vivo	–	EE	SHYAMA and RAHIMAN 1996
PCE	In vivo	–	EE	ZIMMERMANN et al. 1996
PCE	In vivo	+	EE, Cyclodiol, Cyclotriol	HUNDAL et al. 1997
PCE	In vivo	+	E2	DHILLON and DHILLON 1995
PCE	In vivo	+	Mestranol	DHILLON et al. 1994
PCE	In vivo	–	EE	ITOH et al. 1991
PCE	In vivo	–	EE	REIMANN et al. 1996

SHE, Syrian hamster embryo cells; V79, V79 cell line of Chinese hamster lung; PCE, polychromatic erythrocytes of mice; PC/NCE, polychromatic and normochromatic erythrocytes of mice; E2, estradiol; DES, diethylstilbestrol; EE, ethinylestradiol
+[1] By mitotic disturbance

arresting mitosis in metaphase (RAO and EENGELBERG 1967; SAWADA and ISHIDATE 1978; TSUTSUI et al. 1990; ZIMMERMANN et al. 1994; METZLER et al. 1996; REIMANN et al. 1996). This agrees with the finding of numerical chromosomal aberrations by many authors (Table 6).

AIZU-YOKOTA et al. (1995) demonstrated that natural estrogens and their derivatives caused microtubule disruption in V-79 cells in a nongenomic manner at extremely high doses, 2-methoxyestradiol showing the strongest activity. According to REIMANN et al. (1996), the induction of polyploidies and mitotic arrest is a common effect of some endogenous and synthetic steroid hormones under in vitro conditions. However, these findings have to be regarded as irrelevant to the estimation of the mutagenic potential of sex steroids, especially as the observation was made for both endogenous and synthetic sex-steroid hormones. The majority of authors observed no structural chromosomal aberrations in several in vitro and in vivo tests using different estrogens. The reason for the positive findings by some authors remains unclear. In fact, most authors did not reveal positive results in micronucleus tests using different estrogens, or they found the reason for positive findings in numerical aberrations. Findings in sister chromatid-exchange assays are also rather conflicting.

According to HEYWOOD (1989), hormonal agents do not react with DNA and must be classed as nonmutagenic carcinogens. PURCHASE (1994) also grouped estrogens in the group of nongenotoxic carcinogens because they do not damage DNA as their primary biological activity.

HEYWOOD (1989) stated that a positive finding in a gene mutation or chromosome-damage test would most likely represent a false-positive result because of the fact that hormonal agents do not react with DNA. This was

Table 5. Selection of reported special genotoxicity tests with different estrogens

Test	Species/cells		Out-come	Substance	Reference
SCE	hl	In vitro	+	EE, Cyclodiol, Cyclotriol	Hundal et al. 1997
SCE	hl	In vitro	–	E2, E3	Hill and Wolff 1983
SCE	hl	In vitro	+	E2	Dhillon and Dhillon 1995
SCE	hl	In vitro	–	E2	Banduhn and Obe 1985
SCE	hl	In vitro	+	DES	Banduhn and Obe 1985
SCE	ucf	In vitro	–	E2	Hilbertz-Nilsson and Forsberg 1985
SCE	ucf	In vitro	+	DES	Hilbertz-Nilsson and Forsberg 1985
SCE	hl	In vivo	+	OC	Murthy and Prema 1979
SCE	hl	In vivo	–	OC	Husum et al. 1982
SCE	hl	In vitro	+	Mestranol	Dhillon et al. 1994
UDS	SHE	In vitro	–	E2	Tsutsui et al. 1987
DNA adducts	LKS	In vivo	+	DES	Bhat et al. 1994
DNA adducts	h.li.DNA	In vitro	–	E1, E2, E3	Seraj et al. 1996
DNA SSB	V79	In vitro	+	Coumestrol	Kulling and Metzler 1997
DNA SSB	V79	In vitro	–	Daidzein	Kulling and Metzler 1997

SCE, Sister chromatid exchange; UDS, unscheduled DNA synthesis; DNASSB, DNA single strand break; hl, human lymphocytes; SHE, Syrian hamster embryo cells; V79, V79 cell line of Chinese hamster lung; LKS, liver and kidney of Syrian golden hamsters; h.li., DNA human liver DNA; ucf, uterine cervical fibroblasts of mice; E1, E2, E3, estrone, estradiol, estriol; DES, diethylstilbestrol; OC, combined oral contraceptives containing estrogen and progestogen

Table 6. Published data on cytogenetic analysis of different estrogens

Cells		Str. CA	Num. CA	Substance	Reference
CHL	In vitro	–	+	EE	ITOH et al. 1991
CHO	In vitro	+		E2, E1, E3, EE	KOCHHAR 1985
hl	In vitro	–	?	EE	ZIMMERMANN et al. 1994
hl	In vitro	–	+	EE	REIMANN et al. 1996
hl	In vitro	+		EE, Cyclodiol, Cyclotriol	HUNDAL et al. 1997
hl	In vitro	–		EE, E2, Mestranol	STENCHEVER et al. 1969
hl	In vitro	+		E2	DHILLON and DHILLON 1995
hl	In vitro	–	+	E2, DES	BANDUHN and OBE 1985
hl	In vitro	+		Mestranol	DHILLON et al. 1994
SHE	In vitro	–	+	E2, DES	TSUTSUI et al. 1997
SHE	In vitro	–	+	E2	TSUTSUI et al. 1987
bmm	In vivo	+		Mestranol	DHILLON et al. 1994
bmm	In vivo	–		EE	SHYAMA and RAHIMAN 1996
hl	In vivo	–	–	OC	MATTON-VAN LEUVEN et al. 1974
hl	In vivo	–/+[a]		OC	LITTLEFIELD et al. 1975
SHRC	In vivo	+		E2, DES	BANERJEE et al. 1994
SHRC	In vivo	–		17α E2	BANERJEE et al. 1994
SHRC	In vivo	+[b]		EE	BANERJEE et al. 1994

CHL, Chinese hamster lung; CHO, Chinese hamster ovary; hl, human lymphocytes; SHE, Syrian hamster embryo; bmm, bone marrow of mice; SHRC, Syrian hamster renal cortical cells; E1, E2, E3, estrone, estradiol, estriol; DES, diethylstilbestrol; EE, ethinylestradiol; str. CA, structural chromosome aberrations; num. CA, numerical chromosome aberrations

[a] Neg. vs pregnant nonusers, pos. vs. nulligravid nonusers

[b] Only gaps, no breaks

confirmed by SERAJ et al. (1996) who found no DNA adducts after estrone, estradiol and estriol treatment, whereas BHAT et al. (1994) demonstrated DNA adducts with diethylstilbestrol. TSUTSUI et al (1987) failed to demonstrate increased unscheduled DNA repair as response to a DNA damage after estradiol treatment, thus supporting the view that estrogens, at least estradiol, do not react with DNA.

The estradiol metabolite 16α-hydroxy (OH)-estrone binds covalently to DNA in vitro, although binding could not be demonstrated in vivo (YAGER and LIEHR 1996). Increased unscheduled DNA synthesis was observed by SUTO et al. (1993) and TELANG et al. (1992) in mouse mammary epithelial cells, whereas estradiol itself or estriol did not increase unscheduled DNA synthesis (UDS) (TELANG et al. 1992). Their conclusion was that endogenously formed 16α-OH-estrone may function as an initiator and/or promoter of tumorigenesis.

Catecholestrogens and diethylstilbestrol both can undergo redox cycling between hydroquinone (catechol) and quinone forms via semiquinone radical intermediates, thus causing oxidative stress and damage leading to DNA damage. DNA binding was demonstrated in vitro for catecholestrogens and diethylstilbestrol, whereas, in vivo, no adducts of catecholestrogens could be demonstrated. However, diethylstilbestrol adducts were also found in vivo. Whereas the 4-hydroxylated estrogens were carcinogenic in the hamster kidney tumor model, the 2-hydroxylated estrogens were not (YAGER and LIEHR 1996). A possible role of the 4-hydroxylated estrogens in carcinogenic processes is supported by the finding that this metabolite is elevated in hamster kidney, mouse uterus and rat pituitary; these are organs in which estrogens induce tumors (YAGER and LIEHR 1996). However, BRADLOW et al. (1994) favor a causal role of 16α-hydroxylation in mammary cancer.

IV. Carcinogenicity

The most common neoplasms in rodents associated with long-term estrogen treatment are tumors of the pituitary, the mammary gland, the uterus and the liver (HIGHMAN et al. 1980, MCCONNELL 1989a, LEHMANN et al. 1989). Even the lowest tested dose of 1.2 mg/kg estradiol valerate increased the incidence of pituitary adenomas in female rats after 90 weeks (KUHNZ and PUTZ 1989). Ethinyl estradiol increased the incidence of pituitary adenomas at doses of about 60–80 μg/kg in a 2-year study in rats (SCHARDEIN 1980). In this study, doses of about 6–8 μg/kg and 60–80 μg/kg ethinyl estradiol were tested. Food intake and body-weight gain were suppressed in both treatment groups. There was no increase in overall tumor incidence rates in either sex. The incidence of pituitary chromophobe adenomas was increased by a factor of two in the high-dose group, and liver-neoplastic nodules were increased slightly. The incidence of mammary tumors, especially fibroepithelial tumors, was comparable with untreated animals.

Other tumors that can be induced by various estrogens in certain strains of rats and mice include testicular, vaginal, cervical, bone, lymphoid, renal, bladder and subcutaneous (DUNNING et al. 1953; SCHARDEIN 1980; IARC 1979; MCCONNELL 1989a).

A relationship between the occurrence of mammary and pituitary tumors in rodents has long been observed. In addition, increased mammary gland secretory changes, which are known to be associated with increased prolactin activity, were observed in mice and rats with pituitary tumors (MCCONNELL 1989b). Pituitary tumors were shown to develop directly in response to the stimulatory effects of estrogens on pituitary lactotrophs (SOTO and SONNENSCHEIN 1979) and indirectly in response to damaged dopaminergic neurons in the medial-basal hypothalamus which secrete prolactin-inhibiting factor (BRAWER and SONNENSCHEIN 1975). The uninhibited activity of these cells further stimulated by estrogens is thought to be the basis of pituitary tumor development. (MCCONNELL 1989b) The pituitary tumors found in rodents after estrogen administration are adenomas, mainly of the prolactin-secreting cells (ELETREBY et al. 1979). Only SATOH et al. (1997) observed a progression of the adenomas to invasive carcinomas after 7–9 weeks of estradiol dipropionate treatment in Fisher rats. Whether this finding is really due to sampling differences as stated by the authors or caused by different strain susceptibility remains to be clarified.

WELSCH et al. (1977) were able to reduce the estradiol-induced increased incidence of mammary tumors in mice to control levels, when administering the prolactin-inhibiting drug 2-bromo-α-ergocryptine together with estradiol. The antiprolactinemic and antiproliferative pituitary effects of bromocriptine mesylate on estradiol-induced hyperprolactinemia was also demonstrated in rats by RIEBEIRO et al. (1997); the mechanism of the development of pituitary and mammary tumors in rodents appears to have been at least partly explained by this. The above-mentioned estrogen and prolactin-sensitive pathways leading to pituitary and mammary tumors in rodents presently appear to have no counterpart in the human (MCCONNELL 1989b). Although estrogen administration can increase prolactin levels in humans, there has been no evidence that long-term exposure to estrogens is responsible for the development of prolactin-secreting tumors in man (ELETREBY and GUNZEL 1973; ELETREBY et al. 1979). Furthermore, estrogens have not caused mammary cancer in the Rhesus monkey (DRILL 1975), a finding that correlates with the absence of a demonstrable effect in women (COOKSON 1994).

The neoplastic liver changes observed in rodents appear to be a consequence of promotion of spontaneous preneoplastic foci (SCHULTE-HERMANN et al. 1983). A liver tumor-promoting activity of the synthetic estrogens ethinyl estradiol and mestranol was demonstrated by CAMERON et al. (1982) and YAGER and YAGER (1980) in diethylnitrosamine-initiated rats. BOWER et al. (1996) demonstrated promoting activity of the natural estrogen estradiol in ovariectomized rats after initiation with diethylnitrosamine. When ethinyl estradiol was tested for its tumor-initiating potency in the rat liver foci bioassay, according to the protocol of OESTERLE and DEML (1983), no increase in foci number and area was observed (OETTEL et al. 1995). This supports the view that carcinogenicity of estrogens is mainly due to their promoting rather than initiating activity. HALLSTROM et al. (1996) compared the promotion activity of ethinyl estradiol in hypophysectomized, intact and bromocriptine-treated rats after initiation with diethylnitrosamine. They found that the promotive effect of ethinyl estradiol was decreased in hypophysectomized and intact bromocriptine-treated rats. Their conclusion is that estrogens exert at least part of their promotion effects indirectly, by increasing the levels of pituitary prolactin.

A special case is the transplacental carcinogenicity of the nonsteroidal estrogen diethylstilbestrol in laboratory animals and in the human population. It was shown that diethylstilbestrol induced clear-cell adenocarcinoma of the vagina in female offspring of women who received this drug during pregnancy. WALKER and HAVEN (1997) demonstrated a multigenerational carcinogenic effect on F2-generation in mice. Furthermore, male mice exposed prenatally to diethylstilbestrol were shown to transmit a carcinogenic effect to their offspring through their germ cells (WALKER and KURTH 1995).

The mechanisms by which estrogens act on cancer growth are still not fully understood. Many carcinogens were found to be positive in mutagenicity tests; however, no estrogen has thus far been clearly shown to be mutagenic (COOKSON 1994; REIMANN et al. 1996). Estrogens, like other hormones, can clearly affect carcinogenesis by epigenetic mechanisms, such as stimulation of cell proliferation of estrogen-dependent target cells. Cell proliferation can facilitate carcinogenesis by offering spontaneous and carcinogen-induced mutations, a greater chance of fixation or by allowing clonal expansion of preneoplastic cells, i.e., tumor promotion (BARRETT and TSUTSUI 1996). According to REIMANN et al. (1996), estrogens, like other sex steroids, have to be classified as nongenotoxic when tested in validated and generally accepted mutagenicity tests.

Some publications report that there is also some evidence that estrogenic activity is not sufficient to explain the carcinogenic activity in vivo of estrogens in certain target tissues, e.g., renal tumors in hamsters. For instance, ethinyl estradiol shows only a weak carcinogenicity in the hamster kidney, compared with diethylstilbestrol or estradiol, although all three estrogens have a comparable receptor affinity, similar ability to induce renal progesterone receptor and lead to similar, high serum prolactin levels (LI et al. 1983). Based on these findings, the latter authors concluded that estrogenic activity alone may not be sufficient to effect renal tumorigenesis in the hamster. But, as evident from more recent studies, the concomitant administration of androgens, progesterone or antiestrogens completely prevented tumor development (LI and LI 1996). As LI et al. (1995) have shown, the degree of carcinogenicity of a given estrogen correlates with its ability to induce proximal-tubule cell proliferation in the hamster kidney in vitro, whereas the amount of catechol estrogens generated, as it was believed earlier, is not suited as an indicator of carcinogenesis in this model. The weak estrogens 17α-estradiol and 11β-methylestradiol had relatively little effect on tubule-cell proliferation and were not carcinogenic (LI et al. 1995). Furthermore, the poor tumorigenic activity of ethinyl estradiol in this model could be related to unusual pharmacokinetic properties. Therefore, LI and LI (1996) suggested neoplastic transformation in the hamster kidney to be probably nongenotoxic and primarily hormonally driven and hormonally dependent.

According to YAGER and LIEHR (1996), estrogen carcinogenesis is complex and involves both nongenotoxic (cell proliferative) as well as direct and indirect genotoxic effects of diethylstilbestrol itself or the metabolites of estradiol. They hypothesized that any agent that increases the metabolism of endogenous estrogens to their catechol or 16α-OH metabolites will increase the risk of estrogen carcinogenesis in the affected tissues. CAVALIERI et al. (1997) demonstrated that oxidation of 4-OH metabolites of estrone and estradiol to catechol estrogen-3,4-quinones results in electrophilic intermediates that covalently bind to DNA to form depurinating adducts. Their conclusion is that the resultant apurinic sites in critical genes can generate mutations that may initiate various human cancers. In contrast, the 2-OH metabolites did not form depurinating DNA adducts. The formation of adducts was also demonstrated by SHEN et al. (1997) for the Premarin metabolite 4-OH-equilenin. Nevertheless, the implications of all these adducts for in vivo situations, especially for human-risk assessment, are not known, whereas there is general consensus about the promoting effects of estrogens. According to LI and LI (1996), estrogenicity is very likely to be the primary driving force in the development of hormone-induced tumors in animals and humans.

V. Reproductive Toxicity

Estrogens, of course, affect various stages of reproductive processes because of their pharmacodynamic properties. During the reproductive age, estrogens are mainly used as a component of oral contraceptives. Therefore, the possible risks, in principle, relate to disturbed return of fertility after stopping long-term contraceptive use and, only if contraceptives are taken inadvertently during pregnancy, to adverse effects on the developing embryo, especially in the very early gestational phase, but also in the teratogenically sensitive period of organogenesis.

1. Effects on Fertility and Preimplantational Development

An inhibition of ovulation in rats does not occur in pharmacological dosages. Even after extremely high daily doses of 50 μg/kg ethinyl estradiol, all rats had normal ovulation (MISCHLER and GAWLAK 1970). Estrogens can inhibit fertility, although this effect is reversible. A delay of several days in the return of fertile mating was observed by PETERSON and EDGREN (1965). Often, a decrease in the number of implantations and fetuses was observed (VELARDO et al. 1956), indicating lasting endocrine pharmacological effects, resulting in a slightly higher preimplantation loss of ova due to a disturbance of ovum transport. However, the implanted embryos used to develop normally without increased malformations (LEHMANN et al. 1989). When female rats were treated from 5 days before mating until mating had occurred, fertility was totally suppressed with daily intragastric doses of ethinyl estradiol of 50 μg/rat and s.c. doses of estrone amounting to 10 μg/rat, whereas 50% of animals were fertile after a dose of 3 μg estrone s.c. After a recovery period, most of the rats became pregnant (EDGREN et al. 1966).

Estrogen administration interferes with tubular ovum transport. Depending on the species and the administered dose, either an acceleration of egg transport or a retention of ova for longer than the normal period is observed (GREENWALD 1967). The synchronism of the endometrium and the embryo is obviously disturbed by excessive estrogen dosage in the early phase of gestation. In rabbits, after postcoital estrogen treatment, the transformation of the endometrium is retarded by several days (BEIER et al. 1983). Estrogens are also known to prevent implantation in rats when given in early pregnancy (SAUNDERS and ELTON 1967). When estrone was given subcutaneously to rats from day 1 to day 7 of pregnancy, none of the animals was pregnant on day 15 with a dose of 40 μg/kg, whereas nearly 50% were pregnant on day 15 with a dose of 20 μg/kg (SAUNDERS and ELTON 1967). When mestranol was given daily from day 2 to day 4 at a dose of 20 μg/kg, only one of nine animals had fetuses on day 15, whereas all animals had fetuses with a dose of 10 μg/kg, although the number of fetuses per rat was reduced (SAUNDERS and ELTON 1967). HARADA et al. (1991) treated rats with a combination of norethisterone and ethinyl estradiol from 14 days prior to mating until day 7 of pregnancy and examined the development, behavioral performance, reproductive performance and fertility of offspring. At the highest dose of 17.5 μg/kg ethinyl estradiol and 500 μg/kg norethisterone, litter size was reduced, but there were no adverse effects on the offspring. In rabbits, daily subcutaneous treatment from day 1 to day 3 of pregnancy with 2 μg/kg estrone, 20 μg/kg estradiol, 2 μg/kg ethinyl estradiol and 10 μg/kg mestranol fully inhibited implantation (JACOB and MORRIS 1969).

The effects of ethinyl estradiol on male fertility were tested by IWASE et al. (1995). After 4 weeks' daily dosing with 3 mg/kg and 10 mg/kg, all males had lost their reproductive ability. Recovery of fertility was stated after 4 weeks. In animals dosed with 0.1 mg/kg and 0.3 mg/kg, low copulation indices but unchanged-fertility indices were observed. Sperm counts were decreased although sperm motility was not affected. An intramuscular dose of 50 μg estradiol benzoate per rat over 15 days caused a strong decrease in sperm count and resulted in total infertility. Reversibility of this effect was not tested (CHINOY et al. 1984).

2. Embryotoxicity

Generally, embryotoxicity studies in rats and rabbits are performed during the period of organogenesis to detect a possible teratogenic risk of the substances administered for humans. For estrogens, an extrapolation to the human situation is not easy because of the different physiological endogenous hormone concentrations and ratios as well as different feedback mechanisms and endocrine-control mechanisms, as outlined earlier in this chapter in Sect. A. In addition, some basic data on estradiol concentrations and estradiol/progesterone ratios during pregnancy are summarized in Table 7. It is obvious that estradiol levels show species differences with lower values and smaller periodic rises in all other animal species than women. Therefore, it can be concluded from normal endocrine physiology that the human organism is able to handle much higher levels of estradiol properly than the common laboratory animal species (SEIBERT and GUNZEL 1994). Furthermore, it is obvious from the different estradiol/progesterone ratios that animal toxicity tests, using combined contraceptives as they were recommended earlier, are of no predictive value for the human situation.

Estrogens, as well as other sex steroids, cause embryolethality in rats and rabbits. The precise mechanism of this is not fully understood, but it may be

Table 7. Comparison of estradiol plasma concentrations (in pg/ml) and estradiol/progesterone ratios (in brackets) between animal species and humans during pregnancy (data from SEIBERT and GÜNZEL 1994)

Pregnancy phase	Mouse	Rat	Dog	Monkey	Human
Early	16	600	20	170	1000–7000
	(1:1900)	(1:122)	(1:1000)	(1:24)	(1:6)
Middle	20	500	24	550	6000–12,000
	(1:2000)	(1:218)	(1:3600)	(1:7)	(1:3)
Late	30	n.d.	8	550	12,000–29,000
	(1:2700)		(1:1300)	(1:7)	(1:7)

n.d., no data

related to the severe disturbance of the hormonal balance in these species that regulate their pregnancy at far lower estrogen levels than humans and at estrogen/progesterone ratios that are different from humans (GUNZEL et al. 1989). The estrogen sensitivity varies at different stages of gestation in most species. In rabbits, a pronounced estrogen sensitivity exists throughout gestation. In rats and mice, the estrogen sensitivity decreases continuously throughout gestation. For instance, a single s.c. dose of 10 μg estradiol given on day 2 or day 3 of pregnancy was abortive, whereas on the 9th day, the abortive dose was 200 μg. A dose of 500 μg estradiol was not abortive on day 13 (BEIER et al. 1983). A comparable pattern of abortive effects over time was reported for estrone (EDGREN and SHIPLEY 1961).

In classic embryotoxicity studies with administration from day 6 to day 15 (rats) or day 18 (rabbit) postcoitum, increased embryolethality was observed at s.c. estradiol doses of only 1 μg/kg/day; ethinyl estradiol increased embryolethality at oral doses of 100 μg/kg in rats and 10 μg/kg in rabbits. Despite the embryolethal effects of estrogens in rats, rabbits and mice, no or only minor and nonspecific defects of nongenital organs have been reported (OVERBECK et al. 1962; SAUNDERS and ELTON 1967; MORRIS 1970; HEINECKE and KLAUS 1975; JOOSTEN et al. 1978; POGGEL and GUNZEL 1978; HENDRICKX and BINKERD 1989; SEIBERT and GUNZEL 1994). However, especially during late pregnancy, estrogens dose dependently affect the genital system of progeny. Shortened sinus vagina and enlarged Müllerian vagina were found when estrogens were administered from day 15 to day 21 of pregnancy (ELGER 1977). A mild feminization of male mouse fetuses was reported in mice with extremely high doses of 1 mg per animal s.c. or 10 mg per animal p.o. (SEIBERT and GUNZEL 1994). No anomalies were observed in monkeys treated with several estrogens during different pregnancy periods (MORRIS and WAGENUM 1973). In nonhuman primates, ethinyl estradiol in combination with norethisterone acetate, estradiol benzoate in combination with progesterone and estradiol valerate in combination with hydroxyprogesterone caproate caused embryolethality at high doses, but no nongenital teratogenicity (HENDRICKX et al. 1987a,b; SEIBERT and GUNZEL 1994).

3. Perinatal and Postnatal Effects

Ethinyl estradiol at doses of 25 μg/kg given to lactating rats had no effects on the young and the authors suggested that no biologically important amounts of the drug were secreted in the milk (Clancy and Edgren 1968). Estrogens administered in the late-gestational phase and during lactation were able to reduce the fertility of female F1-rats (Saunders 1967). A single s.c. dose of 100 μg estradiol to 3- or 5-day-old female rats caused persistent vaginal estrus in adulthood, whereas most rats that received 10 μg estradiol showed regular 4-day estrus cycles (Kumagai and Shimizu 1982). Neonatal treatment of mice with estradiol resulted in hyperplastic proliferation of the cervicovaginal epithelium, with extensive cornification followed by subsequent neoplastic development (Hajek et al. 1997). These authors have demonstrated the same qualitative effects for the weak estrogen 17α-estradiol and concluded that the estrogenic potency of 17α-estradiol could be age dependent. Another reason for these effects of 17α-estradiol may be the metabolism of s.c.-administered 17α-estradiol to considerable amounts of 17β-estradiol (own unpublished data).

Lehmann et al. (1989) compared the results of different types of reproductive studies using estrogens and/or progestogens with human epidemiological data and arrived at the conclusion that the animal data allowed only a qualitative risk assessment with respect to disturbed return of fertility, absence of teratogenic effects at nonembryolethal doses and adverse effects on the morphological development of genital organs. Furthermore, they suggested that direct quantitative extrapolations to humans seemed to be almost impossible because of the interspecies differences.

VI. Environmental Estrogens

Another field of interest in the toxicology of estrogens that becomes more important with time is that of environmental estrogens which will be discussed here only very briefly. Many naturally occurring and man-made chemicals present in the environment possess estrogenic activity, e.g., plant and fungal products, pesticides and plasticisers (Stancel et al. 1995). The situation with these environmental estrogens is different from that in medicine because everybody can be exposed to these environmental estrogens, and this exposure is difficult to measure compared with that of estrogens used in contraception and medicine. Even if the concentrations of individual compounds that are released into the environment are relatively low, toxicological problems can arise because of the sum of different compounds with estrogenic activity that we are exposed to, especially if some of these substances can accumulate in the environment. Arnold et al. (1996) found extreme synergistic interaction between weak environmental estrogens. Many screening assays for rapid characterization of estrogenic chemicals or estrogenic exposure are in development and/or validation (Soto et al. 1995, 1997; Arnold et al. 1996; O'Connor et al. 1996; Shelby et al. 1996; Klotz et al. 1996).

One of the questions that has arisen over the last few years has been whether there is a causality between exposure to estrogens and increased breast and testicular cancers and decreased male reproductive health. JENSEN et al. (1995) speculate that decreased male fertility may be due to estrogenic exposure during fetal development. According to THIERFELDER et al. (1995) and DASTON et al. (1997), there is little evidence to indicate that exposure to ambient levels of estrogenic xenobiotics affects reproductive health. Therefore, it is concluded that there is a need for research in the field of environmental estrogens. ASHBY et al. (1997) suggest other factors, such as diet and lifestyle, that could have contributed to the observed increase in breast and testicular cancers and decrease in human sperm count and sperm quality. Review articles in the field of environmental estrogens include those by TOPPARI et al. (1996), DASTON et al. (1997), MULLER et al. (1995), KAMRIN et al. (1994) and WILSON and LEIGH (1992).

C. Antiestrogens

When dealing with the toxicology of antiestrogens, one has to consider the same physiological background, including species differences as outlined earlier in this chapter in Sect. A. In addition, the pharmacological profile of the different antiestrogens varies from pure antagonists to mixed antagonists/agonists. Therefore, the impact of antiestrogens on the endogenous-hormonal balance will be different, depending on the special pharmacological profile of the substance used. Unlike the situation with estrogens, there exist only a few reports on the toxicology of antiestrogens, most of them dealing with tamoxifen. Publications on the newer antiestrogens, such as raloxifen are not available.

I. Acute Toxicity

The acute toxicity of the antiestrogens is as low as that of estrogens; some data are summarized in Table 8. Toxic signs of toremifene in rats include irritation, piloerection, spinal curvature and urinary incontinence. Postmortem findings include atony of the gastrointestinal tract, swelling of the adrenal gland and atrophy of the thymus and spleen. In surviving animals, atrophy of the testes, prostate and seminal vesicles was seen, probably due to the estrogenic action of toremifene (SAKAMOTO et al. 1992). The only findings in the monkey were temporary vomiting and decreased food intake 1 week after administration of toremifene (SAKAMOTO et al. 1992).

II. Subchronic and Chronic Toxicity

LOSER et al. (1984) investigated the 4-week toxicity of tamoxifen and droloxifene in rats. Doses administered were 1 mg/kg, 10 mg/kg and 100 mg/kg for

Table 8. LD_{50} values of different antiestrogens in various species

Antiestrogen	Species (sex)	LD_{50} oral (mg/kg)	LD_{50} i.p. (mg/kg)	Reference
TAM	Mice	2150	745	Löser et al. 1984
	Rat	4100	700	Löser et al. 1984
DROL	Mice	>3000	>1500	Löser et al. 1984
	Rat	>4500	>2250	Löser et al. 1984
TOR	Rat (f)	ca. 3000	520	Sakamoto et al. 1992
	Rat (m)	ca. 3000	620	Sakamoto et al. 1992
	Monkey (f)	>2000		Sakamoto et al. 1992

TAM, tamoxifen; DROL, droloxifene; TOR, toremifene; f, female; m, male

both drugs. The highest tamoxifen dose of 100 mg/kg caused a 100% mortality in female rats and an 88% mortality in male rats, whereas with the same dose of droloxifene no deaths were observed. Body-weight gain was reduced in all treatment groups. These effects and the differences observed between the two drugs may be due to their estrogenic activity which is more pronounced with tamoxifen.

Hirsimaki et al. (1993) observed reduced body-weight gain, alopecia and aggressiveness in a 1-year study of female rats with tamoxifen and toremifene. In a 2-year study with toremifene in rats, Karlsson et al. (1996) observed drug-related changes in the genital organs, thyroid, spleen, mammary gland, adrenal, kidney, stomach and lung. All of these changes were due to hormonal disturbances or a result of reduced food consumption.

In a 52-week study of female monkeys, Wood et al. (1992) found endometrial hyperplasia, increased ovary weights, increased liver and adrenal weights, increased glutamate pyruvate transaminase and leucine aminopeptidase activities, decreased hemoglobin and packed cell volume, and transient decrease in body weight. Most of these changes can be attributed to estrogenic and antiestrogenic activities of toremifene.

III. Genotoxicity

Some years ago, tamoxifen was found to form DNA adducts in rat, mouse and hamster liver (Han and Liehr 1992; White et al. 1992; Randerath et al. 1994a,b; Pathak et al. 1995). In a comparative study using tamoxifen and toremifene, only tamoxifen produced DNA adducts in rat liver (Montandon and Williams 1994). After a 7-day administration period, only tamoxifen produced DNA adducts in liver and no adducts were found after droloxifene and toremifene; in other organs no adducts were detectable (White et al. 1992). Cai and Wei (1995) showed that tamoxifen, after topical application, induced DNA adducts in mouse skin in a dose- and time-dependent manner. Tamoxifen induced unscheduled DNA synthesis in rat hepatocytes and micronuclei in the human cell line MCL-5 (Vijayalaxmi and Rai 1996; White et al. 1992).

Tamoxifen was mutagenic at the hypoxanthine phosphoribosyl transferase (HPRT) locus in V79 cells (RAJAH and PENTO 1995).

TSUTSUI et al. (1997) found that tamoxifen, toremifene and ICI 164,384, like estradiol, induced numerical but not structural chromosomal aberrations in Syrian hamster embryo (SHE) cells. STYLES et al. (1994) compared tamoxifen and toremifene in their ability to induce micronuclei in human lymphoblastoid cell lines and found both antiestrogens to be clastogenic, although tamoxifen was more genotoxic than toremifene. They demonstrated that tamoxifen required metabolic activation by specific monooxygenases in order to exert its genotoxic effect. SARGENT et al. (1994) demonstrated that tamoxifen induced numerical and structural chromosomal aberrations in hepatocytes after single-dose administration to rats. Hepatic aneuploidy was induced after single-oral administration of tamoxifen, toremifene and idoxifene in rats (SARGENT et al. 1996). In an in vivo test using mouse bone-marrow cells, tamoxifen induced micronuclei and chromosomal aberrations (VIJAYALAXMI and RAI 1996). STYLES et al. (1997) tested tamoxifen, toremifene, droloxifene, clomiphene and the 4-OH metabolites of tamoxifen and toremifene for their clastogenic and aneugenic effects and found that tamoxifen and toremifene were the only two tested drugs to cause structural and numerical chromosomal aberrations in vitro. In vivo, only tamoxifen was both aneugenic and clastogenic. A review of the genotoxicity of tamoxifen can be found in WHITE et al. (1997). According to PHILLIPS et al. (1994), recent findings strongly suggest that tamoxifen is carcinogenic by a genotoxic mechanism.

IV. Carcinogenicity

Tamoxifen is a liver carcinogen in female rats. HIRSIMAKI et al. (1993), comparing the effects of tamoxifen and toremifene on rat liver in a 1-year study, found hepatocellular carcinomas in the largest-dose tamoxifen group (45 mg/kg), whereas in the lower-dose tamoxifen group (11.3 mg/kg) and in the toremifene groups (45 mg/kg and 12 mg/kg), no liver tumors were observed. Toremifene was tested in a 2-year carcinogenicity study in rats by KARLSSON et al. (1996). Toremifene reduced mortality, mainly due to reduced incidences of pituitary tumors, which is not unexpected for an antiestrogen. Decreases in mammary and testicular tumors and preneoplastic foci of basophilic hepatocytes were also observed. No preneoplastic or neoplastic lesions were induced. Other studies by HARD et al. (1993) and GREAVES et al. (1993) also demonstrated the hepatocarcinogenic effect of tamoxifen in rats, whereas in a mouse study by TUCKER et al. (1984) the only tumor increase was seen in the testes. The difference seen in hepatic carcinogenicity between tamoxifen and toremifene cannot be caused by different hepatic estrogenicity, because both antiestrogens cause comparable estrogenic effects in rat liver as demonstrated by KENDALL and ROSE (1992).

Tamoxifen and toremifene promoted diethylnitrosamine-initiated liver tumors in rats, but toremifene was much less active than tamoxifen (DRAGAN et al. 1995). In another study on promoting activities, Dragan et al. (1994) showed that liver concentrations of tamoxifen, 4-OH tamoxifen and *N*-desmethyl tamoxifen were much higher than serum concentrations.

Despite the findings in rats, there is no evidence for DNA-adduct formation to have occurred in humans (Jordan 1995). There are no reports of hepatocellular carcinomas in women receiving the 20-mg/day dose and only two reports with the higher dose of 40 mg/day (Wiseman and Goa 1997). However, tamoxifen showed carcinogenic potential in increasing the incidence of endometrial carcinoma during long-term therapy (Rutquist 1993).

V. Reproductive Toxicity

Anordrin, an antiestrogen with agonistic activity, caused spermatogenesis arrest by blocking gonadotropin production and/or release by the pituitary, thereby blocking testosterone production by Leydig cells (Vanage et al. 1997).

Antiestrogens were shown to have an antifertility effect on mice and rats when given in early pregnancy; this effect was exerted on the genital tract rather than the blastocyst. No increased incidence of fetal anomalies was reported in surviving fetuses (Lunan and Klopper 1975).

Raloxifene, which showed no estrogenic activity in the mouse uterus, when given to mice from the day of birth for 5 days, resulted in temporary suppression of spermatogenesis in males and in persistent suppression of ovulation in females (Chou et al. 1992). This result underlines the fact that the neonatal phase is very sensitive to hormonal changes and that disturbance of the hormonal balance during this phase has strong consequences on reproductive development.

References

Aizu-Yokota E, Susaki A, Sato Y (1995) Natural estrogens induce modulation of microtubules in Chinese hamster V79 cells in culture. Cancer Res 55:1863–1868

Arnold SF, Klotz DM, Collins BM, Vonier PM, Guillette LJ, McLachlan JA (1996) Synergistic activation of estrogen receptor with combinations of environmental chemicals. Science 272:1489–1492

Ashby J, Houthoff E, Kennedy SJ, Stevens J, Bars R, Jekat FW, Campbell P, Van Miller J, Carpanini FM, Randall GLP (1997) The challenge posed by endocrine-disrupting chemicals. Environ Health Perspect 105:164–169

Attia M, Goldfain F, LeBigot J-F, Zayed I (1996) Comparative 52-week study of 17β oestradiol and ethinyl oestradiol in Beagle dogs and Cynomolgus monkeys. Toxicol Lett 88(suppl 1):107

Banduhn N, Obe G (1985) Mutagenicity of methyl 2-benzimidazole carbamate, diethylstilbestrol and estradiol: structural chromosome aberrations, sister chromatid exchanges, c-mitoses, polyploidies and micronuclei. Mutat Res 156:199–218

Banerjee SK, Banerjee S, Li SA, Li JJ (1994) Induction of chromosome aberrations in Syrian hamster renal cortical cells by various estrogens. Mutat Res 311:191–197

Barrett JC, Tsutsui T (1996) Mechanisms of estrogen-associated carcinogenesis. Prog Clin Biol Res 394:105–111

Beier, S, Düsterberg B, Eletreby MF, Elger W, Neumann F, Nishino Y (1983) Toxicology of Hormonal Fertility-Regulating Agents In: Benagiano G, Diczfalusy E (eds)

Endocrine Mechanisms in fertility Regulation Raven Press, New York, pp 261–346
Bhat HK, Han X, Gladek A, Liehr JG (1994) Regulation of the major diethylstilbestrol-DNA adduct and some evidence of its structure. Carcinogenesis 15:2137–2142
Bojar H, Schütte J, Staib W, Broelsch C (1982) Does human liver contain estrogen receptors? Klin Wschr 60:417–425
Bolt HM (1976) Probleme der Toxikologie von Sexualhormonen. Fortschr Med 94:731–735
Bower N, Mylecraine L, Stein AP, Reuhl KR, Gallo MA (1996) Estradiol promotion of size and number of hepatic enzyme-altered foci in rats in the absence of cell proliferation. In: Li JJ, Li SA, Gustafsson J-A, Nandi S, Sekely LI (eds) Hormonal Carcinogenesis II. Springer, New York, pp 429–433
Bradlow HL, Fishman J, Telang NT et al. (1994) Re: Estrogen metabolism and excretion in oriental and caucasian women. J Natl Cancer Inst 86:1643–1644
Brawer JR, Sonnenschein C (1975) Cytopathological effects of estradiol on the arcuate nucleus of the female rat. A possible mechanism for pituitary tumorigenesis. Am J Anat 144:57–88
Cai Q, Wei H (1995) In vivo formation of DNA-adducts in mouse skin DNA by tamoxifen. Cancer Lett 92:187–192
Cameron RC, Imaida K, Tsuda H, Ito N (1982) Promotive effects of steroids and bile acids on hepatocarcinogenesis initiated by diethylnitrosamine. Cancer Res 42:2426–2428
Capel-Edwards K, Hall DE, Sansom AG (1971) Hematological changes observed in female Beagle dogs given ethinylestradiol. Toxicol Appl Pharmacol 20:319–326
Cavalieri EL, Stack DE, Devanesan PD, Todorovic R, Dwivedy I, Higginbotham S, Johansson SL, Patil KD, Gross ML, Gooden JK, Ramanthan R, Cerny RL, Rogan EG (1997) Molecular origin of cancer: Catechol estrogen-3,4-quinones as endogenous tumor initiators. Proc Natl Acad Sci USA 94:10937–10942
Chinoy MR, Sharma JD, Chinoy NJ (1984) Altered structural and functional integrity of the reproductive tissues in estradiol benzoate-treated intact male albino rats. Int J Fertil 29:98–103
Chou Y-C, Iguchi T, Bern HA (1992) Effects of antiestrogens on adult and neonatal mouse reproductive organs. Reprod Toxicol 6:439–446
Clancy DP, Edgren RA (1968) The effects of norgestrel, ethinyl estradiol, and their combination, Ovral, on lactation and the offspring of rats treated during lactation. Int J Fertil 13: 133–141
Cookson KM (1994) The parenteral toxicity of Cyclofem. Contraception 49:335–345
Daston GP, Gooch JW, Breslin WJ, Shuey DL, Nikiforov AI, Fico TA, Gorsuch JW (1997) Environmental estrogens and reproductive health: a discussion of the human and environmental data. Reprod Toxicol 11:465–481
Dhillon VS, Dhillon IK (1995) Genotoxicity evaluation of estradiol. Mutat Res 345:87–95
Dhillon VS, Singh JR, Singh H, Kler RS (1994) In vitro and in vivo genotoxicity evaluation of hormonal drugs v mestranol. Mutat Res 322:173–183
Dragan YP, Fahey S, Street K, Vaughan J, Jordan VC, Pitot HC (1994) Studies of tamoxifen as a promotor of hepatocarcinogenesis in female Fisher F344 rats. Breast Cancer Res Treat 31:111–25
Dragan YP, Vaughan J, Jordan VC, Pitot HC (1995) Comparison of the effects of tamoxifen and toremifene on liver and kidney tumor promotion in female rats. Carcinogenesis 16:2733–2741
Drevon C, Piccoli C, Montesano R (1981) Mutagenicity assays of estrogenic hormones in mammalian cells. Mutat Res 89:83–90
Drill VA (1975) Oral contraceptives: relation to mammary cancer, benign breast lesions, and cervical cancer. Ann Rev Pharmacol 15:367–385

Dunning WF, Curtis MR, Segaloff A (1953) Strain differences in responses to estrone and the induction of mammary gland, adrenal, and bladder cancer in rats. Cancer Res 13:147–152

Eckert I, Stopper H (1996) Genotoxic effects induced by β-oestradiol in vitro. Toxicology in Vitro 10:637–642

Edgren RA (1994) Issues in animal pharmacology. In: Goldzieher J (ed) Pharmacology of Contraceptive Steroids. Raven Press, New York, pp 81–97

Edgren RA, Peterson BA, Jones RC (1966) Some progestational and antifertility effects of norgestrel. Int J Fertil 11:389–400

Edgren RA, Shipley C (1961) A quantitative study of the termination of pregnancy in rats with estrone. Fertil Steril 12:178–181

Eletreby MF, Gräf KJ, Günzel P, Neumann F (1979) Evaluation of Effects of Sexual Steroids on the Hypothalamic-Pituitary System of Animals and Man. Arch Toxicol (suppl 2):11–39

Eletreby MF, Günzel P (1973) Prolaktinzell-Tumoren im Tierexperiment und beim Menschen. Arzneimittelforsch/Drug Res 23:1768–1790

Elger W (1977) Effects of exogenous estrogens and gestagens on the formation of the vagina in rat and rabbit fetuses. In: Neubert D, Merker HJ, Kwasigrodi TE (eds) Methods in Prenatal Toxicology. Georg Thieme, Stuttgart, pp 371–380

Engel J, Hartmann RW, Schönenberger H (1983) Metahexestrol. Drugs of the Future 8:413–419

Geil RG, Lamar JK (1977) FDA studies of estrogen, progestogens and estrogen/progestogen combinations in the dog and monkey. J Toxicol Environ Health 3:179–193

Gibson JP, Newberne JW, Kuhn WL, Elsea JR (1967) Comparative chronic toxicity of three oral estrogens in rats. Toxicol Appl Pharmacol 11:489–510

Goldzieher JW, Kraemer DC (1972) The metabolism and effects of contraceptive steroids in primates. Acta Endocr 166 (suppl) 71:389–421

Greaves P, Goonetilleke R, Nunn G, Topham J, Orton T (1993) Two-year carcinogenicity study of tamoxifen in Alderley park Wistar-derived rats. Cancer Res 53:3919–3924

Greenwald GS (1967) Species differences in egg transport in response to exogenous estrogen. Anat Rec 157:163–172

Günzel P (1971) Methode der tierexperimentellen Verträglichkeitsprüfung von hormonalen Kontrazeptiva. In: Plotz EJ and Haller J (eds) Methodik der Steroidtoxikologie. Georg Thieme, Stuttgart, pp 1–6

Günzel P, Putz B, Lehmann M, Hasan SH, Hümpel M, Eletreby MF (1989) Steroid Toxicology and the Pill: Comparative Aspects of Experimental Test Systems and the Human In: Dayan AD, Paine AJ (eds) Advances in Applied Toxicology. Taylor and Francis Ltd., London, New York, Philadelphia, pp 19–48

Hajek RA, Robertson AD, Johnston DA, Van NT, Tcholakian RK, Wagner LA, Conti CJ, Meistrich ML, Contreras N, Edwards CL, Jones LA (1997) During development, 17α-estradiol is a potent estrogen and carcinogen. Environ Health Perspect 105 (Suppl 3):577–581

Hallström IP, Liao DZ, Assefaw Redda Y, Ohlson LC, Sahlin L, Eneroth P, Eriksson LC, Gustafsson JA, Blanck A (1996) Role of the pituitary in tumor promotion with ethinyl estradiol in rat liver. Hepatology 24:849–854

Han X, Liehr JG (1992) Induction of covalent DNA adducts in rodents by tamoxifen. Cancer Res 52:1360–1363

Harada S, Takayama S, Shibano T, Chino M, Tanabe A, Amada Y, Yoneyama S (1991) Fertility study of oral contraceptives DT-5061 and DT-5062 (1/35) in rats. Yakuri to Chiryo (Pharmacology and Therapeutics) 19 (suppl):157(S-925)–196(S-964)

Hard GC, Iatropoulos MJ, Jordan K, Radi L, Kaltenberg OP, Imondi AR, Williams GM (1993) Major difference in the hepatocarcinogenicity and DNA adduct forming ability between toremifene and tamoxifen in female Crl:CD(BR) rats. Cancer Res 53:4534–4541

Hart JE (1988) Subacute toxicity of diethylstilboestrol and hexoestrol in the female rat, and the effects of clomiphene pretreatment. Food Chem Toxicol 26:227–232

Heinecke H, Klaus S (1975) Einfluss von Mestranol auf die Gravidität von Mäusen. Pharmazie 30:53–56

Hendrickx AG, Binkerd PE (1989) Teratogenicity studies of sex hormones. In: Michal F (ed) Safety requirements for contraceptive steroids. University Press, Cambridge, pp 289–322

Hendrickx AG, Korte R, Leuschner F, Neumann BW, Poggel A, Binkerd PE, Prahlada S, Guenzel P (1987a) Embryotoxicity of sex steroidal hormone combinations in nonhuman primates: II. Hydroxyprodesterone caproate, estradiol valerate. Teratology 35:129–136

Hendrickx AG, Korte R, Leuschner F, Neumann BW, Prahalada S, Poggel A, Binkerd PE, Guenzel P (1987b) Embryotoxicity of sex steroidal hormone combinations in nonhuman primates: I. Norethisterone acetate + ethinylestradiol and progesterone + estradiol benzoate (*Macaca mulatta, Macaca fascicularis, and Papio cynocephalus*). Teratology 35:119–127

Heywood R (1986) An assessment of the toxicological and carcinogenic hazards of contraceptive steroids In: Gregoire AT and Blye RP (eds) Contraceptive steroids – pharmacology and safety. Plenum Press, New York, pp 231–245

Heywood R (1989) The toxicological assessment of the safety of hormonal contraception. In: Michal F (ed) Safety requirements for contraceptive steroids. University Press, Cambridge, pp 159–165

Heywood R, Wadsworth PF (1980) The experimental toxicology of estrogens. Pharmacol Ther 8:125–142

Highman B, Greenman DL, Norvell MJ (1980) Neoplastic and preneoplastic lesions induced in female C3H mice by diets containing Diethylstilbestrol or 17-*β*-Estradiol. Environ Pathol Toxicol 4:81–95

Hill A, Wolff S (1983) Sister chromatid exchanges and cell division delays induced by diethylstilbestrol, estradiol, and estriol in human lymphocytes. Cancer Res 43:4114–4118

Hillbertz-Nilsson K, Forsberg JG (1985) Estrogen effects on sister chromatid exchanges in mouse uterine, cervical and kidney cells. J Natl Cancer Inst 75: 575–580

Hirsimäki P, Hirsimaki Y, Nieminen L, Payne BJ (1993) Tamoxifen Induces Hepatocellular Carcinoma in Rat Liver – A 1-Year Study with 2 Antiestrogens. Arch Toxicol 67:49–54

Hundal BS, Dhillon VS, Sidhu IS (1997) Genotoxic potential of estrogens. Mutat Res 389:173–181

Husum B, Wulf HC, Niebuhr E (1982) Normal sister-chromatid exchanges in oral contraceptive users. Mutat Res 103:161–164

Hümpel M, Düsterberg B, Beier S, Schuppler J, Gunzel P, Elger W (1986) The role of pharmacokinetics in preclinical safety studies of synthetic sex steroids In: Gregoire AT and Blye RP (eds) Contraceptive steroids – pharmacology and safety. Plenum Press, New York, London, pp 47–65

IARC (International Agency for Research on Cancer) (1979) Evaluation of the Carcinogenic Risk of Chemicals to Humans, Sex Hormones (II). Lyon, IARC, vol. 21

Ishikawa S, Oda T, Sato Y, Mochizuki M (1996) Lack of mutagenicity of diethylstilbestrol metabolite and analog, (+/–)-indenestrols A and B, in bacterial assays. Mutat Res 368:261–265

Itoh S, Matsuura Y, Seki H, Tazawa T, Shimada H (1991) Mutagenicity studies of norethisterone and ethinylestradiol. Yakuri to Chiryo 19(suppl 297):(S-1065)-308(S-1076)

Iwase T, Sano F, Murakami T, Inazawa K (1995) Male reproductive toxicity of ethinylestradiol associated with 4 weeks daily dosing prior to mating in rats. J Toxicol Sci 20:265–279

Jacob D, Morris JM (1969) The estrogenic activity of postcoital antifertility compounds. Fertil Steril 20:211–222
Jensen TK, Toppari J, Keiding N, Skakkebaek NE (1995) Do environmental estrogens contribute to the decline in male reproductive health? Clin Chem 41:1896–1901
Joosten HFP et al. (1978) Interspecies differences in embryolethal activity of steroids. Sixth Conference of the European Teratology Society, Budapest, quoted from LEHMANN et al. (1989)
Jordan A (1992) FDA requirements for nonclinical testing of contraceptive steroids. Contraception 46:499–509
Jordan VC (1995) Tamoxifen: Toxicities and drug resistance during the treatment and prevention of breast cancer. Annu Rev Pharmacol Toxicol 35:195–221
Kamrin MA, Carney EW, Chou K, Cummings A, Dostal LA, Harris C, Henck JW, Lochcaruso R, Miller RK (1994) Female reproductive and developmental toxicology: overview and current approaches. Toxicol Lett 74:99–119
Karlsson S, Hirsimäki Y, Mäntylä E, Nieminen L, Kangas L, Hirsimäki P, Perry CJ, Mulhern M, Millar P, Handa J, Williams GM (1996) A two-year dietary carcinogenicity study of the antiestrogen toremifene in Sprague-Dawley rats. Drug Chem Toxicol 19:245–266
Kendall ME, Rose DP (1992) The effects of diethylstilbestrol, tamoxifen, and toremifene on estrogen-inducible hepatic proteins and estrogen receptor proteins in female rats. Toxicol Appl Pharmacol 114:127–131
Klotz DM, Beckman BS, Hill SM, McLachlan JA, Walters MR, Arnold SF (1996) Identification of environmental chemicals with estrogenic activity using a combination of in vitro assays. Environ Health Perspect 104:1084–1089
Kochhar TS (1985) Inducibility of chromosome aberrations by steroid hormones in cultured chinese hamster ovary cells. Toxicol Lett 29:201–206
Kramer M, Günzel P (1979) The proof of drug effects on endocrine glands or endocrine target organs by means of toxicological investigations. Pharmac Ther 5:287–296
Kuhnz W, Putz B (1989) Pharmacokinetic interpretation of toxicity tests in rats treated wit oestradiol valerate in the diet. Pharmacol Toxicol 65:217–222
Kulling SE, Metzler M (1997) Induction of micronuclei, DNA strand breaks and HPRT mutations in cultured Chinese hamster V79 cells by the phytoestrogen coumoestrol. Food Chem Toxicol 35:605–613
Kumagai S, Shimizu T (1982) Neonatal exposure to zeraleone causes persistent anovulatory estrus in the rat. Arch Toxicol 50:279–286
Lang R, Reimann R (1993) Studies for a genotoxic potential of some endogenous and exogenous sex steroids. I. Communication: Examination for the induction of gene mutations using the Ames salmonella/microsome test and the HPGRT Test in V79 cells. Environ Mol Mutagen 21:272–304
Lehmann M, Putz B, Poggel HA, Günzel P (1989) Experimental toxicity studies with contraceptive steroids and their relevance for human risk estimation In: Dayan AD and Paine AJ (eds) Advances in Applied Toxicology. Taylor and Francis Ltd., London, New York, Philadelphia, pp 51–79
Li JJ, Li SA (1996) Estrogen carcinogenesis in the hamster kidney: a hormone-driven multistep process. Prog Clin Biol Res 394:255–267
Li JJ, Li SA, Klicka JK, Parsons JA, Lam LKT (1983) Relative carcinogenic activity of various synthetic and natural estrogens in the Syrian hamster kidney. Cancer Res 43:5200–5204
Li JJ, Li SA, Oberley TD, Parsons JA (1995) Carcinogenic activities of various steroidal and nonsteroidal estrogens in the hamster kidney: relation to hormonal activity and cell proliferation. Cancer Res 55:4347–4351
Littlefield LG, Lever WE, Miller FL, Kong-OO Goh (1975) Chromosome breakage studies in lymphocytes from normal women, pregnant women, and women taking oral contraceptives. Am J Obstet Gynecol 121:976–980
Löser R, Janiak P, Seibel K (1984) K-21060E. Drugs of the Future 9:186–188
Lunan CB, Klopper A (1975) Antiestrogens: A review. Clin Endocrinol 4:551–572

Makino M, Sakka M, Yamamoto T et al. (1990) Acute toxicity studies of ethinylestradiol in mice and rats. Yakuri to Chiryo 18:2583–2589
Matton-Van Leuven MT, Thiery M, de Bie S (1974) Cytogenetic evaluation of patients in relation to the use of oral contraceptives. Contraception 10:25–38
McConnell RF (1989a) Comparative aspects of contraceptive steroids: effects observed in rats. Toxicol Pathol 17:385–388
McConnell RF (1989b) General observations on the effects of sex steroids in rodents with emphasis on long-term oral contraceptive studies. In: Michal F (ed) Safety requirements for contraceptive steroids. University Press, Cambridge, pp 211–229
Metzler M, Pfeiffer E, Schuler M, Rosenberg B (1996) Effects of estrogens on microtubule assembly: Significance for aneuploidy. In: Li JJ, Li SA, Gustafsson J-A, Nandi S, Sekely LI (eds) Hormonal Carcinogenesis II. Springer, New York, pp 193–199
Minesita T, Muraoka Y, Otori H, Yahara I, Inuta T, Kuramoto Y, Kawaguchi J, Okada T (1970) Toxicity tests of S-3850, and its components, chlormadinone acetate and mestranol in mice and rats. Oyo Yakuri Pharmacometr 4:217–232
Mischler TW Gawlak D (1970) The biological profile of 8-aza quinestrol. J Reprod Fert 22:49–56
Montandon F, Williams GM (1994) Comparison of DNA reactivity of the polyphenylethylene hormonal agents diethylstilbestrol, tamoxifen and toremifene in rat and hamster liver. Arch Toxicol 68:272–275
Morris JM (1970) Postcoital antifertility agents and their teratogenic effect. Contraception 2:85–97
Morris JM, Wagenem Gv (1973) Interception: the use of postovulatory estrogens to prevent implantation. Am J Obstet Gynecol 115:101–106
Murthy PBK, Prema K (1979) Sister-chromatid exchanges in oral contraceptive users. Mutat Res 68:149–152
Müller AMF, Makropoulos V, Bolt HM (1995) Hormonartig wirkende Substanzen in Umwelt und Nahrung. DAZ 135:3923/17–3930/24
Nagae Y, Takahashi S, Takahashi H, Deguchi J, Yasuda S, Kawahara M, Mori T, Miyamoto M (1992) A 3-months subacute toxicity study on 17β-estradiol by continuously subcutaneous infusion in ovariectomized rats. Yakuri to Chiryo 20:37(3899)–51(3913)
Nishino Y, Neumann F (1980) Östrogene Partialwirkung von Gestagenen, speziell von Norethisteronönanthat. Arzneimittelforschung 30:439–452
O'Connor JC, Cook JC, Craven SC, Van Pelt CS, Obourn JD (1996) An in vivo battery for identifying endocrine modulators that are estrogenic or dopamine regulators. Fundam Appl Toxicol 33(2):182–195
Oesterle D, Deml E (1983) Promoting effect of polychlorinated biphenyls on development of enzyme-altered islands in livers of weanling and adult rats. J Cancer Res Clin Oncol 105:141–147
Oettel M, Zimmermann H, Elger W, Böcker R, Ponsold K, Schwarz S, Siemann J (1995) Comparative toxicology of ethinylestradiol (EE) and a new estrogen without 17α-ethinyl group: 14,15α-methylene estradiol 17β (J824). Contracept Fertil Sex, Special n° 1, (suppl au n° 9 Vol. 23):S52–S53
Ohno H, Yamada M, Kudo G, Jindo T, Sakamoto Y, Nomura M (1991) Single dose oral toxicity study of DT-5061 and DT-5062 (1/35) in mice, rats and dogs. Yakuri to Chiryo 19:67–84
Overbeck GA, Madjerek Z, de Visser J (1962) The effect of Lynestrenol on animal reproduction. Endocrinologica 41:351–370
Pathak DN, Pongracz K, Bodell WJ (1995) Microsomal and peroxidase activation of 4-hydroxy-tamoxifen to form DNA adducts: comparison with DNA adducts formed in Sprague-Dawley rats treated with tamoxifen. Carcinogenesis 16:11–15
Peterson DL, Edgren RA (1965) The effect of various steroids on mating behaviour, fertility and fecundity of rats. Int J Fertil 10:327–332

Phillips DH, Potter GA, Horton MN, Hewer A, Crofton-Sleigh C, Jarman M, Venitt S (1994) Reduced genotoxicity of (D_5-ethyl)-tamoxifen implicates α-hydroxylation of the ethyl group as a major pathway of tamoxifen activation to a liver carcinogen. Carcinogenesis 15:1487–1492

Poggel HA., Günzel P (1978) Steroidhormone und Embryotoxizität. In: Schnieders B, Stille G, Grosdanoff P (eds) Embryotoxikologische Probleme in der Arzneimittelforschung. Dietrich Reimer, Berlin, pp 127–132

Purchase IFH (1994) Current knowledge of mechanisms of carcinogenicity: Genotoxins versus non-genotoxins. Hum Exp Toxicol 13:17–28

Rajah TT, Pento JT (1995) The mutagenic potential of antiestrogens at the HPRT locus in V79 cells. Res Commun Mol Pathol Pharmacol 89:85–92

Randerath K, Mabon N, Sriram P, Moorthy B (1994a) Strong intensification of mouse hepatic tamoxifen DNA adduct formation by pretreatment with the sulfotransferase inhibitor and ubiquitous environmental pollutant pentachlorophenol. Carcinogenesis 15:797–800

Randerath K, Moorthy B, Mabon N, Sriram P (1994b) Tamoxifen: evidence by ^{32}P-postlabeling and use of metabolic inhibitors for two distinct pathways leading to mouse hepatic DNA adduct formation and identification of 4-hydroxytamoxifen as a proximate metabolite. Carcinogenesis 15:2087–2094

Rao PN, Engelberg J (1967) Structural specifity of estrogens in the induction of mitotic chromatid non-disjunction in HeLa cells. Experimental Cell Research 48:71–81

Reimann R, Kalweit S, Lang R (1996) Studies for a genotoxic potential of some endogenous and exogenous sex steroids. II. Communication: examination for the induction of cytogenetic damage using the chromosomal aberration assay on human lymphocytes in vitro and the mouse bone marrow micronucleus test in vivo. Environ Mol Mutagen 28:133–144

Ribeiro MF, Spritzer PM, Barbosa-Coutinho LM, Oliveira MC, Pavanato MA, Silva ISB, Reis FM (1997) Effects of bromocriptine on serum prolactin cells in estradiol-treated ovariectomized rats: an experimental model of estrogen-dependent hyperprolactinemia. Braz J Med Biol Res 30:113–117

Rutqvist LE (1993) Long-term toxicity of tamoxifen. Rec Results Cancer Res 127:257–266

Sakamoto M, Ito K, Horikoshi M, Hayashi M, Tsubosaki M, Nakamori K, Lancaster P, Wood JD (1992) Single dose toxicity study of toremifene citrate (NK622) in rats and monkeys. Oyo Yakuri Pharmacometr 44:351–356

Sargent LM, Dragan YP, Bahnub N, Wiley JE, Sattler CA, Schroeder P, Sattler GL, Jordan VC, Pitot HC (1994) Tamoxifen induces hepatic aneuploidy and mitotic spindle disruption after a single in vivo administration to female Sprague-Dawley rats. Cancer Res 54:3357–3360

Sargent LM, Dragan YP, Sattler C, Bahnub N, Sattler G, Martin P, Cisneros A, Mann J, Thorgeirsson S, Jordan VC, Pitot HC (1996) Induction of hepatic aneuploidy in vivo by tamoxifen, toremifene and idoxifene in female Sprague-Dawley rats. Carcinogenesis 17:1051–1056

Satoh H, Kajimura T, Chen C-J, Yamada K, Furuhama K, Nomura M (1997) Invasive pituitary tumors in female F344 rats induced by estradiol dipropionate. Toxicol Pathol 25:462–469

Saunders FJ (1967) Effects of norethynodrel combined with mestranol on the offspring when administered during pregnancy and lactation in rats. Endocrinology 80:447–452

Saunders FJ, Elton RL (1967) Effects of Ethynodiol diacetate and Mestranol in rats and rabbits, on conception, on the outcome of pregnancy and on the offspring. Toxicol Appl Pharmacol 11:229–244

Sawada M, Ishidate M (1978) Colchicine-like effect of diethylstilbestrol (DES) on mammalian cells in vitro. Mutat Res 57:175–182

Schardein JL (1980) Studies of the components of an oral contraceptive agent in albino rats. I. Estrogenic component. J Toxicol Environ Health 6:885–894

Schnitzler R, Foth J, Degen GH, Metzler M (1994) Induction of micronuclei by stilbene-type and steroidal estrogens in Syrian hamster embryo and ovine seminal vesicle cells in vitro. Mutat Res 311:84–93

Schulte-Hermann R, Timmermann-Trosiener I, Schuppler J (1983) Promotion of spontaneous preneoplastic cells in rat liver as a possible explanation of tumor production by non-mutagenic compounds. Cancer Res 43:839–844

Seibert B, Günzel P (1994) Animal toxicity studies performed for risk assessment of the once-a-month injectable contraceptive mesigyna(r). Contraception 49:303–333

Seraj MJ, Umemoto A, Tanaka M, Kajikawa A, Hamada K, Monden Y (1996) DNA adduct formation by hormonal steroids in vitro. Mutat Res 370:49–59

Shelby MD, Newbold RR, Tully DB, Chae K, Davis VL (1996) Assessing environmental chemicals for estrogenicity using a combination of in vitro and in vivo assays. Environ Health Perspect 104:1296–1300

Shen L, Qiu S, van Breemen RB, Zhang F, Chen Y, Bolton JL (1997) Reaction of the Premarin metabolite 4-hydroxyequilenin semiquinone radical with 2′-deoxyguanosine: Formation of unusual cyclic adducts. J Am Chem Soc 119:11126–11127

Shyama SK, Rahiman MA (1996) Genotoxicity of lynoral (ethinyloestradiol, an estrogen) in mouse bone marrow cells, in vivo. Mutat Res 370:175–180

Soto AM, Fernandez MF, Luizzi MF, Oles Karasko AS, Sonnenschein C (1997) Developing a marker of exposure to xenoestrogen mixtures in human serum. Environ Health Perspect 105(suppl 3):647–654

Soto AM, Sonnenschein C (1979) Estrogen receptor levels in estrogen sensitive cells in culture. J Steroid Biochem 11:1185–1190

Soto AM, Sonnenschein C, Chung KL, Fernandez MF, Olea N, Serrano FO (1995) The E-SCREEN assay as a tool to identify estrogens: An update on estrogenic environmental pollutants. Environ Health Perspect 103(suppl 7):113–122

Stancel GM, Boettger Tong HL, Chiappetta C, Hyder SM, Kirkland JL, Murthy L, Loose Mitchell DS (1995) Toxicity of endogenous and environmental estrogens: what is the role of elemental interactions? Environ Health Perspect 103(Suppl) 7:29–33

Stenchever MA, Jarvis JA, Kreger NK (1969) Effect of selected estrogens and progestins on human chromosomes in vitro. Obstet Gynecol 34:249–251

Styles JA, Davies A, Davies R, White INH, Smith LL (1997) Clastogenic and aneugenic effects of tamoxifen and some of its analogues in hepatocytes from dosed rats and in human lymphoblastoid cells transfected with human P450 cDNAs (MCL-5 cells). Carcinogenesis 18:303–313

Styles JA, Davies A, Lim CK, De Matteis F, Stanley LA, White INH, Yan Z-X, Smith LL (1994) Genotoxicity of tamoxifen, tamoxifen epoxide and toremifene in human lymphoblastoid cells containing human cytochrome P450s. Carcinogenesis 15:5–9

Suto A, Bradlow HL, Wong GY, Osborne MP, Telang NT (1993) Experimental downregulation of intermediate biomarkers of carcinogenesis in mouse mammary epithelial cells. Breast Cancer Res Treat 27:193–202

Telang NT, Suto A, Wong GY, Osborne MP, Bradlow HL (1992) Induction by estrogen metabolite 16α-hydroxyestrone of genotoxic damage and aberrant proliferation in mouse mammary epithelial cells. J Natl Cancer Inst 84:634–638

Thierfelder W, Mehnert WH, Laussmann D, Arndt D, Reineke HH (1995) Der Einfluss umweltrelevanter östrogener oder östrogenartiger Substanzen auf das Reproduktionssystem. Bundesgesundhbl 9/95:338–341

Toppari J, Larsen JC, Christiansen P, Giwercman A, Grandjean P, Guillette LJ, Jegou B, Jensen TK, Jouannet P, Keiding N, Leffers H, McLachlan JA, Meyer O, Muller J, Rajpert-De Meyts E, Scheike T, Sharpe R, Sumpter J, Skakkebaek NE (1996) Male reproductive health and environmental xenoestrogens. Environ Health Perspect 104:741–803

Tsutsui T, Suzuki N, Fukuda S, Sato M, Maizumi H, McLachlan JA, Barrett JC (1987) 17ß-estradiol induced cell transformation and aneuploidy of Syrian hamster embryo cells in culture. Carcinogenesis 8:1715–1719

Tsutsui T, Suzuki N, Maizumi H, Barrett JC (1990) Aneuploidy induction in human fibroblasts: comparison with results in Syrian hamster fibroblasts. Mutat Res 240:241–249

Tsutsui T, Taguchi S, Tanaka Y, Barrett JC (1997) 17β-estradiol, diethylstilbestrol, tamoxifen, toremifene and ICI 164,384 induce morphological transformation and aneuploidy in cultured Syrian hamster embryo cells. Int J Cancer 70:188–193

Tucker, MJ, Adam HK, Patterson JS (1984) Tamoxifen In: Laurence DR, McLean DEM, Wetherall M (eds) Safety testing of new drugs: Laboratory Predictions and Clinical Performance. Academic Press, London, pp 125–161

Vanage GR, Dao B, Li X-J, Bardin CW, Koide SS (1997) Effects of Anordiol, an antiestrogen, on the reproductive organs of the male rat. Arch Androl 38:13–21

Velardo JT, Raney N, Smith BG, Sturgis SH (1956) Effect of various steroids on gestation and litter size in rats. Fertil Steril 7:301–311

Vijayalaxmi KK, Rai SP (1996) Studies on the genotoxicity of tamoxifen citrate in mouse bone marrow cells. Mutat Res 368:109–114

Wadsworth PF, Heywood R (1978) Toxicology of ethinyl oestradiol in rhesus monkeys. In: Chivers DJ, Ford EHR (eds) Recent advances in primatology (Volume 4, Medicine) Academic Press, London, pp 179–182

Walker BE, Haven MI (1997) Intensity of multigenerational carcinogenesis from diethylstilbestrol in mice. Carcinogenesis 18:791–793

Walker BE, Kurth LA (1995) Multi-generational carcinogenesis from diethylstilbestrol investigated by blastocyst transfers in mice. Int J Cancer 61:249–252

Weikel JH, Jr, Nelson LW (1977) Problems in evaluating chronic toxicity of contraceptive steroids in dogs. J Toxicol Environ Health 3:167–177

Welsch CW, Adams C, Lambrecht LK, Hassett CC, Brooks CL (1977) 17β-oestradiol and Enovid mammary tumorigenesis in C3H/HeJ female mice: Counteraction by concurrent 2-bromo-α-ergocryptine. Br J Cancer 35:322–328

White IN, Martin EA, Styles J, Lim CK, Carthew P, Smith LL (1997) The metabolism and genotoxicity of tamoxifen. Prog Clin Biol Res 396:257–270

White INH, De Matteis F, Davies A, Smith LL, Crofton-Sleigh C, Venitt S, Hewer A, Phillips DH (1992) Genotoxic potential of tamoxifen and analogues in female Fisher F344/n rats, DBA/2 and C57BL/6 mice and in human MCL-5 cells. Carcinogenesis 13:2197–2203

Wilson CA, Leigh AJ (1992) Endocrine toxicology of the female reproductive system. In: Atterwill CK, Flack JD (eds) Endocrine Toxicology. University Press, Cambridge, pp 313–395

Wiseman LR, Goa KL (1997) Toremifene: A review of its pharmacological properties and clinical efficacy in the management of advanced breast cancer. Drugs 54:141–160

Wood JD, Glaister JR, Goodyer MJ, Yamashita T, Nakamori K (1992) Fifty two-week oral toxicity study of toremifene citrate (NK622) in female monkeys. Oyo Yakuri Pharmacometr 44:375–387

Yager JD, Liehr JG (1996) Molecular mechanisms of estrogen carcinogenesis. Annu Rev Pharmacol Toxicol 36:203–232

Yager JD, Yager R (1980) Oral Contraceptive Steroids as Promotors of Hepatocarcinogenesis in Female Sprague-Dawley Rats. Cancer Res 40:680–3685

Zimmermann H, Elger W, Oettel M, Schwarz S (1996) Comparative toxicology of ethinylestradiol (EE) and a new estrogen without 17-α ethinyl group: 17-β-estradiol-3-sulfamate (J 995). Toxicol Lett 88[Suppl 1]:100

Zimmermann H, Oettel M, Dance CA, Hodson-Walker G (1994) In vitro assessment of the clastogenic and mitogenic activity of ethinyl-estradiol (EE) and a new estrogen, J 824 in human cultured lymphocytes. Toxicol Lett 74[Suppl 1]:96

CHAPTER 37

Estrogens and Sexually Transmitted Diseases

M. Dören

A. Introduction

Communicable diseases are amongst the most common diseases world-wide, a significant part of which is attributable to sexually transmitted diseases (Sciarra 1997). They are caused by a variety of microorganisms, including bacteria, viruses, parasites, yeast, chlamydia, ureaplasma and mycoplasma. They are most often transmitted by penile–vaginal, oral–genital, and genital–anal contact. The physiological presence of *lactobacilli* in the vagina protects, to some extent, against external damage of the epithelium by microorganisms. The presence of cyclic ovarian function ensures that glycogen within the superficial cells of the vaginal epithelium is available as a nutrient for the *lactobacilli* which produce lactic acid to adjust the vaginal pH to 3.8–4. This ensures some protection against various pathogens known to cause (ascending) urogenital infections. However, frequently this barrier is disrupted by microorganisms. Common bacterial communicable diseases include chancroid, caused by the gram-negative bacillus *Hemophilus ducreyi,* to be the most common cause of genital ulcer disease; gonorrhea caused by the gram-negative diplococcus *Neisseria gonorrhea*; and syphilis caused by the spirochete *Treponema pallidum*. Common viral diseases are *herpes genitalis* caused by herpes simplex virus type II and genital warts due to human papillomavirus (HPV) infection. One of the most serious health concerns today is the acquired immune deficiency syndrome (AIDS) caused by type I and II of human immunodeficiency virus (HIV). This affects approximately 17 million women worldwide (Johnson et al. 1994).

Other organisms such as *Chlamydia trachomatis*, parasites such as *Trichomonas vaginalis*, and yeast such as *Candida albicans* are known to cause infections of the female genital tract as well. They are not unanimously regarded as sexually transmitted diseases. However, as these diseases are frequently associated with the bacterial and viral infections mentioned above, their relationship to exogenous and endogenous estrogens will be discussed too.

B. The Impact of Hormonal Contraceptives

There are various mechanisms that probably account for changes in the microenvironment of the vagina in women who use oral contraceptives. First, the progestogenic compounds after long-term use apparently contribute to a thinning of the vaginal epithelium, an increase in the vaginal pH and a reduced amount of cervical mucus in combination with its increased viscosity, whereas estrogens produce the opposite effects. Cervical ectopy is more often seen in women who use oral contraceptives. We may assume that thickened cervical mucus builds a more potent mechanical barrier for pathogens apart from the possibility that immunological reactions towards pathogenic microorganisms may be modified by exogenous steroids. This is important in case microorganisms are attached to spermatozoa or spread directly through the endocervical mucus. If they spread along cell surfaces (intracanalicular infection), then thickened mucus might not effectively prevent ascending infections. Second, oral contraceptives generally reduce menstrual flow. Therefore, it has been suggested that the shortened period of time of endometrial shedding might decrease the possibility of acquiring a genital infection. Third, animal models support the idea that exogenous estrogen increases the number of infected cells and increases the spread to the fallopian tubes; an infection may persist longer after exposure to progesterone. Increased spread and/or clearing of, for example, *chlamydia* may occur because of impaired production of the mucosal-membrane secretory antibodies immunoglobulin (Ig) A and IgG (Arya et al. 1981; Rank and Baron 1987). Few studies examined the impact of various types of hormonal contraception on the acquisition and transmission of sexually transmitted diseases, including HIV infection. Specific data to differentiate between various combined or progestogen-only methods (pills, injections, implants) are missing (Carlin and Boag 1995).

I. Human Immunodeficiency Virus Infection

Information on endogenous ovarian function in women with HIV is very limited. It is possible that secondary amenorrhea may be associated with late-stage infection (Widy-Wirski et al. 1988). However, there is no indication that the disease itself induces menstrual changes due to changes in the secretion of ovarian steroid hormones (Shah et al. 1994). West-African women appear to be more susceptible to HIV-2 infection, known to progress to AIDS in only a proportion of infected individuals, after the age of 40–45 years (Aaby et al. 1996). Whether this phenomenon is related to the ovarian function of that age group is unknown. Various factors have been identified to be significant for transmission of HIV. The risk of HIV transmission is considerably increased in the presence of genital ulcer disease, i.e., herpes simplex virus type-II infection, syphilis, chancroid. Gonorrhea, chlamydial infection and trichomoniasis are associated with significantly increased risks of HIV acquisition as well (Laga et al. 1993).

In infected women who are asymptomatic and not immunosuppressed, combined oral contraceptive pills prevent unwanted pregnancy to the same extent as in women without acquisition of this virus and do not appear to influence the course of an HIV infection. Pregnancy with highest endogenous levels of estrogen and progesterone does not accelerate the development of AIDS (DESCHAMPS et al. 1993). Two prospective studies in African women demonstrated that current use of oral contraceptives is an independent risk factor for HIV infection (CARAEL et al. 1988, PLUMMER et al. 1991). Another prospective study conducted in African women appeared to show the opposite (LAGA et al. 1993). European studies did not demonstrate any association between the use of oral contraceptives and the risk of HIV infection (DE VINCENZI 1994). However, in contrast to a recent overview concluding that there is apparently no overall change of risk for HIV infection among users of combined oral contraceptives (TAITEL and KAFRISSEN 1995), it is not impossible that the transmission of HIV may be different in Caucasian and Afro-Caribbean women. There are no longitudinal data to demonstrate that hormonal contraceptives may increase the risk of HIV transmission to a seronegative partner via cervical ectopia.

Lighter withdrawal bleeding due to oral contraceptives could be seen as a protective factor by a decrease of the exposure time of the endometrium to the HIV antigen. Intrauterine devices may increase the risk of transmission by prolonged or heavier withdrawal bleeding. On the other hand, breakthrough bleeding due to exogenous hormones could enhance the chance of infection as the endometrial layer is disrupted (HICKS 1995). Vaginal sex during menstruation has been shown to be associated with an increased risk of infection in a recent European study (European Study Group on heterosexual transmission of HIV 1992).

II. Human Papillomavirus Infection

More than 60 types of human papillomavirus infection (HPV) are known today; the types 16 and 18 are discussed in conjunction with their increased prevalence in women with cervical carcinoma. HPV infection increases the likelihood of having cervical cancer diagnosed in later life (LEHTINEN et al. 1996). Long-term use of combined oral contraceptives might be associated with an increased risk of cervical dysplasia and cervical cancer (PARAZZINI et al. 1990). Whether this is due to HPV infection or any other (life style) factor is very difficult to establish. A recent cross-sectional case-control study demonstrated that women taking oral contraceptives are at increased risk of having genital warts; however, specific testing for HPV was not performed (ROSS 1996). In vitro experiments show a higher rate of HPV gene expression and cellular transformation by viral DNA in response to increased progesterone levels (PATER et al. 1990).

III. Infection with *Chlamydia trachomatis*, *Neisseria gonorrhea* and Mycoplasma

Infections with *Neisseria gonorrhea* and *Chlamydia trachomatis* have been linked with oral-contraceptive use (LOUV et al. 1989; BLUM et al. 1990). A meta-analysis calculated a twofold increased risk of chlamydial infection in oral-contraceptive users; compared with users of barrier methods, the risk was elevated threefold (Cottingham and Hunter 1992). It is possible that an increased vaginal pH in women who use oral contraceptives is related to a higher incidence of endocervical chlamydia and mycoplasma (Hanna et al. 1985). It is difficult to establish whether the increased frequency of cervical ectopy in oral contraceptive users might lead to easier diagnostic detection. Studies that controlled for the size of the ectropion yielded conflicting results (Arya et al. 1981; Harrison et al. 1985). Endocervical-epithelial cells are the primary sites of attachment and infection for both *Chlamydia trachomatis* and *Neisseria gonorrhea*. The incidence of chlamydial and less-established gonococcal pelvic-inflammatory disease appears to be reduced in oral-contraceptive users (Westrom 1980; Wolner-Hanssen et al. 1990; Cates and Stone 1992), even more when compared with women fitted with intrauterine devices. Current users of oral contraceptives with Chlamydia salpingitis have less perihepatitis, perisplenitis and pericolitis (Fitz-Hugh-Curtis syndrome). The severity of infection, i.e., the degree of tissue inflammation, is apparently reduced in oral contraceptive users. Their geometric mean antibody level is lower than that in women using no hormonal contraception (Wolner-Hanssen et al. 1985). This protection may only be effective if the duration of oral-contraceptive use is at least 1 year (Rubin et al. 1982). Thickened cervical secretion due to the progestogenic compound is thought to be one major reason for this (Grimes and Cates 1990).

IV. *Candida Albicans* Infection

The incidence of *Candida albicans* infection is low before puberty and after menopause, increases during pregnancy, and is reported to peak in the late luteal phase of the menstrual cycle (Sweet 1985). Pregnancy increases both colonization by *Candida albicans* and its conversion to the invasive myceliated form. High-dose estrogen combined oral contraceptives have been implicated in an increased risk of vulvovaginal candidiasis via an increase in vaginal glycogen content, vaginal pH and modified-epithelial cell adherence and receptors for *Candida*. Low-dose estrogen combined oral contraceptives are not thought unequivocally to increase the risk of candidiasis (Avonts 1990; Barbone et al. 1990; Spinillo et al. 1995). Intramuscular depot medroxyprogesterone acetate seems to protect against this infection (Toppozada et al. 1979). In a recent retrospective assessment of urban Californian women, the onset of symptoms of vulvovaginal candidiasis within a woman's menstrual cycle was not associated with the various methods of birth control, including oral contraceptives, injectable progestogens and various non-hormonal methods (Nelson 1997). Previous work suggested that the frequency of a diagnosis of *Candida* infection may change throughout the cycle (Wallin et al. 1974; Segal 1984; Kalo-Klein and Witkin 1989).

V. Trichomoniasis

Trichomoniasis is more common in women of reproductive age and in pregnancy than in postmenopausal women and premenarchal girls. Under the influence of high-maternal estrogens, the vaginal milieu of newborn females has more glycogen deposits and a thicker epithelium than later on in childhood when the onset of cyclic ovarian function begins. It is only during this neonatal period that a young female is susceptible to infection in her childhood (AL-SAHILI et al. 1974). A hypo-estrogenic vaginal epithelium, an alkaline pH and a relative lack of glycogen may constitute a more hostile environment for these organisms. Trichomoniasis appears to be less frequent in users of oral and injectable contraceptives (ROY 1991; DALY et al. 1994). However, estradiol enhances the growth of *Trichomonas vaginalis* organisms in vitro (MARTINOTTI and SAVOIA 1985). Recently a case report suggested that hormonal replacement therapy might increase the susceptibility to infection (SHARMA et al. 1997).

C. The Impact of Hormonal Replacement Therapy

Most studies that examined the prevalence of sexually transmitted disease and urogenital infectious diseases focus on women before menopause and, in particular, young women in their teens and twenties. Only recently have investigations of the effects of hormonal replacement therapy in the detection of HPV and cancer risk been performed. It is unclear what effects hormonal replacement therapy may have on the risk of development of cervical cancer in women infected with HPV (MUNOZ and BOSCH 1992; KJAER et al. 1990).

The only data available today were generated by a randomized, prospective trial, predominantly designed to assess cardiovascular risk factors and endometrial safety data in women assigned to various types of combination therapy and compared with a placebo group [postmenopausal estrogen and progestin intervention trial (PEPI)]. Treatment consisted of either conjugated estrogen 0.625 mg per day alone or in combination with either sequential or continuous medroxyprogesterone acetate or sequential-micronized progesterone. HPV cervical specimens were collected before initiation of treatment and annually thereafter for 3 years. Polymerase chain reaction was used to amplify the viral DNA followed by dot-blot hybridization to detect the presence and types of HPV. Half of all 105 women tested positive for at least one of the subtypes tested (6, 11, 16, 18, 31, 35, 39, 45, 51, 52). At one point during the 3-year interval, only 10.5% tested positive, both at baseline and at follow-up, 40% tested positive at one or the other time. HPV-16 was the most frequent type isolated, followed by HPV-31. HPV-6 and HPV-11, which are associated with benign genital lesions, were detected in 4% of women at baseline or follow-up. HPV-18, which is associated with more aggressive female genital-cancer risk in younger women with adenocarcinoma was identified infrequently (4%). There was no consistent pattern of influence in the HPV status in users of hormonal replacement therapy compared with placebo throughout the study (Smith et al. 1997).

D. Conclusion

It appears to be possible that long-term hormonal contraceptive use may be associated with an increased risk of acquiring various genital infections and sexually transmitted diseases, such as HPV infection. The transmission and course of the most serious health hazard of all sexually transmitted diseases, HIV infection, seems not to be influenced by the use of combined oral contraceptives. Little is known about the impact of hormonal replacement therapy on sexually transmitted diseases and specific genital infections.

References

Aaby P, Ariyoshi K, Buckner M, Jensen H, Berry N, Wilkins A, Richard D, Larsen O, Dias F, Melbye M, Whittle H (1996) Age of wife as a major determinant of male to female transmission of HIV-2 infection: a community study from rural West Africa. AIDS 10:1585–1590

Al-Sahili FL, Curran JP, Wang J (1974) Neonatal *Trichomonas vaginalis*: report of three cases and review of the literature. Pediatrics 53:196–200

Arya LP, Mallinson H, Goddard AD (1981) Epidemiological and clinical correlates of chlamydial infection of the cervix. Br J Vener Dis 57:118–124

Avonts D, Sercu M, Heyerick P, Vandermeeren I, Meheus A, Piot P (1990) Incidence of uncomplicated genital infections in women using oral contraception or an intrauterine device: a prospective study. Sex Transm Dis 17:23–29

Barbone F, Austin H, Louv WC, Alexander WJ (1990) A follow-up study of methods of contraception, sexual activity, and rates of trichomoniasis, candidiasis, and bacterial vaginosis. Am J Obstet Gynecol 163:510–514

Blum M, Gilerovitch M, Benaim J, Appelbaum T (1990) The correlation between chlamydia antigen, antibody, vaginal colonization and contraceptive method in young, unmarried women. Adv Contracept 6:41–45

Carael M, van de Perre PH, Lepage PH, Allen S, Nsengumuremyi F, van Goethem C, Ntahorutaba M, Nzaramba D, Clumeck N (1988) Human immunodeficiency virus transmission among heterosexual couples in Central Africa. AIDS 2:201–205

Carlin EM, Boag FC (1995) Women, contraception and STDs including HIV. Int J STD AIDS 6:373–386

Cates W Jr, Stone KM (1992) Family planning, sexually transmitted diseases and contraceptive choice: a literature update-Part II. Fam Plann Perspect 24:122–128

Cottingham J, Hunter D (1992) *Chlamydia trachomatis* and oral contraceptive use: a quantitative review. Genitourin Med 68:209–216

Daly CC, Maggwa N, Mati JK, Solomon M, Mbugua S, Tukei PM, Hunter DJ (1994) Risk factors for gonorrhoea, syphilis, and trichomonas infections among women attending family planning clinics in Nairobi, Kenya. Genitourin Med 70:155–161

Deschamps M-M, Pape JW, Desvarieux M, Williams-Russo P, Madhavan S, Ho JL, Johnson WD Jr (1993) A prospective study of HIV-seropositive asymptomatic women of childbearing age in a developing country. J Acquired Immune Defic Syndr 6:446–451

De Vincenzi I for The European Study Group on Heterosexual Transmission of HIV (1994) A longitudinal study of human immunodeficiency virus transmission by heterosexual partners. New Engl J Med 331:341–346

European Study Group on Heterosexual Transmission of HIV (1992) Comparison of female to male and male to female transmission in 563 stable couples. Br Med J 304:809–813

Grimes DA, Cates W Jr (1990) Family planning and sexually transmitted diseases. 2nd edn. In: Holmes KK, Mardh PA, Sparling Pf (eds) Sexually transmitted diseases. McGraw-Hill, New York, pp 1087–1099

Hanna NF, Taylor-Robinson D, Kalodiki-Karamanoli M, Harris JRW, McFayden IR (1985) The relation between vaginal pH and the microbiological status in vaginitis. Brit J Obstet Gynaecol 92:1267–1271

Harrison HR, Costin M, Meder JB, Bownds LM, Sim DA, Lewis M, Alexander ER (1985) Cervical *Chlamydia trachomatis* infection in university women: relationship to history, contraception, ectopy, and cervicitis. Am J Obstet Gynecol 153:244–251

Hicks D (1995) Hormonal contraception and HIV. Br J Family Planning 20:103–104

Johnson AM, de Cock KM (1994) What's happening to AIDS. Br Med J 309:1523–1524

Kalo-Klein A, Witkin SS (1989) *Candida albicans*: cellular immune system interactions during different stages of the menstrual cycle. Am J Obstet Gynecol 161:1132–1136

Kjaer SK, Engholm G, Teisen C, Haugaard BJ, Lynge E, Christensen RB, Moller KA, Jensen H, Poll P, Vestergaard BF, de Villiers EM, Jensen OM (1990) Risk factors for cervical human papillomavirus and herpes simplex infections in Greenland and Denmark: a population-based study. Am J Epidemiol 131:169–182

Laga M, Manoka A, Kivuvu M, Malele B, Tuliza M, Nzila N, Goeman J, Behets F, Batter V, Alary M, Heyward WL, Ryder RW, Piot P (1993) Non-ulcerative sexually transmitted diseases as risk factors for HIV-1 transmission in women: results from a cohort study. AIDS 7:95–102

Lehtinen M, Dillner J, Knekt P, Luostarinen T, Aromaa A, Kirnbauer R, Koskela P, Paavonen J, Peto R, Schiller JT, Hakama M (1996) Serologically diagnosed infection with human papillomavirus type 16 and risk for subsequent development of cervical carcinoma: nested case-control study. Brit Med J 312:537–539

Louv WC, Austin H, Perlman J, Alexander WJ (1989) Oral contraceptive use and the risk of chlamydial and gonococcal infections. Am J Obstet Gynecol 160:396–402

Martinotti MG, Savoia D (1985) Effect of some steroid hormones on the growth of *Trichomonas vaginalis*. G Batteriol Virol Immunol 78:52–59

Munoz N, Bosch FX (1992) HPV and cervical neoplasia: review of case-control and cohort studies. IARC Sci Publ 119:251–261

Nelson AL (1997) The impact of contraceptive methods on the onset of symptomatic vulvovaginal candidiasis within the menstrual cycle. Am J Obstet Gynecol 176:1376–1380

Parazzini F, La Vecchia C, Negri E, Maggi R (1990) Oral contraceptive use and invasive cervical cancer. Int J Epidemiol 19:259–263

Pater A, Bayatpour M, Pater MA (1990) Oncogenic transformation by humane papillomavirus type 16 deoxyribonucleic acid in the presence of progesterone or progestins from oral contraceptives. Am J Obstet Gynecol 162:1099–1103

Plummer FA, Simonsen JN, Cameron DW, Ndinya-Achola JO, Kreiss JK, Gakinya MN, Waiyali P, Cheang M, Piot P, Ronald AR, Ngugi EN (1991) Co-factors in male-female sexual transmission of human immunodeficiency virus type 1. J Infect Dis 163:233–239

Rank RG, Barron AL (1987) Specific effect of estradiol on the genital mucosa antibody response in chlamydial ocular and genital infections. Infect Immunol 55:2317–2319

Ross JDC (1996) Is oral contraceptive associated with genital warts? Genitourin Med 72:330–333

Roy S (1991) Nonbarrier contraception and vaginitis and vaginosis. Am J Obstet Gynecol 165:S1240–S1244

Rubin GL, Ory HW, Layde PM (1982) Oral contraceptives and pelvic inflammatory disease. Am J Obstet Gynecol 144:630–635

Sciarra JJ (1997) Sexually transmitted diseases: global importance. Int J Gynecol Obstet 58:107–119

Segal E, Soroka A, Schechter A (1984) Correlative relationship between adherence of *Candida albicans* to human vaginal epithelial cells in vitro and candidal vaginitis. J Med Vet Mycol 22:191–200

Shah PN, Smith JR, Wells C, Barton SE, Kitchen VS, Steer PJ (1994) Menstrual symptoms in women infected by the human immunodeficiency virus. Obstet Gynecol 83:397–400

Sharma R, Pickering J, McCormack WM (1997) Trichomoniasis in a postmenopausal woman cured after discontinuation of estrogen replacement therapy. Sex Transm Dis 24:543–545

Smith EM, Johnson SR, Figuerres EJ, Mendoza M, Fedderson D, Haugen TH, Turek LP (1997) The frequency of human papillomavirus detection in postmenopausal women on hormone replacement therapy. Gynecol Oncol 65:441–446

Spinillo A, Capuzzo E, Nicola S, Baltaro F, Ferrari A, Monaco A (1995) The impact of oral contraception on vulvovaginal candidiasis. Contraception 51:293–297

Sweet RL (1985) Importance of differential diagnosis in acute vaginitis. Am J Obstet Gynecol 152:921–923

Taitel HF, Kafrissen ME (1995) A review of combined pill use and risk of HIV transmission. Br J Fam Plann 20:112–116

Toppozada M, Onsy FA, Fares E, Amir S, Shaala S (1979) The protective influence of progestogen only contraception against vaginal moniliasis. Contraception 20:99–102

Wallin J, Gnarpe H, Forsgren A (1974) Sexually transmitted diseases in women. Br J Vener Dis 50:217–221

Westrom L (1980) Incidence, prevalence, and trends of acute pelvic inflammatory disease and its consequences in industrialized countries. Am J Obstet Gynecol 138:880–892

Widy-Wirski R, Berkley S, Downing R, Okware S, Recine U, Mugerwa R, Lwegaba A, Sempala S (1988) Evaluation of the WHO clinical case definition of AIDS in Uganda. J Am Med Assoc 260:3286–3289

Wolner-Hanssen P, Eschenbach D, Paavonen J, Kiviat N, Stevens LE, Critchlow C, De Rouen T, Holmes KK (1990) Decreased risk of symptomatic chlamydial pelvic inflammatory disease associated with oral contraceptive use. J Am Med Assoc 263:54–59

Wolner-Hanssen P, Svensson L, Mardh PA, Westrom L (1985) Laparoscopic findings and contraceptive use in women with signs and symptoms suggestive of acute salpingitis. Obstet Gynecol 66:233–238

Part 7
Clinical Application and Potential of Estrogens and Antiestrogens

CHAPTER 38
Hormonal Contraception

H. KUHL

A. History

Ovulation inhibition by progestogens was postulated as early as 1921 by HABERLANDT. One year later, a similar action was ascribed to estrogens by FELLNER. In 1944, BICKENBACH and PAULIKOVICS succeeded in suppressing ovulation in a woman by means of daily injections of 20 mg progesterone. Hormonal contraception for all women, however, only became possible after orally active estrogens and progestogens had been developed. The basis for this revolution in birth control was brought about by BUTENANDT, INHOFFEN, HOHLWEG, MARKER, DJERASSI and many other scientists.

The first report of an ovulation-inhibiting effect of the synthetic progestogens, norethisterone and norethynodrel, was published in 1956 by ROCK, PINCUS and GARCIA. In 1958, the results of a large field study demonstrated that a combination of high doses of mestranol (ME) and norethynodrel effectively prevents pregnancy. The first oral contraceptive "Enovid" was approved in the USA in 1960 and contained 150 μg ME and 9.85 mg norethynodrel. "Anovlar" which was introduced in Germany 1 year later, was, at that time, regarded as a very low dose formulation (50 μg ethinylestradiol and 4 mg norethisterone acetate). Until today, ethinylestradiol has remained the estrogen of choice in oral contraceptives.

The results of some large prospective studies, initiated between 1968 and 1970 in England and USA, indicated that oral contraceptives, particularly their estrogen component, may be involved in the development of venous thromboembolic diseases and other complications and disorders. As a consequence, the dose of ethinylestradiol was considerably reduced and new progestogens were developed. The first sequential ovulation inhibitor was marketed in Germany in 1964, and the first triphasic formulation was introduced in 1979, also in Germany. The first estrogen-free preparation, the "mini-pill", which consists of a low-dose progestogen only, was introduced in 1965 in Mexico. In the same country, the first depot-progestogen became available 1 year later. Until this day, many hormonal systems and formulations have been developed that differ in their composition and route of administration, in their efficacy and risk profile. It can, however, be expected that ovulation inhibitors will remain the leading method among the hormonal contraceptives.

B. Types of Hormonal Contraceptives

I. Progestogen-Only Contraceptives

Women with contraindications for ethinylestradiol (EE) or women who do not tolerate estrogen containing formulations may use progestogen-only preparations or non-hormonal methods. The hormonal and non-hormonal methods differ largely in their practicability, efficacy and risks.

1. Mini-pill

Among the progestogen-only preparations, the mini-pill exerts the least impact on metabolism and the least contraceptive efficacy, although it is taken continuously without a hormone-free interval (Table 1). It consists of a progestogen at a very low dose, the effect of which lasts for merely 24h. Therefore, the mini-pill has to be taken regularly at the same time of day. The contraceptive efficiency of the mini-pill is based on its progestogenic effect on cervical mucus, endometrium and tubal function, which impairs or prevents migration and capacitation of sperms, synchronous transport of the zygote and

Table 1. Failure rate during the first year of use, reversibility and health risk of various contraceptive methods. (Modified from MCCANN and POTTER 1994)

Method	PI-1	PI-2	Reversibility	Risk
No method	85	85		
Female sterilization	0.2	0.4	Limited	Very low
Male sterilization	0.1	0.15	Limited	Very low
Ovulation inhibitor	0.1	1	Yes	Low
Mini-pill	0.5	3	Yes	Very low
Depot-progestogen (MPA)	0.3	0.3	Yes	Very low
Depot-progestogen (NETE)		1.5	Yes	Very low
Norplant	0.3	0.3	Yes	Very low
Intrauterine device (Cu)	0.6	0.8	Yes	Low
Intrauterine device (LNG)	0.1	0.1	Yes	Very low
Diaphragm + spermicide	6	18	Yes	None
Cervical cap (parous women)	9	28	Yes	None
Cervical cap (nulliparous women)	6	18	Yes	None
Spermicide	6	21	Yes	None
Sponge (parous women)	20	36	Yes	None
Sponge (nulliparous women)	9	18	Yes	None
Female condom	5	21	Yes	None
Male condom	3	12	Yes	None
Coitus interruptus	4	19	Yes	None
Periodic abstinence		20	Yes	None
Calendar method	9	–	–	–
Basal temperature	3	–	–	–
Sympto-thermal	2	–	–	–

PI-1, Pearl-Index when correctly used; PI-2, Pearl-Index in reality; MPA, medroxyprogesterone acetate; NETE, norethisterone enanthate; Cu, copper; LNG, levonorgestrel implantation (McCann and Potter 1994). In addition, in about one-third of the women, ovulation is inhibited. In young women, the Pearl Index is approximately 3 (Table 1), but is improved in older women (Vessey et al. 1985). The mini-pill is the method of choice for contraception during lactation, as it does not adversely influence infant growth and development (World Health Organization 1994a, b). It should not be used by patients with endometriosis, uterine myoma or mastopathia, as frequently high endogenous estradiol levels can be observed. The relative risk of ectopic pregnancy is increased; therefore, it is contraindicated in women with a history of tubal pregnancy.

2. Depot-Progestogens

The high effectiveness of depot-medroxyprogesterone acetate (depot-MPA), the norplant systems or the levonorgestrel containing intrauterine device (LNG-IUD) is based on the permanent and relatively even influence of a progestogen which is ensured by the hormone depot. The intramuscular injection of depot-MPA (150 mg) effectively suppresses ovulation for at least 18 weeks and, after a transitory period of irregular bleeding, causes amenorrhea during consecutive injections every 3 months. The ovulation-inhibiting action of norethisterone enanthate, however, lasts for only 6–8 weeks after the injection and, thereafter, the contraceptive effect depends on the peripheral progestogenic changes caused in the cervical mucus, tubes and endometrium (Population Reports 1995). In order to increase efficacy, the first injections of norethisterone enanthate (200 mg) are carried out every 2 months and, thereafter, every 3 months.

After the subdermal implantation of Silastic capsules or rods containing levonorgestrel (Norplant), a low but relatively even serum concentration of the progestogen is achieved which causes a reliable contraception for 3–5 years, based on the peripheral effects on cervix, tubes and endometrium (Population Reports 1992). In addition, in half of the cycles, ovulation is inhibited. After insertion of the LNG-IUD, the strong local effect of LNG on the endometrium and cervical mucus warrants a highly effective contraception, while the systemic effect of the progestogen is negligible.

II. Estrogen/Progestogen-Containing Contraceptives

1. Post-coital Pill

In the case of a suspected failure of contraception, an emergency measure may become necessary. Previously it was shown that the use of high doses of estrogens (e.g., 5 mg ethinylestradiol for 5 days) may prevent implantation. As this method was frequently associated with gastrointestinal side effects and bleeding, another regimen was investigated and proved to be effective. The intake of two tablets containing 50 μg EE and 250 μg LNG, not later than 72 h after intercourse, and another two tablets of this high dose ovulation inhibitor (total dose 200 μg EE + 1 mg LNG) 12 h later, prevents implantation in 98% of cases

(YUZPE et al. 1982). This "morning-after" or post-coital pill is not suitable as a normal contraceptive method, as the following cycle may be disturbed. The mechanism of action is not clear, but it is suggested that tubal function and endometrium are profoundly affected, resulting in the prevention of implantation (HASPELS 1994). The side effects, which are less severe than those after use of high-dose estrogens, are nausea, vomiting, breast tenderness, dizziness and headaches. It is also recommended to use an anti-emetic. The patient should be followed up for 3 weeks in order to assess the result. If the time limit of 72 h is passed, the insertion of a copper intrauterine device within 5 days after intercourse may also prevent implantation.

2. Ovulation Inhibitors

Ovulation inhibition can be achieved by means of a prolonged influence of sufficient concentrations of a potent estrogen and/or a potent progestogen. Therefore, the contraceptive efficacy is essentially dependent on correct use and good compliance. Ovulation inhibitors, generally called "oral contraceptives" (OC) represent a combination of an estrogen, EE or ME, with a progestogen which is taken orally, daily, for 21 days or 22 days of the 28-day cycle (Fig. 1). There are monophasic and bi- or triphasic (two- or tri-step formulations) combinations with varying doses of the estrogen and/or the progestogen component. The sequential type of OC consists of a first short phase, with estrogen only, and a second phase with a combination of EE and progestogen. Ovulation inhibitors are not only very reliable and easy-to-use contraceptives, but may influence many organs, tissues and metabolic systems, and may also cause adverse effects.

The modern OC consists of EE at a relatively low dose and a potent progestogen. Formulations containing 35 μg or less EE are called "micro-pills", irrespective of the progestogen component. The doses of the various progestogens used in OC depend on their hormonal potency. The impact on metabolism of different progestogens cannot be compared on the basis of the respective doses, as combinations with a low-dose potent progestogen may be more reliable, although may also cause more adverse effects than those with a less-potent compound at a higher dose.

The monophasic combinations represent the most effective OC, as the progestogen components exert their various contraceptive effects from the first day of the pill cycle. They consist of 21 tablets of the estrogen/progestogen combination at a fixed dose. The most effective preparations are those containing the progestogen at a dose which is about twice that of the respective ovulation inhibition dose (Table 2) (KUHL 1996a). The use of formulations with 23 tablets per cycle, i.e., the shortening of the hormone-free interval to only 5 days, will enhance contraceptive efficacy.

The bi- and triphasic OC consists of an estrogen and a progestogen which are combined at two or three different dosage combinations. Biphasic formulations may contain a constant estrogen dose, while the progestogen dose is

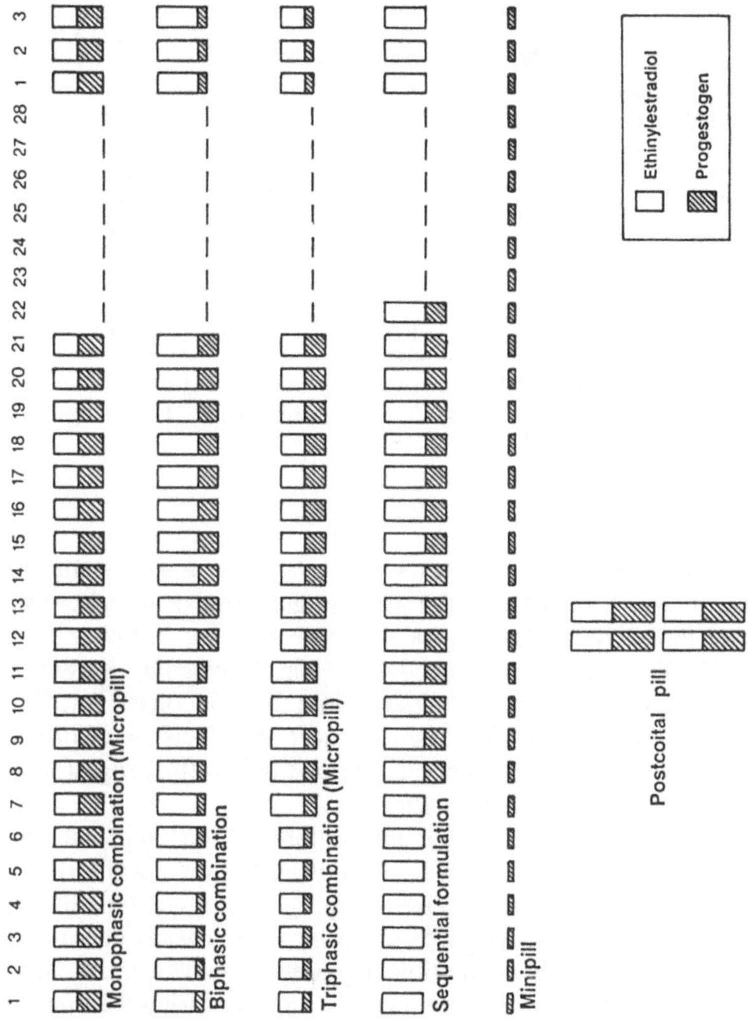

Fig. 1. Graphic representation of various types of hormonal contraceptives

Table 2. Hormonal potency of progestogens

Progestogen	T-Dose mg/cycle	OI-Dose mg/day	Dose (COMB)	Dose (PHAS)	AA (%)
Cyproterone acetate	12	1.0	2.0		100
Chlormadinone acetate	25	1.7	2.0	1.0	30
Dienogest	6	1.0	2.0		33
Norgestimate	7	0.2	0.25	0.18	
Levonorgestrel	4	0.06	0.125	0.05	
Norethisterone	100	0.4	0.5	0.5	
Norethisterone acetate	50	0.5	0.6		
Lynestrenol	70	2.0	0.75		
Desogestrel	2	0.06	0.150	0.125	
Gestoden	3	0.03	0.075	0.05	

T-Dose, transformation dose; OI-Dose, ovulation inhibition dose without estrogen; AA, relative anti-androgenic potency; COMB, combined OC; PHAS, first phase of bi- or triphasic OC

lower during the first phase. Triphasic OCs may consist of a constant estrogen dose and a step-wise increase in the progestogen dose, or a higher estrogen dose during the second phase and a step-wise increase in the progestogen doses (Fig. 1). Obviously, there is no advantage in the bleeding pattern or contraceptive efficacy of phasic pills compared with the monophasic formulations. From a theoretical point of view, the contraceptive efficiency of phasic pills must be less than those of monophasic formulations, as the progestogen dose is lower during the first week of intake.

This also holds true for the sequential formulations, the first phase of which consists of six or seven tablets with 50 μg EE only, while the second phase of 15 tablets represents an estrogen/progestogen combination. As the treatment with EE only at a dose of 50 μg during the first week does not inhibit follicular maturation reliably, ovulation may occur in some women. Moreover, there is no progestogenic influence on cervical mucus during the first week of use. On the other hand, the unopposed action of EE on the endometrium during the first phase allows the full proliferation of the endometrium, while the subsequently ingested estrogen/progestogen combination causes a secretory transformation that is similar to that during the luteal phase of an ovulatory cycle. Women taking sequential formulations show very good cycle control; therefore, this type of OC is the treatment of choice for women who suffer from irregular bleeding when taking combination or phasic pills. The first sequential OC which was introduced in the USA in 1963, was removed from the market in 1976 as a consequence of increased reports of endometrial carcinoma. This adverse effect was due to the short estrogen/progestogen phase of only 5–6 days. The modern sequential formulations which exert a progestogenic effect for 15 days per cycle protect from the development of endometrial hyperplasia and carcinoma.

C. Pharmacology of Contraceptive Steroids

I. Pharmacology of Estrogens

Since the introduction of the first OC, EE has remained the only active estrogen component. ME, which was contained in the first formulations and is still used in only a few OCs, represents the 3-methyl-ether of EE and becomes active after hydrolysis to EE (Fig. 2). As this transformation occurs rapidly, after intake in the intestinal tract and the liver, the pharmacokinetics of ME is similar to that of EE. Some other derivatives of EE, e.g., the 3-cyclopentylether (Quinestrol) or the 3-isopropylsulfonic acid ester (Ethinylestradiolsulfonate) which exert a depot-effect, are used in combination with a progestogen in the so-called once-a-month pill or once-a-week pill, respectively. These compounds also become active after hydrolysis to EE (Fig. 2).

EE is used in OCs, because it is also very potent after oral application and can be used at very low doses. The high potency is based on the ethinyl group at C17α which blocks the oxidative inactivation of estradiol into estrone. The binding affinity of EE for the estrogen receptor is similar to that of estradiol. After oral application, the bioavailability of EE is 38–48%, and the peak serum concentrations of EE are reached within 1–3 h. During daily intake, the EE levels increase up to a steady state which is reached after about 1 week. At a

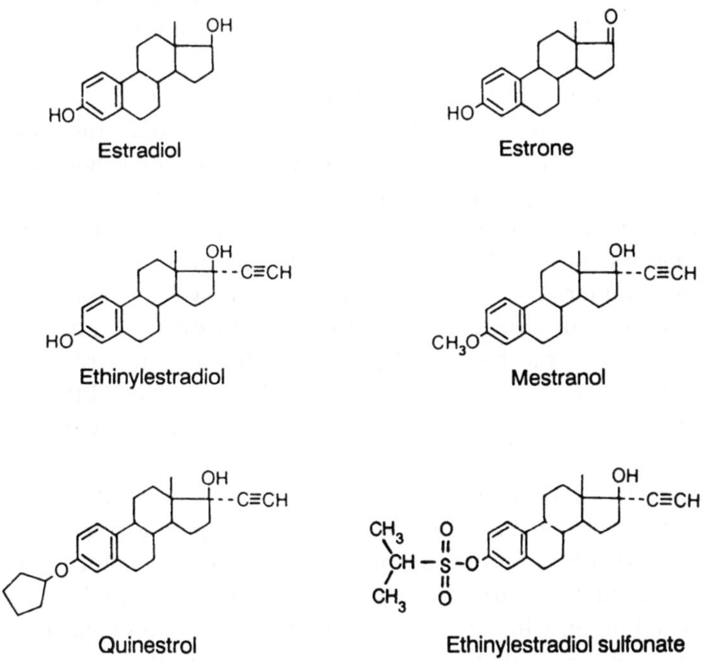

Fig. 2. Structural formulae of estrogens

dose of 30 μg, the maximal serum levels of EE are 100 pg/ml, and with 50 μg, a level of 150 pg/ml may be reached, on average. There are large intra- and interindividual variations both in the EE levels and in the physiological response during treatment with the same EE dose. In the serum, EE which has no affinity for sex hormone-binding globulin (SHBG) is bound to albumin.

EE has a proliferative effect on the epithelium of endometrium, tubes, vagina and urethra, and stimulates the production of cervical mucus. In the breast, it promotes ductal proliferation and alveolar branching. EE causes vasodilation and increases blood flow, stimulates collagen synthesis and water retention and inhibits bone resorption and gonadotropin release. It stimulates hepatic protein synthesis and causes alterations of numerous metabolic serum parameters. The effects of EE are dose-dependent; even at a dose as low as 5 μg, EE may change several serum parameters, but at doses above 20 μg, the increments become continuously smaller (MASHCHAK et al. 1982; MANDEL et al. 1982). Although the hepatic effect of EE is much stronger than that of estradiol, its proliferative effect on the endometrium is less than that of the natural estrogen (BROSENS and PIJNENBORG 1976), provided that no progestogen is present.

II. Pharmacology of Progestogens

The two types of progestogens used in OCs, are the derivatives of progesterone and of 19-nortestosterone. Those compounds which do not possess a 4-ene-3-keto group, represent pro-drugs, which are rapidly transformed in the intestinal tract and the liver into the active progestogen (Fig. 3). The progestogens used in OCs represent potent antagonists of EE that inhibit estrogen-induced proliferation and cause secretory transformation of the endometrium. They increase basal temperature by about 0.5°C and reduce tubal motility and cervical mucus even at low doses. According to their pharmacological properties, the progestogens may influence hepatic metabolism and may counteract the effects of EE. As the various progestogens may bind not only to the progesterone receptor, but also to a different extent to the androgen, glucocorticoid and mineralocorticoid receptor, they may exert some androgenic, anti-androgenic, glucocorticoid or anti-mineralocorticoid effects (Table 3). Therefore, the dose and hormonal pattern of the various progestogens and, consequently, the resulting effects of the combination with different doses of EE determine the tolerability and susceptibility of an individual woman to adverse effects.

Within 1–3 h after intake, the peak serum levels of the active progestogens are reached, which increase during daily intake up to a steady state (KUHL 1990a; STANCZYK 1997). In the serum, most of the nortestosterone derivatives are bound with high affinity/low capacity to SHBG and with low affinity/high capacity to albumin, while the remaining compounds and the progesterone derivatives are bound only to albumin. Similar to EE, the serum levels of the progestogens show large intra- and interindividual variations (KUHL 1990a).

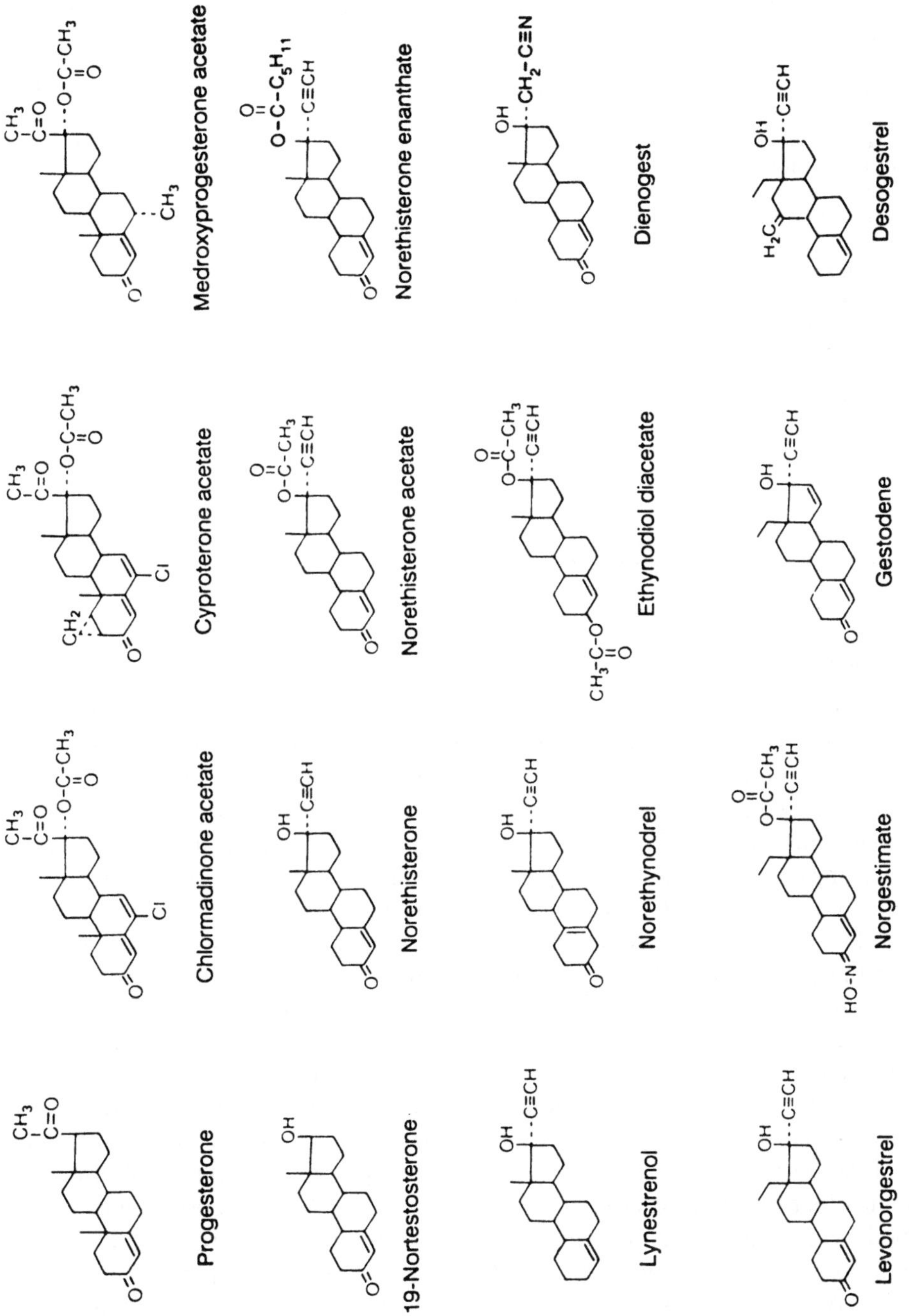

Fig. 3. Structural formulae of progestogens

Table 3. Spectrum of hormonal effects of progestogens (mostly evaluated by animal experiments)

Progestogen	E	AE	A	AA	GC	AMC
Progesterone	–	+	–	(+)	(+)	+
Chlormadinone acetate	–	+	–	+	+	–
Cyproterone acetate	–	+	–	+	+	–
Medroxyprogesterone acetate	–	+	(+)	–	+	–
Dienogest	–	+	–	+	–	–
Norethisterone	(+)	+	+	–	–	–
Norethisterone acetate	(+)	+	+	–	–	–
Lynestrenol	(+)	+	+	–	–	–
Ethynodiol diacetate	(+)	+	+	–	–	–
Norethynodrel	+	+	+	–	–	–
Levonorgestrel	–	+	+	–	–	–
Norgestimate	–	+	+	–		
Desogestrel (DG)	–	+	+	–		–
Gestodene (GSD)	–	+	+	–	(+)	+

E, estrogenic; AE, anti-estrogenic; A, androgenic; AA, anti-androgenic; GC, glucocorticoid; AMC, anti-mineralocorticoid effect

1. Nortestosterone Derivatives

The high oral potency of nortestosterone derivatives depends on the ethinyl group at C17α which may inhibit enzymatic inactivation of the steroid hormones by reductases and monooxidases. Therefore, this type of progestogen, which possesses some androgenic properties, exerts a pronounced effect on hepatic metabolism. The nortestosterone derivatives can be divided into two subgroups. The estranes (13-methyl-gonanes) comprise of norethisterone (NET), lynestrenol (LYN), ethynodiol diacetate (ETY) and norethynodrel (NYD) (Fig. 3). NET was the first orally active progestogen used for hormonal contraception, while LYN, ETY and NYD are pro-drugs, which become active after transformation into NET (Kuhl 1990a). The so-called gonanes (13-ethyl-gonanes) include LNG, desogestrel (DG), norgestimate (NGM), and gestodene (GSD) (Fig. 3). The introduction of the ethyl group at C13 instead of a methyl group led to a pronounced enhancement of hormonal potency, as exemplified by NET and LNG (Table 2). Norgestrel, which consists of LNG and the hormonal inactive stereoisomer dextronorgestrel (1:1), is still used in some older formulations (Stanczyk and Roy 1990). A special position is occupied by dienogest (DNG), which belongs to the nortestosterone derivatives, but has no ethinyl group and no androgenic, but even anti-androgenic, properties. As DNG has a short half-life, the daily dosage must be higher than that of the other nortestosterone derivatives (Kuhl 1996a).

DG and NGM are prodrugs which are transformed into 3-keto-DG and LNG, respectively. Except DG, NGM and DNG, nortestosterone derivatives including 3-keto-DG are bound to SHBG. The main metabolic pathways of nortestosterone derivatives are the reduction of the 4-ene-3-keto group and hydroxylation reactions at various positions of the steroid.

The effects of the progestogens on different organs and tissues may diverge. This is exemplified by DNG which has a pronounced effect on the endometrium comparable with that of NGM or LNG. On the other hand, the ovulation-inhibiting potency of DNG is much less than that of NGM or LNG, and is comparable to that of progesterone derivatives (Table 2).

2. Progesterone Derivatives

As progesterone is rapidly inactivated after oral intake, it is not suitable for oral contraception. By introducing substituents, such as methyl or chloro groups, into the progesterone molecule, the reductive metabolism at ring A and at C20 is slowed down. The lack of an ethinyl group explains not only the lower hormonal potency of progesterone derivatives, but also their lower impact on hepatic metabolism. Among the progesterone derivatives, only chlormadinone acetate (CMA), cyproterone acetate (CPA) and medroxyprogesterone acetate (MPA) are used for hormonal contraception (Fig. 3). Due to their anti-androgenic properties, CMA and particularly CPA are used in OCs for treatment of androgenic symptoms, while MPA is used as a depot-progestogen.

D. Mode of Action of Oral Contraceptives

The high contraceptive efficiency of OCs is based on their profound influence on many target organs involved in the events leading to conception and pregnancy. The estrogen and progestogen component may act synergistically or antagonistically. The effect of the progestogen is largely dependent on the presence of a potent estrogen which induces the synthesis of the progesterone receptor. In contrast, in the endometrium and other organs, progestogens may reduce the binding capacity of estrogen receptors and may decrease the local estrogen concentration by enhancing degradation of estrogens.

The inhibition of ovulation is brought about by an early disturbance of the follicle-stimulating hormone (FSH) and luteinizing hormone (LH) secretion pattern and a direct interaction of EE and the respective progestogen with ovarian processes, which lead to an impairment of follicular development (KUHL 1996a). If this mechanism was not sufficient for the prevention of follicular maturation, the pre-ovulatory LH surge and, hence, ovulation would be prevented by the progestogen component. In most women, the reduction in gonadotropin release results in a suppression of ovarian estradiol and testosterone production. In a certain proportion of women, follicular maturation begins during the hormone-free week and continues during the first days of pill intake (KUHL et al. 1985b). In these women, relatively high serum levels of estradiol can be measured, even though ovulation is reliably suppressed, provided that no omission of tablets has occurred. As follicular development may start during the pill-free week, the first week of pill intake is the most crucial period of time with respect to contraceptive effectiveness. The risk of ovulation and undesired pregnancy is highest, if omission of tablets occurs during the first week of the pill cycle, i.e., if the hormone-free interval is prolonged. Even though, in a normal 28-day cycle, ovulation occurs around day 15, on average, in a certain proportion of women it may take place some days earlier.

Both EE and the synthetic progestogen may impair tubal function, transport of the zygote, composition of tubal secrets and endometrial function, the changes of which are synchronized by the cyclic course of endogenous estradiol and progesterone levels. Using combination formulations, the early action of the progestogens inhibits estrogen-dependent proliferation of the endometrium and causes premature secretory transformation, resulting in an endometrium which is not receptive for implantation. The progestogen also counteracts the effect of the estrogen on cervical mucus and impairs penetration and ascension of sperms. With respect to the effects on the endometrium and cervix, the effect of the progestogen component overweighs that of the estrogen.

In addition to its contribution to the contraceptive effect, the estrogen component plays an important role in the maintenance of regular bleeding. In the presence of a potent progestogen, the intracellular inactivation of estradiol in the endometrium is profoundly enhanced. Therefore, the use of estradiol in oral contraceptives is associated with a high rate of inter-menstrual bleeding. In contrast, EE which is resistant to the action of the 17β-hydroxysteroid dehydrogenase, shows a sufficient endometrial activity and allows relatively stable cycles when combined with potent progestogens.

On the other hand, EE is a potent estrogen which may alter hepatic metabolism and cause changes in many metabolic parameters. In predisposed women, this impact may lead to the development of metabolic disturbances and various diseases.

E. Pharmacokinetics of Oral Contraceptives

I. Factors Influencing Efficacy

The effectiveness of OCs depends on sufficiently high serum concentrations of EE and the respective progestogen. Although there are large inter-individual variations in the serum levels of the synthetic sex steroids, the dosages are sufficient for all women. The levels are not dependent on body weight, but primarily on the metabolic capacity of the intestinal tract and liver. Therefore, the inter-individual differences in the EE and progestogen levels are due to genetic or acquired predisposition. There are also large intra-individual variations from day to day, which may be caused by diet, drugs or environmental factors (Jung-Hoffmann and Kuhl 1990).

During treatment with OCs, the maximal serum concentrations are reached within 1–4 h after intake of the tablet; thereafter, a rapid decline of the serum levels can be observed. During the phase of elimination, the serum concentration of EE is also influenced by the enterohepatic circulation; a certain proportion of EE-conjugates is split by the bacterial flora in the colon, and the reabsorbed EE contributes to the circulating EE. This does not hold true for the progestogens, as their metabolites are hormonally inactive.

The absorption of EE and the progestogens in the small bowel is completed within 2–3 h after intake of a tablet. Therefore, vomiting may impair contraceptive efficacy but only during the first 2–3 h after ingestion. A possible influence of diarrhea depends on the severity of the disease. Generally, only the colon is affected, and this may impair only the enterohepatic circulation of EE. As this mechanism does not concern the progestogen, diarrhea usually does not influence the contraceptive action of formulations with a potent progestogen component, but may cause inter-menstrual bleeding. It should, however, be kept in mind that nausea and illness may lead to an omission of tablets.

In general, the effectiveness of OCs is not impaired in women with inflammatory bowel diseases, e.g., ulcerative colitis, Crohn's disease, cystic fibrosis or celiac disease. While diet may influence the pharmacokinetics of contraceptive steroids, no significant effect of smoking could be demonstrated. Chronic alcohol abuse, but not short-term alcohol drinking, may reduce the contraceptive efficacy of OCs.

II. Interaction of Oral Contraceptives with Drugs

There are numerous reports of irregular bleeding and undesired pregnancies in women who were concomitantly treated with OCs and certain drugs (Back et al. 1988; Szoka and Edgren 1988; Back and Orme 1990). Although in many cases a causal relationship is improbable, an interaction cannot be excluded with certainty, as there are large inter-individual variations in the serum levels of sex steroids and in the metabolic response to treatment with OCs and drugs. A possible interaction with the pharmacokinetics of contraceptive steroids has been investigated only for some drugs, and mostly in small groups of volunteers. As women with chronic diseases often need a lifelong therapy, and as some of the drugs are potentially teratogenic, the knowledge of possible interferences with the contraceptive reliability of OCs is essential.

There are two main pathways leading to a decrease in the serum levels of contraceptive steroids. An enzyme induction induced by drugs may cause an enhancement of hepatic metabolism of both EE and the progestogens. The most important enzyme inducers that may impair efficiency of OCs are rifampicine, griseofulvine, anti-epileptics such as phenobarbital, phenytoin. primidone, carbamazepine and oxcarbazepine, and several other tranquilizers (Table 4). An interruption of the enterohepatic circulation of EE by antibiotics may reduce EE, but not progestogen levels. The latter may be of importance

Table 4. Interaction of drugs with the effectiveness of oral contraceptives

Drug	Effectiveness
Antibiotics, sulphonamides	
Rifampicin	RP, EM
Isoniazid	RP %
Ampicillin	RP
Phenoxymethylpenicillin	RP
Oxacillin	RP
Amoxicillin	RP
Tetracycline	RP
Cotrimoxazole	(RP), IM
Cefalexin	RP, (EM)
Chloramphenicol	RP
Neomycin	RP
Eythromycin	(RP), (IM)
Fusidic acid	RP
Sulphamethoxazole	RP
Sulphmethoxypyridazine	RP
Sulfisoxazole	RP
Dapsone	(RP), IM
Nitrofurantoin	RP
Metronidazole	RP
Clarithromycin	%
Roxithromycin	%
Doxycycline	%
Temafloxazin	%
Ciprofloxazin	%
Dirithromycin	EM
Triacetyloleandomycin	IM
Primaquine	%
Chloroquine	%
Ofloxacin	%
Vitamins	
Vitamin C	%
Isotretinoin	%
Tretinoin derivatives	%
Anti-epileptics	
Phenobarbital	RP, EM
Methylphenobarbital	RP, EM
Primidone	RP, EM
Phenytoin	RP, EM
Carbamazepine	RP, EM
Ethosuximide	(RP) %
Methosuximide	RP
Sodium valproate	(RP) %
Oxcarbazepine	%
Analgesics, anti-inflammatory agents	
Phenylbutazone	RP, EM
Oxyphenbutazone	RP
Aspirin	RP
Phenazone (Anti-pyrine)	RP
Aminophenazone	RP
Phenacetin	RP

Table 4. (*Continued*)

Drug	Effectiveness
Aminopyrine	RP
Paracetamol	(RP), %
Ibuprofen	%
Anti-histamines, H2-Antagonists	
Cimetidine	%
Roxatidine	%
Ranitidine	%
Diphenhydramine	(EM)
Adsorbents, antacids	
Aluminium hydroxide	%
Magnesium trisilicate	%
Kaolin	%
Activated charcoal	%
Tranquillisers, neuroleptics, sedatives	
Barbital	RP, EM
Phenobarbital	RP, EM
Promethazine	RP, EM
Chlorpromazine	RP, EM
Diazepam (Valium)	RP, EM
Chlordiazepoxide	RP, EM
Clorazepate	RP, EM
Meprobamate	RP, EM
Triflupromazine	(EM)
Prazepam	EM
Alprazolam	EM
Nordazepam	EM
Temazepam	EM
Oxazepam	%
Lorazepam	%
Diphenhydramine	(EM)
Anti-depressants	
Imipramine	(EM)
Tolbutamide	%
Carbutamide	%
Antidiabetic agents	
Tolbutamide	%
Carbutamide	%
Immunosuppressants	
Cyclosporin	IM
Anti-fungal agents	
Griseofulvin	RP, EM
Ketoconazole	IM
Terbinafine	IM
Diuretics	
Metyrapone	(IM)

RP, reported pregnancies; EM, enhancement of metabolism and attenuation of efficacy; IM, inhibition of metabolism and enhancement of efficacy; %, no influence on hepatic metabolism of contraceptive steroids when OC formulations are used that contain a less potent progestogen component, or a very low or no progestogen dose during the first week of intake, e.g., triphasic or sequential OCs. The problem of antibiotic interaction is complicated by the fact that within several days of antibiotic therapy, resistant bacterial flora develops in the colon, which normalizes enterohepatic circulation. A large proportion of the population might already have an intestinal bacterial flora that is resistant to the commonly used antibiotics. There are some drugs that may modulate conjugation and metabolism of steroids in the small bowel, and there are others that may inhibit hepatic metabolism and, hence, enhance the effects and side effects of OCs (Table 4).

However, OCs may alter the effects of drugs by inhibition of drug metabolism, or by enhancing conjugation of drugs. As the therapeutic window of most drugs is large, this effect is of minor clinical relevance. It has, however, been shown that OCs may considerably increase the extent of side effects of imipramine due to the action of EE. The estrogen component may also enhance the conjugation of some drugs and reduce their effectiveness. This is exemplified by the inhibition of warfarin metabolism and the enhancement of phenprocoumone conjugation (Table 5). In such cases, the effectiveness of the drugs may be adapted by changing the drug dose.

Table 5. Interaction of oral contraceptives with the efficacy of drugs

Drug	Efficacy
Antibiotics	
Ampicillin	%
Tetracycline	%
Metronidazole	%
Isoniazid	%
Triacetyloleandomycin	IM
Quinine	%
Mefloquine	%
Analgesics, anti-inflammtory agents	
Phenylbutazone	IM
Phenazone (Anti-pyrine)	IM
Aminopyrine	IM
Metamizol	IM
Phenacetin	IM
Meperidine (Pethidine)	IM
Paracetamol	EC
Diflunisal	EC
Aspirin	EC
Salicylic acid	EC
Morphine	EC
Ethylmorphine	IM
Corticosteroids	
Cortisol	IM
Prednisolone	IM
Fluocortolone	%

Table 5. (*Continued*)

Drug	Efficacy
Tranquilizers, neuroleptics, sedatives	
Diazepam	IM
Alprazolam	IM
Triazolam	IM
Nitrazepam	IM
Chlordiazepoxide	IM
Bromazepam	%
Clotiazepam	%
Lorazepam	EC
Temazepam	EC
Midazolam	%
Doxylamine	%
Diphenhydramine	%
Anti-arrhythmic agents, β-adrenoceptor blocking agents	
Chinidine	(IM)
Lidocaine	(IM)
Propranolol	IM,EC
Oxprenolol	(IM)
Metoprolol	(IM)
Calcium antagonists, cardiac inotropic agents	
Nifedipine	(IM)
Digitoxin	(IM)
Anti-epileptics	
Phenytoin	IM
Anti-depressants	
Imipramine	IM
Clomipramine	%
Xanthines	
Caffeine	IM
Theophylline	IM
Immunosuppressants	
Cyclosporin	IM
Anti-coagulants	
Warfarin	IM
Phenprocoumon	EC
Anti-diabetic agents	
Tolbutamide	IM
H_2-Antagonists	
Cimetidine	%
Doxylamine	%
Lansoprazol	%
Diphenhydramin	%
Vitamins	
Vitamin A	IM
Lipid regulating agents	
Clofibrate	EC

IM, inhibition of metabolism and enhancement of efficacy; EC, enhancement of conjugation and attenuation of efficacy; %, no influence on metabolism of drugs

F. Use of Oral Contraceptives

I. Choice of Oral Contraceptives

Before an OC is prescribed, an internal and gynecological examination must be carried out. A careful anamnesis and family history may help to recognize risks or predisposition for the development of thromboembolic or other diseases. Possible absolute and relative contraindications must be taken into consideration (Table 6). Hormone determinations are worthless except in women

Table 6. Absolute and relative contraindications to the use of oral contraceptives

Absolute contraindications
Acute and chronic progressive liver disease
Impairment of bile secretion, cholestatic jaundice (also a history of)
Hemolytic uremic syndrome
Previous or current thrombo-embolic disease (venous thrombosis, stroke, myocardial infarction)
Micro- or macroangiopathia
Hereditary thrombophilia
Lupus erythematosus
Vasculitis
Anti-phospholipid antibodies
Circulatory disorders
Severe hypertension
Diabetes mellitus with angiopathia
Hyperhomocysteinemia
Severe hypertriglyceridemia
Breast cancer
Undiagnosed vaginal bleeding
Pregnancy
Relative contraindications
Liver disease (e.g., porphyria)
Gall-bladder disease
Disorders of lipid metabolism
Diabetes mellitus
Disorders of hemostasis
Vascular lesions
Cardiac or renal dysfunction, edema
Cardiac surgery
Angina pectoris
Previous or current thrombophlebitis
Smoking
Hypertension
Obesity
Lactation
Mastopathia with epithelial atypia
Uterine myoma
Elective surgery with major thrombo-embolic risk
Long-term immobilization
Endometrial cancer
Cervical cancer
Migraine

with androgenic symptoms. Biochemical laboratory parameters should be measured only when indicated. First time users should be reassessed after 3–4 months of use. Thereafter, monitoring is necessary only every 12 months and, in the case of any risk factors, every 6 months. During monitoring, it is important to look for the onset of new risk factors or warning signs. Regular measurement of blood pressure, cervical cytology and breast examination is recommended.

The choice of the formulation depends on the individual disposition and situation. As most side effects are dependent on the EE dose, a low-dose combination formulation should be prescribed for first-time users, i.e., preparations containing 35 μg EE or less. The doses of the various progestogens cannot be compared with respect to the metabolic impact, as their potencies differ largely. A comparison of the dose used with the ovulation-inhibition dose of the respective progestogen may give a reference point for efficacy. Oral contraception should be started on the first day of menses in order to increase contraceptive effectiveness. Thereafter, the pills must be taken according to instructions for 21–22 days, followed by a hormone-free interval of 6–7 days. During this time, withdrawal bleeding takes place. The next package of pills should be started immediately after this time. In preparations with 28 pills, the last seven tablets contain a placebo in order to facilitate correct use.

Omission of tablets may impair contraceptive efficacy and lead to undesired pregnancies, particularly during the first week of intake. If one tablet is missed, the woman should take that pill as soon as possible and take the following tablet as usual. If two tablets are missed, two pills on each of the following 2 days should be taken. If more than two tablets are missed, alternative contraceptive measures, e.g., condom or diaphragm plus spermicides, should be used additionally.

It is important to inform the patient of the possible occurrence of intermenstrual bleeding and other side effects which may occur more frequently during the first treatment cycles. If, thereafter, side effects continue and cannot be tolerated, change to another preparation should be taken into consideration. In the case of certain symptoms indicating serious diseases, pill intake must be stopped immediately (Table 7).

II. Metabolic Effects of Oral Contraceptives

OCs may exert a pronounced influence on many metabolic and clinical laboratory parameters. In most cases, the estrogen component acts in a stimulatory manner, i.e., it increases the serum concentrations of most parameters, while the progestogen components show mostly a modulatory or antagonistic effect. In general, the effects are dose- and time-dependent. Some of the alterations induced by OCs are highest during the first weeks of intake and are attenuated during long-term treatment, while other parameters increase continuously during several months before reaching a steady state. Therefore, the estimation of long-term effects and risks, on the basis of the results of short-term studies, might be questionable.

Table 7. Reasons for immediate discontinuation of oral contraceptives

Pregnancy
First-time occurrence or aggravation of migraine or severe headaches
Transient cerebral attacks (e.g., speech disorder, numb feeling)
Acute blurred vision
Severe pain in the chest which increases at breathing (myocardial infarction)
Unilateral severe leg pain (thrombosis)
Pain in the chest, unclear difficulty of breathing, hemoptysis (pulmonary embolism)
Thrombophlebitis
Cholestatic jaundice
Epigastric pain (liver disease, gall-stones, thrombosis)
Marked rise of blood pressure
Severe generalized cutaneous eruption (erythema multiforme)
Growth of existing uterine myoma
4–6 Weeks before an elective surgery
Long-term immobilization (e.g., after an accident)

1. The Liver

Orally taken estrogens and progestogens exert a considerably stronger effect on the liver than parenterally applied sex steroids. This is due to the high steroid concentration of serum during the first liver passage, which is about 4-fold that of the peripheral levels (BACK et al. 1982). The liver also differs from other organs, as the higher permeability of hepatic vascular beds and the larger surface area of the liver's microvasculature cause a stronger influx of sex steroids than other organs (STEINGOLD et al. 1986). Binding to albumin, which is the only carrier of EE in serum, does not or only marginally influence uptake of the synthetic estrogens by hepatocytes (STEINGOLD et al. 1986). Moreover, synthetic sex steroids used in OCs elicit a much stronger effect on hepatic metabolism than natural sex steroids, as their inactivation is retarded and the local concentration in the hepatocytes is probably higher. Among the progestogens, nortestosterone derivatives containing an ethinyl group probably have a more pronounced impact on hepatic metabolism than dienogest and progesterone derivatives. Therefore, the effect of OCs on hepatic factors is much more pronounced than that on non-hepatic parameters, and the ratio between hepatic and peripheral activity of EE is much higher than that of natural estrogens (MASHCHAK et al. 1982).

The liver is one of the most important target organs of estrogens, which not only affect the synthesis and activity of many enzymes and other proteins, but may also alter the composition of gall-bladder bile and impair the transport of biliary components. The EE component increases biliary concentration of cholesterol and cholic acid and reduces that of deoxycholic acid and chenodeoxycholic acid. In predisposed women, this may lead to gall-bladder disease, intra-hepatic cholestasis, jaundice and pruritus during treatment with OCs.

During the first three cycles of treatment with high-dose OCs, a transitory rise of various liver function parameters may occur, e.g., of aspartate aminotransferase (ASAT), alanin aminotransferase (ALAT), γ-glutamyltransferase (γ-GT), glutamate dehydrogenase, or β-glucuronidase. Such effects are rarely observed when low-dose OCs are used. A decrease in the activity of alkaline phosphatase and cholinesterase and an increase in lactate dehydrogenase can, however, be measured also when low-dose OCs are used. In general, alterations of liver function parameters remain in the normal range and are reversed after some treatment cycles. Therefore, screening of liver function is justified only on suspicion of liver disease or a past history of these diseases.

2. Lipid Metabolism

Due to their strong hepatic effect, EE and – to a different extent – the progestogens may alter lipid metabolism. The estrogen increases the synthesis of triglycerides (TG), apolipoproteins A and B, and high-density lipoprotein (HDL); it also stimulates the receptor-mediated uptake of remnants and low-density lipoprotein (LDL) in the liver and inhibits hepatic lipoprotein lipase (Krauss and Burkman 1992). Moreover, EE increases the proportion of unsaturated fatty acids and lecithin in phospholipids. Progestogens, particularly those with androgenic properties, may counteract the estrogen-induced effects by reducing TG synthesis and increasing the activity of hepatic lipoprotein lipase and the clearance of TGs.

Therefore, according to the composition of the OC, treatment with estrogen-dominant formulations results in a rise of HDL-cholesterol (HDL-CH), very low density lipoprotein (VLDL) and total TG, while LDL-CH remains unchanged (Kuhl et al. 1990b). The levels of lipoprotein (a) are reduced during treatment with OCs (Kuhl et al. 1993). The estrogen-induced rise in TG cannot be regarded as being deleterious, as it is not the result of a disturbed lipolysis, but is caused by an elevated synthesis of VLDL. At the same time, the hepatic elimination of LDL and TG-rich remnants is enhanced, which leads to a shortening of the residence time of atherogenic lipoproteins in the circulation. It has been suggested that the anti-oxidative effect of EE prevents LDL oxidation and accumulation in the arterial wall. Moreover, estrogens show a vasodilatory effect on the arteries and inhibit proliferation of smooth muscle cells. Therefore, even during use of OCs, which may change lipid metabolism in an unfavorable manner, there seems to be no acceleration of the development of atherosclerosis. The increased risk of myocardial infarction in pill users who smoke, may be due to the occurrence of vasospasms in atherosclerotic coronary arteries, which might be triggered by the vasoconstrictory action of the progestogen component. In these cases, the manifestation of an ischemic cardiovascular disease may also be facilitated by the pro-coagulatory action of OCs.

In women with hyperlipoproteinemia type II or III, OCs are contraindicated if vascular lesions have developed. If no additional risk factors are present, OCs may be used (Knopp et al. 1993). In women with hypertriglyceridemia, due to an impaired lipolysis (type VI or V), OCs are contraindicated.

3. Renin-Angiotensin-Aldosterone System (RAA System)

Due to the strong influence of EE on the liver, OCs may increase the serum concentration of angiotensinogen 3- to 5-fold which, after discontinuation of treatment, is rapidly reversed (DERKX et al. 1986). Although the renal release of renin is reduced, the high renin substrate level leads to an enhanced production of angiotensin I which is transformed by the angiotensin converting enzyme (ACE) to angiotensin II. The 2- to 3-fold increase in angiotensin II causes a corresponding rise in aldosterone secretion which reduces the release of renin by 50%. The EE-induced increase in renin activity which is dose-dependent, is less pronounced during treatment with low-dose OCs (DERKX et al. 1986) While aldosterone may increase the retention of sodium and water and cause the development of edema, the elevated angiotensin-II levels may contribute to a slight increase in blood pressure. There is, however, no correlation between the alterations of blood pressure and plasma concentrations of renin, angiotensin II or aldosterone, or plasma volume (WEIR et al. 1975). The changes observed during treatment with OCs are similar in normotensive and hypertensive women. Therefore, the development of hypertension is not directly caused by the stimulation of the RAA system. In contrast to OCs, no influence of progestogen-only preparations on RAA system is observed.

4. Carbohydrate Metabolism

Low-dose OCs have only a marginal influence on fasting glucose levels, but may cause a slight hyperinsulinemia. This is the result of a decrease in insulin sensitivity, which is caused by a synergistic action of EE and the progestogen. The elevated C-peptide levels indicate an enhancement of both the secretion and clearance of insulin. Long-term treatment with OCs is associated with a slight insulin resistance and impairment of glucose tolerance which is compensated by the rise in insulin levels (CROOK et al. 1988; KRAUSS and BURKMAN 1992; GODSLAND and CROOK 1994). There is, however, no change in the levels of glycated proteins or HbA1. The glucagon levels are increased by EE and decreased by progestogens, and the resulting effect is dependent on the composition of the OC.

In most women, the changes in carbohydrate metabolism are reversible and are without clinical relevance; only in predisposed women, i.e., in 4–5% of OC users, are pathological alterations observed. In women with impaired glucose tolerance, OCs may cause a slight deterioration which may be enhanced during long-term treatment (Wynn 1982). The progression of a disturbed glucose tolerance to a manifest diabetes mellitus is, however, independent of OC treatment. Therefore, there is no significant influence of OCs on the incidence of non-insulin-dependent diabetes mellitus (NIDDM, type II) (RIMM et al. 1992). In women with diabetes mellitus, treatment with OCs might, however, increase the risk of cardiovascular diseases. The role of an OC-induced increase in the serum levels of growth hormone and TG is not yet clarified.

5. Hemostasis

Due to the marked action of EE on the liver, there is a rise of many coagulation and fibrinolysis factors. Progestogens, particularly those with androgenic properties, may slightly antagonize these effects (KUHL 1996b). During treatment with OCs, an increase in the serum levels of fibrinogen, Willebrand factor, prothrombin, and factors VII, VIII, IX, X, and XII can be observed (COHEN et al. 1988; BALL et al. 1990; DALY and BONNAR 1990; PETERSEN et al. 1993; TAUBERT and KUHL 1995). While there is a decrease in antithrombin-III activity, Protein C is increased and protein S mostly unchanged. The enhancement of fibrinolytic activity is mainly based on the rise of the plasminogen level and the activity of tissue-plasminogen activator (t-PA). The elevated turnover of fibrin is reflected by a marked increase in the serum levels of fibrinopeptide A, fragments 1 + 2, fibrin monomers, D-Dimer and fibrin degradation products. The changes of the various clotting and fibrinolytic factors induced by EE are time- and dose-dependent, but there are differences between the various factors with respect to the time-course and extent of changes. During OC use, a steady state of the hemostasis/fibrinolysis system at a higher level is reached within a few weeks. After termination of treatment, the return to baseline of the altered concentrations and activities of hemostasis factors which have varying half-lives, needs different periods of time.

Although the rise in coagulation factors is generally compensated by the increase in fibrinolytic activity, the hypercoagulable state induced by OCs may contribute to the development of thromboembolic diseases in predisposed women. The risk of deep vein thromboses and pulmonary emboli is particularly increased in women with resistance to activated protein C (APC), or with a deficiency of anti-thrombin III, protein C or protein S. Recent investigations indicate that OCs may impair the effect of activated protein C and induce APC-resistance (OLIVIERI et al. 1996; ROSING et al. 1997).

The development of venous thromboses is a local phenomenon. As the activation of the coagulation cascade is triggered by the interaction of thrombocytes and endothelium, it appears probable that OCs may influence these local mechanisms. It is known that OCs may influence platelet aggregation and the metabolic activity of the endothelium.

6. Serum Proteins

According to their composition, oral contraceptives may alter the serum levels of various hepatic proteins, particularly serum binding globulins. These binding proteins play a role in the regulation of various hormones including sex steroids. It is assumed that only the free proportion of a hormone may be bound at the cell membrane or penetrate through the membrane, while the protein-bound hormone is hormonally inert and may serve as a circulating reservoir. This may particularly hold true for the role of serum binding globulins, which may bind sex steroids, corticosteroids or thyroid hormones with high affinity and may influence the biological activity of the hormones.

Treatment with EE increases dose-dependently the serum concentrations of SHBG, corticosteroid-binding globulin (CBG) and thyroxine-binding globulin (TBG).

In post-menopausal women, a dose of only 5 µg EE increases SHBG levels by 100% and a dose of 20 µg EE by 200% (MANDEL et al. 1982). While progesterone derivatives do not influence SHBG, nortestosterone derivatives may decrease SHBG levels according to their androgenic properties. Therefore, the composition of the OC is crucial to the resulting effect of the formulation: treatment with 20 µg EE + 250 g LNG reduces SHBG by 50%; 30 µg EE + 250 µg LNG has no influence; 30 µg EE + 150 µg LNG increases SHBG by 30%, and the EE/LNG containing triphasic OC causes a rise in SHBG by 100–150%. The combination of 30 µg EE with 150 µg DG increases SHBG levels by 200% and 35 µg EE + 2 mg cyproterone acetate by 400%. An excessive EE-induced rise in serum SHBG is, however, not associated with a corresponding decrease in free testosterone (JUNG-HOFFMANN and KUHL 1987; VAN DER VANGE et al. 1990).

CBG responds to a lesser extent than SHBG to treatment with EE. In post-menopausal women, a dose of 10 µg EE increases CBG levels by 50% and 20 µg EE by 100% (MANDEL et al. 1982). Progesterone derivatives have no effect and nortestosterone derivatives only a weak effect on CBG. Therefore, treatment with low dose OCs may cause a rise in CBG by 100–150% (VAN DER VANGE et al. 1990).

In post-menopausal women, 5 µg EE increase TBG by 40% and 20 µg EE by 60%. Progesterone derivatives do not influence TBG, while nortestosterone derivatives with androgenic properties may antagonize the estrogen-induced rise. During treatment with 30 µg EE + 1 mg NET, TBG levels rise by 50–70%, and with 30 µg EE + 150 µg DG by 100%.

OCs do not or only marginally affect the serum levels of albumin or total protein. The EE component of OCs causes a rapid increase in the serum concentrations of transferrin, ferritin and ceruloplasmin, while haptoglobin is reduced. In many women, a rise in C-reactive protein and a transitory increase in α-fetoprotein is observed during treatment with OCs. Immunglobulins A, G and M may be slightly elevated by OCs, while α1- and α2-globulin may increase and β-globulin decrease.

7. Hormones

Intake of OCs leads to a time-dependent decrease in FSH and LH secretion and, consequently, to a suppression of ovarian steroid synthesis. Therefore, the serum levels of estradiol are suppressed, whereby a direct inhibition of ovarian steroid production by EE and the synthetic progestogen might play a role. In a certain proportion of women, follicular maturation begins during the hormone-free interval of 7 days and may continue during the first week of OC intake, resulting in elevated estradiol levels. In the case of profound suppression of endogenous estradiol, no symptoms of estrogen deficiency occur, as the

EE levels are sufficient for replacement. OC treatment leads to a decrease in 17α-hydroxyprogesterone by 70–80% and in DHEA-S by 30–50% (KUHL et al. 1985b; JUNG-HOFFMANN et al. 1988) which reflects an inhibition of adrenal steroid synthesis, presumably by a direct interaction of ethinylated steroids (FERN et al. 1978). Accordingly, adrenal testosterone and androstendione production is reduced by OC. The action of OC on both ovarian and adrenal steroid synthesis leads to a reduction in testosterone levels by 30%.

Treatment with OCs does not significantly influence insulin-like growth factor-1 (IGF-1) or its binding protein IGFBP-1, while the level of growth hormone is increased by the estrogen component by about 50%. As a result of a rise in CBG levels, clearance of cortisol is reduced. Consequently, the serum concentration of total cortisol is elevated. There is, however, also a slight increase in free cortisol. High-dose OCs may cause a rise in adrenocorticotropic hormone (ACTH), while low-dose OCs are less effective.

OCs do not or only slightly influence thyroid function: the EE-induced rise in TBG is associated with a significant increase in the serum concentrations of T3 and T4; there is, however, no change or only a marginal change in the level of TSH, free T3, free T4, and the effective thyroxine ratio (KUHL et al. 1985a; JUNG-HOFFMANN et al. 1988).

Treatment with OCs causes a rise in the serum levels of vasopressin, oxytocin and atrial natriuretic peptide (ANP) and a fall in those of somatostatin and cholecystokinin, while leptin, gastrin and the vasoactive intestinal peptide (VIP) are unaffected.

8. Other Biochemical Laboratory Parameters

The EE component of an OC causes a rapid increase in the levels of ferritin, transferrin and total iron-binding capacity, which leads to a rise in serum iron. Similarly, the increase in the serum level of copper is caused by the EE-induced increase in the copper-binding protein ceruloplasmin which may be antagonized by progestogens with androgenic properties. In general, treatment with OCs results in a rise of serum ceruloplasmin by 100–140%. Most of the electrolytes and trace elements are not altered, except for a reduction in magnesium and nickel, and a slight but significant increase in sodium (Table 8). The latter may be due to the EE-dependent rise in aldosterone and vasopressin. The levels of hemoglobin and hematocrit, erythrocyte count and platelet count are not altered. There is, however, an increase in blood viscosity. OCs do not alter homocysteine concentrations. During treatment with high-dose formulations, the levels of several amino acids, e.g., glutaminic acid, alanin, glycin, glutamin and taurin, may be reduced.

During treatment with OCs, the serum concentrations of vitamin A increase by 50%, whereby the rise in retinol is accompanied by a decrease in β-carotin, particularly in smokers. There is also an elevation of calcitriol levels. In some of the women who use OCs, the serum concentrations of vitamin B1, B2, B6 and B12 may be – mostly transitorily – reduced. No change occurs with

Table 8. Effect of low-dose oral contraceptives on various clinical chemical laboratory parameters

Blood sedimentation rate	Slightly increased
Hematocrit	Unchanged
Erythrocyte count	Unchanged
Leukocyte count	Slightly increased
Platelet count	Unchanged
Plasma volume	Increased
Blood viscosity	Increased
Hemoglobin	Unchanged
Total protein	Unchanged
Albumin	Unchanged
SHBG	Increased
CBG	Increased
TBG	Increased
α_2-Macroglobulin	Unchanged
α_1-Anti-trypsin	Increased
α-fetoprotein	Unchanged
C-Reactive protein	Increased
Orosomucoid	Unchanged
Haptoglobin	Reduced
Ceruloplasmin	Increased
Transferrin	Increased
Ferritin	Increased
Iron	Increased
Copper	Increased
Magnesium	Reduced
Zinc	Unchanged
Calcium	Unchanged
Nickel	Reduced
Manganese	Unchanged
Selenium	Unchanged
Mercury	Unchanged
Aluminium	Unchanged
Lead	Unchanged
Cadmium	Unchanged
Sodium	Increased
Oxytocin	Increased
Vasopressin	Increased
Angiotensin I and II	Increased
Growth hormone	Increased
IGF-1 (somatomedin)	Unchanged
Somatostatin	Reduced
Cholecystokinin	Reduced
Gastrin	Unchanged
Leptin	Unchanged
Melatonin	Unchanged
Vasoactive intestinal peptide	Unchanged
Atrial natriuretic peptide	Increased
ACTH	Unchanged
Cortisol	Increased
Aldosterone	Slightly increased
TSH	Unchanged
T_3	Increased
FT_3	Unchanged

Table 8. (*Continued*)

T_4	Increased
FT_4	Slightly increased
Effective thyroxine ratio	Slightly increased
Reverse trijodothyronine	Unchanged
Vitamin A	Increased
Vitamin B_1	Unchanged
Vitamin B_2	Unchanged
Vitamin B_6	Unchanged
Vitamin B_{12}	Reduced
Vitamin C	Unchanged
25-OH-Vitamin D_3	Unchanged
1,25-$(OH)_2$-Vitamin D_3	Increased
Vitamin E	Unchanged
Vitamin H	Unchanged
Vitamin K	Unchanged
Folic acid	Unchanged
Panthotenic acid	Unchanged
Lactate	Increased
Pyruvate	Unchanged
Free fatty acids	Unchanged
Uric acid	Slightly reduced
Creatinine	Increased
Bilirubin	Reduced
Alkaline phosphatase	Reduced
ASAT (SGOT)	(Increased)
ALAT (SGPT)	(Increased)
Gamma-GT	(Increased)
Glutamate dehydrogenase	(Increased)
Lactate dehydrogenase	(Increased)
β-glucuronidase	(Increased)
Cholinesterase	(Increased)
Xanthuric acid excretion	Increased
Sulfobromophthalein retention	Increased

ACTH, adrenocorticotropic hormone;TSH, thyroid stimulating hormone; SHBG, sex hormone-binding globulin; CBG, corticosteroid-binding globulin; TBG, thyroxine binding globulin; ASAT, aspartate aminotransferase; SGOT, serum glutamic oxaloacetic transferase; ALAT, alanine aminotransferase; SGPT, serum glutamic pyruvate transferase; GT, glutamyl transferase; IGF-1, insulin-like growth factor-1

respect to other vitamins or folic acid (Table 8). If there is an effect, the alteration is probably caused by the estrogen component. In the case of vitamin A and D, the increase is probably due to an elevated serum level of binding proteins.

III. Beneficial Effects of Oral Contraceptives

The use of OCs is not only associated with an increased risk of various diseases (Table 9), but also with many benefits (Mishell 1982) (Table 10). The

Table 9. Adverse effects of oral contraceptives; increase in the relative risk of various diseases

Disease	Relative risk
Cardiovascular diseases (total)	1.5
Myocardial infarction (total)	3.3
Myocardial infarction (non-smokers)	1.0
Myocardial infarction (light smokers)	3.5
Myocardial infarction (heavy smokers)	20.0
Cerebrovascular diseases (total)	1.4
Cerebral thromboses	2.5
Subarachnoidal bleeding (heavy smokers)	10.0
Pulmonary embolism	3.0
Deep venous thromboses	2.5
Gall-bladder diseases	3.0
Benign liver tumors	50.0
Hepatocellular carcinoma	3.0
Erythema nodosum et multiforme	3.0
Pruritus	2.0
Photosensitive eczema	4.0
Irritant agent eczema	2.0
Dermatitis	2.0
Chloasma	1.5
Cervicitis (6 years of use)	3.0
Chlamydial infections	2.5

Table 10. Beneficial effects of oral contraceptives; reduction in the relative risk of various disorders or diseases

Disorders or diseases	Relative risk
Iron-deficiency anemia	0.58
Menorrhagia	0.52
Irregular cycles	0.65
Inter-menstrual bleeding	0.72
Dysmenorrhea	0.37
Pelvic inflammatory disease	0.50
Benign breast disease	0.69
Rheumatoid arthritis	0.49
Endometrial cancer	0.50
Ovarian cancer	0.37

most important advantage of OC use is the significant reduction in the risk of ovarian cancer and endometrium cancer, benign breast disease and pelvic inflammatory disease. There are also less ovarian cysts, probably less uterine myoma and endometriosis, less rheumatoid arthritis, and – due to the effective inhibition of ovulation – less ectopic pregnancies and abortions. Treatment with OCs may also prevent bone loss in women with oligo- or amenorrhea.

Although in pill starters, inter-menstrual bleeding is frequently observed during the first to third treatment cycles, use of OCs leads to regular cycles, a reduction of menstrual blood loss and an amelioration of dysmenorrhea. Therefore, OCs may be used for treatment of various bleeding disorders (MISHELL 1982).

Therapeutic use of OCs is indicated in women with irregular bleeding, menorrhagia and anemia, dysmenorrhea, and mittelschmerz. If use of combination preparations remains unsuccessful in women with intermenstrual bleeding or irregular cycles, treatment with a sequential formulation will often be effective. It is also possible to postpone withdrawal bleeding if the time of expected menses is inconvenient. By omitting the hormone-free interval of 7 days, i.e., by the uninterrupted intake of the pill for some days or weeks subsequently to the 21-day pill cycle, withdrawal bleeding is prevented. If intake is stopped, bleeding takes places within 2–4 days. By means of this procedure, the day of bleeding may be timed, although during prolonged intake, breakthrough bleeding or spotting may occur in some women.

OCs effectively reduce the serum levels of testosterone and, consequently, those of free testosterone. Therefore, some androgen-dependent symptoms, such as seborrhea, acne, hirsutism or alopecia, may be improved (LEMAY et al. 1990). The therapeutic effect may be enhanced using estrogen-dominant formulations which increase the levels of SHBG, resulting in a more pronounced suppression of free testosterone levels. In more severe cases of acne and hirsutism, the use of OCs containing progestogens with anti-androgenic properties is indicated, particularly formulations with CPA. In cases with an insufficient therapeutic effect, an elevation of the CPA dose may lead to an improvement of symptoms (MOLTZ et al. 1980).

In women with functional ovarian cysts, treatment with OCs containing a potent progestogen component may improve complaints and prevent a relapse. Mastopathia of stage I or II may also be treated with combination OCs containing low-dose EE and a potent progestogen. Intake of this type of OC may also cause an improvement of symptoms of endometriosis.

IV. Risks and Side Effects of Oral Contraceptives

Many complaints, disorders and diseases have been ascribed to the use of OCs. Most of the epidemiological data on adverse effects of OCs have been gained from retrospective studies and refer to high-dose formulations. Nevertheless, severe diseases may also be caused by low-dose preparations. They occur, however, mostly in predisposed women; the events are rare, and a large proportion of them might be prevented by a careful examination, anamnesis and family history. Most of the disorders associated with pill use are reversed after discontinuation of OCs, and most of the subjective complaints which are known to occur during the first weeks of treatment do not differ from the effect of taking placebos.

1. Minor Complaints During Intake of Oral Contraceptives

Inter-menstrual bleeding and spotting, the incidence of which is high during the first cycle of intake and declines thereafter, may lead to a premature cessation of treatment. This type of bleeding is frequently the consequence of tablet omission and may indicate a loss of contraceptive efficacy, but is generally not a sign of disease. In most cases, inter-menstrual bleeding disappears after the third pill cycle; if not, change to a sequential formulation may lead to regular cycles, while switching to high-dose combination preparations will rarely be effective. If inter-menstrual or irregular bleeding persists, the cause must be clarified.

Amenorrhea during use of OCs ("silent menstruation") may be a consequence of high progestogenic and low estrogenic effectiveness, which results in a suppression of endometrial proliferation. It might, however, also be the result of an undesired pregnancy and has to be investigated. If amenorrhea occurs repeatedly, change to another formulation should be considered.

During the first cycle of treatment with OCs, particularly with high-dose formulations, nausea, vomiting, intestinal disorders or vertigo may occur. In most cases, the complaints disappear during further intake, but irrespective of improvement, high-dose preparations should be replaced by low-dose pills. Most subjective complaints, such as loss of libido, headaches, abdominal aches, backaches etc., are also observed during an ovulatory cycle. A placebo-controlled, blind study revealed that the incidence of such symptoms is obviously not due to an effect of the contraceptive steroids (AZNAR–RAMOS et al. 1969). In this study, only one third of the women were without side effects during intake of the placebo.

2. Fertility, Pregnancy and Lactation

In general, discontinuation of oral contraception is rapidly followed by ovulatory cycles. In nulliparous women aged more than 30 years, fertility may be slightly impaired during the first 2 years. Thereafter, no difference exists between women who had been treated with OCs and those who had used another contraceptive method. In cases of the so-called post-pill amenorrhea, i.e., long-term anovulation after cessation of OC use, a causal relationship is questionable. The rate of 1–3% in ex-users is similar to that of spontaneous secondary amenorrhea (VESSEY et al. 1978). In most of these cases, there had been irregular cycles before starting treatment with OCs. This also holds true for treatment of adolescents with OCs. There is no influence on the incidence of abortion in pregnancies after discontinuation of OC and no influence on the risk of malformations, pregnancy outcome or sex ratio of fetuses.

In a small proportion of women taking OCs, withdrawal bleeding is lacking. The uncertainty regarding pregnancy or not may lead to a continuation of pill intake early in pregnancy. As organogenesis takes place between the third and eighth weeks of fetal development (5–10 weeks since last menstruation), the question arises whether or not the use of an OC increases the risk of congenital malformations. A *meta*-analysis revealed that intake of OCs during pregnancy does not increase the risk of malformations, the general rate of which is 2–3% (Bracken 1990).

In postpartum women, OCs may impair lactation and reduce milk volume. Moreover, the uptake by the infant of contraceptive steroids with the milk is not negligible, although no adverse effects have been observed. As the contraceptive effect of breast feeding is not reliable in all women, the use of the mini-pill is recommended.

3. Immune System

The results of some studies indicated that OCs may suppress the cellular and partly the humoral immune system, resulting in a reduction of autoimmune diseases. Treatment with an OC appears to decrease the incidence of thyroid disease and rheumatoid arthritis, particularly chronic polyarthritis. On the other hand, some data indicate that viral infections may occur more frequently during intake of OCs (Taubert and Kuhl 1995).

4. Genital Tract Infections

It is known that pelvic inflammatory disease is associated with a high risk of tubal infertility. Therefore, the protection from upper genital tract infection is an important beneficial effect of OCs. After at least 12 months of intake, the risk is reduced by 50% and, if a pelvic infection occurs, the severity of a salpingitis is decreased (Rubin et al. 1982). However, OC use increases the risk of chlamydia infection and cervicitis (Roy 1991). The latter correlates with the duration of treatment and the potency of the progestogen component.

There is no association between OC use and sexually transmitted viral disease. As OCs do not protect from human immunodeficiency virus (HIV), human papilloma virus (HPV), hepatitis B (HBV) or herpes simplex virus (HSV), the additional use of barrier methods (e.g., condom) is recommended if the risk of an infection is increased.

5. Respiratory Tract and Gingiva

OCs do not impair pulmonary function or diseases except for infections that might occur more frequently during treatment. In patients with chronic diseases, it should be kept in mind that certain drugs may interfere with the pharmacokinetics of contraceptive steroids and impair the efficacy of the OC. The EE component of OCs may stimulate proliferation of nasal and gingival mucosa (Taubert and Kuhl 1995).

The female voice is very susceptible to the influence of androgens. Therefore, treatment with OCs containing progestogens with androgenic properties may cause lowering of vocal pitch (Lembke and Freund 1990). On the other hand, estrogen-dominant OCs may stimulate edema and hyperplasia of vocal fold.

6. Eyes and Ears

There are reports of impaired vision and development of amaurosis with OC use, caused by occlusion of retinal veins or arteries. Moreover, lens opacities and ocular symptoms such as keratoconjunctivitis sicca have been observed. This may lead to contact lens wear problems. The alteration of tear composition induced by OCs may cause the formation of deposits on the contact lenses (TAUBERT and KUHL 1995).

OCs have no significant influence on hearing or on the incidence and course of otosclerosis. There are, however, some reports of sudden deafness due to an impairment of microcirculation.

7. Skin

The EE component may influence the pigment system of melanocytes. Therefore, in synergism with ultra violet (UV) light, OCs may cause changes in pigmentation. The most frequent disorder is chloasma which occurs in 20% of the women using OCs. After discontinuation, the changes are to a certain extent reversible.

The EE component may modify immunological and autoimmunological mechanisms and promote the development of erythema multiforme or nodosum and rosacea. Allergic reactions may also be induced by coloring matter contained in the tablets. The enhancement of susceptibility of the skin to exogenous agents by OCs may facilitate the development of eczema, dermatitis or urticaria. OCs may cause the manifestation or aggravation of porphyria, on the one hand, while, on the other, preventing the onset of acute intermittent porphyria during the luteal phase (TAUBERT and KUHL 1995).

8. Gastrointestinal Tract Disease

In general, there is no association between OC use and gastritis. Treatment with an OC increases, however, the risk of Crohn's disease and colitis ulcerosa (GODET et al. 1995). On the other hand, gastrointestinal bleeding may be reduced or stopped during intake of OCs.

9. Urinary Tract

OCs increase creatinine clearance and sodium and potassium excretion. There are a few casuistics of hemolytic–uremic syndrome during OC use which were due to intimal alterations and thromboses of renal arterioles or to glomerular necroses (TAUBERT and KUHL 1995). Obviously, low dose OCs do not increase infections of urinary tract. There is no information on the effect of dialysis on pharmacokinetics and efficacy of contraceptive steroids. In any case, the pill should be ingested after dialysis, and if hypertension develops, the pill has to be discontinued.

10. Endocrine Effects

In contrast to high-dose OCs, the low-dose formulations do not or only slightly increase the serum level of prolactin. In most women with an elevated prolactin concentration, the rise occurs sporadically and transitorily. In women with hyperprolactinemia or amenorrhea/galactorrhea, the use of OCs may cause a 30% rise in prolactin levels during the first cycle which remains at this range throughout further treatment.

The EE component of OCs may increase CBG levels, resulting in a rise in the serum concentrations of total cortisol. There is, however, a slight increase in free cortisol. The adrenal response to ACTH is not affected. The EE-induced rise in TBG causes an increase in T3 and T4. Thyroid function is, however, not influenced as reflected by unchanged levels of FT3, FT4 and TSH.

11. Neurological Diseases

Sex steroids may effectively influence the functional and morphological organization of the central nervous system and modulate psyche, mood and well-being. In general, estrogens may enhance neuronal activity, while progestogens may act in an antagonistic way. Reports on libido changes could not be confirmed by placebo-controlled double-blind studies. Treatment with OCs may exert a beneficial effect on the symptoms of pre-menstrual syndrome, but the success does not exceed that of a placebo.

Migraine headaches may occur during intake of OCs or during the hormone-free interval. As severe migraine or visual symptoms may be a prodrome of stroke, occurrence for the first time or an unusual worsening of the symptoms is an indication for immediate cessation of intake; although one third of patients suffering from migraine, report on an improvement during use of OCs (Mattson and Rebar 1993).

Estrogens may enhance the excitability and seizure frequency while progesterone has a sedative effect. Use of depot-MPA in addition to anti-convulsants may reduce the rate of seizures by 30%, while treatment with OCs does not influence the course of the disease. As anti-epileptics may exert a teratogenic effect, a reliable contraception is necessary. Most anti-convulsants may cause an enzyme induction and enhancement of hepatic inactivation of contraceptive steroids. Continuous use of OCs, without the hormone-free interval, or of depot-MPA may provide an increase in efficacy and prevent undesired pregnancies.

No adverse effects of OCs have been recorded with respect to multiple sclerosis or myasthenia gravis (Villard-Mackintosh and Vessey 1993). There are casuistics of other neurological diseases that developed during treatment with OCs and mostly improved or reversed after discontinuation. Some neurological diseases that may be associated with a deficiency of folic acid or vitamin B may be improved by a corresponding replacement.

12. Psychiatric Diseases

OCs have no significant influence on depression, psychosis, neurosis or phobia or other psychical diseases. Improvement, but also deterioration have been observed in some cases of menses-associated psychoses (TAUBERT and KUHL 1995). There are no data on the effect of OCs in patients with schizophrenia.

13. Physical Condition and Sports

About 10–20% of women observe an increase in body-weight during the first treatment cycles. As in double-blind studies the same effect was recorded during intake of placebos, this is probably a psychological phenomenon. In general, estrogens do not increase water retention in young women, and the weak anabolic effect of nortestosterone derivatives is without clinical relevance at the doses used in OCs. Using a high-dose OC containing 50 μg EE and 0.5 mg NG, no change in nitrogen or electrolyte balance and fat body mass could be observed (KUDZMA et al. 1972). Despite this, individual predisposed women may, however, experience an increased water retention and, consequently, a rapid weight gain. Most cases of an increase in body weight appear to be caused by an elevated caloric intake.

Low dose OCs do not decrease bone mineral content even in women with a profound suppression of ovarian estrogen synthesis, as a daily dose of 20 μg EE is sufficient for the prevention of loss of bone mass. In many high-performance athletes, irregular cycles or amenorrhea may develop resulting in an estrogen deficiency and an increased risk of stress fractures. Treatment with OCs may prevent bone loss and stress fractures.

14. Venous Complaints

In veins, EE exert a dilatory effect and increase capillary permeability, while progestogens may enhance capacitance and distensibility (FAWER et al. 1978). An increase in interstitial fluid volume in predisposed women may lead to the development of edema, “heavy legs”, leg pain and cramps. The complaints depend on the EE dose and the progestogenic potency (VIN et al. 1992). In the case of severe pain and swelling in only one leg, a physician must be consulted for exclusion of a venous thrombosis.

15. Venous Thromboembolic Diseases

Treatment with OCs increase the relative risk of venous thromboembolic disease 3- to 4-fold. The incidence of deep vein thromboses and pulmonary emboli is highest during the first year of treatment, indicating a crucial role of thrombophilia, and is reversed within 4–6 weeks after cessation of OC intake. Nevertheless, the absolute risk is very low, as the annual incidence is about 4 of every 10,000 OC users. The mortality from venous thromboses and pulmonary emboli is estimated at about 5 in 1 million OC users. The most important risk factors are genetic coagulation disorders, such as resistance to activated protein C (APC-resistance), antithrombin III-, protein C- and protein S-deficiency, anti-phospholipid-antibodies, dysfibrinogenemia, hyperhomocysteinemia, disorders of fibrinolysis and obesity. Routine screening is not recommended because of the unfavorable cost-effectiveness, but in women with a positive personal or family history, a selective thrombophilia screening may help to estimate the individual risk associated with OC use. As major surgery is also a risk factor, OC use should be stopped 4–6 weeks before an elective surgery carrying an increased risk of post-operative thrombosis.

Although epidemiological data are inconsistent due to the influence of many bias, it can be assumed that the risk increases with the EE dose of OC (GERSTMAN et al. 1991; KOSTER et al. 1995). Recent studies revealed that the use of formulations containing DG or GSD is associated with a slightly higher risk than with OCs containing LNG or NET (WORLD HEALTH ORGANIZATION 1995; JICK et al. 1995; BLOEMENKAMP et al. 1995), even though the findings have been called in question. The difference was, however, most striking in first-time users, indicating that the results are not biased by a "healthy-user effect". The mechanism of action is not clarified, but it cannot be excluded that formulations containing DG and GSD may impair the anti-coagulatory system more than the other OCs (ROSING et al. 1997).

16. Stroke

Stroke is a rare event in young women. Use of OCs increases the risk of ischemic stroke 3-fold, whereby it is less in younger, normotensive and nonsmoking women (WORLD HEALTH ORGANIZATION 1996a). The relative risk is 1.5 with low-dose OCs and 5.3 with high-dose OCs, but is reversed after discontinuation of treatment. Hypertension is a strong risk factor (WORLD HEALTH ORGANIZATION 1996a). The relative risk of hemorrhagic stroke is not increased in young women, but is doubled in women above the age of 35 years (WORLD HEALTH ORGANIZATION 1996b). In women with a history of hypertension, treatment with an OC leads to a more than 10-fold rise in risk, while in smokers it is increased 3-fold. It is important to know that an unusual onset of visual symptoms or severe headaches may precede a stroke and is an indication to discontinue use of the pill immediately.

17. Hypertension

Treatment of women with low-dose OCs may cause a slight elevation in systolic (by 5 mmHg) and diastolic (by 2–3 mmHg) blood pressure, which is mostly reversed within 3 months after discontinuation (GODSLAND et al. 1995). The risk of hypertension is, however, doubled to an incidence of between 2% and 4%. The development of a high blood pressure may occur rapidly or slowly and correlates with the duration of OC treatment and age. Further risk factors are smoking, obesity, diabetes mellitus and hyperlipidemia. If blood pressure is normalized by anti-hypertensive therapy, use of OCs is possible under careful supervision. If, in this case, blood pressure rises once more, OC use has to be stopped immediately. It has been suggested that EE is the main factor in the development of hypertension, although the progestogen component may enhance this effect. As there is no difference between hypertensive and normotensive women in the alterations of the RAA system induced by OCs, a local mechanism at the arterial wall has been proposed.

18. Myocardial Infarction

In young women, myocardial infarction is a very rare event, and the annual mortality rate is 2 in 100,000 women. In total, the relative risk is increased 5-fold by OC use; however, young women who do not smoke and have no other risk factors are not affected (WORLD HEALTH ORGANIZATION 1997). In heavy smokers, the relative risk increases 20-fold during OC treatment. Further risk factors are age, hypertension, diabetes mellitus, obesity, hyperlipoproteinemia, hyperfibrinogenemia and elevated blood viscosity (CROFT and HANNAFORD 1989). Epidemiological data indicate an involvement of the progestogen component in the development of arterial diseases (KAY 1982; WINGRAVE 1982; MEADE 1988). There is, however, no evidence that the risk is higher when OCs containing LNG or NET are used. Although formulations with an overweigh of the androgenic effect of the progestogen may cause deleterious alterations in lipid metabolism, no accelerated development of atherosclerosis has been observed. This might be due to the anti-oxidative effect of EE which may act as a free radical scavenger and prevent LDL oxidation. It is known that estrogens exert a dilatory, and progestogens a constrictor effect on the arterial wall. At sites of arterial lesions, e.g., in smokers, the progestogen might enhance vasospasms and trigger an arterial thrombosis.

19. Raynaud's Syndrome

Treatment with OCs appears to increase to risk of development of Raynaud's syndrome. Although there is no evidence of a deterioration of the symptoms of Raynaud's syndrome with OC use, the pill should not be used in patients with disturbed peripheral circulation. In cases of occlusion of retinal vessels or changes in renal microcirculation, an improvement may be observed after discontinuation of OCs.

20. Diabetes Mellitus

During intake of OCs, a slight insulin resistance and impairment of glucose tolerance is observed that is compensated by a small increase in insulin levels and reversed after discontinuation (GODSLAND and CROOK 1994; GODSLAND et al. 1990). The mechanism is not clarified, but it is suggested that EE reduces insulin clearance, while the progestogen component impairs peripheral glucose consumption (TAUBERT and KUHL 1995). There are no epidemiological data indicating that OCs may cause diabetes mellitus, neither in women with normal glucose tolerance nor in patients with pathological values. The proportion of women with an impaired glucose tolerance is, however, doubled by treatment with OCs. In contrast, the progression to diabetes mellitus is independent of OC use (DUFFY and RAY 1984).

Patients with manifest type-I diabetes mellitus (IDDM) who need a reliable contraception may use OCs under careful supervision, provided that no macro- or microangiopathies exist. It is also important to pay attention to additional risk factors. Use of OCs in young women with insulin-dependent diabetes mellitus probably does not additionally impair endothelial function (PETERSEN et al. 1994; GARG et al. 1994).

21. Liver

Except for rare liver tumors, the relative risk of which is elevated during long-term OC use, the incidence of serious liver diseases among current or former pill users is not significantly influenced. There was, however, a modestly increased risk of mild liver disease, which declined after 4 years of use and after discontinuation (HANNAFORD et al. 1997).

It is known that synthetic sex steroids containing an ethinyl group, particularly EE, may impair hepatic functions and cause morphological alterations of the liver. In 1% of women, a rise in serum liver-enzyme parameters is observed during the first cycles of treatment with low-dose OCs, which mostly normalizes thereafter. In patients with severe chronic liver disease, all synthetic steroids are contraindicated.

EE and nortestosterone derivatives may impair the excretory function of the liver and change the composition of bile. In women with a reduced excretory capacity, an intra-hepatic cholestasis may develop during the first six cycles of treatment with OCs, leading to pruritus and jaundice (RANNEVIK et al. 1972). In most cases, it is reversed within 2 months after discontinuation.

In predisposed women, treatment with OCs may lead to an excessive increase in porphyrin precursors, resulting in the manifestation of porphyria (DOSS 1984). In most cases, the symptoms are rapidly reversible after stopping pill intake. In some patients with acute intermittent porphyria, who frequently suffer from an outburst during the luteal phase, treatment with OCs may even stabilize the latent sub-clinical phase of the disease.

OCs do not influence the development or course of hepatitis (SCHWEITZER et al. 1975). In women with a history of hepatitis, the liver function parameters may increase markedly during the first cycles of OC treatment (SHAABAN et al. 1982). If the elevated values still persist after 6 months, the pill should be discontinued.

The prevalence of gall bladder disease in women is about 4% and is 4- to 5-fold higher in patients with a positive family history, indicating the role of predisposition. OCs may cause a premature manifestation of cholelithiasis or cholecystitis, but the total risk of gall bladder diseases is not influenced by the pill (VESSEY and PAINTER 1994; GRODSTEIN et al. 1994).

In women with a latent hyperlipoproteinemia type IV or V, EE may cause an excessive rise in TGs. Therefore, during treatment with estrogen-dominant OCs, predisposed women may develop a pancreatitis within three cycles that is reversible after cessation of intake.

22. Liver Tumors

Benign liver tumors are an extremely rare event, the relative risk of which is, however, increased by OC use (ROOKS et al. 1979). The effect depends on the duration of treatment and the dose and potency of EE and the progestogen. Moreover, a genetic predisposition appears to play a role. Even though most liver adenomas regress after discontinuation of the pill, their importance lies in the risk of rupture and intraperitoneal hemorrhage, which is higher for liver cell adenomas than for focal nodular hyperplasias. Regular palpation of the liver may help to recognize liver tumors.

The development of hepatocellular carcinomas, which is also a very rare event, is mostly associated with liver cirrhosis due to hepatitis B infection or alcohol abuse. Long-term treatment with an OC appears to increase the relative risk (NEUBERGER et al. 1986; ROSENBERG 1991).

23. Breast

In 2–5% of women taking OCs, an enhanced fluid retention may lead to an increase in breast volume, which may cause mastodynia. Treatment with combination formulations with low EE and a potent progestogen or with depot-progestogens may improve the condition. During use of OCs, the risk of benign breast disease is reduced. This beneficial effect correlates with the duration of treatment and potency of the progestogen component (ROYAL COLLEGE OF GENERAL PRACTITIONERS 1977). Therefore, in women with mastopathia grade I or II, the use of formulations with a potent progestogen is recommended.

24. Breast Cancer

Nearly 30% of all cancers in women concerns breast cancer, and about 10% of the women will develop breast cancer during their lifetime. In young women, the disease is rare, but increases markedly with age. Numerous case-control and prospective studies have been carried out regarding the effect of OCs on breast-cancer risk. Due to the long latency of the disease and the relative small case numbers in younger women, the results were inconsistent, particularly with respect to sub-groups.

A re-analysis of the individual data of more than 50,000 women with breast cancer and 100,000 controls from 54 studies was published in 1996 (COLLABORATIVE GROUP ON HORMONAL FACTORS IN BREAST CANCER 1996). The results clearly show that during current use of OCs, there is a small increase in the relative risk of a breast cancer diagnosis of 24%, which is reversed during the first 10 years after discontinuation. The additional cases diagnosed in women who had used OC are less advanced and have a better prognosis than those diagnosed in women who had never used OCs. The extremely small risk associated with OC use is exemplified by the small number of additional breast cancers. Among 10,000 women, between the age of 25 years and 29 years, who take OCs for 5 years, there will be seven additional localized breast cancers diagnosed before age 50, while there will be a deficit of four breast cancers spread beyond the breast. The cumulative number of breast cancers additionally diagnosed in 10,000 women in the period between starting OC use and 10 years after stopping are 0.5 in women aged 16–19 years, 1.5 in those aged 20–24 years, 4.7 in those aged 25–29 years and 11.1 in those aged 30–34 years. The relative risk associated with OCs was not influenced by dose or composition of the pill, duration of treatment, parity, time of first childbirth, family history, body mass index or alcohol. The effect appears to be due to an increase in the proportion of proliferative epithelial cells (WILLIAMS et al. 1991) and a stimulation of growth of that proportion of the tumor which still is controlled by estrogens.

25. Uterine Tumors

Treatment with OCs containing a low EE dose and a potent progestogen component may reduce rather than increase the incidence of uterine myoma. As growth of uterine myoma is estrogen-dependent, depot-progestogens should be preferred. Progestogen-dominant OCs may also improve symptoms of endometriosis, but cannot cause atrophy of endometriotic tissue.

Long-term unopposed influence of endogenous estrogens as observed in persisting anovulatory cycles or polycystic-ovary syndrome, may cause endometrial hyperplasia and cancer. The regular influence of the progestogen contained in combined OCs protects from the development of endometrial hyperplasia and reduces the relative risk of endometrial cancer by 50% (SCHLESSELMAN 1991; STANFORD et al. 1993). The beneficial effect is greatest after 3 years of intake and persists after discontinuation of the pill for many years. It can be assumed that the use of sequential formulations with 15 days of progestogenic influence per cycle also protects from endometrial carcinoma.

As OCs do not stimulate growth of trophoblastic cells, the incidence of hydatidiform mole or chorionic carcinoma is not increased.

26. Cervical Neoplasia

During use of OCs, endocervical glandular hypersecretion and proliferation of endocervical glands may be observed, which are suggested to be induced primarily by the progestogen component. Similarly, the occurrence of metaplasia and dysplasia of the uterine cervix have been ascribed to the action of progestogens.

Epidemiological findings indicate that OCs increase the risk of carcinoma-in-situ and invasive carcinoma of the uterine cervix, time-dependently, particularly in adolescents (World Health Organization 1985; Brinton et al. 1986). The results are, however, not definitive, as the role of other important risk factors, such as smoking and sexual behavior, have not been clarified. It is known that infection with human papilloma virus is involved in the development of cervical neoplasia (Bosch et al. 1992), and contraceptive barrier methods may reduce the risk. Regular examination and *Pap* smear in women who use OCs may also contribute to preventing invasive cervical cancer (Parazzini et al. 1997).

27. Ovarian Tumors

The incidence of functional ovarian cysts is reduced by the use of combined OCs. Moreover, existing functional ovarian cysts may regress during treatment with OCs. Ovulation inhibitors also reduce the relative risk of epithelial ovarian cancer by 40%. This important protective effect correlates with the duration of treatment and persists for at least 10 years after discontinuation of the pill (Whittemore et al. 1992).

28. Other Cancers

Similar to cervical cancer, the risk of vulvar cancer correlates with the number of sexual partners and is decreased by the use of contraceptive barrier methods. No influence on the relative risk of vulvar cancer with OC use could, however, be observed.

OCs do not influence the incidence of prolactinoma and, in patients with microprolactinoma, OCs may be used after normalization of prolactin levels by treatment with dopamine agonists.

Although estrogens may influence melanocytes in synergism with UV light and may cause hyperpigmentation, there is no evidence that OCs may influence the incidence of malignant melanoma (Palmer et al. 1992). Therefore, a history of melanoma is no contraindication for hormonal contraception.

Even though estrogen and progesterone receptors have been demonstrated to be present in nearly every tissue, no other malignant tumors have been shown to be promoted by the use of OCs.

References

Aznar-Ramos R, Giner-Velasquez J, Lara-Ricalde R, Martinez-Manautou J (1969) Incidence of side-effects with contraceptive placebo. Am J Obstet Gynecol 105: 1144–1149

Back DJ, Barkfeldt JO, Breckenridge AM, Odlind V, Orme M, Park BK, Purba H, Tjia J, Victor A (1982) The enzyme inducing effect of rifampicin in the rhesus monkey and its lack of interaction with oral contraceptive steroids. Contraception 25:307–316

Back DJ, Grimmer SFM, Orme ML'E, Proudlove C, Mann RD, Breckenridge AM (1988) Evaluation of committee on safety of medicines yellow card reports on oral

contraceptive-drug interactions with anticonvulsants and antibiotics. Br J Clin Pharmacol 25:527–532
Back DJ, Orme MLE (1990) Pharmacokinetic drug interactions with oral contraceptives. Clin Pharmacokinet 18:472–484
Ball MJ, Ashwell E, Jackson M, Gillmer MDG (1990) Comparison of two triphasic contraceptives with different progestogens: effects on metabolism and coagulation proteins. Contraception 41:363–376
Bloemenkamp KWM, Rosendaal FR, Helmerhorst FM, Büller HR, Vandenbroucke JP (1995) Enhancement by factor V Leiden mutation of risk of deep-vein thrombosis associated with oral contraceptives containing a third-generation progestagen. Lancet 346:1593–1596
Bosch FX, Munoz N, de Sanjose S, Izarzugaza I, Gili M, Viladiu P et al. (1992) Risk factors for cervical cancer in Colombia and Spain. Int J Cancer 52:750–758
Bracken MB (1990) Oral contraception and congenital malformations in offspring: A review and meta-analysis of the prospective studies. Obstet Gynecol 76:552–557
Brinton LA, Huggins GR, Lehman HF, Mallin K, Savitz DA, Trapido E, Rosenthal J, Hoover R (1986) Long-term use of oral contraceptives and risk of invasive cervical cancer. Int J Cancer 38:339–344
Brosens IA, Pijnenborg R (1976) Comparative study of the estrogenic effect of ethinylestradiol and mestranol on the endometrium. Contraception 14:679–685
Cohen H, Mackie IJ, Walshe K, Gillmer MDG, Machin SJ (1988) A comparison of the effects of two triphasic oral contraceptives on haemostasis. Br J Haematol 69:259–263
Collaborative Group on Hormonal Factors in Breast Cancer (1996) Breast cancer and hormonal contraceptives: collaborative reanalysis of individual data on 53297 women with breast cancer and 100 239 women without breast cancer from 54 epidemiological studies. Lancet 347:1713–1727
Croft P, Hannaford PC (1989) Risk factors for acute myocardial infarction in women: evidence from the Royal College of General Practitioners' oral contraception study. Br Med J 298:165–168
Crook D, Godsland IF, Wynn V (1988) Oral contraceptives and coronary heart disease: modulation of glucose tolerance and plasma lipid risk factors by progestins. Am J Obstet Gynecol 158:1612–1620
Daly L, Bonnar J (1990) Comparative studies of 30μg ethinylestradiol combined with gestodene and desogestrel on blood coagulation, fibrinolysis, and platelets. Am J Obstet Gynecol 163:430–437
Derkx FHM, Stünkel C, Schalekamp MPA, Visser W, Huisveld IH, Schalekamp MADH (1986) Immunoreactive renin, prorenin, and enzymatically active renin in plasma during pregnancy and in women taking oral contraceptives. J Clin Endocrinol Metab 63:1008–1015
Doss M (1984) Porphyrie und hormonale Kontrazeptiva. Dtsch Med Wschr 109:1701–1702
Duffy TJ, Ray R (1984) Oral contraceptive use: prospective follow-up of women with suspected glucose intolerance. Contraception 40:197–208
Fawer R, Dettling A, Weihs D, Welti H, Schelling JL (1978) Effect of the menstrual cycle, oral contraception and pregnancy on forearm blood flow, venous distensibility and clotting factors. Eur J Clin Pharmacol 13:251–257
Fern M, Rose DP, Fern EB (1978) Effect of oral contraceptives on plasma androgenic steroids and their precursors. Obstet Gynecol 51:541–544
Fisch IR, Frank J (1977) Oral contraceptives and blood pressure. J Am Med Ass 237:2499–2503
Garg SK, Chase HP, Marshall G, Hoops SL, Holmes DL, Jackson WE (1994) Oral contraceptives and renal and retinal complications in young women with insulin-dependent diabetes mellitus. J Am Med Ass 271:1099–1102

Gerstman BB, Piper JM, Tmita DK, Ferguson WJ (1991) Oral contraceptive estrogen dose and the risk of deep venous thromboembolic disease. Am J Epidemiol 133:32–36

Godet PG, May GR, Sutherland LR (1995) Meta-analysis of the role of oral contraceptive agents in inflammatory bowel disease. Gut 37:668–673

Godsland IF, Crook D (1994) Update on the metabolic effects of steroidal contraceptives and their relationship to cardiovascular disease risk. Am J Obstet Gynecol 170:1528–1536

Godsland IF, Crook D, Davenport M, Wynn V (1995) Relationships between blood pressure, oral contraceptive use and metabolic risk markers for cardiovascular disease. Contraception 52:143–149

Godsland IF, Crook D, Wynn V (1990) Low-dose oral contraceptives and carbohydrate metabolism. Am J Obstet Gynecol 163:348–353

Grodstein F, Colditz GA, Hunter DJ, Manson JE, Willett WC, Stampfer MJ (1994) A prospective study of symptomatic gallstones in women: relation with oral contraceptives and other risk factors. Obstet Gynecol 84:207–214

Hannaford PC, Kay CR, Vessey MP, Painter R, Mant J (1997) Combined oral contraceptives and liver disease. Contraception 55:145–151

Haspels AA (1994) Emergency contraception: a review. Contraception 50:101–108

Jick H, Jick SS, Gurewich V, Myeres MW, Vasilakis C (1995) Risk of idiopathic cardiovascular death and non-fatal venous thromboembolism in women using oral contraceptives with differing progestagen components. Lancet 346:1589–1593

Jung-Hoffmann C, Kuhl H (1987) Divergent effects of two low-dose oral contraceptives on sex hormone-binding globulin and free testosterone. Am J Obstet Gynecol 156:199–203

Jung-Hoffmann C, Kuhl H (1990) Intra- and interindividual variations in contraceptive steroid levels during 12 treatment cycles: no relation to irregular bleedings. Contraception 42:423–438

Jung-Hoffmann C, Heidt F, Kuhl H (1988) Effect of two oral contraceptives containing 30μg ethinylestradiol and 75μg gestodene or 150μg desogestrel upon various hormonal parameters. Contraception 38:593–603

Kay CR (1982) Progestogens and arterial disease – evidence from the Royal College of General Practitioners' study. Am J Obstet Gynecol 142:762–765

Knopp RH, LaRosa JC, Burkman RT (1993) Contraception and dyslipidemia. Am J Obstet Gynecol 168:1994–2005

Koster T, Small RA, Rosendaal FR, Helmerhorst FM (1995) Oral contraceptives and venous thromboembolism: a quantitative discussion of the uncertainties. J Int Med 238:31–37

Krauss RM, Burkman RT (1992) The metabolic impact of oral contraceptives. Am J Obstet Gynecol 167:1177–1184

Kudzma DJ, Bradley EM, Goldzieher JW (1972) A metabolic balance study of the effects of an oral steroid contraceptive on weight and body composition. Contraception 6:31–37

Kuhl H (1990a) Pharmacokinetics of estrogens and progestogens. Maturitas 12:171–197

Kuhl H (1994a) Wie Darmerkrankungen, Ernährung, Rauchen und Alkohol die Wirkung von oralen Kontrazeptiva beeinflussen. Geburtsh Frauenheilk 54:M1–M10

Kuhl H (1994b) Wie sich orale Kontrazeptiva und Medikamente in ihrer Wirkung beeinflussen. Geburtsh Frauenheilk 54:M23–M30

Kuhl H (1996a) Comparative pharmacology of newer progestogens. Drugs 51:188–215

Kuhl H (1996b) Effects of progestogens on haemostasis. Maturitas 24:1–19

Kuhl H, Gahn G, Romberg C, Althoff PH, Taubert HD (1985a) A randomized cross-over comparison of two low-dose oral contraceptives upon hormonal and metabolic parameters: II.Effects on thyroid function, gastrin, STH, and glucose tolerance. Contraception 32:97–107

Kuhl H, Gahn G, Romberg C, März W, Taubert HD (1985b) A randomized cross-over comparison of two low-dose oral contraceptives upon hormonal and metabolic parameters: I. Effects upon sexual hormone levels. Contraception 31:583–593
Kuhl H, März W, Jung-Hoffmann C, Heidt F, Gross W (1990b) Time-dependent alterations and lipid metabolism during treatment with low-dose oral contraceptives. Am J Obstet Gynecol 163:363–369
Kuhl H, März W, Jung-Hoffmann C, Weber J, Siekmeier R, Gross W (1993) Effect on lipid metabolism of a biphasic desogestrel-containing oral contraceptive: divergent changes in apolipoprotein B and E and transitory decrease in Lp (a) levels. Contraception 47:69–83
Lemay A, Dodin Dewailly S, Grenier R, Huard J (1990) Attenuation of mild hyperandrogenic activity in postpubertal acne by a triphasic oral contraceptive containing low doses of ethynyl estradiol and d,l-norgestrel. J Clin Endocrinol Metab 71:8–14
Lembke S, Freund H (1990) Einfluß hormonaler Kontrazeptiva auf die Stimme. Z Ärztl Fortb 84:47–49
Mandel FP, Geola FL, Lu JKH, Eggena P, Sambhi MP, Hershman JM, Judd HL (1982) Biologic effects of various doses of ethinyl estradiol in postmenopausal women. Obstet Gynecol 59:673–679
Mashchak CA, Lobo RA, Dozono-Takano R, Eggena P, Nakamura RM, Brenner PF, Mishell DR (1982) Comparison of pharmacodynamic properties of various estrogen formulations. Am J Obstet Gynecol 144:511–518
Mattson RH, Rebar RW (1993) Contraceptive methods for women with neurologic disorders. Am J Obstet Gynecol 168:2027–2032
McCann MF, Potter LS (1994) Progestin-only oral contraception: a comprehensive review. Contraception 50(Suppl.1):S1–S198
Meade TW ((1988) Risks and mechanisms of cardiovascular events in users of oral contraceptives. Am J Obstet Gynecol 158:1646–1652
Mishell DR (1982) Noncontraceptive health benefits of oral steroidal contraceptives. Am J Obstet Gynecol 142:809–816
Moltz L, Schwartz U, Hammerstein J (1980) Die klinische Anwendung von Antiandrogenen bei der Frau. Gynäkologe 13:1–17
Neuberger J, Forman D, Doll R, Williams R (1986) Oral contraceptives and hepatocellular carcinoma. Br Med J 292:1355–1357
Olivieri O, Friso S, Manzato F, Grazioli S, Bernardi F, Lunghi B, Girelli D, Azzini M, Brocco G, Russo C, Corrocher R (1996) Resistance to activated protein C, associated with oral contraceptives use; effect of formulations, duration of assumption, and doses of oestro-progestins. Contraception 54:149–152
Palmer JR, Rosenberg L, Strom BL, Harlap S, Zauber AG, Warshauer ME, Shapiro S (1992) Oral contraceptive use and risk of cutaneous malignant melanoma. Cancer Causes Control 3:547–554
Parazzini F, LaVecchia C, Negri E, Franceschi S, Moroni S, Chatenoud L, Bolis G (1997) Case-control study of estrogen replacement therapy and risk of cervical cancer. Br Med J 315:85–88
Petersen KR, Skouby SO, Sidelmann J, Jespersen J (1994) Assessment of endothelial function during oral contraception in women with insulin-dependent diabetes mellitus. Metabolism 43:1379–1383
Petersen KR, Sidelmann J, Skouby SO, Jespersen J (1993) Effects of monophasic low-dose oral contraceptives on fibrin formation and resolution in young women. Am J Obstet Gynecol 168:32–38
Population Reports (1995) Injectables and implants – new era for injectables. Series K:1–31
Population Reports (1992) Decisions for Norplant programs. Series K:1–31
Rannevik G, Jeppson S, Kullander S (1972) Effect of oral contraceptives on the liver in women with recurrent cholestasis (hepatosis) during previous pregnancies. J Obstet Gynaecol Br Cmwlth 79:1128–1136

Rimm EB, Manson JE, Stampfer MJ, Colditz GA, Willett WC, Rosner B, Hennekens CH, Speizer FE (1992) Oral contraceptive use and risk of type 2 (non-insulin-dependent) diabetes mellitus in a large prospective study of women. Diabetologia 35:967–972

Rooks JB, Ory HW, Ishak K, Strauss LT, Greenspan JR, Paganini-Hill A, Tyler CW (1979) Epidemiology of hepatocellular adenoma. The role of oral contraceptive use. J Am Med Ass 242:644–648

Rosenberg L (1991) The risk of liver neoplasia in relation to combined oral contraceptive use. Contraception 43:643–652

Rosing J, Tans G, Nicolaes GAF, Thomassen MCLGD, van Oerle R, van der Ploeg PMEN, Heijnen P, Hamulyak K, Hemker HC (1997) Oral contraceptives and venous thrombosis: different sensitivities to activated protein C in women using second- and third- generation oral contraceptives. Br J Haematol 97:233–238

Roy S (1991) Nonbarrier contraceptives and vaginitis and vaginosis. Am J Obstet Gynecol 165:1240–1244

Royal College of General Practitioners' Oral Contraceptive Study (1977) Effect on hypertension and benign breast disease of progestagen component in combined oral contraceptives. Lancet I:624

Rubin GL, Ory HW, Layde PM (1982) Oral contraceptives and pelvic inflammatory disease. Am J Obstet Gynecol 144:630–635

Schlesselman JJ (1991) Oral contraceptives and neoplasia of the uterine corpus. Contraception 43:557–580

Schweitzer IL, Weiner JM, McDeak CM, Thursby MW (1975) Oral contraceptives in acute viral hepatitis. J Am Med Ass 233:979–980

Shaaban MM, Hammad WA, Fathalla MF, Ghaneimah SA, El-Sharkawy MM, Salim TH, Ali MY, Liao WC, Smith SC (1982) Effects of oral contraception on liver function tests and serum proteins in women with past viral hepatitis. Contraception 26:65–74

Stanczyk FZ (1997) Pharmacokinetics of the new progestogens and influence of gestodene and desogestrel on ethinylestradiol metabolism. Contraception 55:273–282

Stanczyk FZ, Roy S (1990) Metabolism of levonorgestrel, norethindrone, and structurally related contraceptive steroids. Contraception 42:67–96

Stanford JL, Brinton LA, Berman R, Mortel R, Twiggs LB, Barrett RJ, Wilbanks GD, Hoover RN (1993) Oral contraceptives and endometrial cancer: do other risk factors modify the association? Int J Cancer 54:243–248

Steingold KA, Cefalu W, Pardridge W, Judd HL, Chaudhuri G (1986) Enhanced hepatic extraction of estrogens used for replacement therapy. J Clin Endocrinol Metab 62:761–766

Szoka PR, Edgren RA (1988) Drug interactions with oral contraceptives: compilation and analysis of an adverse experience report database. Fertil Steril (Suppl) 49:31–38

Taubert HD, Kuhl H (1995) Kontrazeption mit Hormonen, 2nd edn, Thieme, Stuttgart New York

van der Vange N, Blankenstein MA, Kloosterboer HJ, Haspels AA, Thijssen JHH (1990) Effects of seven low-dose combined oral contraceptives on sex hormone binding globulin, corticosteroid binding globulin, total and free testosterone. Contraception 41:345–352

Vessey MP, Lawless M, McPherson K, Yeates D (1985) Progestogen-only oral contraceptives. Findings in a large prospective study with special reference to effectiveness. Br J Family Plann 10:117–121

Vessey MP, Painter R (1994) Oral contraceptive use and benign gallbladder disease; revisited. Contraception 50:167–173

Vessey MP, Wright NH, McPherson K, Wiggins P (1978) Fertility after stopping different methods of contraception. Br Med J I:265–267

Villard-Mackintosh L, Vessey MP (1993) Oral contraceptives and reproductive factors in multiple sclerosis incidence. Contraception 47:161–168

Vin F, Allaert FA, Levardon M (1992) Influence of estrogens and progesterone on the venous system of the lower limbs in women. Phlebology 18:888–892
Weir RJ, Davies DL, Fraser R, Morton JJ, Tree M, Wilson A (1975) Contraceptive steroids and hypertension. J Steroid Biochem 6:961–964
Whittemore AS, Harris R, Intyre J, and the Collaborative Ovarian Cancer Group (1992) Characteristics relating to ovarian cancer risk: collaborative analysis of 12 US case-control studies. Am J Epidemiol 136:1184–1203
Williams G, Anderson E, Howell A, Watson R, Coyne J, Roberts SA, Potten CS (1991) Oral contraceptive (OCP) use increases proliferation and decreases estrogen receptor content of epithelial cells in the normal human breast. Int J Cancer 48:206–210
Wingrave SJ (1982) Progestogen effects and their relationship to lipoprotein changes. Acta Obstet Gynecol Scand (Suppl) 105:33–36
World Health Organisation Collaborative Study of Neoplasia and Steroid Contraceptives (1985) Invasive cervical cancer and combined oral contraceptives. Br Med J 190:961–965
World Health Organisation Task Force for Epidemiological Research on Reproductive Health (1994a) Progestogen-only contraceptives during lactation: I. Infant growth. Contraception 50:35–53
World Health Organisation Task Force for Epidemiological Research on Reproductive Health (1994b) Progestogen-only contraceptives during lactation: II. Infant development. Contraception 50:55–68
World Health Organisation Collaborative Study of Cardiovascular Disease and Steroid Hormone Contraception (1995) Effect of different progestagens in low estrogen oral contraceptives on venous thromboembolic disease. Lancet 346:1582–1588
World Health Organisation Collaborative Study of Cardiovascular Disease and Steroid hormone Contraception (1996a) Ischaemic stroke and combined oral contraceptives: results of an international, multicentre, case-control study. Lancet 348:498–504
World Health Organisation Collaborative Study of Cardiovascular Disease and Steroid Hormone Contraception (1996b) Haemorrhagic stroke, overall stroke risk, and combined oral contraceptives: results of an international, multicentre, case-control study Lancet 348:505–510
World Health Organisation Collaborative Study of Cardiovascular Disease and Steroid Hormone Contraception (1997) Acute myocardial infarction and combined oral contraceptives: results of an international multicentre case-control study, Lancet 349:1202–1209
Wynn V (1982) Effect of duration of low-dose oral contraceptive administration on carbohydrate metabolism. Am J Obstet Gynecol 142:739–746
Yuzpe AA, Smith RP, Rademaker AW (1982) A multicentre clinical investigation employing ethinylestradiol combined with dl-norgestrel as a postcoital contraceptive agent, Fertil Steril 37:508–513

CHAPTER 39

Hormone Replacement Including Osteoporosis

H.L. JØRGENSEN and B. WINDING

A. The Menopause

There has been a steady increase in female life expectancy in developed countries over the past century. Since the age of onset of menopause has remained relatively constant (in western countries, it occurs at about 50 years of age (NACHTIGALL 1988), this has resulted in a marked increase of women living beyond the menopause and, thus, potentially being in a state of hormone deficiency. In fact, women can now expect to live one-third of their lives in this phase (Fig. 1).

The onset of menopause is caused by a decline in the number of oocytes in the ovaries from several million in the fetal state to a few hundred at the time of the menopausal transition. The resulting lower estrogen level has consequences in a host of tissues, in which estrogen receptors are present, giving rise to both diverse immediate symptoms (climacteric complaints) and changes that have long-term effects on organ systems, such as bone, the cardiovascular system and the brain.

I. Climacteric Complaints

Apart from an irregular bleeding pattern, the most prominent and immediate sign of the menopausal transition is a vasomotor disorder causing the phenomenon of hot flashes. Although this symptom has, of course, been recognized for many years and KUPPERMAN et al. (1953) tried to quantify the whole range of climacteric complaints, including hot flashes, it was not until 1975 that a study of the actual physiological changes was published (MOLNAR 1975). Hot flashes appear suddenly and cause a sensation of warmth or even intense heat propagating throughout the body. Often, the women experience perspiration and flushing at the same time. The prevalence of hot flashes varies considerably among studies, ranging from around 30% to 65% in perimenopausal women and 60% to 90% in the first two postmenopausal years (KRONENBERG 1990). Hot flashes can be severe and impair normal daily activities of the afflicted women. This symptom is the primary reason why menopausal women seek out their general practitioner or gynecologist for treatment. As summarized in Table 1, there are several other symptoms that may accompany the menopausal transition.

Table 1. Kuppermann's index scoring system for evaluating climacteric symptoms. Each symptom is graded on a scale from 0 to 3. This grade is multiplied by a factor (ranging from 1 to 4), thus yielding a weighted total. The highest possible sum of the weighted totals is 51

Symptom	Kuppermann's index score				Factor	Total
	0	1	2	3		
Hot flushes	None	One per day	2–5 per day	More than five per day	4	
Sweats	None	Up to one per week	1–4 per week	Five or more per week	2	
Sleep	Good	Moderate	Poor	Very poor	2	
Nervousness	None	Mild	Moderate	Severe	2	
Headaches	None	Rare	Frequent	Very frequent	1	
Depression	None	Slight	Moderate	Severe	1	
Vertigo	None	Rare	Frequent	Very frequent	1	
Asthenia	None	Slight	Moderate	Severe	1	
Arthralgia	None	Slight	Moderate	Severe	1	
Palpitations	None	Up to one per week	2–5 per week	More than five per week	1	
Vaginal dryness	None	Mildly uncomfortable	Uncomfortable	Very uncomfortable	1	
Sum						

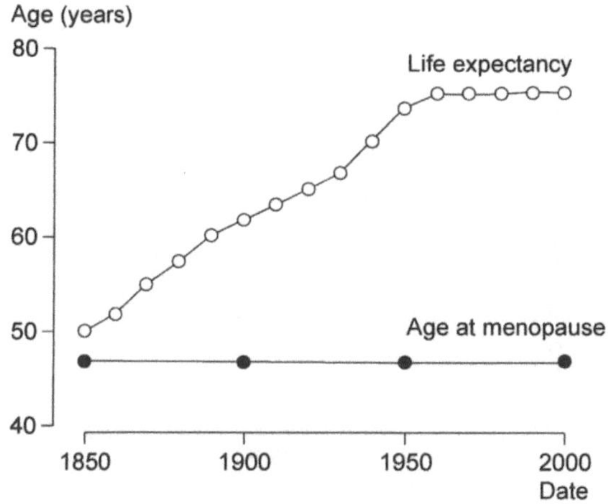

Fig. 1. Life expectancy and age at menopause from 1850 to 2000. There has been a steady increase in life expectancy during this period, whereas the age of menopause has remained constant. Adapted from (NACHTIGALL 1988)

The coupling of these symptoms with declining ovarian function, has been realized for almost a century, leading physicians to the obvious conclusion that replacing the missing sex hormones would alleviate the subjective evidence of physical disturbance. However, it is only in the latter part of the 20th century, with the introduction of scoring systems, such as Kupperman's, that systematic placebo controlled, randomized studies have been able to quantify the efficacy of hormone replacement therapy (HRT) on the symptoms. These scoring systems are now widely used in clinical studies in various modified forms. From the overwhelming number of studies examining the effect of HRT on climacteric symptoms, the positive effect of these treatments has now been fully established.

Until the mid-1970s, HRT was almost exclusively used for the relief of climacteric symptoms. Today, the picture is much more complicated. On one hand, side effects, for example a probable increase in the risk of breast cancer, have come to light. On the other hand, unexpected positive long-term effects on the skeletal system and, possibly, also on the cardiovascular and central nervous system have emerged.

II. Estrogens and the Skeleton

As described in Chap. 26, estrogen is an extremely important regulator of skeletal metabolism. With the dramatic decline of estrogen levels occurring at menopause, an imbalance in the activity of the two most important bone cells, osteoblasts and osteoclasts, occurs. The activity of both of these cell populations increases significantly, accelerating the process of bone turnover. As seen in Fig. 2, the activity of the osteoclasts (bone-resorbing cells) increases more

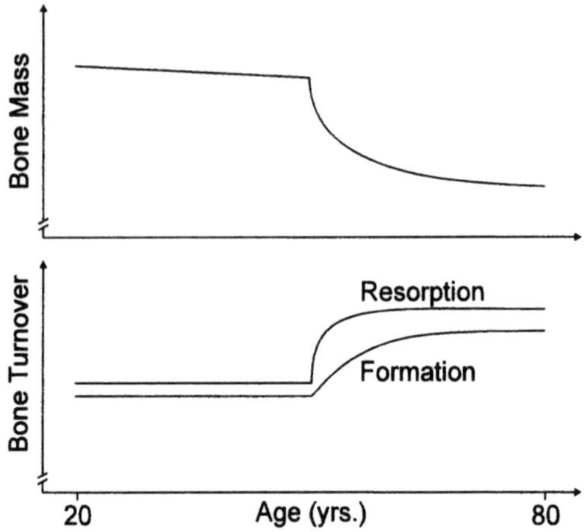

Fig. 2. Bone turnover and bone mass in normal women with increasing age

than that of the osteoblasts (bone-forming cells), leading to a net deficiency of newly formed bone during the remodeling process. In all women, this will lead to fragilization of the skeleton. However, the clinical significance of this depends on two crucial factors: first, the amount of bone mass attained during adolescent growth, which is called the peak bone mass and, second, the rate of bone loss occurring after the menopause. The clinical condition resulting from the loss of bone is called osteoporosis. Although, osteoporosis under certain conditions also occurs in men, the prevalence is much higher in women due the fact that men obtain a much higher peak bone mass and do not have an accelerated bone loss comparable with that seen in women after the menopause (Fig. 3).

Osteoporosis is defined as a systemic skeletal disease characterized by low bone mass and microarchitectural deterioration of bone tissue, with a subsequent increase in bone fragility and susceptibility to fractures (CONSENSUS DEVELOPMENT CONFERENCE 1993). Since this definition is somewhat difficult to apply to a clinical setting, a study group under the World Health Organization (KANIS et al. 1994) has proposed the following guidelines for the diagnosis of osteoporosis in adult women, based on a bone mineral density (BMD) measurement:

A. Normal. BMD not more than one standard deviation (SD) below the mean value of peak bone mass in young normal women.
B. Low bone mass (osteopenia). BMD between 1 SD and 2.5 SD below the mean value of peak bone mass in young normal women.
C. Osteoporosis. BMD more than 2.5 SD below the mean value of peak bone mass in young normal women.
D. Established osteoporosis. BMD more than 2.5 SD below the mean value of peak bone mass in young normal women and the presence of fractures.

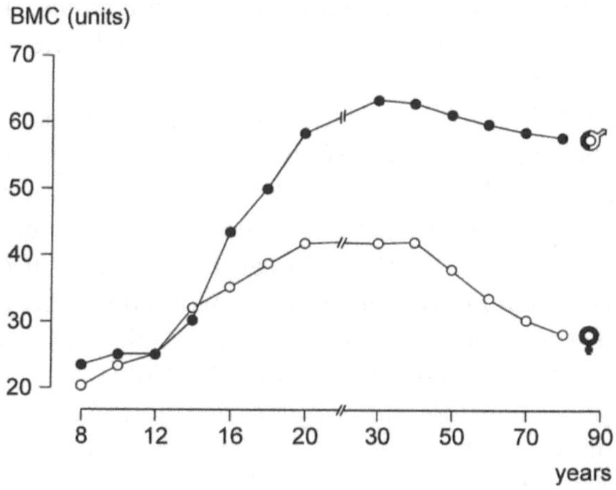

Fig. 3. Bone mineral content (*BMC*) at the lower forearm in men and women. At maturity, BMC is 30–50% lower in women then in men. Furthermore, an accelerated rate of bone loss occurs in women at the time of menopause. These two factors lead to osteoporosis being much more common in women than in men

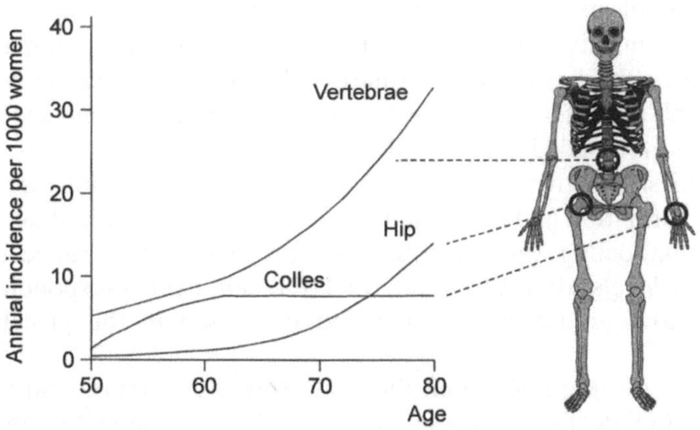

Fig. 4. Incidence of osteoporotic fractures in women with increasing age

Based on these definitions, 54% of postmenopausal white women in the United States have osteopenia and another 30% have osteoporosis. About half of the osteoporotic women are estimated to have established osteoporosis according to the above definition (MELTON 1995).

As can be seen in Fig. 4, the must common sites of osteoporotic fractures are the vertebral column, the hip, and the lower forearm. The incidence of osteoporotic fractures is growing due to the rise in life expectancy, and a threefold increase in worldwide fracture incidence is expected over the next 60 years (MELTON et al. 1992). Osteoporosis is thus a major health problem both in terms of the cost to society and to the afflicted individuals. This is especially true of patients with hip fractures, who experience a high rate of mortality and morbidity (MILLER 1978).

Identifying individuals at risk is therefore of prime importance since prevention is possible, e.g., using HRT (CONSENSUS DEVELOPMENT CONFERENCE 1993). Measurement of BMD by bone densitometry is currently considered to be the most important predictor of osteoporotic fractures (CONSENSUS DEVELOPMENT CONFERENCE 1993). BMD can be measured at different skeletal sites by a variety of non-invasive methods. Most commonly used are single-energy X-ray absorptiometry (SEXA) (KELLY et al. 1994) and dual-energy X-ray absorptiometry (DEXA) (HANSEN et al. 1990).

Although, bone densitometry is by far the most widely used diagnostic tool in osteoporosis at the moment, several other methods are currently emerging. The rate of bone turnover can be indirectly assessed by use of a number of biochemical markers. The major part of the organic bone matrix consists of type-I collagen. When bone is resorbed, this collagen is degraded into smaller fragments and released into the extracellular fluid. These fragments are excreted into the urine and can be measured by a variety of assays as markers of bone resorption. However, the specificity and clinical utility vary considerably among the different markers.

The osteoblasts produce small amounts of osteocalcin for incorporation into the bone matrix. A proportion of this escapes into the serum and can be measured as a marker of bone formation. Alkaline phosphatase, believed to reflect osteoblast differentiation, is less specific, but newer bone-specific assays have been developed (ROSALKI and FOO 1984).

Figure 5 shows that the estrogen deficiency after menopause increases both bone formation and resorption, while HRT reverses this effect. The level of bone remodeling (as measured by the markers) is reduced to premenopausal levels after 12 months of HRT (Fig. 6). Correspondingly, bone mass increased in the HRT group and decreased in the placebo-treated group.

When instituting HRT, both the rate of bone formation and the rate of bone resorption decline. However, in the initial phase, the rate of bone resorption decreases faster than that of bone formation (Fig. 7), leading to a gain in bone mass in the first months of treatment (Fig. 8). After a while, this difference disappears and bone mass will remain relatively constant for the duration of antiresorptive therapy. As can also be seen from Fig. 8, the rate of bone loss will resume at the same rate when the treatment is stopped (CHRISTIANSEN et al. 1981).

Recently, several alternatives to HRT for the prevention of osteoporosis have been developed. The so-called selective estrogen receptor modulators, SERMs (see below), and the group of compounds called bisphosphonates, such as the recently launched Fosamax (Alendronate), are now good alternatives for the treatment of women who cannot tolerate HRT or who are reluctant to use HRT for fear of side effects.

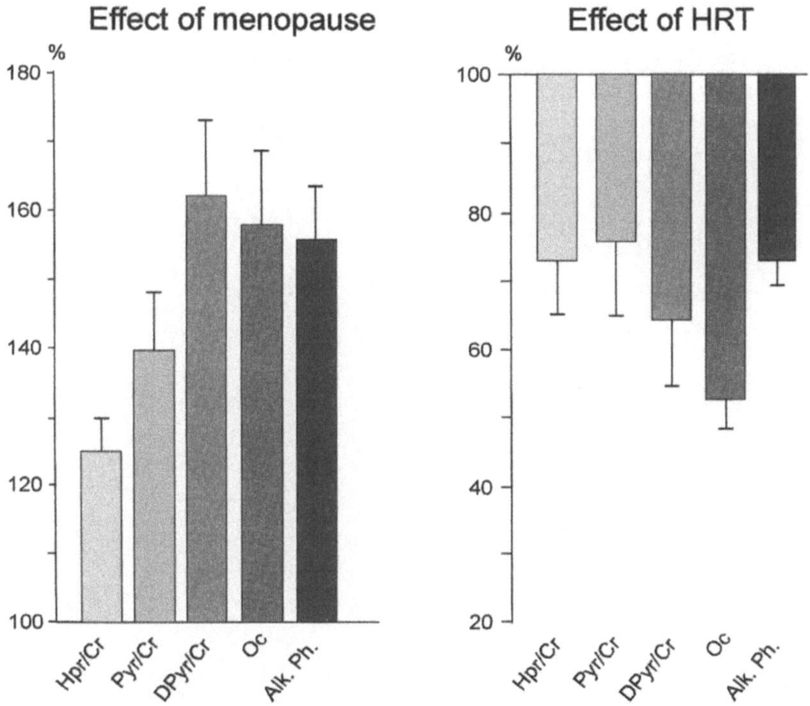

Fig. 5. The effects of menopause and hormone replacement therapy (HRT), respectively, on three urinary markers of bone resorption (corrected for urinary creatinine excretion): hydroxyproline (*Hpr/Cr*); pyridinoline (*Pyr/Cr*); deoxypyridinoline (*DPyr/Cr*); and on two serum markers of bone formation: osteocalcin (*Oc*) and alkaline phosphatase (*AP*)

III. HRT, Serum Lipids and the Risk of Cardiovascular Disease

Popular belief has held for many years that coronary heart disease is much more prevalent in men than in women. In fact, this is not true. Cardiovascular disease (CVD) remains the leading cause of death among postmenopausal women in the developed world. It is, however, incontrovertible that the mean age of the afflicted women is 8–10 years higher than the mean age for men. Indeed, CVD is rarely seen in premenopausal women, whereas it is by no means uncommon that men suffer their first myocardial infarction in their forties.

The increased incidence of CVD after the menopause and the accompanying rise in cholesterol levels (Fig. 9) have formed the basis of a hypothesis that female sex hormones could have a cardioprotective effect, possibly mediated through changes in the lipid profile. In support of this hypothesis, numerous prospective studies have shown that HRT has a favorable effect on total cholesterol and induces beneficial changes in the lipoprotein subfractions, i.e., elevating high-density lipoprotein (HDL) levels and decreasing low-density

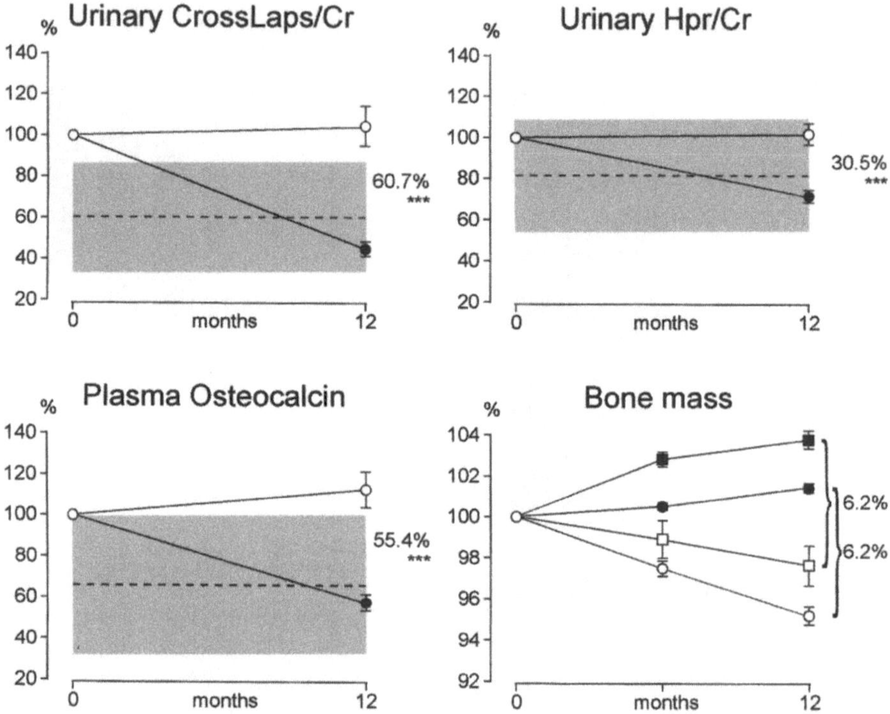

Fig. 6. Mean changes in the biochemical markers and forearm (*squares*) and spinal (*circles*) bone mass after 12 months of hormone replacement therapy (*HRT*) ($n = 80$, *filled squares* and *circles*) or placebo ($n = 35$, *open squares* and *circles*), expressed as a percentage of baseline values. The *hatched areas* represent the variation in premenopausal women. CrossLaps is a relatively new marker of bone resorption, based on the measurement of degradation products of type-I collagen crosslinks

lipoprotein (LDL) levels (MATTHEWS et al. 1989; JENSEN et al. 1990; STEVENSON et al. 1993).

Furthermore, several investigations, in both animals and humans, have found the prevalence of coronary calcification to be lower among HRT users. An example of this was demonstrated in an interesting recent Israeli case control study (SHEMESH et al. 1997), involving 41 postmenopausal women on HRT and 37 age-matched controls who had never taken HRT. This study used computed tomography (CT) to assess the prevalence and extent of coronary calcification. Using stepwise logistical regression, HRT was found to be the only independent variable determining the presence of coronary calcium with an odds ratio of 0.2 [95% confidence interval (CI), 0.06–0.63, $P = 0.006$]. Until recently, the only evidence that HRT decreases the actual risk of coronary heart disease by 35–50% came from meta analyses of observational studies (STAMPFER and COLDITZ 1991; GRADY et al. 1992).

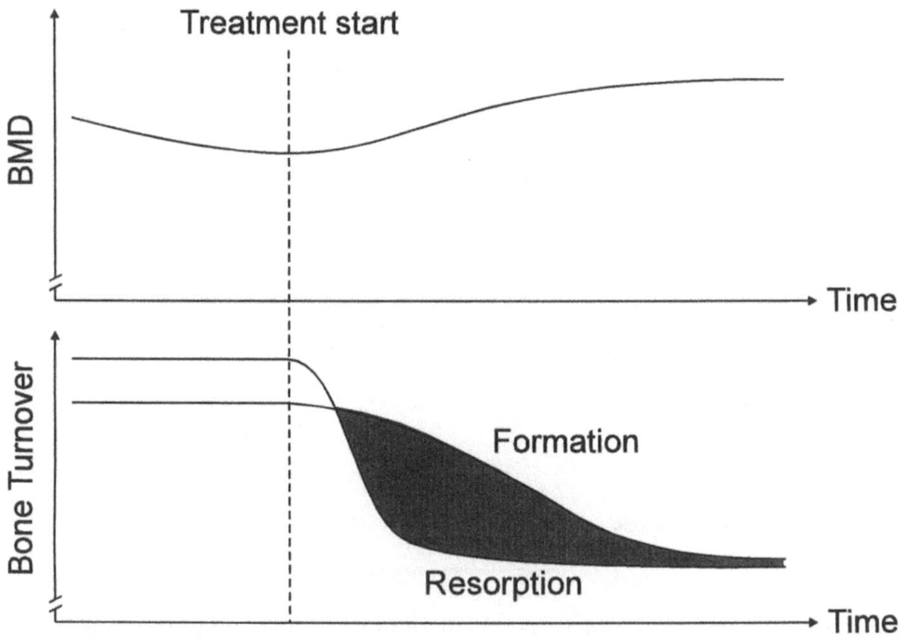

Fig. 7. The effect of antiresorptive therapy on bone mineral density (*BMD*) and bone turnover

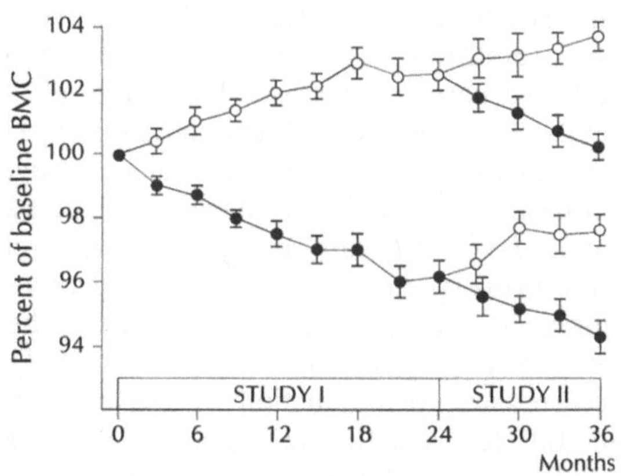

Fig. 8. Bone mineral content (*BMC*) at the lower forearm in early postmenopausal women (study I), receiving either combined estrogen/progestogen treatment (*open circles*) or placebo (*filled circles*). The women were re-randomized after 2 years (study II); 50% of the hormone replacement therapy (*HRT*)-treated women were changed to placebo and 50% of the placebo-treated women were changed to active treatment

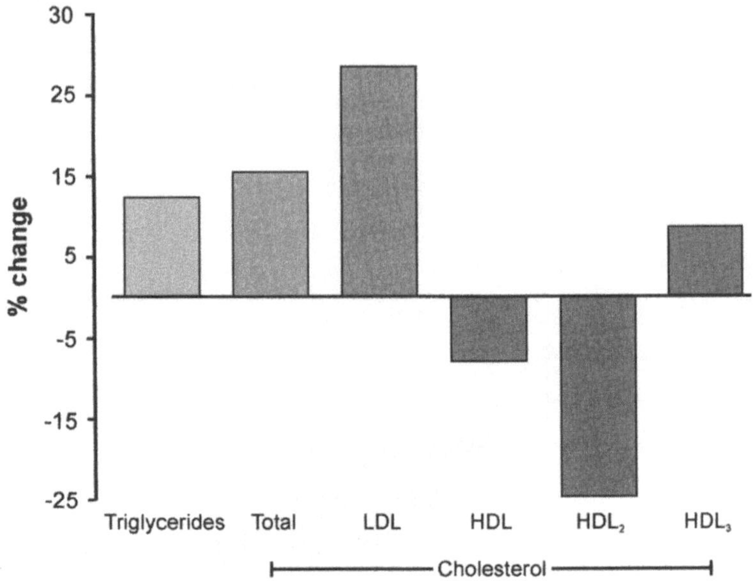

Fig. 9. Changes in serum lipids and lipoproteins resulting from the menopause. Adapted from (STEVENSON et al. 1993)

All the above evidence is circumstantial, but weighty, in favor of a positive effect of HRT on CVD. However, the only randomized, placebo-controlled trial (the HERS study) published to date, very surprisingly, showed no reduction in the overall rate of cardiovascular events in postmenopausal women with established coronary disease (HULLEY et al. 1998). The study included a total of 2763 women with a mean age of 66.7 years; 1380 women were treated with 0.625 mg of conjugated equine estrogens plus 2.5 mg medroxyprogesterone acetate and 1383 women were treated with placebo. The mean follow-up period was 4.1 years. In the hormone treated group, 172 women suffered the primary end point of combined non-fatal myocardial infarction or cardiovascular death versus 176 women in the placebo group. This result was clearly not significant [relative risk (RR) 0.99 (95% CI, 0.80–1.20)] and occurred despite a reduction in LDL levels of 11% ($P < 0.001$) and an increase in HDL levels of 10% ($P < 0.001$) in the hormone group. The study even suggested an initial detrimental effect with a trend toward more cardiovascular deaths in the HRT group than placebo during the first year. Furthermore, the HRT group experienced significantly higher levels of venous thrombolic events [RR 2.89 (95% CI, 1.50–5.58)] and gallbladder disease [RR 1.38 (95% CI, 1.00–1.92)]. On this basis, the authors of the HERS study do not recommend initiating HRT for the purpose of secondary prevention of CVD.

Although this study is disappointing, it cannot be completely excluded that HRT has a role in primary prevention of CVD. The women in the HERS study all had established CVD and, therefore, it can be speculated that they constitute a subset of women with a lower responsiveness to endogenous estrogen. This could explain the lack of effect of adding exogenous estrogen to this group. There is clearly a need for large randomized placebo-controlled primary prevention studies to determine the future role of HRT for this indication.

IV. Estrogens and Neurodegenerative Diseases

HRT has been proposed to be preventive of the development of neurodegenerative diseases, though this has been based merely on retrospective and short-term, low-numbered prospective trials and animal studies. In a recent paper, in which a retrospective study and a low-numbered, open-label prospective trial were combined (Resnick et al. 1997), it was found that women receiving HRT, when compared with a group of women who had never received HRT, had higher scores in a test measuring short-term visual memory, visual perception and constructional skills. This study, though, suffers from the unavoidable selection bias inherent in all observational studies. Several other retrospective analyses have supported the idea of a protective effect of HRT on dementia.

During the last year, two groups have questioned this concept based on both a meta analysis of the available data (Yaffe et al. 1998) and the, to date, only prospective randomized placebo-controlled study (Polo-Kantola et al. 1998), which showed no effect of HRT. The meta analysis found a 29% decrease in the development of dementia among HRT users. However, the studies were heterogeneous and, although plausible biological mechanisms exist for the action of HRT on the central nervous system, the authors concluded that large-scale randomized, placebo-controlled, long-term clinical trials must be conducted in order to definitively establish the efficacy of estrogen in the prevention of dementia, Alzheimer's disease and other neurodegenerative diseases. The study by Polo-Kantola and co-workers (1998) does not, though, meet this criteria since the treatment duration was short and the study only included 70 women.

It should be reinforced that the general belief that there should be a beneficial effect of HRT on neurodegenerative diseases in human is merely speculative. Consequently, when women are receiving information about HRT, the effect on neurodegenerative diseases should probably not be discussed before the appropriate clinical studies have been conducted and evaluated.

V. Selective Estrogen Receptor Modulators

A great deal of interest has recently been focused on the class of drugs known as selective estrogen receptor modulators (SERMs). For many years, the only SERM on the market was tamoxifen, widely used as adjuvant treatment of breast cancer. However, another treatment indication namely prevention of osteoporosis, has come to light with the introduction of raloxifene (Delmas et al. 1997). This study, which randomized 601 postmenopausal women to either different doses of raloxifene or placebo with a treatment duration of 2 years, concluded that daily therapy with raloxifene increases BMD, lowers serum concentrations of total and LDL cholesterol, and does not stimulate the endometrium. Several similar SERMs are under clinical trials and can be expected to reach the market soon. Although, tamoxifen also prevents bone loss to a certain extent, this drug is clearly unsuitable for use in this indication due to the well-known side effect of increased risk of endometrial cancer. This side effect is not seen with the newer SERMs.

In addition to the positive effect of SERMs on bone, another very interesting effect has recently been shown. In a study conducted by Fisher et al. (1998), 13,388 healthy women at increased risk of breast cancer were randomized to receive either tamoxifen or placebo. The women were followed for 5 years and all cases of breast cancer were documented. In this study, tamoxifen reduced the risk of invasive breast cancer by 49% ($P < 0.00001$), with a cumulative incidence of 43.4 vs 22.0 per 1000 women in the placebo and the tamoxifen group, respectively. However, in two similar tamoxifen chemopreventive trials, Powles et al. (1998) and Veronesi et al. (1998), which were of considerably smaller size, the preventive effect was not reproduced. Furthermore, pooled data from large, randomized double-blind trials with raloxifene has shown a relative risk for the development of breast cancer of 0.42 (CI, 0.25–0.73) (Jordan et al. 1998).

While the evidence for the protective effect of SERMs on the development of breast cancer seem to be accumulating rapidly, its still seems to be too soon to advocate widespread use of such drugs solely for a preventive indication. Several questions remain unanswered, such as whether the reduction in breast cancer incidence will be reflected in a comparable reduction in mortality.

References

Christiansen C, Christensen MS, Transbol I (1981) Bone mass in postmenopausal women after withdrawal of estrogen/gestagen replacement therapy. Lancet 1:459–61

Consensus Development Conference (1993) Diagnosis, prophylaxis, and treatment of osteoporosis. JAMA 94:646–650

Delmas PD, Bjarnason NH, Mitlak BH, Ravoux AC, Shah AS, Huster WJ, Draper M, Christiansen C (1997) Effects of raloxifene on bone mineral density, serum cholesterol concentrations, and uterine endometrium in postmenopausal women. N Engl J Med 337:1641–1647

Fisher B, Costantino JP, Wickerham DL, Redmond CK, Kavanah M, Cronin WM, Vogel V, Robidoux A, Dimitrov N, Atkins J, Daly M, Wieand S, Tan-Chiu E, Ford L, Wolmark N (1998) Tamoxifen for prevention of breast cancer: report of the National Surgical Adjuvant Breast and Bowel Project P-1 Study. J Natl Cancer Inst 90:1371–1388

Grady D, Rubin SM, Petitti DB, Fox CS, Black D, Ettinger B, Ernster VL, Cummings SR (1992) Hormone therapy to prevent disease and prolong life in postmenopausal women. Ann Intern Med 117:1016–1037

Hansen MA, Hassager C, Overgaard K, Marslew U, Riis BJ, Christiansen C (1990) Dual-energy X-ray absorptiometry, a precise method of measuring bone mineral density in the lumbar spine. J Nucl Med 31:1156–1162

Hulley S, Grady D, Bush T, Furberg C, Herrington D, Riggs B, Vittinghoff E (1998) Randomized trial of estrogen plus progestin for secondary prevention of coronary heart disease in postmenopausal women. JAMA 280:605–613

Jensen J, Nilas L, Christiansen C (1990) Influence of menopause on serum lipids and lipoprotein. Maturitas 12:321–331

Jordan VC, Glusman JE, Eckert S, Lippman M, Powles T, Costa A, Morrow M, Norton L (1998) Incident primary breast cancers are reduced by raloxifene: integrated data from multicenter double-blind, randomized trials in 12,000 postmenopausal women (abstract). Proceedings of the 34th Annual Meeting of the American Society of Clinical Oncology, 16–19 May, Philadelphia, USA

Kanis JA, Melton LJ III, Christiansen C, Johnston CC, Khaltaev N (1994) The diagnosis of osteoporosis. J Bone Miner Res 9:1137–1141

Kelly TL, Crane G, Baran DT (1994) Single X-ray absorptiometry of the forearm, precision, correlation and reference data. Calcif Tissue Int 54:212–218

Kronenberg F (1990) Hot flashes: epidemiology and physiology. Ann NY Acad Sci 592:52–86

Kupperman HS, Blatt MHG, Wiesbader (1953) Comparative clinical evaluation of estrogenic preparations by the menopause and amenorrheal indices. Endocrinology 13:688–703

Matthews KA, Meilahn E, Kuller LH, Helsey SF, Caggiula AW, Wing RR (1989) Menopause and risk factors for coronary heart diseases. N Engl J Med 321:641–646

Melton LJ III (1995) How many women have osteoporosis now? J Bone Miner Res 10:175–177

Melton LJ, Chrischilles EA, Cooper C, Lane AW, Riggs BL (1992) Perspective: how many women have osteoporosis. J Bone Miner Res 7:1005–1010

Miller CW (1978) Survival and ambulation following hip fracture. J Bone Joint Surg Am 60:930–934

Molnar GW (1975) Body temperatures during menopausal hot flashes. J Appl Physiol 38:499–503

Nachtigall LE (1988) Female population changes. Managing the climacteries. CIBA symposium, San Francisco, USA

Polo-Kantola P, Portin R, Polo O, Helenius H, Irjala K, Erkkola R (1998) The effect of short-term estrogen replacement therapy on cognition: a randomized, double-blind, cross-over trial in postmenopausal women. Obstet Gynecol 91:459–466

Powles T, Eeles R, Ashley S, Easton D, Chang J, Dowsett M, Tidy A, Viggers J, Davey J (1998) Interim analysis of the incidence of breast cancer in the Royal Marsden Hospital tamoxifen randomised chemoprevention trial. Lancet 352:98–101

Resnick SM, Metter EJ, Zonderman AB (1997) Estrogen replacement therapy and longitudinal decline in visual memory. A possible protective effect? Neurology 49:1491–1497

Rosalki SB, Foo AY (1984) Two new methods for separating and quantifying bone and liver alkaline phosphatase isoenzymes in plasma. Clin Chem 30:1182–1186

Shemesh J, Frenkel Y, Leibovitch L, Grossman E, Pines A, Motro M (1997) Does hormone replacement therapy inhibit coronary artery calcification? Obstet Gynecol 89:989–92

Stampfer MJ, Colditz GA (1991) Estrogen replacement therapy and coronary heart disease: a quantitative assessment of the epidemiologic evidence. Prev Med 20:47–63

Stevenson JC, Crook D, Godsland IF (1993) Influence of age and menopause on serum lipids and lipoproteins in healthy women. Atherosclerosis 98:83–90

Veronesi U, Maisonneuve P, Costa A, Sacchini V, Maltoni C, Robertson C, Rotmensz N, Boyle P (1998) Prevention of breast cancer with tamoxifen: preliminary findings from the Italian randomised trial among hysterectomised women. Italian Tamoxifen Prevention Study. Lancet 352:93–97
Yaffe K, Sawaya G, Lieberburg I, Grady D (1998) Estrogen therapy in postmenopausal women: effects on cognitive function and dementia. JAMA 279:688–695

CHAPTER 40
Gynaecological Disorders

G. SAMSIOE

A. Introduction

Estrogen therapy is indicated whenever estrogen deficiency is at hand. Such deficiency could be either general or local. The major indications for systemic therapy would be climacteric symptoms, such as sweats and hot flushes etc., and for inducing or maintaining female sex characteristics in primary and secondary amenorrhoea. However, estrogen monotherapy is seldom used for lengthier periods due to its stimulatory effect on the endometrium. Prolonged estrogen treatment may lead to endometrial cancer, but much more often to various types of bleeding disorders. A progestogen is therefore usually added when administering estrogen to women with intact uteri. In this chapter, a number of clinical conditions will be mentioned, in which estrogen treatment has been shown to be useful. However, it should be remembered that, in most cases, a progestogen co-medication is necessary in women with intact uteri.

The classical indications for anti-estrogens are adjuvant therapy to breast cancer (tamoxifen) and ovulation induction (clomiphene). However, these molecules may also possess estrogenic agonism. Novel molecules, so-called selective estrogen-receptor modulators (SERMs), have been developed which possess beneficial estrogen agonistic effects on bone and lipid metabolism, as well as anti-estrogen effects on, e.g. the breast and the endometrium.

B. Primary Amenorrhoea

In young women with primary amenorrhoea, it is essential to establish that no congenital anomalies are present of essence to the establishment of a menstrual cycle. Such anomalies include the absence of ovaries, uterus or vagina. An impermeable hymen could result in a haematocolpic condition with no visualisation of blood. Furthermore, the functional capacity of the organs needs to be outlined. In other words, do the ovaries contain follicles? Does the uterus contain an endometrium? Does the cervical canal open into the vagina?

It is beyond the scope of this chapter to outline the differential diagnostic procedures used to establish the cause behind a primary amenorrhoea. Women with congenital anomalies and women with various gonadal dysfunctions, such as those with the chromosomal abnormality Turner's syndrome, may need estrogen replacement therapy. This is particularly prudent in young girls in whom pubertal development has been impeded by the lack of estrogens. Treatment with estrogen monotherapy is usually indicated to enhance the overall female appearance, especially to increase the volume of the breast tissue. The time and dose of the monotherapy is dependent on the clinical outcome. A time frame of 2–3 years is often advocated. In order to mimic the natural pubertal development, estrogen treatment, e.g. oral oestradiol, should start at a fairly low dose, which is then gradually increased, aiming at reaching concentrations similar to those in normal menstruating women.

Commonly, oral oestradiol is given in doses between 1–2 mg and 3 mg and corresponding doses for transdermal oestradiol. A stepwise increase is further recommended to minimise the side effects, especially for oral preparations. Transdermal preparations and conjugated equine estrogens could also be used. Uterine monotherapy also induces uterine growth and an increase of uterine volume, and endometrial development should be monitored preferably by ultrasound. If the endometrium becomes too thick or estrogen breakthrough bleedings occur, a progestogen should be added to ensure a normal cyclical bleed. Oral contraceptives could also eventually be used.

Higher doses of estrogen, usually in the form of ethinyloestradiol, are used when inducing femininity in transsexuals; usually 50–100 μg per day is required to induce breast development. Also the maintenance dose in transsexuals is commonly higher than in female phenotypes.

C. Secondary Amenorrhoea

The reason behind secondary amenorrhoea should be outlined prior to administration of any type of treatment. Hormone replacement therapy (HRT) is only indicated in women who do not desire a pregnancy, and estrogen monotherapy only in women who do not have a uterus. The most common indication for estrogen monotherapy is women who have undergone surgery for various reasons. Untill recently, cervical cancer was treated by radical hysterectomy, including bilateral oophorectomy, also in young women. The most common form of cervical cancer is the squamous type, which is not estrogen dependent. Hence, estrogen monotherapy could well be used in these cases, in doses similar to or somewhat higher than in HRT in the climacteric. Ovarian tumours, which require bilateral oophorectomy and hysterectomy, may also fall into this category. Prior to the last 5 or 10 years, hysterectomy and bilateral oophorectomy were also common in conjunction with severe pelvic inflammatory diseases or extensive endometrioses.

In other cases of secondary amenorrhoea, with the uterus left in situ, estrogen or anti-estrogen therapy are seldom indicated as monotherapy. Oral contraceptives could well be the drug of choice in women at risk of pregnancy. Women with androgen-producing tumours or other signs of androgen over activity could be given HRT after proper diagnostic procedures had been performed with removal of the excessive source of androgens (if possible). Again, if the uterus is present, HRT should be in the combined form, i.e. a combination of a progestogen and an estrogen. In this context, progestogens, which do not impair the estrogen induction of sex hormone-binding globulin (SHBG) are preferred. Progestogens with a low displacement activity on SHBG are also preferred. The type of estrogen is of little or no importance. Women undergoing this treatment should be informed that hair may now grow slower and become less coarse. However, hairs will not disappear. Virilizing signs, such as a lowered voice, are difficult to reverse, but at least further progress will cease.

Other indications exist for estrogen monotherapy. This includes the so-called low estrogen dose concept for urogenital disorders in women past the climacteric. Traditionally, estriol in the form of pessaries, cream or, even, as tablets is used for this indication. During recent years, vaginal preparations with a slow-release form of oestradiol have been introduced. Oestradiol-releasing vaginal rings, releasing 7.5 μg oestradiol per day, and a so-called vaginal tablet, containing 25 μg oestradiol per day, to be used twice a week, yield similar clinical results. Hence, the restoration of vaginal atrophy calls for a delivery of about 50 μg per week, whereas treatment of climacteric symptoms requires some 50 μg per day.

This low dose estrogen therapy is capable of restoring the glands of the vagina, inducing lubrication and recurrence of lactobacillae, with a consequent drop of pH from between 6 and 7 to around 4. In addition, parts of the urethra and bladder neck are also stimulated by estrogens and the atrophy may also be counterbalanced. There are data to show that recurrent urinary-tract infections (RAZ and STAMM 1993) and various forms of incontinence, especially of the urgency form and of the mixed form (SAMSIOE et al. 1985), could be treated with this type of low dose estrogen monotherapy. Vaginal preparations have a very small systemic absorption, once the vaginal mucosa is matured (NILSSON and HEIMER 1994). For this reason, progestogen co-medication is not necessary. However, these types of preparations could not be anticipated to have any major systemic effects, although a protective effect on osteoporosis has been demonstrated in one recent study. Due to its high safety and lack of such effects, vaginal preparations containing estriol and indeed dienoestrol were made as over-the-counter preparations in 1991. This decision increased sales. It seems prudent to believe that elderly women are still embarrassed to consult their doctors for urogenital problems and that the over-the-counter accessibility was a great relief to these women.

In postmenopausal women with signs of atrophy of the urogenital tract, vaginal surgery may sometimes be difficult. A surgical procedure per se is often facilitated by pre-operative treatment with estrogens. This could be done as an acute treatment, using for instance 50 μg ethinyloestradiol for 2–3 days. More efficaciously, oestradiol, in the oral dose of 2–4 mg could be given for a week. The true surgical procedures are hence facilitated by the stimulated tissue, although capillary bleeds are more common in an estrogen-stimulated tissue.

There is now information to suggest that estrogen deficiency may facilitate HIV infections and estrogen treatment may well affect the local immune defence in a positive way. In addition, estrogen therapy, local or systemic, increases the vaginal lubrication and may sometimes positively affect chronic cervicitis and vulva itching and burning.

Cervical glands are also stimulated by estrogens to produce a cervical mucus with high spinnbarkeit and good ferning properties. This type of treatment may be helpful in chronic cervicitis and in treating cervical mucus in permeable sperms, which is sometimes a cause of infertility.

Newer pharmaceutical concepts have, nowadays, replaced estrogen therapy for women, in whom it is desirable to stop lactation, women with premenstrual tension etc. Recent advances in the effects of estrogens also include a decrease of colonic cancer and an improvement in asthmatic cases (CHANDLER et al. 1997; GRODSTEIN et al. 1997).

Estrogen addition to a given hormonal regimen may also be attempted when vaginal dryness occurs as a result of progestogen influence. In women using continuous combined regimens, the progestogenic dose that is required to induce endometrial atrophy may also reduce vaginal atrophy, resulting in a feeling of dryness. A supplementation of a (low dose) local estrogen is often needed in some 20–25% users of continuous combined HRT, especially at ages over 60.

Another indication where estrogens have been used is in adolescent girls in order to prevent them from becoming too tall. Usually, oral contraceptives are used, but it is the estrogen component that facilitates the closure of the epiphyseal lining.

D. Bleeding Problems

Anti-estrogens (tamoxifen) can also act as estrogen agonists or have little or no effect (raloxifene) on the endometrium. For this reason, the so-called anti-estrogens have not been used in clinical series to treat bleeding dis-orders in women.

A long-standing experience relates the use of progestogens in the treatment of various bleeding disorders during fertile life. The reason for the progestogenic effect is the induced transformation of the endometrium into a secretory phase, coupled to suppression of its growth and, thereby, control of bleeding. However, the exact mechanism as to how women bleed and what initiates bleeding is still not known in detail. It has also been established that the concurrent endometrial histology bears no relation to the bleeding pattern (STURDEE et al. 1994). When using progestogen-only contraception, especially if ovulation inhibition is also induced, the endometrium usually becomes atrophic.

An atrophic endometrium is also the therapeutic goal in the so-called continuous combined therapy for climacteric complaints. Bleeding problems, usually in the form of spotting, are common during the first few months of continuous combined therapy (ARCHER et al. 1994). After 12 months of treatment, some 70–80% of women report amenorrhoea. Also in users of depot medroxyprogesterone acetate, irregular bleeds and spottings may occur. These types of bleeds, that emanate from an atrophic endometrium, are different in character and background than the so-called withdrawal bleeds and other bleeds incurred during fertile life. The exact mechanism behind the atrophic bleed is not known, but increased capillary activity and proliferation of capillaries have been brought forward as possible candidates for explanation. Of particular interest, in this context, is the activity of the so-called vascular endothelial growth factor (VEGF) which is stimulated, not only by hypoxia, but also by progestogens (SMITH 1993). Atrophic bleeds could be diminished by decreasing the progestogenic influence.

In the World Health Organization program on Depo-Provera, an addition of estrogen is advocated as a therapeutic possibility when irritating bleeds occur during progestogen-only contraception. However, it should only be used when ovulation inhibition is induced, as during Depo-Provera treatment. In other instances of progestogen-only contraception, an addition of estrogen may jeopardise the contraceptive purpose. Prior to initiating estrogen therapy in these cases, ultrasonography is mandatory and an endometrial biopsy is recommended. It should be borne in mind that bleeding problems are more common in women with uterine fibroids, irrespective of whether they use some kind of hormonal treatment or not. Adenomyosis is another diagnosis that is sometimes the culprit of intractable bleeds.

The estrogen component of the oral contraceptive and HRT are both linked to an increased incidence of clotting. The reason for this is unknown, but estrogen has been used to treat bleeds, especially from the urological tract. Case reports have been presented by MILLER et al. (1994) and by VIGANÓ et al. (1988) in which idiopathic haemorrhagic cystitis is treated successfully by conjugated equine estrogens. This effect can be achieved by oral conjugated equine estrogens, but also by intravenously injected conjugated estrogens. The estrogens are esterified at the 3-hydroxy position and, thereby, are made water soluble. After intravenous injection, this type of ester is not hydrolysed, but is rapidly cleared by the kidneys, thereby inducing a high concentration in the urological tract. Most cases described are also from urological bleeds, although effects have been reported when treating bleeds from the lungs, the gastrointestinal tract and the nose.

Estrogen has also been used to prevent and even treat the so-called Asherman's syndrome. It was also recently reported that 35% of women with asthma noticed a deterioration of their disease pre-menstrually. Exogenously administered oestradiol had a significant effect on pre-menstrual asthma and dyspnoea inclusive of an increase of PEF. It is unclear how oestradiol works in this particular setting, but it seems not to be via the so-called $\beta 2$ receptors (Chandler et al. 1997).

Other indications for estrogens include depression, especially postpartum, but newer treatment modalities have limited this indication. Recent data imply that estrogen treatment may modulate the effects of modern anti-depressions, but the role of estrogen in combination with these drugs needs to be further evaluated before any specific recommendation can be given. In addition, an attempt with estrogen treatment could be instituted in men with senile dementia showing signs of aggression and in men with hyper-sexuality. The only persistent indication for estrogen use in men is the treatment of prostatic carcinoma, but other possibilities, inclusive of positive effects on the cardiovascular system, may well be at hand, although this issue is still controversial.

E. Anti-Estrogens

The classical use of anti-estrogens is firstly for the induction of ovulation in secondary amenorrhoea, when a hypothalamic disorder is suspected. Commonly, a dose of 50 mg/day for 5 days is used. As much as 100 mg could be required if there is no response after the first cycle. In certain states of ovulatory dysfunction with and without amenorrhoea, such as the PCO syndrome, patients may be extremely sensitive to clomiphene and a lower dose (25 mg/day) in the PCO syndrome is often advocated to avoid hyper-stimulation of the ovaries. Also, in several in vitro fertilisation (IVF) programs, use of clomiphene could be advocated to time the ovulatory luteinising hormone (LH) surge better. More recently, clomiphene therapy has been replaced by other methods such as the sequential treatments with gonadotrophin releasing hormone (GnRH) analogues and gonadotrophin stimulation.

The other classical use of the so-called anti-estrogens is the treatment of breast cancer, especially with tamoxifen. There is substantial evidence to suggest a reduction of recurrence of breast cancer in tamoxifen users, especially in women with positive estrogen receptors. However, during these studies, it was found that anti-estrogens could also be estrogen agonists on certain bodily systems, such as bone and possibly also the cardiovascular system. Tamoxifen itself is a weak agonist on the endometrium. Newer anti-estrogens, SERMs, display an estrogen antagonism also on the endometrium. Raloxifene, which is a SERM under clinical development, has been used in well over 1000 women for long periods. When results of these studies were revealed, it was shown that raloxifene seemed to reduce the incidence of breast cancer. However, these data should be interpreted with care, as the lag phase for breast cancer may be longer and the use of an anti-estrogen may well disguise breast tumours that may later be detected by the routine mammography programme, which was part of the clinical trials on raloxifene. There are several other SERMs under development, and it seems to be an interesting pharmacological concept to have estrogen antagonism and estrogen agonism in the same molecule. It has been shown convincingly that raloxifene inhibits endometrial proliferation. Whether it could also be used for the treatment of endometrial hyperplasia or endometrial cancer remains to be clarified.

In addition, estrogens have been used to correct the male type of fat distribution, which is probably achieved via an increase in SHBG and a decrease in insulin resistance and hyperinsulinaemia. By doing so, the free androgen fraction can be decreased. Estrogen treatment has been used extensively, with a significant clinical effect, in hirsutism and in other signs of increased endogenisity in the female, such as body fat distribution, hair distribution etc.

It has also been discussed whether estrogens possess some anti-inflammatory properties. The clinical results are not without difficulties to interpret. As far as rheumatoid arthritis is concerned, there is data to suggest that an improvement may occur (Spector and Hochberg 1990; Hall et al. 1994); other studies found no such effect (Koepsell et al. 1994). There are no data to suggest that a deterioration of the disease should occur during estrogen treatment. Some data have suggested beneficial effects in diseases like systemic lupus erythematosus (SLE) or Crohn's disease (Boumpas et al. 1995); oral contraceptives containing up to 15 mg of ethinyloestradiol and postmenopausal HRT have been used safely in SLE (Julkunen et al. 1993) with the important exceptions of the presence of high levels of anti-phospholipid antibodies (Arden et al. 1994).

References

Archer DF, Pickar JH, Bottiglioni F (1994) For the Menopause Study Group: Bleeding patterns in postmenopausal women taking continuous combined or sequential regimens of conjugated estrogens with medroxy-progesterone acetate. Obstet Gynecol 83:686–692

Arden NK, Lloyd ME, Spector TD, Hughes GR (1994) Safety of hormone replacement therapy (HRT) in systemic *lupus erythematosus* (SLE). Lupus 3:11–13

Boumpas DR, Fessler BJ, Austin HA et al. (1995) Systemic lupus erythematosus: emerging concepts. Part 2: Dermatologic and joint disease, the anti-phospholipid antibody syndrome, pregnancy and hormonal therapy, morbidity and mortality, and pathogenesis. Ann Intern Med 123:42–53

Chandler M, Schuldheisz S, Phillips BA, Muse KN (1997) Premenstrual Asthma: The effect of estrogen on symptoms, pulmonary function, and $\beta 2$-receptors. Pharmacotherapy 17:224–234

Grodstein F, Stampfer MJ, Colditz GA et al. (1997) Postmenopausal hormone therapy and mortality. N Engl J Med 336:1769–1775

Hall GM, Daniels M, Muskisson EC, Spector TD (1994) A randomised controlled trial of the effect of hormone replacement therapy on disease activity in postmenopausal rheumatoid arthritis. Am Rheum Dis 3:112–116

Julkunen HA, Kaaja R, Friman C (1993) Contraceptive practice in women with systemic lupus erythematosus. Br J Rheumatol 32:227–230

Koepsell TD, Duguwson CE, Nelson JL et al. (1994) Non-contraceptive hormones and the risk of rheumatoid arthritis in menopausal women. In J Epidemiol 23:1248–1255

Miller J, Burfield GD, Moretti KL (1994) Oral conjugated estrogen therapy for treatment of hemorrhagic cystitis. J Urol 51:1349–1350
Nilsson K, Heimer G (1994) Ultralow dose transdermal estrogen therapy in postmenopausal urogenital estrogen deficiency, a placebo controlled study. Menopause. The Journal of the North American Menopause Society 4:191–197
Raz R, Stamm W (1993) A controlled trial of intravaginal estriol in postmenopausal women with recurrent urinary tract infections. N Engl J Med 329:753–756
Samsioe G, Jansson I, Mellström D, Svanborg A (1985) The occurrence, nature and treatment of urinary incontinence in a 70 year old population. Maturitas 7:335–342
Smith SK (1993) Endometrial changes during the perimenopausal years. In: Berg G, Hammar M (eds) The Modern Management of the Menopause. Parthenon Publishing, London, pp 201–206
Spector TD, Hochberg MC (1990) The protective effect of the oral contraceptive pill on rheumatoid arthritis: an overview of the analytic epidemiological studies using metaanalysis. J Clin Epidemiol 43:1221–1230
Sturdee DW, Barlow DH, Ulrich LG et al. (1994) Is the timing of withdrawal bleeding a guide to endometrial safety during sequential estrogen-progestogen replacement therapy? Lancet 344:979–982
Viganó G, Gaspari F, Locatelli M et al. (1988) Dose-effect and pharmaco-kinetics of estrogens given to correct bleeding time in uremia. Kidney Int 34:853–858

CHAPTER 41

Oncology

R.Q. WHARTON and S.K. JONAS

A. Breast Cancer

In the western world, approximately one woman in twelve will develop breast cancer during her life; of those, 30% will ultimately die of the disease. The peak incidence occurs at age 50–70 years and only 0.5–1.0% of cases will arise in men (MILLER and BULBROOK 1986). Risk factors include a strong family history, early menarche, late menopause, nulliparity and irradiation. Increasing parity, young age at first pregnancy, breast feeding and artificial onset of menopause are protective factors. Management depends on multiple factors, including stage of disease, timing with respect to the menopause, age and condition of the patient, and variations in clinical practice. Mainstays of treatment include surgical resection, chemotherapy, radiotherapy and hormonal manipulation.

I. Systemic Hormonal Treatment of Breast Cancer

Two distinct entities of treatment of breast cancer utilise the pharmacology of estrogens and anti-estrogens: that of adjuvant therapy to loco-regional surgical resection and that of therapy for advanced breast cancer. One further utilisation that has been studied, but not yet universally adopted, is that of prophylaxis in healthy women who fall into high-risk groups. At present, the drug of choice in the first-line treatment of breast cancer by hormonal manipulation is tamoxifen.

II. Tamoxifen

Tamoxifen is a non-steroidal substituted triphenylethylene estrogen receptor (ER) antagonist which also has limited estrogen agonist activity. This dual activity is attributed to the configuration of the ER complex produced in any particular cell (GALLO and KAUFMAN 1997). In the body, it forms two major metabolites, *des*-methyl tamoxifen (d-Me T) and 4-hydroxytamoxifen (4-OH T). Tamoxifen was first introduced in the 1970s – it was approved for use in the United States in 1977 – for the management of metastatic breast cancer and only later became accepted as standard adjuvant therapy. Since 1973, tamoxifen has been prescribed to over one million women. Although a small percentage of women report minor side effects, such as headache, gastrointestinal disturbance and fluid retention, these are not significantly increased over placebo (NOLVADEX ADJUVANT TRIAL ORGANISATION 1985). Menopausal symptoms such as vaginal dryness and hot flushes are often reported by patients, but usually disappear within a few months. More problematic is deep vein thrombosis, which occurred in 1 of every 800 patients in a recent study (SAPHER et al. 1991). However, the emergence of tamoxifen as the leader in hormonal manipulation was due to its ability to increase the disease-free period and increase overall survival.

In the early 1980s, tamoxifen was being compared to established hormonal therapy, such as aminoglutethimide and chemotherapy (CLAVEL et al. 1982; PREECE et al. 1982) with favourable results. It was also shown that there was seemingly no advantage to the addition of chemotherapy to tamoxifen therapy in postmenopausal advanced breast cancer (BEZWODA et al. 1982), although this has recently been refuted and modern current practice is to offer combined hormonotherapy and chemotherapy to women who are more at risk of recurrence (ALLUM and SMITH 1995). Investigators had also compared tamoxifen alone to multiple hormonotherapy and shown no advantage for those patients receiving the multiple therapy; conversely, the multiple therapies had the disadvantage of higher side-effect profiles (POWLES et al. 1982; SMITH et al. 1982).

In an overview of the anti-estrogen treatment of breast cancer (PEARSON et al. 1982), it was reported that tamoxifen had been shown to be as effective as surgical hypophysectomy in postmenopausal women with breast cancer with metastases. Additionally, it was shown to be effective in selected patients after other ablative procedures, such as ovariectomy and adrenalectomy. The writers suggested that tamoxifen treatment be considered as the drug of choice in postmenopausal women with advanced breast cancer, in view of its effectiveness, minimal side effects and acceptability to patients.

However in 1989, a retrospective study reported that tamoxifen therapy was associated with an increased risk of endometrial cancer (FORNANDER et al. 1989). Since then, tamoxifen therapy has also been implicated in an increased risk of hepatocellular and gastrointestinal cancer (JORDAN 1995). Because of the difficulties in setting up a clinical trial, most analysis has been modelled on meta-analysis and, therefore, absolute values for risk are still not available.

III. Present Strategies

It was decided, through the overview of clinical studies, that hormonal therapy was indicated in postmenopausal women: premenopausal women without metastatic disease were to receive combination chemotherapy if indicated (NATIONAL INSTITUTE OF HEALTH 1985). Tamoxifen does show benefit in premenopausal women – 6% reduction in annual mortality and 12% reduction in recurrence annually – but its use is usually confined to women with metastatic disease in view of its associated endocrine effects.

From the onset of use of tamoxifen, questions began to be asked regarding its indications, relative indications, dosage and duration. One of the most influential studies that attempted to answer these questions was a meta-analysis of 133 trials (THE EARLY CANCER TRIALISTS COLLABORATION GROUP [ECTC] 1992a, b). Guidelines extrapolated from this work and others of similar consequence are used in the clinical practice of most breast specialists. The ECTC group's overview showed, in postmenopausal women, a 20% reduction in mortality annually and a reduction of 29% in the annual risk of recurrence when tamoxifen 20mg was taken once a day. Additionally, risk of contralateral cancer was reduced by up to 40%.

The most recent overviews and completed trials all conclude that hormonotherapy has a primary role in adjuvant treatment of breast cancer in postmenopausal women and that chemotherapy has a far less established place in these patients (PLOWMAN 1996).

1. Adjuvant Therapy

Adjuvant therapy intrinsically is administered in addition to a surgical procedure, which in the treatment of breast cancer, means either mastectomy, segmental mastectomy or wide local excision, with or without axillary clearance (note that, at present, neoadjuvant therapy – preoperative – is restricted to chemotherapy). Three main aims are apparent in adjuvant hormonal therapy: the therapy should increase survival, increase the disease-free period and increase the probability of a cure. The disease-free period is doubly important in that, firstly, it is related to improved overall survival and, secondly, it increases quality of life.

In clinical practice, tamoxifen has been prescribed as adjuvant therapy for 2 years. However, many groups have been looking at 5-year therapy. Two major recent trials have suggested that 5 years is more advantageous than 2 years (CAMPAIGN BREAST CANCER TRIALS GROUP 1996; SWEDISH BREAST CANCER COOPERATIVE GROUP 1996), although both groups agree that more data are needed. Whether more than 5 years is advisable is again debatable. One group investigated postmenopausal patients who were node negative and ER positive and found no advantage in using more than 5 years of tamoxifen therapy (FISHER et al. 1996).

2. Metastatic Therapy

Hormonal therapy, especially tamoxifen, is the mainstay of the palliative non-surgical treatment of breast cancer, largely because of its low side-effect profile, easy compliance and efficacy. Tamoxifen is the drug of choice in the initial treatment in both premenopausal and postmenopausal women; ovariectomy and luteinizing hormone releasing hormone analogues are often substituted in premenopausal patients (VOGEL 1996). Second-line treatment increasingly involves the use of the aromatase inhibitors. Tamoxifen analogues with less estrogen agonism are being developed which may replace tamoxifen as first-line treatment in the near future. Both of these groups are discussed later in Sect. A.VI.1.

3. Prophylaxis

It has become apparent that, in high risk patients, tamoxifen could be used to prevent breast cancer, with the additional health benefits of tamoxifen's limited estrogen agonism, although in one particular model looking at potential health benefits, the predicted advantages, other than breast-cancer prevention, were modest (NEASE and ROSS 1995). However, the unacceptability of prophylactic bilateral mastectomy in genetically susceptible women has increased the interest in hormonotherapy. Three major studies – one in the USA, one in the UK and one in Italy – are being conducted, all of which are double blind and involve women without breast cancer taking tamoxifen for 5 years. The results of these trials are still awaited. As well as tamoxifen, luteinizing hormone releasing hormone agonists and moderated oral contraceptives are being investigated (LOVE 1994).

4. Male Breast Cancer

Male breast cancer differs from that of females in that the patients are, on average, much older, have more extensive disease on presentation and are more likely to be estrogen-receptor positive (CULUTI et al. 1995). Both the old age of the patients – making chemotherapy largely unacceptable – and the receptor status mean that tamoxifen is often used as adjuvant therapy and in the treatment of metastatic disease. The most common side effect is, not surprisingly, loss of libido which occurs in approximately 30% of these patients on tamoxifen (ANELLI et al. 1994); others include depression and weight gain. Unfortunately, this has an adverse effect on compliance, which is much worse than in women. However, in one trial, tamoxifen has delivered an increase in 5-year survival from 44% to 61%, and a disease-free period from 28% to 56% at 5 years (RIBEIRO et al. 1992).

5. Desmoid Tumours

In view of tamoxifen's efficacy in breast cancer, some clinicians have used the drug to treat a certain rare mesenchymal tumour called a desmoid, which is associated with *Polyposis coli*. Interestingly, a 65% response was reported in one series of 20 patients (BROOKS et al. 1992). In a tumour with low levels of ERs, but consisting mainly of stroma, such as that found in the breast, may point to a stromal intermediate that may be useful in the management of breast cancer.

IV. Estrogen Hormone Receptors in Breast Cancer

Approximately 50% of all primary breast tumours are so-called ER positive (ALLEGRA et al. 1980). ER positivity is defined as the measurement of the binding capacity of ^{3}H-estradiol above a set point. This point is still open to debate, but is commonly set at a value above 10 fmol ^{3}H-estradiol per milligram of cytosolic protein; higher values have been suggested (HEISE and GORLICH 1993). A value below 3 fmol/mg is considered ER negative and between 3 fmol/mg and 10 fmol/mg borderline. The standard assay for the receptor is the dextran-coated charcoal method. ER positivity is related to histological subtype, menopause and whether the sample originates from the primary lymph node or visceral metastases. For instance, in the most common form of breast carcinoma, that of infiltrating ductal carcinoma, the ER is deemed positive in 60–70% of cases, whereas in tumours with high lymphocytic infiltration, the ER is positive in 30–40%. Premenopausal women are positive in 30% of cases compared with 20% of perimenopausal women and 60% of postmenopausal women (KIANG et al. 1977). Additionally, within the same patient who has advanced disease, the ER value has been shown to be different in the primary, the affected lymph nodes and the liver metastases (HEISE and GORLICH 1993).

The response of patients with advanced breast carcinoma to hormonal manipulation closely correlates with the value of ER positivity: ER-positive tumours have a 60% response rate to hormonal manipulation compared with ER-negative tumours which have a response rate of only 10% (DOWSETT 1996). These figures show, therefore, that approximately 40% of ER-positive patients fail to produce any response. Furthermore, virtually all patients with advanced breast cancer eventually relapse on hormonal manipulation. This de novo resistance to hormonal manipulation appears to be dependent on the tumour's switching to other growth pathways. Second-line hormonal manipulation is therefore less effective: fewer than 50% of patients respond.

Because of previous clinical investigations showing that the treatment of premenopausal women with anti-neoplastic agents influenced the commencement of menopause, it was thought that there was a connection between hormone receptors and anti-neoplastic chemotherapy, and that this might be useful in treating advanced disease. GORLICH et al. 1997 showed that the mechanism of action of chemotherapeutic drugs is not directly related to the presence or absence of steroid-hormone receptors, although it was clear that ER positivity was still useful in identifying high- and low-risk patients undergoing chemotherapy.

V. Second-Line Therapy

Unfortunately, the vast majority of patients with metastatic breast cancer will relapse even if, as about 50% do, they initially responded to first-line hormonal manipulation with tamoxifen; the de novo resistance seemingly arising from dependence on intermediates other than the ER and population selection of ER-negative cell lines within the tumour.

Half of the patients in relapse respond, to varying degrees, to second-line therapy. The group of drugs generating most interest with respect to second-line therapy are the aromatase inhibitors. These drugs have largely replaced progestins, such as megestrol acetate and medroxyprogesterone, which were troubled by a high side-effect profile.

1. Aromatase Inhibitors

It is suggested that estrogen is active in driving tumour growth via pathways other than via the ER and, therefore, in patients with de novo resistance to tamoxifen, reduction in circulating estrogens may be of benefit. This can be achieved by preventing estrogen synthesis by inhibiting the aromatase cytochrome P450 enzyme complex. It can be noted that, although estrogen synthesis occurs principally in the ovaries, the adipose tissue and extragonadal sites, such as placenta and brain, also produce estrogen. Thus, even after ovariectomy estrogens are present in the circulation. The pathophysiological importance of these extragonadal sites is therefore apparent.

Aromatase inhibitors can be steroidal or non-steroidal, but as the non-steroidals are more specific biologically, they have undergone most investigation. The agents can produce competitive, irreversible or mechanism-based (enzyme-activated) inhibition (BRUEGGEMEIER 1994). Mechanism-based inhibition is preferred in that it is more specific, requires less frequent dosing and exhibits less toxicity.

2. Evolution of Aromatase Inhibitors

The first popular aromatase inhibitor, aminoglutethimide – a non-steroidal – is relatively non-specific and suffers from poor tolerance due to its side-effect profile, notably sedation, fever and rash. It requires that corticosteroid replacement therapy be administered concurrently because of its non-specific action (GOSS and GWYN 1994). Trilostane was also developed, but again because of its non-specific action, required corticosteroid replacement therapy. In view of these inadequacies, analogues have been sought and investigated over the last decade.

The second-generation agent formestane – a mechanism-based inhibitor with few side effects – was a newer contender, but has a poor bioavailability when taken orally making intramuscular injection mandatory (BAJETTA 1996). However, it was intensively investigated and workers concluded that it was effective with satisfactory tolerability and the additional benefit of insignificant effects on the levels of cortisol and aldosterone (BAJETTA 1994a; DOWSETT 1994). In a phase-II trial of patients with advanced breast cancer, formestane produced disease stabilisation in 29% and complete or partial response in 23% (POSSINGER 1994). However, the duration of response, as in first-line hormonotherapy, is limited: in the case of formestane it is limited to approximately 12 months (WISEMAN and GOA 1996).

Since 1994, third- and fourth-generation agents have been investigated – notably letrozole, fadrozole, vorozole and anastrozole. They have been shown to have greater selectively and potency in vitro (BAJETTA 1996b). Fadrozole and letrozole (non-steroidal imidazole derivatives) were compared in a phase-

I trial: letrozole was found to be more potent in reducing circulating estrogens and more selective; fadrozole had a measurable effect on cortisol and aldosterone levels (DEMERS 1994). More recent research showed letrozole to be 10,000 times as potent in vivo as aminoglutethimide and to be well tolerated (TRUMET et al. 1996). Phase-III trials are still outstanding in comparing its clinical efficacy to the older agents. Likewise, phase-III trials are underway with anastrozole and vorozole which have a similar profile to letrozole (PLOURDE et al. 1994; JOHNSTON et al. 1994).

VI. Future Developments

Although tamoxifen is currently the endocrine therapy of choice, attempts have been made to improve its efficacy and also to develop new groups of agents.

1. Tamoxifen Analogues

In general, this has involved altering the triphenylethylene ring structure to form new non-steroidal ring structures with anti-estrogen activity, producing agents such as droloxiphene and toremiphene. A number of these newer agents have shown in pre-clinical trials increased activity with regards to estrogen antagonism, ER binding affinity and anti-tumour activity when compared to tamoxifen (HOWELL 1997).

Although clomiphene has anti-estrogen activity in mammary tissue, it is far less potent than tamoxifen and, therefore, is not used in standard practice in the management of breast cancer (LERNER and JORDAN 1990). Its clinical use is dealt with elsewhere.

Droloxiphene (3-hydroxy tamoxifen) was assessed in a in a phase-II clinical trial (RAUSCHNING and PRITCHARD 1994). It was concluded that droloxiphene has a higher affinity for the ER, higher anti-estrogenic to estrogenic ratio and more effective inhibition of cell growth and division in ER-positive cell lines; they also reported less toxicity. In their animal studies, there was significantly less carcinogenicity when compared to tamoxifen. However, droloxiphene reaches peak levels and is eliminated far more quickly than tamoxifen and, therefore, may require a different dosing regime.

Response to toremifene, again a triphenylethylene derivative, was also assessed in a phase-II clinical trial (STENBYGAARD et al. 1993). The workers showed that response rates and survival were not statistically different between toremifene and tamoxifen. It was also noted that toxicity was comparable. In addition, the study concluded that tamoxifen and toremifene are clinically cross-resistant in patients with advanced breast cancer.

2. Benzothiophene Derivatives

Raloxifene is the major representative of this class of compounds, which has the benefit of estrogen agonist-like actions on bone tissues and serum lipids, but is a potent estrogen antagonist when acting on the breast and uterus (BLACK et al. 1994). These types of compounds are designated as selective ER modulators (SERMs). It is their reduced ability to affect uterine tissue that has caused interest. However, some authorities (HOWELL et al. 1996) believe that raloxifene, if used as endocrine therapy in breast cancer, will need an extra agent co-administered to attain the efficacy of tamoxifen in prevention of osteoporosis and cardiovascular disease. Raloxifene has been modified recently (PALKOWITZ et al. 1997) to increase its efficacy in in vitro trials on human mammary cells and in in vivo studies in the ovariectomized rat model with encouraging results. At the time of writing this chapter, the majority of investigations involve assessment of ramoxifene as a possible replacement to current hormone replacement therapies rather than an anti-neoplastic agent (FROLICH et al. 1996).

3. Steroidal "Pure" Anti-estrogens

The two main contenders for this title are ICI 182,780 and ICI 164,384. They are chemically 7α-alkylamine derivatives of 17β-estradiol and differ from tamoxifen pharmacodynamically in that they have no partial estrogen agonist activity. Although they both bind strongly to the estrogen receptor, ICI 182,780 has the greater anti-estrogenic properties (WAKELING and BOWLER 1992). Workers have also investigated the "pure" anti-estrogen RU 58668 (VAN DE VELDE et al. 1995) and shown in in vivo animal models that it has promise for the development of a new therapy for advanced breast cancer.

However, ICI 182,780 has drawn the most attention. Cell-culture studies have shown that breast cancer lines insensitive to tamoxifen are still sensitive to ICI 182,780, producing growth inhibition of those cells. Its effects on patients with advanced breast cancer resistant to tamoxifen have been investigated in pilot trials. One such trial showed a response rate of 69% (HOWELL et al. 1996) when the drug was delivered by monthly intramuscular depot injection. In addition, no significant changes occurred in serum levels of lipids, the pituitary-hypothalamic axis was unaffected and side effects were infrequent. There was no apparent effect on the endometrium.

The "pure" anti-estrogens' limitation is the lack of the cardioprotective ability ascribed to tamoxifen, which may limit its use to second-line treatment only. Conversely, it has been argued that "pure" anti-estrogens should be used as first-line treatment for the following reason (WAKELING 1993): a breast cancer consists of two populations – one reacting to the anti-estrogen effect of tamoxifen and one reacting, not in the desired way, to the estrogen agonism. This second population, therefore, has an evolutional advantage and brings about the effect of de novo resistance. However, treatment with "pure" anti-estrogens should suppress both populations from the beginning of treatment and, thus, provide a greater duration of response.

B. Ovarian Cancer

The leading cause of gynaecological cancer mortality in the United States is ovarian cancer with 21,000 new cases and 13,000 deaths per year.

I. Hormonal Involvement

The role of hormonal involvement in the risk of ovarian cancer is supported by several epidemiological studies. These include hormonally regulated events, such as parity, length of reproductive life and oral contraceptives. The reported risk progressively decreases for women who have more children (CRAMER et al. 1983; CASANGRANDE et al. 1979). The prevalence of hormone receptors in ovarian cancer has been reviewed in more than 50 studies, and the incidence of ERs in more than 1600 ovarian cancers was reported at 63% compared with 48% of progesterone receptors and 69% of androgen receptors (SLOTMAN et al. 1988). Several studies have linked the presence of ERs and progesterone receptors to survival, and demonstrated that high ER levels correlated with improved survival (KIELBUCK et al. 1993).

Studies on ER expression by means of immunohistochemical methods in different types of ovarian cancer has recently demonstrated that the greatest expression is present in serous borderline tumours. ABU-JAWDEH et al. (1996) and RISCH et al. (1996) studied risk factors for epithelial ovarian cancer by histological type and found that child bearing and use of oral contraceptives were associated with a decreasing risk of ovarian cancer in all but the mucinous-type tumours. On the basis of their and other studies, they concluded that the mucinous ovarian tumours should be considered separately in studies of ovarian cancer.

1. Hormone Replacement Therapy

Some inconsistencies exist in the literature regarding estrogen replacement therapy (ERT) and the risk of ovarian cancer, particularly in postmenopausal women. This has been addressed by taking into account differences in histological groups. Non-mucinous ovarian cancer, as a proportion of all ovarian cancers, was increased by 5.1% for each year of ERT use. It is therefore suggested that ERT usage may contribute to the development of non-mucinous types of ovarian cancer (RISCH 1996). Epidemiological studies suggest an increased risk of fatal ovarian cancer with long-term use of ERT. This was concluded from a large prospective mortality study of women with no prior history of cancer. The increased rate ratio was reported at 1.15 and increased with duration of use (RODRIGNEZ et al. 1995).

2. Oral Contraceptives

Oral contraceptives have been implicated in decreased risk of ovarian cancer (ROSENBLATT et al. 1992). They have, in the past, received some attention in terms of their protective role against ovarian and endometrial cancer. This has been repeatedly shown with high-dose oral contraceptives, but not to date with low doses. Due to the disproportionate attention that has been given to the thrombotic risks induced by high-dose contraceptives, the protective role has been largely undermined (GOLDZIEHER 1994).

3. Tamoxifen

The first report of a response to tamoxifen by two patients with ovarian cancer appeared in 1981 (MYRES et al. 1981). Subsequent studies reported only up to 10% complete responses to tamoxifen with stage-III/IV disease. Although this appears to be a low response rate, it is thought to be clinically useful if the response in patients can first be identified. These same authors studied the efficacy of three different hormonal manipulations in the treatment of chemoresistant cancer. Three phase-II trials were performed using high-dose megestrol acetate, high-dose tamoxifen or aminoglutethimide. The results demonstrated five responses in 29 patients (17%) treated with tamoxifen and no response with either of the other two drugs. It was suggested that anti-estrogen therapy may offer long-term palliation of refractory epithelial ovarian cancer with little toxicity. This is further supported by the demonstration that significantly higher estrogen binding sites could be found in membrane fractions of ovarian tumours than in normal membrane fractions, suggesting that tamoxifen alone and in combination with chemotherapy could be used to an advantage in ovarian cancer (BATRA and IOSIF 1996). Tamoxifen has also been used in other malignancies (GELMAN 1997). In ovarian cancer, the action of tamoxifen is suggested to occur through competitive binding to the estrogen receptor, which is the primary mechanism of inhibiting cell replication.

C. Endometrial Cancer

Endometrial cancer is the most commonly diagnosed gynaecological malignancy seen in the United States.

I. Hormonal Involvement

Amongst the risk factors identified for endometrial cancer are continuous exposure to estrogen, obesity, nulliparity and late menopause. A 5-fold increase in risk is apparent when all these factors are evident. An increased incidence of endometrial cancer is also observed in patients with hormone-secreting tumours, particularly of the ovary.

Further association of estrogen and endometrial cancer comes from the observation that a greater conversion of androstenedione to estrone occurs in these patients than in healthy postmenopausal patients. Adipose tissue also stores precursors of estrone, hence the association with obesity (CEASMAN 1997).

The single etiological pathway, i.e., exposure to high levels of estrogen has been questioned to account for all the risk of endometrial cancer (POTISCHMAN et al. 1996). Involvement of steroid hormones has also been implicated. One of the strongest associations with risk in this study was found to be the level of androstenedione. This androgen is a major substrate for the

Table 1. Endometrial cancer and hormone replacement therapy: annual incidence per 1000 patient

	Untreated	Estrogen	Estrogen/progesterone
NACHTIGALL et al.	1.2	–	0
HAMMOND et al.	0.5	–	0
GAMBRELL	2.5	6.5	0.5
PERSSON et al.	1.4	1.8	0.9

aromatase enzyme responsible for estrogen synthesis. Normal endometrium is not thought to have aromatase activity, but there is evidence of increased activity in malignant endometrial tumours. It has been suggested that early in the neoplastic process, abnormal endometrial cells gain the ability to produce estrone from androstenedione and have a growth advantage, independent of circulating estrogen levels (MILLER and LAYDON 1997).

1. Hormone Replacement Therapy

In view of this suggested relationship between estrogen and endometrial cancer, several studies have been undertaken to compare the incidence of endometrial cancer and hormone replacement therapy (HRT). These studies demonstrated a much lower incidence in patients receiving both estrogen and progesterone than in patients receiving estrogen alone (Table 1). Based on these studies, treatment now includes both estrogen and progesterone.

There has been a recent report on the continuous low-dose combined HRT and the risk of endometrial cancer (COMERCI et al. 1997), which concluded that there may be an increased risk.

2. Tamoxifen

There are conflicting reports on the use of tamoxifen and endometrial cancer. One of the largest studies to date, the National Surgical Adjuvant Breast and Bowel Project (NSABP), reported by FISHER et al. (1994) suggests an increased annual rate of endometrial cancer per 1000 patients on tamoxifen of 1.7 compared with those not receiving the agent. Other studies showed no relationship.

References

Abu-Jawdeh GM, Jacobs TW, Niloff J, Cannistra SA (1996) Estrogen receptor expression is a common feature of ovarian borderline tumours. Gynecol Oncol 60(2):301–307

Allegra JC, Barlock A, Huff KK et al. (1980) Changes in multiple or sequential estrogen receptor determinations in breast cancer. Cancer 46:362–367

Allum WH, Smith IE (1995) Carcinoma of the breast. Br J Hosp Med 54(6):255–258

Anelli TF, Anelli A, Tran KN, Lebwohl DE, Borgen PI (1994) Tamoxifen administration is associated with a high rate of treatment-limiting symptoms in male breast cancer. Cancer 74(1):74–77

Bajetta E, Celio L, Buzzoni R, Bichisao E (1996) Novel non-steroidal aromatase inhibitors: are there new perspectives in the treatment of breast cancer? Tumori 82(5):417–422
Bajetta E, Zilembo N, Buzzoni R, Noberasco C, Martinetti A, Ferrari L, Bartoli C, Sacchini V, Attili A, Lepera P (1994) Formestane: effective therapy in postmenopausal women with advanced breast cancer. Ann Oncol 5(Suppl 7):15–17
Batra S, Iosif CS (1996) Elevated concentrations of antiestrogen binding sites in membrane fractions of human ovarian tumours. Gynecol Oncol 60(2):228–232
Bezwoda WR, Derman D, De Moor NG, Lange M, Levin J (1982) Treatment of metastatic breast cancer in estrogen receptor positive patients. A randomized trial comparing tamoxifen alone versus tamoxifen plus CMF. Cancer 50(12):2747–2750
Black LJ, Sato M, Rowley ER, Magee DE, Bekele A, William DC, Cullinan GJ, Bendele R, Kauffman RF, Bensch WR et al. (1994) Raloxifene (LY139481 HCl) prevents bone loss and reduces serum cholesterol without causing uterine hypertrophy in ovariectomized rats. J Clin Invest 93(1):63–69
Brooks MD, Ebbs SR, Colletta AA, Baum M (1992) Desmoid tumours treated with triphenylethylenes. Eur J Cancer 28 A(6–7):1014–1018
Brueggemeier RW (1994) Aromatase inhibitors – mechanisms of steroidal inhibitors. Breast Cancer Res Treat 30(1):31–42
Casangrande JT, Louie EW, Pike MC, Roy S, Ross RK, Henderson BE (1979) Incessant ovulation and ovarian cancer. Lancet II:170–173
Clavel B, Cappelaere JP, Guerin J, Klein T, Pommatau E, Berlie J (1982) Management of advanced breast cancer in postmenopausal women. A comparative trial of hormonal therapy, chemotherapy, and a combination of both. Sem Hop 58(34):1919–1923
Comerci JT, Fields AL, Runowicz CD, Goldberg GL (1997) Steroid hormones and cancer: (III) Observations from human subjects. Eur J Surgical Oncol 23:163–183
Current Trials Working Party of the Cancer Research Campaign Breast Cancer Trials Group (1996) Preliminary results from the cancer research campaign trial evaluating tamoxifen duration in women aged fifty years or older with breast cancer. J Natl Cancer Inst (88924):1834–1839
Cutuli B, Lacroze M, Dilhuydy JM, Velten M, De Lafontan B, Marchal C, Resbeut M, Graic Y, Campana F, Moncho-Bernier V et al. (1995) Male breast cancer: results of the treatments and prognostic factors in 397 cases. Eu J Cancer 31 A(12):1960–1964
Cramer DW, Hutchinson GB, Welch WR, Scully RE, Ryan KJ (1983) Determinants of ovarian cancer risk. I. Reproductive experiences and family history. J Natl Cancer Inst 71:711–716
Creasman WT (1997) Endometrial cancer. Incidence,Prognostic Factors, Diagnosis and Treatment. Seminars in Oncology 24 (1) (Suppl 1):1–140 – 1–150
Demers LM (1994) Effects of Fadrozole (CGS 16949 A) and Letrozole (CGS 20267) on the inhibition of aromatase activity in breast cancer. Breast Cancer Res Treat 30(1):95–102
Dowsett M (1994) Aromatase inhibition: basic concepts, and the pharmacodynamics of formestane. Ann Oncol 5(Suppl 7):3–5
Dowsett M (1996) Endocrine resistance in advanced breast cancer. Acta Oncol 35(Suppl 5):91–95
Fisher B, Costantino JP, Redmond CK, Fisher ER, Wickerham DL, Cronin WM (1994) Endometrial cancer in tamoxifen-treated breast cancer patients: findings from the National Surgical Adjuvant Breast and Bowel Project (NSABP) B–14. J Natl Cancer Inst 86(7):527–537
Fisher B, Dignam J, Bryant J, DeCillis A, Wickerham DL, Wolmark N, Costantino J, Redmond C, Fisher ER, Bowman DM, Deschenes L, Dimitrov NV, Margolese RG, Robidoux A, Shibata H, Terz J, Paterson AH, Feldman MI, Farrar W, Evans J, Lickley HL (1996) Two versus more that five years of tamoxifen therapy for breast

cancer patients with negative lymph nodes and estrogen receptor-positive tumours. J Natl Cancer Inst 88(21):1529–1542
Fornander T, Rutquist LE, Cedermark B et al. (1989) Adjuvant tamoxifen in early breast cancer: occurrence of new primary cancers. Lancet 1(8630):117–120
Frolich CA, Brant HU, Black EC, Magee DE, Chandrasekbar S (1996) Time-dependent changes in biochemical bone markers and serum cholesterol in ovariectomized rats: effect of raloxifene HCl, tamoxifen, estrogen and alendronate. Bone 18(6):621–67
Gallo MA, Kaufman D (1997) Antagonistic and agonistic effects of tamoxifen: significance in human cancer. Semin Oncol 24 (Suppl 1):1–71–1–80
Gambrell RD Jr.(1982) The menopause: Benefits and risks of estrogen-progestogen replacement therapy. Fertil Steril 37:457
Goldzieher JW (1995) Are low-dose oral contraceptives safer and better? Comment in: Am J Obstet Gynecol 172:1948–1950
Goss PE, Gwyn KM (1994) Current perspectives on aromatase inhibitors in breast cancer. J Clin Oncol 12(11):2460–2470
Hammond CB, Jelovsek FR, Lee KL, Creasman WT, Parker RT (1979) Effects of long term estrogen replacement therapy. I. Metabolic effects. Am J Obstet Gynecol 133(5):525–536
Heise E, Gorlich M (1993) Estradiol receceptor and prognosis in human breast cancer and its metastases. Neoplasia 40(1):55–57
Howell A (1997) Antiestrogens: future prospects. Oncology-Huntingt 11(Suppl 2): 59–64
Howell A, Downey S, Anderson E (1996) New endocrine therapies for breast cancer. Eur J Cancer 32 A(4):576–588
Jensen EV, Jacobsen HI (1962) Basic guides to the mechanism of estrogen action. Recent Progress Hormone Res 18:387–414
Johnston SR, Smith IE, Doody D, Jacobs S, Robertshaw H, Dowsett M (1994) Clinical and endocrine effects of the oral aromatase inhibitor vorozole in postmenopausal patients with advanced breast cancer. Cancer Res 54(22):5875–5881
Jordan VC (1995) Tamoxifen for breast cancer prevention. Proc Soc Exp Biol Med 208(2):144–149
Kiang DT, Kennedy BJ (1977) Factors affecting estrogen receptors in breast cancer. Cancer 40:1571–1576
Kielback DG, McCamant SK, Press MF, Atkinson EN, Gallager HS, Edwards CL, Hajek RA, Jones LA (1993) Cancer Res 53(21):5188–5192
Lerner LJ, Jordan VC (1990) Development of antiestrogens and their use in breast cancer: Eighth Cain Memorial Award Lecture. Cancer Res 50:4177–4189
Love RR (1994) Prevention of breast cancer in premenopausal women. J Natl Cancer Inst Monogr 16:61–65
McQuire WL, Horowitz KB, Zava DT et al. (1978) Progress endrocrinology and metabolism – ormones in breast cancer. Metabolism 27:487–501
Miller AB, Bulbrook RD (1986) UICC multidisciplinary project on breast cancer: the epidemiology and prevention of breast cancer. Int J Cancer 37:173–177
Miller WR, Langdon SP (1997) Steroid hormones and cancer: (III) Observations from human subjects. Eur J Surg Oncol 23:163–13
Myres AM, Moore GE, Major FJ (1981) Advanced ovarian carcinoma: response to antiestrogen therapy. Cancer 48:2368–2370
Nachtigall LE, Nachtigall R, Nachtigall RD, Beckman EM (1979) Estrogen replacement therapy II: a prospective study in the relationship to carcinoma and cardiovascular and metabolic problems. Obstet Gynecol 54(1):74–7
National Institute of Health (1985) Consensus development conference statement: adjuvant chemotherapy for breast cancer. JAMA 254:3661–363
Nease RF Jr, Ross JM (1995) The decision to enter a randomized trial of tamoxifen for the prevention of breast cancer in healthy women: an analysis of the tradeoffs. Am J Med 99(2):180–189

Nolvadex Adjuvant Trial Organisation (1985) Controlled trial of tamoxifen as a single adjuvant agent in the management of early breast cancer. Lancet 1:836–840
Palkowitz AD, Glasebrook AL, Thrasher KJ, Hauser KL, Short LL, Phillips DL, Muehl BS, Sato M, Shelter PK, Cullenoin GJ, Pell TR,Bryant HU (1997) Discovery and synthesis of [6-hydroxy-3-[2-(1-piperidinyl)ethoxy]phenoxy]-2-(4-hydroxyphenyl)]benzo[b]thiophene: a novel, highly potent selective estrogen receptor modulator. J Med Chem 40(10):1407–1416
Pearson OH, Manni A, Arafah BM (1982) Antiestrogen treatment of breast cancer: an overview. Cancer Res 42(Suppl 8):3424–3429
Persson I, Adami HO, Bergkvist L, Lindgren A, Pettersson B, Hoover R, Schairer C (1989) Risk of endometrial cancer after treatment with estrogens alone or in conjunction with progestogens: results of a prospective study. BMJ 298(6667): 147–151
Plourde PV, Dyroff M, Dukes M (1994) Arimidex: a potent and selective fourth-generation aromatase inhibitor. Breast Cancer Res Treat 30(1):103–111
Plowman PN (1996) Adjuvant therapy in breast cancer. Drugs Aging 9(3):185–190
Possinger K, Jonat W, Hoffken K (1994) Formestane in the treatment of advanced post-menopausal breast cancer. Ann Oncol 5(Suppl 7):7–10
Potischman N, Hoover RN, Brinton LA, Siiteri P, Dorgan JF, Swanson CA, Berman ML, Mortal R, Twiggs LB, Barrett RJ, Wilbanks GD, Persky V, Lurain JR (1996) Case-control study of endogenous steroid hormones and endometrial cancer. J Natl Cancer Inst 88(16):1127–1135
Powles TJ, Gordon C, Coombes RC (1982) Clinical trial of multiple endocrine therapy for metastatic and locally advanced breast cancer with tamoxifen-aminoglutethimide-danazol compared to tamoxifen used alone. Cancer Res 42(Suppl 8):3458–3460
Preece PE, Wood RA, Mackie CR, Cuschieri A (1982) Tamoxifen as the initial sole treatment of localised breast cancer in elderly women: a pilot study. Br Med J Clin Res Ed 284(6319):869–870
Rauschning W, Pritchard KI (1994) Droloxifene, a new antiestrogen: its role in metastatic breast cancer. Breast Cancer Res Treat 31(1):83–94
Ribeiro G, Swindell R (1992) Adjuvant tamoxifen for male breast cancer. Br J Cancer 65(2):252–254
Risch HA (1996) Estrogen replacement therapy and risk of epithelial ovarian cancer. Gynecol Oncol 63:254–257
Rodriguez C, Calle EE, Coates RJ, Miracle MC, Mahill HL, Thun, MJ, Heath CW (1995) Estrogen replacement therapy and fatal ovarian cancer. Am J Epidemiology 141:828–835
Rosenblatt KA, Thomas DB, Nooman EA (1992) High dose and low dose combined oral contraceptives: protection against epithelial ovarian cancer and the length of the protective effect. Eur J Cancer 28 A:1872–1875
Sapher T, Torrney C, Gray R (1991) Venous and arterial thrombosis in patients who received adjuvant therapy for breast cancer. J Clin Oncol 9:286–29
Slotman BJ, Rao BR (1988) Ovarian cancer (review): etiology, diagnosis, prognosis, surgery, radiotherapy, chemotherapy and endocrine therapy. Anticancer Res 8:417–434
Smith IE, Harris AL, Morgan M, Gazet JC, McKinna JA (1982) Tamoxifen versus aminoglutethimide versus combined tamoxifen and aminoglutethimide in the treatment of advanced breast carcinoma. Cancer Res 42(Suppl 8):3430–3433
Stenbygaard LE, Herrstedt J, Thomsen JF, Svendsen KR, Engelholm SA, Dombernowsky P (1993) Toremifene and tamoxifen in advanced breast cancer – a double-blind cross-over trial. Breast Cancer Res Treat 25(1):57–63
Swedish Breast Cancer Cooperative Group (1996) Randomized trial of two versus five years of adjuvant tamoxifen for postmenopausal early stage breast cancer. J Natl Cancer Inst 88(21):1543–1549

Trunet PF, Bhatnagar AS, Chaudri HA, Hornberger U (1996) Letrozole (CGS 20267), a new oral aromatase inhibitor for the treatment of advanced breast cancer in postmenopausal patients. Acta Oncol 35(Suppl 5):15–18

Van de Velde P, Nique F, Planchon P, Prevost G, Bremaud J, Hameau MC, Magnien V, Philibert D, Teutsch G (1996) RU 58668: further in vitro and in vivo pharmacological data related to its antitumoral activity. J Steroid Biochem Mol Biol 59(5–6):449–457

Vogel CI (1996) Hormonal approaches to breast cancer treatment and prevention: an overview. Semin Oncol 23(Suppl 9):2–9

Wakeling AE (1993) Are breast tumours resistant to tamoxifen also resistant to pure antiestrogens? J Steroid Biochem Mol Biol 47:107–114

Wiseman LR, Goa KL (1996) Formestane. A review of its pharmacological properties and clinical efficacy in the treatment of postmenopausal breast cancer. Drugs Aging 9:292–306

CHAPTER 42
Cardiology

E.F. MAMMEN

A. Introduction

Soon after the introduction of the first combined oral contraceptive (OC), Enovid 10[R], in 1961, there appeared a case report that seemed to suggest a possible link between OC use and thromboembolic disease (JORDAN 1961). Thromboembolism is a part of the broader term "cardiovascular disease." This chapter will exclusively discuss the relationship between OC use and thromboembolic diseases; the references cited are by necessity selective.

B. Epidemiological Association Between Oral Contraceptive Use and Thrombosis

The first report by JORDAN (1961) was followed by additional case reports that ultimately prompted several prospective epidemiological studies in the early 1970s (INMAN and VESSEY 1968; INMAN et al. 1970; ROYAL COLLEGE OF GENERAL PRACTITIONERS 1974). These investigations assessed the possible associations between OC use and deep vein thrombosis (DVT), pulmonary embolism (PE), acute myocardial infarction (MI) and cerebrovascular accidents (CVA) either thrombotic or hemorrhagic. In the following years, several additional epidemiological studies were published that were summarized and critically reviewed by REALINI and GOLDZIEHER (1985). The observations suggested a relative increased risk of 2–4 of acute MI, 2–6 of CVA and 8 of fatal PE (INMAN and VESSEY 1968; MANN et al. 1975; ROYAL COLLEGE OF GENERAL PRACTITIONERS 1978). An increased risk of DVT was difficult to establish because of the unreliability of diagnosis. THOROGOOD and VESSEY (1990) drew four conclusions from these epidemiological observations:

1. The absolute risk of fatal cardiovascular diseases seemed to increase with age of the OC user.
2. The increased risk of acute MI seemed to be influenced by co-existing risk factors, most notably cigarette smoking.
3. The risk seemed to be confined to present OC users.
4. The risk seemed to be little influenced by the duration of the OC use.

Early on, the estrogen component of the OC was causally implicated by DVT and PE (INMAN et al. 1970), while the progestogens were thought to be primarily related to changes in lipids and lipoproteins and, thus, potentially to the development of atherosclerosis and changes in carbohydrate metabolism (MEADE 1988).

These observations prompted two major formulary changes for OCs:

1. A reduction of estrogen content from greater than 50 μg to 30–35 μg and, more recently, to 20 μg
2. A reduction of progestogen dose because of concerns about the lipid changes (ROBINSON 1994) and the introduction of three new progestogens (desogestrel, gestodene and norgestimate); the estrogen component in OCs is usually ethinyl estradiol (EE)

The OC formulations are presently classified as three generations:

First generation:	Estrogen (50 μg) with or without synthetic progestogens
Second generation:	Estrogen (<50 μg) with progestogens other than gestodene or desogestrel
Third generation:	Estrogen (<50 μg) plus gestodene or desogestrel

Each change from first to third generation was intended to increase safety, and first-generation OCs are practically no longer in clinical use (LEWIS et al. 1996a). Follow-up epidemiological studies on users of second-generation OCs demonstrated a decreased incidence of thromboembolic events, in general, and of DVT and PE, in particular (GERSTMAN et al. 1990). Also, the risk of CVA seems to have disappeared (PETITTI et al. 1996).

Recently, a number of case-control studies have suggested that the third-generation OCs carry a greater risk, especially for DVT and PE, than the second generation (FARLEY et al. 1995; POULTER et al. 1985; JICK et al. 1995; BLOEMENKAMP et al. 1995; SPITZER et al. 1996). It was suggested that this might be due to desogestrel or gestoden. More recently, this difference could not be confirmed (FARMER et al. 1997, LIDEGARD and MILSOM 1996). Two studies even suggested a cardioprotective effect of third-generation OCs, at least on acute MI (LEWIS et al. 1996b; JICK et al. 1996). This benefit seems to outweigh the risk of DVT, especially in older women (35–40 years) (SCHWINGL and SHELTON 1997).

These conflicting results strongly indicate that co-existing factors or bias must exist (LEWIS et al. 1996a; DOUKETIS et al. 1997), and they reiterate the cautions expressed by REALINI and GOLDZIEHER (1985) when interpreting epidemiological studies. There is presently no evidence that progestogens of the third-generation OCs exert a greater thrombophilic activity, as will be discussed below. It should be noted that "the large benefits of appropriate contraceptive practices, such as use of oral contraceptive agents, are often obscured by legitimate concerns about small absolute risks of rare but serious events" (BURING 1996). It seems that pregnancy represents a greater risk for thromboembolic disorders than current OC formulations (BURING 1996; FOTHERBY 1996). General agreement seems to exist, however, that whatever risk might still exist, it is only present during the time of OC ingestion and is influenced by age and smoking.

A plausible explanation for these controversial data would be that those few women who encounter a thromboembolic event during OC use would have an underlying prethrombotic disorder that predisposes to these complications (FARAG et al. 1988; MAMMEN 1992a).

C. Oral Contraceptives and Hemostasis

Thrombosis is a pathological event related to an activation of the hemostasis system. There is, however, a difference in the pathogenesis of arterial and of venous thrombosis. Arterial thromboses (acute MI, thrombotic CVA) are predominantly platelet driven and mediated; platelets respond to endothelial lesions, most commonly atherosclerotic disease. Also congenital hyperaggregability of platelets, the "sticky-platelet syndrome", can lead to arterial thrombotic occlusions (MAMMEN 1994). In contrast, venous thrombosis is greatly influenced by disturbed venous blood flow patterns and by hypercoagulability, i.e., changes in the hemostasis system that lead to a thrombophilic state (FUJII and MAMMEN 1992b).

The observed epidemiological association between OC use and thromboembolic events has obviously led to intensive investigations of the clotting and the fibrinolytic systems, since an activated clotting system, and an inhibited fibrinolytic system, can lead to thrombophilic states.

I. Oral Contraceptives and the Vessel Wall

The potential effects of OCs on the vessel wall, on platelets and on the hemostasis system have been extensively reviewed (MAMMEN 1992a; NOTOLOVITZ 1985; BELLER and EBERT 1985) and details may be obtained from these reviews. There is, at this time, no convincing evidence that OCs adversely affect endothelium. There is also no evidence that OCs increase atherosclerosis in humans (TIKKANEN 1990) or in monkeys (MANNING et al. 1997; BELLINGER et al. 1998).

II. Oral Contraceptives and Platelets

The publications on OCs and platelet function, especially platelet hyperaggregability, are controversial, which is likely due to technical difficulties. Studies on low-dose estrogen OCs, triphasic and monophasic, have shown no changes (see MAMMEN 1992a). More recent studies performed on women using different formulations of OC and Norplant again revealed no signs of hyperaggregability (SALEH et al. 1995; WEINGES et al. 1995). Platelet factor 4 and β-thromboglobulin, two platelet-released proteins, were measured in OC users and normal levels seem to be the predominant finding (see MAMMEN 1992a).

There is, at this time, no convincing evidence that OCs of any type activate platelets and thereby contribute to thromboembolic events, especially arterial.

III. Oral Contraceptives and Coagulation

The coagulation system comprises a mechanism that leads to the formation of fibrin, i.e., the clotting system, and a mechanism that dissolves fibrin, i.e., the fibrinolytic system. Both systems are in a constant but subliminal state of activation. This implies that a balanced equilibrium has to exist between the two systems. It is well recognized that an activation of the clotting system or an impaired fibrinolytic system may lead to a hypercoagulable state (COMP 1986; MAMMEN and FUJII 1989).

1. Clotting System

The clotting system centers around the conversion of prothrombin to thrombin, with the subsequent formation of fibrin from fibrinogen. A large number of clotting factors drive this process. For details, see recent reviews (HARKER and MANN 1992; BICK and MURANO 1993; COLMAN et al. 1994; MAMMEN and WILSON 1996).

This activation sequence is regulated by two major inhibitor devices, antithrombin (AT) and the protein C (PC) and protein S (PS) system. AT is a serine protease inhibitor that neutralizes all enzymes of the clotting system. PC in its enzymatic form, PC_a, proteolytically destroys two co-factors of the clotting system, factors V and VIII. Through this mechanism, the amounts of thrombin and factor X_a generated are regulated. For details, see ESMON (1984) and BROZE and TOLLEFSEN (1994).

It is well recognized that decreased plasma levels of AT, PC and PS, congenital or acquired, are associated with thromboembolic diseases. Most patients with congenital AT or PC defects are heterozygotes; homozygosity is barely compatible with life. In cases of PS defect, both heterozygotes and homozygotes are seen, whereby the latter have a greater risk of thrombosis. More recently, "activated PC resistance" (APCR) has been discovered (DAHLBÄCK et al. 1993) that seems to be widely distributed in the Caucasian population (see REES 1996; ZÖLLER et al. 1997). The cause of APCR is a single base change, G $\rightarrow$ A, at nucleotide 1691 of the factor-V gene. An arginine residue at position 506 is replaced by a glutamine (BERTINA et al. 1994). This abnormal factor-V mutation is called "factor-V Leiden." It is found in 40–60% of patients with a history of DVT and in 1–15% of healthy Caucasians (ZÖLLER et al. 1997). In contrast, AT, PC and PS defects are only encountered in 5–10% of thrombosis patients. More recently, also combination defects such as APCR plus PC or PS defects and others have been reported.

A large volume of literature is available on the effect of OCs on the clotting system, some with considerable controversy. Lack of standardization and technical differences account for some of the findings in the older literature. More recent studies, using standardized technology and in many instances automation, have yielded more consistent and more reliable data. The screening tests of the clotting system, prothrombin time (PT) and activated partial thromboplastin time (APTT) may be shorter in OC users than non-users, especially in patients who use high-dose estrogen OCs (>50 μg); the shortening is expressed less with the second- and third-generation OCs. In these preparations, only PT may be shorter, likely due to elevated factor VII levels (Lox 1996; Creatsas et al. 1997; see Mammen 1992a).

Clotting-factor levels increase in OC users and, again, the increases seem to be related to the estrogen dose. During the use of first-generation OCs, practically all procoagulant clotting factors were elevated (see Mammen 1992a). In more recent studies, only a few factors are affected, most notably the vitamin K-dependent factors VII and X, and fibrinogen (Poller et al. 1991; Scarabin et al. 1995; Norris and Bonnar 1996). It appears that OCs containing 20 μg estrogen or OCs containing levonorgestrel and 30 μg EE exert the least effect on clotting-factor levels (Poller et al. 1991; Norris and Bonnar 1996). For reasons outlined before (Mammen 1982), there is no evidence that elevated factor levels suggest hypercoagulability. The positive correlation between elevated fibrinogen and factor-VII levels and subsequent cardiovascular events, such as acute MI, CVA and sudden death, as reviewed by Hultin (1991), relates apparently to arterial events and not to DVT, PE or increased atherosclerotic vessel disease. As pointed out before, there is no evidence that OC users have an increased risk of atherosclerosis.

In contrast, decreases in inhibitor levels, especially AT, PC and PS, are known to predispose to thrombosis, but levels must be about half that of normal for this association; these inhibitor levels have been extensively studied in OC users. Again, due to technical problems, data are inconsistent, especially in the older literature (see Mammen 1992a). In general, dependent on the estrogen levels, AT and PS levels seem to decrease during OC use, and PC levels increase. Third-generation OC users only demonstrate a minor change or none at all in AT, with similar findings for PC and PS (Poller et al. 1991, Norris and Bonnar 1996; Creatsas et al. 1997). Progesterone-only OCs showed no effect on hemostasis, especially AT, PC and PS (Blombäck et al. 1997). Measurement of clotting factors and inhibitors gave only very limited and indirect information on the actual status of activation of the systems in vivo, and is not likely clinically relevant (Mammen 1982, 1992a; Comp 1997).

2. Fibrinolytic System

During fibrinolysis plasminogen is converted to plasmin, principally by tissue-type plasminogen activator (t-PA). It is primarily regulated by plasminogen-activator inhibitor 1 (PAI-1) and α_2-antiplasmin (for details, see Bick and Murano 1993; Colman et al. 1994; Mammen and Wilson 1996). There is strong evidence in the literature that decreased plasma levels of plasminogen and t-PA and elevated levels of PAI-1 predispose to thrombosis (LIJNEN and COLLEN 1989).

a) Oral Contraceptives and Fibrinolysis

Many studies have been conducted on components of the fibrinolytic system in OC users (see MAMMEN 1992a); most of these relate to second- and third-generation OCs. In general, plasminogen levels seem to be elevated, t-PA levels unchanged and PAI-1 levels decreased. Newer studies reveal very little changes in these parameters (POLLER et al. 1991; NORRIS and BONNAR 1996; SCARABIN et al. 1995; JESPERSEN and GRAM 1996). Measurement of the end-products of fibrinolysis, fibrin(ogen) split products (FSP) and D-dimer demonstrated elevations (NORRIS and BONNAR 1996; JESPERSEN and GRAM 1996), suggesting a mild activation of the fibrinolytic system during OC use.

As with measurement of clotting components, fibrinolytic parameters also only give limited information on the status of activation of this system. Elevated FSP and D-dimer levels, however, allow a more accurate assessment. If one were to assume, based on the data presented so far, that there is a greater than normal activation of the clotting system in vivo induced by OCs, there seems to be an equally greater activation of fibrinolysis. This would suggest that a balance between the two systems remains (NORRIS and BONNAR 1996; POLLER et al. 1991; GEVERS LEUVEN et al. 1996).

IV. Oral Contraceptives and Molecular Markers of Hemostasis Activation

In the last few years, new technologies have been developed that yield more direct information on an activated hemostasis system. Antibodies have been raised against either endproducts or intermediate products of hemostasis activation (FAREED 1984; BAUER and WEITZ 1994). Table 1 summarizes the presently available markers for clotting and fibrinolysis activation.

Several investigators have applied this technology to users of OCs. In a pilot study, we looked at 56 women on three different low-estrogen dose OCs and found no difference in TAT and F 1 + 2 levels between them and 47 non-users, matched for age and race (SALEH et al. 1994). Also PETERSEN et al. (1993) observed no changes in TAT and D-dimer levels in women taking OCs containing either 20 μg EE and 150 μg desogestrel, or 30 μg EE and 75 μg gestodene. PÄRTAN et al. (1991) reported normal TAT levels in users of OCs containing 30 μg EE and 75 μg gestodene. In contrast, GRAM et al. (1990) reported elevated TAT levels in older (>30 years) users.

FpA levels were reported to be elevated in women ingesting OCs containing greater than 50 μg EE, but were normal in lower dose OCs (MELIS et al. 1990; ABBATE et al. 1990). The greatest experiences with these markers were published by WINKLER et al. (1991, 1995, 1996a, b, c). These investigators compared OCs with identical progestogen, but different EE content (35 μg and

Table 1. Available molecular markers of in vivo hemostatic activation

Clotting system	
Fibrinopeptide A (FpA):	Generated from fibrinogen by thrombin
Thrombin/antithrombin (TAT) complex:	Generated when thrombin is inactivated by AT
Prothrombin fragment 1 + 2 (F 1 + 2):	Generated from prothrombin by factor Xa
Fibrinolysis system	
Fibrin(ogen) split products (FSP):	Generated when plasmin cleaves fibrinogen or fibrin
Plasmin/antiplasmin (PAP) complex:	Generated when plasmin is inactivated by α_2-antiplasmin
Both systems	
D-dimer:	Generated when fibrin is formed by thrombin and subsequently proteolytically digested by fibrin, i.e., a *fibrin* degradation product

50 μg), and found elevated TAT and D-dimer levels in women on the higher dose EE pills, but no major changes in the group taking 35 μg EE (WINKLER et al. 1991). They also compared women taking OCs containing 35 μg EE and 250 μg norgestimate with those taking 30 μg EE and 75 μg gestodene (WINKLER et al. 1995). They reported significant elevations in TAT and F 1 + 2 with both OCs, but the F 1 + 2 increase was expressed less ($P < 0.05$) in the gestodene group. Also, PAP and D-dimer levels were markedly higher in both groups. In another study (WINKLER et al. 1996a), the effects of 20 μg EE and 75 μg gestodene were compared with 30 μg EE and 75 μg gestodene; F 1 + 2 and PAP levels were elevated in both groups. Also the effects of two monophasic gestodene OCs, containing 30 μg and 20 μg EE were reported (WINKLER et al. 1996b). In both patient groups, F 1 + 2, PAP, D-dimer and FSP levels were elevated, but TAT concentrations were non-significantly altered. Lastly, WINKLER et al. (1996c) compared OCs containing 150 μg desogestrel and 20 μg EE with those containing 30 μg EE in a multicenter trial. Both groups of women had elevated F 1 + 2 and D-dimer levels, but the levels were significantly lower in the group taking 20 μg EE.

Recently QUEHENBERG et al. (1997) reported elevated PAP and D-dimer levels in women on three different OCs, all containing 30 μg EE but different norgestimate or gestodene concentrations. All three groups had elevated levels without significant differences between them. JESPERSEN and GRAM (1996) described elevated PAP and D-dimer levels in women over 30 years who also smoked. Users of OCs containing 30–50 μg EE had the highest D-dimer and PAP levels, and non-users who smoked had higher levels than non-smoking non-users.

From these data, inconsistent as they may be for the moment, several conclusions may be drawn:

1. During OC use there seems to be a greater than normal activation of the hemostasis system.
2. Both the clotting and the fibrinolytic system seem to be activated with no indication of an imbalance between the two systems.
3. The effects seem to be related to the estrogen and not to the progestogen component.
4. Age and smoking seem to exercabate the response.

All of these data still do not give a satisfactory answer to the question why only a few women encounter a thromboembolic event, when all have an activated hemostasis system with no evidence for an imbalance between clotting and fibrinolysis.

D. Oral Contraceptives and Thrombophilia

A possible answer to the above question might be that those women on OCs who develop a thrombotic event might have a predisposing disorder, i.e., congenital or acquired thrombophilia (FARAG et al. 1988; MAMMEN 1992a). It is known that patients with congenital AT, PC and PS defects or APCR have a strong predisposition to thromboembolic problems. Unless there are combined defects or a precipitating factor is introduced, a single defect rarely results in an event. The precipitating factor may be surgery, trauma, pregnancy or, potentially, OC use (see COMP 1986).

Of all known congenital defects, APCR is the most frequently encountered abnormality in the Caucasian population (see REES 1996; ZÖLLER et al. 1997; COMP 1997). There are a number of reports that have measured the presence of APRC (factor-V Leiden) in women who developed DVT while using OCs (BOKAREWA et al. 1995; HELLGREN et al. 1995; SVENSSON and DAHLBÄCK 1994; see COMP 1997) and between 14% and 40% of these patients had this defect. The incidence of factor-V Leiden in the normal population is about 5% and 20–40% in patients with DVT (VANDENBROUCKE et al. 1994; SVENSSEN and DAHLBÄCK 1994; KOSTER et al. 1993). VANDENBROUCKE et al. (1994) calculated that OC use in women with congenital APCR increases the risk 34.7-fold in comparison with healthy non-users. Similar incidences were found in women who developed DVT during pregnancy (HELLGREN et al. 1995, CUMMING et al. 1995). In contrast to other congenital inhibitor defects, APCR-related thromboses might develop rather late (after approximately 6 cycles) in OC-users (GIROLAMI et al. 1995).

These findings suggest that women with congenital or acquired thrombophilia might be at risk of encountering thrombotic events when placed on OCs. Table 2 lists the presently known and suspected defects associated with thrombophilia. In addition to factor-V Leiden, hyperhomocysteinemia seems to become a rather frequent finding. With rare exceptions, most patients with these defects have a positive family history of thromboembolic events. Based on present knowledge it may, thus, be prudent to exercise caution when pre-

Table 2. Congenital and acquired conditions possibly associated with thrombophilia

Congenital[a]
Activated protein C resistance (APCR) or factor-V Leiden
Hyperhomocysteinemia
Protein C defects
Protein S defects
Antithrombin defects
Plasminogen defect (?)
Impaired release of t-PA (?)
Elevated PAI-1 levels (?)
Factor XII defect (?)
Heparin co-factor II defect (?)
Dysfibrinogenemia
Dysprothrombinemia
Acquired
Antiphospholipid antibody syndrome
Lupus anticoagulant
Anticardiolipin antibodies

t-PA, tissue-type plasminogen activator; PAI-1, plasminogen-activator inhibitor 1; (?), relationship possible but not proven
[a] Defects can also be acquired

scribing OCs to patients with a positive history and, if at all possible, the nature of the defect should be determined first. It must be remembered, however, that at least 50% of patients with congenital thrombophilia can, at present, not yet be classified into any one of the defects listed in Table 2. Acquired thrombophilic disorders, most notably the antiphospholipid-antibody syndrome, may not have a positive family history.

It has been debated whether APCR should be routinely determined in all women about to start OC treatment. Based on statistics and theoretical projections, this may not be cost-effective (Comp 1997, Winkler et al. 1996d); others are of a different opinion (Schambeck et al. 1997) and still others support a work-up on all women with a positive history before prescribing OCs (Hellgren et al. 1995; Bokarewa et al. 1995).

Several recent studies have found that APRC can be acquired, especially during OC use and pregnancy (Cumming et al. 1995; Olivieri et al. 1995, 1996; Comp 1997). This resistance is different from the factor-V Leiden defect, i.e., these patients are negative for the mutation. It is not related, in OC users, to dose and composition of the formulations and is independent of time of use (Olivieri et al. 1996). There is, at this time, no evidence that this form of APCR poses a thrombotic risk (Comp 1997).

E. Summary

Many epidemiological studies have reported a link between OC use and thromboembolic disorders, arterial and venous. A reduction in estrogen content has markedly reduced the observed risk and the newly developed low-dose estrogen OC seem to be safe; risks seem to persist only for older women (>30 years) and for smokers. The recently reported higher incidence of events with third-generation OCs than with second generation may reflect bias and lacks a physiological basis. Any risk ceases with discontinuation of OC use.

There is no convincing evidence that presently used OCs adversely impact vessel wall, platelet function and the coagulation system. The slightly increased activation of the clotting system seems to be balanced by a similarly activated fibrinolytic system. The use of molecular markers of in vivo hemostasis activation seems to confirm this.

Evidence accumulates that women who encounter a thrombotic event, likely have a pre-existing thrombophilic state, and APCR or factor-V Leiden seems to markedly increase the risk. All other known defects must also be considered. Since most patients with these defects have a history of thromboembolism, routine screening of all women prior to OC use is likely not economical.

References

Abbate R et al. (1990) Effects of long–term gestodene–containing oral contraceptive administration on hemostasis. Am J Obstet Gynecol 163:424–429

Bauer KA, Weitz JI (1994) Laboratory markers of coagulation and fibrinolysis. In: Colman RW, Hirsh J, Marder VJ, Salzman EW (eds) Hemostasis and thrombosis. Basic principles and clinical practice. 3rd edn. Lippincott, Philadelphia, PA, p 1197

Beller FK, Ebert C (1985) Effects of oral contraceptives on blood coagulation. A review. Obstet Gynecol Surv 40:425–436

Bellinger DA et al. (1998) Oral contraceptives and hormone replacement therapy do not increase the incidence of arterial thrombosis in a nonhuman primate model. Arterioscler Thromb Vasc Biol 18:92–99

Bertina RM et al. (1994) Mutation in blood coagulation factor V associated with resistance to activated protein C. Nature 369:64–67

Bick RL, Murano G (1993) Physiology of hemostasis. In: Bick RL (ed) Hematology. Clinical and laboratory practice. Mosby, St. Louis, MO, p 84

Bloemenkamp KWM (1995) Enhancement of factor V Leiden mutation of risk of deep-vein thrombosis associated with oral contraceptives containing third generation progestagen. Lancet 346:1589–1593

Blombäck M et al. (1997) The effect of progesterone on the haemostatic mechanism. Thromb Haemostas 77:105–108

Bokarewa MI et al. (1995) Thrombotic risk factors and oral contraceptives. J Lab Clin Med 126:294–298

Buring JE (1996) Low–dose oral contraceptives and stroke. N Engl J Med 335:53–55

Broze GJ Jr, Tollefsen DM (1994) Regulation of blood coagulation by protease inhibitors. In: Stamatoyannopoulos G, Nienhuis AW, Majerus PW, Varmus H (eds) The molecular basis of blood diseases. 2nd edn. Saunders, Philadelphia, PA, p 629

Coata G et al. (1995) Effect of low–dose oral triphasic contraceptives on blood viscosity, coagulation and lipid metabolism. Contraception 52:151–157

Colman RW et al. (1994) Overview of hemostasis. In: Colman RW, Hirsh J, Marder VJ, Salzman EW (eds) Hemostasis and thrombosis. Basic principles and clinical practice. 3rd edn. Lippincott, Philadelphia, PA, p 3

Comp PC (1986) Hereditary disorders predisposing to thrombosis. Progr Hemostas Thromb 8:71–102

Comp PC (1997) Thrombophilic mechanisms of OCs. Int J Fertil 42 (Suppl 1):170–176
Creatsas G et al. (1997) Effects of two combined monophasic and triphasic ethinylestradiol/gestodene oral contraceptives on natural inhibitors and other hemostatic variables. Eur J Contracept Reprod Health Care 2:31–38
Cumming AM et al. (1995) Development of resistance to activated protein C in pregnancy. Br J Haematol 90:725–727
Dahlbäck B et al. (1993) Familial thrombophilia due to a previously unrecognized mechanism characterized by poor anticoagulant response of activated protein C: prediction of a cofactor to activated protein C. Proc Natl Acad Sci USA 90:1004–1008
Douketis JD et al. (1997) A reevaluation of the risk for venous thromboembolism with the use of oral contraceptives and hormone replacement therapy. Arch Intern Med 157:1522–1530
Esmon CT (1984) Protein C. Semin Thromb Hemostas 10:109–166
Farag AM et al. (1988) Oral contraceptives and the hemostasis system. Obstet Gynecol 71:584–588
Fareed J (1984) Molecular markers of hemostatic disorders. Semin Thromb Hemostas 10:215–332
Farley TMM et al. (1995) Effect of different progestagens in low estrogen oral contraceptives on venous thromboembolism. Lancet 346:1582–1588
Farmer RTD et al. (1997) Population-based study of risk of venous thromboembolism associated with various oral contraceptives. Lancet 349:83–88
Fotherby K (1996) Thrombosis and the pill (guest editorial). J Drug Dev Clin Pract 8:1–5
Gerstman et al. (1990) Oral contraceptive estrogen and progestin potencies and the incidence of deep venous thrombosis. Int J Epidemiol 19:931–936
Gevers Leuven JA et al. (1996) Effects of low-dose ethinylestradiol oral contraceptives differing in progestogen compound on coagulation and fibrinolytic–risk variables for venous and arterial thromboembolic diseases. In: Glas–Greenwalt P (ed) Fibrinolysis diseases. CRC, Boca Raton, FL, p 226
Girolami A et al. (1995) Patients with APC resistance compared with those with other clotting inhibitor deficiencies show late onset of venous thrombosis during oral contraception. Clin Appl Thromb/Hemostas 1:274–276
Gram J et al. (1990) Enhanced generation and resolution of fibrin in women above the age of 30 years using oral contraceptives low in estrogen. Am J Obstet Gynecol 163:438–442
Harker LA, Mann KG (1992) Thrombosis and fibrinolysis. In: Fuster V, Verstraete M (eds) Thrombosis in cardiovascular disorders. Saunders, Philadelphia, PA, p 1
Hellgren M et al. (1995) Resistance to activated protein C as a basis for venous thromboembolism associated with pregnancy and oral contraceptives. Am J Obstet Gynecol 173:210–213
Hultin MB (1991) Fibrinogen and factor VII as risk factors in vascular disease. Progr Hemostas Thromb 10:215–241
Inman WHW, Vessey MP (1968) Investigation of deaths from pulmonary, coronary and cerebral thrombosis and embolism in women of childbearing age. Br Med J 2:193–199
Inman WHW et al. (1970) Thromboembolic disease and the steroidal content of oral contraceptives: A report to the Committee on Safety of Drugs. Br Med J 2:203–209
Jespersen J, Gram J (1996) Increased fibrin formation in blood of women above the age of 30 who are both oral contraceptive users and smokers. Fibrinolysis 10:9–13
Jick H et al. (1995) Risk of idiopathic cardiovascular death and nonfatal venous thromboembolism in women using oral contraceptives with differing progestogen components. Lancet 346:1589–1593
Jick H et al. (1996) Risk of acute myocardial infarction and low-dose combined oral contraceptives (Letter). Br Med J 347:627–628
Jordan WM (1961) Pulmonary embolism. Lancet II:1146–1147

Koster T et al. (1993) Venous thrombosis due to poor anticoagulant responses to activated protein C: Leiden thrombophilia study. Lancet 342:1503–1506
Lewis MA et al. (1996a) The increased risk of venous thromboembolism and the use of third generation progestagens: Risk of bias in observational research. Contraception 54:5–13
Lewis MA et al. (1996b) Third generation oral contraceptives and risk of myocardial infarction: An international case-control study. Br Med J 312:88–90
Lidegaard O, Milson I (1996) Oral contraceptives and thrombotic diseases: Impact of new epidemiological studies. Contraception 53:135–139
Lijnen HR, Collen D (1989) Congenital and acquired deficiencies of components of the fibrinolytic syndrome and their relation to bleeding and thrombosis. Fibrinolysis 3:62–77
Lox CD (1996) Biochemical effects in women following one year's exposure to a new triphasic oral contraceptive. II. Coagulation profiles. Gen Pharmacol 27:371–374
Mammen EF (1982) Oral contraceptives and blood coagulation: A critical review. Am J Obstet Gynecol 142:781–790
Mammen EF (1992a) Oral contraceptives and thrombotic risk: A critical overview. In: Ramwell P, Rubanyi G, Schillinger E (eds) Sex steroids and the cardiovascular system. Springer, Berlin Heidelberg New York, p 65
Mammen EF (1992b) Pathogenesis of venous thrombosis. Chest 102:640S–644S
Mammen EF (1994) Ten years experience with the "sticky platelet syndrome". Clin Appl Thromb/Hemostas 1:66–72
Mammen EF, Fujii Y (1989) Hypercoagulable states. Lab Med 20:611–615
Mammen EF, Wilson RF (1996) Coagulation abnormalities in trauma. In: Wilson RF, Walt AJ (eds). Management of trauma. Pitfalls and practice. 2nd edn. Williams and Wilkins, Baltimore, MD, p 1012
Mann JI et al. (1975) Oral contraceptives and myocardial infarction in young women: A further report. Br Med J 3:631–632
Manning JM et al. (1997) Effects of contraceptive estrogen and progestin on the atherogenic potential of plasma LDLs in cynomolgus monkeys. Arterioscler Thromb Vasc Biol 17:1216–1223
Meade TW (1988) Risks and mechanisms of cardiovascular events in users of oral contraceptives. Am J Obstet Gynecol 158:1646–1652
Melis GB et al. (1990) A comparative study on the effects of monophasic pill containing desogestrel plus 20 micrograms ethinylestradiol, a triphasic combination containing levonorgestrel and a monophasic combination containing gestodene on coagulation factors. Contraception 43:23–31
Norris LA, Bonnar J (1996) The effect of estrogen dose and progestogen type on haemostatic changes in women taking low dose oral contraceptives. Br J Obstet Gynecol 103:261–267
Notolovitz M (1985) Oral contraception and coagulation. Clin Obstet Gynecol 28:73–83
Olivieri O et al. (1995) Resistance to activated protein C in healthy women taking oral contraceptives. Br J Haematol 91:465–470
Olivieri O et al. (1996) Resistance to activated protein C, associated with oral contraceptives use; effect of formulations, duration of assumption, and doses of oestro-progestins. Contraception 54:149–152
Pärtan C et al. (1991) Effect of contraceptive pill containing 0.03 mg ethinylestradiol and 0.075 mg gestogen on blood coagulation and fibrinolysis. Thromb Haemostas 65:894 (Abst)
Petersen KR et al. (1993) Effects of monophasic low-dose oral contraceptives on fibrin formation and resolution in young women. Am J Obstet Gynecol 168:32–38
Petitti DB et al. (1996) Stroke in users of low-dose oral contraceptives. N Engl J Med 335:8–15
Poller L for the Task Force on Oral Contraceptives–WHO (1991) A multicentre study of coagulation and haemostatic variables during oral contraception: Variations with four formulations. Br J Obstet Gynecol 98:1117–1128

Poulter NR et al. (1995) Venous thromboembolic disease and combined oral contraceptives: Results of international multicentre case–control study. Lancet 346:1575–1582

Quehenberg P et al. (1997) Effects of third generation oral contraceptives containing newly developed progestagens on fibrinolytic parameters. Fibrinolysis Proteolysis 11:97–101

Realini JP, Goldzieher JN (1985) Oral contraceptives and cardiovascular disease: A critique of the epidemiological studies. Am J Obstet Gynecol 152:729–798

Rees DC (1996) The population genetics of factor V Leiden (Arg 506 Gln). Br J Haematol 95:579–586

Robinson GE (1994) Low–dose combined oral contraceptives. Br J Obstet Gynaecol 101:1036–1041

Royal College of General Practitioners (1974) Oral contraceptives and health. Pitman, London

Royal College of General Practitioners (1978) Oral contraceptives, venous thrombosis, and varicose veins. J Royal Coll Gen Pract 28:393–399

Saleh AA et al. (1994) TAT complexes and prothrombin fragment 1+2 in oral contraceptive users. Thromb Res 73:137–142

Saleh AA et al. (1995) Hormonal contraception and platelet function. Thromb Res 78:363–367

Scarabin P-Y et al. (1995) Changes in haemostatic variables induced by oral contraceptives containing 50μg or 30μg estrogen: Absence of dose–dependent effect on PAI–1 activity. Thromb Haemostas 74:928–932

Schambeck CM et al. (1997) Selective screening for the factor V Leiden mutation: Is it advisable prior to the prescription of oral contraceptives? Thromb Haemostas 78:1480–1483

Schwingl PJ, Shelton J (1997) Modeled estimates of myocardial infarction and venous thromboembolic disease in users of second and third generation oral contraceptives. Contraception 55:125–129

Spitzer WO et al. (1996) Third generation oral contraceptives and risk of venous thromboembolic disorders: An international case-control study. Br Med J 312:83–88

Svensson PJ, Dahlbäck B (1994) Resistance to activated protein C as a basis for venous thrombosis. N Engl J Med 330:517–522

Thorogood M, Vessey MP (1990) An epidemiological survey of cardiovascular disease in women taking oral contraceptives. Am J Obstet Gynecol 163:274–281

Tikkanen MJ (1990) Role of plasma lipoproteins in the pathogenesis of atherosclerotic disease, with special reference to sex hormone effects. Am J Obstet Gynecol 163:296–304

Vandenbroucke et al. (1994) Increased risk of venous thrombosis in oral-contraceptive users who are carriers of factor V Leiden mutation. Lancet 344:1453–1457

Weinges KF et al. (1995) The effects of two phasic oral contraceptives on hemostasis and platelet function. Adv Contracept 11:227–237

Winkler UH et al. (1991) Changes of the dynamic equilibrium of hemostasis associated with the use of low–dose oral contraceptives: A controlled study of cyproterone acetate containing oral contraceptives combined with either 35 or 50μg ethinyl estradiol. Adv Contraception 7 (Suppl 3):273–284

Winkler UH et al. (1995) Hemostatic effects of two oral contraceptives containing low doses of ethinyl estradiol and either gestodene or norgestimate: An open, randomized, parallel–group study. Int J Fertil 40:260–268

Winkler UH et al. (1996a) The influence of a low-dose oral contraceptive containing 20μg ethinylestradiol and 75μg gestodene on lipid and carbohydrate metabolism and hemostasis. In: Lopes P, Killick SR (eds) New option low–dose oral contraceptive. Parthenon, Carnfouth, p 49

Winkler UH et al. (1996b) A comparative study of the effects on the hemostatic system of two monophasic gestodene oral contraceptives containing 20μg and 30μg ethinylestradiol. Contraception 53:75–84

Winkler UH et al. (1996c) Ethinylestradiol 20 versus 30μg combined with 150μg desogestrel: A large comparative study of the effects of two low-dose oral contraceptives on the hemostatic system. Gynecol Endocrinol 10:265–271

Winkler UH et al. (1996d) Routine screening for coagulation inhibitors prior to prescribing the pill: Prevalence data from a large cohort of German pill starters. Eur J Contraception Reprod Health Care 1:47–52

Zöller B et al. (1997) Activated protein C resistance: Clinical implications. Clin Appl Thromb/Hemostas 3:25–32

CHAPTER 43

Urogenital Ageing and Dermatology

M. DÖREN

A. Urogenital Aging

The cessation of ovarian function after menopause is associated with a variety of symptoms from the urogenital tract in many postmenopausal women such as urinary incontinence, urgency, recurrent urinary tract infection, and dyspareunia (OLDENHAVE et al. 1993). The underlying condition is the development of local atrophy of the urethral and vaginal epithelium and their supportive tissues. The gradual increase of these symptoms is age-dependent. Thus, a substantial number of women at very old age are affected. One specific feature attached to this complex set of symptoms frequently referred to as urogenital aging is the influence of personal attitudes and cultural background towards hygiene and sexuality by both menopausal women and health care providers.

I. Specific Functional Aspects of the Urogenital System

The urinary and genital systems share a common embryology. The urethra, bladder, and the distal third of the vagina arise from the urogenital sinus. Specific estrogen receptors are present in the vagina, the urethra, the periurethral and pelvic floor muscles, ligaments, and the vesical triangle of the bladder (IOSIF et al. 1981). The estrogen receptors of the vaginal epithelium are located mainly in the basal and parabasal layers, but also in stromal cells and smooth muscle fibres. The density of estrogen receptors in the urethral wall and the surrounding muscular and connective tissue is similar to the density of estrogen receptors in the vagina, the cytology of urethral and vaginal cells is similar.

Urethral closure pressure is maintained by muscular, vascular, and connective tissue and the epithelium. The submucosal network of arteriovenous anastomoses is important for the sphincteric function of the urethra. The urethral epithelium and this highly vascularized submucosa constitute an intentionally easily deformable layer which should ensure occlusion of the urethral lumen and thus continence under external compression of the urethral smooth and striated muscles. It appears that estrogen modifies the response of the urethra and bladder neck to α-adrenergic stimulation, thus smooth muscle contraction is enhanced.

Vaginal estrogen receptors are thought to be significantly involved in the continuous production of vaginal secretion, the proliferation and vacuolization of the vaginal epithelium and the deposition of glycogen. The latter is a nutritive substrate for the vaginal flora lactobacilli. Production of lactic acid by these bacteria ensures a low vaginal pH in women with endogenous estrogen secretion. The presence of these bacteria significantly contributes to the creation of an acidic barrier against various pathogens responsible for urogenital infections.

II. Functional Changes Due to Age and Menopause

Estrogen deficiency causes atrophic changes of the (sub)urethral tissues and dimished arterial blood supply. This may lead to a variety of symptoms including inadequate urethral closure with an abnormal urinary flow pattern (Table 1). Therefore, urethral atrophy predisposes to ascending bacterial infections. Atrophy itself, without involvement of exogenous pathogenic bacteria, leads to an inflammatory reaction which may lead to a sensation of bladder filling. Dysuria, urgency, urge incontinence, urinary frequency and nocturia increase, urethral pressure and functional urethral length decrease with age (RUD 1980). However, it is difficult to differentiate changes due to hypoestrogenism and those due to increasing age. Many postmenopausal women with urinary symptoms relate the onset of their problems to the menopause. Although the etiology of incontinence is multifactorial, it is likely that the menopause contributes to the onset of symptoms, at least in urge incontinence due to detrusor instability. In postmenopausal women with urinary urge incontinence, estrogen deficiency appears to affect the sensitivity threshold of the lower urinary tract. It is felt that the event of the menopause plays only a minor role in the development of genuine stress incontinence.

Urinary incontinence is frequently associated with a severe impairment of a formerly pleasant sexual relationship. A study of incontinent women demonstrated a link of incontinence and decreased sexual activity due to secondary

Table 1. Urogenital estrogen deficiency: symptoms

Symptoms related to micturition
Itching or soreness of the external urethral orifice
Recurrent urinary tract infection
Dysuria (painful micturition)
Frequent micturition, urgency
Urge incontinence – involuntary loss of urine due to detrusor instability
Mixed incontinence – combination of urge and stress incontinence
(Stress incontinence – urethral sphincter incompetence)
Symptoms related to vaginal function
Dryness, itching or soreness of the vagina
Dyspareunia (Painful intercourse)
Vaginal bleeding

dyspareunia and reduced libido in many women (SUTHERST 1979). With estrogen deficiency the vaginal epithelium becomes thin, fragile, dry and less elastic. The degree of maturation of the vaginal epithelial cells decreases, fragmentation of elastic and hyalinization of collagen fibres in the underlying tissues reflect the general loss of function. Therefore (petechial) bleeding is common, in particular after intercourse. The glycogen deposits disappear, the glands atrophy and vaginal blood flow decreases (SARREL 1990). The colonization with lactic acid bacteria is diminished in favour of pathogenic intestinal bacteria because the epithelial cells store less glycogen reflective of normal cell function. Thus, an increased incidence of vaginal and urinary tract infection is quite common as the pH of the vagina increases. This can result in vaginitis with discharge, minor hemorrhage, very unpleasant if not very painful local sensations, and dyspareunia.

III. Epidemiology of Urogenital Symptoms

Epidemiological data on the prevalence of urogenital symptoms among postmenopausal are scarce compared to other estrogen-deficiency related diseases i.e. osteoporosis (Table 2). One possible reason is that, due to the particular mutual reluctance of women to give information and of physicians to specifically ask for symptoms, our knowledge about the scope of urogenital symptoms in postmenopausal women is fragmentary. Urinary incontinence has been reported to be prevalent in some 25% of postmenopausal women, recurrent urinary tract infection in some 10%, vaginal dryness and dyspareunia in

Table 2. Prevalence of urogenital symptoms

	IOSIF et al. 1984 (Sweden)	MOLANDER et al. 1990 (Sweden)	OLDENHAVE et al. 1993 (Netherlands)	VAN GEELEN et al. 1996 (Netherlands)	BARLOW et al. 1997 (Denmark, France, Germany, Italy, Netherlands, UK)
Sample size (n)	1206	4206	5213	2159	3000
Age (years)	61	65–84	39–60	50–75	55–75
Prevalence (%) of symptoms					
Urinary incontinence	29	16.9	25	25	5.5
Dysuria	–	–	25[a]	6	3.2
Recurrent UTI	13	17.1	8	–	2.2
Urinary frequency	–	11.0[b]	–	–	8.4
Dyspareunia/ dryness	38	–	25	9	2.2

UTI, urinary tract infection
[a] Combined percentage provided for discharge / pain + urinary incontinence + dysuria
[b] Combined percentage provided for pain, burning, pruritus, or vaginal discharge

some 30%. Interestingly, many of the studies and surveys originate from few countries, in particular Sweden. In contrast, a recent survey in six European countries showed a much lower prevalence for these symptoms (BARLOW et al. 1997). Furthermore, other studies did not find a relationship between incontinence and menopause (BURGIO et al. 1991).

IV. Clinical Evaluation

The assessment of urogenital signs of estrogen deficiency includes, apart from the medical history, a gynaecological examination which may show atrophic changes of the urethral orifice and the vagina besides gross anatomical changes such as prolapse of the urethral orifice, the anterior and/or the posterior vaginal wall. The mere inspection of the epithelium may show a change of colour and a loss of moisture of epithelial structures indicative of estrogen deficiency, often associated with petechial bleeding. Cytology may confirm the loss of superficial cells. Standardized questionnaires and urodynamic assessments may be necessary to record the individual pattern of micturition and to differentiate the various forms of urinary symptoms denoted as irritable bladder syndrome.

V. Treatment

Various systemic and local estrogen preparations have been used for decades in the management of lower urinary tract symptoms, including incontinence. Reviews and one meta-analysis show, however, that few controlled clinical trials are available to evaluate the actual efficacy. As diagnostic criteria of how to assess the various forms of urinary incontinence and how to evaluate the closely linked various vaginal symptoms differ substantially and are not universally agreed upon, an assessment of the efficacy of treatment with estrogens is most difficult to perform. Generally it is felt that urge incontinence may improve with estrogen therapy. In the few, mostly short-term studies conducted – all of them with small numbers of patients – oral estradiol 2 mg or estriol 3 mg or a combination of both were administered (for review see FANTL et al. 1994; SULTANA and WALTERS 1994; Table 3). Whether local treatment is superior to systemic has not convincingly been shown yet. α-Adrenergic drugs have been suggested as adjunct treatment together with estrogens. Controlled studies failed to show a beneficial effect in genuine stress incontinence as regards to objective assessment. However, it is important to note that many treated women stated an subjective improvement of all types of incontinence. It should be kept in mind that, although estrogen therapy was effective in some studies, placebo treatment had a good effect on various urogenital symptoms, too.

Estrogen administration leads to improvement of vaginal blood flow, reversal of atrophic changes in vaginal cytology and recolonization of lactobacilli. It may take months before effects can be seen; however, the optimal

Table 3. Some controlled trials of estrogen treatment for incontinence in postmenopausal women

Type of incontinence	Estrogen therapy	Route, duration, sample size	Efficacy	Authors
Urge	Estradiol 2 mg + estriol 1 mg (placebo controlled, double-blind, randomized)	Oral, 4 × 20 days; $n = 15$	Subjective improvement (less frequency, urgency, urge incontinence)	Walter et al. 1978
	Estriol 3 mg (placebo controlled, double-blind, randomized)	Oral, 3 months; $n = 12$	Subjective improvement	Samsioe et al. 1985
Mixed	Estriol 3 mg (design as above)	Oral, 3 months; $n = 8$	Subjective improvement	Samsioe et al. 1985
Stress	Estriol 3 mg (design as above)	Oral, 3 months; $n = 11$	No improvement	Samsioe et al. 1985
	Piperazine oestrone sulphate 3 mg (placebo controlled, double-blind, randomized)	Oral, 3 months, n = 18	Similar subjective improvement in both groups	Wilson et al. 1987

dose, route of administration and duration of treatment with estrogens or estrogen-medicated devices still needs to be determined (Barentsen et al. 1997). Whether nonhormonal treatment with a vaginal moisturizer is indeed as useful as local estrogen treatment (Bygdeman and Swahn 1996) remains to be confirmed.

VI. Conclusion

Evidence-based medicine indicates a role for estrogen replacement therapy in estrogen-deprived women to alleviate lower urinary tract symptoms, in particular urge incontinence and vaginal atrophy. However, convincing data from appropriate clinical trials are necessary to substantiate the long-term indication for estrogens in the prevention and treatment of both conditions.

B. Dermatology

The skin is one of the largest and most exposed connective tissue organs of the human body and may be regarded as an endocrine organ. The alteration of connective tissue's metabolism throughout the body is age-dependent. All connective tissues consist of cells embedded in an extracellular matrix composed of proteins, mainly collagens of various types dependent on the body compartment, and the polysaccharide glycosaminoglycans. There are 12 different types of collagen to constitute one-third of the total body mass; some 80% are located in the dermis of the skin.

Ageing of the skin translates into atrophic, gradual thinning due to increased collagen breakdown accompanied by a diminished local blood flow. The distribution of hair changes with increasing age: the numbers of hairs on the face increase, while in the pubic, axillary and scalp areas they decrease. To what extent estrogen deficiency after the menopause is a factor independent of age involved in these mechanisms is yet to be determined.

I. Skin Composition

The human skin may be divided into two main compartments. The outer layer, the epidermis, consists mainly of keratinocytes and melanocytes. The inner layer, the dermis, is composed almost exclusively of collagens produced by fibroblasts and polymerized to cross-linked peptide chains, mainly type I and, to a minor extent, type III, and a small amount of elastin fibres. The predominant amino acids of these collagens are glycine, proline and hydroxyproline. Capillary networks in the dermal papillae are responsible for the blood supply of both the dermis, with sweat glands, sebaceous glands and hair follicles, and the epidermis (Fig. 1). The gel-like ground substance fills the space between fibrous and cellular elements of the dermis. It contains mainly polysaccharide

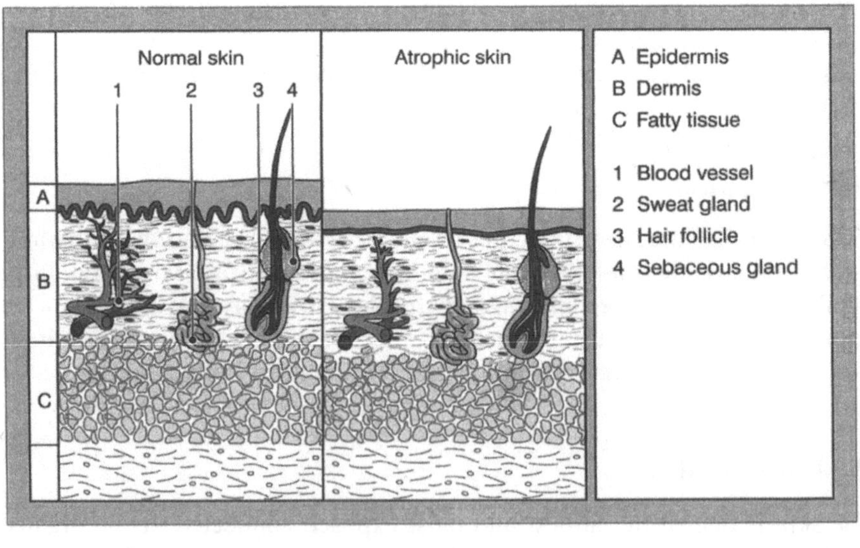

Fig. 1. Sections of normal and atrophic skin (Source Hormone Replacement Therapy and the Menopause, SCHERING AG, SBU FC/HT, International Product Management, RA Henry, 2nd edn, JUNI 1997, Figs. 2–16, p 53

glycosaminoglycans and non-collagenous proteins. Thus, it forms a depot that regulates the diffusion of water and electrolytes and determines the turgor of the skin within the dermis.

It is well recognised that the human skin, hair and subcutaneous tissue are target organs for both endogenous estrogens and androgens. The skin fibroblasts in the germinative basal cell layer and hair follicles contain estrogen and androgen receptors, receptors for growth hormone and for insulin-like growth factor I. Fibroblasts are capable of aromatizing androgens to estrogens. Thyroid hormones, glucocorticoids and insulin may influence fibroblast IGF-I synthesis.

II. Effects of Age and Estrogen Deficiency on the Skin

1. Epidermis

Atrophy of the skin begins as early as in young adulthood, in both women and men, and many years before the cessation of ovarian function. The junctions between epidermis and dermis begin to flatten, the rate of epidermal basal cell proliferation decreases and elastic properties deteriorate. Therefore, healing times after skin lesions are longer. Estrogens are believed to be involved together with other hormones and growth factors in the control of epidermal cell function via yet-unknown mechanisms. It has been demonstrated that bilateral oophorectomy will lead to a decrease of epidermal thickness by more than 10% after half a year, as assessed by planimetric measurements as early as 4 months later (PUNNONEN 1972).

2. Dermis

Skin collagen content of the dermis and the number of fibroblasts and mast cells decline in an age-dependent fashion. Various techniques have been assessed to estimate the collagen content by skin thickness measurement (TAN et al. 1982; SMEETS et al. 1994; VARILA et al. 1995). The average loss was calculated to be 2% per year after menopause (BRINCAT et al. 1987a). The skin of women has considerably less collagen content than that of men. Some 30% of skin collagen is apparently lost in the first 5 years after the menopause. Women with hirsutism are believed to have a higher skin collagen content, which may be related to the excess of androgens. It appears that a positive correlation between skin thickness and collagen content exists and relates to menopausal, not chronological age (BRINCAT et al. 1987b). In young women with premature ovarian failure, atrophic changes otherwise seen in postmenopausal women have been demonstrated by skin biopsies. Degenerative changes in the dermal elastic fibres, a reduced vascularisation, and a reduced reactivity to temperature changes have been demonstrated in postmenopausal women (BOLOGNIA et al. 1989). Most of the wrinkling of ageing skin is assumed to be the cumulative result of chronic injury, such as sun exposure which causes a loss of texture of the intradermal collagens and elastic fibres.

Estrogens do not affect the sebaceous glands directly, androgens stimulate mitoses and increase sebum production. After menopause, the relative predominance of androgens may stimulate the growth and may change the appearance of hair, in particular facial hair. Finer texture and the change from terminal to vellus hair may be noticed after menopause.

III. Effects of Estrogen Replacement Therapy

One of the first reports on the effects of exogenous estrogens on the skin of estrogen-deficient women dates back to RAURAMO and PUNNONEN, (1969). Since then, most of the clinical studies to focus on skin thickness have been conducted in postmenopausal and bilateral oophorectomized women who used estradiol implants alone or in combination with testosterone. Treatment with estradiol implants 50–100 mg, administered twice per year, sometimes combined with testosterone implants (100 mg), apparently prevented the decrease in collagen content, as assessed by thigh skin biopsies with hydroxyproline determination as an indirect estimate of collagen content (BRINCAT et al. 1984; BRINCAT et al. 1987a–c). This particular therapy used to induce supraphysiological levels of both hormones appeared to be most effective in women with a low collagen content before initiation of treatment (BRINCAT et al. 1987c) with a specific increase of collagen type III (SAVVAS et al. 1993). However, other the authors failed to demonstrate an increased content of mature collagen cross-links (HOLLAND et al. 1994).

The administration of various oral and transdermal estrogen and sequential progestogen-replacement therapies may be associated with an increased skin thickness. Results from the first National Health and Nutrition Examination Survey suggest that estrogens prevent dry skin and wrinkling on postmenopausal women (DUNN et al. 1997). Clinical trials show, for example, that estradiol valerate 2 mg per day alone or in combination with sequential norgestrel 0.5 mg per day and estriol succinate 2 mg each increased skinfold thickness, as measured with callipers on the dorsal side of the hands (PUNNONEN et al. 1984). Transdermal estradiol, 50 μg per day, in combination with medroxyprogesterone acetate, 10 mg per day, sequentially increased skin thickness determined by a radiological method in a placebo-controlled study within a period of 1 year (MESCHIA et al. 1994). B-mode ultrasound (7.5 MHz) measurements of skin thickness demonstrated that conjugated estrogens, 0.625 mg per day administered for 1 year, increased thigh skin thickness; a similar trend was demonstrated for the skin above the thyroid gland and below the umbilicus. In addition, thigh skin biopsies confirmed the beneficial effect of estrogen therapy as dermal thickness was increased by 30% compared with baseline in this randomized, double-blind, placebo-controlled trial (MAHEUX et al. 1994). However, treatment with oral estradiol 2 mg, with and without norethisterone acetate 1mg daily for 1 year, did not yield any effect on skin thickness measured by ultrasound, amount and rate of collagen synthesis, elastic fibre content and thickness of the epidermis assessed by biochemical and immunohistochemical methods of skin biopsies compared with controls (Haapasaari et al. 1997).

The decrease of skin collagen content has been correlated to the decrease of bone mass, another organ that contains a significant amount of collagen (McConkey et al. 1963). The outcomes of clinical trials comparing skin thickness assessed by radiography, manual callipers measurement or various skin ultrasound techniques and bone-mass measurements differ substantially (Maheux et al. 1996). At present, there are no consistent data to confirm whether skin thickness may be of additional information to bone-mass respective density measurements to assess the risk of an osteoporotic fracture. Skin colour is not an established risk factor for osteoporosis (Nelson et al. 1993).

Other biophysical properties of the skin – the water-holding capacity of the skin and, in particular, the cornified external cell layer and skin extensibility – seem to improve after treatment, with either transdermal estradiol replacement therapy, 50 μg per day, or conjugated estrogens, 0.625 mg, in combination with sequential medroxyprogesterone acetate, 5 mg per day (Piérard-Franchimont et al. 1995; Piérard et al. 1995). Topical facial treatment with estradiol or estriol creme suggested flattening of wrinkles and an improvement of skin elasticity in an uncontrolled study (Schmidt et al. 1994). Whether the local administration of estradiol or estriol has any specific, beneficial effects on the skin of postmenopausal women compared with oral treatment has not been studied. Apart from local skin irritations, caused by transdermal estradiol reservoir patches, for example, which are extremely unlikely to be due to the administration of 17β-estradiol itself (Cheang et al. 1993), it remains to be confirmed that endogenous estrogens may cause skin irritations in some premenopausal women (Shelley et al. 1995).

IV. Conclusion

Today, atrophic skin changes are not generally a recognized indication for estrogen replacement therapy (Kenemans et al. 1996). Convincing data are yet to be presented. However, the improvement of skin quality seems possible and could be regarded as a beneficial side-effect in some postmenopausal women.

References

Barentsen R, van der Weijer PHM, Schram JHN (1997) Continuous low dose estradiol released from a vaginal ring versus estriol vaginal cream for urogenital atrophy, Eur J Obstet Gynecol 71:73–80

Barlow DH, Samsioe G, van Geelen JM (1997) A study of European women's experience of the problems of urogenital aging and its management. Maturitas 27:239–247

Bolognia JL, Bravermann IM, Rousseau ME, Sarrel PM (1989) Skin changes in menopause. Maturitas 11:295–304

Brincat M, Studd JWW, Moniz CF, Parson V, Darby AJ (1984) Skin thickness measurement. A simple screening method for determining patients at risk of developing postmenopausal osteoporosis. In: Christiansen C, Arnaud CD, Nordin BEC, Parfitt AM, Peck WA, Riggs BL (eds) Osteoporosis, Proceedings of the Copenhagen international symposium on osteoporosis. Aalborg Styftsbogtrykkeri, Aalborg, pp 323–326

Brincat M, Kabalan S, Studd JWW, Moniz CF, de Trafford J, Montgomery J (1987a) A study of the decrease of skin collagen content, skin thickness, and bone mass in the postmenopausal woman. Obstet Gynecol 70:840–845

Brincat M, Moniz CF, Kabalan S, Versi E, O'Dowd T, Magos AL, Montgomery J, Studd JWW (1987b) Decline in skin collagen content and metacarpal index after the menopause and its prevention with sex hormone replacement. Br J Obstet Gynaecol 94:126–129

Brincat M, Versi E, Moniz CF, Magos A, de Trafford J, Studd JWW (1987c) Skin collagen changes in postmenopausal women receiving different regimens of estrogen therapy. Obstet Gynecol 70:123–127

Burgio KL, Matthews KA, Engle BT (1991) Prevalence, incidence and correlates of urinary incontinence in healthy, middle aged women. J Urol 146:1255–1259

Bygdeman M, Swahn ML (1996) Replens versus dienoestrol cream in the symptomatic treatment of vaginal atrophy in postmenopausal women. Maturitas 23:259–263

Cheang A, Sitruk-Ware R, Utian W (1993) A risk-benefit appraisal of transdermal estradiol. Drug Safety 9:365–379

Dunn LB, Damesyn M, Moore AA, Reuben DB, Greendale GA (1997) Does estrogen prevent skin aging? Results from the First National Health and Nutrition Examination Survey (NHANES 1) Arch Dermatol 133:339–342

Fantl JY, Cardozo L, McClish DK, and The Hormones and Urogenital Therapy Committee (1994) Estrogen therapy in the management of urinary incontinence in postmenopausal women: a meta-analysis. First report of the hormones and urogenital therapy committee. Obstet Gynecol 83:12–8

Haapasaari K-M, Raudaskoski T, Kallionen M, Suvanto-Luukkonen E, Kauppila A, Läära E, Risteli J, Oikarinen A (1997) Systemic therapy with estrogen or estrogen with progestin has no effect on skin collagen in postmenopausal women. Maturitas 27:153–162

Holland EF, Studd JWW, Mansell JP, Leather AT, Bailey AJ (1994) Changes in collagen composition and cross-links in bone and skin of osteoporotic postmenopausal women treated with percutaneous estradiol implants. Obstet Gynecol 83:180–183

Iosif CS, Batra S, Ek A, Åstedt B (1981) Estrogen receptors in the human female lower urinary tract. Am J Obstet Gynecol 141:817–820

Iosif C, Bekassy Z (1984) Prevalence of genito-urinary symptoms in the late menopause. Acta Obstet Gynecol Scand 63:257–260

Kenemans P, Barentsen R, van der Weijer P (1996) Other climacteric complaints and symptoms. In: Kenemans P, Barentsen R, van der Weijer P (eds) Practical HRT, 2nd edn. Medical Forum International, Zeist, The Netherlands, p 53

Maheux R, Naud F, Rioux M, Grenier R, Lémay A, Guy J, Langevin M (1994) A randomized, double-blind, placebo-controlled study on the effect of conjugated estrogens on skin thickness. Am J Obstet Gynecol 170:642–649

Maheux R, Guy J, Dumont M, Maesse B (1996) Correlation between skin thickness and bone mass in women. Menopause 3:197–200

McConkey B, Fraser GM, Bligh AS, Whitely H (1963) Transparent skin and osteoporosis. Lancet I:693–695

Meschia M, Bruschi F, Amicarelli F, Barbacini P, Monza GC, Crosignani PG (1994) Transdermal hormone replacement therapy and skin in post-menopausal women: a placebo controlled study. Menopause 1:79–82

Molander U, Milsom I, Ekelund P, Mellström D (1990) An epidemiological study of urinary incontinence and related urogenital symptoms in elderly women. Maturitas 12:51–60

Nelson DA, Kleerekoper M, Peterson E, Parfitt AM (1993) Skin color and body size as risk factors for osteoporosis. Osteoporosis Int 3:18–23

Oldenhave A, Jaszman LJB, Haspels AA, Everaerd WTAM (1993) Impact of climacteric on well-being: a survey based on 5213 women aged 39 to 60 years old. Am J Obstet Gynecol 168:772–780
Piérard GE, Letawe C, Dowlati A, Piérard-Franchimont C (1995) Effect of hormone replacement therapy for menopause on the mechanical properties of skin. J Am Geriatr Soc 43:662–665
Piérard-Franchimont C, Letawe C, Goffin V, Piérard GE (1995) Skin water-holding capacity and transdermal estrogen therapy for menopause: a pilot study. Maturitas 22:155–161
Punnonen R (1972) Effect of castration and peroral estrogen therapy on the skin. Acta Obstet Gynecol Scand 21:S3–S44
Punnonen R, Vilska S, Rauramo L (1984) Skinfold thickness and long-term postmenopausal hormone therapy. Maturitas 5:259–262
Rauramo L and Punnonen R (1969) Wirkung einer oralen Therapie mit Östriolsuccinat auf die Haut kastrierter Frauen. Z Haut Geschl Kr 44:463–470
Rud T (1980) Urethral pressure profile in continent women from childhood to old age. Acta Obstet Gynecol Scand 59:331–335
Samsioe G, Jansson I, Mellström D, Svanborg A (1985) Occurence, nature and treatment of urinary incontinence in a 70-year-old female population. Maturitas 7:335–342
Sarrel PM (1990) Sexuality and menopause. Obstet Gynecol 75:26S–32S
Savvas M, Bishop J, Laurent G, Watson N, Studd J (1993) Type III collagen content in the skin of postmenopausal women receiving oestradiol and testosterone implants. Br J Obstet Gynaecol 100:154–156
Schmidt JB, Binder M, Macheiner W, Kainz Ch, Gitsch G, Bieglmayer Ch (1994) Treatment of skin ageing symptoms in perimenopausal females with estrogen compounds. A pilot study. Maturitas 20:25–30
Shelley WB, Shelley ED, Talanin NY, Santoso-Pham J, Toledo DO (1995) Estrogen dermatitis. J Am Acad Dermatol 32:25–31
Smeets AJ, Kujper JW, van Kuijk C, Berning B, Zwamborn AW (1994) Skin thickness does not reflect bone mineral density in postmenopausal women. Osteoporosis Int 4:32–35
Sultana CJ, Walters MD (1994) Estrogen and urinary incontinence. Maturitas 20: 129–138
Sutherst JR (1979) Sexual dysfunction and urinary incontinence. Br J Obstet Gynecol 86:387–388
Tan CY, Statham B, Marks R, Payne PA (1982) Skin thickness measurement by pulsed ultrasound: its reproducibility, validation and variability. Br J Dermatol 106: 657–667
van Geelen JM, van der Weijer PHM, Arnolds HT (1996) Urogenital verschijnselen en hinder daarvan bij thuiswonende Nederlandse vrouwen van 50–75 jaar (Urogenital symptoms and resulting discomfort in non-institutionalized Dutch women aged 50 to 75 years). Ned Tijdschr Genesk 140:713–716
Varila E, Rantala I, Oikarinen A, Risteli J, Reunala T, Oksanen H, Punnonen R (1995) The effect of topical oestriol on skin collagen of postmenopausal women. Br J Obstet Gynaecol 102:985–989
Walter S, Wolf H, Barlebo H, Jensen HK (1978) Urinary incontinence in postmenopausal women treated with estrogens: A double-blind clinical trial. Urol Int 33:135–143
Wilson PD, Faragher B, Butler B, Bu'Lock D, Robinson EL, Brown ADG (1987) Treatment with oral piperazine oestrone sulphate for genuine stress incontinence in postmenopausal women. Br J Obstet Gynecol 94:568–574

CHAPTER 44

Geriatric Neurology and Psychiatry

V.W. HENDERSON

A. Introduction

For both women and men, estrogen and other gonadal steroids have widespread effects on many organ systems. Estrogen affects target tissues directly through interactions with nuclear and non-nuclear receptors and indirectly through actions at more distant sites. This chapter considers the clinical impact of estrogen on the brain, emphasizing effects clinically relevant to neurological and psychiatric disorders of older women. So-called organizational effects of estrogen are viewed as irreversible in nature and are presumed to occur during prenatal, perinatal, and pubescent stages of development. Of greater relevance to geriatric health and disease are the more transient, activational effects of estrogen on central nervous system function.

Among the changes that accompany aging, none are more pervasive or important than those affecting behavior and cognition. In its most dreaded guise, age-associated mental deficits are manifest as dementia, a leading cause of health care expenditures (SCHNEIDER and GURALNIK 1990), morbidity, and even mortality (BOWEN et al. 1996). Other common neuropsychiatric conditions are also considered within this chapter. These disorders were selected to illustrate the range of illnesses that may be influenced by estrogen.

B. Estrogen and the Brain

Actions of estrogen on central nervous system function are of great potential importance for geriatric neurology and geriatric psychiatry. Throughout the life span, ovarian steroids play crucial roles in the formation, maintenance, and remodeling of neuronal circuits in the mammalian brain; and hormonal changes that occur during the aging process impact brain function. In responsive neurons of the central nervous system, estrogens mediate neuronal plasticity, manifest in part through the growth of nerve processes (TORAN-ALLERAND 1991; BRINTON 1993) and the formation of synapses (CHUNG et al. 1988; MATSUMOTO 1991).

Estrogen is not synthesized within the nervous system itself, although neurons that contain the enzyme aromatase can convert circulating testosterone to estradiol in situ (NAFTOLIN and RYAN 1975). There are various means by which estrogen can modulate brain function at the genomic level

(KATZENELLENBOGEN et al. 1996): cell-specific effects and regional selectivity depend in part on the focal distribution of classical estrogen receptors, which are confined to specific subsets of neurons (WOOD and NEWMAN 1995) and whose numbers vary as a function of gender and developmental age. Like other steroid hormones, estrogen acts by first binding to its intracellular receptor (EVANS 1988). Within the nucleus, the ligand–receptor complex then binds to a DNA sequence known as the estrogen response element, resulting in the transcription of target genes and the synthesis of specific protein products. The presence of different isoforms of the estrogen receptor differentially expressed in different tissues (FRIEND et al. 1995) and the discovery of different estrogen response elements (YANG et al. 1996) suggest that other mechanisms exist by which focal estrogen effects on brain might be obtained. To add to this modulatory complexity, the recently described beta estrogen receptor is regionally distributed in a pattern different from that of the classic alpha receptor (SHUGHRUE et al. 1996).

Some estrogen actions, such as changes in neuronal excitability or stimulation of neurotransmitter release, occur too rapidly to be attributed to genomic activation and are presumed to be mediated by receptors on the cell membrane (WONG et al. 1996). Membrane receptors have been characterized pharmacologically, but not biochemically. Thus, the distribution within different brain areas of intracellular estrogen receptors will provide an imprecise guide to the regional specificity of estrogen's actions. In the adult hippocampus, a brain region critical for learning and memory, only a small number of neurons possess estrogen receptors (LOY et al. 1988), and yet there are prominent estrogen effects on the morphology and physiological properties of specific populations of hippocampal neurons (WEILAND 1992; WOOLLEY and MCEWEN 1993; WARREN et al. 1995; GIBBS et al. 1996; WOOLLEY et al. 1997). Changes in hippocampal neurons induced by estrogen are sex specific and are modified by early developmental exposures (LEWIS et al. 1995).

Other estrogen influences on brain function occur indirectly. Receptor-mediated effects in one neuronal population can influence the function of neurons in a distant but interconnected brain region. Estrogen actions on glial cells also influence neurons. Glia can express estrogen receptors, and estrogen promotes the extension of astrocytic processes (GARCIA-SEGURA et al. 1994; SANTAGATI et al. 1994). Other indirect effects can occur through the enhancement of cerebral blood flow (BELFORT et al. 1995) or the augmentation of glucose utilization by the brain (BISHOP and SIMPKINS 1992).

C. Estrogen, Mood, and Behavior

Estrogen may influence different aspects of mood and behavior, perhaps mediated through monoamine neurotransmitter systems of the brain. The sense of well-being is heightened in the late follicular stage of the menstrual cycle, a time when estradiol levels are high but progesterone levels are still low

(Hampson 1990a). Similarly, estrogen replacement given during the perimenopausal or postmenopausal period is reported to diminish anxiety and to enhance mood and subjective well-being (Fedor-Freybergh 1977; Schneider et al. 1977; Montgomery et al. 1987; Sherwin 1988a; Ditkoff et al. 1991; Best et al. 1992). Estrogen might also reduce aggressive behaviors, as suggested by case reports of elderly men with agitated dementia who responded to estrogen treatment (Kyomen et al. 1991).

Women have higher rates of depression than men (Weissman et al. 1993). The postpartum period is a time of particular vulnerability. Low mood is also common during the premenstrual phase of the menstrual cycle and during the climacteric. Among older postmenopausal women, those who use estrogens typically have fewer depressive symptoms than non-users (Palinkas and Barrett-Connor 1992). In general, effects of estrogen replacement on mood have been demonstrated in healthy women who did not have diagnosed depression or other psychopathology (Schneider et al. 1977). Among asymptomatic postmenopausal women in randomized studies of oral (Fedor-Freybergh 1977; Ditkoff et al. 1991) or intramuscular (Sherwin and Gelfand 1985) estrogen, active treatment reduced depressive scores.

More severe depression may also be influenced by hormonal therapy. A 3-month study that used very high oral doses of conjugated estrogens to treat women with severe unipolar depression found significant abatement of depressive symptoms (Klaiber et al. 1979). In this randomized placebo-controlled trial, women in the estrogen group showed clinically meaningful improvement, although residual depressive symptoms persisted. In another double-blind study, transdermal estradiol was found to be effective against symptoms of severe postnatal depression (Gregoire et al. 1996).

The manner in which estrogen might influence mood is uncertain. Speculatively, however, estrogen interactions with noradrenergic (Ball et al. 1972; Greengrass and Tonge 1974) or serotonergic (Biegon et al. 1983; Cohen and Wise 1988; Sumner and Fink 1995) neurotransmitters of the brain may be critical, as pharmacological potentiation of these monoamines is a mainstay in the treatment of clinical depression. Although mood can impact scores on tests of cognitive abilities, estrogen effects on cognition appear to be independent of mood (Phillips and Sherwin 1992a; Kimura 1995).

Estrogen may also influence symptoms of schizophrenia, a chronic psychotic disorder characterized by delusions, auditory hallucinations, disorganized thought processes, affective blunting, and difficulty in sustaining goal-directed activity. The incidence of schizophrenia is similar in both sexes. Onset is often in the third decade of life, but averages about 3–5 years later for women than men (Loranger 1984; Hafner et al. 1993). Late-onset schizophrenia is also more common in women than men (Lindamer et al. 1997), but there is no strong evidence that the menopause represents a time of heightened risk (Loranger 1984).

Agents that block specific central dopamine receptors and atypical antipsychotic medications that bind serotonin receptors are effective antipsychotic agents. It is speculated that estrogen effects on schizophrenia may occur through actions on dopaminergic (Chiodo and Caggiula 1980; Hruska and Silbergeld 1980; Roy et al. 1990) or serotonergic (Biegon et al. 1983; Cohen and Wise 1988; Sumner and Fink 1995) systems of the brain. Consistent with this view, higher serum estrogen levels in ovulating women are associated with milder psychopathology (Riecher-Rössler et al. 1994) and reduced requirements of neuroleptic medications. In a small open-label trial of women with schizophrenia, estrogen added to standard antipsychotic drugs led to more rapid amelioration of psychotic symptoms (Kulkarni et al. 1996).

D. Estrogen and Cognition

Only limited numbers of animal studies have examined estrogen effects on non-reproductive behaviors. In adult ovariectomized rats, estrogen replacement is reported to enhance sensorimotor skills (Becker et al. 1987). It also may improve certain types of memory, as suggested by improved performances on a water-escape (O'Neal et al. 1996) and a passive-avoidance (Singh et al. 1994) task, but not on a radial-maze spatial-memory task (Luine and Rodriguez 1994).

Estrogen appears to influence the pattern of brain activation during the performance of cognitive tasks, as inferred from measures of regional cerebral flow (Berman et al. 1997). The effect of estrogen on cognitive abilities of healthy adult women has been studied in a variety of clinical settings, but findings across studies are not fully consistent (Hampson and Kimura 1988; Hampson 1990a, b; Phillips and Sherwin 1992b; Gordon and Lee 1993; Krug et al. 1994). The most common interpretation of these various investigations is that estrogen maintains or enhances skills at which women tend to excel, but impairs other skills conceptualized as *male advantaged*. For example, spontaneously cycling women may show better verbal fluency, manual speed, articulatory agility, or creativity during the late follicular or midluteal phase of the menstrual cycle (when estradiol levels are elevated) than during menses. Conversely, performance on spatial tasks may be enhanced during the menses. Other evidence supporting this view comes from the evaluation of trans-sexual men prior to surgery for sex reassignment. In these men, verbal fluency improved after the administration of estrogen given with an antiandrogen (van Goozen et al. 1995).

Small observational studies of healthy community-dwelling older women suggest that women who use hormone replacement may perform better on a broad spectrum of cognitive skills (Kimura 1995) or may outperform non-users on certain aspects of verbal memory (Kampen and Sherwin 1994; Robinson et al. 1994). In a well-characterized aging cohort, women receiving estrogens performed significantly better than women who had never used estrogens on a measure of short-term non-verbal memory and drawing skills (Resnick et al. 1997). A population-based case-control study from Austria also found that postmenopausal women who used estrogen performed better than non-users on several psychometric measures (SCHMIDT et al. 1996). Studies in two other large cohorts, however, failed to discern appreciable differences between users and non-users of postmenopausal estrogen on a variety of cognitive measures (BARRETT-CONNOR and KRITZ-SILVERSTEIN 1993; SZKLO et al. 1996).

Several randomized controlled trials have examined cognitive effects of estrogen given after surgical or non-surgical menopause. Preliminary analyses from one interventional trial indicated no effect of estrogen replacement on measures of attention, reaction time, or manual dexterity (DOODY et al. 1995). However, in other studies, women given estrogen outperformed women given only placebo on a variety of psychometric measures (SHERWIN 1988b; PHILLIPS and SHERWIN 1992a). Improvement on memory tasks may be more apparent when verbal, as opposed to non-verbal, memory is assessed (SHERWIN 1988b; PHILLIPS and SHERWIN 1992a). Verbal memory enhancement was also reported in a controlled study of women, whose ovarian function had been suppressed with a gonadotrophin releasing hormone agonist prior to "add-back" treatment with estrogen (SHERWIN and TULANDI 1996). The magnitude of the estrogen effect in healthy adults appears to be modest, although differences as high as one standard deviation are sometimes reported on some measures of verbal memory.

E. Stroke

Stroke refers to brain disease caused by abnormalities in the cerebrovascular system. In most developed countries, stroke is the third leading cause of death, after heart disease and cancer, and its incidence rises dramatically with age. At all ages, rates are lower in women than men, but rates are less divergent after the menopause.

Stroke is etiologically and pathogenetically heterogeneous. Cerebral damage is usually due to vascular occlusion or rupture, but changes in blood flow – in the absence of frank infarction or hemorrhage – occasionally affect brain tissue. Vascular occlusion can affect large arteries, small arteries, or veins. Hemorrhage is most often associated with the intracerebral rupture of a small penetrating artery or with bleeding into the subarachnoid space after rupture of an aneurysm or vascular malformation.

In the older woman, estrogen replacement protects against cardiovascular disease (GRODSTEIN and STAMPFER 1995). Cerebrovascular and cardiovascular disease share a number of common risk factors, including age, hypertension, smoking, diabetes, and lipid abnormalities. Atherosclerotic changes within the coronary arteries and the extracranial carotid arteries predispose to ischemic heart disease and to cerebral infarction, respectively. Angiographic studies of the coronary arteries clearly link estrogen use to significant reductions in atherosclerosis (SULLIVAN 1995). Similarly, estrogen replacement also appears to prevent or delay atherosclerotic narrowing of the carotid arteries, as inferred by ultrasonography (MANOLIO et al. 1993; ESPELAND et al. 1995). Favorable effects on serum levels of specific plasma lipoproteins, lipid peroxidation, fibrinolysis, and vascular endothelial function may contribute to observed atherosclerotic reductions. Estrogen also increases cerebral blood flow (BELFORT et al. 1995; OHKURA et al. 1995b), another action of possible relevance to cerebrovascular disease.

A review of epidemiological studies on postmenopausal estrogen therapy and cerebral vascular disease found the risk of fatal stroke among estrogen users to be reduced by 20–60% (PAGANINI-HILL 1995). A protective effect of estrogen on stroke mortality was evident across studies. However, estrogen effects on overall stroke incidence are inconsistent, and most studies show no significant benefit (PAGANINI-HILL 1995).

In part, discrepancies between cardioprotection and neuroprotection reflect a failure or inability to distinguish among different forms of stroke. For example, risk factors for hemorrhagic and ischemic strokes would be expected to differ substantially. However, even for ischemic disease, mechanisms and risk factors are not identical for heart and brain (PUDDU et al. 1995). Coronary occlusion leading to myocardial infarction is most often due to ulceration of an atherosclerotic plaque followed by acute thrombosis (FALK 1983). Although a similar process occurring near the bifurcation of the common carotid artery is an important cause of cerebral infarction, the majority of ischemic strokes are not due to atherosclerotic thrombosis (ADAMS et al. 1997). Thus, even studies that have considered cerebral infarction separately – where similarities to cardiovascular disease are most evident – have not demonstrated the degree of risk reduction anticipated on the basis of studies of heart disease. For example, among the women in the Nurses' Health Study, current estrogen users compared with never-users demonstrated a significantly lower risk of non-fatal myocardial infarction or death from coronary disease (odds ratio [OR] = 0.60, 95% confidence interval [CI] = 0.47–0.76), whereas the risk of ischemic stroke in this group was elevated (OR = 1.40, 95% CI = 1.02–1.92) (GRODSTEIN et al. 1996).

At present, therefore, effects of estrogen on overall stroke risk remain uncertain. This issue is likely to be resolved only with additional data from well-designed case-control and cohort studies, or from randomized primary and secondary prevention trials of estrogen and stroke.

F. Dementia

The term dementia refers to cognitive loss severe enough to interfere with usual daily activities. The most common cause of dementia is Alzheimer's disease, accounting for about 55–65% of cases (SCHOENBERG et al. 1987; BACHMAN et al. 1992). Dementia attributed to cerebrovascular disease represents less than 10% of cases. Other causes individually account for only small proportions.

I. Alzheimer's Disease

In Alzheimer's disease, symptom onset is insidious, and there is a progressive loss of memory and other cognitive abilities. Pathological characteristics include neurofibrillary tangles within vulnerable neurons of the cerebrum and neuritic plaques within the neuropil between nerve cell bodies. Plaques typically contain dystrophic nerve processes and a central core of β-amyloid protein. An inflammatory process is suggested by the presence of reactive astrocytes and microglia, as well as by deposition within the plaque of complement proteins, inflammatory cytokines, and acute phase reactants (McGeer and McGeer 1995).

So-called "early-onset" Alzheimer's disease, in which symptoms are apparent before about age 60 years, is usually inherited as an autosomal dominant disorder due to point mutations in specific genes identified on chromosomes 14, 1, or 21 (Pericak-Vance and Haines 1995). Late-onset illness, which has not been linked to specific genetic mutations, is far more common, and the prevalence of Alzheimer's disease doubles about every 5 years between the ages of 65 years and 90 years (Jorm et al. 1987). Although most genetic and environmental risk factors for late-onset disease remain to be identified, Alzheimer's risk is known to be strongly influenced by polymorphisms of apolipoprotein E (Strittmatter et al. 1993), a glycoprotein involved in lipid transport and lipid metabolism which is synthesized by various tissues, including astrocytes and microglia. Apolipoprotein-E expression within the nervous system is increased in the setting of neuronal injury, presumably reflecting a reparative role in lipid redistribution during neurite regrowth and synapse formation (Mahley 1988; Poirier 1994). A single genetic locus on the long arm of chromosome 19 encodes one of three common apolipoprotein-E alleles, with increased risk associated with possession of the so-called $\varepsilon 4$ allele (Seshadri et al. 1995). The risk conferred by the $\varepsilon 4$ allele is more evident among women than men (Poirier et al. 1993; Payami et al. 1996).

Gender is one of several factors postulated to influence the risk of developing late-onset Alzheimer's disease (Breteler et al. 1992; van Duijn and Hofman 1992; Graves and Kukull 1994). Most observations (Mölsä et al. 1982; Rorsman et al. 1986; Katzman et al. 1989; Payami et al. 1996; Fratiglioni et al. 1997), but not all (Bachman et al. 1993) indicate that the incidence of late-onset illness is elevated among women. Alzheimer's disease also appears to affect women differently than men. Certain cognitive symptoms differ between men and women with Alzheimer's disease, with demented women showing greater deficits on semantic memory (naming) tasks than men (Henderson and Buckwalter 1994; Ripich et al. 1995; Buckwalter et al. 1996).

1. Estrogen Impact on Alzheimer's Disease: Possible Mechanisms

There are several mechanisms by which estrogen deprivation after the menopause might influence the development or manifestations of Alzheimer's disease (Table 1). Declining estrogen levels that accompany the menopause have the capacity to impact brain function through effects on neuronal plasticity (CHUNG et al. 1988; MATSUMOTO 1991; TORAN-ALLERAND 1991) or through other actions that affect neurons.

Axons of cholinergic, noradrenergic, and serotonergic neurons originate in discrete nuclei and project widely over great distances. These neurotransmitter systems, among others, are influenced by estrogen (SAR and STUMPF 1981; COHEN and WISE 1988; TORAN-ALLERAND et al. 1992). Cell bodies of origin are heavily affected by neurofibrillary tangle formation in Alzheimer's disease (KEMPER 1994). Estrogen effects on acetylcholine and estrogen interactions with the neurotrophins may be especially important. Manipulations that interfere with cholinergic systems of the brain impede memory (BARTUS et al. 1981), and some cognitive deficits of Alzheimer's disease are attributed to prominent pathological alterations of basal forebrain neurons that use this neurotransmitter (COYLE et al. 1983). In adult ovariectomized female rats, estradiol replacement increases cholinergic markers in the basal forebrain and its projection target areas (LUINE 1985; GIBBS and PFAFF 1992; GIBBS et al. 1996).

The neurotrophins are a group of related proteins that play key roles in maintaining neuronal viability and promoting growth and differentiation. Two particular neurotrophins, nerve growth factor and brain-derived neurotrophic factor, promote the survival of cholinergic neurons after experimental lesions (TUSZYNSKI et al. 1991; KNUSEL et al. 1992). Nerve growth factor has been advocated for the treatment of Alzheimer's symptoms (OLSON et al. 1992). Cholinergic neurons within the basal forebrain have receptors for estrogen and for neurotrophins, and estrogen may modulate neurotrophin effects in this region (TORAN-ALLERAND et al. 1992; MIRANDA et al. 1993; SALEHI et al. 1996).

Estrogens might also act to reduce the formation of β-amyloid, a key biochemical abnormality of Alzheimer's disease. The large amyloid precursor protein can be proteolytically processed at alternative sites to yield smaller

Table 1. Estrogen effects in Alzheimer's disease: possible mechanisms (adapted from HENDERSON 1997c)

Estrogens can influence
Neuronal plasticity
Neurotrophins
Acetylcholine, noradrenaline, serotonin, and other neurotransmitters
β-Amyloid protein
Apolipoprotein-E levels
Inflammation
Oxidative damage
Stress response

degradation products; estradiol promotes metabolism of the amyloid precursor protein that yields fragments less likely to accumulate as β-amyloid (JAFFE et al. 1994).

Several observations imply an important link between estrogens and apolipoprotein E. Although estrogens reduce circulating levels of apolipoprotein E (APPLEBAUM-BOWDEN et al. 1989; MUESING et al. 1992), the situation appears to differ in the brain, where, experimentally, estrogen increases the expression of apolipoprotein E (STONE et al. 1997). Lesions in rat brain that disrupt cholinergic input to the hippocampus elevate apolipoprotein E in denervated areas (POIRIER et al. 1991). Although the severe cholinergic deficits that typify Alzheimer's disease might be expected to cause a similar increase, apolipoprotein-E protein levels in Alzheimer's disease are reported to be reduced both in cerebrospinal fluid (BLENNOW et al. 1994) and the hippocampus (BERTRAND et al. 1995). Interestingly, hippocampal reductions of apolipoprotein E are greater for $\varepsilon 4$ homozygotes than for other apolipoprotein E genotypes (BERTRAND et al. 1995). It remains speculative as to whether estrogen, by increasing brain apolipoprotein-E expression, might also facilitate neuronal repair processes and, thereby, retard the progression of Alzheimer's disease, and whether such a putative effect might vary as a function of apolipoprotein-E genotype.

Other estrogen properties might improve or preserve brain function in patients with Alzheimer's disease. These include effects on inflammation, free radicals, and cortisol. Inflammatory responses are implicated in neuritic plaque formation (BAUER et al. 1992; MCGEER and MCGEER 1995), and estrogen may moderate some aspects of this process (JOSEFSSON et al. 1992; GILMORE et al. 1997; HASHIMOTO et al. 1997). Oxidative damage caused by free oxygen radicals is prominent in brains of Alzheimer's patients (SMITH et al. 1997). The β-amyloid protein is toxic to neurons, and β-amyloid damage may be mediated or potentiated by free radicals (SAGARA et al. 1996; MCDONALD et al. 1997). At physiological concentrations, estrogens act as antioxidants (NIKI and NAKANO 1990; MOORADIAN 1993; SACK et al. 1994). Finally, basal cortisol levels are elevated in Alzheimer patients (DAVIS et al. 1986); stress-induced corticosteroid increases are toxic to specific populations of hippocampal neurons (MCEWEN and SAPOLSKY 1995); and estrogen replacement after the menopause may mitigate deleterious consequences of stress (LINDHEIM et al. 1992).

2. Estrogen and Alzheimer's Disease Risk

Initial case-control studies failed to demonstrate an association between a woman's use of estrogen after the menopause and the risk of Alzheimer's disease (HEYMAN et al. 1984; AMADUCCI et al. 1986; BROE et al. 1990; GRAVES et al. 1990) (Table 2). However, the number of estrogen users in these studies was small, limiting the power to detect any possible association. A 1994 analysis of volunteers in a longitudinal study of aging and dementia found current estrogen use to be significantly more common among elderly non-demented

Table 2. Epidemiological studies of estrogen replacement therapy and Alzheimer's disease risk[a]

Study	Number of Alzheimer's cases	Number of controls	Type of study	Relative risk	95% Confidence interval
Heyman et al. 1984	40	80	Case-control	2.38	–
Amaducci et al. 1986	60	60[b]	Case-control	0.71	–
		50[b]		1.67	–
Broe et al. 1990	170	170	Case-control	0.78	0.39–1.56
Graves et al. 1990	130	130	Case-control	1.15	0.50–2.64
Henderson et al. 1994	143	92	Case-control	0.33	0.15–0.74
Paganini-Hill and Henderson 1994[c]	138	550	Case-control	0.69	0.46–1.03
Brenner et al. 1994[c]	107	120	Case-control	1.1	0.6–1.8
Mortel and Meyer 1995	93	148	Case-control	0.55	0.26–1.16
Paganini-Hill and Henderson 1996[c,d]	248	1198	Case-control	0.65	0.49–0.88
Tang et al. 1996[c]	167	957	Cohort	0.5	0.25–0.9
van Duijn et al. 1996[e]	124	124	Case-control	0.40	0.19–0.91
Lerner et al. 1997	88	176	Case-control	0.58	0.25–0.91
Kawas et al. 1997[c]	34	438	Cohort	0.46	0.209–0.997
Waring et al. 1997[c]	222	222	Case-control	0.4	0.2–0.8
Baldereschi et al. 1998	92	1476	Case-control	0.28	0.8–0.98

[a] Estrogen use included unopposed estrogen and estrogen opposed by progestogen. Relative risks of less than 1 imply a protective effect of estrogen
[b] Both hospital and population controls were used in comparison with Alzheimer's cases
[c] Studies in which information on estrogen use was collected prospectively, before the onset of dementia symptoms
[d] Includes cases from the earlier analysis of Paganini-Hill and Henderson, (1994)
[e] Analyses, restricted to early-onset cases, indicated that the estrogen effect was modified by apolipoprotein-E genotype; see text for details

women (18%) than among women with AD (7%) (Henderson et al. 1994). In this study, cases and controls did not differ with respect to total numbers of prescription medications or prior gynecological procedures that might have influenced the use of hormone therapy. Other analyses based on current hormone use or derived from data on hormone use that was obtained retrospectively also imply that estrogen may be protective (Birge 1994; Mortel and Meyer 1995; Baldereschi et al. 1998; van Duijn et al. 1996; Lerner et al. 1997) (Table 2).

Several recent case-control and cohort studies provide stronger evidence for a protective effect of estrogen (Table 2). In these studies, estrogen use was documented prospectively, before some women developed symptoms of dementia (Brenner et al. 1994; Paganini-Hill and Henderson 1994; Paganini-Hill and Henderson 1996; Tang et al. 1996; Kawas et al. 1997; Waring et al. 1997). The largest such study is from the Leisure World retirement community cohort in southern California, where 56% of women self-reported estrogen use after the menopause (Paganini-Hill and Henderson 1996). In this nested case-control study, 248 cases identified on the basis of death certificate records were matched with 1198 controls. Forty-eight percent of controls but only 39% of cases had used estrogen, and estrogen users had a 35% lower risk of Alzheimer's disease. Similar risk reductions were found when analyses in this cohort were restricted to oral estrogen use (OR = 0.70, 95% CI = 0.50–0.98).

Protective effects are also reported in analyses from Baltimore, New York City, and Rochester, Minnesota. In these studies, estrogen users had about half the risk of Alzheimer's disease as non-users (Table 2). The Baltimore Longitudinal Study on aging included 472 perimenopausal and postmenopausal women who had provided information on oral and transdermal estrogen use (Kawas et al. 1997). Among 34 women who developed Alzheimer's disease over a follow-up period of up to 16 years, 29% had used estrogen; 47% of women who did not develop Alzheimer's had used estrogen. In the New York City cohort, the use of oral estrogens had been reported by 15% of 957 healthy elderly women, but by only 5% of the 167 women who developed Alzheimer's disease during a 1–5 year follow-up interval (Tang et al. 1996). Moreover, the age of dementia onset was later in women who had taken estrogen than in non-users. A population-based analysis from Rochester involved 222 women diagnosed with Alzheimer's disease matched to an equal number of controls. Medical records were used to document the use of any estrogen preparation for at least 6 months; 12% of controls but only 5% of cases were estrogen users (Waring et al. 1997).

Different conclusions were reached in a health maintenance organization population in Seattle (Brenner et al. 1994). Computerized pharmacy records were available for 107 incident Alzheimer's disease cases and 120 matched controls, about half of whom had used estrogens. As in the Leisure World cohort, the risk of an Alzheimer diagnosis was reduced by nearly one-third among women who had previously used oral estrogens (OR = 0.7, 95%

CI = 0.4–1.5), but no protective effect was evident when all types of estrogen preparations were considered (Table 2).

Negative results are also reported in preliminary analyses from the Framingham Heart Study cohort, where 181 women with at least three years of continuous estrogen use by age 65 years were compared with 1358 other women (McNULTY et al. 1997). Members of the estrogen group had higher scores on a cognitive screening task, but there was no evidence that estrogen influenced the risk of dementia, where the type of dementia was not specified.

a) *Alzheimer's Risk: Strength of Association*

If estrogen helps to protect against Alzheimer's disease, then it is likely that greater estrogen exposure would be more protective than lesser exposure. Indeed, a significant dose effect was observed in the Leisure World study (PAGANINI-HILL and HENDERSON 1994; PAGANINI-HILL and HENDERSON 1996). Here, the risk of Alzheimer's disease decreased significantly with increasing duration of postmenopausal estrogen use, with increasing duration of oral estrogen use, and with higher dosages of the most often prescribed oral estrogen. Similarly, women in the New York City study who had used oral estrogens for longer than 1 year showed a greater reduction in risk than those who had used estrogens for 1 year or less (TANG et al. 1996), and in Rochester, risk estimates among estrogen–users decreased with increasing duration of estrogen use and with increasing cumulative estrogen dose (WARING et al. 1997). In the Baltimore study, however, the duration of estrogen use did not appear to influence risk estimates, although the number of Alzheimer's cases was relatively small (KAWAS et al. 1997).

b) *Alzheimer's Risk: Estrogen and Apolipoprotein E*

Two studies have considered a possible interaction between apolipoprotein E genotype and estrogen use. In the New York City study, a protective effect of estrogen replacement was evident for women heterozygous for the $\varepsilon 4$ allele (OR = 0.13, 95% CI = 0.02–0.95) and for women without the $\varepsilon 4$ allele (OR = 0.4, 95% CI = 0.2–0.9) (TANG et al. 1996). However, in a population-based study of early-onset Alzheimer's disease, a protective effect of estrogen replacement was apparent only among women who possessed the $\varepsilon 4$ allele (OR = 0.14, 95% CI = 0.02–0.87) (VAN DUIJN et al. 1996).

c) *Estrogen and Alzheimer's Risk: Caveats*

In aggregate, recent epidemiological studies imply that estrogen use could reduce Alzheimer's risk by about one-third to one-half (Table 2). However, women who use estrogen differ from those who do not (HEMMINKI et al. 1993; DERBY et al. 1995), and bias or unrecognized confounding cannot be fully excluded in some studies. Any causative relationship could be more firmly established in the future by findings from randomized clinical trials.

Case-control and cohort studies of Alzheimer's disease have understandably included women who used various regimens of opposed and unopposed estrogen. Higher daily doses may confer greater protection than lower doses (PAGANINI-HILL and HENDERSON 1996), and postmenopausal estrogens used for longer periods of time may confer greater protection than short-term use (PAGANINI-HILL and HENDERSON 1996; TANG et al. 1996; WARING et al. 1997). However, these preliminary inferences remain to be confirmed, and no convincing data indicate whether or not putative benefits of estrogen might be modified by concomitant use of a progestogen.

3. Estrogen and Alzheimer's Disease Symptoms

Women with Alzheimer's disease taking estrogen show better cognitive skills than women with Alzheimer's disease not taking estrogen (HENDERSON et al. 1994; HENDERSON et al. 1996). In an observational study, where subjects were matched for age, education, and duration of dementia symptoms, estrogen users outperformed non-users on a variety of psychometric tasks (HENDERSON et al. 1996) (Fig. 1). Differences favoring estrogen users were greatest on a naming (semantic memory) task, the same type of skill that impacts women with Alzheimer's disease more severely than men (HENDERSON and BUCKWALTER 1994; RIPICH et al. 1995; BUCKWALTER et al. 1996).

Recent intervention trials have examined estrogen effects in women, although some of these are reported only in a preliminary manner (FILLIT et al. 1986; HONJO et al. 1989, 1993; FILLIT 1994; OHKURA et al. 1994a, b, c, 1995a; ASTHANA et al. 1996; BIRGE 1997). Despite small sample sizes, overall results suggest a benefit of treatment (HENDERSON 1997b); however, only two trials (OHKURA et al. 1994a; BIRGE 1997) followed subjects for longer than 2 months, and most have been conducted as open-label trials.

The only randomized trial of estrogen in Alzheimer's disease reported in full is a 3-week study conducted in Japan (HONJO et al. 1993). Fourteen

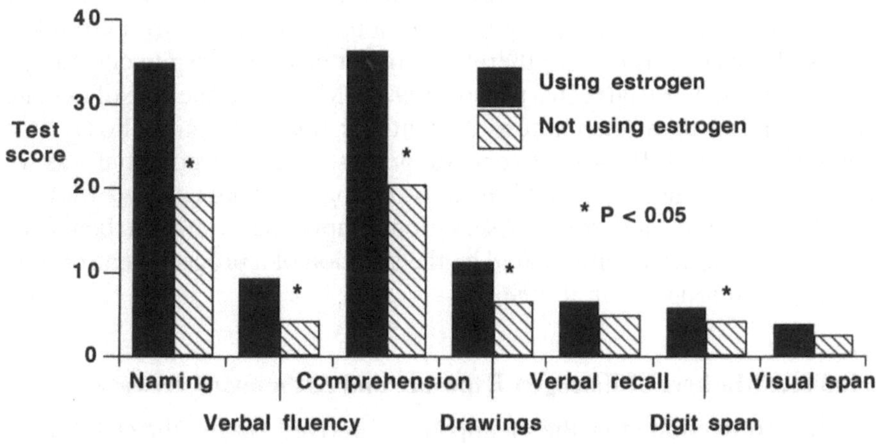

Fig. 1. Neuropsychological test scores for women with Alzheimer's disease who were either using estrogen therapy ($n = 9$) or not ($n = 27$). Data are from HENDERSON et al. (1996)

Alzheimer's subjects (mean age 84 years) received either 1.25 mg per day oral conjugated estrogens or placebo for 3 weeks. Women in the active arm improved significantly over baseline on each of three brief psychometric measures. In a second double-blind trial, which has been reported in a preliminary manner, ten demented women were randomized to 50 mg per day transdermal estradiol or to placebo (ASTHANA et al. 1996). After 2 months, treated subjects improved significantly over baseline on two verbal memory tasks. Interim analyses from an ongoing 9-month double-blind study involving 20 women treated with 0.625 mg per day conjugated estrogens or placebo showed significant improvement in the active treatment group on a standardized clinician interview-based impression of change (BIRGE 1997).

Some of the postulated estrogen effects in women with Alzheimer's disease may be mediated through cholinergic mechanisms. This contention is supported by a study of ovariectomized rats on a T-maze alternation task (DOHANICH et al. 1994). After the administration of scopolamine, a high-affinity blocker of cholinergic muscarinic receptors, maze performance deteriorated. However, estradiol reversed deleterious effects of scopolamine on this task. In Alzheimer's disease, inhibitors of acetylcholinesterase (e.g., tacrine and donepezil), which act to increase brain levels of acetylcholine, are of modest therapeutic benefit (KNAPP et al. 1994; ROGERS et al. 1996). In a retrospective analysis of data from a multi-center cholinesterase-inhibitor trial in patients with Alzheimer's disease, women using hormone replacement at the time of study enrolment and subsequently randomized to receive the anticholinesterase agent performed significantly better on outcome measures than women randomized to the placebo arm (SCHNEIDER et al. 1996). Demented women in the active treatment arm who were not using estrogen performed similarly to demented women in the placebo group. Findings must be interpreted cautiously, however, as estrogen use in this study was not randomized, and the number of women using hormone replacement was small.

No large placebo-controlled, double-blind trial of estrogen has been conducted over a substantial time period, although such trials are currently in progress. Firm conclusions are therefore not yet possible, despite encouraging results from smaller trials (HENDERSON 1997b). Moreover, the role of estrogen may be limited to women whose dementia is relatively mild (FILLIT et al. 1986; OHKURA et al. 1994c). Of concern, progesterone can potentiate, as well as downregulate, neurotrophic effects of estrogen (WOOLLEY and MCEWEN 1993; GIBBS 1996), and some observations imply that estrogen benefits in Alzheimer's disease are attenuated by the addition of a progestogen (OHKURA et al. 1995a; SCHNEIDER et al. 1996).

4. Indirect Markers of Estrogen Exposure and Alzheimer's Disease

Estrogen production after the menopause is derived largely through the aromatization of androstenedione and other androgen precursors. These steroids are produced by the ovarian stroma and adrenal cortex and are converted principally in adipose tissue. Among older women, body weight therefore correlates with circulating levels of estrogen (MELDRUM et al. 1981). Because Alzheimer's patients tend to be thin (BERLINGER and POTTER 1991), it is suggested that body weight after the menopause might influence the risk of Alzheimer's disease. In Leisure World, higher body weight at the time of cohort enrolment predicted a lower risk of subsequent Alzheimer's disease (PAGANINI-HILL and HENDERSON 1996). For women weighing at least 63 kg, the estimated risk of developing Alzheimer's disease was about 30% lower than women weighing less than 56 kg.

Among women with Alzheimer's disease, body weight is also associated with cognitive skills. An analysis of psychometric scores from 347 women with Alzheimer's disease confirmed that greater weight was significantly associated with better cognitive performance on two tasks (the Mini-Mental State examination and the cognitive sub-score of the Alzheimer's Disease Assessment Scale), in analyses that adjusted for height, symptom duration, age, and education (BUCKWALTER et al. 1997). When two weight extremes (40 kg vs 80 kg) were considered, it was estimated that scores on the two psychometric tasks examined would differ by about 12% and 25%, respectively (BUCKWALTER et al. 1997).

Reproductive variables can modify disease risk, when the disease in question is hormonally sensitive. Consistent with the hypothesis that estrogen exposure is protective for Alzheimer's disease, in the Leisure World cohort there was a slight tendency ($P = 0.11$) for earlier age at menarche to be associated with a lower risk of this disorder (PAGANINI-HILL and HENDERSON 1996). However, in this and a second cohort of predominately late-onset Alzheimer's cases (TANG et al. 1996), the predicted association with later age at menopause was not observed. In Leisure World, parity was also unrelated to the risk of Alzheimer's disease (PAGANINI-HILL and HENDERSON 1996). However, among early-onset cases with a documented family history of dementia, an early age at menopause was significantly associated with Alzheimer's disease risk (VAN DUIJN et al. 1996), possibly implying an interaction between genetic factors and a protective effect of estrogen in early-onset illness.

II. Multi-infarct Dementia

The second most prevalent cause of dementia is attributed to cerebrovascular disease. In its most common form, this form of dementia is referred to as multi-infarct dementia. In the older woman, estrogen replacement protects against cardiovascular disease (GRODSTEIN and STAMPFER 1995), but there is no clear association between estrogen use and stroke incidence (PAGANINI-HILL 1995). More specifically, it is unknown whether estrogen might have a preventive role in multi-infarct dementia. One study compared 65 women with a clinical diagnosis of vascular dementia (mean age 74 years) with 148 non-demented control subjects (mean age 72 years) (MORTEL and MEYER 1995). In this analysis of current estrogen use, patients were somewhat less likely to be receiving hormone-replacement therapy than controls (OR = 0.50, 95% CI = 0.20–1.20).

G. Other Neuropsychiatric Disorders

Potential estrogen effects on migraine, epilepsy, Parkinson's disease, and multiple sclerosis are discussed in this section. These four illnesses are briefly presented as illustrative examples of common neuropsychiatric disorders, in which estrogen might play some role in disease pathogenesis or the symptomatic manifestations.

I. Migraine Headache

Migraine is a common form of headache characterized by recurrent throbbing head pain lasting from hours to days. The pain is often unilateral and often accompanied by nausea or vomiting. The term "classic" migraine is used when pain is preceded by scintillating visual field defects or other focal neurological symptoms. The neurological aura lasts about 20 min, and deficits tend to occur contralateral to the side of the ensuing head pain. "Common" migraine is not preceded by neurological disturbances.

Migraine pain is pulsatile in nature, and vascular factors are strongly implicated in migraine pathogenesis. One common view is that pain is due to distention and high-amplitude pulsation of the external carotid artery, whereas the preceding neurological aura reflects a period of intense vasoconstriction of intracranial vessels. However, this theory does not fully explain the pathophysiology of migraine. A reduction in regional cerebral blood flow is a consistent early event of classic migraine, although the pattern of blood flow during the ensuing headache phase is more variable, and regional flow does not necessarily increase at the onset of pain (LAURITZEN and OLESEN 1984; OLESEN et al. 1990). The primary event of classic migraine may be a process of depressed electrical activity that spreads slowly through the cerebral cortex and is accompanied by reductions in cerebral blood flow (LAURITZEN and OLESEN 1984; WOODS et al. 1994). Overactivity of the neurotransmitter serotonin is also implicated in migraine pathogenesis.

Estrogen has the potential to affect migraine headache via its actions on cerebral blood flow (BELFORT et al. 1995; OHKURA et al. 1995b) or serotonergic neurons (BIEGON et al. 1983; COHEN and WISE 1988; SUMNER and FINK 1995). Indeed, a possible role for sex hormones is suggested by changes in the frequency of headaches during the menstrual cycle, with pregnancy, and with the use of ovarian hormones.

The prevalence of migraine is three times higher in women than men (STEWART et al. 1992). Migraine attacks are more common preceding or during the menses (WATERS and O'CONNOR 1971; DALTON 1973) and, in some migraineurs, pain occurs only during the premenstrual days, where it is attributed primarily to estrogen withdrawal (SOMERVILLE 1972a; LICHTEN et al. 1996). In some patients, falling progesterone levels could also play a role (WHITTY et al. 1966). Estrogen treatment can postpone or prevent attacks of premenstrual migraine (MAGOS et al. 1983; DE LIGNIÈRES et al. 1986). It is also suggested that the frequency of migraine attacks can worsen in women who use estrogen-containing oral contraceptives or who receive estrogen replacement after menopause (KUDROW 1975), perhaps attributable to fluctuations in estrogen levels.

During pregnancy, when estrogen levels are elevated, migraine frequency is typically reduced (SOMERVILLE 1972b), particularly for women with a history of menstrual migraine (LANCE and ANTHONY 1966). Some women, however, experience their first migraine headache while pregnant (SOMERVILLE 1972b). Headache frequency is often reduced in the third trimester of pregnancy, compared with early pregnancy (CHEN and LEVITON 1994). Headaches are commonly reported in the puerperium (STEIN 1981).

Migraine frequency tends to decline with age (WATERS and O'CONNOR 1971), but is usually unaffected by the menopause (WHITTY and HOCKADAY 1968). Effects of estrogen therapy on migraine symptoms are variable, but estrogen may contribute to migraine symptoms in some women (KUDROW 1975). Oophorectomy is ineffective in the treatment of migraine (ALVAREZ 1940).

II. Epilepsy

A seizure is the paroxysmal disruption of brain function due to the abrupt discharge of neurons. The abnormal discharge results in involuntary muscle contractions, loss of consciousness, or other neurological symptoms. Epilepsy, the tendency to experience recurrent seizures, is a common and important neurological disorder.

In female rats, the threshold for seizures induced by electrical stimulation of the brain varies during the estrus cycle. Ovariectomy abolishes cyclical changes in this threshold (TERASAWA and TIMIRAS 1968). The seizure threshold is lowest during proestrus (TERASAWA and TIMIRAS 1968), when estradiol levels are at their peak. Synaptic excitability (WONG and MOSS 1992) and spontaneous neuronal activity in the hippocampus are also greatest during this time (KAWAKAMI et al. 1970). Estradiol replacement in ovariectomized rats transiently lowers baseline thresholds in the hippocampus, whereas the effect of progesterone is to decrease the likelihood of seizures (WOOLLEY and TIMIRAS 1962; TERASAWA and TIMIRAS 1968; LANDGREN et al. 1978). Consistent with these observations, estrogen administration induces abnormal electrical activity in the cerebral cortex (LOGOTHETIS and HARNER 1960) and can potentiate experimentally induced seizures in animals of either sex (WOOLLEY and TIMIRAS 1962; NICOLETTI et al. 1985; BUTERBAUGH and HUDSON 1991).

Morphological changes in individual neurons parallel hormonally induced changes in hippocampal excitability. Within the CA1 region of the hippocampus, synaptic density fluctuates during the rat estrus cycle, reaching a peak during proestrus (WOOLLEY and MCEWEN 1992). Hippocampal synaptic plasticity, as measured by long-term potentiation, is similarly increased in female rats during the afternoon of proestrus (WARREN et al. 1995).

In humans, as in laboratory animals, estrogen is postulated to lower seizure threshold (LOGOTHETIS et al. 1959). Within seconds to minutes after the intravenous administration of conjugated estrogens in one study, 11 of 16 epileptic patients showed epileptiform discharges on their electroencephalographic records (LOGOTHETIS et al. 1959). Epilepsy often first develops at about the time of menarche (LOGOTHETIS et al. 1959).

Although slight changes in electroencephalographic rhythms are detectable during the menstrual cycle (CREUTZFELDT et al. 1976), most epileptic women show no relationship between menstruation and seizures (DICKERSON 1941). However, in so-called catamenial epilepsy, recurrent seizures are in fact linked to the menstrual cycle. In these patients, seizures are most apt to occur immediately preceding or during menstruation (LAIDLAW 1956; LOGOTHETIS et al. 1959). Catamenial seizures are thought to be triggered primarily by fluctuations in the concentrations of ovarian hormones (LAIDLAW 1956; BÄCKSTRÖM 1976), but cyclical changes in serum concentrations of anticonvulsant medications might play an additional role (ROSCISZEWSKA et al. 1986).

Epileptic manifestations may be altered during other times of hormonal change. Seizure frequency commonly increases during pregnancy, particularly in women whose seizures were poorly controlled before they became pregnant (KNIGHT and RHIND 1975). However, reasons for this exacerbation are myriad and are not necessarily related to hormonal effects on the seizure threshold. During pregnancy, declining serum concentrations of anticonvulsants, sleep deprivation, hormonally responsive neurological illnesses (e.g., cerebral thrombosis, enlargement of a cerebral meningioma), and pregnancy complications, such as eclampsia, can all cause seizures. The effect of menopause on seizure frequency has not been well studied.

Progesterone secretion may protect against seizures (LAIDLAW 1956; BÄCKSTRÖM 1976). The ratio between estrogen and progesterone may also be important in ovulating women, with a higher ratio associated with an increased frequency of seizures (BÄCKSTRÖM 1976). Progestogen administration may reduce seizure frequency in epileptic women (MATTSON et al. 1984; HERZOG 1986). Although observational reports suggest that progesterone also reduces seizure frequency in catamenial epilepsy, this hypothesis has not been tested in randomized controlled trials.

III. Parkinson's Disease and Other Movement Disorders

An association between ovarian hormones and movement disorders has long been suspected in younger women. Both pregnancy (chorea gravidarum) and oral-contraceptive use are associated with reversible chorea (WILLSON and

Preece 1932; Lewis and Parsons 1966; Nausieda et al. 1979). Although the substantia nigra and striatum in adult brains do not appear to contain substantial numbers of estrogen receptors (Pfaff and Keiner 1973), a large body of evidence indicates that estrogen modifies dopamine receptors, dopaminergic neuronal activity, and motor behaviors mediated by dopamine (Chiodo and Caggiula 1980; Hruska and Silbergeld 1980; Joseph et al. 1989; Roy et al. 1990).

Among older persons, Parkinson's disease is a common progressive neurodegenerative disorder of the basal ganglia characterized by resting tremor, rigidity of the limbs and trunk, the paucity of voluntary movement (bradykinesia), and impaired postural reflexes. Biochemical abnormalities involve deficiencies of the catecholamine neurotransmitters. Most Parkinsonian symptoms are attributed to the prominent loss of dopamine-containing neurons in the substantia nigra region of the midbrain; these neurons project to the striatum. This nigro–striatal pathway is an important part of the extrapyramidal system that controls movement, and drugs that increase brain levels of dopamine ameliorate symptoms of Parkinson's disease. In contrast, chorea and certain other hyperkinetic movement disorders are typically exacerbated by dopaminergic drugs and may benefit from treatment with neuroleptic agents or other dopamine antagonists.

Estrogen clearly modulates dopaminergic activity within the nigro–striatal system, but there is only weak evidence that estrogen affects symptoms of patients with movement disorders. Younger women with Parkinson's disease sometimes complain of increased Parkinsonian symptoms preceding or concurrent with menstruation (Quinn and Marsden 1986). Case reports imply that Parkinsonian symptoms may be precipitated by estrogen in women receiving neuroleptic drugs (Gratton 1960) and that estrogen might aggravate symptoms of Parkinson's disease (Bedard et al. 1977; Koller et al. 1982). For the treatment of hyperkinetic dyskinesias, however, estrogen may have at least some limited efficacy. An open-label study of 20 men with neuroleptic-induced tardive dyskinesia, a hyperkinetic movement disorder, found a significant decrease in dyskinesias after 6 weeks of treatment with oral conjugated estrogens (Villeneuve et al. 1980). In another study of conjugated estrogens, patients with hyperkinesias of various etiologies failed to show significant change (Koller et al. 1982). However, it was observed that estrogen treatment led to a noticeable decrease in dyskinesia scores in two of ten Huntington's disease patients and four of ten patients with tardive dyskinesia. Eight dystonic patients in the same study showed no effect of therapy (Koller et al. 1982).

Experimentally, estradiol protects against the loss of dopamine neurons induced by specific neurotoxins (Dluzen et al. 1996), but estrogen use after the menopause does not appear to affect the risk of uncomplicated Parkinson's disease. A cohort study that compared 87 women with idiopathic Parkinson's disease with 989 women without stroke or dementia found no association with estrogen therapy (Mayeux et al. 1997). In contrast, 80 other women in the same cohort had both Parkinson's disease and dementia, and estrogen use was associated with a significantly reduced risk of dual symptoms. Parkinsonian symptoms, however, are common in Alzheimer's disease patients, and demented patients with pathological features of Parkinson's disease often have additional neuropathological changes of Alzheimer's disease (CLARK et al. 1997; HANSEN and SAMUEL 1997). It is possible, therefore, that the favorable effects of estrogen on Parkinsonian dementia in this study were due to putative protective effects in Alzheimer's disease (HENDERSON 1997a).

IV. Multiple Sclerosis

Multiple sclerosis is a chronic neurological disorder affecting myelinated nerve fibers of the central nervous system. Protean symptoms of this common illness depend on the distribution of demyelinated plaques within the cerebrum, cerebellum, brain stem, and optic nerves. The onset typically occurs during the second and third decades of life, but sometimes it appears later. The clinical course can be punctuated by remissions and relapses, or symptoms might progress unremittingly.

Like many illness thought to have an autoimmune pathogenesis (MARTIN and MCFARLAND 1995), multiple sclerosis affects women more often than men. In this disorder, the target of the inflammatory response is believed to be specific components of the myelin sheath. Estrogen effects on inflammation could be important in this regard. Estrogens inhibit leukocyte production in the bone marrow, suppress delayed hypersensitivity, enhance antibody production, and modulate cytokine production by T-cell lymphocytes (CARLSTEN et al. 1989; PACIFICI et al. 1991; JOSEFSSON et al. 1992; GILMORE et al. 1997). Nevertheless, the clinical relevance of estrogen in multiple sclerosis is poorly established. Pregnancy does not appear to affect the course of multiple sclerosis deleteriously, although neurological relapses might occur less frequently during pregnancy and more frequently in the puerperium (KORN-LUBETZKI et al. 1984; BERNARDI et al. 1991; SADOVNICK et al. 1994). Effects of menopause or postmenopausal hormone replacement on symptoms or the long-term course of multiple sclerosis are unknown.

H. Conclusions

There is a complex and still elusive relationship between estrogens and other gonadal steroids and the brain, cognition, and behavior. As considered above, an abundant literature testifies to the biological plausibility of estrogen effects on the maintenance of psychological and neurological well-being. Data on Alzheimer's disease are perhaps the best developed, so far, but firm conclusions are still lacking. In other specific neuropsychiatric illnesses, experimental data are sparse, and findings of observational studies are not completely convincing; the clinical relevance is for this reason obscure. Undoubtedly, continual advances in our understanding of mechanisms of hormonal actions on neural tissues coupled with new observations on hormone effects in health and disease will spur well-designed treatment and prevention trials.

References

Adams RD, Victor M, Ropper AH (1997) Principles of Neurology. McGraw–Hill, New York

Alvarez WC (1940) Can one cure migraine in women by inducing menopause? Report on forty-two cases. Mayo Clin Proc 15:380–382

Amaducci LA, Fratiglioni L, Rocca WA, Fieschi C, Livrea P, Pedone D, Bracco L, Lippi A, Grandolfo C, Bino G, Prencipe M, Bonatti ML, Girotti F, Carella F, Tavolato B, Ferla S, Lenzi GL, Carolei A, Gambi A, Grigoletto F, Schoenberg BS (1986) Risk factors for clinically diagnosed Alzheimer's disease: a case-control study of an Italian population. Neurology 36:922–931

Applebaum-Bowden D, McLean P, Steinmetz A, Fontana D, Matthys C, Warnick GR, Cheung M, Albers JJ, Hazzard WR (1989) Lipoprotein, apolipoprotein, and lipolytic enzyme changes following estrogen administration in postmenopausal women. J Lipid Res 30:1895–1906

Asthana S, Craft S, Baker LD, Raskind MA, Avery E, Lofgreen C, Wilkinson CW, Falzgraf S, Veith RC, Plymate SR (1996) Transdermal estrogen improves memory in women with Alzheimer's disease [abstract]. Soc Neurosci Abstr 22:200

Bachman DL, Wolf PA, Linn R, Knoefel JE, Cobb J, Belanger A, D'Agostino RB, White LR (1992) Prevalence of dementia and probable senile dementia of the Alzheimer type in the Framingham Study. Neurology 42:115–119

Bachman DL, Wolf PA, Linn RT, Knoefel JE, Cobb JL, Belanger AJ, White LR (1993) Incidence of dementia and probable Alzheimer's disease in a general population: the Framingham study. Neurology 43:515–519

Bäckström T (1976) Epileptic seizurers in women related to plasma estrogen and progesterone during the menstrual cycle. Acta Neurol Scand 54:321–347

Baldereschi M, Di Carol A, Lepore V, Bracco L, Maggi S, Grigoletto F, Scarlato G, Amaducci L (1998) Estrogen-replacement therapy and Alzheimer's disease in the Italian longitudinal study on aging. Neurol 50:996–1002

Ball P, Knuppen R, Haupt M, Breuer H (1972) Interactions between estrogens and catechol amines. II. Studies on the methylation of catechol estrogens, catechol amines and other catechols by the catechol–O–methyltransferases of human liver. J Clin Endocrinol Metab 34:736–746

Barrett-Connor E, Kritz-Silverstein D (1993) Estrogen replacement therapy and cognitive function in older women. JAMA 269:2637–2641

Bartus RT, Dean RL, Beer B, Lippa AD (1981) The cholinergic hypothesis of geriatric memory dysfunction. Science 217:208–217

Bauer J, Ganter U, Strauss S, Stadtmüller G, Frommberger U, Bauer H, Volk B, Berger M (1992) The participation of interleukin-6 in the pathogenesis of Alzheimer's disease. Res Immunol 143:650–657

Becker JB, Snyder PJ, Miller MM, Westgate SA, Jenuwine MJ (1987) The influence of estrous cycle and intrastriatal estradiol on sensorimotor performance in the female rat. Pharmacol Biochem Behav 27:53–59

Bedard P, Langelier P, Villeneuve A (1977) Estrogens and extrapyramidal system [letter]. Lancet 2:1367–1368

Belfort MA, Saade GR, Snabes M, Dunn R, Moise KJ, Jr, Cruz A, Young R (1995) Hormonal status affects the reactivity of the cerebral vasculature. Am J Obstet Gynecol 172:1273–1278

Berlinger WG, Potter JF (1991) Low body mass index in demented outpatients. J Am Geriatr Soc 39:973–978

Berman KF, Schmidt PJ, Rubinow DR, Danaceau MA, Van Horn JD, Esposito G, Ostrem JL, Weinberger DR (1997) Modulation of cognition-specific cortical activity by gonadal steroids: a positron-emission tomography study in women. Proc Natl Acad Sci USA 94:8836–8841

Bernardi S, Grasso MG, Bertollini R, Orzi F, Fieschi C (1991) The influence of pregnancy on relapses in multiple sclerosis: a cohort study. Acta Neurol Scand 84:403–406

Bertrand P, Poirier J, Oda T, Finch CE, Pasinetti GM (1995) Association of apolipoprotein E genotype with brain levels of apolipoprotein E and apolipoprotein J (clusterin) in Alzheimer disease. Mol Brain Res 33:174–178

Best NR, Rees MP, Barlow DH, Cowen PJ (1992) Effect of estradiol implant on noradrenergic function and mood in menopausal subjects. Psychoneuroendocrinology 17:87–93

Biegon A, Reches A, Snyder L, McEwen BS (1983) Serotonergic and noradrenergic receptors in the rat brain: modulation by chronic exposure to ovarian hormones. Life Sci 32:2015–2021

Birge SJ (1994) The role of estrogen deficiency in the aging central nervous system. In: Lobo RA (ed) Treatment of the Postmenopausal Woman: Basic and Clinical Aspects. Raven Press, New York, pp 153–157

Birge SJ (1997) The role of estrogen in the treatment of Alzheimer's disease. Neurology 48[suppl 7]:S36-S41

Bishop J, Simpkins JW (1992) Role of estrogens in peripheral and cerebral glucose utilization. Rev Neurosci 3:121–137

Blennow K, Hesse C, Fredman P (1994) Cerebrospinal fluid apolipoprotein E is reduced in Alzheimer's disease. NeuroReport 5:2534–2536

Bowen JD, Malter AD, Sheppard L, Kukull WA, McCormick WC, Teri L, Larson EB (1996) Predictors of mortality in patients diagnosed with probable Alzheimer's disease. Neurology 47:433–439

Brenner DE, Kukull WA, Stergachis A, van Belle G, Bowen JD, McCormick WC, Teri L, Larson EB (1994) Postmenopausal estrogen replacement therapy and the risk of Alzheimer's disease: a population-based case-control study. Am J Epidemiol 140:262–267

Breteler MMB, Claus JJ, van Duijn CM, Launer LJ, Hofman A (1992) Epidemiology of Alzheimer's disease. Epidemiol Rev 14:59–82

Brinton RD (1993) 17β-estradiol induction of filopodial growth in cultured hippocampal neurons within minutes of exposure. Mol Cell Neurosci 4:36–46

Broe GA, Henderson AS, Creasey H, McCusker E, Korten AE, Jorm AF, Longley W, Anthony JC (1990) A case-control study of Alzheimer's disease in Australia. Neurology 40:1698–1707

Buckwalter JG, Rizzo AA, McCleary R, Shankle R, Dick M, Henderson VW (1996) Gender comparisons of cognitive performances among vascular dementia, Alzheimer disease, and older adults without dementia. Arch Neurol 53:436–439

Buckwalter JG, Schneider LS, Wilshire TW, Dunn ME, Henderson VW (1997) Body weight, estrogen and cognitive functioning in Alzheimer's disease: an analysis of the tacrine study group. Arch Gerontol Geriatr 24:261–267

Buterbaugh GG, Hudson GM (1991) Estradiol replacement to female rats facilitates dorsal hippocampal but not ventral hippocampal kindled seizure acquisition. Exp Neurol 111:55–64

Carlsten H, Holmdahl R, Tarkowski A, Nilsson L-A (1989) Oestradiol- and testosterone-mediated effects on the immune system in normal and autoimmune mice are genetically linked and inherited as dominant traits. Immunology 68:209–214

Chen T-C, Leviton A (1994) Headache recurrence in pregnant women with migraine. Headache 34:107–110

Chiodo LA, Caggiula AR (1980) Alterations in basal firing rate and autoreceptor sensitivity of dopamine neurons in the substantia nigra following acute and extended exposure to estrogen. Eur J Pharmacol 67:165–166

Chung SK, Pfaff DW, Cohen RS (1988) Estrogen-induced alterations in synaptic morphology in the midbrain central gray. Exp Brain Res 69:522–530

Clark CM, Ewbank D, Lerner A, Doody R, Henderson VW, Panisset M, Morris JC, Fillenbaum GG, Heyman A (1997) The relationship between extrapyramidal signs and cognitive performance in patients with Alzheimer's disease enrolled in the CERAD study. Neurology 49:70–75

Cohen IR, Wise PM (1988) Effects of estradiol on the diurnal rhythm of serotonin activity in microdissected brain areas of ovariectomized rats. Endocrinol 122: 2619–2625

Coyle JT, Price DL, DeLong MR (1983) Alzheimer's disease: a disorder of cortical cholinergic innervation. Science 219:1184–1190

Creutzfeldt OD, Arnold P-M, Becker D, Langenstein S, Tirsch W, Wilhelm H, Wuttke W (1976) EEG changes during spontaneous and controlled menstrual cycles and their correlation with psychological performance. Electroenceph Clin Neurophysiol 40:113–131

Dalton K (1973) Progesterone suppositories and pessaries in the treatment of menstrual migraine. Headache 12:151–159

Davis KL, Davis BM, Greenwald BS, Mohs RC, Mathé AA, Johns CA, Horvath TB (1986) Cortisol and Alzheimer's disease. Am J Psychiatry 143:442–446

de Lignières B, Vincens M, Mauvais-Jarvis P, Mas JL, Touboul PJ, Bousser MG (1986) Prevention of menstrual migraine by percutaneous oestradiol. Br Med J 293:1540

Derby CA, Hume AL, McPhillips JB, Barbour MM, Carleton RA (1995) Prior and current health characteristics of postmenopausal estrogen replacement therapy users compared with nonusers. Am J Obstet Gynecol 173:544–550

Dickerson WW (1941) The effect of menstruation on seizure incidence. J Nerv Ment Dis 94:160–169

Ditkoff EC, Crary WG, Cristo M, Lobo RA (1991) Estrogen improves psychological function in asymptomatic postmenopausal women. Obstet Gynecol 78:991–995

Dluzen DE, McDermott JL, Liu B (1996) Estrogen as a neuroprotectant against MPTP-induced neurotoxicity in C57/B1 mice. Neurotoxicol Teratol 18:603–606

Dohanich GP, Fader AJ, Javorsky DJ (1994) Estrogen and estrogen-progesterone treatments counteract the effect of scopolamine on reinforced T-maze alternation in female rats. Behav Neurosci 108:988–992

Doody RS, Young R, Lynn P, Cooke N, Rosenfeld JE (1995) Estrogen and psychomotor performance [abstract]. J Neuropsychiatr Clin Neurosci 7:410

Espeland MA, Applegate W, Furberg CD, Lefkowitz D, Rice L, Hunninghake D (1995) Estrogen replacement therapy and progression of intimal–medial thickness in the carotid arteries of postmenopausal women. Am J Epidemiol 142:1011–1019

Evans RM (1988) The steroid and thyroid hormone receptor superfamily. Science 249:889–895

Falk E (1983) Plaque rupture with severe pre-existing stenosis precipitating coronary thrombosis: characteristics of coronary atherosclerotic plaques underlying fatal occlusive thrombi. Br Heart J 50:127–134

Fedor-Freybergh P (1977) The influence of estrogens on the wellbeing and mental performance in climacteric and postmenopausal women. Acta Obstet Gynecol Scand [suppl 64]:1–99

Fillit H (1994) Estrogens in the pathogenesis and treatment of Alzheimer's disease in postmenopausal women. Ann NY Acad Sci 743:233–238

Fillit H, Weinreb H, Cholst I, Luine V, McEwen B, Amador R, Zabriskie J (1986) Observations in a preliminary open trial of estradiol therapy for senile dementia–Alzheimer's type. Psychoneuroendocrinology 11:337–345

Fratiglioni L, Viitanen M, von Strauss E, Tontodonati V, Herlitz A, Winblad B (1997) Very old women at highest risk of dementia and Alzheimer's disease: incidence data from the Kungsholmen Project, Stockholm. Neurology 48:132–138

Friend KE, Ang LW, Shupnik MA (1995) Estrogen regulates the expression of several different estrogen receptor mRNA isoforms in rat pituitary. Proc Natl Acad Sci USA 92:4367–4371
Garcia-Segura LM, Chowen JA, Dueñas M, Torres-Aleman I, Naftolin F (1994) Gonadal steroids as promoters of neuro–glial plasticity. Psychoneuroendocrinology 19:445–453
Gibbs RB (1996) Fluctuations in relative levels of choline acetyltransferase mRNA in different regions of the rat basal forebrain across the estrus cycle: effects of estrogen and progestrone. J Neurosci 16:1049–1055
Gibbs RB, Hashash A, Johnson DA (1996) Effects of estrogen on potassium-stimulated acetylcholine release in the hippocampus and overlying cortex of adult rats. Brain Res 749:143–146
Gibbs RB, Pfaff DW (1992) Effects of estrogen and fimbria/fornix transection on $p75^{NGFR}$ and ChAT expression in the medial septum and diagonal band of Broca. Exp Neurol 116:23–39
Gilmore W, Weiner LP, Correale J (1997) Effect of estradiol on cytokine secretion by proteolipid protein-specific T cell clones isolated from multiple sclerosis patients and normal control subjects. J Immunol 158:446–451
Gordon HW, Lee PA (1993) No difference in cognitive performance between phases of the menstrual cycle. Psychoneuroendocrinology 18:521–531
Gratton L (1960) Neuroleptiques, parkinsonnisme et schizophrénie. Union Med Canada 89:679–694
Graves AB, Kukull WA (1994) The epidemiology of dementia. In: Morris JC (ed) Handbook of Dementing Illnesses. Marcel Dekker, New York, pp 23–69
Graves AB, White E, Koepsell TD, Reifler BV, van Belle G, Larson EB, Raskind M (1990) A case-control study of Alzheimer's disease. Ann Neurol 28:766–774
Greengrass PM, Tonge SR (1974) The accumulation of noradrenaline and 5–hydroxytryptamine in three regions of mouse brain after tetrabenazine and iproniazid: effects of ethinyloestradiol and progesterone. Psychopharmacologia 39:187–191
Gregoire AJP, Kumar R, Everitt B, Henderson AF, Studd JWW (1996) Transdermal estrogen for treatment of severe postnatal depression. Lancet 347:930–933
Grodstein F, Stampfer M (1995) The epidemiology of coronary heart disease and estrogen replacement in postmenopausal women. Progress in Cardiovascular Disease 38:199–210
Grodstein F, Stampfer MJ, Manson JE, Colditz GA, Willett WC, Rosner B, Speizer FE, Hennekens CH (1996) Postmenopausal estrogen and progestin use and the risk of cardiovascular disease. N Engl J Med 335:453–461
Hafner H, Riecher-Rössler A, An Der Heiden W, Maurer K, Fatkenheuer B, Loffler W (1993) Generating and testing a causal explanation of the gender difference in age at first onset of schizophrenia. Psychol Med 23:925–940
Hampson E (1990a) Estrogen–related variations in human spatial and articulatory–motor skills. Psychoneuroendocrinology 15:97–111
Hampson E (1990b) Variations in sex-related cognitive abilities across the menstrual cycle. Brain Cognit 14:26–43
Hampson E, Kimura D (1988) Reciprocal effects of hormonal fluctuations on human motor and preceptual–spatial skills. Behav Neurosci 102:456–459
Hansen LA, Samuel W (1997) Criteria for Alzheimer's disease and the nosology of dementia with Lewy bodies. Neurology 48:126–132
Hashimoto S, Katou M, Dong Y, Murakami K, Terada S, Inoue M (1997) Effects of hormone replacement therapy on serum amyloid P component in postmenopausal women. Maturitas 26:113–119
Hemminki E, Malin M, Topo P (1993) Selection to postmenopausal therapy by women's characteristics. J Clin Epidemiol 46:211–219
Henderson VW (1997a) The epidemiology of estrogen replacement therapy and Alzheimer's disease. Neurology 48[suppl 7]:27–35
Henderson VW (1997b) Estrogen replacement therapy for the prevention and treatment of Alzheimer's disease. CNS Drugs 8:343–351

Henderson VW (1997c) Estrogen, cognition, and a woman's risk of Alzheimer's disease. Am J Med 103[suppl 3 A]:11–18
Henderson VW, Buckwalter JG (1994) Cognitive deficits of men and women with Alzheimer's disease. Neurology 44:90–96
Henderson VW, Paganini-Hill A, Emanuel CK, Dunn ME, Buckwalter JG (1994) Estrogen replacement therapy in older women: comparisons between Alzheimer's disease cases and nondemented control subjects. Arch Neurol 51:896–900
Henderson VW, Watt L, Buckwalter JG (1996) Cognitive skills associated with estrogen replacement in women with Alzheimer's disease. Psychoneuroendocrinology 21:421–430
Herzog AG (1986) Intermittent progesterone therapy and frequency of complex partial siezures in women with menstrual disorders. Neurology 36:1607–1610
Heyman A, Wilkinson WE, Stafford JA, Helms JJ, Sigmon AH, Weinberg T (1984) Alzheimer's disease: a study of epidemiological aspects. Ann Neurol 15:335–341
Honjo H, Ogino Y, Naitoh K, Urabe M, Kitawaki J, Yasuda J, Yamamoto T, Ishihara S, Okada H, Yonezawa T, Hayashi K, Nambara T (1989) In vivo effects by estrone sulfate on the central nervous system – senile dementia (Alzheimer's type). J Steroid Biochem 34:521–525
Honjo H, Ogino Y, Tanaka K, Urabe M, Kashiwagi T, Ishihara S, Okada H, Araki K, Fushiki S, Nakajima K, Hayashi K, Hayashi M, Sakaki T (1993) An effect of conjugated estrogen to cognitive impairment in women with senile dementia – Alzheimer's type: a placebo-controlled double blind study. J Jpn Menopause Soc 1:167–171
Hruska RE, Silbergeld EK (1980) Increased dopamine receptor sensitivity after estrogen treatment using the rat rotation model. Science 208:1466–1468
Jaffe AB, Toran-Allerand CD, Greengard P, Gandy SE (1994) Estrogen regulates metabolism of Alzheimer amyloid β precursor protein. J Biol Chem 269: 13065–13068
Jorm AF, Korten AE, Henderson AS (1987) The prevalence of dementia: a quantitative integration of the literature. Acta Psychiatr Scand 76:465–479
Josefsson E, Tarkowski A, Carlsten H (1992) Anti-inflammatory properties of estrogen. Cell Immunol 142:67–78
Joseph JA, Kochman K, Roth GS (1989) Reduction of motor behavioural deficits in senescence via chronic prolactin or estrogen administration: time course and putative mechanisms of action. Brain Res 505:195–202
Kampen DL, Sherwin BB (1994) Estrogen use and verbal memory in healthy postmenopausal women. Obstet Gynecol 83:979–983
Katzenellenbogen JA, O'Malley BW, Katzenellenbogen BS (1996) Tripartite steroid hormone receptor pharmacology: interaction with multiple effector sites as a basis for the cell- and promotor-specific action of these hormones. Mol Endocrinol 10:119–131
Katzman R, Aronson M, Fuld P, Kawas C, Brown T, Morgenstern H, Frishman W, Gidez L, Eder H, Ooi WL (1989) Development of dementing illnesses in an 80-year-old volunteer cohort. Ann Neurol 25:317–324
Kawakami M, Terasawa E, Ibuki T (1970) Changes in multiple unit activity in the brain during the estrous cycle. Neuroendocrinology 6:30–48
Kawas C, Resnick S, Morrison A, Brookmeyer R, Corrada M, Zonderman A, Bacal C, Donnell Lingle D, Metter E (1997) A prospective study of estrogen replacement therapy and the risk of developing Alzheimer's disease: the Baltimore Longitudinal Study of Aging. Neurology 48:1517–1521
Kemper TL (1994) Neuroanatomical and neuropathological changes during aging and dementia. In: Albert ML, Knoefel JE (eds) Clinical Neurology of Aging 2nd ed. Oxford University Press, New York, pp 3–67
Kimura D (1995) Estrogen replacement therapy may protect against intellectual decline in postmenopausal women. Hormon Behav 29:312–321
Klaiber EL, Broverman DM, Vogel W, Kobayashi Y (1979) Estrogen therapy for severe persistent depressions in women. Arch Gen Psychiatry 36:550–554

Knapp MJ, Knopman DS, Solomon PR, Pendlebury WW, Davis CS, Garcon SI (1994) A 30-week randomized controlled trial of high-dose tacrine in patients with Alzheimer's disease. JAMA 271:985–991
Knight AH, Rhind EG (1975) Epilepsy and pregnancy: a study of 153 pregnancies in 59 patients. Epilepsia 16:99–110
Knusel B, Beck KD, Winslow JW, Rosenthal A, Burton LE, Widmer HR, Nikolics K, Hefti F (1992) Brain-derived neurotrophic factor administration protects basal forebrain cholinergic but not nigral dopaminergic neurons from degenerative changes after axotomy in the adult rat brain. J Neurosci 12:4391–4402
Koller WC, Barr A, Biary N (1982) Estrogen treatment of dyskinetic disorders. Neurology 32:547–549
Korn-Lubetzki I, Kahana E, Cooper G, Abramsky O (1984) Activity of multiple sclerosis during pregnancy and puerperium. Ann Neurol 16:229–231
Krug R, Stamm U, Pietrowsky R, Fehm HL, Born J (1994) Effects of menstrual cycle on creativity. Psychoneuroendocrinology 19:21–31
Kudrow L (1975) The relationship of headache frequency to hormone use in migraine. Headache 15:36–40
Kulkarni J, de Castella A, Smith D, Taffe J, Keks N, Copolov D (1996) A clinical trial of the effects of estrogen in acutely psychotic women. Schizophr Res 20:247–252
Kyomen HH, Nobel KW, Wei JY (1991) The use of estrogen to decrease aggressive physical behavior in elderly men with dementia. J Am Geriatr Soc 39:1110–1112
Laidlaw J (1956) Catamenial epilepsy. Lancet 2:1235–1237
Lance JW, Anthony M (1966) Some clinical aspects of migraine: a prospective study of 500 patients. Arch Neurol 15:356–361
Landgren S, Bäckström T, Kalistratov G (1978) The effect of progesterone on the spontaneous interictal spike evoked by the application of penicillin to the cat's cerebral cortex. J Neurol Sci 36:119–133
Lauritzen M, Olesen J (1984) Regional cerebral blood flow during migraine attacks by xenon-133 inhalation and emission tomography. Brain 107:447–461
Lerner A, Koss E, Debanne S, Rowland D, Smyth K, Friedland R (1997) Smoking and estrogen–replacement therapy as protective factors for Alzheimer's disease [letter]. Lancet 349: 403–404
Lewis BV, Parsons M (1966) Chorea gravidarum. Lancet 1:284–286
Lewis C, McEwen BS, Frankfurt M (1995) Estrogen–induction of dendritic spines in ventromedial hypothalamus and hippocampus: effects of neonatal aromatase blockade and adult GDX. Dev Brain Res 87:91–95
Lichten EM, Lichten JB, Whitty A, Pieper D (1996) The confirmation of a biochemical marker for women's hormonal migraine: the depo-estradiol challenge test. Headache 36:367–371
Lindamer LA, Lohr JB, Harris MJ, Jeste DV (1997) Gender, estrogen, and schizophrenia. Psychopharmacol Bull 33:221–228
Lindheim SR, Legro RS, Bernstein L, Stanczyk FZ, Vijod MA, Presser SC, Lobo RA (1992) Behavioral stress responses in premenopausal and postmenopausal women and the effects of estrogen. Am J Obstet Gynecol 167:1831–1836
Logothetis J, Harner R (1960) Electrocortical activation by estrogens. Arch Neurol 3:290–297
Logothetis J, Harner R, Morrell F, Torres T (1959) The role of estrogens in catamenial exacerbation of epilepsy. Neurology 9:352–360
Loranger AW (1984) Sex differences in age at onset of schizophrenia. Arch Gen Psychiatry 41:157–161
Loy R, Gerlach JL, McEwen BS (1988) Autoradiographic localization of estradiol-binding neurons in the rat hippocampal formation and entorhinal cortex. Dev Brain Res 39:245–251
Luine V (1985) Estradiol increases choline acetyltransferase activity in specific basal forebrain nuclei and projection areas of female rats. Exp Neurol 89:484–490

Luine V, Rodriguez M (1994) Effects of estradiol on radial arm maze performance of young and aged rats. Behav Neural Biol 62:230–236

Magos A, Zilkha KJ, Studd JWW (1983) Treatment of menstrual migraine by oestradiol implants. J Neurol Neurosurg Psychiatry 46:1044–1046

Mahley RW (1988) Apolipoprotein E: cholesterol transport protein with expanding role in cell biology. Science 240:622–630

Manolio TA, Furberg CD, Shemanski L, Psaty BM, O'Leary DH, Tracy RP, Bush TL (1993) Associations of postmenopausal estrogen use with cardiovascular disease and its risk factors in older women. Circulation 88:2163–2171

Martin R, McFarland HF (1995) Immunological aspects of experimental allergic encephalomyelitis and multiple sclerosis. Crit Rev Clin Lab Sci 32:121–182

Matsumoto A (1991) Synaptogenic action of sex steroids in developing and adult neuroendocrine brain. Psychoneuroendocrinology 16:25–40

Mattson RH, Cramer JA, Caldwell BV, Siconolfi BC (1984) Treatment of seizures with medroxyprogesterone acetate: preliminary report. Neurology 34:1255–1258

Mayeux R, Tang M-X, Marder K, Cote LJ, Jacobs DM, Stern Y (1997) Postmenopausal estrogen use and Parkinson's disease with and without dementia [abstract]. Neurology 48[Suppl 2]:A79

McDonald DR, Brunden KR, Landreth GE (1997) Amyloid fibrils activate tyrosine kinase-dependent signaling and superoxide production in microglia. J Neurosci 17:2284–2294

McEwen BS, Sapolsky RM (1995) Stress and cognitive function. Curr Opin Neurobiol 5:205–216

McGeer PL, McGeer EG (1995) The inflammatory response system of brain: implications for therapy of Alzheimer and other neurodegenerative diseases. Brain Res Rev 21:195–218

McNulty K, Au R, White RF, Myers R, Seshadri S, Knoeffel J, Beiser A, D'Agostino RB, Wolf PA (1997) Estrogen replacement therapy in association with dementia in the Framingham Study [abstract]. J Intern Neuropsychol Soc 3:35–36

Meldrum DR, Davidson BJ, Tataryn IV, Judd HL (1981) Changes in circulating steroids with aging in postmenopausal women. Obstet Gynecol 57:624–628

Miranda RC, Sohrabji F, Toran-Allerand CD (1993) Presumptive estrogen target neurons express mRNAs for both the neurotrophins and neurotrophin receptors: a basis for potential developmental interactions of estrogen with neurotrophins. Mol Cell Neurosci 4:510–525

Mölsä PK, Marttila RJ, Rinne UK (1982) Epidemiology of dementia in a Finnish population. Acta Neurol Scand 65:541–552

Montgomery JC, Appleby L, Brincat M, Versi E, Tapp A, Fenwick PBC, Studd JWW (1987) Effect of estrogen and testosterone implants on psychological disorders in the climacteric. Lancet 1:297–299

Mooradian AD (1993) Antioxidant properties of steroids. J Steroid Biochem Mol Biol 45:509–511

Mortel KF, Meyer JS (1995) Lack of postmenopausal estrogen replacement therapy and the risk of dementia. J Neuropsychiatr Clin Neurosci 7:334–337

Muesing RA, Miller VT, LaRosa JC, Stoy DB, Phillips EA (1992) Effects of unopposed conjugated equine estrogen on lipoprotein composition and apolipoprotein-E distribution. J Clin Endocrinol Metab 75:1250–1254

Naftolin F, Ryan KJ (1975) The metabolism of androgens in central neuroendocrine tissues. J Steroid Biochem 6:993–997

Nausieda PA, Koller WC, Weiner WJ, Klawans HL (1979) Chorea induced by oral contraceptives. Neurology 29:1605–1609

Nicoletti F, Speciale C, Sortino MA, Summa G, Caruso G, Patti F, Canonico PL (1985) Comparative effects of estradiol benzoate, the antiestrogen clomiphen citrate, and the progestin medroxyprogesterone acetate on kainic acid–induced seizures in male and female rats. Epilepsia 26:252–257

Niki E, Nakano M (1990) Estrogens as antioxidants. Method Enzymol 186:330–333

O'Neal MF, Means LW, Poole MC, Hamm RJ (1996) Estrogen affects performance of ovariectomized rats in a two-choice water-escape working memory task. Psychoneuroendocrinology 21:51–65

Ohkura T, Isse K, Akazawa K, Hamamoto M, Yaoi Y, Hagino N (1994a) Evaluation of estrogen treatment in female patients with dementia of the Alzheimer type. Endocrine J 41:361–371

Ohkura T, Isse K, Akazawa K, Hamamoto M, Yaoi Y, Hagino N (1994b) Low-dose estrogen replacement therapy for Alzheimer disease in women. Menopause 1:125–130

Ohkura T, Isse K, Akazawa K, Hamamoto M, Yaoi Y, Hagino N (1994c) An open trial of estrogen therapy for dementia of the Alzheimer type in women. In: Berg G, Hammar M (eds) The Modern Management of the Menopause: A Perspective for the 21st Century. Parthenon, Carnforth, England, pp 315–333

Ohkura T, Isse K, Akazawa K, Hamamoto M, Yaoi Y, Hagino N (1995a) Long-term estrogen replacement therapy in female patients with dementia of the Alzheimer type: 7 case reports. Dementia 6:99–107

Ohkura T, Teshima Y, Isse K, Matsuda H, Inoue T, Sakai Y, Iwasaki N, Yoshimasa Y (1995b) Estrogen increases cerebral and cerebellar blood flows in postmenopausal women. Menopause 2:13–18

Olesen J, Friberg L, Olsen TS, Iversen HK, Lassen NA, Andersen AR, Karle A (1990) Timing and topography of cerebral blood flow, aura, and headache during migraine attacks. Ann Neurol 28:791–798

Olson L, Nordberg A, von Holst H, Backman L, Ebendal T, Alafuzoff I, Amberla K, Hartvig P, Hertlitz A, Lilja A, Lundquist H, Langstron B, Meyerson B, Persson A, Vitanen M, Winblad B, Seiger A (1992) Nerve growth factor affects 11C-nicotine binding, blood flow, EEG and verbal episodic memory in an Alzheimer patient. J Neural Trans 4:79–95

Pacifici R, Brown C, Puscheck E, Friedrich E, Slatopolsky E, Maggio D, McCracken R, Avioli LV (1991) Effect of surgical menopause and estrogen replacement on cytokine release from human blood mononuclear cells. Proc Natl Acad Sci USA 88:5134–5138

Paganini-Hill A (1995) Estrogen replacement therapy and stroke. Prog Cardiovascular Dis 38: 223–242

Paganini-Hill A, Henderson VW (1994) Estrogen deficiency and risk of Alzheimer's disease in women. Am J Epidemiol 140:256–261

Paganini-Hill A, Henderson VW (1996) Estrogen replacement therapy and risk of Alzheimer's disease. Arch Intern Med 156:2213–2217

Palinkas LA, Barrett-Connor E (1992) Estrogen use and depressive symptoms in postmenopausal women. Obstet Gynecol 80:30–36

Payami H, Zareparsi S, Montee KR, Sexton GJ, Kaye JA, Bird TD, Yu CE, Wijsman EM, Heston LL, Litt M, Schellenberg GD (1996) Gender difference in apolipoprotein E-associated risk for familial Alzheimer disease: a possible clue to the higher incidence of Alzheimer disease in women. Am J Hum Genet 58:803–811

Pericak-Vance MA, Haines JL (1995) Genetic susceptibility to Alzheimer disease. Trends Genet 11:504–508

Pfaff D, Keiner M (1973) Atlas of estradiol-concentrating cells in the central nervous system of the female rat. J Comp Neurol 151:121–158

Phillips SM, Sherwin BB (1992a) Effects of estrogen on memory function in surgically menopausal women. Psychoneuroendocrinology 17:485–495

Phillips SM, Sherwin BB (1992b) Variations in memory function and sex steroid hormones across the menstrual cycle. Psychoneuroendocrinology 17:497–506

Poirier J (1994) Apolipoprotein E in animal models of CNS injury and in Alzheimer's disease. Trends Neurosci 17:525–530

Poirier J, Davignon J, Bouthillier D, Kogan S, Bertrand P, Gauthier S (1993) Apolipoprotein E polymorphism and Alzheimer's disease. Lancet 342:697–699

Poirier J, Hess M, May PC, Finch CE (1991) Astrocytic apolipoprotein E mRNA and GFAP mRNA in hippocampus after entorhinal cortex lesioning. Mol Brain Res 11:97–106

Puddu P, Puddu GM, Bastagli L, Massarelli G, Muscari A (1995) Coronary and cerebrovascular atherosclerosis: two aspects of the same disease or two different pathologies? Arch Gerontol Geriatr 20:15–22

Quinn NP, Marsden CD (1986) Menstrual-related fluctuations in Parkinson's disease. Movement Dis 1:85–87

Resnick SM, Metter EJ, Zonderman AB (1997) Estrogen replacement therapy and longitudinal decline in visual memory. A possible protective effect? Neurology 49:1491–1497

Riecher-Rössler A, Häfner H, Stumbaum M, Maurer K, Schmidt R (1994) Can estradiol modulate schizophrenic symptomatology? Schizophr Bull 20:203–214

Ripich DN, Petrill SA, Whitehouse PJ, Ziol EW (1995) Gender differences in language of AD patients: a longitudinal study. Neurology 45:299–302

Robinson D, Friedman L, Marcus R, Tinklenberg J, Yesavage J (1994) Estrogen replacement therapy and memory in older women. J Am Geriatr Soc 42:919–922

Rogers SL, Friedhoff LT, Group DS (1996) The efficacy and safety of donepezil in patients with Alzheimer's disease: results of a US multicentre, randomized, double-blind, placebo-controlled trial. Dementia 7:293–303

Rorsman B, Hagnell O, Lanke J (1986) Prevalence and incidence of senile and multi-infarct dementia in the Lundby study: a comparison between the time periods 1947–1957 and 1957–1972. Neuropsychobiology 15:122–129

Rosciszewska D, Buntner B, Guz I, Zawisza L (1986) Ovarian hormones, anticonvulsant drugs, and seizures during the menstrual cycle in women with epilepsy. J Neurol Neurosurg Psychiatry 49:47–51

Roy EJ, Buyer DR, Licari VA (1990) Estradiol in the striatum: effects on behavior and dopamine receptors but no evidence for membrane steroid receptors. Brain Res Bull 25:221–227

Sack MN, Rader DJ, Cannon ROI (1994) Estrogen and inhibition of oxidation of low-density lipoproteins in postmenopausal women. Lancet 343:269–270

Sadovnick AD, Eisen K, Hashimoto SA, Farquhar R, Yee IML, Hooge J, Kastrukoff L, Oger JJ-F, Paty DW (1994) Pregnancy and multiple sclerosis: a prospective study. Arch Neurol 51:1120–1124

Sagara Y, Dargusch R, Klier FG, Schubert D, Behl C (1996) Increased antioxidant enzyme activity in amyloid β protein-resistant cells. J Neurosci 16:497–505

Salehi A, Verhaagen J, Dijkhuizen PA, Swaab DF (1996) Co-localization of high-affinity neurotrophin receptors in nucleus basalis of Meynert neurons and their differential reduction in Alzheimer's disease. Neurosci 75:373–387

Santagati S, Melcangi RC, Celotti F, Martini L, Maggi A (1994) Estrogen receptor is expressed in different types of glial cells in culture. J Neurochem 63:2058–2064

Sar M, Stumpf WE (1981) Central noradrenergic neurones concentrate ^{3}H-oestradiol. Nature 289:500–502

Schmidt R, Fazekas F, Reinhart B, Kapeller P, Fazekas G, Offenbacher H, Eber B, Schumacher M, Freidl W (1996) Estrogen replacement therapy in older women: a neuropsychological and brain MRI study. J Am Geriatr Soc 44:1307–1313

Schneider EL, Guralnik JM (1990) The aging of America: impact on health care costs. JAMA 263:2335–2340

Schneider LS, Farlow MR, Henderson VW, Pogoda JM (1996) Effects of estrogen replacement therapy on response to tacrine in patients with Alzheimer's disease. Neurology 46:1580–1584

Schneider MA, Brotherton PL, Hailes J (1977) The effect of exogenous estrogens on depression in menopausal women. Med J Australia 2:162–163

Schoenberg BS, Kokmen E, Okazaki H (1987) Alzheimer's disease and other dementing illnesses in a defined United States population: incidence rates and clinical features. Ann Neurol 22:724–729

Seshadri S, Drachman DA, Lippa CF (1995) Apolipoprotein E $\varepsilon 4$ allele and the lifetime risk of Alzheimer's disease. Arch Neurol 52:1074–1079
Sherwin BB (1988a) Affective changes with estrogen and androgen replacement therapy in surgically menopausal women. J Affect Disord 14:177–187
Sherwin BB (1988b) Estrogen and/or androgen replacement therapy and cognitive functioning in surgically menopausal women. Psychoneuroendocrinology 13: 345–357
Sherwin BB, Gelfand MM (1985) Sex steroids and affect in the surgical menopause: a double-blind cross-over study. Psychoneuroendocrinology 10:325–335
Sherwin BB, Tulandi T (1996) "Add–back" estrogen reverses cognitive deficits induced by a gonadotropin–releasing hormone agonist in women with leiomyomata uteri. J Clin Endocrinol Metab 81:2545–2549
Shughrue PJ, Komm B, Merchenthaler I (1996) The distribution of estrogen receptor-β mRNA in the rat hypothalamus. Steroids 61:678–681
Singh M, Meyer EM, Millard WJ, Simpkins JW (1994) Ovarian steroid deprivation results in a reversible learning impairment and compromised cholinergic function in female Sprague–Dawley rats. Brain Res 644:305–312
Smith MA, Harris PLR, Sayre LM, Beckman JS, Perry G (1997) Widespread peroxynitrite-mediated damage in Alzheimer's disease. J Neurosci 17:2653–2657
Somerville BW (1972a) The role of estradiol withdrawal in the etiology of menstrual migraine. Neurology 22:355–365
Somerville BW (1972b) A study of migraine in pregnancy. Neurology 22:824–828
Stein GS (1981) Headaches in the first post partum week and their relationship to migraine. Headache 21:201–205
Stewart WF, Lipton RB, Celentano DD, Reed ML (1992) Prevalence of migraine headache in the United States. JAMA 267:64–69
Stone DJ, Rozovsky I, Morgan TE, Anderson CP, Hajian H, Finch CE (1997) Astrocytes and microglia respond to estrogen with increased apoE mRNA *in vivo* and *in vitro*. Exp Neurol 143:313–318
Strittmatter WJ, Saunders AM, Schmechel D, Pericak-Vance M, Enghild J, Salvesen GS, Roses AD (1993) Apolipoprotein E: high-avidity binding to β-amyloid and increased frequency of type 4 allele in late-onset familial Alzheimer disease. Proc Natl Acad Sci USA 90:1977–1981
Sullivan JM (1995) Coronary arteriography in estrogen–treated postmenopausal women. Prog Cardiovascular Dis 38:211–222
Sumner BEH, Fink G (1995) Estrogen increases the density of 5-hydroxytryptamine$_{2A}$ receptors in cerebral cortex and nucleus accumbens in the female rat. J Steroid Biochem Mol Biol 54:15–20
Szklo M, Cerhan J, Diez-Roux AV, Chambless L, Cooper L, Folsom AR, Fried LP, Knopman D, Nieto FJ (1996) Estrogen replacement therapy and cognitive functioning in the Atherosclerosis Risk in Communities (ARIC) study. Am J Epidemiol 144:1048–1057
Tang M-X, Jacobs D, Stern Y, Marder K, Schofield P, Gurland B, Andrews H, Mayeux R (1996) Effect of estrogen during menopause on risk and age at onset of Alzheimer's disease. Lancet 348:429–432
Terasawa E, Timiras PS (1968) Electrical activity during the estrous cycle of the rat: cyclic changes in limbic structures. Endocrinol 83:207–216
Toran-Allerand CD (1991) Organotypic culture of the developing cerebral cortex and hypothalamus: relevance to sexual differentiation. Psychoneuroendocrinology 16:7–24
Toran-Allerand CD, Miranda RC, Bentham WDL, Sohrabji F, Brown TJ, Hochberg RB, MacLusky NJ (1992) Estrogen receptors colocalize with low-affinity nerve growth factor receptors in cholinergic neurons of the basal forebrain. Proc Natl Acad Sci USA 89:4668–4672
Tuszynski MH, U HS, Yoshida K, Gage FH (1991) Recombinant human nerve growth factor infusions prevent cholinergic neuronal degeneraton in the adult primate brain. Ann Neurol 30:625–636

van Duijn C, Meijer H, Witteman JCM, Havekes LM, de Knijff P, Van Broeckhoven C, Hofman A (1996) Estrogen, apolipoprotein E and the risk of Alzheimer's disease [abstract]. Neurobiol Aging 16[suppl]:S79–S80
van Duijn CM, Hofman A (1992) Risk factors for Alzheimer's disease: the EURODEM collaborative re-analysis of case–control studies. Neuroepidemiol 11[suppl 1]: 106–113
Van Goozen SHM, Cohen-Kettenis PT, Gooren LJG, Frijda NH, Van de Poll NE (1995) Gender differences in behaviour: activating effects of cross-sex hormones. Psychoneuroendocrinology 20:343–363
Villeneuve A, Cazejust T, Côté M (1980) Estrogens in tardive dyskinesia in male psychiatric patients. Neuropsychobiology 6:145–151
Waring SC, Rocca WA, Petersen RC, Kokmen E (1997) Postmenopausal estrogen replacement therapy and Alzheimer's disease: a population-based study in Rochester, Minnesota [abstract]. Neurology 48[suppl 2]:A79
Warren SG, Humphreys AG, Juraska JM, Greenough WT (1995) LTP varies across the estrous cycle: enhanced synaptic plasticity in proestrus rats. Brain Res 703:26–30
Waters WE, O'Connor PJ (1971) Epidemiology of headache and migraine in women. J Neurol Neurosurg Psychiatry 34:148–153
Weiland NG (1992) Estradiol selectively regulates agonist binding sites on the N-methyl-D-aspartate receptor complex in the CA1 region of the hippocampus. Endocrinol 131:662–668
Weissman MM, Bland R, Joyce PR, Newman S, Wells JH, Wittchen H-U (1993) Sex differences in rates of depression: cross-national perspectives. J Affect Disord 29:77–84
Whitty CWM, Hockaday JM (1968) Migraine: a follow-up study of 92 patients. Br Med J 1:735–736
Whitty CWM, Hockaday JM, Whitty MM (1966) The effect of oral contraceptives on migraine. Lancet 1:856–859
Willson P, Preece AA (1932) Chorea gravidarum. Arch Intern Med 49:471–533, 671–697
Wong M, Moss RL (1992) Long-term and short-term electrophysiological effects of estrogen on the synaptic properties of hippocampal CA1 neurons. J Neurosci 12:3217–3225
Wong M, Thompson TL, Moss RL (1996) Nongenomic actions of estrogen in the brain: physiological significance and cellular mechanisms. Crit Rev Neurobiol 10:189–203
Wood RI, Newman SW (1995) Androgen and estrogen receptors coexist within individual neurons in the brain of the Syrian hamster. Neuroendocrinology 62:487–497
Woods RP, Iacoboni M, Mazziotta JC (1994) Bilateral spreading cerebral hypoperfusion during spontaneous migraine headache. N Engl J Med 331:1689–1692
Woolley CS, McEwen BS (1992) Estradiol mediates fluctuation in hippocampal synapse density during the estrous cycle in the adult rat. J Neurosci 12:2549–2554
Woolley CS, McEwen BS (1993) Roles of estradiol and progesterone in regulation of hippocampal dendritic spine density during the estrous cycle in the rat. J Comp Neurol 336:293–306
Woolley CS, Weiland NG, McEwen BS, Schwartzkroin PA (1997) Estradiol increases the sensitivity of hippocampal CA1 pyramidal cells to NMDA receptor–mediated synaptic input: correlation with dendritic spine density. J Neurosci 17:1848–1859
Woolley DE, Timiras PS (1962) The gonad–brain relationship: effects of female sex hormones on electroshock convulsions in the rat. Endocrinol 70:196–209
Yang NN, Venugopalan M, Hardikar S, Glasebrook A (1996) Identification of an estrogen response element activated by metabolites of 17β-estadiol and raloxifene. Science 273:1222–1225

CHAPTER 45
Estrogens and Antiestrogens in the Male

M. OETTEL

A. Introduction

Estrogens and men – this is a topic which has been neglected for quite some time and it is only relatively recently that discussion of it has intensified remarkably. The discussion has not been restricted to the scientific community. In former times, the role of estrogens in male physiology received little attention since androgens clearly have a dominant role. It is now becoming more and more evident that estrogens are far more important in the male than we had supposed (HABENICHT 1998).

But no area of hormone replacement in elderly males has been discussed so much and is so little known as the usefulness of estrogen therapy. Does a male estrogen deficit, or even an extensive full or partial aromatase defect (with normal testosterone levels and reduced estrogen serum concentrations), exist at all, or do males have lifelong resources (at least of brain estrogen) through the intracerebral conversion of testosterone to estrogen, thus avoiding late-life estrogen deficiency (BIRGE 1996)? Research has been extraordinarily stimulated by the availability of transgenic mice with disruption of the estrogen-receptor gene and by reports of estrogen-receptor deficiencies in males (DIXON et al. 1997). Certain conclusions of this research will be discussed in this chapter.

In fact, men have much higher 17β-estradiol levels than postmenopausal women. In the testicles and by peripheral aromatization of androgens, the human male produces considerable amounts of estrogens whose function has not yet been sufficiently clarified. The fact that the presence of estrogen receptors (ERs) have been demonstrated in many tissues suggests a physiological role of estrogens, even in males. Because the circulating levels of testosterone in the male are similar to the K_m of aromatase (20–30 nmol/l), it is likely that circulating testosterone can be converted efficiently in extragonadal sites to give rise to local concentrations of estradiol sufficient to transactivate both α- and β-ERs (K_D~1 nmol/l; SIMPSON and DAVIS 1998).

Recent clinical reports (SMITH et al. 1994; FOREST et al. 1996; SUDHIR et al. 1997) of disruptive mutations of the genes for ERs or for cytochrome P-450 aromatase have shed new light on the role of estrogen. A mutation in the estrogen-receptor gene in a 28-year-old man led to elevated estradiol and estrone serum concentrations, and serum testosterone concentrations (T) were normal. Serum follicle-stimulating hormone (FSH) and luteinizing hormone (LH) levels were increased. Glucose tolerance was impaired, and hyperinsulinemia was present. The bone mineral density of the lumbar spine (LS) was 0.745 g/cm^2, 3.1 SD below the mean for age-matched normal women; there was also biochemical evidence of increased bone turnover. The patient was tall (204 cm) and had incomplete epiphyseal closure, with a history of continued linear growth into adulthood despite otherwise normal pubertal development. The patient had no detectable response to estrogen administration despite a tenfold increase in the serum free-estradiol concentration. Contrary to previous view, disruption of the ER in humans need not be lethal (Smith et al. 1994).

Males with an aromatase defect have an absolute estrogen deficiency, with increased FSH and LH. The skeletal age appears to be retarded, with unclosed epiphyses. The clinical picture essentially corresponds to osteoporosis. Moreover, increased basal insulin levels and a reduced high-density to low-density lipoprotein (HDL–LDL) quotient are observed (Bulun 1996). In females, the lack of estrogen due to aromatase deficiency leads to pseudohermaphroditism and progressive virilization at puberty whereas, in males, pubertal development is normal. In members of both sexes, epiphyseal closure is delayed, resulting in an eunuchoid habitus, and osteopenia is present (Morishima et al. 1995). These findings suggest a crucial role of estrogen in skeletal maturation. Carani et al. (1997) describe a therapeutic response to estrogen therapy, but not to androgen therapy, in a man with aromatase deficiency. However, the questions of when to initiate the estrogen treatment, at what doses, and for how long remain to be investigated in further studies.

B. Age-Related Changes of Estrogen Secretion

Young men have estrogen levels between 10 pg/ml and 100 pg/ml (mean: 40 ± 15 pg/ml; Kuhl 1997). In another communication, the serum 17β-estradiol levels (E2) are 51.6 pg/ml on average [standard error (SE): 5.3, range: 20–90; Ambrosi et al. 1981]. They correspond to the levels found in women in the early follicular phase (Kuhl 1997) and are sufficient to develop a female phenotype (testicular feminization) in the case of an androgen-receptor defect, e.g., absence of androgenic actions. Elderly men have higher estradiol or estrone levels than postmenopausal women (Janssen et al. 1998).

Andersson et al. (1997), in a cross-sectional study, determined the E2 and T in 400 healthy Danish prepubertal, pubertal, and adolescent males aged 6–25 years. While the highest mean concentrations of about 22 ± 5 nmol/l for T were reached at age 19, for 17β-estradiol the highest mean concentrations (±SD) of about 100 ± 25 pmol/l were reached at about age 24 years. This means that the estradiol rise is flatter in young males, with peak values being reached about 6 years after the secretion peak of testosterone. In pubertal boys, the serum estrogen concentrations were established mainly through aromatization of testosterone (McDonald et al. 1979).

Comparing the ratio of T to E2 in younger and older men, we can at first assume that relatively and/or absolutely more 17β-estradiol is present in the elderly (PIRKE and DOERR 1973; BAKER et al. 1976; MURONO et al. 1982; DESLYPÈRE et al. 1987). Estradiol has been reported to increase or remain unchanged with age (KLEY et al. 1974; STEARNS et al. 1974; RUBENS et al. 1974; ZUMOFF et al. 1982; KAISER et al. 1988). In this context, fatty tissue is an important site of aromatization, and obesity is associated with increased estrone concentrations in aged men (KLEY et al. 1979; STANIK et al. 1981). Therefore, it is not surprising to find some data suggesting an increase in estrogenic concentration with age, as the percentage of body fat in normal individuals increases as an age-related phenomenon. Additionally, the metabolic clearance of estradiol decreases with age (BAKER et al. 1976).

In our own study (WINKELMANN 1998), measuring different hormone serum levels of 698 male patients aged 21–88 years with or with suspicion of coronary heart disease, we found the following results:

- A significant increase of FSH and LH (Fig. 1); that means a hypogonadotropic or – at least in some cases – an eugonadotropic state
- A significant decrease of total testosterone (Fig. 2). Of the voluteers, 59.6% are eugonadal, 15.5% hypogonadal (total T less than 10 nmol/l), and 24.9% of the males are in the borderline range. That means partial androgen deficiency of the aging male (PADAM)
- No significant changes of E2 and significant decrease of the T/E2 ratio (Fig. 3)

Of the patients studied, 52.9% showed physiological total T (>14 nmol/l) and E2 (>20 pg/ml); 37.0% were hypogonadal (T < 14 nmol/l) but normo- or hyperestrogenic (E2 > 20 pg/ml). Of the patients studied, 3.4% were simultaneously in a hypoandrogenic and a hypoestrogenic state. Interestingly enough, 6.7% of the men showed at least a partial aromatase defect. That means normal total T over 14 nmol/l but extremely low E2 (<20 pg/ml). This aromatase defect doesn't show any age-related changes (Fig. 4).

However, these data are not at all consistent with other reports of a decrease in E2 with age. In contrast to the above-mentioned studies indicating an age-related increase or constancy in E2 levels in men, more and more papers show an age-related decline in estrogen secretion. In an additional small pilot study with healthy men, we found 31 (27%) of a total of 116 men (aged 43–71 years) in an estrogen-deficient state (E2 < 30 pmol/l). Among the 47 hypogonadal men (T < 14 nmol/l), 32 (68%) showed normal or elevated E2 (hypoandrogenic and normo- or hyperestrogenic state) whereas 15 men (32%) showed E2 decreased below 30 pmol/l (hypoandrogenic and hypoestrogenic states). However, there were 16 (14%) eugonadal men (T > 14 nmol/l) with estrogen deficiency or with an aromatase defect. Despite the presence of enough substrate for the aromatase there was not enough estrogen in the blood (HEINEMANN 1998). This shows that T and E2 do not correlate, and partial aromatase defect in men should be given greater attention in the future.

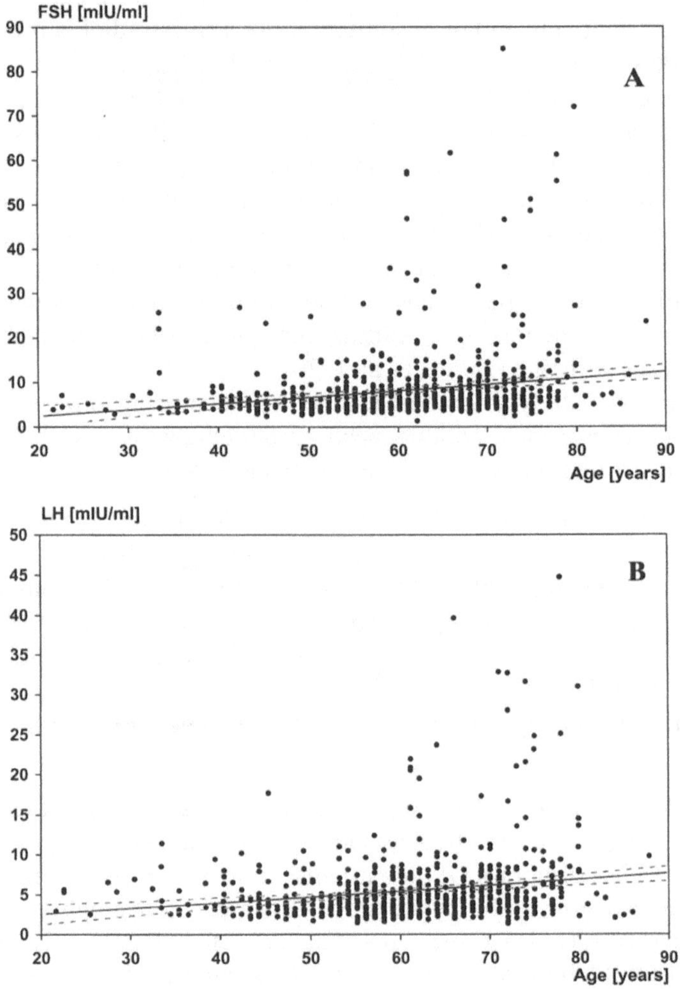

Fig. 1A,B. Serum follicle-stimulating hormone (*FSH*; **A**) and luteinizing hormone (*LH*; **B**) concentrations in 698 men aged 21–88 years. The age-related increase of mean gonadotropin concentrations is significantly ($P > 0.01$) (WINKELMANN 1998)

FERRINI and BARRETT-CONNOR (1998) found that testosterone and bioavailable estradiol levels in plasma samples obtained from 810 men aged 24–90 years (Rancho Bernardo Study) decreased significantly with age, independently of covariates. In contrast to the weaker association between age and total estradiol, there was a strong association of bioavailable estradiol with age. The age-associated decrease in bioavailable estradiol among these men may be partially explained by decreasing levels of testosterone, the primary substrate for male estradiol production, coupled with the higher levels of sex-hormone-binding globulin (SHBG) in older adults (FIELD et al. 1994).

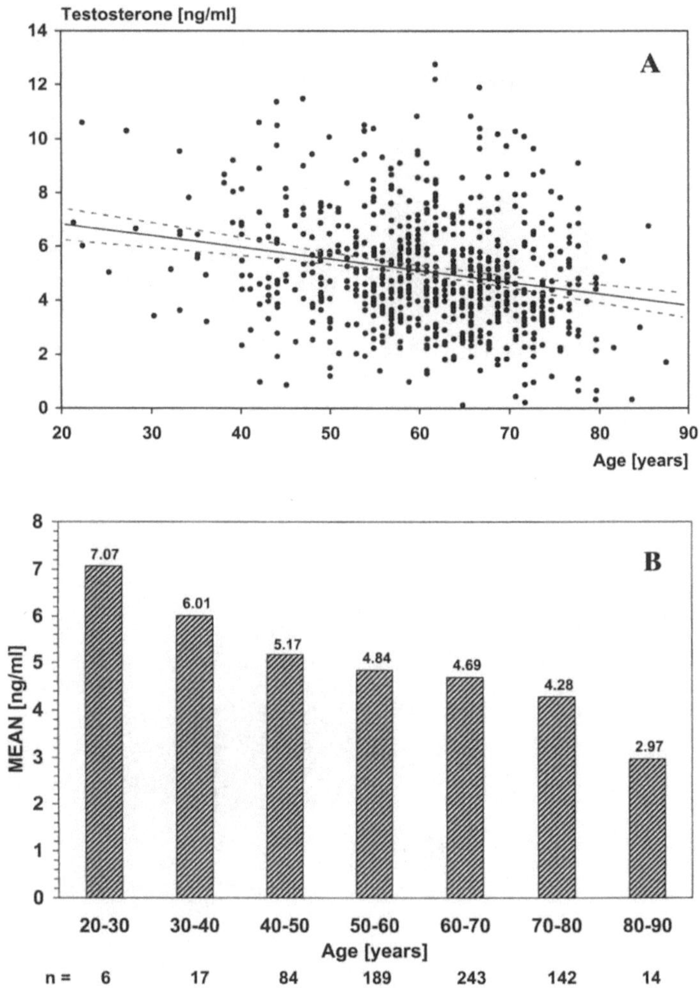

Fig. 2A,B. Total testosterone serum concentrations (**A**) in 698 men aged 21–88 years including the mean levels. The mean testosterone concentrations (**B**) are significantly age-related decreased ($P > 0.001$) (Winkelmann 1998)

In 466 male subjects, ranging in age from 2 years to 101 years, Baker et al. (1976) found that free testosterone levels, as well as estradiol levels, fell in old age. The metabolic clearance rates (MCRs) of testosterone and estradiol also fell in old age, while the conversion of testosterone to estradiol was increased. Simon et al. (1992) found that both total testosterone and estradiol levels showed a significant stepwise decrease with age ($P < 0.001$) starting in the early adult years, while estrone levels did not vary. This decline in testosterone and estradiol with age remained significant after adjustment for body-mass index, subscapular skinfold, and tobacco and alcohol consumptions, and they were

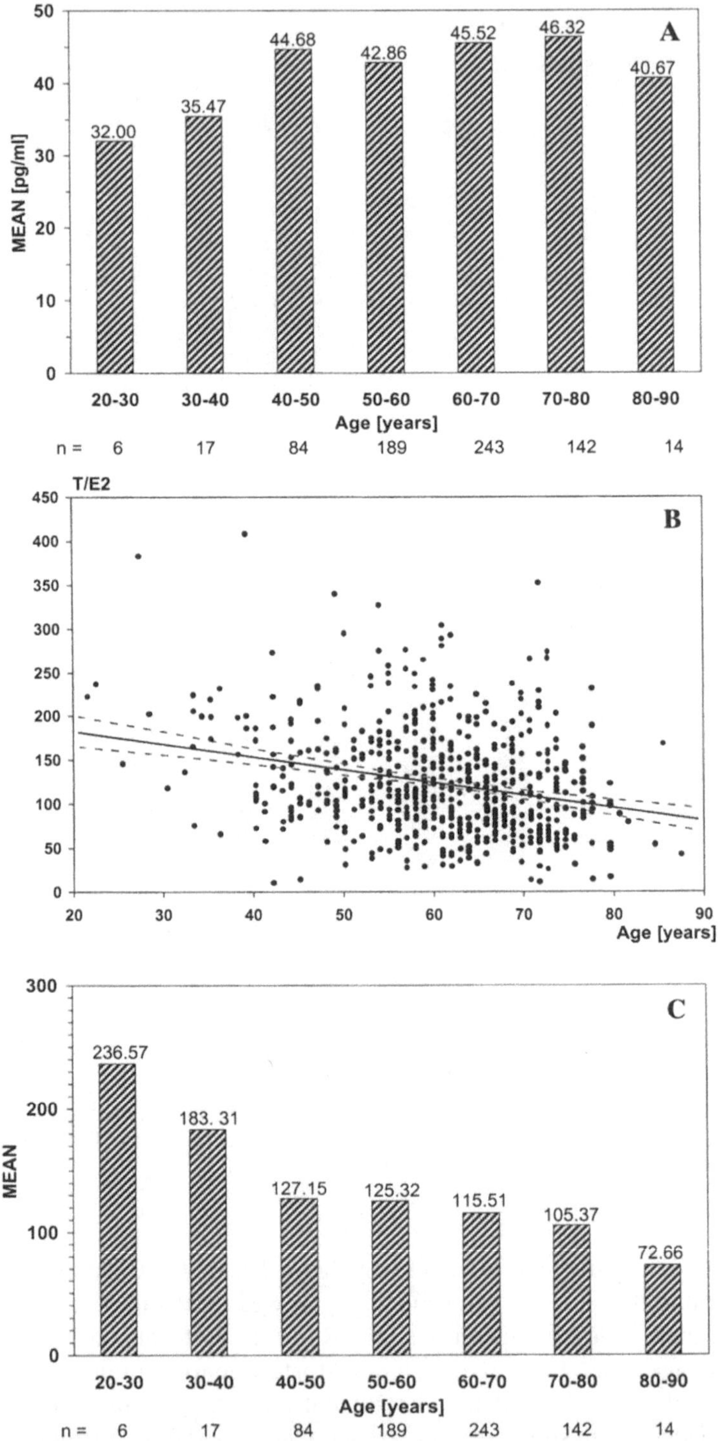

Fig. 3A–C. Mean serum-17β estradiol levels (**A**) in 698 men aged 21–88 years. There are no age-related changes. Additionally, the serum testosterone/estradiol-ratio (**B**) and the mean of the testosterone/estradiol-ratio (**C**) is shown. The mean testosterone/estradiol-ratio exhibits a significant age-related decrease ($P < 0.001$) (WINKELMANN 1998)

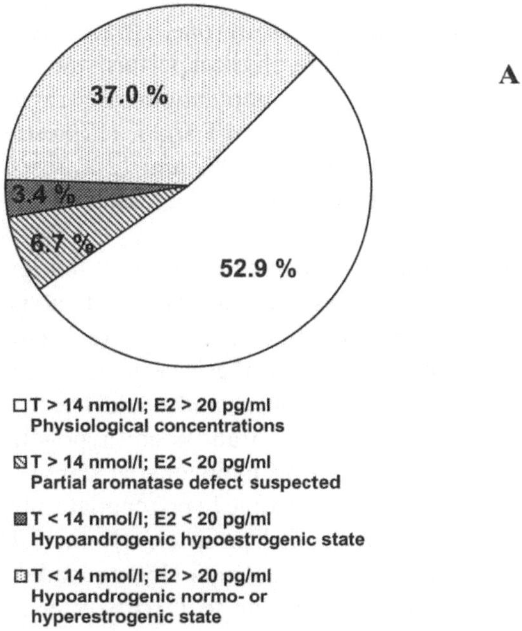

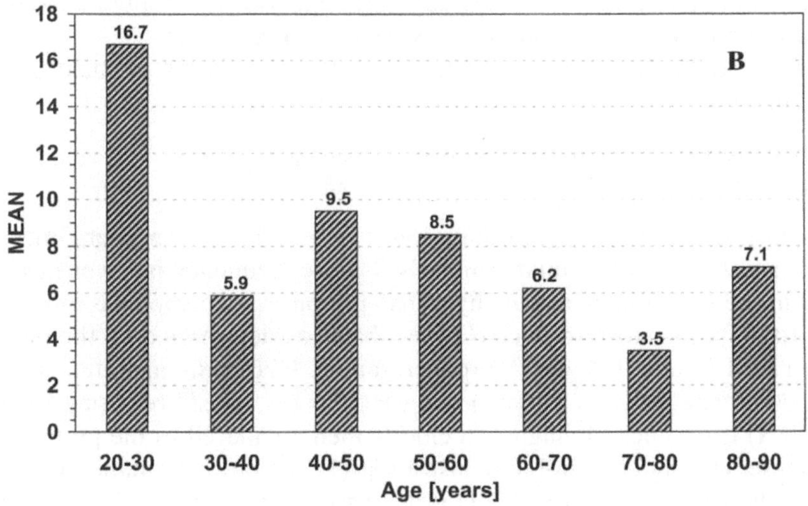

Fig. 4A,B. Testosterone and 17β-estradiol serum concentrations in 698 men aged 21–88 years (**A**). Note the assumed partial aromatase defect in 6.7% of the volunteers. There are no significant age-related changes in the portion of the assumed partial aromatase defect (**B**) (WINKELMANN 1998)

not modified by exclusion of the men who reported chronic disease. Also, VAN DEN BELD et al. (1998) found that estradiol significantly decreased with age in elderly men. In contrast to testosterone, estradiol seems to be an independent determinant of quality of life, physical performance, and proximal femur bone-mineral density (BMD). In a population-based, age-stratified sample of 346 men, aged 23–90 years, serum total T and E (17β-estradiol plus estrone) levels decreased over the life span by 30% and 12%, respectively, but bioavailable (non-SHBG bound) T and E levels decreased by 64% and 47%, respectively (KHOSLA et al. 1998).

As testosterone levels decrease and estradiol levels increase, the ratio of free testosterone to estradiol reaches a critical point and the estrogenic gonadotropin-suppressive effects predominate. The suppression of LH inhibits endogenous testosterone biosynthesis in the Leydig cell. This ratio may signal the biological point of no return and could become one of the criteria for defining the separation of the transitional hypogonadal state from the final "end stage" hypogonadotropic hypogonadal state (COHEN 1998).

However, a hypogonadotropic hypogonadal state represents the relatively infrequent form of hypogonadism in elderly men. The other form observed more often is a hypergonadotropic hypogonadal state (TENNEKOON and KARUNANAYAKE 1993). According to DAVIDSON et al. (1983), the age-related decrease in free testosterone is greater than that in total testosterone. Prolactin and 17β-estradiol did not change with age. There is a simultaneous age-related increase in the gonadotropins FSH and LH.

One could therefore speculate about the two forms of male hypogonadism:

- Hypergonadotropic hypogonadism: T $\downarrow$ and E2 $\rightarrow$ or $\downarrow$
- Hypogonadotropic hypogonadism: T $\downarrow$ and E2 $\uparrow$

LH-pulse frequency is significantly lower in elderly than in young men. The pulse amplitudes are similar. Similarly, T-pulse frequency is lower in elderly than in the young men and the hypothalamo-pituitary complex is more sensitive to dihydrotestosterone (DHT) feedback, as determined by the decrease in serum LH and T levels. Moreover, during DHT administration (125 mg/day, percutaneously, for 10 days), the LH response to LH-releasing hormone (LHRH) is significantly higher in elderly men compared to the pretreatment response. During estradiol administration (1.5 mg/day, percutaneously for 10 days), the LH response to LHRH is decreased in elderly men but unchanged in young men, suggesting greater responsiveness to estradiol in elderly men (DESLYPÈRE et al. 1987).

According to UMBREIT (1996), there is an estrogen deficit if the level is below 20 pg/ml, and estrogen deficit is said to be frequently accompanied by high androstenedione concentrations (>3.0 ng/ml). However, exact data in a low serum-concentration range should be interpreted with some caution, since most commercially available radio- or enzyme immunoassays do not work precisely in this concentration range. Recently, however, ultrasensitive bioassays for estradiol having sufficient sensitivity for monitoring estrogen-deficient men have been developed (KLEIN et al. 1998).

In contrast to testosterone, no diurnal rhythm was found for estradiol secretion in aged men (MURONO et al. 1982). According to JANSSEN et al. (1998), free, non-SHBG-bound estradiol is positively correlated with serum insulin-like growth factor-1 (IGF-1) levels in men. O'CONNOR et al. (1998) found that serum IGF-I levels declined with age in both men and women. In men, the decline was linear, whereas IGF-1 levels decreased faster in women less than 45 years of age than in older women or in men.

In males suffering from systemic lupus erythematosus, the estradiol levels were significantly lower than in controls (VILARINHO and COSTALLAT 1998). Chronic exercise training – in contrast to testosterone – does not cause a decline in E2 in men (HACKNEY et al. 1997).

C. Metabolism and Pharmacokinetics of the Estrogens in the Male

The main source of estrogens is the adipose tissue. In addition, however, there is a variety of other sources, such as the adrenal gland, brain, breast, hair, liver, and testis. The daily production rate of estrogens in men is comparable to that in postmenopausal women. Of the total estrogen production, the testes contribute 5% in the form of 17β-estradiol and 25% in the form of estrone production (HABENICHT 1998).

According to ZUMOFF et al. (1968), the urinary elimination of estrogen is faster in young women than in elderly women and men. However, the biliary excretion is larger in men than in women. Moreover, the conjugation of the metabolites differs between men and women.

The urinary concentrations of estrone, estradiol, and estriol in men range between 20% and 30% of those in cyclic women. However, the 2-hydroxy estrone or the 2-methoxy estrone metabolites in male urine amount to 40–90% of the quantities measured in women. These observations show that, in males, the metabolism of estrone on the ring A (in contrast to ring C) is preferred (FISHMAN 1980). Also, in blood plasma the percentage of the 2-hydroxy estrone fraction is clearly higher than that of the estrone, estradiol or estriol fractions. This is important, inasmuch as the 2-hydroxy or 2-methoxy metabolites are counted among the non-dangerous estrogens because these compounds fail to show up as mutagens in cell-culture studies or as carcinogens in animals (SERVICE 1998). So, in men, the "good" metabolic pathway of the estrogens is preferred. Hydroxyestrone shows antiproliferative activities in vitro as well as in vivo (BRADLOW et al. 1996). In every experimental model in which 2-hydroxylation was increased, protection against tumors was achieved. Correspondingly, when 2-hydroxylation was decreased, an increase in cancer risk was observed. 2-Hydroxyestradiol in men is preferably metabolized to 2-methoxyestradiol (BANGER et al. 1996). 2-Methoxyestradiol is the only metabolite of estrogen devoid of uterotropic, vaginotropic, or tumorigenic activity in vivo and is now emerging as a potential therapeutic agent for the treatment of angiogenically based diseases. Recent studies have shown 2-methoxyestradiol to be a potent nonspecific antimitotic agent in vitro and an effective oral anti-angiogenic and antitumor agent in vivo. It may be the first endogenous chemotherapeutic compound that is a physiological metabolite in humans (PRIBLUDA and GREEN 1998). In addition, it is a particularly interesting fact that the phytochemical indole-3-carbinol, a constituent of cabbage-family vegetables, strongly increases the "good" estrogen 2-hydroxylation (TELANG et al. 1997; MICHNOVICZ 1998). WONG et al. (1997), in a dose-ranging study of 60 women at increased risk of breast cancer, demonstrated a significantly increased urinary 2-hydroxyestrone-to-16α-hydroxyestrone ratio after giving subjects 300 mg indole-3-carbinol per day.

In men, 2-hydroxy estrone suppresses gonadotropins when given in doses high enough to compensate for their rapid clearance and degradation (MERRIAM et al. 1983). The metabolism of 2-hydroxyestradiol in men is dominated by methyl ether formation, that of 4-hydroxy estrone by direct conjugation (EMONS et al. 1987). However, it should be taken into account that under physiological conditions the plasma value of 2-hydroxy estrone (approximately 15 pg/ml) is below the detection limit for most of the assays (KONO et al. 1980, 1983).

ZIMMERMANN et al. (1994) reported on gender-specific differences in the pharmacokinetics of ethinyl estradiol (EE). In a study using an oral combination of EE (0.06 mg/day) and levonorgestrel (0.250 mg/day) in male and female volunteers, the area-under-the-curve values for EE were 2.850 pg/ml/h in males and 6.216 pg/ml/h in females. The C_{max} values were 251 pg/ml in males and 495 pg/ml in females. The elimination half-lives were 10.2 h in the males volunteers and 16.5 h in the women. These results show, obviously, that the EE absorption in males is lower or the first-pass effect greater in males than in females. Additionally, the excretion of the estrogen is faster in males. Is this phenomenon part of a "protection mechanism" of the male organism e.g. against environmental estrogens?

D. Genitourinary System

I. What Can We Learn from the ERα Knock-Out (ERKO) Mice?

Both sexes produce estrogens, but at present relatively little is known about the physiological role of the "female" hormone in males. HESS et al. (1997) showed that estrogen-receptor activation is necessary for normal male fertility, and indicated that the effects of estradiol are important for reproduction in males (BERG 1998). In a case report of a man with dysfunctional ERs, the sperm count was normal but the viability of the sperm was low (SMITH et al. 1994). A low level of estrogen caused by aromatase deficiency was seen in a

24-year-old man with increased testicular volume (Morishima et al. 1995), whereas a 38-year-old man with the same deficiency had small testicles and severe oligozoospermia (Carani et al. 1997).

The first ER was cloned and sequenced in 1986. It was denoted ERα (Green et al. 1986) after the identification of an additional receptor (ERβ) ten years later (Kuiper et al. 1996). In 1993, a knock-out mouse model lacking a functional ERα was created (ERKO mice; Lubahn et al. 1993). The phenotype of male ERKO mice initially appeared normal and excluded a role for ERα in the fetal development of genitalia. Adult male mice became infertile, with atrophy of the testes and dysmorphic seminiferous tubules. Sperm from the mice were abnormal, and the sperm concentration in the epididymis was low (Eddy et al. 1996). Testicles of male ERKO mice developed normally until puberty, after which they started to degenerate. A transient increase in weight occurred between 32 and 81 days of age, when the rete testis became dilated and the efferent ductules were swollen, with increased luminal area. This was caused by an increased secretion of fluid by the testis or a defective removal of the fluid secreted. After 185 days of age, the observed atrophy and decreased weight of the testes appeared to result from long-standing increased luminal-fluid pressure (Berg 1998).

The efferent ductules conduct sperm from the testis to the epididymis (for review see Ilio and Hess 1994). They are the first site of epithelial ER expression in the developing male reproductive organ in mice, and both the male rat and marmoset monkey show a pronounced immunoexpression of ERα in the efferent ductules (Cooke et al. 1991; Fisher et al. 1997). The concentration of estrogens in fluid from the testes is high and comparable to the E2 level in females of reproductive age. Interestingly, spermatids express aromatase and may convert androgens to estrogens. The number of spermatozoa in transit to the epididymis may determine the estrogenic stimulation of the cells lining the efferent ductules. These cells have a well-developed apparatus for the reabsorption of fluid, and more than 90% of the fluid continuously secreted by the seminiferous epithelium of the testes is reabsorbed in the efferent ductules. By occluding the ductal system at different levels, Hess et al. (1997) demonstrated that the apparent increase in luminal pressure in mice lacking ERα was caused by a defect in the reabsorption of testicular fluid secretion. Experiments with cultured pieces of epididymis corroborated these findings. Thus, ERα seems to be important for fluid reabsorption and normal adult function of the efferent ductules. However, wild-type efferent ductules treated with the "pure" antiestrogen ICI 182.780 in vitro did not swell like the ductules from the ERKO mice. The presence and effects of ERβ in efferent ductules and the epididymis may explain this discrepancy (Hess et al. 1997).

II. Gonads

It is well known that androgens as well as estrogens are secreted from the testis, and the Sertoli cells contain ERs (Panno et al. 1996). Estrogens play a pivotal role in regulating spermatogenesis, e.g. by potentiating the stimulatory effect of FSH (MACCALMAN et al. 1997).

As shown above, in the estrogen receptor α gene-knock out (ERKαO) mice, estrogens are necessary for mating frequency, sperm number, and several sperm functions (EDDY et al. 1996). While the necessity for ERα-mediated function in male fertility has meanwhile been well supported by the ERαKO experiments, there are still some open questions about the physiological role of ERβ. This ER subtype is expressed in the Leydig cells, elongated spermatids, efferent ductules, and the initial segment of the epididymides of ERαKO mice, but the presence of ERβ is not able to compensate for the absence of ERα in male reproductive function (ROSENFELD et al. 1998).

For a number of years, estrogens have been discussed as one of the factors regulating testicular function locally. As in the 1970s, DORRINGTON and ARMSTRONG (1975), PAYNE et al. (1976, 1987) and ROMMERTS et al. (1982) demonstrated aromatase activity to be present in the testis. In the context of this investigation, an age-dependent shift in the localization of this enzyme system was shown to exist. Originally, aromatase activity was demonstrated in the Sertoli cells of immature animals. Later on, it became evident that estrogens are produced in the Sertoli cells of immature animals only whereas, in the adult ones, Leydig cells are the major source (KMICIKIEWICZ et al. 1997). Most of the studies were performed in rats, but there is now good evidence that the same is true for human beings (ROMMERTS et al. 1982; INSKER et al. 1995). This has not been fully clarified yet as far as the exact function of estrogens in the male gonads is concerned, but there are a number of suggestions that estrogens has a mitogenic action on Sertoli cells and Leydig cells (PAYNE et al. 1987). Furthermore, an inhibitory effect on 17α-hydroxylase/c17–20-lyase has been demonstrated, indicating a local paracrine/autocrine regulatory effect on testosterone production. Recently, LIO et al. (1997) reported on a direct stimulatory effect of estrogens on rat gonocyte proliferation mediated via the plateled-derived growth factor (PDGF) pathway, indicating an essential role of estrogens in supplying the adult testis with a sufficient number of germ cells. Furthermore, data are available indicating a direct inhibitory effect of pachytene spermatocytes and early spermatides on Sertoli cell estradiol production (DUPAIX et al. 1996), a finding speaking clearly in favor of a possibly important role of estrogens in the context of a cross-talk between germ cells and Sertoli cells.

However, estradiol treatment can increase germ cell apoptosis mainly at stages IV-X of the spermatogenic cycle, rather than at stage VII when apoptotic germ cell death is mainly triggered by gonadotropin withdrawal e.g. caused by hypophysectomy. Therefore, estradiol plays a specific local role in the modulation of germ cell death in the adult testis (BLANCO-RODRIGUEZ and MARTINEZ-GARCIA 1997). SAH (1998) treated 14 men with oligospermia with daily 0.044 mg EE plus 3.6 mg methyl testosterone orally over 4 months. In 9 of the 14 men (64.2%), treatment was effective; there was some improvement in the remaining 5 men. The wives of three patients became pregnant within

6 months of therapy initiation. None of the patients had any side effects. The mean post-treatment semen index was 18.6–8.4× larger than the pre-treatment level.

Spermatozoa also express ERs (DURKEE et al. 1998). Interestingly, sperm themselves are sources of estrogen. HESS et al. (1995) and JANULIS et al. (1996) demonstrated that developing spermatids in several species contain aromatase. This observation is the basis for the hypothesis that estrogen, synthesized by sperm, plays a role in the regulation of epididymal function proportional to the number of sperm being transported. Human epididymides of fertile men aged 14 years to 64 years express ERs whereas, in the epididymis in boys aged from neonatal to 12 years and in men over 65 years, ER was not detected (MISAO et al. 1997). Furthermore, the sperm penetration into oocytes is enhanced by estradiol (CHIAN et al. 1996).

III. Epididymis

Another target where estrogens might play a role is the epididymis, as this tissue is particularly rich in ERs. Furthermore, there is some evidence of a relationship between estrogens and sperm function (see above). There are new signs indicating that estrogens have a function in the reproductive system as well as in non-reproductive organs. As shown above, these new signs originate mainly from results obtained with the ERαKO mice (LUBAHN et al. 1993; EDDY et al. 1996) and first descriptions of a naturally appearing defect either in the aromatase gene and/or the ER in men (MORISHIMA et al. 1995; SMITH et al. 1994). Unexpectedly, male ERKO mice were infertile (EDDY et al. 1996). Altough spermatogenesis was mainly normal, epididymal sperm were not able to fertilize an egg. These findings further support the data of HABENICHT (1998) on the role of estrogens for male physiology. Treatment of rats with antiestrogens resulted in an inhibition of motility as well as alterations of acrosomal status. SMITH et al. (1994) described a man with mutation on the aromatase gene and another man with mutation on the ER gene. In these cases, a disturbance of sperm motility was evident.

IV. Prostate

ERα is found almost exclusively in stromal components, with isolated reports indicating basal epithelial cell localization as well (EHARA et al. 1995; PRINS and BIRCH 1997). However, it is widely believed that estrogen exerts its influence on the prostate mainly via the epithelium (FARNSWORTH 1996). From this point of view, it is remarkable that ERα mRNA was not detected in any epithelial samples in dysplastic epithelial tissues of Noble rats. In contrast to this, both testosterone and estradiol (or both combined) can induce large amounts of ERβ in the epithelium (KUIPER et al. 1997; LAU et al. 1998). On the basis of this finding one might speculate that estrogens or selective estrogen modulators (SERMs) with preferred binding to ERβ have stronger dysplastic effects on the prostate than ligands showing stronger binding to ERα. Also, Prins et al. (1998) found that ERβ-mRNA was localized to rat prostatic epithelial cells, which contrasts with the normal stromal localization of ERα in the rat prostate.

Estrogens are known to be involved in the etiology of benign prostatic hyperplasia (BPH), probably because of their ability to increase androgen receptors (ARs; Moore et al. 1979). We know that estrogens can be produced in the prostate by aromatization of testosterone. Aromatase is present in the prostatic stroma, and aromatase activity has been reported to increase with age (Hemsell 1974).

What is new is that estradiol (but not diethylstilbestrol) can act as a natural ligand for the AR in DU145 human prostate cancer cells. While all three known AR coactivators – ARA_{70}, steroid receptor coactivator 1, and RAC3/ACTR – can enhance AR transcriptional activity at 1 nM dihydrotestosterone, Yeh et al. (1998) demonstrated that ARA_{70} can induce AR transcriptional activity less than 30-fold only in the presence of 17β-estradiol.

It is known that unliganded SHBG binds to a receptor (R_{SHBG}) on prostate membranes. The R_{SHBG}–SHBG complex is rapidly activated by estradiol to stimulate adenylate cyclase, with a resultant increase in intracellular cyclic adenosine monophosphate (cAMP). Nakhla et al. (1997) found that dihydrotestosterone (DHT) caused an increase in prostate-specific-antigen (PSA) secretion in serum-free organ cultures of human prostates. This event was blocked by the anti-androgens cyproterone acetate and hydroxyflutamide. In the absence of androgens, estradiol added to prostate tissue whose R_{SHBG} was occupied by SHBG reproduced the results seen with DHT. The estradiol-SHBG-induced increase in PSA was not blocked by anti-estrogens, but was blocked both by anti-androgens and 2-methoxyestradiol, which prevents the binding of estradiol to SHBG. Furthermore, an inhibitor of protein kinase A (PKCA) prevented the estradiol-SHBG-induced increase in PSA, but not that which followed DHT. The conclusion of the authors is that there is a signaling system that amalgamates steroid-initiated intracellular events with steroid-dependent occurrences generated at the prostate cell membrane and that the latter signaling system proceeds by a pathway that involves PKCA.

Concerning the metabolism of 17β-estradiol, Lane et al. (1997) found that the formation of catechol estrogens is much lower in the rat prostate than in the liver. Thus, these estrogen metabolites are unlikely to be involved in the hormonally induced prostatic dysplasia.

E. Mammary Gland

The mammogenic actions of estrogen consist of indirect actions via pituitary intermediaries such as prolactin (PRL) and growth hormone (GH) and direct effects on the mammary gland. The two direct estrogenic, mitogenic pathways concern ductal elongation (ductal growth) and ductal maintenance (ductal branching) (SILBERSTEIN et al. 1994).

The connection between estrogen replacement and the development of breast cancer is the subject of controversial discussions. Existing evidence supports a causal relationship between use of estrogens and progestins, levels of endogenous estrogens, and breast cancer incidence in postmenopausal women. Hormones may act to promote the late stages of carcinogenesis among postmenopausal women and to facilitate the proliferation of malignant cells (COLDITZ 1998), but what about the situation in men?

In the male ERαKO mouse the mammary glands were undeveloped, with only vestigial ducts present at the nipples, indicating a causal role of estrogens for the development of the male breast (KORACH 1994). Gynecomastia is common in adolescent and adult men, and reflects in some – but not in all cases – an underlying imbalance in hormonal physiology in which there is an increase in estrogen action relative to androgen action at the breast-tissue level (WILSON et al. 1980; KORENMAN 1985; MATHUR and BRAUNSTEIN 1997). According to BAUDUCEAU et al. (1993), gynecomastia is present in almost 40% of young men and is linked with reduced T and hypogonadism. This suggests an increased aromatase activity. However, SASANO et al. (1996) found relatively strong aromatase immunoreactivity in only 11 of 30 cases (37%) of gynecomastia; in contrast to this, they found such immunoreactivity in all 15 cases (100%) of mammary carcinoma. ER and progesterone-receptor (PR) expression was observed in the nuclei of ductal cells in all cases of gynecomastia.

Carcinoma of the male breast is rare and represents only 1% of the female mammary cancer rate (RIBEIRO 1985). There is strong circumstantial evidence to implicate hormonal factors in the development of male breast carcinoma (ROGERS et al.1993). However, some investigators assume that there has been no consistent evidence of increased serum estrogen or progesterone concentrations in men with breast malignancy. However, ERs and PRs are found even more commonly in breast carcinomas of men than in breast carcinomas of women (ROGERS et al. 1993; PICH et al. 1994). Altered androgen metabolism has also been proposed to be involved in the development of male breast carcinoma (LABACCARO et al. 1993). According to THOMAS (1993), it is now suggested that men with breast carcinoma may have endocrine profiles different from those of normal male subjects. Increased aromatase expression in the malignant stromal cells is considered to contribute to the increment in the in situ estrogen concentration and the development of male breast carcinoma (SASANO et al. 1996). Male breast carcinoma patients showed a positive correlation between ERs and PRs; however, there was a lack of correlation between heat-shock protein 27 and ERs or PRs. This lack of correlation differs from the results known for female breast carcinoma, suggesting that male breast carcinoma and female breast carcinoma are biologically different tumors (MUNOZ DE TORO and LUQUE 1997).

There is some evidence that the cardiac glycoside digoxin may share intriguing similarities to SERMs. This digitalis glycoside shares a structural homology with steroid hormones, and digoxin-induced gynecomastia in men is well known (SCHUSSHEIM and SCHUSSHEIM 1998).

F. Liver

I. Estrogens

Relatively little is known about gender-specific estrogen effects on the liver. In male rats, cholestasis can be induced by EE or estradiol-17β-glucuronide (ALVARO et al. 1997;TAKIKAWA 1997). Gallstone disease is known in men during estrogen treatment of prostatic carcinoma (ANGELIN et al. 1992). Hepatic calcium-binding protein regucalcin mRNA was expressed in male rats more than in females (UEOKA and YAMAGUCHI 1998), and there are profound gender-dependent differences in the regulation of sulfotransferase mRNA in female and male rats (KLAASSEN et al. 1998).

In men, LDL-receptor expression under estrogen treatment is 3× higher than in untreated controls, and the 3-hydroxy-3-methylglutaryl-coenzyme A reductase was increased two-fold (ANGELIN et al. 1992). Oral estrogen treatment induces a decrease in expression of sialyl Lewis X and α_1-acid glycoprotein in females and male-to-female transsexuals (BRINKMAN-VAN DER LINDEN et al. 1996).

There is a great deal of controversy regarding hormonal changes in hepatic cirrhosis patients. GLUUD et al. (1983, 1987), on behalf of the Copenhagen Study Group for Liver Diseases, claim that T levels in cirrhotic patients could be low, normal, or subnormal, and that estrone and E2 levels should be high. However, in most of the reports in the literature it is maintained that T levels are lower in cirrhotic patients, while estrogens are normal or a little higher in severe cases (VAN THIEL et al. 1981; WANG et al. 1991). Still other reports declare that estrogen levels are found to be lower in cirrhotic patients (WANG et al. 1993). It has been reported by GORDON et al. (1975) that hepatic extraction of testosterone decreases in cirrhosis, because testosterone circulates towards extrasplenic tissues (mainly in fatty tissues and muscles), peripheral aromatization increases, and finally the level of estrogen increases. VAN THIEL et al. (1980, 1983) have shown in rats that after diverting portal blood flow, T decreases, testicular atrophy develops, estradiol formation increases, and the hypothalamo-pituitary-gonadal axis is disrupted. It is generally accepted that the primary mechanism causing hormonal changes in cirrhosis is blocked entero-hepatic circulation and portal hypertension, disrupting the hypothalamic-pituitary-gonadal axis (VAN THIEL et al. 1974; GAVALER and VAN THIEL 1988). Nevertheless, a significant increase in the level of SHBG and a decrease in the level of free T because of the high affinity of SHBG have been reported in cirrhotic patients. CETINKAYA et al. (1998) investigated 60 patients with postnecrotic cirrhosis and alcoholic cirrhosis at age 40 and over, and 20 voluntary subjects in the same age group with normal hepatic functions. Although the mean E2 was a little higher in cirrhotic patients compared to the controls, the difference was not significant. However, the higher ratio of E2 to free T in the cirrhotic group was significant.

GANNE-CARRIÈ et al. (1997) summarize that for liver carcinoma, while high SHBG levels have an independent predictive value for the occurrence of hepatocellular carcinoma, the serum estrogen levels were without significance in this respect.

II. Antiestrogens

The effects of antiestrogens on the liver are particularly worthy of notice. Covalent binding of [C^{14}] tamoxifen (45 μM) in rats was 3.8-fold higher, and 17-fold higher in mice, than in human liver microsomal preparations (WHITE et al. 1995). Treatment of rats with tamoxifen increases the mRNA levels of many phase-I and phase-II drug-metabolizing enzymes, and this pleiotropic response to the antiestrogen has implications for the carcinogenicity and/or therapeutic activity of the drug (HELLRIEGEL et al. 1996; NUWAYSIR et al. 1996). While both antiestrogens, tamoxifen and toremifene, are effective promoting agents for diethylnitrosamine-initiated lesions, tamoxifen is more potent than toremifene in the induction of rat hepatocarcinogenesis (DRAGAN et al. 1994, 1995; KIM et al. 1996). Tamoxifen is clearly DNA-reactive and carcinogenic in rat liver. Since most human carcinogens are DNA-reactive carcinogens in rodents, such findings with the antiestrogen raise grave concerns. Tamoxifen administration to rats results in a shift toward growth of diploid hepatocytes, thus contributing to its carcinogenic action in the rat liver (DRAGAN et al. 1998).

The observations that tamoxifen is also DNA-reactive in hamster liver and is activated by human liver microsomes to form DNA adducts provide evidence that its toxicity is not confined to the rat. Accordingly, tamoxifen can be presumed as a potential human cancer hazard unless some mechanistic basis for non-susceptibility of humans is discovered. But it is important in this regard to recognize that the target site for human carcinogens is not always the same as in rodents. The presumption of hazard is reinforced by the finding of p53 mutations in the tamoxifen-induced liver tumors. The p53 gene is the most commonly mutated gene in human neoplasms (WILLIAMS 1995). Interestingly enough, tamoxifen is also useful for the treatment of advanced hepatocellular carcinoma in men, namely in combination with recombinant interferon-α2a or etoposide (VP-16) (CHENG et al. 1996; WERNER et al. 1996).

G. Bone

More than half of all women and about one-third of men will develop fractures related to osteoporosis (ROSS 1998). The consequences of fractures are often severe, and can lead to persistent declines in quality of life and increased mortality rates.

The male risk of osteoporosis is often underestimated. However, men will be 5 years older on average at the time of reaching the critical bone density. Therefore, men and women of the same age and the same bone density are at the same fracture risk (DELAET et al. 1997). Bone loss accelerates with age, as seen more clearly in men than in women (BURGER et al. 1998). Compared with women, elderly men presenting with hip fracture have a higher mortality and have more risk factors for osteoporosis (DIAMOND et al. 1997). Adjusting for age, body-mass index (BMI), and BMD at the trochanter in grams per square centimeter, men had a two-fold higher risk of deformity than women (LUNT et al. 1997). The bone density of the ERαKO male mice was 20 to 25% lower than in wild-type mice, suggesting a direct role for estrogen in male bone physiology (KORACH 1994).

Concerning the race- and gender-specific incidence rate of hip fractures in the United States, JACOBSEN et al. (1990) found a higher risk for white men than for black women. Since black women have a high aromatase activity, this study provides the first epidemiological indication of the importance of estrogens for the etiopathogenesis of male osteoporosis.

Reports of severe osteopenia in several men with estrogen deficieny (ER abnormality or absent aromatase activity) have raised questions concerning the role of estrogens in maintaining bone mass (SMITH et al. 1994; MORISHIMA et al. 1995). Aromatase activity is present in bone (SASANO et al. 1997), and probably the local production of estrogen may not be mirrored by serum estrogen concentrations (ORWOLL 1998). In transsexuals, estrogens can maintain bone mass in the presence of low concentrations of androgens after orchiectomy (VAN KESTEREN et al. 1996), and some reports suggest that estrogen levels are closely associated with bone mass in older men (SLEMENDA et al. 1987; GREENDALE et al. 1997). Although studies in experimental animals and in osteoblastic cells in vitro indicate that nonaromatizable androgens are potent modulators of skeletal homeostasis (WAKLEY et al. 1991; GOULDING and GOLD 1993; MASON and MORRIS 1997; WIREN et al. 1997), the important roles of estrogens – including aromatization of androgens to estrogens – is beyond question. For example, 17β-estradiol reduces IL-1β-induced IL-6 mRNA production and IL-6 secretion by human osteoblast-like cells (KOKA et al. 1998).

Summarizing the preclinical and clinical arguments favoring a role of estrogens on skeletal homeostasis in the male, VANDERSCHUEREN (1996) stressed the following points:

- Both androgens (BELLIDO et al. 1995) and estrogens (GIRASOLE et al. 1992) are able to prevent osteoclastogenesis in vitro via interleukin-6 after stimulation of their specific receptors in bone marrow cells
- Male-derived osteoblasts not only possess ERs (COLVARD et al. 1989) but are also able to aromatize androgens into estrogens (BRUCH et al. 1992)
- Both androgens and estrogens prevent bone loss in the aged male rat by a similar mechanism (FRANCIS et al. 1986)

- Androgen deficiency in the aged male rat (VANDERSCHUEREN et al. 1992) has skeletal effects very similar to estrogen deficiency in the aged female rat (KALU et al. 1984).
- Both ERKO mice (LUBAHN et al. 1993) and ER deficiency in man (SMITH et al. 1994) were reported to be associated with osteopenia
- Both inhibition of aromatization in the rat (VANDERSCHUEREN et al. 1995) and aromatase deficiency in man (KENAN et al. 1995) are also associated with osteopenia

Men with a defect in the aromatase gene or in the ER gene showed incomplete epiphyseal closure, resulting in continued growth and marked osteopenia, clearly indicating an important role of estrogens for bone physiology in men (SMITH et al. 1994). Treatment of a man with aromatase deficiency with daily 0.3–0.75 mg conjugated estrogens ceased non-physiological linear growth and increased bone mass (BILEZIKIAN et al. 1998). Although androgens have direct growth-stimulating actions on the epiphysis, the final phase of skeletal maturation associated with epiphyseal closure seems to be primarily an estrogen-dependent phenomenon in men (BACHRACH and SMITH 1996). In healthy men, ERα mRNA was consistently identified in osteoblasts and chondrocytes (KUSEC et al. 1998). As early as 1994, LIESEGANG et al. succeeded in detecting two ER types in human osteoblast-like cells. Both high-affinity, low-capacity and low-affinity, high-capacity specific [^{3}H]17β-estradiol binding were demonstrable.

17β-Estradiol inhibits bone resorption by directly inducing apoptosis of the bone-resorbing osteoclasts (KAMEDA et al. 1997). In 346 men aged 23–90 years, univariate analyses showed that serum bioavailable T and E levels correlated positively with BMD at the total body, spine, proximal femur, and distal radius and negatively with urinary N-telopeptide of type-I collagen (NTx) excretion. Urinary Ntx excretion was also negatively associated with BMD. By multivariate analyses, however, the serum bioavailable E level was the consistent, independent predictor of BMD in men. These studies suggest that, in contrast to traditional belief, age-related bone loss may be the result of E deficiency not just in postmenopausal women, but also in men (KHOSLA et al. 1998).

In a case-control study, 56 men with vertebral fractures had E2 levels 30% (SD 5%, $P < 0.0005$) lower than controls (BERNECKER et al. 1995). In a cross-sectional study in 37 healthy older men with no history of bone disease, ANDERSON et al. (1996) found that BMD at the LS and hip correlated more closely with E2 ($R = 0.383$, $P < 0.03$) than with T ($R = 0.245$, $P > 0.15$). In a prospective study of 93 healthy men aged over 65 years, E2 levels were positively associated with initial BMD values at all sites (SLEMENDA et al. 1995) and were associated with significantly lower rates of bone loss at the radius and hip on twice-yearly BMD measurement over a mean of 2 years ($P < 0.05$, test for trend), whereas T levels were not predictive (SLEMENDA et al. 1996). In men with vertebral osteoporosis, E2 levels were found to be positively correlated with BMD at the femoral neck ($R = 0.29$, $P > 0.03$; BERNECKER et al. 1995) and negatively correlated with markers of bone resorption such as hydroxyproline ($R = -0.57$, $P < 0.05$). On bone morphometry, T and E2 levels were associated with differing and complementary features such as trabecular thickness and trabecular number (SELBY et al. 1995). In a therapeutic trial of testosterone supplementation in men with vertebral osteoporosis (ANDERSON et al. 1997), pharmacological doses of testosterone were associated with proportionate rises in E2 levels. Increases in LS-BMD during testosterone therapy correlated more closely with changes in E2 than changes in T.

The essential role of estrogens for development and maintenance of the male skeleton is proved by a variety of further clinical studies. Bone density and sex steroids were measured in 93 healthy men aged over 65 years at 6-month intervals for an average of 2.1 years. Bone density was significantly positively associated with greater E2 ($R = 0.21$; $P < 0.05$) at all skeletal sites. There were weak negative correlations between T and bone density at the spine and the hip. SHBG was negatively associated only with bone density in the greater trochanter. Greater body weight was associated with lower T and SHBG and greater E2. These data indicate that, within the normal range, lower T is not associated with low bone density in men (SLEMENDA et al. 1997). In 534 men aged 68.6 years on average (Rancho Bernardo Study) a statistically significant positive relation was seen between bioavailable estradiol and BMD at all sites (GREENDALE et al. 1997). In age-related (type-II) femoral neck osteoporosis in men, the free estradiol index in 40 patients was significantly lower (1.3) than in 40 control subjects (BOONEN et al. 1997).

403 healthy men living independently (aged 73–94 years) were randomly selected from a population-based sample. E2 decreased significantly with age. In contrast to this, T did not change with age. E2 was significantly related to BMD at all sites. This was independent of T (VAN DEN BELD et al. 1998; JANSSEN et al. 1998). This also raises the question whether or not men may have a partial aromatase deficiency. The hypothesis is supported by the study of KAPS and GIRG (1998). After all, out of 302 male patients with osteoporosis, 56.7% had estrogen deficiency (E2 less than 20 pg/ml; that means practically non-detectable concentrations for conventional assays). However, only 20.5% of the patients showed testosterone deficiency ($T < 2.5$ ng/ml). Finally, BERNECKER et al. (1995) found that mean levels of serum estradiol but not testosterone were significantly reduced in men with established idiopathic osteoporosis. Collectively, these data support the hypothesis that estrogen deficiency plays a major role in involutional bone loss in men (RIGGS et al. 1998) and that the characterization of an – at least partial – aromatase defect will certainly stimulate future clinical research.

Has this knowledge already led to therapeutic consequences in the prevention and therapy of male osteoporosis? So far, unfortunately not. In view of the unknown side effects of estrogen replacement in osteoporotic men, many clinicians may prefer to use other agents, such as calcitonin, bisphosphonates, or perhaps fluorides. Nevertheless, carefully designed trials with suitable estrogenic drugs offer the prospect of better understanding the role of testosterone-independent estrogen deficiency in men (ANDERSON et al. 1998). Here, too, clinical studies including male-to-female transsexuals may be useful. In these patients, estrogen therapy definitely prevents bone loss after testosterone deprivation (VAN KESTEREN et al. 1998). According to SCHLATTERER et al. (1998), for male-to-female transsexual patients undergoing cross-gender hormone replacement therapy, the risk of developing osteoporosis is very low.

However, not only bone loss but also the regulation of chondrocyte metabolism is estrogen-dependent. 17β-estradiol inhibits cyclooxygenase-2 mRNA expression in chondrocytes, suggesting that estrogens could be implicated in the control of cartilage metabolism (MORISSET et al. 1998).

H. Cardiovascular System

Estrogen receptors α (ERα) were found in either cell fraction of aorta samples in men. Therefore, the male blood vessels may be considered as estrogen-sensitive (CAMPISI et al. 1993). In male Sprague-Dawley rats, ERα as well as estrogen receptor β (ERβ) were expressed in the aorta. The level of expression of ERα and ERβ-mRNA in male-rat aortas was examined before and after vascular injury using *en face* (HÄUTCHEN) preparations and in situ hybridization. Little or no change in ERα expression was observed after vascular injury in either vascular endothelial or smooth muscle cells at any time point. In contrast, ERβ-mRNA was found to be expressed markedly after balloon injury. In endothelial cells, ERβ was increased by 2 days after injury, and high levels of expression were maintained at 8 days and 14 days. Furthermore, ERβ expression was high in luminal smooth muscle cells at 8 days and 14 days after injury and had decreased to low levels by 28 days after injury. These data demonstrate the presence of ERβ in male vascular tissues and the induction of ERβ-mRNA expression after vascular injury, supporting a role for ERβ in the direct vascular effects of estrogen (LINDNER et al. 1998).

Interestingly, the story of the cardiovascular estrogen effects in men started in 1952 with the paper of BARR et al. (1952). In 16 men with advanced atherosclerosis, they found that treatment with extremely high dosages of 1 mg EE or 15 mg conjugated estrogens/day over 9 weeks decreased the plasma concentrations of total cholesterol and changed the cholesterol subfractions.

But the continuation of the estrogen story was not very promising. Estrogen supplementation in men with proven cardiovascular disease and previous myocardial infarction (MI) has been less successful (THE CORONARY DRUG PROJECT RESEARCH GROUP 1970). Unfortunately, oral estrogen dosing in this group was approximately 5–10× that used in replacement estrogen doses for postmenopausal women. At a dose of 5.0 mg of conjugated equine estrogens, a significant increase in recurrent MI, venous thrombosis, and pulmonary embolism occurred in men with established cardiovascular disease. However, at lower doses (2.5 mg of conjugated equine estrogens) an insignificant increase in these clinical complications was noted.

Recently, there was bad news: the mean E2 level in men who had had a MI was higher ($P = 0.002$) than in men who had not had MI (PHILLIPS et al. 1996). Estradiol supplementation augmented the flow-mediated vasodilatation and serum level of nitrite/nitrate (metabolites of NO) in women but not in men (KAWANO et al. 1997).

It is well known that estrogens have also been used in the treatment of prostate cancer in men. Again, high doses of oral estrogen have been associated with an increased risk of death from cardiovascular disease. High-dose estrogens may be prothrombotic in elderly men (BYAR and CORLE 1988). On the other hand, this risk has not been associated with lower doses of estrogen (0.2–1.0 mg of estradiol). In addition, many of these studies in men have evaluated the cardiovascular risk in the setting of metastatic prostatic cancer. These are obviously patients with an increased risk of thromboembolic disease, as often seen in malignancies (HENRIKSSON et al. 1990).

Fortunately, the estrogen story in men took a turn for the better. Estrogen treatment limits transplant-associated atherosclerosis in male animals by mechanisms thar may require receptor activation and transcriptional regulation of growth factors and expression of major-histocompatibility-complex class-II antigens (CHENG et al. 1991; SAITO et al. 1997). Whether activation of specific ERs (α and/or β) is required for the vascular effects of estrogen to be mediated is unclear. Probably, estrogen receptor α is not required, as estrogen reduces proliferation after arterial injury and increases endothelium-mediated vasodilatation in male mice and humans deficient in this receptor (IAFRATI et al. 1997; RUBANYI et al. 1997; SUDHIR et al. 1997).

The cardiovascular protective effects of estrogen in men are known to be mediated by its beneficial effects on lipid metabolism as well as by its direct actions on the vessel wall (KNOPP and ZHU 1997; SELZMAN et al. 1998). The latter can be mediated also by a specific ER present on smooth muscle cells and endothelial cells (MATSUBARA et al. 1997). NEVALA et al. (1998) found no gender differences concerning the estrogen-induced relaxation of mesenteric artery in vitro.

Meanwhile, we know the clinical-molecular-biological basis for this phenomenon. A man with a disruptive mutation in the ER gene showed premature coronary-artery disease (SUDHIR et al. 1997a). The sublingual administration of estradiol increased the brachial-artery diameter in the same manner as sublingual nitroglycerin, indicating a nongenomic mode of estrogen action (SUDHIR et al. 1997b). KOMESAROFF et al. (1998) describe a different nongenomic action of estrogen on the cardiovascular system (CVS). To examine the time course and mechanisms of action of single doses of estrogen on skin microvasculature, two double-blind placebo-controlled cross-over studies were conducted in healthy young men using the noninvasive technique of laser Doppler velocimetry with iontophoretic application of vasodilator substances. Estradiol (2 mg sublingually) produced a significant increase in the response to the endothelial vasodilator acetylcholine (Ach) after 15min, but not after 20min or 30min. An intravenous (iv) bolus of 25mg conjugated equine estrogens produced significant increases in the response to Ach at 15min and 20min but not at 30min. There was no change in responses to the non-endothelial vasodilators sodium nitroprusside or nicotine, and administration of placebo produced no change in Ach responses at any time point. These experiments show that, at E2s within the physiological range for premenstrual women, estrogens act directly on the cutaneous microvasculature through a rapid onset – rapid offset, nongenomic mechanism that is specific to the endothelium. In addition, these experiments support the view that estrogens can act on the male CVS in a manner that is potentially clinically beneficial.

Iv conjugated estrogens (0.625mg once) favorably modulate acetylcholine-induced changes in coronary hemodynamics in men. This suggests that estrogens may have anti-ischemic effects in men (Blumenthal et al. 1997). The results of iv conjugated estrogens (1.25mg) in men with cardiac allografts point in the same direction. The acutely given estrogens abolished abnormal cold pressor test (CPT)-induced coronary constriction. This favorable vasomotor effect suggests that estrogen may prevent inappropriate coronary-artery constriction in men with cardiac transplants (Reis et al. 1998).

However, Collins et al. (1995) found that 20min after intracoronary administration of 2.5 μg of 17β-estradiol into atherosclerotic, nonstenotic coronary arteries of nine postmenopausal women and seven men 52 ± 4 years old, the estrogen modulated the acetylcholine-induced coronary responses of female but not male atherosclerotic arteries in vivo.

There were significant negative associations of urinary total estradiol excretion with total cholesterol, LDL cholesterol, and apo B levels in 46 healthy Chinese men (Ooi et al. 1996). In 313 Japanese men aged 50–54 years, HDL and HDL_2 cholesterol were positively associated with E2, and E2 levels were also higher among current alcohol drinkers. Obesity, especially waist-to-hip ratio, was a strong correlate of both total and free testosterone, but not of estradiol (Handa et al. 1997). The administration of exogenous estrogen increases HDL-cholesterol levels in men (Wallentin and Varenhorst 1978).

Because the estrogen levels are rather low, they were not regarded as physiologically important until recently, when epidemiological research into heart-disease risk suggested a protective effect of endogenous estrogens in men (Khaw and Barrett-Connor 1991). This is partly explained by the fact that a lower proportion of circulating estradiol is bound to SHBG in men than in women due to competition with the higher levels of androgens (Anderson 1974). In subsequent prospective controlled trials, Bagatell et al. (1994) treated men receiving gonadotropin-releasing hormone (GnRH) agonist and testosterone replacement therapy with an aromatase inhibitor and showed that the estrogen-deficient men had a significant 8% fall in HDL cholesterol compared with the other group or normal controls, demonstrating that the levels of estradiol normally found in men are sufficient to alter lipid profiles favorably. Zmuda et al. (1993) showed that adding an aromatase inhibitor to supraphysiological testosterone treatment in weightlifters caused a 38% increase in lipoprotein-lipase activity, with worsening HDL levels.

It has been clearly demonstrated that the treatment of men with non-aromatizable androgens (anabolics, DHT derivatives) leads to an increase in LDL and triglyceride and a decrease in HDL, whereas treatment with aromatizable androgens such as testosterone enanthate only marginally affects the lipid system. However, when these men were treated with an aromatase inhibitor simultaneously, a decrease in HDL was observed to change the HDL/LDL ratio in favor of LDL. In summary, there were no significant effects of testosterone esters and other aromatizable androgens on HDL, but major effects on lipid levels by anabolic steroids which cannot be converted to estrogens (Sudhir et al. 1997; Friedl et al. 1990; Bagatell et al. 1994). Consequently, the administration of 17β-estradiol to a man with an ER defect shows the same benefits as estrogen replacement in postmenopausal women (Sudhir et al. 1997). This shows that, in men, nongenomic estrogen actions are more important for the extragenital estrogen benefits. According to Yeung (1997), the development of an estrogenic molecule that can be vascular protective yet devoid of feminizing properties could be very promising for estrogen replacement in men.

A combination of oral and parenteral estrogens administered to males with prostatic cancer profoundly lowers LDL cholesterol, apolipoprotein B (apo B), and lipoprotein (a) [Lp(a)] levels (Eriksson et al. 1989; Henriksson et al. 1992). Surprisingly, when estrogens were given only via the parenteral route, no significant changes in serum lipid levels were observed (Berglund et al. 1996). This suggests that the regulatory role of estrogens on serum lipoproteins, including Lp(a) levels in men, may depend on their capability of influencing hepatic metabolic pathways.

Twenty-two healthy elderly men (age 74 ± 3 years, mean ± SD) received 0.5, 1 or 2 mg/day of oral micronized 17β-estradiol over 9 weeks. LDL-C (–6%), apo B (–9%), triglyceride (–5%) and homocysteine (–11%) concentrations decreased, whereas HDL-C (+14%) increased. Intermediate-size very-low-density lipid (VLDL) subclass concentrations were lowered, and LDL and HDL subclass levels altered, in such a way as to cause average LDL and HDL particle size to increase. Lp(a) did not change. Fibrinogen (–13%) and plasminogen activator inhibitor-1 (PAI-1) concentrations (–26%) decreased, but there were no changes in thrombotic markers including thrombin-antithrombin III complex, prothrombin fragment 1.2, D-dimer, antithrombin activity, proteins-C and S, and von Willebrand-factor antigen. Breast tenderness occurred in four men and heartburn in five, but did not require discontinuation of treatment. The conclusion of the authors is that oral estrogen in men reduces homocysteine, fibrinogen, and PAI-1 concentrations and favorably influences VLDL, LDL and HDL subclass levels without increasing markers of thrombotic risk (Giri et al. 1998).

As in the case of osteoporosis, in the assessment of the benefits and risks of estrogen treatment on the CVS clinical experience with cross-sex hormone administration in transsexuals is very helpful. New et al. (1997) and McCrohon et al. (1997) independently studied two populations of male-to-female transsexuals in Melbourne (New et al. 1997) and Sydney (McCrohon et al. 1997) maintained on high-dose estrogen with and without antiandrogen therapy. The Melbourne group studied 14 male-to-female transsexuals, 14 age-matched men and 14 age-matched premenopausal women. The transsexuals had been receiving estrogen therapy for 61 ± 70 months. The daily mean oral dose of EE was 118 ± 60 μg, and the daily mean oral dose of Premarin was 1.03 ± 0.35 mg. Of the 14 transsexuals, 21% had had an orchidectomy and the rest received antiandrogen therapy. Six transsexuals had coronary risk factors (smoking, family history, diabetes, hypercholesterolemia). The flow-mediated response in the transsexuals, women, and men was 11.5 ± 1.3% vs 9.4 ± 1.1% vs 5.2 ± 1.0%, with baseline diameters of 4.1, 3.6, and 4.9 mm, respectively. The nitroglycerin response was also significantly different: 21.6 ± 1.7% vs 21.0 ± 0.9% vs 14.5 ± 1.2%. The duration of the therapy was not predictive, and only the estrogen therapy and vessel size were predictive of the brachial response. The Sydney group studied 15 male-to-female transsexuals and 15 men, well matched for age, smoking history, and rest vessel size (4.0 ± 4.2 mm). The transsexuals had been receiving estrogen for between 6 months to 21 years, with concomitant orchidectomy (50%) and antiandrogen therapy (50%). The flow-mediated response was markedly better in the transsexuals than in the control subjects (7.1 ± 3.1% vs 3.2 ± 2.8%), with an equally impressive improvement in vasodilation response to nitroglycerin (21.2 ± 6.7% vs 14.6 ± 3.3%). Both of these improvements were not related to the duration of estrogen treatment. In summary, both studies showed that male-to-female transsexuals had better arterial function (endothelial-dependent and -independent) as measured by brachial ultrasound.

Seventeen male-to-female transsexuals were treated with 0.1 mg EE and 100 mg cyproterone acetate (antiandrogen) daily. This treatment of the male subjects increased median serum leptin levels from 1.9 ng/ml before to 4.8 ng/ml after 4 months and 5.5 ng/ml after 12 months of treatment ($P < 0.0001$; Elbers et al. 1997). Beyond this, the plasma total homocysteine level decreased from a geometric mean of 8.2 μmol/l to 5.7 μmol/l ($P < 0.001$; Giltay et al. 1998). The same group reported that in male-to-female transsexuals mean plasma levels of tissue plasminogen activator (–4.4 ng/ml), big endothelin-1 (–0.8 pg/ml), urokinase-type plasminogen activator (–0.5 ng/ml) and PAI-1 (–26 ng/ml) decreased (all *Ps* < 0.02). The level of von Willebrand factor (vWF) decreased (+24%; $P = 0.005$), while vWF:Ag(II) did not change (van Kesteren et al. 1998). A favorable endothelium-dependent dilatation in the brachial artery was seen in estrogen-treated male-to-female transsexuals (McCrohon et al. 1997).

The mortality in male-to-female transsexuals is not increased during estrogen treatment. However, transdermal application of estradiol is recommended, particularly in the population over 40 years, in whom a high incidence of venous thromboembolism could be expected with enhanced liver estrogenicity, which is typical of oral administration of estrogens (VAN KESTEREN et al. 1997).

These findings bring us to the following question: will only women who will benefit from the cardiovascular or gene-therapy-related benefits of estrogens? At present, we don't know the answer, since men unfortunately appear reluctant to take estrogens for prolonged periods, despite the marked salubrious cardiovascular effects of the estrogens (LUFT 1998). However, this situation can be changed in the near future.

Should we use estrogen to treat men with coronary artery disease? For now, we should say probably not. This would be too early, because the cardioprotective effects in men would be hard to demonstrate in the clinical setting without significant feminization and other side effects such as prothrombotic states (YEUNG 1997). However, our current knowledge encourages us to develop non-feminizing estrogens (see below).

I. Central Nervous System

It is sometimes believed that our fate could be influenced by our endocrine glands. Obviously, this hypothesis has never been fully tested. However, there is abundant experimental evidence suggesting that endogenous steroid hormones, mainly though not exclusively from the adrenal and gonadal glands, may exert powerful effects on the central and peripheral nervous system. These include effects on neuronal cell development and differentiation, plastic changes in the organization of synaptic connections, and modulation of the efficiency of neuronal signal-transduction events. Consequently, it is not surprising that physiological, pharmacological, or pathological changes in circulating levels of steroid hormones may induce important modifications of neuroendocrine responses related to the maintenance of general body homeostasis and appearance, behavior, mood states, and even memory (Chap. 15, 27 and 44; MIRANDA and SOHRABJI 1996; ALONSO and LÓPEZ-COVIELLA 1998).

The role of estrogens in imprinting of the brain in males is an important aspect of an involvement of estrogens in male physiology which has been known for many years. It is 17β-estradiol which mediates the male priming of the brain in terms of behavior, at least in the rat (PLAPINGER and MCEWEN 1978; MCEWEN and WOOLLEY 1994). There is no doubt that, in this species, the priming phenomenon is essentially dependent on a conversion of testosterone into estradiol in the brain itself. Interestingly, even in the adult male rat brain, aromatase activity is three times higher than in females; therefore, this aromatase activity is dependent on estrogens (ROSELLI and RESKO 1987). However, SASANO et al. (1998) found no difference between men and women in the post-mortem expression of aromatase in various regions of the brain.

17β-Estradiol, which is formed in the central nervous system (CNS) by local aromatization of testosterone, is involved in the regulation of male gonadotropin production, becoming active mainly on the hypophyseal level. Estrogens are also assumed to modulate other central nervous processes, including the psyche. The clinical proofs of the concept are the signs and symptoms of an aromatase defect. A 24-year-old man with aromatase deficiency due to mutation in exon 9 of the P450 aromatase gene (cyp19) presented normal secondary sex characteristics, macroorchidism and elevated FSH, LH, and testosterone levels associated with very low estradiol and estrone concentrations (MORISHIMA et al. 1995). This observation gives prominence to the fact that estrogens are also important regulators of gonadotropin secretion in the male. Relatively low doses of estrogen and progestin are sufficient in men to influence gonadotropin secretion and thus to realize the concept of an orally active contraceptive (BRIGGS and BRIGGS 1974; see also section K.III).

OGAWA et al. (1997) have determined the role of ER activation by endogenous estrogen in the development of male-typic behaviors by the use of transgenic male ERKO mice. Surprisingly, in spite of the fact that they are infertile, ERKO mice showed normal motivation to mount females, but they achieved less intromissions and virtually no ejaculations. Aggressive behaviors were dramatically reduced and male-typical offensive attacks were rarely displayed by ERKO males. Moreover, ER-gene disruption demasculinized open-field behaviors. It should be noted that these changes of sexual and aggressive behaviors were not due to the reduced levels of testosterone in these mice. In the brain, despite the evident loss of functional ER protein, the androgen-dependent system appears to be normally present in ERKO mice. Together, these findings indicate thet ER-gene expression during development plays a major role in the organization of male-typical aggressive and emotional behaviors in addition to simple sexual behaviors (for review: COUSE and KORACH 1998).

The distribution of ERα and ERβ mRNA in the CNS is asymmetrical both in female wild-type mice and in female ERα KO mice, indicating that activation of ERα as well as ERβ may have different organizational and activational effects in the CNS (SHUGRUE et al. 1997a, b; LAFLAMME et al. 1998). Using adult male ERαKO mice, WERSINGER et al. (1997) showed that – in contrast to wild-type male mice – little masculine sexual behavior was displayed during testosterone replacement after gonadectomy. Therefore, it seems clear that ERα plays a key role in the expression of masculine sexual behavior. In earlier experiments using isolation-induced male agressive behavior in mice, we were able to show that pheromone-induced agressiveness was also dependent on the aromatization of testosterone to estrogens (KURISCHKO and OETTEL 1977; OETTEL and KURISCHKO 1978).

Can all estrogen effects in the brain be explained by transcriptional activation of either ER-α or ER-β? In this context, SHUGRUE et al. (1997b) conducted an interesting experiment using ERα-disrupted mice. There is only a weak affinity of ^{125}I-estradiol for ER-β. Surprisingly, the induction of PR mRNA by 17β-estradiol in ERαKO mice was much greater than expected. Based on these observations, estrogen appears to regulate some genes in the developing and adult ERαKO brain via an unknown, non-classical ER.

Besides the PR, another estrogen-driven gene in the CNS is the oxytocin receptor (OTR). Using ERαKO mice, YOUNG et al. (1998) demonstrated that ERα is not necessary for basal OTR synthesis, but is absolutely necessary for the induction of OTR binding in the brain by estrogen.

I. What Do We Know About Gender Differences Concerning Cerebral Blood Flow?

MATHEW et al. (1986) have reported that women had higher cerebral flow than men, and the differences were most obvious in the frontal regions. This finding is confirmed by many reports of higher cerebral blood flow in females compared with males (RODRIQUEZ et al. 1988; DANIEL et al. 1988). Studies of brain metabolism have also indicated that there are sex-related CNS glucose metabolic rate differences in humans and rats. BAXTER et al. (1987) have shown that women have whole-brain glucose metabolic rates that are 19% higher than those of men. The 19% difference between sexes for the brain as a whole is close to that found in the cerebral blood-flow study. ÖZTAS (1998) hypothesize that the permeability of the blood-brain barrier differs between males and females since brain function, glucose utilization, and blood flow ehibit sex-related differences.

II. What Do We Know About the Connection Between the Serum Levels of Endogenous Estrogen and Certain CNS Functions in Men?

Regarding sexual behavior, circulating levels of estradiol – in contrast to testosterone – play only a limited role (BAGATELL et al. 1994b). However, VAN DEN BELD et al. (1998) found a significant positive association between E2 and the outcome of a quality of life questionnaire in men aged 73–94 years. Low estradiol levels have been associated with decreases in verbal skills in older men (CHERRIER et al. 1998), but men retain verbal skills better than women do during the initial stages of Alzheimer's disease (AD; HENDERSON and BUCKWALTER 1994).

In another study, the influence of endogenous testosterone and estrogen on memory was investigated in 33 healthy young men. Tests of visual memory, visuospatial ability, verbal memory, and attention were administered, and circulating levels of estradiol and free testosterone were measured. Participants with high levels of estradiol performed better on two measurements of visual memory than did those with normal but lower levels. There were no differences between individuals with high and low levels of testosterone on any cognitive measurement. These results support the contention that estradiol influences memory in men (KAMPEN and SHERWIN 1996).

Finally, higher serum levels of estradiol – but not of testosterone – in women were associated with improved sequential movement as a sign of better fine-motor performance. In contrast with this, hormone levels in men's blood were unrelated to key-pressing performance (JENNINGS et al. 1998).

III. What Do We Know about the Influence of Exogenous Estrogen on CNS Functions in Men?

Treatment of four men with idiopathic hypothalamic hypogonadism with testosterone or 17β-estradiol resulted in decreased mean levels of biologically and immunologically active LH and FSH, whereas administration of DHT did not alter gonadotropin secretion. These data suggest that some of the direct effects of testosterone at the pituitary level in men are mediated by estradiol, whereas peripherally formed DHT may not play an important role in this process (BAGATELL et al. 1994c). In pubertal boys, the administration of testosterone, but not of the non-aromatizable androgen DHT, significantly influenced the secretion pattern of GH, indicating that aromatization of testosterone to estradiol in boys is the proximate sex-steroid stimulus amplifying secretory activity of the GH axis at puberty (VELDHUIS et al. 1997).

A sample of 16 young men received a patch delivering 0.1 mg of 17β-estradiol per day transdermally. They were exposed to a brief psychosocial stressor (free speech and mental arithmetic in front of an audience) 24–48 h later. The estradiol-treated subjects showed exaggerated peak adrenocorticotropic hormone ($P < 0.001$) and cortisol ($P < 0.002$) responses compared to the placebo group (KIRSCHBAUM et al. 1996). Studies by KUDIELKA et al. (1998) could support this finding. Elderly males (probably with higher estradiol levels) show larger hypothalamic-pituitary-adrenal responses to psychosocial stress than postmenopausal women (with lower estradiol levels).

In contrast to this, percutaneous admistration of estradiol (100 μg) reduced the mental arithmetic stress in 20 normal young men. The increase in epinephrine was significantly lower in the estradiol-treated groups (DEL RIO et al. 1994).

It is well known that hot flushes experienced during GnRH therapy or after orchidectomy in men e.g. with cancer of prostate can be stopped with different estrogens. Estrogens reduce adrenergic stimulation and therefore the occurrence of hot flushes. The parenteral use of estrogen preparations (e.g. patches) in men has scarcely been studied so far. However, based on the data available, it can be assumed that their use should be associated with much fewer side effects than the use of orally active estrogen drugs (KLIESCH et al. 1997).

Now, considering the fact that the involvement of exuberant oxidative mechanisms in AD is very likely (SMITH and PERRY 1995) and that meanwhile the delaying effect of 17β-estradiol replacement on the progression of senile dementia of Alzheimer's type (SDAT) can be assumed to be proven (PAGANINI-HILL and HENDERSON 1994; PAGANINI-HILL 1996), attractive possibilities emerge for utilization of this advantage in men with the help of the so-called non-feminizing scavestrogens (Sect. M). The BEYREUTHER group in Heidelberg found that in AD, the reduction of Cu(II) to Cu(I) by Alzheimer's amyloid precursor protein (APP) involves an electron-transfer reaction and could lead to production of hydroxyl radicals. Thus, copper-mediated toxicity of APP-Cu(II)/(I) complexes may contribute to neurodegeneration in AD (MULTHAUPT et al. 1997).

Women have more strokes than men do but are more likely to recover from them. In this context, estrogen might be able to reduce brain damage from a stroke in both women and men. After ischemia, estrogen may provide protection against cellular injury related to antioxidant properties of the estrogenic molecules (HALL et al. 1991; KEANEY et al. 1994; CAULIN-GLASTER et al. 1997). TOUNG et al. (1998) gave estrogen to 36 rats and plain salt water to another 21. They then stimulated the effects of a stroke by cutting blood flow to the brain for 2h. Female as well male rats that got the estrogen had half the brain damage. These findings demonstrate that the benefit of estrogen can be extended to the male brain, reducing tissue injury. HAWK et al. (1998) found that testosterone increases and estradiol decreases middle cerebral artery occlusion lesion size in male rats. A strong positive correlation ($R = 0.922$) between plasma testosterone concentrations and ischemic lesion size was observed. Estradiol treatment reduced the ischemic brain area of the male rats significantly.

J. The Influence of Environmental Estrogens (Xenoestrogens) on the Fertility of Men

The hormone-like activity of environmental chemicals and natural plant constituents has evoked a controversial debate in both the public and the scientific community (GREIM 1998). For the last 40 years, substantial evidence has surfaced on the hormone-like effects of environmental chemicals such as pesticides and industrial chemicals in wildlife and humans. The endocrine and reproductive effects of these chemicals are believed to be due to their ability to:

1. Mimic the effect of endogenous hormones
2. Antagonize the effect of endogenous hormones
3. Disrupt the synthesis and metabolism of endogenous hormones
4. Disrupt the synthesis and function of hormone receptors (SAFE et al. 1997; NIEMANN et al. 1998; SONNENSCHEIN and SOTO 1998)

The discovery of the hormone-like activity of these chemicals occurred long after they were released into the environment. Aviation crop dusters handling dichloro-diphenyl-trichloroethane were found to have reduced sperm counts (SINGER 1949), and workers at a plant producing the insecticide kepone were reported to have lost their libido, become impotent and have low sperm counts

(GUZELIAN 1982). Subsequently, experiments conducted in lab animals demonstrated unambiguously the estrogenic activity of these pesticides (SONNENSCHEIN and SOTO 1998). Man-made compounds used in the manufacture of plastics were accidentally found to be estrogenic because they fouled experiments conducted in laboratories studying natural estrogens (SOTO et al. 1991). For example, polystyrene tubes released nonylphenol, and polycarbonate flasks released bisphenol-A (KRISHNAN et al. 1993). Alkylphenols are used in the synthesis of detergents (alkylphenol polyethoxylates) and as antioxidants. These detergents are not estrogenic; however, upon degradation during sewage treatment they may release estrogenic alkylphenols. The surfactant nonoxynol is used as an intravaginal spermicide and condom lubricant. When administered to laboratory animals, it is metabolized to free nonylphenol. Bisphenol-A was found to contaminate the contents of canned foods; these tin cans are lined with lacquers such as polycarbonate (BROTONS et al. 1994). Bisphenol-A is also used in dental sealants and composites; up to 950 μg of this compound were retrieved from saliva collected during the first hour of polymerization. Other xenoestrogens recently identified among chemicals used in large volumes are the plastizicers benzylbutylphthalate, dibutylphthalate, the antioxidant butylhydroxyanisole, the rubber additive *p*-phenylphenol, and the disinfectant *o*-phenylphenol. Feminized male fish were found near sewage outlets in several rivers in the U.K. (for review see JOHNSON et al. 1998; SONNENSCHEIN and SOTO 1998).

Male reproductive toxicology has made some important discoveries in the past few decades. It is likely that man will increase his deliberate and incidental exposure to old and new chemicals, so it would not be surprising if, in the future, the chemicals we have discussed are replaced by others which will be a greater concern for male reproductive health (MORRIS et al. 1996).

However, it should be stated that synthetic, natural and hormonally active steroids are easily degradable in the environment by aerobic and anaerobic microorganisms, including fungi, and therefore are not an ecotoxicological risk (HÖRHOLD-SCHUBERT et al. 1995). In particular, this applies to the unjustified fear that the elimination of steroids after using oral contraceptives or hormone replacement might cause environmental pollution.

A discussion of environmental estrogens (the so-called hormonal disruptors as well as the "good" phytoestrogens) means discussing very different compounds like phytoestrogens (isoflavones, coumestane, soya), estrogenic chemicals (tetrachlorobenzodioxin, 4-octylphenol) or estrogens such as diethylstilbestrol (DES). To demonstrate a role of DES or other estrogenic chemicals, pregnant animals, including rats, were treated with these compounds and the effects on the pups investigated (MAJDIC et al. 1996). Under these circumstances, a disturbance of testicular development and an imbalance of somatic sexual organ development were demonstrated. However, these data are in very good agreement with what was shown more than 50 years ago (RAYNAUD 1940), and has since been confirmed by many investigators. Thus, Raynaud described the same effect after treatment of pregnant mice with large doses of 17β-estradiol. However, one aspect which is of utmost importance in this context has been neglected to a great extent, and this is the relevance of the presented data in terms of fertility in context with the dose used. Negative effects on the development of the male sexual system can only be observed if extremely high doses of so-called environmental estrogens are used, and there is good evidence that the concentration necessary to induce adverse testicular effects during the time of sexual differentiation is hardly reached in men. Furthermore, it should be taken into account, and it should not be forgotten that men are used to certain amounts of estrogens during their own lives, and even more during uterine life before birth.

MÜLLER et al. (1998) estimated the estrogenic potency of 4-nonylphenol (NP) and a risk calculation for non-occupationally exposed humans was performed. The daily intake of non-occupationally exposed persons was estimated to be less than 0.16 mg/day. Risk estimates were based on this daily intake and the relative potency of NP to 17β-estradiol. Comparison of this intake with the no-observed-adverse-effect level derived from a 90-day subchronic toxicity study in animals, results in a safety factor of about 20,000. A safety margin of 3000 can be derived when comparing the resulting NP blood concentrations with 17β-estradiol levels in the blood of adult males.

In this context, estrogens have been discussed in the last few years as a possible factor for declining fertility in men (SHARPE 1993; SHARPE and SKAKKEBAEK 1993). This phenomenon captured the attention of scientists, physicians, and the media. In spite of clear evidence of a major role of estrogens in the normal male physiology, estrogens have been discussed as risk factors in men, particularly in the context of male fertility. Specifically, exposure to environmental estrogens has been connected to a constant fall in sperm count over the last decades (COMHAIRE et al. 1996; MENCHINI-FABRIS et al. 1996; GIWERCMAN and SKAKKEBAEK 1992; JOFFE 1996). At present, there are the following reports concerning the decline in male reproductive health: LETO and FRENSILLI 1981; BOSTOFEE et al. 1983; BRAKE and KRAUSE 1992; CARLSEN et al. 1992, 1995; AUGER et al. 1995; IRVINE et al. 1996; VAN WAELEGHEM et al. 1996; YOUNGLAI et al. 1998.

However, it has been seriously questioned whether there is a decline in sperm count at all (NIESCHLAG and LERCHL 1996). These and other authors came to the conclusion that the re-analysis of the data, showing that male sperm counts decreased by over 40% between 1940 and 1990, indicated that inadequate statistical methods were used and that the presented data did not support a significant decline in sperm count (NIESCHLAG and LERCHL 1996; JOUANNET and AUGER 1996). All of the studies published to date were flawed in one way or another because of subject selection bias, choice of statistical model for analysis, methodological inconsistencies, inconsistencies with abstinence, quality of data (i.e., means used for skewed distributions of data not normally distributed), potential regional and ethnic variations in semen quality, or study design. These significant flaws make it imposible to conclude anything regarding potential changes in semen quality over time (BERMAN et al. 1996; for review see LAMB 1997).

A very witty comment, of course, comes from a veterinarian: SETCHELL (1997) compared the data from 137 studies on sperm counts of bulls, boars, and rams in the literature from the early 1930s to 1995. The bull data showed no correlation of sperm count with year of publication, for the boars there was a slight but insignificant positive correlation, and for the sheep there was a slight, but significant, rise in sperm counts with time. He concluded: it would appear that, if the fall in human sperm counts is real, then it must be due to something which is not affecting farm animals.

In contrast to the controversial discussions about whether or not there has actually been a decline in sperm quality during the last five decades, there is no doubt about an increase in testicular cancer in young men (FORMAN and MOLLER 1994). The evaluation of cancer registries in several countries in north and central Europe, Australia, New Zealand, and the U.S.A. showed consistently increasing incidences, e.g. by 2–4% per year in men under the age of 50 years. Those at highest risk are men aged between 20 and 45 years. So far, it is still unclear whether the apparent increase in this type of cancer found in many countries is due to hormonally active substances (MCLACHLAN et al. 1998), changed ways of life, or other causes. For instance, the fact that there is a definite increase in testicular cancer in the U.S. white population, while in the U.S. population of Asian and African origins there is not was cited by the Danish health authority in their report of 1995 as a reason to doubt a relation between testicular cancer and the presence of environmental hormonally active substances (GREIM 1998).

We might argue further here. If environmental estrogens were to have such a negative impact on fertility and health in men, another question would have to be raised: namely, are estrogens bad for women? In terms of the reproductive system, it has also been known for decades that the treatment of pregnant animals with 17β-estradiol results in a paradoxical virilizing effect in female pups (RAYNAUD 1940; NEUMANN et al. 1974). But there are no reports about an increased virilizing phenomenon in girls or women. This is just one example of what should be expected if any relevant concentration of environmental estrogens exists (HABENICHT 1998).

K. Regimens for Estrogen and Antiestrogen Treatment in Men

I. Treatment of Prostate Cancer with Estrogens

Androgens are known to be very important for the development of the prostate gland. However, the level of testosterone decreases with age, indicating that factors other than androgens could be important for the development of prostate hyperplasia or prostate cancer. Therefore, it is only logical to ask, what do we know about the role of estrogens in the pathophysiology of the prostate?

Long-term administration of estrogens or aromatizable androgens to either intact or castrated males of various animal species results in altered prostatic growth (SANTTI et al. 1994). The most striking morphological change caused by estrogen treatment of intact or castrated male animals is squamous epithelial metaplasia (OETTEL 1974) extending from the prostatic utricle into the epithelium covering the seminal colliculus (PYLKKÄNEN et al. 1993). Neonatal estrogenization of male mice promotes epithelial hyperplasia and dysplasia based on altered *c-fos* expression (SAITO et al. 1997). In addition, the chronic co-administration of estrogens and androgens to inbred Noble rats causes dysplastic lesions, and in some cases adenocarcinomas, in the dorsolateral lobes of the prostate and the coagulating glands (NOBLE 1977, 1980, 1982; DRAGO 1984; LEAV et al. 1988). These findings suggest that estrogens may play an important role in the control of prostatic growth and development.

Estrogens also have direct cytotoxic effects on prostatic carcinoma cells in vitro, possibly by an effect on the cell membrane, stimulating a rapid Ca^{2+} influx (WIDMARK et al. 1995; AUDY and DUFY 1996). Another interesting contribution to the nongenomic action of estrogens on the prostate comes from the Rosner group (NAKHLA and ROSNER 1996; NAKHLA et al. 1997). SHBG not only regulates the free – and therefore biologically active – concentrations of certain steroid sex hormones in plasma but is involved in a nongenomic mechanism of steroid-hormone action. It binds to a receptor on prostatic cell membranes and is activated by an appropriate steroid to initiate the generation of intracellular cAMP. In serum-free medium, both dihydrotestosterone and estradiol increase growth in the presence, but not in the absence, of SHBG.

HIRAMATSU et al. (1996) found that ER expression was only observed in the stromal cells of six out of 26 (23%) patients with prostatic carcinoma, but in none of the prostatic cells of 19 patients with benign prostatic hyperplasia (BPH). They conclude that estrogens do not have a direct genomic effect on the biological behavior of BPH or prostatic carcinoma. On the other hand, androgen deprivation leads to an upregulation of stromal ER expression in the human prostate (KRUITHOF-DEKKER et al. 1996).

In a case-controlled pair study based on 320 men who developed surgically treated BPH, GANN et al. (1995) found a positive correlation between increased E2 and risk of BPH. The excess risk associated with 17β-estradiol was confined to men with relatively low androgen levels. Contrasting with this are the findings of CETINKAYA et al. (1998). They found that, whereas E2 and the ratio of estradiol to free testosterone were increased in patients with hepatic cirrhosis, serum PSA levels and mean prostatic volume were significantly higher in the healthy control group. LAGIOU et al. (1997), too, found that E2 was not significantly related to the risk of BPH. GANN et al. (1996) also performed a case-controlled pair study based on 222 men who developed prostate cancer. High levels of circulating testosterone and low levels of SHBG – both within normal endogenous ranges – were associated with increased risks of prostate cancer. Low levels of circulating estradiol were assumed to represent an additional risk factor. Circulating levels of DHT and 3α-androstanediol glucuronide did not appear to be strongly related to prostate-cancer risk.

What knowledge comes from the estrogen treatment of male-to-female transsexuals? In nine cases of estrogen-treated males who had undergone orchiectomy, the prostates were small only after long-term estrogens. No malignancies were found (VAN KESTEREN et al. 1996). JIN et al. (1996) measured the prostate sizes in 11 estrogen-treated male-to-female transsexuals. Compared with age-matched controls, total prostate volume, central prostate volume (CPV), ratio of CPV to peripheral prostate volume, PSA levels, and total and free testosterone were significantly reduced. There was no influence of the estrogen treatment on prostatic acid phosphatase. SHBG levels were elevated by nearly 500% ($P < 0.001$), indicating the same well-known situation of estrogen dependence of SHBG induction as in women.

The basis of palliative therapy with estrogens in advanced prostatic cancer is the systemic blockade of the androgenic stimulation of the tumor. This is achieved, among other things, by using estrogens (HUGGINS and HODGES 1941; HUGGINS et al. 1941a, b), which inhibit the pituitary secretion of LH, and therefore indirectly the testicular secretion of testosterone. Further observations include a rise in SHBG, induction of pituitary prolactin secretion, and decreased DNA synthesis in carcinoma cells of the prostate.

Unfortunately, estrogen therapy has fallen into disrepute, since numerous side effects are allegedly due to estrogens, such as cardiovascular and hepatic toxicity, salt and water retention, hyperprolactinemia, endocrine-determined psychological disturbances, immunosuppression, etc. (MELLINGER et al. 1967; BYAR 1973).

It is still the general consensus that if estrogens are to be used, contraindications, especially thrombophilia and a history of embolism, must be excluded. A general thrombosis prophylaxis with low-dose heparin is recommended, particularly in the case of iv injection. In addition, in order to prevent gynecomastia, mamillary radiation with 10–20 Gy is necessary before the beginning of therapy (ZINGG and HEINZEL 1968). On the other hand, MAUERMAYR et al. (1978) recommended pretherapeutic andromastectomy.

DES is the only estrogenic drug still used in primary therapy in the USA, with recommended doses from 0.2 mg up to 1 mg/day maximally. In previous studies, even 5 mg DES was used (THE VETERANS ADMINISTRATION COOPERATIVE UROLOGICAL RESEARCH GROUP, VACURG 1967). High-dose oral estrogens were associated with an increased risk of cardiovascular death. A dose of 1 mg DES daily is not associated with an increased risk of cardiovascular death (COX and CRAWFORD 1995). Besides the suppression of LH, a direct, local antitumor effect has also been demonstrated for DES in rats (CUI et al. 1998).

The rationale behind the use of DES diphosphate in doses of 200 mg each day, or sometimes greater, was that the free, unconjugated, presumably active

DES was released directly into the prostate by the high levels of phosphatases present in the tissue. But there is little evidence that DES diphosphate accumulates in the prostate or that DES is released in any substantial amount to elicit a cytotoxic intraprostatic effect (GRIFFITHS et al. 1994). However, in 36 patients with advanced prostate cancer there was a stronger suppression of serum testosterone and FSH levels by DES diphosphate than by castration or a GnRH agonist (KITAHARA et al. 1997). Premarin (2.5mg t.d.s.) is also clinically effective (BOYNS et al. 1974). Synthetic EE (0.15 to 1.0mg/day), too, has had its proponents (GRIFFITHS et al. 1994).

The long-acting, intramuscular polyestradiol phosphate (PEP) is popular in Scandinavia and does not show an increased rate of cardiovascular complications in patients suffering from cancer of the prostate (HAAPIAINEN et al. 1990; STEGE et al. 1995). Patients with prostatic carcinoma and who are on oral estrogen therapy presumably have an altered coagulation system and suffer cardiovascular side effects. The reason is that estrogens – especially oral estrogens – are potent inducers of liver-synthesized proteins, including coagulation factors. HENRIKSSON et al. (1990) have assessed the effect of non-oral estrogen on the coagulation system in patients with prostatic carcinoma. Monthly intramuscular injections of 320mg PEP were given to 12 patients. No change was found in any of the coagulation factors, including factor VII, with the exception of a significant decrease in antithrombin III. No patient, including another 38 patients treated with PEP, had any cardiovascular complications after an average follow-up period of 12.9 ± 0.7 months; 76% of the patients responded to treatment.

A distinct optimization of oral estrogen therapy was achieved with the development of ethinyl estradiol sulfonate (EES; Turisteron; SCHWARZ and WEBER 1970; SCHWARZ et al. 1975; CHEMNITIUS and ONKEN 1976; OETTEL et al. 1981). EES is an orally active depot estrogen and is administered once per week in a dose of 1–2mg (GUDDAT et al. 1979; BUSCHMANN and GUDDAT 1997; HÖFLING and HEYNEMANN 1998). The good tolerability of EES compared to the mother compound EE is remarkable. That is why a 3-sulfamate was substituted for the sulfonate in position 3 of the estrogen molecule. This resulted in the so-called estrogen sulfamates, with distinct, low hepatic estrogenicity (ELGER et al. 1995, 1998). However, the excellent preclinical results need to be confirmed in patients with prostatic cancer.

Numerous studies have been performed to investigate estramustine phosphate after tumor progression, but also in primary therapy. Estramustine, a conjugate of nor-nitrogen mustard and estradiol, is an effective chemotherapeutic agent that is presently mainly used to treat metastatic, hormone-refractory prostate cancer (HUDES et al. 1992). Apart from inhibiting LH, this drug shows a direct cytotoxic activity based on its ability to attack the microtubular system and on the tau expression of the cell (TEW and STEARUS 1987; LAING et al. 1998; LIDSTRÖM et al. 1998; SANGRAJRANG et al. 1998). Its activity is dose-dependent (BENSON and GILL 1986). Infusion of 300–450mg daily over 10days is recommended; for oral use, up to 840mg per day is recommended.

The combination of the vinca alkaloids vinblastine or vinorelbine with estramustine phosphate is active and well tolerated in hormone-resistant prostate cancer and supports the therapeutic strategy of combining agents that impair microtubule function (BATRA et al. 1996; CARLES et al. 1998). The metabolites of estramustine phosphate perform as androgen antagonists of ARs, an additional mechanism involved in the therapeutic effect of the parent compound in patients with prostate cancer (WANG et al. 1998). Fosfestrol 1000–1200 mg daily is now only used for alleviation of pain in tumor progression, mostly as 10-day infusion therapy.

Disillusioned about the marginal therapeutic progress achieved with the expensive complete androgen blockage (GnRH analogs), we turn to Scandinavia and the UK again, where estrogen therapy is still among the standard treatments of metastatic prostate cancer. So far, the side effects which have become known since the publication of the VACURG study in 1970 have deterred us from estrogen therapy. These side effects mainly include edema, cardiovascular toxicity, thrombosis, and the development of male breast (PRESTI 1996). They can be controlled symptomatically, the cardiovascular effects may possibly be reduced or prevented by ASS dosage or the systemic use of the estrogens with circumventing of the enterohepatic circulation, or the modern development of estrogen sulfamates. If we succeed in freeing a modern estrogen therapy of prostate cancer from the stigma of threatening cardiovascular death, it would be easy to predict a revival of the use of estrogens, since their influence on prostate cancer is at least equal to that of all other therapeutic methods. The distinctly lower economic load this therapeutic alternative involves should not be underestimated (MOHREN 1998).

Prostate-cancer patients who underwent orchiectomy or GnRH treatment experience hot flushes. Estrogens, owing to their binding to ERs while simultaneously showing antitumor activity, seem to reduce adrenergic stimulation and therefore the occurrence of hot flushes. However, unlike oral administration, the parenteral use of estrogen preparations in men has altogether been scarcely investigated so far. The data available suggests a definitely lower rate of side effects for the parenteral than for the conventional, orally active estrogen drugs (KLIESCH et al. 1997).

II. Estrogen Replacement in Men

No area of hormone replacement in elderly males is so speculated on and so little understood as the sense and usefulness of estrogen therapy in males. As mentioned above, previous negative experiences in terms of adverse effects of estrogen therapy of prostate cancer are totally useless. Estrogen is used here without preceding measurement of endogenous estrogens. What is more, synthetic products, such as DES or EE are used in unphysiologically high doses. These, of course, must have adverse effects on breast tissue, the clotting system, and the vascular system. A good example is the comparison of the rate of side effects of the synthetic estrogen EE in oral contraceptives with the natural estrogens in the different hormone-replacement regimens used in postmenopausal women. The risk of deep venous thromboembolism in postmenopausal estrogen replacement using natural estrogens appears to be smaller compared with the risk from contraception using EE. Mimicking natural estrogen sulfation by sulfonation of EE in position 3 (Turisteron) considerably reduces the cardiovascular risk of large-dose estrogen treatment in men suffering from cancer of the prostate (Buschmann and Guddat 1997). There is an exogenous supply of aromatizable androgens, e.g. testosterone esters; at the same time, aromatization will always lead to increased endogenous serum concentrations of 17β-estradiol and estrone (Sih et al. 1997).

It is crucial to know whether there is an estrogen deficit in men (section B). Men with a complete aromatase defect are devoid of any estrone and 17β-estradiol in serum (absolute estrogen deficiency), although FSH and LH are increased. The skeletal age seems retarded, with incomplete epiphyseal closure. The clinical picture corresponds to osteoporosis. Moreover, increased basal insulin levels and a decreased HDL/LDL ratio are detected (Bulun 1996). At present, only speculations about the frequency of complete or partial aromatase defect in men (normal testosterone concentrations and lacking or very low estrogen concentrations) are possible.

Remarkably, there are regional differences in fatty-tissue ER concentrations of men and women. While the numbers of estrogen receptors in visceral fat are the same, the number of ERs in the abdominal subcutaneous fatty tissue in men is double that of women (Pedersen et al. 1996).

An interesting proposal is made by Kuhl et al. (1994). The daily oral intake of 2–10 mg estriol (t.i.d.) is said to be appropriate for the prevention and treatment of arterial diseases in men. It being well known that, due to its short half-life, estriol has little or no "classical", i.e. genomic (nucleus-mediated) estrogenic actions; the estriol effect must be based on other mechanisms of action. As we have found out, estriol has distinct antioxidative (i.e. nongenomic) effects which, with regard to the inhibition of lipid peroxidation in synaptosomal membranes, are about as strong as vitamin E (Römer et al. 1993). Generally, nongenomic actions of estrogens on the vascular system (Farhat et al. 1996) or in the CNS (Ramirez et al. 1996) are a very interesting starting point for the development of novel drugs (Sect. M).

III. Male Contraception with Estrogens

The disruption of spermatogenesis by the estrogen treatment of experimental animals has been widely used as a useful approach in the study of spermatogenesis regulation (Blanco-Rodríguez and Martínez-García 1996). Steroidal estrogens and nonsteroidal estrogenic compounds such as DES can suppress spermatogenesis in man so that, in the end, only Sertoli cells and spermatogonia remain in the thickened seminiferous tubules (Jackson and Jones 1972). Nevertheless, the small number of publications so far which have dealt with the use of estrogens for male contraception is surprising. Too little is known to definitely either recommend or reject the use of a low-dose estrogen component (besides androgens or progestins). The cautiousness in using or recommending estrogens certainly results from the negative experience gained with previous estrogen therapy of prostate cancer, where it should be considered that the estrogen doses chosen were much too high (Sect. D.IV.). While it is possible that the diminished libido and sexual potency could be improved by concomitant androgen replacement, the occurrence of symptomatic gynecomastia and other metabolic side effects, such as those demonstrated in women, make it unlikely that high-dose estrogens will be accepted in formulations designed to interrupt male fertility (DE KRETSER 1974). An example of previous, extremely high estrogen dosage is reported by HELLER et al. (cited by DE KRETSER 1974). Mestranol (0.45 mg/day) led to azoospermia, but this was accompanied by decreased libido and potency together with the universal occurrence of painful gynecomastia.

In men, the role of testosterone and estradiol in a feedback regulation has been established. The very low daily dose of 15 μg EE obviously is the borderline amount for decreasing plasma LH and testosterone levels in normal males. LH and testosterone are reliably reduced by 30 μg EE whereas, in patients suffering from Klinefelter's syndrome, the EE dose must be approximately five times higher (SMALS et al. 1974). On the other hand, KULIN and REITER (1997) found that daily 32 μg and 42 μg suppressed FSH in urine and blood, respectively. A significant suppression of LH was not obtained. A more powerful decline in T was seen with daily 50 μg EE plus 0.5 mg norgestrel over 9 days (KJELD et al. 1977). However, after checking the individual components, it became evident that the stronger effect regarding the decrease in testosterone emanated from norgestrel/levonorgstrel (KJELD et al. 1979). Using a combination of 0.02 mg EE and 10 mg methyltestosterone per day, KULIN and REITER (1972) and BRIGGS and BRIGGS (1974) found a suppression of pituitary secretion of FSH sufficient to stop the sperm production in normal adult men. Curiously, these experiments have not been carried on.

High-dose sex-steroid administration had marked effects on adrenal androgen levels, which decreased by 27–48% in males treated daily with 0.1 mg EE orally and increased by 23–70% in females treated with 250 mg testosterone enanthate intramuscularly every 2 weeks (POLDERMAN et al. 1995).

IV. Treatment of Male-to-Female Transsexuals with Estrogens

Transsexualism belongs to the group of gender dysphoric disorders. These have been known from antiquity onward across national and cultural boundaries. Formally, the prevalence of transsexualism has increased over the past years. This – in our current opinion – is due to the improvement of therapies and to an increased public acceptance of this phenomenon (SCHLATTERER et al. 1996). Cross-gender hormone treatment of transsexual patients is carried out to suppress secondary sex characteristics of the original sex and to induce those of the opposite sex. This kind of therapy is life-long. Prescription of feminizing drugs bears some risks. A continuous monitoring has to be performed by the experienced endocrinologist during cross-gender hormone-replacement therapy.

Initial treatment of male-to-female transsexuals consists of judicious administration of an estrogen such as 100 μg EE or conjugated estrogens 1.25–2.5 mg daily, or high-dose intramuscular estradiol valerate (100 mg) or polyestradiol phosphate (80 mg) every 2 weeks, mostly in combination with oral antiandrogens (e.g. cyproterone acetate 100 mg daily or spironolactone administered in a divided daily dosage of 100–200 mg; SCHLATTERER et al. 1996; FUTTERWEIT 1998). The addition of the antiandrogen is indicated in patients who may benefit from possible enhanced breast development and the clinical benefits of a likely further reduction of facial-hair regrowth. The treatment is often given in conjunction with near-physiological doses of estrogens (PRIOR et al. 1989). With decreasing testosterone levels, the antiandrogen dose can be reduced stepwise until it is zero. If no further development of secondary sex characteristics is observed, normally after one year of pharmacological estrogen doses, an oral therapy with estrogens, such as estrogen replacement for regular postmenopausal women (e.g. 2–4 mg estradiol or estradiol valerate according to serum estrogen and LH levels), is recommended. Prior to sex-reversal surgery, estrogen therapy should be discontinued for 3–4 weeks to avoid a hypercoagulable state (FUTTERWEIT 1998). Surgical sex reassignment completes the multidisciplinary treatment concept and should not be performed until 6–9 months following the start of cross-gender hormone therapy (SCHLATTERER et al. 1998).

One major clinical complication of estrogen administration is the expected thromboembolic event. In one large series of 303 male-to-female transsexuals a 45-fold increase in deep venous thrombosis and/or pulmonary embolism above expectation for age has been observed (ASSCHEMAN et al. 1989). In the same study with estrogen- and cyproterone-acetate-treated male-to-female transsexuals, a 400-fold increase in hyperprolactinemia incidence was shown. But interestingly enough, only single prolactinoma case reports have been cited for estrogen-treated male-to-female transsexuals (KOVACS et al. 1994). The remaining biochemical and endocrinological effects of the estrogen treatment of male-to-female transsexuals are discussed in other sections of this chapter.

L. The Therapeutic Value of Antiestrogens and Aromatase Inhibitors in Men

In male mammals, including men, testicular androgen production is dependent on the tropic hypophyseal effect of LH, which in turn is synthesized and secreted under the tropic control of GnRH. Testosterone and estradiol have negative feedback effects on the levels of these tropic hormones (GOOREN et al. 1984; WINTERS and TROEN 1985; CROWLEY et al. 1991). The feedback is mediated by androgens and/or ERs (HABENICHT and EL ETREBY 1992). As a consequence, the treatment with either an ER and/or AR antagonist does not only inhibit the effect of androgens or estrogens at the corresponding peripheral target organ but also at the brain, inducing a counter-regulatory increase to compensate for a supposedly deficient hormone synthesis. The same happens under treatment with inhibitors of androgens or estrogens, such as, for instance, aromatase inhibitors.

In the rat, feedback regulation is dependent almost exclusively on androgens. Thus, treatment with an AR antagonist such as flutamide results in an immediate and permanent increase in LH and testosterone levels. Conversely, treatment with an aromatase inhibitor leads only to a transient increase in serum testosterone and LH concentration. Anordrin is an antiestrogen with agonist activity and is used as a postcoital contraceptive in China. Anordrin and its metabolite anordriol act in rats like fully active estrogen agonists. A significant decrease in serum LH, FSH, and testosterone levels occurred in treated animals as compared to controls. The compound adversely affected spermatogenesis and induced moderate to severe degenerative changes in the epididymis and prostate, and atrophy of the glands of the seminal vesicles (VANAGE et al. 1997).

The SERM raloxifene did not compete for binding of the androgen [^{3}H]-R1881 (methyltrienolone) in cytosolic extracts of the ventral prostate in rats. It produced a significant dose-dependent regression of the ventral prostate and seminal vesicles. This was associated with a decline in T. Raloxifene antagonizes testosterone-induced increases in the prostate weight of castrated rats, but does not bind to AR or affect prostatic 5α-reductase or testicular steroid 17α-hydroxylase activity (BUELKE-SAM et al. 1998). Administration of estradiol to castrated male rats stimulated four-fold increases in in vitro ventral prostatic binding of [^{3}H]-R1881. When raloxifene was co-administered with estradiol, the compound markedly antagonized the estrogen-induced increase in prostatic binding of [^{3}H]-R1881, confirming its antiestrogenic properties in male rats. Raloxifene increased serum prolactin and decreased serum FSH, while leaving LH unaffected (NEUBAUER et al. 1993). There was no indication in either male rat study that raloxifene caused important changes in sperm production, sperm quality, or male reproductive performance at doses as high as 100 mg/kg/day orally (HOYT et al. 1998).

In dogs, feedback regulation is also exclusively dependent on estrogens (WINTER et al. 1983). Therefore, in this species, treatment with an aromatase inhibitor results in immediate and permanent increases in LH and testosterone, whereas treatment with an antiandrogen such as flutamide does not induce counter-regulatory increases in LH and testosterone secretion. Subhuman primates, as well as men, have a position somewhere in between these two extremes. Thus, in the monkey, treatment with the antiandrogen flutamide induced a slight increase in T, while combined treatment with an aromatase inhibitor plus an antiandrogen resulted in complete opening of the feedback loop (HABENICHT and EL ETREBY 1992; HABENICHT et al. 1992).

For the assessment of antiestrogenic activities in the brain, it is important that while certain antiestrogens do overcome the blood-brain barrier, others do not. For example, tamoxifen, clomiphene, RU 58668, and UC 23469 act on the brain after systemic use (VAGELL and MCGINNIS 1997; ORTEGA et al. 1993), while the "pure" antagonists, e.g. ZM 182780 (formerly named ICI 182780), ICI 182780, and ZK 164015, do not (WADE et al. 1993; FUHRMANN et al. 1998).

How about the therapeutic value of antiestrogens in men? Certain forms of male gynecomastia seem to respond to antiestrogens (EVERSMANN et al. 1984; GLASS 1993; VIZNER et al. 1994). The use of a tamoxifen therapy in oligospermia is being discussed controversially. The mechanism of action in male infertility treatment is presumably based on a competitive displacement of estrogens from the ERs in the hypothalamus and pituitary. Since estrogens inhibit pituitary gonadotropin secretion via a negative feedback mechanism, blockage of the estrogen effect may cause an increase in gonadotropins. Increase in LH and FSH is assumed to improve spermatogenesis (NIESCHLAG et al. 1993).

ROLF et al. (1996) analyzed clinical studies including a total of 1586 infertile men treated with tamoxifen. Six studies were randomized and placebo-controlled. The randomized study with the largest sample size provided no data about the frequency of pregnancies but showed a significant increase in sperm concentration and the number of viable sperm in ejaculate (KOTULAS et al. 1994). No significant changes in the sperm parameters examined were stated in the remaining randomized studies. ROLF et al. (1996) conclude that there is no proof of any improvement in fertility chances by tamoxifen. According to STERZIK et al. (1993), there is a lack of correlation between a rise in hormone levels and improvement of sperm quality, which suggests that tamoxifen is of questionable value in men with idiopathic oligozoospermia (BREZNIK and BORKO 1993).

On the other hand, more favorable results have been reported with clomiphene citrate (HAMMAMI 1996) or with the combination of tamoxifen citrate plus clomiphene citrate (SUGINAMI et al. 1993). The prospective randomized clinical study with 80 oligozoospermic men by ADAMOPOULOS et al. (1997) yielded a similar result. The authors found that the combination of testosterone undecanoate with tamoxifen citrate enhanced the effects of each agent given independently on seminal parameters in men with idiopathic oligozoospermia.

The therapeutic value of aromatase inhibitors in infertile men, at present, cannot yet be assessed. The aromatase inhibitors CGS 16949A as well as CGS 20267 suppress the E2 and increase FSH, LH, and T dose-dependently in men (BHATNAGAR et al. 1992; TRUNET et al. 1993).

Another interesting aspect is changes in estrogen biosynthesis by concomitant administration of other drugs. HERZOG et al. (1991) found total and non-SHBG-bound E2 levels to be significantly higher among phenytoin-treated men with epilepsy than among untreated epileptic men or normal controls. Consequently, treatment using a combination of testosterone and the aromatase inhibitor testolactone may have significantly better effects on sexual function and seizure frequency than testosterone alone (HERZOG et al. 1998).

M. The Concept of Non-Feminizing Estrogens

The utilization of the benefits of estrogen on bone (Sect. G), CVS (Sect. H), and CNS (Sect. I) without feminizing effects on the genitourinary system (Sect. D) or mammary gland (Sect. E) is doubtless a big challenge and might be a great gain for hormone therapy in men. The story began with the phytoestrogens (MURKIES et al. 1998; PRZYREMBEL 1998). Phytoestrogens emerged from their esoteric role in animal husbandry following the hypothesis that the western human diet is relatively deficient in these substances compared with societies where large amounts of plant foods and legumes are eaten. Evidence is beginning to accrue that they may begin to offer protection against a wide range of human conditions, including breast, bowel, prostate and other cancers, cardiovascular disease, impaired brain function, alcohol abuse, osteoporosis, and menopausal symptoms. Of the two main classes of these weak estrogens, the isoflavones are under intensive investigation due their high levels in soyabean. Like the "antiestrogen" tamoxifen, these seem to have estrogenic effects in human subjects in the CVS and bone. Although previously only available from food, isoflavones are now being marketed in health-food supplements or drinks, and tablets may soon be available over the counter as "natural" hormone-replacement therapy. In cancer, antiestrogenic effects are thought to be important, although genistein especially has been shown to induce wide-ranging anticancer effects on cell lines independent of any hormone-related influence. There are few indications of harmful effects at present, although possible proliferative effects have been reported. In infants, the effects of high levels of isoflavones in soya-milk formulas are uncertain. The second group of phytoestrogens are the lignans. They have been less investigated despite their known antiestrogenic effects and more widespread occurrence in foods (ADLERCREUTZ and MAZUR 1997; BINGHAM et al. 1998; THAM et al. 1998).

Epidemiological studies, as well as in vitro and in vivo experiments, provide evidence suggesting that isoflavonoids and lignans are protective and lower the risk of prostate cancer. The protective effect seems to occur mainly during the promotional phase of the proliferative disease. However, as long as the role of estrogens in prostate cancer is unclear (Sect. D.IV.), the potential beneficial effects of phytoestrogens in lowering prostate-cancer risk remain hypothetical (ADLERCREUTZ and MAZUR 1997).

Investigation into the possible benefits of phytoestrogens is hampered by lack of analytical standards and, hence, inadequate methods for the measurement of low levels in most foods (BINGHAM et al. 1998). This has induced scientists to search, *inter alia*, for other non-feminizing estrogens for the treatment of men. As a first example, SERMs may be cited (BRYANT and DERE 1998; GRESE and DODGE 1998; HOYT et al. 1998).

Recent advances in our understanding of the molecular mechanism of action of ER agonists and antagonists have indicated that the pharmacology of SERMs is complex, and have highlighted the fact that not all estrogens are the same. These advantages have led to the conclusion that ERs do not merely change from inactive to an active forms upon binding a ligand but are quite malleable and can exist in several different induced conformations. Importantly, the transcription apparatus within a cell is configured so as to be able to distinguish between these different complexes (MCDONELL 1998).

In cholesterol-fed male rabbits, the oral administration of estradiol or the SERM levormeloxifene reduced cholesterol accumulation in the aortic arch by approximately 50%. The other SERM, *d*-ormeloxifene, was ineffective in this respect (HOLM et al. 1997).

We found that 17α-estradiol and analogs of this epimer of the "classical" estrogen, 17β-estradiol and a derivative of 17α-estradiol – J861 –, show the same or more powerful antioxidative potential as 17β-estradiol, but possess only marginal proliferative activities on endometrium and uterus (Fig. 5). This observation correlates very well with the low binding affinities of 17α-estradiol to ER. The daily administration of 2 mg 17α-estradiol to elderly men prolonged significantly the lag time for ex vivo oxidation of LDL by copper. No signs of "classical" estrogenicity (no changes in serum levels of 17β-estradiol, estrone, SHBG, prolactin, or testosterone or subjectively recorded breast tenderness) were seen (OETTEL et al. 1995, 1996a and b). WASHBURN et al. (1996) apparently followed the same strategy. Using a different 17β-estradiol epimer, 17α-dihydroequilenin, in male rhesus monkeys, they were successful in preventing arteriosclerosis without feminizing effects. In contrast to female rhesus monkeys, the serum insulin concentrations were decreased by 17α-dihydroequilenin, and an insulin-receptor resistance was overcome.

Reactive oxygen species play an important role in both the physiology and pathology of the human spermatozoon (AITKEN 1995). Physiological action of LH in the testis causes lipid peroxidation and maintains high activities of peroxide-metabolizing enzymes in the gonadal interstitial tissue. The steroidogenic steps regulated by P450 enzymes are the most likely sites of free-radical generation (PELTOLA et al. 1996).

All estrogens have an antioxidative potential. This means that they protect, e.g., the cell membrane against the oxidative attack of free-radical oxygen species (ROS). No influence on ROS-generating enzyme systems has become known so far. The synthesis of $\Delta^{8,9}$-dehydro derivatives of 17α-estradiol has made it possible that the formation of ROS is blocked too (RÖMER et al. 1997). The so-called scavestrogens are novel derivatives of 17α-estradiol and combine nongenomic (antioxidative, radical-scavenging) effects (Fig. 5) with classical genomic (ER-mediated) activities in the brain while having a low (and clinically non-relevant) estrogenic effect outside the CNS. These scavestrogens distinguish themselves by a CNS-targeting

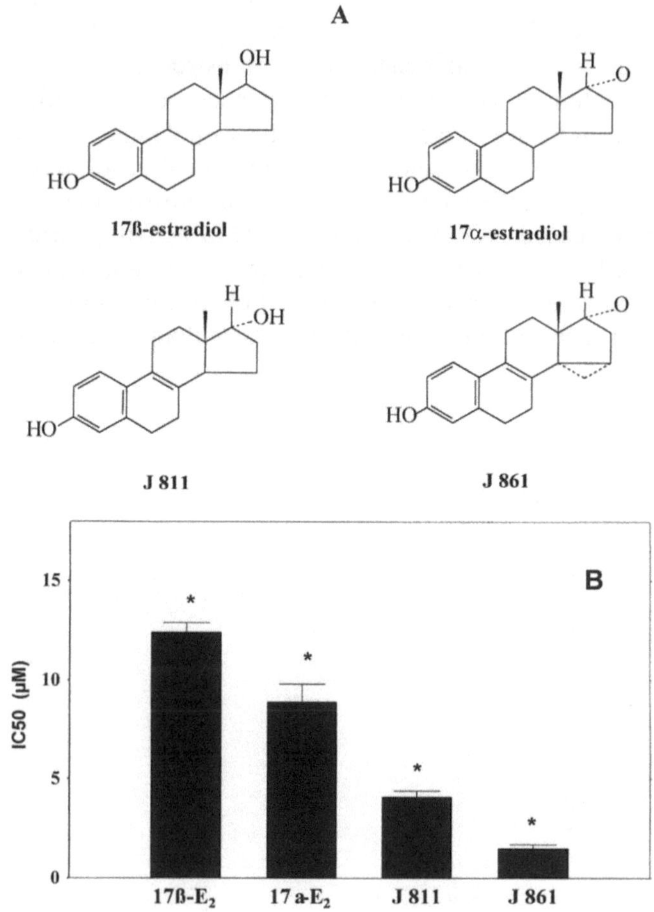

Fig. 5A,B. Chemical structure of the so-called scavestrogens J 811 and J 861 in comparison to 17β- and 17α-estradiol (**A**) and inhibition of lipid peroxidation in synaptosomal membranes from rats (**B**) (RÖMER et al. 1997)

behavior and are a completely new alternative for estrogen replacement with non-feminizing estrogens in elderly men.

In several in vitro as well as in vivo studies, we compared the estrogen-like effects of J 861 in the genital tract and the CNS with those of 17β-estradiol. The scavestrogen displayed strongly reduced affinity for, and marginal transactivational efficacy at, the "classic" ERs (ERα as well as ERβ). However, while their uterotropic efficacy in vivo was 100–1000 times lower than that of 17β-estradiol, several observations following chronic administration in ovariectomized rats are indicative of CNS-selective estrogen-like activity. Thus, J 861 regulated the secretory activity of the pituitary-adrenal axis in an estrogen-like fashion and significantly stimulated the transcription of estrogen-dependent genes in the CNS (corticotropin-releasing hormone, oxytocin, Bcl-2; see Fig. 6) at doses which were unable to promote uterine enlargement, unlike 17β-estradiol. Chronic treatment with J861 in doses which barely affected the genital tract produced estrogen-like anxiolysis and improved the retention of conditioned avoidance behavior. Moreover, pretreatment with J861 was able to attenuate scopolamine-induced learning impairment in an estrogen-resembling manner. Finally, J861 was as potent as 17β-estradiol in protecting nerve cells from oxidative stress in vitro, and significantly decreased NO production and NOS activity in rat brain upon administration in vivo. These results indicate that non-feminizing estrogens like J861 act as selective neurotropic and, probably, neuroprotective agents, combining estrogen-like genomic effects with enhanced radical-scavenging

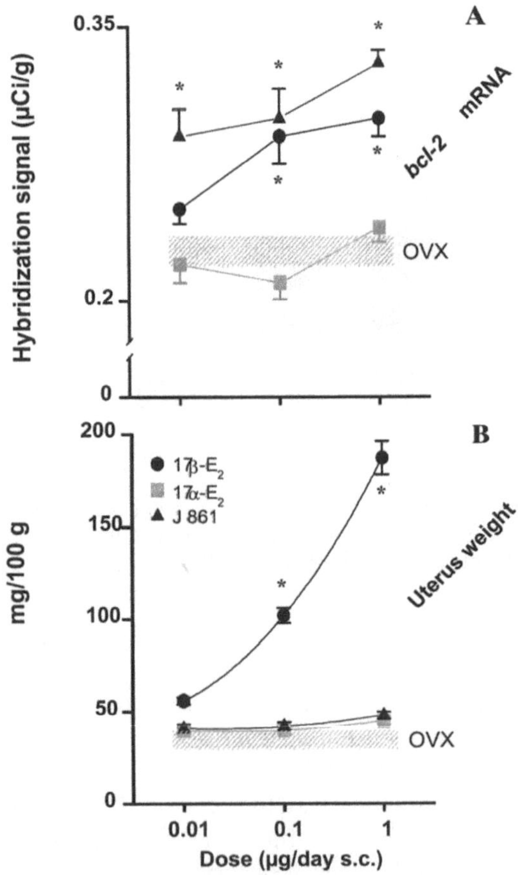

Fig. 6A,B. Non-feminizing estrogen J 861 enhances the expression of the estrogen-driven anti-apoptotic gene *Bcl-2* in the rat hippocampus (**A**), whereas J 861, like the parent compound 17α-estradiol, doesn't show any estrogenic activity on the uteri of the same animals (**B**) (PATCHEV et al. 1998)

capacity (Patchev et al. 1998). Thus, the non-feminizing estrogens emerge as prime candidates for gender-independent estrogen-replacement therapy.

Thus, all the pharmaceutical tasks of a specific brain-enhanced delivery of estradiol become more or less superfluous (Simpkins et al. 1998). A goal for the future is to find an answer to the question of whether scavestrogens can effectively exert their effects even in the presence of normal estradiol levels in men, and thus expand the therapeutic applications of these compounds.

References

Adamopoulos DA, Nicopoulou S, Kapolla N, Karamertzanis M, Andreou E (1997) The combination of testosterone undecanoate with tamoxifen citrate enhances the effects of each agent given independently on seminal parameters in men with idiopathic oligozoospermia. Fertil Steril 67:756–762

Adlercreutz H, Mazur W (1997) Phyto-estrogens and western diseases. Ann Med 29:95–120

Aitken RJ (1995) Free radicals, lipid peroxidation and sperm function. Reprod Fertil Dev 7:659–668

Alonso R, López-Coviella I (1998) Gonadal steroids and neuronal function. Neurochem Res 23:675–688

Alvaro D, Gigliozzi A, Piat C, Carli L, Fraioala F, Romeo R, Francia C, Attili AF, Capocaccia L (1997) Inhibition of biliary bicarbonate secretion in ethinyl estradiol-induced cholestasis is not associated with impaired activity of the Cl^-/HCO_3^--exchanger in the rat. J Hepatol 26:146–157

Ambrosi B, Gaggini M, Travaglini P, Moriondo P, Elli R, Faglia G (1981) Hypothalamic-pituitary-testicular function in men with PRL-secreting tumors. J Endocrinol Invest 4:309–315

Anderson DC (1974) Sex hormone binding globulin. Clin Endocrinol 3:69–96

Anderson FH, Francis RM, Hindmarsh P, Fall C, Cooper C (1996) Serum oestradiol in osteoporotic and normal men is related to bone mineral density. In: Papapoulos SE, Lips P, Pols HAP, Johnston CC, Delmas PD (eds) Osteoporosis 1996. Proceedings of the 1996 World Congress on Osteoporosis. Elsevier, Amsterdam, pp 377–381

Anderson FH, Francis RM, Peaston RT, Wastell HJ (1997) Androgen therapy in eugonadal men with osteoporosis – effects of six months' treatment on markers of bone formation and resorption. J Bone Miner Res 12:472–478

Anderson FH, Francis RM, Selby PL, Cooper C (1998) Sex hormones and osteoporosis in men. Calcif Tissue Int 62:185–188

Andersson AM, Juul A, Petersen JH, Müller J, Groome NP, Skakkebaek NE (1997) Serum inhibin B in healthy pubertal and adolescent boys: regulation to age, stage of puberty, and follicle-stimulating hormone, luteinizing hormone, testosterone, and estradiol levels. J Clin Endocrinol Metab 82:3976–3981

Angelin B, Olivecrona H, Reihnér E, Rudling M, Stahlberg D, Eriksson M, Ewerth S, Henriksson P, Einarsson K (1992) Hepatic cholesterol metabolism in estrogen-treated men. Gastroenterology 103:1657–1663

Asscheman H, Gooren LJG, Eklund PLE (1989) Mortality and morbidity in transsexual patients with cross-gender hormone treatment. Metabolism 38:869–873

Audy MC, Dufy B (1996) 17β-Estradiol stimulates a rapid Ca^{2+} influx in LNCaP human prostate cancer cells. Eur J Endocrinol 135:367–373

Auger J, Kunstmann JM, Czyglik F, Jouannet P (1995) Decline in semen quality among fertile men in Paris during the past 20 years. New Engl J Med 332:281–285

Bachrach BE, Smith EP (1996) The role of sex steroids in bone growth and development: evolving new concepts. Endocrinologist 6:362–368

Bagatell CJ, Knopp RH, Rivier JE, Bremner WJ (1994a) Physiological levels of estradiol stimulate plasma high density lipoprotein cholesterol levels in normal men. J Clin Endocrinol Metab 78:855–861

Bagatell CJ, Heiman JR, Rivier JE, Bremner WJ (1994b) Effects of endogenous testosterone and estradiol on sexual behavior in normal young men. J Clin Endocrinol Metab 78:711–716

Bagatell CJ, Dahl KD, Bremner WJ (1994c) The direct pituitary effect of testosterone to inhibit gonadotropin secretion in men is partially mediated by aromatization to estradiol. J Androl 15:15–21

Banger M, Hiemke C, Haupt M, Knuppen R (1996) Excretion of 2- and 3-monomethyl ethers of 2-hydroxyestrogens in healthy male volunteers. Eur J Endocrinol 135:193–197

Baker HWG, Burger HG, de Kretser DM, Hudson B, O'Connor S, Wang C, Mirovics A, Court J, Dunlop M, Rennie GC (1976) Changes in the pituitary-testicular system with age. Clin Endocrinol 5:349–372

Barr DP, Russ EM, Eder HA (1952) Influence of estrogens on lipoproteins in atherosclerosis. Trans Ass Am Phys 65:102–113

Batra S, Karlsson R, Witt L (1996) Potentiation by estramustine of the cytotoxic effect of vinblastine and doxorubicin in prostatic tumor cells. Int J Cancer 68:644–649

Bauduceau B, Reboul P, Le Guyadec T, Legrelle M, Mayaudon H, Gautier D (1993) Profil hormonal des gynécomasties idiopathiques de l'adulte jeune. Ann d'Endocrinologie 54:163–167

Baxter LR, Mazziotta JC, Phelps ME, Selin CE, Guze BH, Fairbanks I (1987) Cerebral glucose metabolic rates in normal human females vs. normal males. Psychiatry Res 21:237–245

Bellido T, Jelka RJ, Boyce BF, Grasole G, Broxmeyer H, Dabrynyle SA, Murray R, Managolas S (1995) Regulation of interleukin-6, osteoclastogenesis and bone mass by androgens. J Clin Invest 95:2886–2895

Benson RC Jr, Gill GM (1986) Estramustine phosphate compared with diethylstilbestrol. Am J Clin Oncol 9:341–351

Berglund L, Carlström K, Stege R, Gottlieb C, Eriksson M, Angelin B, Henriksson P (1996) Hormonal regulation of serum lipoprotein (a) levels: effects of parenteral administration of estrogen or testosterone in males. J Clin Endocrinol Metab 81:2633–2637

Berman NG, Wang C, Paulsen CA (1996) Methodological issues in the analysis of human sperm concentration data. J Androl 17:68–73

Bernecker PM, Willvonseder R, Resch H (1995) Decreased estrogen levels in patients with primary osteoporosis. J Bone Miner Res 10[Suppl 1]:T364

Bhatnagar AS, Müller P, Schenkel L, Trunet PF, Beh I, Schieweck K (1992) Inhibition of estrogen biosynthesis and its consequences on gonadotrophin secretion in the male. J Steroid Biochem Mol Biol 41:437–443

Bilezikian JP, Morishima A, Bell J, Grumbach MM (1998) Increased bone mass as a result of estrogen therapy in a man with aromatase deficiency. N Engl J Med 339:599–603

Bingham SA, Atkinson C, Liggins J, Bluck L, Coward A (1998) Phyto-estrogens: where are we now? Br J Nutr 79:393–406

Birge SJ (1996) Is there a role for estrogen replacement therapy in the prevention and treatment of dementia? J Am Geriatr Soc 44:865–870

Blanco-Rodríguez J, Martínez-García C (1996) Further observations on the early events that contribute to establishing the morphological pattern shown by the oestradiol suppressed testis. Tissue Cell 28:387–399

Blanco-Rodríguez J, Martínez-García C (1997) Apoptosis pattern elicited by oestradiol treatment of the seminiferous epithelium of the adult rat. J Reprod Fert 110:61–70

Blumenthal RS, Heldman AW, Brinker JA, Resar JR, Coombs VJ, Gloth ST, Gerstenblith G, Reis SE (1997) Acute effects of conjugated estrogens on coronary blood flow response to acetylcholine in men. Am J Cardiol 80:1021–1024

Boonen S, Vanderschueren D, Cheng XG, Verbeke G, Dequeker J, Geusens P, Broos P, Bouillon R (1997) Age-related (type II) femoral neck osteoporosis in men: Biochemical evidence for both hypovitaminosis D – and androgen deficiency -induced bone resorption. J Bone Miner Res 12:2119–2126

Bostofee E, Serup J, Rebbe H (1983) Has the fertility of Danish men declined through the years in terms of semen quality? A comparison of semen qualities between 1952 and 1972. Int J Fertil 28:91–95

Boyns AR, Cole EN, Phillips MEA (1974) Plasma prolactin, GH, LH, FSH, TSH and testosterone during treatment of prostatic carcinoma with estrogens. Eur J Cancer 10:445–449

Bradlow HL, Telang NT, Sepkovic DW, Osborne MP (1996) 2-Hydroxyestrone: the "good" estrogen. J Endocrinol 150:S259–S265

Brake A, Krause W (1992) Decreasing quality of semen. BMJ 305:1498

Breznik R, Borko E (1993) Effectiveness of antiestrogens in infertile men. Arch Androl 31:43–48

Briggs M, Briggs M (1974) Oral contraceptive for men. Nature 252:585–586

Brinkman-Van der Linden ECM, Havenaar EC, Van Ommen ECR, Van Kamp GJ, Gooren LJG, Van Dijk WV (1996) Oral estrogen treatment induces a decrease in expression of sialyl Lewis x on α_1 -acid glycoprotein in females and male-to-female transsexuals. Glycobiology 6:407–412

Bromwhich P, Cohen P, Stewart I, Walker A (1994) Decline in sperm counts: an artefact of changed reference range of "normal"? BMJ 309:19–22

Brotons JA, Olea-Serrano MF, Villalobos M, Olea N (1994) Xenoestrogens released from lacquer coating in food cans. Environ Health Perspect 103:608–612

Bruch HB, Wolf L, Budde R, Romalo G, Schweikert HU (1992) Androstenedione metabolism in cultured human osteoblast-like cells. J Clin Endocrinol Metab 75:101–105

Bryant HU, Dere WH (1998) Selective estrogen receptor modulators: an alternative to hormone replacement therapy. Proc Soc Exp Biol Med 217:45–52

Buelke-Sam J, Bryant HU, Francis PC (1998) The selective estrogen receptor modulator, raloxifene: an overview of nonclinical pharmacology and reproductive and developmental testing. Reprod Toxicol 12:217–221

Bujan L, Mansat A, Pontonnier F, Mieusset R (1996) Time series analysis of sperm concentration in fertile men in Toulouse, France, between 1977 and 1992. BMJ 305:471–472

Bulun SE (1996) Clinical review 78. Aromatase deficiency in women and men: would you have predicted the phenotypes? J Clin Endocrinol Metab 56:1278–1281

Burger H, de Laet CEDH, van Daele PLA, Weel AEAM, Witteman JCM, Hofman A, Pols HAP (1998) Risk factors of increased bone loss in an elderly population – the Rotterdam study. Am J Epidemiol 147:871–879

Buschmann I, Guddat HM (1997) Langzeitergebnisse der Therapie des Prostatakarzinoms mit Turisteron®. Beilage Aktuelle Urologie 28: Heft 7

Byar DP (1973) VACURG studies of prostate cancer. Cancer 32:1126

Byar DP, Corle DK (1988) Hormone therapy for prostate cancer: results of the Veterans Administration Cooperative Urological Research Group studies. NCI Monogr 7:165–170

Campisi D, Cutolo M, Carruba G, Lo Casto M, Comito L, Granata OM, Valentino B, King RJB, Castagnetta L (1993) Evidence for soluble and nuclear site I binding of estrogens in human aorta. Atherosclerosis 103:267–277

Carani C, Qin K, Simoni M, Faustini-Fustini M, Serpente S, Boyd J, Korach KS, Simpson ER (1997) Effect of testosterone and estradiol in a man with aromatase deficiency. N Engl J Med 337:91–95

Carles J, Domenech M, Gelabert-Mas A, Nogue M, Tabernero JM, Arcusa A, Guasch I, Miguel A, Ballesteros JJ, Fabregat X (1998) Phase II study of estramustine and vinorelbine in hormone-refractory prostate carcinoma patients. Acta Oncologica 37:187–191

Carlsen E, Giwercman A, Keiding N, Skakkebaek NE (1992) Evidence for decreasing quality of semen during past 50 years. Br J Med 305:609–613

Carlsen E, Giwercman A, Keiding N, Skakkebaek NE (1995) Declining semen quality and increasing incidence of testicular cancer: is there a common cause? Environ Health Perspect 103[Suppl 7]:137–139

Caulin-Glaster T, Garcia-Cardena G, Sarrel P, Sessa WC, Bender JR (1997) 17β-estradiol regulation of human endothelial cell basal nitric oxide release, independent of cytosolic Ca^{2+} mobilization. Circ Res 81:885–892

Cetinkaya M, Cetinkaya H, Ulusoy E, Baz S, Memis A, Yasa H, Yanik B, Öztürk B, Uzunalimoglu Ö (1998) Effect of postnecrotic and alcoholic hepatic cirrhosis on development of benign prostatic hyperplasia. Prostate 36:80–84

Chemnitius KH, Onken D (1976) Deposiston. Germed-Informationen, Jenapharm

Cheng AL, Chen YC, Yeh KH, Chuang SE, Chen BR, Chen DS (1996) Chronic oral etoposide and tamoxifen in the treatment of far-advanced hepatocellular carcinoma. Cancer 77:872–877

Cheng LP, Kuwahara M, Jacobsson J, Foegh ML (1991) Inhibition of myointimal hyperplasia and macrophage infiltration by estradiol in aorta allografts. Transplantation 52:967–972

Cherrier M, Craft S, Plymate S, Asthana S, Matsumoto A, Bremner B (1998) Effects of testosterone on cognition in healthy older. The 80th Annual Meeting of the Endocrine Society, June 24–27, 1998, New Orleans, pp 2–643

Chian RC, Blondin P, Sirard MA (1996) Effect of progesterone and/or estradiol-17β on sperm penetration in vitro of bovine oocytes. Theriogenology 46:459–469

Cui L, Mori T, Takahashi S, Imaida K, Akagi K, Yada H, Yaono M, Shirai T (1998) Slight promotion effects of intermittent administration of testosterone propionate and/or diethylstilbestrol on 3′,2′-dimethyl-4-aminobiphenyl-initiated rat prostate carcinogenesis. Cancer Lett 122:195–199

Cohen PG (1998) The role of estradiol in the maintenance of secondary hypogonadism in males in erectile dysfunction. Med Hypotheses 50:331–333

Colditz GA (1998) Relationship between estrogen levels, use of hormone replacement therapy, and breast cancer. J Natl Cancer Inst 90:814–823

Collins P, Rosano GMC, Sarrel PM, Ulrich L, Adamopoulos S, Beale CM, McNeill JG, Poole-Wilson PA (1995) 17β-Estradiol attenuates acetylcholine-induced coronary arterial constriction in women but not in men with coronary heart disease. Circulation 92:24–30

Colvard DS, Eriksen EF, Keeting PE, Wilson EM, Lubahn DB, French FS, Riggs BL, Spelsberg TC (1989) Identification of androgen receptors in normal human osteoblast-like cells. Proc Natl Acad Sci USA 86:854–857

Comhaire F, Waeleghem KV, de Clercq N, Schoonjans F (1996) Declining sperm quality in European men. Andrologia 28:300–301

Cooke PS, Young P, Hess RA, Cunha GR (1991) Estrogen receptor expression in developing epididymis, efferent ductules and other male reproductive organs. Endocrinology 128:2874–2879

Couse JF, Korach KS (1998) Exploring the role of sex steroids through studies of receptor deficient mice. J Mol Med 76:497–511

Cox RL, Crawford ED (1995) Estrogens in the treatment of prostate cancer. J Urol 154:1991–1998

Crowley WFJ, Whitcomb RW, Jameson LJL, Weiss J, Finkelstein JS, O'Dea LSL (1991) Neuroendocrine control of human reproduction in the male. Rec Prog Horm Res 47:27–67

Daniel DG, Mathew RJ, Wilson WH (1988) Sex roles and regional cerebral blood flow. Psychiatry Res 27:55–64

Davidson JM, Chen JJ, Crapo L, Gray GD, Greenleaf WJ, Catania JA (1983) Hormonal changes and sexual function in aging men. J Clin Endocrinol Metab 57:71–77

DeLaet CEDH, van Hout LB, Burger H, Hofman A, Pols HAP (1997) Bone density of hip fractures in men and women: cross sectional analysis. BMJ 315:221–225

Del Rio G, Velardo A, Zizzo G, Avogaro A, Cipolli C, Della Casa L, Marrama P, Macdonald IA (1994) Effect of estradiol on the sympathoadrenal response to mental stress in normal men. J Clin Endocrinol Metab 79:836–840
De Kretser DM (1974) The regulation of male fertility. The state of the art and future possibilities. Contraception 9:561–583
Deslypère JP, Kaufman JM, Vermeulen T, Vogelaers D, Vandalem JL, Vermeulen A (1987) Influence of age on pulsatile luteinizing hormone release and responsiveness of the gonadotrophs to sex hormone feedback in men. J Clin Endocrinol Metab 64:68–73
Diamond TH, Thornley SW, Sekel R, Smerdely P (1997) Hip fractures in elderly men: prognostic factors and outcomes. Med J Aust 167:412–415
Dixon D, Couse JF, Korach KS (1997) Disruption of the estrogen receptor gene in mice. Toxicol Pathol 25:518–520
Dorrington JH, Armstrong DT (1975) Follicle-stimulating hormone stimulated oestradiol-17β synthesis in cultured Sertoli cells. Proc Natl Acad Sci USA 72:2677–2681
Dragan YP, Fahey S, Street K, Vaughan J, Jordan VC, Pitot HC (1994) Studies of tamoxifen as a promoter of hepatocarcinogenesis in female Fischer F344 rats. Breast Cancer Res Treat 31:11–25
Dragan YP, Vaughan J, Jordan VC, Pitot HC (1995) Comparison of the effects of tamoxifen and toremifene on liver and kidney tumor promotion in female rats. Carcinogenesis 16:2733–2741
Dragan YP, Shimel RJ, Sattler G, Vaughan JR, Jordan VC, Pitot HC (1998) Effect of chronic administration of mestranol, tamoxifen, and toremifene on hepatic ploidy in rats. Toxicol Sci 43:129–138
Drago JR (1984) The induction of Nb rat prostatic carcinomas. Anticancer Res 4:255–256
Dupaix A, Pineau C, Piquet-Pellorce C, Jégou B (1996) Paracrine and autocrine regulations of spermatogenesis. In: Hamamah S, Mieusset R (eds) Male gametes production and quality. INSERM, Paris, pp 47–63
Durkee TJ, Mueller M, Zinaman M (1998) Identification of estrogen receptor protein and messenger ribonucleic acid in human spermatozoa. Am J Obstet Gynecol 178:1288–1297
Eddy EM, Washburn TF, Bunch DO, Goulding EH, Gladen BC, Lubahn DB, Korach KS (1996) Targeted disruption of the estrogen receptor gene in male mice causes alteration of spermatogenesis and infertility. Endocrinology 137:4796–4805
Ehara H, Koji T, Deguchi T, Yoshii A, Nakano M, Nakane PK, Kawada Y (1995) Expression of estrogen receptor in diseased human prostate assessed by non-radioactive in situ hybridization and immunohistochemistry. Prostate 27:304–313
Elbers JMH, Asscheman H, Seidell JC, Frolich M, Meinders AE, Gooren LJG (1997) Reversal of the sex difference in serum leptin levels upon cross-sex hormone administration in transsexuals. J Clin Endocrinol Metab 82:3267–3270
Elger W, Schwarz S, Hedden A, Reddersen G, Schneider B (1995) Sulfamates of various estrogens are prodrugs with increased systemic and reduced hepatic estrogenicity at oral application. J Steroid Biochem Mol Biol 55:395–403
Elger W, Palme HJ, Schwarz S (1998) Novel estrogen sulfamates: a new approach to oral hormone therapy. Exp Opin Invest Drugs 7:575–589
Emons G, Merriam GR, Pfeiffer D, Loriaux DL, Ball P, Knuppen R (1987) Metabolism of exogenous 4- and 2-hydroxyestradiol in the human male. J Steroid Biochem 28:499–504
Eversmann T, Moito J, von Werder K (1984) Testosteron- und Oestradiolspiegel bei der Gynäkomastie des Mannes. Klinische und endokrine Befunde bei Behandlung mit Tamoxifen. Dtsch Med Wochenschr 109:1678–1682
Farhat MY, Abi-Younes S, Ramwell PW (1996) Non-genomic effects of estrogen and the vessel wall. Biochem Pharmacol 51:571–576

Farnsworth WE (1996) Roles of estrogen and SHBG in prostate physiology. Prostate 28:17–23
Ferrini RL, Barrett-Connor E (1998) Sex hormones and age: a cross-sectional study of testosterone and estradiol and their bioavailable fractions in community-dwelling men. Am J Epidemiol 147:750–754
Field AE, Colditz GA, Willett WC (1994) The relation of smoking, age, relative weight, and dietary intake to serum adrenal steroids, sex hormones, and sex hormone-binding globulin in middle-aged men. J Clin Endocrinol Metab 79:1310–1316
Fisch H, Goluboff ET (1996) Geographic variations in sperm counts: a potential cause of bias in studies on semen quality. Fertil Steril 65:1044–1046
Fisch H, Goluboff ET, Olson JH, Feldshuh J, Broder SJ, Barad DH (1996) Semen analyses in 1,283 men from the United States over 25-year period: no decline in quality. Fertil Steril 65:1009–1014
Fisher JS, Millar MR, Majdic G, Saunders PTK, Fraser HM, Sharpe RM (1997) Immunolocalisation of estrogen receptor-α within the testis and excurrent ducts of the rat and marmoset monkey from perinatal life to adulthood. J Endocrinol 153:485–495
Fishman J (1980) Estrogens in the environment. In: Proceedings of the Symposium on Estrogens in the Environment, Raleigh, North Carolina, U.S.A., September 10–12, 1979. Elsevier/North Holland
Forest MG, Portrat-Doyen S, Nicolino M, Morel Y, Chatelain PC (1996) Proceedings of the IV International Aromatase Conference, Tahoe City, Calif., p 22
Friedl KE, Hannan CJ, Jones RE, Kettler TM, Plymate SR (1990) High-density lipoprotein is not decreased if an aromatizable androgen is administered. Metabolism 39:69–77
Fuhrmann U, Parczyk K, Klotzbücher M, Klocker H, Cato ACB (1998) Recent developments in molecular action of antihormones. J Mol Med 76:512–524
Futterweit W (1998) Endocrine therapy of transsexualism and potential complications of long-term treatment. Arch Sex Behav 27:209–226
Gann PH, Hennekens CH, Longcope C, Verhoek-Oftedahl W, Grodstein F, Stampfer MJ (1995) A prospective study of plasma hormone levels, nonhormonal factors, and development of benign prostatic hyperplasia. Prostate 26:40–49
Gann PH, Hennekens CH, Ma J, Loncope C, Stampfer MJ (1996) Prospective study of sex hormone levels and risk of prostate cancer. J Natl Cancer Inst 88:1118–1126
Ganne-Carrié N, Chastang C, Uzzan B, Pateron D, Trinchet JC, Perret G, Beaugrand M (1997) Predictive value of serum sex hormone binding globulin for the occurrence of hepatocellular carcinoma in male patients with cirrhosis. J Hepatol 26:96–102
Gavaler JS, van Thiel DH (1988) Gonadal dysfunction and inadequate sexual performance in alcoholic cirrhotic men. Gastroenterology 5:1680–1683
Giltay EJ, Hoogeveen EK, Elbers JMH, Gooren LJG, Asscheman H, Stehouwer CDA (1998) Effects of sex steroids on plasma total homocysteine levels: a study in transsexual males and females. J Clin Endocrinol Metab 83:550–553
Girasole G, Jilka RL, Passeri G, Scott B, Boder G, Williams DC, Manolagas SC (1992) Estradiol inhibits interleukin-6 production by bone marrow-derived stromal cells and osteoblasts in vitro. A potential mechanism for the antiosteoporotic effect of estrogens. J Clin Invest 89:883–891
Giri S, Thompson PD, Taxel P, Contois JH, Otvos J, Allen R, Ens G, Wu AHB, Waters DD (1998) Oral estrogen improves serum lipids, homocysteine and fibrinolysis in elderly men. Atherosclerosis 137:359–366
Giwercman A, Skakkebaek NE (1992) The human testis – an organ at risk? Int J Androl 15:373–375
Glass AR (1994) Gynecomastia. Endocrinol Metab Clin North Am 23:825–837
Gluud C, Bahnsen M, Bennett P, Brodthagen UA, Dietrichson O, Johnsen SG, Nielsen J, Micic S, Svendsen LB, Svenstrup B (1983) Hypothalamic-pituitary-gonadal function in relation to liver function in men with alcoholic cirrhosis. Scand J Gastroenterol 18:939–944

Gluud C, Copenhagen Study Group for Liver Diseases (1987) Serum concentration in men with alcoholic cirrhosis: background for variation. Metabolism 36: 373–378
Gooren LJG, van der Veen EA, van Kessel H, harmsen-Louman W (1984) Estrogens in the feedback regulation of gonadotropin secretion in men: effects of administration of estrogen to agonadal subjects and the antiestrogen tamoxifen and the aromatase inhibitor Δ′-testolactone to eugonadal subjects. Andrologia 16: 568–577
Gordon CG, Olivo J, Rafil F, Southern AL (1975) Conversion of androgens to estrogens in cirrhosis of the liver. J Clin Endocrinol Metab 40:1018–1026
Goulding A, Gold E (1993) Flutamide-mediated androgen blockade evokes osteopenia in the female rat. J Bone Miner Res 8:763–769
Green S, Walter P, Kumar V, Krust A, Bornert JM, Argos P, Chambon P (1986) Human estrogen receptor cDNA: sequence, expression and homology to v-erb-A. Nature 320:134–139
Greendale GA, Edelstein S, Barret-Connor E (1997) Endogenous sex steroids and bone mineral density in older women and men: the Rancho Bernardo Study. J Bone Miner Res 12:1833–1843
Grese TA, Dodge JA (1998) Selective estrogen receptor modulators (SERMs). Curr Pharm Design 4:71–92
Greim H (1998) Hormonähnlich wirkende Stoffe in der Umwelt – Einführung und Sachstand. Bundesgesundhbl 8:326–329
Griffiths K, Eaton CL, Harper ME, Turkes A, Peeling WB (1994) Hormonal treatment of advanced disease: some newer aspects. Semin Oncol 21:672–687
Guddat HM, Schnorr D, Dörner G, Stahl F, Rohde W (1979) Erste Erfahrungen mit Äthinylöstradiolsulfonat (J 96) bei der Therapie des Prostatakarzinoms. Z Urol Nephrol 73:153–157
Guzelian PS (1982) Comparative toxicology of chlorodecone (kepone) in humans and experimental animals. Annu Rev Pharmacol Toxicol 22:89–113
Haapiainen R, Rannikko S, Alfthan O, and the Finnprostate Group (1990) Comparison of primary orchidectomy and polyestradiol phosphate in the treatment of advanced prostatic cancer. Br J Urol 66:94–97
Habenicht UF (1998) Estrogens for men: good or bad news. Aging Male 1:73–79
Habenicht UF, El Etreby MF (1992) Role of estrogens and androgens for the negative feedback control of gonadotropin secretion in different species. In: Rossmanith WG, Scherbaum WA (eds) Neuroendocrinology of Sex Steroids. de Gruyter, Berlin, pp 135–147
Habenicht UF, El Etreby MF, Lewis R, Ghoniem G, Roberts J (1992) Long term effect of different types of androgen deprivation in combination with the aromatase inhibitor atamestane on the prostate of intact male cynomolgus monkeys. Presented at the 36th Symposium of the German Society of Endocrinology, Erlangen, March 11–14, pp 57
Hackney AC, Fahrner CL, Stupnicki R (1997) Reproductive hormonal responses to maximal exercise in endurance-trained men with low resting testosterone levels. Exp Clin Endocrinol Diabetes 105:291–295
Hall ED, Pazara KE, Linseman KL (1991) Sex differences in postischemic neuronal necrosis in gerbils. J Cereb Blood Flow Metab 11:292–298
Hammami MM (1996) Hormonal evaluation in idiopathic oligozoospermia: correlation with response to clomiphene citrate therapy and sperm motility. Arch Androl 36:225–232
Handa K, Ishii H, Kono S, Shinchi K, Imanishi K, Mihara H, Tanaka K (1997) Behavioral correlates of plasma sex hormones and their relationships with plasma lipids and lipoproteins in Japanese men. Atherosclerosis 130:37–44
Hawk T, Zhang YQ, Rajakumar G, Day AL, Simpkins JW (1998) Testosterone increases and estradiol decreases middle cerebral artery occlusion lesion size in male rats. Brain Res 796:296–298
Heinemann L (1998) Report to Jenapharm

Hellriegel ET, Matwyshyn GA, Fei P, Dragnev KH, Nims RW, Lubet RA, Kong ANT (1996) Regulation of gene expression of various phase I and phase II drug-metabolizing enzymes by tamoxifen in rat liver. Biochem Pharmacol 52:1561–1568

Henderson VW, Buckwalter JG (1994) Cognitive deficits of men and women with Alzheimer's disease. Neurology 44:90–96

Henriksson P, Blombäck M, Eriksson A, Stege R, Carlström K (1990) Effect of parenteral estrogen on the coagulation system in patients with prostatic carcinoma. Br J Urol 65:282–285

Herzog AG, Levesque L, Drislane F, Ronthal M, Schomer D (1991) Phenytoin-induced elevation of serum estradiol and reproductive dysfunction in men with epilepsy. Epilepsia 32:550–553

Herzog AG, Klein P, Jacobs AR (1998) Testosterone versus testosterone and testolactone in treating reproductive and sexual dysfunction in men with epilepsy and hypogonadism. Neurology 50:782–784

Hess RA, Bunick D, Bahr JM (1995) Sperm, a source of estrogen. Environ Health Perspect 103[Suppl 7]:59–62

Hess RA, Bunick D, Lee KH, Bahr J, Taylor JA, Korach KS, Lubahn DB (1997a) A role for estrogens in the male reproductive system. Nature 390:509–512

Hess RA, Gist DH, Bunick D, Lubahn DB, Farrell A, Bahr J, Cooke PS, Greene GL (1997b) Estrogen receptor (α and β) expression in the excurrent ducts of the adult male rat reproductive tract. J Androl 18:602–611

Hiramatsu M, Maehara I, Orikasa S, Sasano H (1996) Immunolocalization of estrogen and progesterone receptors in prostatic hyperplasia and carcinoma. Histopathology 28:163–168

Höfling G, Heynemann H (1998) Die orale Östrogentherapie des fortgeschrittenen Prostatakarzinoms – Anlaß für eine Neubewertung? Urologe (B) 2:165–170

Holm P, Shalmi M, Korsgaard N, Guldhammer B, Skouby SO, Stender S (1997) A partial estrogen receptor agonist with strong antiatherogenic properties without noticeable effect on reproductive tissue in cholesterol-fed female and male rabbits. Arterioscler Thromb Vasc Biol 17:2264–2272

Hörhold-Schubert C, Hobe G, Kaufmann G, Schumann G (1995) Zur Biotransformation von Dienogest durch Mikroorganismen. In: Teichman AT (ed) Dienogest – Präklinik und Klinik eines Gestagens, 2nd edn. Walter de Gruyter Berlin, pp

Hoyt JA, Fisher LF, Buelke-Sam J, Francis PC (1998a) The selective estrogen receptor modulator, raloxifene: reproductive assessments following premating exposure in female rats. Reprod Toxicol 12:233–245

Hoyt JA, Fisher LF, Swisher DK, Byrd RA, Francis PC (1998b) The selective estrogen receptor modulator, raloxifene: reproductive assessments in adult male rats. Reprod Toxicol 12:223–232

Huggins C, Hodges CV (1941) Studies of prostatic cancer: I. effect of castration, estrogen and androgen injection on serum phosphatase in metastatic carcinoma of the prostate. Cancer Res 1:293–297

Huggins C, Stevens RE, Hodges CV (1941a) Studies of prostatic carcinoma: II. the effect of castration on advanced carcinoma of the prostate gland. Arch Surg 43:209–223

Huggins C, Scott W, Hodges CV (1941b) Studies of prostatic cancer: III. the effects of favour of desoxicorticosterone and of estrogen on clinical patients with metastatic carcinoma of the prostate. J Urol 46:997–1006

Iafrati MD, Karas RH, Aronovitz M, Kim S, Sullivan TR, Lubahn DB, O'Donell TF Jr, Korach KS, Mendelsohn ME (1997) Estrogen inhibits the vascular injury response in estrogen receptor α-deficient mice. Nat Med 3:545–548

Ilio KY, Hess RA (1994) Structure and function of the ductuli efferentes: a review. Microsc Res Tech 29:432–467

Insker S, Yue W, Brodie A (1995) Human testicular aromatase: immunocytochemical and biochemical studies. J Clin Endocrinol Metab 80:1941–1947

Irvine S, Cawood E, Richardson D, MacDonald E, Aitken J (1996) Evidence of deteriorating semen quality in the United Kingdom: birth cohort study in 577 men in Scotland over 11 years. BMJ 312:467–471
Jackson H, Jones AR (1972) The effects of steroids and their antagonists on spermatogenesis. Adv Steroid Biochem Pharmacol 3:167–174
Jacobsen SJ, Goldberg J, Miles TP, Brody JA, Stiers W, Rimm AA (1990) Hip fracture incidence among the old and very old: a population-based study of 745,435 cases. Am J Public Health 80:871–873
Janssen JAMJL, Stolk RP, Pols HAP, Grobbee DE, de Jong FH, Lamberts SWJ (1998a) Serum free IGF-I, total IGF-I, IGFBP-1 and IGFBP-3 levels in an elderly population: relation to age and sex steroid levels. Clin Endocrinol 48:471–478
Janssen JAML, Burger H, Stolk RP, Grobbee DE, de Jong FH, Lamberts SWJ, Pols HAP (1998b) Gender specific relationship between serum free and total IGF-1 and bone mineral density in elderly men and women. Eur J Endocrinol 138:627–632
Janulis L, Hess RA, Bunick D, Nitta H, Janssen S, Asawa Y, Bahr JM (1996) Mouse epididymal sperm contain P450 aromatase which decreases as sperm traverse the epididymis. J Androl 17:111–116
Jennings PJ, Janowsky JS, Orwoll E (1998) Estrogen and sequential movement. Behav Neurosci 112:154–159
Jin B, Turner L, Walters WAW, Handelsman DJ (1996) Androgen or estrogen effects on human prostate. J Clin Endocrinol Metab 81:4290–4295
Joffe M (1996) Decreased fertility in Britain compared with Finland. Lancet 347:1519–1522
Johnson ML, Salveson A, Holmes L, Denison MS, Fry DM (1998) Environmental estrogens in agricultural drain water from the central valley of California. Bull Environ Contam Toxicol 60:609–614
Jouannet P, Auger J (1996) Declining sperm counts? More research is needed. Andrologia 28:302–303
Kaiser FE, Viosca SP, Morley JE, Mooradian AD, Davis SS, Korenman SG (1988) Impotence and aging, clinical and hormonal factors. J Am Geriatr Soc 36:511–516
Kalu DN, Hardin RR, Cockermam R (1984) Evaluation of the pathogenesis of skeletal changes in ovariectomized rats. Endocrinology 115:507–512
Kameda T, Mano H, Yuasa T, Mori Y, Miyazawa K, Shiokawa M, Nakamuru Y, Hiroi E, Hiura K, Kameda A, Yang NN, Hakeda Y, Kumegawa M (1997) Estrogen inhibits bone resorption by directly inducing apoptosis of the bone-resorbing osteoclasts. J Exp Med 186:489–495
Kampen DL, Sherwin B (1996) Estradiol is related to visual memory in healthy young men. Behav Neurosci 110:613–617
Kaps P, Girg F (1998) Die beherrschende Rolle des Östradiol (!) in der hypogonadalen Genese der Osteoporose – auch des Mannes. Osteologie 7[Suppl 1]:91–92
Kawano H, Motoyama T, Kugiyama K, Hirashima O, Ohgushi M, Fujii H, Ogawa H, Yasue H (1997) Gender difference in improvement of endothelium-dependent vasodilation after estrogen supplementation. J Am Coll Cardiol 30:914–919
Keaney JF Jr, Shwaery GT, Xu A, Nicolosi RJ, Loscalzo J, Foxall TL, Vita JA (1994) 17β-Estradiol preserves endothelial vasodilator function and limits low-density lipoprotein oxidation in hypercholesterolemic swine. Circulation 89:2251–2259
Kenan Q, Fisher CR, Grumbach MM, Morishima A, Simpson ER (1995) Aromatase deficiency in a male subject: characterization of a mutation in the CYP 19 gene in an affected family (abstract). In: The premier event in endocrinology. 77th Annual Meeting of, P3–27, p 475
Khaw KT, Barrett-Connor E (1991) Endogenous sex hormones, high density lipoprotein cholesterol and other lipoprotein fractions in men. Arterioscler Thromb 11:489–494
Khosla S, Melton LJ, Atkinson EJ, O'Fallon WM, Klee GG, Riggs BL (1998) Relationship of serum sex steroid levels and bone turnover markers with bone mineral

density in men and women: a key role for bioavailable estrogen. J Clin Endocrinol Metab 83:2266–2274
Kim DJ, Han BS, Ahn B, Lee KK, Kang JS, Tsuda H (1996) Promotion potential of tamoxifen on hepatocarcinogenesis in female SD or F344 rats initiated with diethylnitrosamine. Cancer Lett 104:13–19
Kirschbaum C, Schommer N, Federenko I, Gaab J, Neumann O, Oellers M, Rohleder N, Untiedt A, Hanker J, Pirke KM, Hellhammer DH (1996) Short-term estradiol treatment enhances pituitary-adrenal axis and sympathetic responses to psychosocial stress in healthy young men. J Clin Endocrinol Metab 81:36639–36643
Kitahara S, Yoshida KI, Ishizaka K, Kageyama Y, Kawakami S, Tsujii T, Oshima H (1997) Stronger suppression of serum testosterone and FSH levels by a synthetic estrogen than by castration or an LH-RH agonist. Endocr J 44:527–532
Kjeld JM, Puah CM, Kaufman B, Loizu S, Vlotides J, Joplin GF (1977) Suppression of serum testosterone concentrations in men by an oral contraceptive preparation. BMJ 2:1261
Kjeld JM, Puah CM, Kaufman B, Loizu S, Vlotides J, Gwee HM, Kahn R, Sood R, Joplin GF (1979) Effects of norgestrel and ethinyloestradiol ingestion on serum levels of sex hormones and gonadotrophins in men. Clin Endocrinol 11:497–504
Klaassen CD, Liu L, Dunn II RT (1998) Regulation of sulfotransferase mRNA expression in male and female rats of various ages. Chem Biol Interact 109:299–313
Klein KO, Baron J, Barness KM, Pescovitz OH, Cutler GB (1998) Use of an ultrasensitive recombinant cell bioassay to determine estrogen levels in girls with precocious puberty treated with a luteinizing hormone-releasing hormone agonist. J Clin Endocrinol Metab 83:2387–2389
Kley HK, Nieschlag E, Bidlingmaier F, Kruskemper HL (1974) Possible age-dependent influence of estrogens on the binding of testosterone in plasma of adult men. Horm Metab Res 6:213–217
Kley HK, Solbach HG, McKinnan JC, Kruskemper HL (1979) Testosterone decrease and estrogen increase in male patients with obesity. Acta Endocrinol 91:553–558
Kliesch S, Behre HM, Roth St (1997) Rationale Therapie der Hitzewallungen unter Hormonentzugsbehandlung bei Patienten mit fortgeschrittenem Prostatakarzinom. Dtsch med Wochenschr 122:940–945
Kmicikiewicz I, Krezolek A, Bilinska B (1997) The effect of aromatase inhibitor on basal and testosterone-supplemented estradiol secretion by Leydig cells in vitro. Endocrinol Diabetes 105:113–118
Knopp RH, Zhu X (1997) Multiple beneficial effects of estrogen on lipoprotein metabolism (editorial). J Clin Endocrinol Metab 82:3952–3954
Koka S, Petro TM, Reinhart RA (1998) Estrogen inhibits interleukin-1β-induced interleukin-6 production by human osteoblast-like cells. J Interferon Cytokine Res 18:479–483
Komesaroff PA, Black CVS, Westerman RA (1998) A novel, nongenomic action of estrogen on the cardiovascular system. J Clin Endocrinol Metab 83:2313–2316
Kono S, Brandon D, Merriam GR, Loriaux DL, Lipsett MB (1980) Low plasma levels of 2-hydroxyestrone are consistent with its rapid metabolic clearance. Steroids 36:463–472
Kono S, Merriam GR, Brandon DD, Loriaux DL, Lipsett MB, Fujino T (1983) Radioimmunoassay and metabolic clearance rate of catecholestrogens, 2-hydroxyestrone and 2-hydroxyestradiol in man. J Steroid Biochem 19:627–633
Korach KS (1994) Insights from the study of animals lacking functional estrogen receptor. Science 266:1524–1526
Korenman SG (1985) The endocrinology of the abnormal male breast. Ann N Y Acad Sci 65:400–408
Kotulas IG, Cardamakis E, Michopoulos J, Mitropoulos D, Dounis A (1994) Tamoxifen treatment in male infertility. I. Effect on spermatozoa. Fertil Steril 61:911–914
Kovacs K, Stefaneanu L, Ezzat S, Smyth HS (1994) Prolactin-producing pituitary producing adenoma in a male-to-female transsexual patient with protracted estrogen administration. Arch Pathol Lab Med 118:562–565

Krishnan AV, Starhis P, Permuth SF, Tokes L, Feldman D (1993) Bisphenol-A: an estrogenic substance is released from polycarbonate flasks during autoclaving. Endocrinology 132:2279–2286

Kruithof-Dekker IG, Têtu B, Janssen PJA, van der Kwast TH (1996) Elevated estrogen receptor expression in human prostatic stromal cells by androgen ablation therapy. J Urol 156:1194–1197

Kudielka B, Hellhammer J, Hellhammer DH, Wolf OT, Pirke KM, Varadi E, Pilz J, Kirschbaum C (1998) Sex differences in endocrine and psychological responses to psychosocial stress in healthy elderly subjects and the impact of a 2-week dehydroepiandrosterone treatment. J Clin Endocrinol Metab 83:1756–1761

Kuhl H, Jung-Hoffmann C, Ehrlich M (1994) Oestriol-containing hormonal agent for the prophylaxis and treatment of arterial conditions in humans, method of preparing it and its use. Patent no. WO 94/28905

Kuiper GJM, Enmark E, Pelto-Huikko M, Nilsson S, Gustafsson JA (1996) Cloning of a novel estrogen receptor expressed in rat prostate and ovary. Proc Natl Acad Sci USA 93:5925–5930

Kuiper GJM, Carlsson B, Grandien K, Enmark E, Häggblad J, Nilsson S, Gustafsson JA (1997) Comparison of the ligand binding specificity and transcript tissue distribution of estrogen receptors α and β. Endocrinology 138:863–870

Kulin HE, Reiter EO (1972) Gonadotropin suppression by low dose estrogen in men: evidence for differential effects upon FSH and LH. J Clin Endocr Metab 35:836–839

Kurischko A, Oettel M (1977) Androgen-dependent fighting behaviour in male mice. Endokrinologie 70:1–5

Kusec V, Virdi AS, Prince R, Triffit JT (1998) Localization of estrogen receptor-α in human and rabbit skeletal tissues. J Clin Endocrinol Metab 83:2421–2428

Labaccaro JM, Lumbroso S, Belon C (1993) Androgen receptor gene mutation in male breast carcinoma. Hum Mol Genet 2:1799–1802

Laflamme N, Nappi RE, Drolet G, Labrie C, Rivest S (1998) Expression and neuropeptidergic characterization of estrogen receptors (ERα and ERβ) throughout the rat brain: anatomical evidence of distinct roles of each subtype. J Neurobiol 36:357–378

Lagiou P, Mantzoros CS, Tzonou A, Signorello LB, Lipworth L, Trichopoulos D (1997) Serum steroids in relation to benign prostatic hyperplasia. Lab Invest 54:497–501

Laing NM, Belinsky MG, Kruh GD, Bell DW, Boyd JT, Barone L, Testa JR, Tew KD (1998) Amplification of the ATP-binding cassette 2 transporter gene is functionally linked with enhanced efflux of estramustine in ovarian carcinoma cells. Cancer Res 58:1332–1337

Lamb DJ (1997) Hormonal disrupters and male infertility: are men at serious risk? Regul Toxicol Pharmacol 26:30–33

Lane KE, Ricci MJ, Ho SM (1997) Effect of combined testosterone and estradiol -17β treatment on the metabolism of E_2 in the prostate and liver of Noble rats. Prostate 30:256–262

Lau KM, Leav I, Ho SM (1998) Rat estrogen receptor-α and -β, and progesterone receptor mRNA expression in various prostatic lobes and microdissected normal and dysplastic epithelial tissues of the Noble rats. Endocrinology 139:424–427

Leav I, Ho SM, Ofner P, Merk FB, Kwan PWL, Damassa D (1988) Biochemical alterations in sex hormone-induced hyperplasia and dysplasia of the dorsolateral prostates of Noble rats. J Natl Cancer Inst 80:1045–1053

Leto S, Frensilli FJ (1981) Changing parameters of donor semen. Fertil Steril 36:766–770

Lidström P, Bonasera TA, Marquez-M M, Nilsson S, Bergström M, Langström B (1998) Synthesis and in vitro evaluation of [carbonyl-^{11}C]estramustine and [carbonyl-^{11}C]estramustine phosphate. Steroids 63:228–234

Liesegang P, Romalo G, Sudmann M, Wolf L, Schweikert HU (1994) Human osteoblast-like cells contain specific, saturable, high-affinity glucocorticoid, androgen, estrogen, and 1α,25-dihydroxycholecalciferol receptors. J Androl 15:194–199

Lindner V, Kim SK, Karas RH, Kuiper GGJM, Gustafsson JA, Mendelsohn ME (1998) Increased expression of estrogen receptor-β mRNA in male blood vessels after vascular injury. Circ Res 83:224–229
Lio H, Papadopoulos V, Vidic B, Dym M, Culty M (1997) Regulation of rat testis gonocyte proliferation by platelet-derived growth factor and estradiol: identification of signaling mechanisms involved. Endocrinology 138:1289–1298
Lubahn DB, Moyer JS, Golding TS, Couse JF, Korach KS, Smithies O (1993) Alteration of reproductive function but not prenatal sexual development after insertional disruption of the mouse estrogen receptor gene. Proc Natl Acad Sci USA 90: 11162–11166
Luft FC (1998) Estrogens and the myth of male privilege. J Mol Med 76:657–658
Lunt M, Felsenberg D, Reeve J, Benevolenskaya L, Cannata J, Dequeker J, Dodenhof C, Falch JA, Masaryk P, Pols HAP, Poor G, Reid DM, Scheidt-Nave C, Weber K, Varlow J, Kanis JA, O'Neill TW, Silman AJ (1997) Bone density variation and its effects on risk of vertebral deformity in men and women studied in thirteen European centers: the EVOS study. J Bone Miner Res 12:1883–1894
MacCalman CD, Getsios S, Farookhi R, Blaschuk OW (1997) Estrogens potentiate the stimulatory effects of follicle-stimulating hormone on N-cadherin messenger ribonucleic acid levels in cultured mouse sertoli cells. Endocrinology 138:41–48
MacLeod J, Wang J (1979) Male fertility potential in terms of semen quality: a review of the past, a study of the present. Fertil Steril 31:103–116
Majdic G, Sharpe RM, O'Shaughnessy PJ, Saunders PTK (1996) Expression of cytochrome P450 17α-hydroxylase/C17–20lyase in the fetal rat testis is reduced by maternal exposure to exogenous estrogens. Endocrinology 137:1063
Mason RA, Morris HA (1997) Effects of dihydrotestosterone on bone biochemical markers in sham and oophorectomized rats. J Bone Miner Res 12:1431–1437
Mathew RJ, Wilson WH, Stephen RT (1986) Determination of resting regional cerebral blood flow in normal subjects. Biol Psychiatry 21:907–914
Mathur R, Braunstein GD (1997) Gynecomastia: pathomechanisms and treatment strategies. Horm Res 48:95–102
Matsubara Y, Murata M, Kawano K, Zama T, Aoki N, Yoshino H, Watanabe G, Ishikawa K, Ikeda Y (1997) Genotype distribution of estrogen receptor polymorphisms in men and postmenopausal women from healthy and coronary populations and its relation to serum lipid levels. Arterioscler Thromb Vasc Biol 17:3006–3012
Mauermayr WRR, Sintermann R, Olbricht R (1978) Die Andromastektomie zur Verhinderung der Gynäkomastie bei der Behandlung des Prostatakarzinoms. Urologe A17:123–124
McCrohon JA, Walters WAW, Robinson JTC, McCredie RJ, Turner L, Adams MR, Handelsman DJ, Celermajer DS (1997) Arterial reactivity is enhanced in genetic males taking high dose estrogens. J Am Coll Cardiol 29:1432–1436
McDonald PC, Madden JP, Brenner PF, Wilson JD, Siiteri PK (1979) Origin of estrogen in normal men and in women with testicular feminization. J Clin Endocrinol Metab 49:905–916
McDonell DP (1998) Definition of the molecular mechanism of action of tissue-selective estrogen-receptor modulators. Recomb Antibodies Receptors Reagents Drugs 26:54–60
McEwen BS, Woolley CS (1994) Estradiol and progesterone regulate neuronal structure and synaptic connectivity in adult as well as developing brain. Exp Gerontol 29:431–436
McLachlan JA, Newbold RR, Li S, Negishi M (1998) Are estrogens carcinogenic during development of testis? APMIS 106:240–244
Mellinger GT, Gleason D, Bailar J (1967) The histology and prognosis of prostatic cancer. J Urol 97:331–337
Menchini-Fabris F, Rossi P, Palego P, Simi S, Turchi P (1996) Declining sperm counts in Italy during the past 20 years. Andrologia 28:373–375

Merriam GR, Pfeiffer DG, Loriaux DL, Lipsett MB (1983) Catechol estrogens and the control of gonadotropin and prolactin secretion in man. J Steroid Biochem 19:619–625

Michnovicz JJ (1998) Increased estrogen 2-hydroxylation in obese women using oral indole-3-carbinol. Int J Obesity 22:227–229

Miranda RC, Sohrabji F (1996) Gonadal steroid receptors: possible roles in the etiology and therapy of cognitive and neurological disorders. Ann Rep Med Chem 39:11–20

Misao R, Fujimoto J, Niwa K, Morishita S, Nakanishi Y, Tamaya T (1997) Immunohistochemical expressions of estrogen and progesterone receptors in human epididymis at different ages – a preliminary study. Int J Fertil 42:39–42

Mohren J (1998) Die Behandlung des Prostatakarzinoms (III). T&E Urol Nephrol 10:119–122

Morishima A, Gumbach MM, Simpson ER, Fischer C, Qin K (1995) Aromatase deficiency in male and female siblings caused by a novel mutation and the physiological role of estrogens. J Clin Endocrinol Metab 80:3689–3698

Morisset S, Patry C, Lora M, Brum-Fernandes AJ (1998) Regulation of cyclooxygenase-2 expression in bovine chondrocytes in culture by interleukin 1α, tumor necrosis factor–α, glucocorticoids, and 17β-estradiol. J Rheumatol 25:1146–1153

Morris ID, Hoyes KP, Taylor MF, Woolveridge I (1996) Male reproductive toxicology. A review with special consideration of hazards to men. In: Hamamah S, Mieusset R (eds) Research in male gametes: production and quality. INSERM, France, pp 135–150

Müller S, Schmid P, Schlatter C (1998) Evaluation of the estrogenic potency of nonylphenol in non-occupationally exposed humans. Environ Toxicol Pharmacol 6:27–33

Munoz de Toro MM, Luque EH (1997) Lack of relationship between the expression of Hsp27 heat shock estrogen receptor-associated protein and estrogen receptor or progesterone receptor status in male breast carcinoma. J Steroid Biochem Mol Biol 60:277–284

Murkies AL, Wilcox G, Davis SR (1998) Phytoestrogens. J Clin Endocrinol Metab 83:297–303

Murono EP, Nankin HR, Lin T, Osterman J (1982) The aging Leydig cell V. Diurnal rhythms in aged men. Acta Endocrinologica 99:619–623

Nakhla AM, Rosner W (1996) Stimulation of prostate cancer growth by androgens and estrogens through the intermediacy of sex hormone-binding globulin. Endocrinology 137:4126–4129

Nakhla AM, Romas NA, Rosner W (1997) Estradiol activates the prostate androgen receptor and prostate-specific antigen secretion through the intermediacy of sex hormone-binding globulin. J Biol Chem 272:6838–6841

Neubauer BL, Best KL, Clemens JA, Gates CA, Goode RL, Jones CD, Laughlin ME, Shaar CJ, Toomey RE, Hoover DM (1993) Endocrine and antiprostatic effects of raloxifene (LY 156758) in the male rat. Prostate 23:245–262

Neumann F, Gräf KJ, Elger W (1974) Hormones and embryonic development: hormone-induced disturbances in sexual differentiation. Adv Biosci 13:71

Nevala R, Korpela R, Vapaatalo H (1998) Plant derived estrogens relax rat mesenteric artery in vitro. Life Sci 63:95–100

New G, Timmins KL, Duffy SJ, Tran BT, O'Brien RC, Harper RW, Meredith IT (1997) Long-term estrogen therapy improves vascular function in male to female transsexuals. J Am Coll Cardiol 29:1437–1444

Niemann L, Hilbig V, Pfeil R (1998) Pflanzenschutzmittel und Hormonsystem – Möglichkeiten gesundheitlicher Störungen und ihre Manifestation im Tierversuch. Bundesgesundhbl 8:330–335

Nieschlag E, Lerchl A (1996) Declining sperm counts in European men – fact or fiction? Andrologia 28:305–306

Nieschlag E, Behre HM, Keck C, Kliesch S (1993) Treatment of male infertility. In: Hillier SG (ed) Gonadal development and function. Raven Press, New York, pp257–272
Noble RL (1977) The development of prostatic adenocarcinoma on the Nb rats following prolonged sex hormone administration. Cancer Res 37:1929–1933
Noble RL (1980) Production of Nb rat carcinoma of the dorsal prostate and response of estrogen-dependent transplants to sex hormones and tamoxifen. Cancer Res 40:3574–3550
Noble RL (1982) Prostate carcinoma in the Nb rat in relation to hormones. Int Rev Exp Pathol 23:113–159
Nuwaysir EF, Daggett DA, Jordan VC, Pitot HC (1996) Phase II enzyme expression in rat liver in response to the antiestrogen tamoxifen. Cancer Res 56:3704–3710
O'Connor KG, Tobin JD, Harman SM, Plato CC, Roy TA, Sherman SS, Blackman MR (1998) Serum levels of insulin-like growth factor-I are related to age and not to body composition in healthy women and men. J Gerontol 53A:M176–M182
Oettel M (1974) Untersuchungen über die Verwendungsmöglichkeiten des Hundes bei der pharmakologisch-endokrinologischen und toxikologischen Prüfung von Sexualwirkstoffen (Dissertation zur Promotion B). Universität Leipzig, Germany
Oettel M, Kurischko A (1978) Maintenance of aggressive behaviour in castrated mice by sex steroids: modification by neonatal injections of gonadal hormones. In: Dörner G, Kawakami (eds) Hormones and brain development. Elsevier, Holland, pp 49–56
Oettel M, Chemnitius KH, Claußen C, Stölzner W (1981) Ethinylestradiolsulfonate – new compound for the treatment of carcinoma of the prostate. Acta Endocr 97[Suppl 243]:223
Oettel M, Dören M, Hübler D, Römer W, Schröder J, Schumann I, Schwarz S, Stelzner A (1995) Freie Radikale und Sexualhormone. J Menopause 2:21–28
Oettel M, Römer W, Heller R (1996a) The therapeutic potential of scavestrogens. Eur J Obstet Gynecol Reprod Biol 65:153
Oettel M, Dören M, Heller R, Hübler D, Römer W, Schröder J, Schumann I, Schwarz S, Stelzner A (1996b) Estrogens and antioxidative capacity. In: Römer T, Straube W (eds) Klimakterium und Hormonsubstitution. Klaus Pia Verlagsgesellschaft GmbH, Nürnberg, pp 109–118
Öztas B (1998) Sex and blood-brain barrier. Pharmacol Res 37:165–167
Ogawa S, Lubahn DB, Korach KS, Pfaff DW (1997) Behavioral effects of estrogen receptor gene disruption in male mice. Proc Natl Acad Sci USA 94:1476–1481
Olsen GW, Bodner KM, Ramlow JM, Ross CE, Lipshultz LI (1995) Have sperm counts been reduced 50 percent in 50 years? A statistical model revisited. Fertil Steril 63:887–893
Ooi LSM, Panesar NS, Masarei JRL (1996) Urinary excretion of testosterone and estradiol in Chinese men and relationships with serum lipoprotein concentrations. Metabolism 45:279–284
Ortega CB, Garcia BG, Esparza AN, Ponce MH, Valencia SA, Villanueva TT, Gallegos CA (1993) Antiestrogen U23,469 induced alterations of catecholamine levels on plasma and central nervous system. Arch Med Res 24:27–31
Orwoll ES (1998) Osteoporosis in men. Endocrinol Metab Clin North Am 27:349–367
Panno ML, Sisci D, Salerno M, Lanzino M, Mauro L, Morrone EG, Pezzi V, Palmero S, Fugassa E, Andó S (1996) Effect of triiodothyronine administration on estrogen receptor contents in peripuberal Sertoli cells. Eur J Endocrinol 134:633–638
Patchev V, Römer W, Schwarz S, Mitev Y, Blum-Degen D, Riederer P, Elger W, Oettel M (1998) Non-feminizing radical-scavenging estrogens: evidence for selective neurotropic action in vivo and implications in neuroprotection (abstract). Presented at the Xth International Congress on Hormonal Steroids, Québec City, June 17–21. p 198

Paulsen CA, Berman NG, Wang C (1996) Data from men in the greater Seattle area reveal no downward trend in semen quality: further evidence that deterioration of semen quality is not geographically uniform. Fertil Steril 65:1015–1020
Payne AH, Kelch RP, Musich S, Halpern ME (1976) Intratesticular site of aromatization in the human. J Clin Endocrinol Metab 42:1081–1087
Payne AH, Perkins LM, Georgiou M, Quinn PG (1987) Intratesticular site of aromatase activity and possible function of testicular estradiol. Steroids 50:437–448
Pedersen SB, Hansen PS, Lund S, Andersen PH, Odgaard A, Richelsen B (1996) Identification of estrogen receptors and estrogen receptor mRNA in human adipose tissue. Eur J Clin Invest 26:262–269
Peltola V, Huhtaniemi I, Metsa-Ketela T, Ahotupa M (1996) Induction of lipid peroxidation during steroidogenesis in the rat testis. Endocrinology 137:105–112
Phillips GB, Pinkernell BH, Jing TY (1996) The association of hyperestrogenemia with coronary thrombosis in men. Arterioscler Thromb Vasc Biol 16:1383–1387
Pich A, Margaria E, Chiusa L (1994) Proliferative activity is a significant prognostic factor in male breast carcinoma. Am J Pathol 145:481–489
Pirke KM, Doerr P (1973) Age related changes and interrelationships between plasma testosterone estradiol and testosterone-binding globulin in normal adult males. Acta Endocrinologica 74:792–800
Plapinger L, McEwen BS (1978) Gonadal steroid-brain interactions in sexual differentiation. In: Hutchinson J (ed) Biological deterimnation of sexual behavior. Wiley and Sons, New York, pp 193–218
Polderman KH, Gooren LJG, van der Veen EA (1995) Effects of gonadal androgens and estrogens on adrenal androgen levels. Clin Endocrinol 43:415–421
Presti JC (1996) Estrogen therapy for prostate carcinoma. JAMA 275:1153
Pribluda VS, Green SJ (1998) A good estrogen. Science 280:987–988
Prins GS, Birch L (1997) Neonatal estrogen exposure upregulates estrogen receptor expression in the developing and adult rat prostate lobes. Endocrinology 138:1801–1809
Prins GS, Marmer M, Woodham C, Chang W, Kuiper G, Gustafsson JA, Birch L (1998) Estrogen receptor-β messenger ribonucleic acid ontogeny in the prostate of normal and neonatally estrogenized rats. Endocrinology 139:874–883
Prior JC, Vigna YM, Watson D (1989) Spironolactone with physiological female steroids for presurgical therapy of male-to-female transsexualism. Arch Sex Behav 18:49–57
Przyrembel H (1998) Natürliche Pflanzeninhaltsstoffe mit Wirkung auf das Hormonsystem. Bundesgesundhbl 8:335–340
Pylkkänen L, Mäkelä S, Valve E, Härkönen P, Santti R (1993) Prostatic dysplasia associated with increased expression of *c-myc* in neonatally estrogenized mice. J Urol 149:1593–1601
Ramirez VD, Zheng J, Siddiqui KM (1996) Membrane receptors for estrogen, progesterone, and testosterone in the rat brain: fantasy or reality. Cell Mol Neurobiol 16:175–198
Rasmussen PE, Erb K, Westergaard LG, Laursen SB (1997) No evidence for decreasing semen quality in four birth cohorts of 1,055 Danish men born between 1950 and 1970. Fert Steril 68:1059–1064
Raynaud A (1940) Effets sur la différenciation sexuelle des embryons, d´un mélange de dipropionate d´oestradiol et de testostérone injecté á la souris en gestation. Comptes Rendus Séances Acad Sci 211:572
Reis SE, Bhoopalam V, Zell KA, Counihan PJ, Smith AJC, Pham S, Murali S (1998) Conjugated estrogens acutely abolish abnormal cold-induced coronary vasoconstriction in male cardiac allografts. Circulation 97:23–25
Ribeiro G (1985) Male breast cancer: review of 301 cases from Christ Hospital and Holt Radium Institute, Manchester. Br J Cancer 51:115–119
Riggs BL, Khosla S, Melton LJ (1998) A unitary model for involutional osteoporosis: estrogen deficiency causes both type I and type II osteoporosis in postmenopausal

women and contributes to bone loss in aging men. J Bone Miner Res 13:763–773
Rodriquez G, Warkentin S, Risberg J, Rosadin G (1988) Sex differences in regional cerebral blood flow. J Cereb Blood Flow Metab 8:783–789
Rogers S, Day CA, Fox SB (1993) Expression of cathepsin D and estrogen receptor in male breast carcinoma. Hum Pathol 24:148–151
Rolf C, Behre HM, Nieschlag E (1996) Tamoxifen bei männlicher Infertilität. Dtsch Med Wochenschr 121:33–39
Römer W, Schröder J, Oettel M (1993) Influence of estrogens and progestins on lipid peroxidation and LDL oxidation. Arch Pharmazie 326:700
Römer W, Oettel M, Droescher P, Schwarz S (1997) Novel "scavestrogens" and their radical scavenging effects, iron-chelating, and total antioxidative activities: $\Delta^{8,9}$-dehydro derivatives of 17α-estradiol and 17β-estradiol. Steroids 62:304–310
Rommerts FFG, de Jong FH, Brinkmann AO, van der Molen HJ (1982) Development and cellular localization of rat testicular aromatase activity. J Reprod Fertil 65:281–288
Roselli CE, Resko JA (1987) The distribution and regulation of aromatase activity in the central nervous system. Steroids 50:495–508
Rosenfeld CS, Ganjam VK, Taylor JA, Yuan X, Stiehr JR, Hardy MP, Lubahn DB (1998) Endocrinology 139:2982–2987
Ross PD (1998) Osteoporosis: epidemiology and risk assessment. In: Kaiser FE, Nourhashemi F, Betiere MC, Ouchi YF (eds) Women, aging and health. Serdi, Paris, pp 189–200
Rubanyi GM, Freay AD, Kauser K, Sukovich D, Burton G, Lubahn DB, Couse JF, Curtis SW, Korach KS (1997) Vascular estrogen receptors and endothelium-derived nitric oxide production in the mouse aorta: gender difference and the effect of estrogen receptor gene disruption. J Clin Invest 99:2429–2437
Rubens R, Dhont M, Vermeulen A (1974) Further studies on Leydig cell function in old age. J Clin Endocrinol Metab 39:40–44
Safe S, Connor K, Ramamoorthy K, Gaido K, Maness S (1997) Human exposure to endocrine-active chemicals: hazard assessment problems. Regul Toxicol Pharmacol 26:52–58
Sah P (1998) Role of low-dose estrogen-testosterone combination therapy in men with oligospermia. Fert Steril 70:780–781
Saito S, Motomura N, Lou H, Ramwell PW, Foegh ML (1997) Specific effects of estrogen on growth factor and major histocompatibility complex class II antigen expression in rat aortic allograft. J Thorac Cardiovasc Surg 114:803–809
Sangrajrang S, Denoulet P, Millot G, Tatoud R, Podgorniak MP, Tew KD, Clavo F, Fellous A (1998) Estramustine resistance correlates with tau over-expression in human prostatic carcinoma cells. Int J Cancer 77:626–631
Santti R, Newbold R, Mäkelä S, Pylkkänen L, McLachlan JA (1994) Developmental estrogenization and prostatic neoplasia. Prostate 24:67–78
Sasano H, Kimura M, Shizawa S, Kimura N, Nagura H (1996) Aromatase and steroid receptors in gynecomastia and male breast carcinoma: an immunohistochemical study. J Clin Endocrinol Metab 81:3063–3067
Sasano H, Uzuki M, Sawai T (1997) Aromatase in human bone tissue. J Bone Miner Res 12:1416–1423
Sasano H, Takahashi K, Satoh F, Nagura H, Harada N (1998) Aromatase in the human central nervous system. Clin Endocrinol 48:325–329
Schlatterer K, von Werder K, Stalla GK (1996) Multistep treatment concept of transsexual patients. Expl Clin Endocrinol Diabetes 104:413–419
Schlatterer K, Auer DP, Yassouridis A, von Werder K, Stalla GK (1998) Transsexualism and osteoporosis. Exp Clin Endocrinol Diabetes 106:365–368
Schussheim DH, Schussheim AE (1998) Is digoxin a designer estrogen? Lancet 351:1734
Schwarz S, Weber G (1970) Alkan- und Cycloalkansulfonate des 17α-Äthinyloestradiols. J Prakt Chemie 312:653–659

Schwarz S, Weber G, Schreiber M (1975) Steroide. Pharmazie 30:17–21
Selby PL, Braidman IP, Freemont AJ, Mawer EB (1995) Is estrogen an important regulator of bone turnover in the male? Preliminary evidence from osteoporosis. J Endocrinol [Suppl 1]:O 54
Selby PL, Braidman IP, Mawer EB, Freemont AJ (1996) Hormonal influences in male osteoporosis. Osteoporosis Int 6[Suppl 1]:279
Selzman CH, Whitehill TA, Shames BD, Pulido EJ, Cain BC, Harken AH (1998) The biology of estrogen-mediated repair of cardiovascular injury. Ann Thorac Surg 65:868–874
Service RF (1998) New role for estrogen in cancer? Science 279:1631–1633
Setchell BP (1997) Sperm counts in semen of farm animals 1932–1995. Int J Androl 20:209–214
Sharpe EM (1993) Declining sperm counts in men – is there an endocrine cause? J Endocrinol 136:357–360
Sharpe RM, Skakkebaek NE (1993) Are estrogens involved in falling sperm counts and disorders of the male reproductive tract? Lancet 341:1392–1395
Shugrue P, Scrimo P, Lane M, Askew R, Merchenthaler I (1997a) The distribution of estrogen receptor-β mRNA in forebrain regions of the estrogen receptor-α knock-out mouse. Endocrinology 138:5649–5652
Shugrue PJ, Lane MV, Merchenthaler I (1997b) Comparative distribution of estrogen receptor-α and -β mRNA in the rat central nervous system. J Comp Neurol 388:507–525
Shugrue PJ, Lubahn DB, Negro-Vilar A, Korach KS, Merchenthaler I (1997c) Responses in the brain of estrogen receptor α-disrupted mice. Proc Natl Acad Sci USA 94:11008–11012
Sih R, Morley JE, Kaiser FE, Perry III HM, Patrick P, Ross C (1997) Testosterone replacement in older hypogonadal men: a 12-month randomized controlled trial. J Clin Endocrinol Metab 82:1661–1667
Silberstein GB, van Horn K, Shyamalia G, Daniel CW (1994) Essential role of endogenous estrogen in directly stimulating mammary growth demonstrated by implants containing pure antiestrogens. Endocrinology 134:84–90
Simon D, Preziosi P, Barrett-Connor E, Roger M, Saint-Paul M, Nahoul K, Papoz L (1992) The influence of aging on plasma sex hormones in men: the telecom study. Am J Epidemiol 135:783–791
Simpkins JW, Rabbani O, Shi J, Panickar KS, Green PS, Day AL (1998) A system for the brain-enhanced delivery of estradiol: an assessment of its potential for the treatment of Alzheimer's disease and stroke. Pharmazie 53:505–511
Simpson E, Davis S (1998) Why do the clinical sequelae of estrogen deficiency affect women more than men? J Clin Endocrinol Metab 83:2214
Singer PL (1949) Occupational oligospermia. JAMA 140:1249
Slemenda C, Hui SL, Longcope C (1987) Sex steroids and bone mass. Am Soc Clin Invest 80:1261–1269
Slemenda C, Zhou L, Longcope C, Hui S, Johnston CC (1995) Sex steroids, bone mass and bone loss in older men: estrogens or androgens? J Bone Miner Res 10[Suppl 1]:S440
Slemenda C, Longcope C, Hui S, Zhou L, Johnston CC (1996) Estrogens but not androgens are positively associated with bone mass in older men. Osteoporosis Int 6[Suppl 1]:139
Slemenda CW, Longcope C, Zhou L, Hui SL, Peacock M, Johnston CC (1997) Sex steroids and bone mass in older men. J Clin Invest 100:1755–1759
Smals AGH, Kloppenborg PWC, Lequin RM, Benraad TJ (1974) The effect of estrogen administration on plasma testosterone, FSH and LH levels in patients with Klinefelter's syndrome and normal men. Acta Endocr 77:765–783
Smith EP, Boyd J, Frank GR, Takahashi H, Cohen RM, Specker B, Williams TC, Lubahn DB, Korach KS (1994) Estrogen resistance caused by a mutation in the estrogen-receptor gene in a man. N Engl J Med 331:1056–1061

Smith KD, Steinberger E (1977) What is oligospermia? In: Troen P, Nankin HR (eds) The testis in normal and infertile men. Raven Press, New York, pp 489–503
Sonnenschein C, Soto AM (1998) An updated review of environmental estrogen and androgen mimics and antagonists. J Steroid Biochem Mol Biol 65:143–150
Soto AM, Justicia H, Wray JW, Sonnenschein C (1991) p-Nonyl-phenol: an estrogenic xenobiotic released from "modified" polystyrene. Environ Health Perspect 92:167–173
Stanik S, Dornfeld LP, Maxwell MH, Viosca SP, Korenman SG (1981) The effect of weight loss on reproductive hormones in obese men. J Clin Endocrinol Metab 53:828–831
Stearns EL, MacDonell JA, Kaufman BJ, Padua R, Lucman TS, Winter JSD, Faiman C (1991) Declining testicular function with age: hormonal and clinical correlates. Am J Med 57:761–765
Stege R, Carlström K, Hedlund PO, Pousette A, von Schoultz B, Henriksson P (1995) Intramuskuläres Depotöstrogen (Estradurin®) in der Behandlung von Patienten mit Prostatakarzinom. Urologe A 34:398–403
Sterzik K, Rosenbusch B, Mogck J, Heyden M, Lichtenberger K (1993) Tamoxifen treatment of oligozoospermia: a re-evaluation of its effects including additional sperm function tests. Arch Gynecol Obstet 252:143–147
Sudhir K, Chou TM, Chatterjee K, Smith EP, Williams TC, Kane JP, Malloy MJ, Korach KS, Rubanyi G (1997a) Premature coronary disease associated with a disruptive mutation in the estrogen receptor gene in a man. Circulation 96:3774–3777
Sudhir K, Chou TM, Messina LM, Hutchison SJ, Korach KS, Chatterjee K, Rubanyi G (1997b) Endothelial dysfunction in a man with disruptive mutation in estrogen-receptor gene. Lancet 349:1146–1147
Suginami H, Kitagawa H, Nakahashi N, Yano K, Matsubara K (1993) A clomiphene citrate and tamoxifen citrate combination therapy: a novel therapy for ovulation induction. Fertil Steril 59:976–979
Takikawa H, Sano N, Aiso M, Takamori Y, Yamanaka M (1997) Effect of tauro-α-muricholate and tauro-β-muricholate on oestradiol-17β-glucuronide-induced cholestasis in rats. J Gastroenterol Hepatol 12:84–86
Telang NT, Katdare M, Bradlow HL, Osborne MP, Fishman J (1997) Inhibition of proliferation and modulation of estradiol metabolism: novel mechanisms for breast cancer prevention by the phytochemical indole-3-carbinol. Proc Soc Exp Biol Med 216:246–252
Tennekoon KH, Karunanayake EH (1993) Serum FSH, LH, and testosterone concentrations in presumably fertile men: Effect of age. Int J Fertil 38:108–112
Tew KD, Stearus ME (1987) Hormone-independent, non-alkylating mechanism of cytotoxicity for Estramustine. Urol Res 15:155–160
Tham DM, Gardner CD, Haskell WL (1998) Potential health benefits of dietary phytoestrogens: a review of the clinical, epidemiological, and mechanistic evidence. J Clin Endocrinol Metab 83:2223–2235
The Coronary Drug Project Research Group (1970) The coronary drug project. JAMA 214:1303–1313
The Veterans Administration Cooperative Urological Research Group, VACURG (1967) Carcinoma of the prostate: treatment comparisons. J Urol 98:516–522
Thomas DB (1993) Breast cancer in men. Epidemiol Rev 15:220–231
Toung TJK, Traystman RJ, Hurn PD (1998) Estrogen-mediated neuroprotection after experimental stroke in male rats. Stroke 29:1666–1670
Trunet PF, Mueller P, Bhatnagar AS, Dickes I, Monnet G, White G (1993) Open dose-finding study of a new potent and selective nonsteroidal aromatase inhibitor, CGS 20 267, in healthy male subjects. J Clin Endocrinol Metab 77:319–323
Ueoka S, Yamaguchi M (1998) Sexual difference with hepatic calcium-binding protein regucalcin mRNA expression in rats with different ages: effect of ovarian hormone. Biol Pharm Bull 21:405–407

Umbreit K (1995) Klimakterium virile. Altersbeschwerden: Wunderwaffe in Sicht. TW Urol Nephrol 7:175–182

Vagell ME, McGinnis MY (1997) Inhibition of brain estrogen receptors by RU 58668. J Neuroendocrinol 9:797–800

Vanage GR, Dao B, Li XJ, Bardin CW, Koide SS (1997) Effects of anordriol, an antiestrogen, on the reproductive organs of the male rat. Arch Androl 38:13–21

Vanderschueren D (1996) Androgens and their role in skeletal homeostasis. Horm Res 46:95–98

Vanderschueren D, Van Herck E, Suiker AMH, Visser WJ, Schot LPC, Bouillon R (1992) Bone and mineral metabolism in aged male rats: short- and long-term effects of androgen deficiency. 130:2906–2916

Vanderschueren D, Van Herck E, DeCoster R, Bouillon R (1995) Androgen action is partially mediated by aromatisation (abstract). 22nd International Conference on Calcium Regulatory Hormones, Melbourne. Bone 16[Suppl]:294 (159 S)

Vanderschueren D, van Herck E, de Coster R, Bouillon R (1996) Aromatization of androgens is important for skeletal maintenance of aged male rats. Calcif Tissue Int 59:179–183

Van den Beld AW, Grobee DE, Pols HAP, Lamberts SWJ (1998) The role of estrogens in physical and psychological well-being in elderly men (abstract). Abstracts of the First World Congress on the Aging Male. Aging Male 1[Suppl 1]:54

Van Kesteren P, Lips P, Deville W (1996a) The effect of one-year cross-sex hormonal treatment on bone metabolism and serum insulin-like growth factor-1 in transsexuals. J Clin Endocrinol Metab 81:2227–2232

Van Kesteren P, Meinhardt W, van der Valk P, Geldof A, Megens J, Gooren L (1996b) Effects of estrogens only on the prostate of aging men. J Urol 156:1349–1353

Van Kesteren PJM, Asscheman H, Megens JAJ, Gooren LJG (1997) Mortality and morbidity in transsexual subjects treated with cross-sex hormones. Clin Endocrinol 47:337–342

Van Kesteren PJM, Kooistra T, Lansink M, van Kamp GJ, Asscheman H, Gooren LJG, Emeis JJ, Vischer UM, Stehouwer CDA (1998a) Thromb Haemost 79:1029–1033

Van Kesteren PJM, Lips P, Gooren LJG, Asscheman H, Megens JAJ (1998b) Long-term follow-up of bone mineral density and bone metabolism in transsexuals treated with cross-sex hormones. Clin Endocrinol 48:347–354

Van Thiel DH, Lester R, Sherins RJ (1974) Hypogonadism in alcoholic liver disease. Gastroenterology 67:1188–1199

Van Thiel DH, Gavaler JS, Slone FL, Cobb CF, Smith WI, Bron KM Jr, Lester R (1980) Is feminization in alcoholic men due in part to portal hypertension: a rat model. Gastroenterology 78:81–91

Van Thiel DH, Gavaler JS, Spero JA, Eggler KM, Wight C, Sanghvi AT, Hasiba U, Lewis JH (1981) Patterns of hypothalamic-pituitary-gonadal dysfunction in men with liver disease due to differing etiologies. Hepatology 1:39–46

Van Thiel DH, Gavaler JS, Cobb CF, McClain CJ (1983) An evaluation of the respective roles of portosystemic shunting and portal hypertension in rats upon the production of gonadal dysfunction in cirrhosis. Gastroenterology 85:154–159

Van Waeleghem K, Declercq N, Vermeulen L, Schoonjans F, Comhaire F (1996) Deterioration of sperm quality in young healthy Belgian men. Hum Reprod 11:325–329

Veldhuis JD, Metzger DL, Martha PM, Mauras N, Kerrigan JR, Keenan B, Rogol AD, Pincus SM (1997) Estrogen and testosterone, but not a nonaromatizable androgen, direct network integration of the hypothalamo-somatotrope (growth hormone)-insulin-like growth factor I axis in the human: Evidence from pubertal pathophysiology and sex-steroid hormone replacement. J Clin Endocrinol Metab 82:3414–3420

Vilarinho ST, Costallat LTV (1998) Evaluation of the hypothalamic-pituitary-gonadal axis in males with systemic lupus erythematosus. J Rheumatol 25:1097–1103

Vizner B, Vilibic T, Brkic K, Vrkljan M, Smircic L, Sekso M (1994) Gynecomastia – clinical and therapeutic aspects. Acta Clinica Croatica 33:205–212
Wade GN, Blaustein JD, Gray JM, Meredith JM (1993) ICI 182,780: a pure antiestrogen that affects behaviors and energy balance in rats without acting in the brain. Am J Physiol 265:R1392–R1398
Wakley GK, Schutte HDJ Hannon KS (1991) Androgen treatment prevents loss of cancellous bone in the orchiectomized rat. J Bone Miner Res 6:325–330
Wallentin L, Varenhorst E (1978) Changes of plasma lipid metabolism in males during estrogen treatment of prostatic carcinoma. J Clin Endocrinol Metab 47:596–601
Wang LG, Liu XM, Kreis W, Budman DR (1998) Androgen antagonistic effect of estramustine phosphate (EMP) metabolites on wild-type and mutated androgen receptor. Biochem Pharmacol 55:1427–1433
Wang YJ, Wu JC, Lee SD, Tsai YT, Ko KJ (1991) Gonadal dysfunction and changes in sex hormones in postnecrotic cirrhotic men: a matched study with alcoholic cirrhotic men. Hepatogastroenterology 38:531–534
Wang YJ, Lee SD, Lin HC, Hsia HC, Lee FY, Tsai YT, Lo KJ (1993) Changes of sex hormone levels in patients with hepatitis B virus-related postnecrotic cirrhosis: relationship to the severity of portal hypertension. J Hepatol 18:101–105
Washburn SA, Honoré EK, Cline JM, Helman M, Wagner JD, Adelman SJ, Clarkson TB (1996) Effects of 17α-dihydroequilenin sulfate on atherosclerotic male and female rhesus monkeys. Am J Obstet Gynecol 173:341–351
Werner A, Bender E, Mahaffey W, McKeating J, Marrangoni A, Katoh A (1996) Inhibition of experimental liver metastasis by combined treatment with tamoxifen and interferon. Anti-Cancer Drugs 7:307–311
Wersinger SR, Sannen K, Villalba C, Lubahn DB, Rissman EF, De Vries GJ (1997) Masculine sexual behavior is disrupted in male and female mice lacking a functional estrogen receptor α gene. Horm Behav 32:176–183
White IN, de Matteis F, Gibbs AH, Lim CK, Wolf CR, Henderson C, Smith LL (1995) Species differences in the covalent binding of [^{14}C] tamoxifen to liver microsomes and the forms of cytochrome P450 involved. Biochem Pharmacol 49:1035–1042
Widmark A, Grankvist K, Bergh A, Henriksson R, Damber JE (1995) Effects of estrogens and progestogens on the membrane permeability and growth of human prostatic carcinoma cells (PC-3) in vitro. Prostate 26:5–11
Williams GM (1995) Tamoxifen experimental carcinogenicity studies: Implications for human effects. Proc Soc Exp Biol Med 208:141–143
Winkelmann BR (1998) LURIC/Jenapharm study: report to Jenapharm,
Winter M, Falvo RE, Schambacher BD, Verholtz S (1983) Regulation of gonadotropin secretion in the male dog. J Androl 4:319–323
Winters SJ, Troen P (1985) Evidence for a role of endogenous estrogen in the hypothalamic control of gonadotropin secretion in men. J Clin Endocrinol Metab 61:842–845
Wiren KM, Zhang X, Chang C (1997) Transcriptional up-regulation of the human androgen receptor by androgen in bone cells. Endocrinology 138:2291–2300
Wong GYC, Bradlow L, Sepkovic D, Mehl S, Mailman J, Osborne MP (1997) Dose-ranging study of indole-3-carbinol for breast cancer prevention. J Cell Biochem Suppl 28/29:111–116
Yeh S, Miyamoto H, Shima H, Chang C (1998) From estrogen to androgen receptor: a new pathway for sex hormones in prostate. Proc Natl Acad Sci USA 95:5527–5532
Yeung AC (1997) Estrogen for men: Reversal of cardiovascular misfortune? J Am Coll Cardiol 29:1445–1446
Young LJ, Wang Z, Donaldson R, Rissman EF (1998) Estrogen receptor α is essential for induction of oxytocin receptor by estrogen. Neuroreport 9:933–936
Younglai EV, Collins JA, Foster WG (1998) Canadian semen quality: an analysis of sperm density among eleven academic fertility centers. Fertil Steril 70:76–80

Zimmermann H, Puri C, Elger W, Hobe G (1994) Comparative pharmacokinetics of selected steroids (abstract). 2nd German-Chinese (R.O.C.) Symposium on "Biotechnical Drugs", May 31 to June 4, Jena
Zingg E, Heinzel F (1968) Verhütung der Gynäkomastie beim hormonbehandelten Prostatakarzinom-Patienten durch Röntgenbestrahlung. Urologe Ausg A 7:96–103
Zmuda JM, Fahrenbach MC, Younkin BT, Bausserman LL, Terry RB, Catlin DH, Thompson PD (1993) The effect of testosterone aromatization on high density lipoprotein cholesterol level and postheparin lipolytic activity. Metab Clin Exp 42:446–450
Zumoff B, Fishman J, Cassouto J, Gallagher TF, Hellman L (1968) Influence of age and sex on normal estradiol metabolism. J Clin Endocrinol 28:937–941
Zumoff B, Strain GW, Kream J, O'Connor J, Rosenfeld RS, Levin J, Fukushima DK (1982) Age variation of the 24-hour mean plasma concentration of androgens, estrogens, and gonadotropins in normal adult men. J Clin Endocrinol Metab 61:705–711

Part 8
Comparative Endocrinology

CHAPTER 46

Comparative Aspects of Estrogen Biosynthesis and Metabolism and the Endocrinological Consequences in Different Animal Species

H.H.D. MEYER

A. Evolution of Estrogens

Steroidal estrogens have been isolated from marine and terrestrial animals representative for all major classes of vertebrates including fish, amphibians, reptiles, birds and mammals. In general, estrogens are responsible for most features characteristic of the female sex of a species, such as metabolic, behavioral and morphological changes during the steps of reproduction; they also support several events in the males. Other steroids are very common in all living organisms, with the exception of bacteria (DORFMAN and UNGAR 1965; FIESER and FIESER 1967). Although quite a number of functions of the various steroids was evolved, it is generally accepted, nowadays, that the endocrine functions of estrogens are a phenomenon of all vertebrates and, usually, estradiol-17β is the most potent endogenous substance. However, there are indications that estrogens are involved in the ovarian activities of non-vertebrate deuterostomic and even protostomic animals, but the function of estrogens in invertebrates is unclear. More recent literature considering these animals is very limited and, according to our knowledge, the presence of aromatase or the estradiol receptor is not yet documented for invertebrates. It is also generally accepted that, at least in plants, estrogens have no "estradiol-like" endocrine function during plant reproduction and the numerous phytoestrogens (HESSE et al. 1981) – which interact with mammalian estrogen receptors due to their structural similarities with estradiol – seem to have many other functions in the plant, such as protection against herbivory or oxygen stress similar to other phenols, condensed tannins or further secondary plant compounds (MCARTHUR et al. 1993, JUNG 1977, FEENY 1976).

The evolution of a hormonal system always involves both the ligand and its sites of interaction. In the case of estrogens, the steroid producing enzymes, mainly the aromatase complex, and the estrogen receptor belong together within their co-evolution. Whereas the literature from a few decades ago concentrated on the steroids as such, nowadays, the focus of interest has moved to the molecular modes of steroid-synthesis regulation and the signal transduction via the receptor. Nevertheless, the earlier findings of estrogenic steroids and the more-recent identification of aromatase and receptor genes and their expression patterns match together. They confirm the importance of estrogens for all vertebrates; the physiological functions of estradiol seem to be less conserved. In early vertebrates, estrogens are known to support the metabolic needs of the extensive oocyte production, whereas in the Eutherian placentalia the function of estrogens has focused on the physiological phenomena of estrus.

B. Comparative Biochemistry

I. Estrogen Biosynthesis

Testosterone is the major substrate within the biosynthesis of estradiol, but also a large number of other substrates or intermediates have been tested within in vivo and in vitro systems, showing quite various conversion rates (DORFMAN and UNGAR 1965). The required elimination of C19 and desaturation of the A-ring is catalyzed by the microsomal enzyme aromatase cytochrome P450 (P450arom, the product of the CYP19 gene) in association with a reduced nicotinamide adenine dinucleotide phosphate (NADPH)-reductase (SIMPSON et al. 1994, SIMMONS et al. 1985). The aromatase reaction utilizes 3 mol NADPH plus 3 mol oxygen per mol C_{19}steroid and the C_{19} methyl group is the site of oxygen attack by a standard hydroxylation mechanism, producing the 19-hydroxyl-C_{19} steroids (AKHTAR et al. 1982, 1993). The formation of the 19-oxo-C_{19} steroid also involves a hydroxylation, whereas for the third oxidative reaction a peroxidative attack is postulated (COLE and ROBINSON 1988).

The evolution of aromatase structure–function relationships was reviewed recently by SIMPSON et al. (1994). The identities of the derived amino-acid sequences of aromatase from mouse, rat, chicken and rainbow trout to the human are 81%, 77%, 73% and 52%, respectively. The variation between two teleost fish – trout and gold fish – is almost as great as that between trout and human, indicating the great evolutionary distance among the teleosts. The highly conserved regions – the heme binding site and the steroid pocket – are nearly identical within the investigated mammals and showed only few amino-acid replacements when compared with birds and fish. This conservation of the CYP19 gene has been used for reinvestigation of the molecular phylogeny of the tribe bovini (PITRA et al. 1997), which enabled a more precise dating of the respective last common ancestors and a new placement of the *anoa depressicornis*.

For the human aromatase, rather good models for the putative mechanism of enzyme catalysis are available (GRAHAM–LORENCE et al. 1991), and they support the major pathway via the 19-oxo-C19-steroid and removal of formic acid. However, these models do not yet explain all the variations in biochemical pathways (Fig. 1) that were postulated on the basis of the different intermediates that had been discovered much earlier (DORFMAN and UNGAR 1965). In addition to testosterone, the intermediates, the respective 17-keto forms and several other synthetic modifications can also become aromatized. In this respect, it is remarkable that the potent androgens 19-nortestosterone and

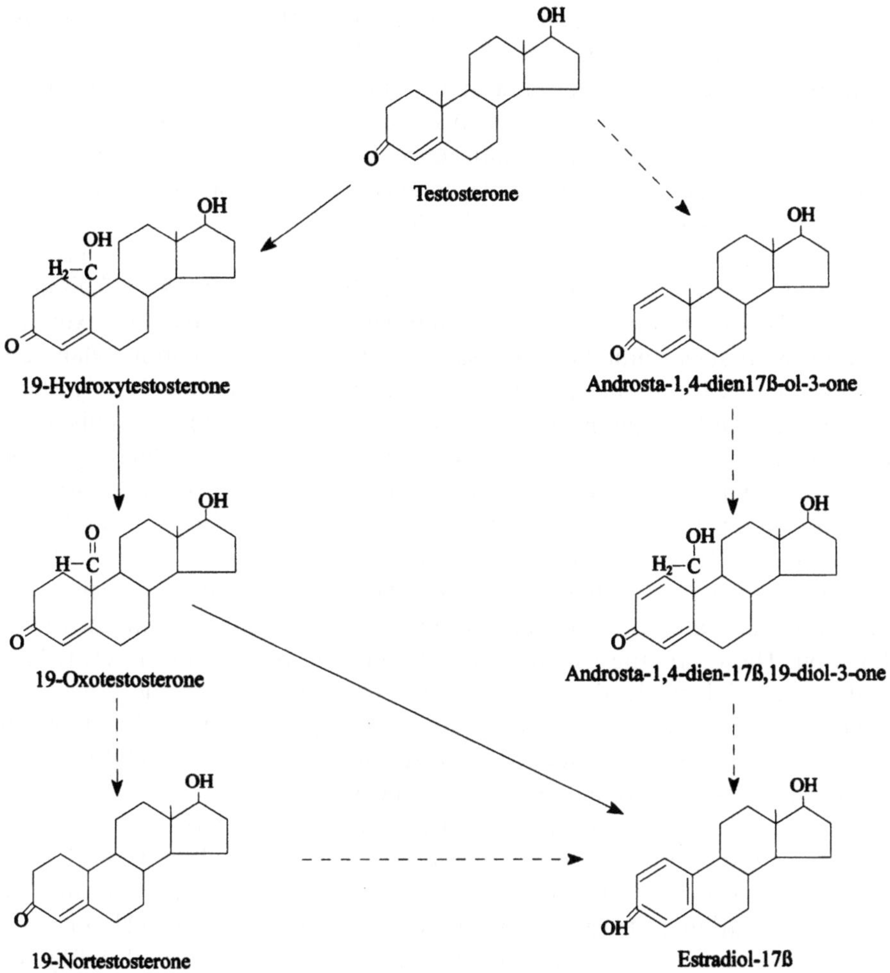

Fig. 1. Identified intermediates and possible biochemical pathways of estrogen synthesis

boldenone (androsta-1,4-dien-17β-ol-3-one) can be produced on the different routes from testosterone to estradiol. The endogenous production of 19-nortestosterone in the placenta of the pregnant cow (MEYER et al. 1992); in males, such as stallions (BENOIT et al. 1985) or boars (MAGHUIN–ROGISTER et al. 1988; VAN GINKEL et al. 1989); and in human (DEHENNIN et al. 1984) or equine (SILBERZAHN et al. 1985) follicular fluid raised problems in discrimination of xenobiotic and physiological origin of these anabolic agents during food or doping control. Recently, boldenone was also identified in urine from untreated calves (ARTS et al. 1996).

At present, their are no explanations for the selection of possible biochemical pathways in different animals and different tissues. Only a single copy of the gene was identified, at least in the cattle and the human genome, and the very high conservation of the active site of P450 arom indicates similar reaction mechanisms. Different splicing of mRNA was documented for aromatizing tissues, such as placenta and granulosa cells (HINSHELWOOD et al. 1995) from three ungulate species with placental estrogen production (cow, horse and pig), but tissue specific differences were limited to promoter regions. In contrast, different ovarian and placental isoforms of functional porcine aromatase were described (CORBIN et al. 1995).

The tissue-specific expression of the aromatase-encoding gene has been investigated most intensively in cattle (VANSELOW and FÜRBASS 1995; FÜRBASS et al. 1997). Six different transcription start sites and alternative first exons are used in the tissues studied (granulosa cells, placenta, testis, adrenal gland and hippocampus). The tissue-specific pattern of transcript variants is only due to different promoter regions and alternative splicing (Table 1) – possibly allowing selective regulations of enzyme production. From the coding region, no evidence emerged for the existence of different genes or enzyme isoforms. Despite these findings, the bovine genome contains a truncated, nonfunctional CYP19 pseudogene that is transcribed in the placenta without providing an enzyme (FÜRBASS and VANSELOW 1995). The pseudogene was obviously generated by gene duplication and later mutations resulting in malfunction. Therefore, the production and secretion of different intermediates within estradiol production – such as 19-nortestosterone in the bovine placenta or boldenone in calf urine – cannot be explained by modified enzyme activities and might be due to tissue-specific substrate availability or yet unknown post-translational enzyme modifications.

Aromatase is expressed in ovaries of all the studied vertebrates and ovarian estradiol production is probably older than 350 million years. Expression in the ovary preferentially utilizes a promoter just proximal to the start site of translation, at least in humans (MEANS et al. 1991), rats (HICKEY et al. 1990), cows (FÜRBASS et al. 1997) and chickens (MATSUMINE et al. 1991), with

Table 1. Relative abundance of tissue-specific CYP 19 transcripts in cattle (according to FÜRBASS et al. 1997)

	Total RNA	Exon 1.1	Exon 1.2	Exon 1.3	Exon 1.4	Exon 2
Granulosa cells	+++	++	+	+	+	+++
Placenta	+++	+++	++	+++	+	+
Testis	++	+	–	+	+	+
Adrenal gland	+	–	–	–	–	+
Hippocampus	++	–	+	+	++	–

+++, Intense expression
++, Intermediate expression
+, Minor expression
–, No transcript detectable

a high degree of conservation in the 5′-untranslated end (77% identity between human and cow).

Placental estrogen production evolved only 50–100 million years ago and it is still doubtful whether it has a common origin or whether there were independent beginnings followed by convergent evolutions in artiodactylids, perissodactylids and primates. The low degree of conservation in the 5′-untranslated ends of the major CYP19 mRNA from placenta (39% identity in human and cow) supports the latter hypothesis. As known from other mammals – such as carnivores (HOFFMANN et al. 1994) – placental or ovarian estrogen production is not essential for maintenance of pregnancy, mammogenesis or parturition. This suggests that placental estrogen production evolved later and independently in several mammalian orders to support the metabolic and further physiological requirements during gestation.

The complete picture of comparative estrogen biosynthesis will also need the parallel elaboration of knowledge on CYP19 mRNA translation and regulation of aromatase enzyme activity. However, such studies are still limited to several cell culture systems or laboratory animal models which provide little insight into the diversity within the animal kingdom. Even biosynthesis of estradiol-17β fatty acyl esters by bovine liver microsomes has been described (PARIS and RAO 1989) and storage of such silent esters or physiological pro-drugs in hepatic fat is suggested. The discovered enzyme activity – a fatty acyl coenzyme A: estradiol-17β acyl transferase – produced the respective arachidonate, linoleate, oleate, palmitate and stearate esters. Estradiol was more susceptible to esterification with polyunsaturated fatty acids than saturated ones.

In horse plasma and urine, four further derivatives were identified – namely 17β-dihydroequilin, 17β-dihydroequilenin, equilin and equilenin (Fig. 2). These substances seem to originate from alternative pathways in sterol biosynthesis of the mare placenta, resulting in intermediates with the respec-

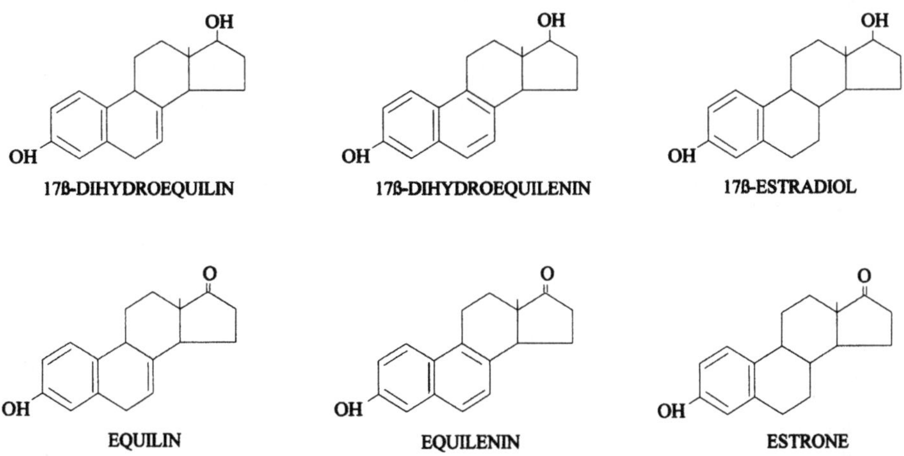

Fig. 2. Estrogens identified in plasma and urine from pregnant horses

tive double bonds in the B-ring, and these estrogens are not a product of steroid-eliminating enzymes (BHAVNANI 1988). To test this hypothesis, various ^{14}C-precursors were injected into the jugular vein of pregnant mares. Only ^{14}C-acetate but not ^{14}C-cholesterol or ^{14}C-androgens gave ^{14}C-equilin and ^{14}C-equilenin (HEARD et al. 1954). The biological activity of B-ring unsaturated estrogens was investigated by Allen–Doisy tests, uterotrophic assay, chick-oviduct assays and relative receptor binding; a bioactivity was documented for all compounds and receptor binding affinities indicated the following order of activity: 17β-dihydroequilin > 17β-estradiol > estrone > 17β-dihydroequilenin > equilin > 17α-dihydroequilin > 17α-estradiol > 17α-dihydroequilenin > equilenin (BHAVNANI 1988).

II. Estrogen Receptor and Signal Transduction

At present the estrogen receptor (ERα) sequence is known for more than 20 species (EMBL DATA BANK INFORMATION SERVICE 1998), representing six different mammalian orders (primates, lagomorphes, rodents, carnivores, artiodactylids, perissodactylids) and all other major vertebrate classes, namely birds, reptiles, amphibians and teleost fish (GAUB et al. 1990; LAUDET et al. 1992; LEROUX et al. 1993; TAN et al. 1995; ING et al. 1996). The shared sequence homologies differ for the single exons depending on their functions, and homologies are maximal in the DNA-binding regions and the steroid pocket (HAGEN–MANN 1994). In general, the degrees of identity are in line with the evolution disparity (Table 2).

A second ER (named ERβ) has been cloned from rat prostate cDNA. The differential ligand activation suggests, that ERα and ERβ may play different roles in gene regulation (PAECH et al. 1997). The molecular characteristics of the ERα protein are quite similar (MW~60 kDa, pI~5.0–5.4) within various species, but rather different affinities to estradiol have been published ranging from 10^{-9}–10^{-11} (apparent K_d). These studies were carried out with different procedures and preparations, in the presence or absence of binding globulins and at varying temperatures. Therefore, these results are not suitable for comparative evaluations. In general, they agree that estradiol shows the highest affinity among the endogenous agonists, whereas relative binding of estrone

Table 2. Homology (expressed as a percentage) of human estrogen receptor alpha (ERα) cDNA exons with cattle, mouse, rat, chicken, xenopus and trout

	Exon 1	Exon 2	Exon 3	Exon 4	Exon 5
Cattle	91.3	98.9	92.2	97.8	99.8
Mouse	85.8	83.6	86.9	85.6	90.6
Rat	82.3	83.6	85.0	85.7	88.5
Chicken	60.0	79.1	86.1	73.8	81.3
Xenopus	59.7	73.6	84.3	67.3	76.3
Trout	47.4	51.5	54.2	45.4	44.9

is about ten times less – as shown, for example, for cattle (WAGNER et al. 1972) or elephants (GREYLING et al. 1997). In any case, estradiol and the synthetic estrogens diethylstilbestrol and ethinyl estradiol exhibited almost identical affinities, and their different in vivo activities must be attributed to different inactivation and elimination pharmacokinetics. There are no indications that relative binding activities of estrogens vary considerably among the species, unlike the progestin receptor for which a co-evolution of receptor and ligand within mammalian species was found (JEWGENOW and MEYER 1998). This states that the key features of the estradiol binding pocket of the estrogen receptor are quite constant, and this seems to be unimpaired by evolved primary structure modifications.

A good indicator of the absolute affinity of the steroid–receptor complex is the physiological range of endogenous estradiol levels in the respective species, because the dissociation constant (K_d) must be in the same order of magnitude as that to get a functional cybernetic system in which the formation of hormone–receptor complexes responds to changing ligand levels. In this respect, it is obvious that an increasing affinity evolved from fish to birds and mammals. In trout, estradiol of ovarian origin ranges from 1 ng/ml to 10 ng/ml in blood plasma (BROMAGE et al. 1982), in chickens 0.1–1 ng/ml (JOHNSON et al. 1980), in horses 5–20 pg/ml (DAELS et al. 1991) and in cows (MEYER et al. 1990) or elephants (TAYA et al. 1991) it does not exceed 10 pg/ml during estrus. Also, in women and other primates, the available "free estradiol" in blood is probably in the same range, as most of the circulating estradiol (range 50–500 pg/ml) is bound to sex hormone binding globulin (SHBG). During evolution of large Eutherian mammals (placentalia), the affinity of the estrogen receptor obviously improved by a factor of almost 1000. This contrasts with the values for androgens, progestins and corticoids, which are mostly regulated in the nanogram range in large mammals, including the elephant (MEYER et al. 1997a). This amplified sensitivity to estrogens may have created the potential of some secondary plant compounds – namely the phytoestrogens – to interact with the estrogen receptors and impair normal reproduction (SHEMESH et al. 1972; FARNSWORTH et al. 1975).

III. Elimination

Removal of hormones from the circulation is part of activity regulation. In the case of the lipophilic steroid hormones, the elimination involves, in the first step, an inactivation or formation of hydroxyl groups by redox reactions, hydroxylations or epimerizations if necessary and, in the second step, a conjugation to a sulfate ester or a glycosidic glucuronide to become water soluble before excretion via the bile or urine. The active estradiol-17β can be oxidized to estrone, hydroxylated to estriol or catechol estrogens and epimerized to estradiol-17α and, in the second step, a large number of inactive mono- or diconjugates can be formed. It seems of little significance to the organism what kind of product is finally excreted; even within one species, great individual

Table 3. Studies of the excretion of radiolabeled estrogens by Eutherian mammals (according to SCHWARZENBERGER et al. 1996)

Species	Steroid	Reference
Sheep	E_2	ADAMS et al. 1994
	E_1	PALME et al. 1996
Pony	E_1	"
Pig	E_1	"
Domestic cat	E_2	SHILLE et al. 1990
	E_2	BROWN et al. 1994
Sibirian pole cat	E_2	GROSS 1992
Otter	E_2	"
White rhinoceros	E_2	HINDLE and HODGES 1990
African elephant	E_2	WASSER et al. 1996
Asian elephant	E_2	CZEKALA et al. 1992
Slow lori	E_2	PEREZ et al. 1988
Ring-tailed lemur	E_2	"
Cotton-top lemur	E_2	ZIEGLER et al. 1989
	E_2	"
Macaque	E_2	SHIDELER et al. 1993
Yellow baboon	E_2	WASSER et al. 1994

deviations were monitored. The bovids predominantly excrete the epimer estradiol-17α (DUNN et al. 1977); in horse and swine, estrone is the major product, but significant amounts of estradiol-17β are also excreted (CHOI 1987); primates produce estriol; and elephants preferentially excrete estradiol-17β conjugates (CZEKALA et al. 1992).

PALME et al. (1996) have carried out very detailed excretion studies of infused radio-labeled ^{14}C-estrone in ponies, pigs and sheep. They concluded, that sheep excrete via feces preferentially (89%), whereas urine contains most of the estrogens in ponies (98%) and pigs (96%). However, if testosterone, progesterone or cortisol was infused, rather different distributions were found. In any case, the peak levels in urine appeared a few hours after infusion and elimination via urine was almost complete after 1 day. Fecal excretion reflected the known passage rates of digesta in the intestine. Peak concentrations of fecal steroids were reached most rapidly in sheep (0.5 days), followed by ponies (1 day) and pigs (2 days). Urine steroids, therefore, reflect quite well the endocrine status of the animal with only a few hours delay; whereas, for fecal estrogens, a longer delay that depends on the species and the food has to be considered. Due to all the found differences concerning the route of excretion, the major metabolites and the time lags, any extrapolation to further species or uninvestigated circumstances are hardly possible. Table 3 summarizes the excretion studies with radiolabeled estrogens in different species.

C. Evolution of Estrogen Functions

Estrogens regulate numerous functions that are strategically important for reproduction. Almost each tissue has estrogen receptors and responds more or less to estrogens. In early vertebrates, the strategy of reproduction is based on the production of a very large number of oocytes, and the metabolic needs for the voluminous fish-roe production are of major importance within the circannual life cycles of fish. This changed after development of both internal fertilization and the cleidoic egg in reptiles. It enabled more investment for internal "brood care" and the production of a much lower number of eggs, later, in birds. The metabolic needs moved from the pre-ovulatory to the post-ovulatory time span. The parallel development of internal semen conservation did not limit matings to ovulations with estrus.

The outstanding importance of estrus is a characteristic feature of placentalia, in which ovulation and mating must be well co-ordinated. Estrogen production by the developing follicle mediates all signals of the nearing optimal mating time, preparing the female and the male for this event of most vital importance within the life of both genders. The induction of estrus symptoms seems to be mediated by estradiol-17β in each mammalian species – at least there is no known Eutherian female animal that uses another compound for this purpose. The word estrus, originating from the Greek word *οιστροσ* (oistros) and meaning a gadfly, goes back to similar behavior of cows that toss their tails into the air when pursued by gadflies or when they are in heat (SHORT 1984). The type of estrus symptoms – whether mediated by olfactory, acoustic, tactile or optic signals – is rather variable and is well adapted to the performances of senses in a species.

Following activities of estrogens on implantation, pregnancy recognition, pregnancy maintenance, parturition and secondary features such as the physiological complexes of lactation or any kind of metabolic effect either directly or indirectly affecting minerals, carbohydrates, amino acids or lipids is most variable and is an important part of the developed reproduction strategy of a species. This provides more and more insight into the evolution of hormonal activities within and between mammalian orders.

I. Oviparous Vertebrates

The processes of oogonial multiplication, oogenesis, oocyte maturation, ovulation and oviposition constitute the seasonal reproductive cycle in fish and require co-ordination in order to adjust spawning for that season of the year most favorable for survival. It is now clear that a variety of geophysical parameters, such as photoperiod, temperature, rainfall and salinity, and biological factors, such as nutrition, play important roles in the timing of reproduction of fish (WHITEHEAD et al. 1978; LAM 1983). For salmonids, photoperiod appears to be the major controlling factor (BROMAGE et al. 1984; POHL–BRANSCHEID and HOLTZ 1990) and, as in mammals, it is melatonin that co-ordinates the circannual rhythms. Estrogen synthesis in the gonads is controlled by neuroendocrine changes involving the hypothalamus and the pituitary gland. The circannual pattern of gonadal estrogen secretion in the rainbow trout – a teleost fish – is well related to the steps of oocyte formation and is reduced to basal levels around spawning.

In fish, estrogens are most important regulators of metabolic requirements for reproduction and oocyte maturation, including yolk production which needs considerable amounts of nutrients; this is obvious in view of the high nutritional value of fish roe. Estrogens in fish are not related to spawning that coincides with external fertilization. Estrus – as typical in mammals at the peak of estrogen synthesis and most important for timing of mating – seems not to be developed yet in fish. A number of proteins are induced by estrogens and vitellogenin is the one most investigated. Vitellogenins are common, but quite variable, proteins (400–600 kDa) of all egg-laying animals and they are produced under the specific control of estrogens in the liver of fish, amphibians, reptiles and birds. After secretion into the blood they are absorbed by the oocytes for egg yolk production. The vitellogenins are glyco-lipo-phosphoproteins (12% lipids, 1.3% protein-bound phosphor and 1% carbohydrates in the frog *Xenopus laevis*) that are composed of several subunits and are heavily processed for yolk production (HARA and HIRAI 1978; ROACH and DAVIES 1980; SCHUBIGER and WAHLI 1986; WALLACE 1978; WILEY and WALLACE 1981).

Males and immature females do not produce vitellogenins, but exogenous estrogens can induce its production. The in vivo stimulation of vitellogenin synthesis in juvenile fish has been used to compare the hormonal activities of xenoestrogens, such as alcylphenols (*p*-octylphenol) or the 1,1,1-trichlor-2,2-bis (*p*-chlor-phenyl) ethane (DDT)-derivatives, with estradiol-17β or ethinylestradiol (PURDOM et al. 1994; SHEAHAN et al. 1994; REN et al. 1996; LECH et al. 1996). These systems have been used for eco-toxicological investigations of sewage-water contaminations. Alternatively, hepatocyte cultures from fish or amphibians have been developed for studying estrogen-induced vitellogenin synthesis and much simpler screening of water quality (PELISSERO et al. 1993; JOBLING and SUMPTER 1993; WHITE et al. 1994). The relative activities of the tested physiological and xenobiotic estrogens deviated considerably within in vivo and in vitro systems – presumably due to the much lower metabolic capacities of the cell-culture systems.

Estrogen-related ovulation mechanisms in female birds are well studied (FARNER and WINGFIELD 1980; DAWSON 1983; AKESSON and RAVELING 1984; BLUHM 1988) and the patterns obtained agree that estrogens are maximal prior to laying and remain elevated throughout laying. Recently developed non-invasive test systems using fecal steroid-hormone analysis (HIRSCHENHAUSER et al. 1997) allow unimpaired hormone monitoring of wild birds, as shown for greylag geese (Fig. 3). Estrogens play a major role in yolk-protein synthesis (vitellogenin), egg-shell formation and female reproductive behavior, including nest building (DONHAM 1979; HARVEY et al. 1981). The egg-white proteins (ovalbumin, conalbumin, avidin) seem to be controlled by estrogens and progestins. Estrogens also seem to be involved in avian fat deposition, winter-flock formation and courtship, which is indicated by the second, lower peak in the winter months. In chickens, estrogens support the retention of calcium and phosphor and their deposition in bone tissues for rapid availability of these minerals during egg-shell formation (KARG 1969).

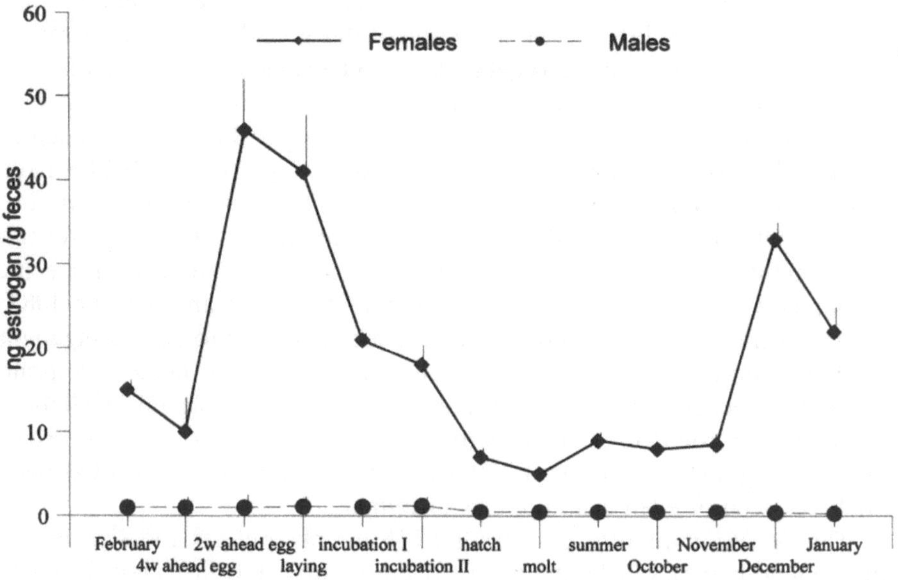

Fig. 3. Fecal estrogen pattern in greylag geese (HIRSCHENHAUSER et al. 1997)

Pre-natal endocrine differences and steroidogenic mechanisms of sexual differentiation have been widely investigated and well documented for avian species (GALLIEN and LE FOULGOC 1957; GALLI and WASSERMANN 1972; TANABE et al. 1979; BERCOVITZ et al. 1985), turtles (WIBBELS and CREWS 1992) and fish (REINBOTH and BECKER 1984). Estradiol and estrone were isolated from ovarian tissue of 10- to 21-day-old chicken embryos and plasma estradiol concentrations in female embryos are consistently greater than in males. For this reason, sexual differentiation of oviparous animals is sensitive to environmental estrogens, which is most problematic for fish. This feature has also been used for estrogen testing (BERGERON et al. 1994) and for production purposes in fish farms, as males have better weight gains which are not impaired by roe production. In this context, it should be mentioned that in addition to direct effects of xeno-estrogens, indirect eco-toxicological effects on estrogen production have also been described. Tributyltin and related tin derivatives seem to inhibit aromatase activity, and androgens instead of estrogens are produced (OEHLMANN et al. 1992; FENT and BUCHELI 1994; BETTIN et al. 1994). Such compounds originate from ship paintings and seem to be active in the pg/ml range.

II. Viviparous Eutherian Mammalia/Placentalia

1. Estrus

Estradiol production by the enlarging follicles is under the strict control of the gonadotropic hormones, follicle-stimulating hormone (FSH) and luteinizing hormone (LH), which are controlled by the hypothalamic gonadotropin-releasing hormone (GnRH). The induction of this gonadotropic axis is rather variable among species and is regulated by biological and geophysical factors – as in non-mammalian vertebrates. Biological factors include nutrition, e.g., in cows (ROCHE and DISKIN 1995); lactation, e.g., in sows; the presence of males recognized by pheromones, e.g., in sheep, goats or pig (CLAUS 1979; CLAUS et al. 1990), or acoustic signs, e.g., deer roaring. The best investigated geophysical factor is photoperiodism, in which melatonin provides the information of night length. The females of sensitive species respond in a specific manner (LEGAN and WINANS 1981; REITER 1978; CLAUS and WEILER 1985) either to increasing night length, e.g., roe deer/red deer, or to decreasing night length, e.g., european brown hare, fox and wild pigs. Males need to respond differently, because the whole process of spermatogenesis takes about 3 months and, in roe deer, gonadal activity has to start earlier, during decreasing night length (BLOTTNER et al. 1996). Seasonal induction of estrus has been widely reduced or even lost in domestic animals.

The secretion pattern of estradiol-17β mainly reflects the size of the enlarging dominant follicle, and the estrogen rise may only last a few days, e.g., in cows (MEYER et al. 1990) or up to 2 weeks, e.g., in horses (DAELS et al. 1991). Absolute levels of estradiol are extremely low (range 1–10 pg/ml in blood plasma) in some large mammals, e.g., in the Indian elephant (TAYA et al. 1991) and the cow (Fig. 4), and the reliable measurement of circulating estradiol-17β levels requires advanced assay systems. Animals can be classified as monoestrus or polyestrus species. In monoestrus animals, the estradiol peak is mostly followed by progestin production, lasting until parturition, as in dogs (CONCANNON 1986), by an intermediate embryonic diapause, as in the giant

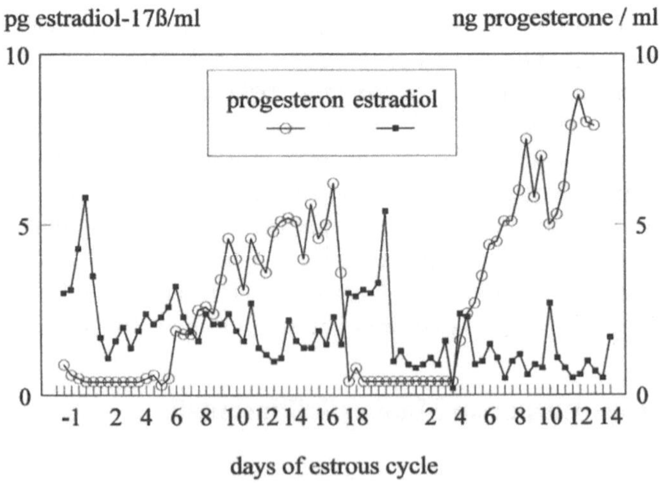

Fig. 4. Estradiol-17β and progesterone in blood plasma of cows (MEYER et al. 1990)

panda (MEYER et al. 1997b), or retarded embryo development, as in roe deer (LENGWINAT and MEYER 1996).

The polyestrus species (most ungulates and primates) have developed various pregnancy detection and estrus-cycle termination systems (MEYER 1994) to provide a new chance of reproduction if it fails during the first estrus. Induced ovulation, e.g., in camel, cat or rabbit, enables alternative polyestrus, because ovulation and corpus luteum (CL) formation depend on copulation-induced mechanisms. Without mating, estradiol peaks follow each other, without a LH peak, no intermediate progestin production and a cycle length of 2 weeks in the cat (SHILLE et al. 1979). Similar non-luteal estrus cycles of 3-week duration, detectable by estradiol-17β peaks and anovulatory LH peaks, were reported for Asian (HESS et al. 1983; BROWN et al. 1991) and African elephants (KAPUSTIN et al. 1996). One or two non-luteal cycles alternated with a luteal estrus cycle lasting 10–11 weeks and resulted in a "total cycle length" of about 16 weeks.

The estradiol-induced effects on primary sex organs (endometrium proliferation, cervical mucus) hardly vary among species, but the estradiol-induced signs to the male are most variable. There is clear evidence for a multitude of unidentified, necessarily species-specific pheromones that can be detected, e.g., by the male dog, even if kilometers away from a female in estrus. Male ruminants, when smelling a female in estrus, raise the head and curl the upper lip back (called flehmen), which in some way allows odors to reach the vomeronasal organ, which seems to be specifically used for savoring sexual smells. The identification of the large variety of estradiol-induced pheromones is still a rather undeveloped field in biochemical sciences, although there is great commercial interest, e.g., for detection of silent estrus in dairy cows. Recently, attempts were undertaken to purify the estrus-signaling pheromone from cow urine and characterize it (DEHNHARD and CLAUS 1996). The neutral pheromone has a molecular weight between 50 Da and 130 Da and is highly volatile, but the molecular structure remains to be identified. In other studies, a few estrus-associated pheromones were found: dimethyldisufide in hamster (O'CONNELL et al. 1979); isobutyric plus isovaleric acid in Rhesus monkey (CURTIS et al. 19971); methyl-*p*-hydroxybenzoate in dog (GOODWIN et al. 1979); and *Z*-7-dodecen-1-yl acetate in the Asian elephant (RASMUSSEN et al. 1997).

Cats call unequivocally when in estrus, but there are many other behavioral changes, such as male tolerance, "bulling" or sit down of female camels. Estrous sows becoming immobile depend on the boar's pheromones, androst-16-en-3-ol and androst-16-en-3-one. Any kind of estrus sign obviously depends on the presence of estradiol and the absence of progesterone, and the application of exogenous estrogens will induce a false estrus with all physiological and behavioral effects.

In primates, the CL also produces estradiol and New-World primates have maximal blood and urine levels of estrogens during the luteal phase of the estrus cycle, when estrogens are 10-fold higher than during the follicular phase and estrus. Fecal estrogen studies in Goeldi's monkeys, marmoset and tamarins proved this unusual estrogen secretory pattern of the corpus luteum (HEISTERMANN et al. 1993; PRYCE et al. 1994; ZIEGLER 1989).

2. Embryonic Pregnancy Signaling

Most ruminants, suides and equides have estrus cycles of about 2–4 weeks, and luteolysis is induced by endometrial release of prostaglandin $F_2\alpha$, ($PGF_2\alpha$) during late diestrus (MCCRACKEN et al. 1984; GADSBY et al. 1990; BAZER 1989). Luteal sensitivity to $PGF_2\alpha$ seems to be induced by increased oxytocin sensitivity; oxytocin receptors are controlled by the increasing estrogens from the next follicular wave (MEYER et al. 1986). Pregnancy-recognition signals have been extensively investigated in artiodactylids. Flushing embryos from the uteri of gilts before day 11 did not extend the estrus cycle. The presence of embryos in the uteri at day 12 or later resulted in an extension of the interestrus interval, suggesting the initiation of embryonic signals around day 11–12 (DHINDSA and DZIUK 1968). These findings initiated a long series of studies that documented the ability of day 11 and older conceptuses to produce estrogens (PERRY et al. 1973; FISCHER et al. 1985). By utilizing immunocytochemistry, the aromatase was localized in the trophectoderm surrounding the embryonic disc (BATE and KING 1988) and the yolk sac (KING and ACKERLY 1985) of pig conceptuses. In the meantime, there is a large body of literature indicating that maternal estrogens control $PGF_2\alpha$ synthesis and secretion (BAZER et al. 1986; BAZER et al. 1989; GEISERT et al. 1990). While conceptus estrogen synthesis is a critical event, maternal recognition of early pregnancy in the pig involves a number of biochemical and cellular interactions set in motion by the initial rise of estrogens. In addition to the early events that ultimately block luteolysis, estrogens are also key regulators of conceptus–endometrial interactions, which control vascular permeability, blood flow, placental attachment and immunological protection (KNIGHT 1994). The estimation of circulating estrogens is used for early pregnancy detection in pigs.

The ability of the embryonic tissues of the dromedary camel to synthesize estrogens was studied in 15- to33-day-old conceptuses (SKIDMORE et al. 1994). There was good evidence for strong aromatase activity and, surprisingly, estradiol was synthesized predominantly. The authors suggested analogous mechanisms in achieving $PGF_2\alpha$ control, luteal maintenance and maternal pregnancy recognition in camelids. They also have a diffuse, epithelio-chorial placenta like that of pigs (VAN LENNEP 1961). Camelids are induced ovulators, but in the case of unsuccessful matings, $PGF_2\alpha$ causes luteolysis after 2–3 weeks as in other artiodactylids.

In ruminants, interferon tau (IFNτ) has become the anti-luteolytic signal produced by conceptuses for taking over all functions, in blocking $PGF_{2\alpha}$ release and maintaining luteal function (ROBERTS et al. 1992); IFNτ was identified in all ruminant families (cervids, giraffids, bovids). As it is absent in camelids and suides, it must have evolved approximately 30–25 million years ago from embryonic type-1 IFN. Therefore, it seems obvious, that estradiol-

17β is the older phylogenic pregnancy recognition signal of conceptuses, which was already used by the common ancestors of all artiodactylids and has maintained its function until nowadays in pigs and camels. Even in horses, estradiol-17β is produced as early as by the 12-day-old conceptus and seems to be involved in pregnancy signaling as well as other factors (BAZER 1989; SHARP et al. 1989).

3. Placental Estrogens

The dynamics of placental estrogen production have been well investigated in four species: human (ALBRECHT and PEPE 1990), horse (AINSWORTH and RYAN 1969; BHAVNANI 1988), pig (KNIGHT 1994) and cow (HOFFMANN 1983). The very high estrogen levels during pregnancy already enabled the first determination with early biological assays (KÜST 1934). Further studies in quite a number of different species suggest that placental estrogens are common in most perissodactylids, artiodactylids and primates. Circulating estrogen levels in dams increase continuously over the whole pregnancy period. In cows, estradiol concentrations on day 20 are similar to those during estrus, and a few weeks before parturition they reach 200–600 pg/ml, which is almost 100 times more than around estrus (THEYERL–ABELE et al. 1990). Horses develop further follicles, producing estrogens after the fifth week of pregnancy; after luteinization, the formed accessory CL produce gestagens that support pregnancy maintenance (TERQUI and PALMER 1979). After 3–4 months, the placenta takes over estrogen and progestin production. This pattern of estrogen production is also found in other equides, tapirs (BROWN et al. 1994; KASMAN et al. 1985) and rhinoceroses (HINDLE et al. 1992; KASSAM et al. 1981). However, it is unclear whether accessory follicles and CL are produced in all perissodactylids. Elephants are also supposed to produce estrogens during pregnancy (MCNEILLY et al. 1983; HODGES et al. 1983), but the secretion pattern shows only a modest elevation in the last 12 months of pregnancy (HODGES et al. 1987). Estradiol-17β seems to dominate in plasma at any time, but levels are much lower than in other ungulates or primates. The origin of pregnancy estrogens in elephants is unknown. Dogs do not produce placental estrogens (HOFFMANN et al. 1994) and there is no indication of production in other carnivores.

The functions of estrogens during pregnancy may be of a multitude, but it is not yet possible to study estradiol "knock-out" models and, therefore, this discussion remains speculative. It was suggested, that in the pig, estradiol supports placental and endometrial vascularization and proliferation (KNIGHT 1994). In cattle, estrogens are essential for mammogenesis and lactogenesis (SCHAMS et al. 1984), but estrogens will also suppress galactopoesis before the next parturition; the amount of estrogen seems to be also paternally controlled. In domestic ungulates, estrogens are very effective anabolic agents and they support retention of proteins and other nutrients; it may be a further function to store protein and minerals for the coming lactation or to direct it to the growing fetus. Moreover, estrogens are involved in oxytocin-receptor

Table 4. Zoo animal species excreting fecal estrogens, indicating pregnancy

Species	Reference
Yak	SAFAR-HERMANN et al. 1987
Grevy's zebra	
Red Buffalo	
Nubian Ibex	
Sable antelope	CHAPEAU et al. 1993
Gorilla	
Malaysian Tapir	
Muskox	DESAULNIERS et al. 1989
Caribou	MESSIER et al. 1990
Orang-utan	BAMBERG et al. 1991
Mhorr gazelle	
Przewalski Mare	MÖSTL et al. 1988
Yellow baboon	WASSER et al. 1991

induction and parturition (MAGGI et al. 1991). Aromatase deficiency in human placenta resulted in maternal and fetal virilization and low levels of estrogen excretion (HARADA 1993).

The outstanding estrogen production capacity of the placenta has provided a means of non-invasive pregnancy diagnosis; first, via skimmed milk in cattle (HEAP and HAMON 1979) and, later, via feces in many species. This feature was found most useful for zoo animals. Elevated estrogens indicating pregnancy were present in bovids, equides and several cervids and primates (Table 4), but were not detectable in the hippopotamus, the black rhinoceros, the giraffe, the okapi or carnivores (SAFAR–HERMANN et al. 1987; SCHWARZENBERGER et al. 1996). Maximal levels were found in the Nubian ibex (2 μg/g).

4. Estrogens in Males

The daily estrogen production of boar and stallion testes amounts to 20–100 mg (RAESIDE 1978; MAHESH et al. 1987), and the resulting mean levels of unconjugated estradiol-17β in plasma (200–500 pg/ml in boars of different breeds and wild boars; 40 pg/ml in stallions) are 2- to 5-fold higher than the estrus values of the respective female (WEILER et al. 1998; CLAUS et al. 1992; SCHOPPER et al. 1984). Means of total estrogens in blood plasma were 17 ng/ml in boars and 2.4 ng/ml in stallions. The in vitro production of estrogens was documented for both species (SMITH et al. 1987; RAESIDE et al. 1989); in addition to the C19 androgens, 19-norandrogens also become aromatized. In the bull, circulating estradiol-17β levels increase when the animal becomes older and up to 30 pg/ml was identified, which is 3- to 5-fold higher than in cows around estrus (MEYER et al. 1990). Fecal estrogen determination was applied to stallions and boars (BAMBERG et al. 1986). The estrogen content in stallion feces can reach values comparable to those of pregnant mares and its determination provides a reliable indicator of cryptorchidism in horses (PALME et al. 1994).

The functions of estrogens in males are less clear. Seminal estrogens will induce lordosis of ovulation, endometrial $PGF_2\alpha$ secretion and myometrial contractions in the sow (CLAUS et al. 1987; CLAUS 1990). Boar ejaculates contained up to 15 μg/ejaculate. After insemination, up to 5 ng E_2/ml blood plasma was found in the uterine vein. In peripheral circulation, exogenous estradiol from the boar exceeds the endogenous estrogens from the sow and semen estrogens support the LH secretion. As well as the estrogens, other factors also seem to be involved in the induction of ovulation.

The continuously present estrogens in the male will support the anabolic activity of androgens. Therefore, estrogens are also responsible for allometric growth and the resulting sexual dimorphism, as obvious in the investigated domestic species.

Male estradiol-receptor "knock-out" mice were found to be infertile, which proved that estradiol is essential in male fertility (HESS et al. 1997). Estrogens were shown to be responsible for testicular fluid production and, hence, for intra-testicular sperm transport. In the epididymis, estrogens regulate luminal fluid resorption and sperm concentration. In this respect, comparative studies will be most interesting in future.

The presence and the functions of estrogens in males support the hypothesis that sex steroids can no longer be strictly divided into male and female ones. This questions the idea, that low levels of xenoestrogens in our environment might impair the fertility of men or terrestrial male animals.

5. Metabolic, Anabolic and Other Secondary Features

Farm animals with low endogenous estrogen production (calves, lambs, heifers, steers) respond well to exogenous estrogens and they enhance their growth rates by 5–15% (MEISSONNIER 1983). Protein deposition in skeletal muscle is improved and nitrogen excretion is reduced. The anabolic activity of estrogens is much superior to that of androgens, at least in cattle and sheep. The xenoestrogen diethylstilbestrol was licensed in the USA until 1979, when it became forbidden to use it for growth promotion in all farm animals. It has been replaced by estradiol-17β and zeranol implants, which are now registered in the USA and many other countries. In contrast, in the EU, the registration of estrogens and other anabolic hormones is not allowed. When given parenterally estradiol-17β and zeranol are almost equipotent and only 50 μg per day per heifer or steer is required for a steady and optimal growth promotion (WAGNER 1983). Growth is effected by estrogens in a dose dependent manner. As is the case during puberty, low doses will stimulate growth, but high estrogen doses will enhance calcium retention in bones and, thereby, causing cessation of overall growth (KARG 1969; BRÜGGEMANN and KARG 1960). Many studies have shown, that meat quality is not impaired and residues do not provide any harm to the consumer after correct use of the drug within good animal husbandry (MEISSONNIER 1983; KOSSILA 1983; FARBER et al. 1983; LAMMING et al. 1987). However, any kind of misuse may provoke adverse effects in consumers, especially small children (KARG et al. 1990; MEYER 1991).

The anabolic mode of action has been well investigated over the past two decades and includes direct and indirect effects (MEYER and KARG 1989). Estrogen receptor in the bovine skeletal muscle was identified by means of several independent methods, and the physicochemical and biochemical characteristics match with those of the uterine estrogen receptor (MEYER and RAPP 1985). The estrogen receptor concentrations amount to only 0.5–2.0 fmol/mg cytosolic protein, which is less than 1% in comparison with the uterus. The relative concentrations in individual muscles is well in line with the estrogen-induced allometric growth impetus of the respective muscle and there was no difference in male and female calves. The concentrations of androgen receptor in the same muscles were only one-fourth on average (SAUERWEIN and MEYER 1989), corresponding to minor anabolic activity of androgens in cattle (Table 5). In addition, estrogens act indirectly via stimulating growth-hormone secretion (DAVIS et al. 1977), growth-hormone receptors in liver and insulin-like growth factor 1 (IGF1) levels in blood (BREIER et al. 1988a, b; SAUERWEIN et al. 1992). Estrogens seem to act in all strategically important tissues to improve protein anabolism and mineral retention (Fig. 5), including the rumen and intestine epithelia (MEYER et al. 1995; SAUERWEIN et al. 1995).

The biological functions of estrogen-induced anabolism have scarcely been discussed. However, on the one hand, estrogens together with androgens will be important in the male, in preparing the animal for dominance fights in the mating season. On the other hand, high estrogens during pregnancy strongly support protein, calcium and phosphor accretion in the mother to

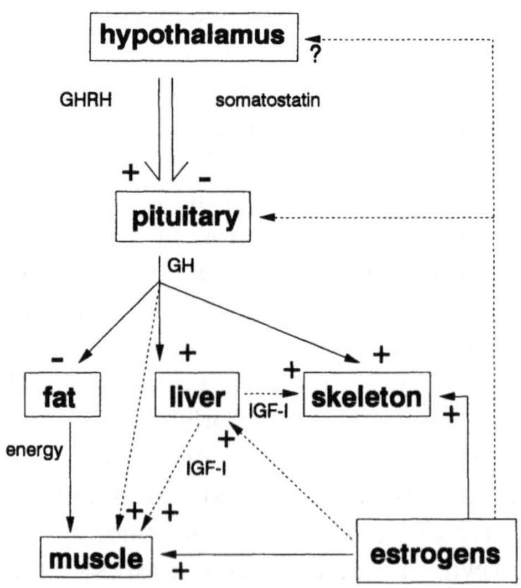

Fig. 5. Synopsis of direct and indirect metabolic effects of estrogens in cattle

Table 5. Estrogen and androgen receptors in different skeletal muscles of female calves

	Estrogen receptor	Androgen receptor
Neck	2.03 ± 0.58	0.56 ± 0.14
Shoulder	1.18 ± 0.32	0.50 ± 0.11
Abdomen	0.98 ± 0.86	0.23 ± 0.13
Hind leg	1.63 ± 0.44	0.52 ± 0.17

create a store for lactation (Lenkeit 1972). The removal of estrogens after parturition stops the "super-retention" and allows the mobilization of proteins and minerals for availability within milk production.

References

Adams NR, Abordi JA, Briegel JR, Sanders MR (1994) Effect of diet on the clearance of estradiol-17β in the ewe. Biol Reprod 51:668–674

Ainsworth L, Ryan KJ (1969) Steroid hormone transformations by endocrine organs from pregnant mammals. III Biosynthesis and metabolism of progesterone by the mare placenta in vitro. Endocrinology 84:91–97

Akesson TR, Raveling DG (1984) Endocrine and behavioral correlates of nesting in Canada Geese. Can J Zool 62:845–850

Akhtar M, Calder MR, Corina DL, Wright JN (1982) Mechanistic studies on C19 demethylation in estrogen biosynthesis. Biochem J 201:569–580

Akhtar M, Njar VCO, Wright JN (1993) Mechanistic studies on aromatase and related C–C bond cleaving P450 enzymes. J Steroid Biochem 44:375–387

Albrecht ED, Pepe GJ (1990) Placental steroid hormone biosynthesis in primate pregnancy. Endocr Rev 11:124–150

Arts CJM, Schilt R, Schreurs M, van Ginkel LA (1996) Boldenone is a naturally occurring (anabolic) steroid in cattle. In: Haagsma N, Ruiter A (eds) Proceedings EuroResidue III, Conference on Residues of Veterinary Drugs in Food, FECS Event no. 213, pp 212–217

Bamberg E, Choi HS, Möstl E (1986) Östrogenbestimmung im Kot zur Trächtigkeitsdiagnose bei Pferd, Rind, Schwein, Schaf und Ziege. Tierärztl Umschau 41:406–408

Bamberg E, Möstl E, Patzl M, King GJ (1991) Pregnancy diagnosis by enzyme immunoassay of estrogens in feces from nondomestic species. J Zoo Wildlife Med 22:73–77

Bate LA, King GJ (1988) Production of oestrone and oestradiol-17β by different regions of the filamentous blastocyst. J Reprod Fertil 84:163–169

Bazer FW, Vallet JL, Roberts RM, Sharp DC, Thatcher WW (1986) Role of conceptus secretory products in establishment of pregnancy. J Reprod Fertil 76:841–850

Bazer FW (1989) Mediators of maternal recognition of pregnancy in mammals. Proc Soc Exp Biol Med 199:373–384

Bazer FW, Vallet JL, Harney JP, Gross TS, Thatcher WW (1989) Comparative aspects of maternal recognition of pregnancy between sheep and pigs. J Reprod Fertil (Suppl 37):85–89

Benoit E, Garnier F, Courtot D, Delatour P (1985) Radioimmunoassay of 19-nortestosterone. Evidence of its secretion by the testis of the stallion. Ann Rech Vet 16:379–383

Bercovitz AB, Mirsky A, Frye F (1985) Non-invasive assessment of endocrine differences in day-old chicks (*Gallus domesticus*) by analysis of the immunoreactive estrogen excreted in the egg. J Reprod Fertil 74:681–686

Bergeron JM, Crews D, McLachlan JA, (1994) PCBs as environmental estrogens: turtle sex determination as a biomarker of environmental contamination. Environ Health Persp 102:80–781

Bettin C, Oehlmann J, Stroben E (1996) TBT–induced imposex in marine neogastropods is mediated by an increasing androgen level. Helgoländer Meeresun 50:299–317

Bhavnani BR (1988) The saga of the ring B unsaturated equine estrogens. Endocr Rev 9:396–416

Blottner S, Hingst O, Meyer HHD (1996) Seasonal spermatogenesis and testosterone production in roe deer (*Capreolus capreolus*). J Reprod Fertil 108:299–305

Bluhm C (1988) Temporal patterns of pair formation and reproduction in annual cycles and associated endocrinology in waterfowl. In: Johnson RF (ed) Current Ornithology, Vol. 5, Chapter 3, pp 123–185

Breier BH, Gluckman PD, Bass JJ (1988a) The somatotrophic axis in young steers: influence of nutritional status and oestradiol–17β on hepatic high– and low–affinity somatotrophic binding sites. J Endocrinol 116:169–177

Breier BH, Gluckman PD, Bass JJ (1988b) Influence of nutritional status and oestradiol-17β on plasma growth hormone, insulin–like growth factors–I and –II and the response to exogenous growth hormone in young steers. J Endocrinol 118:243–250

Bromage NR, Whitehead C, Breton B (1982) Relationships between serum levels of gonadotropin, oestradiol-17β, and vitellogenin in the control of ovarian development in the rainbow trout. Gen Comp Endocr 47:366–376

Bromage NR, Elliot JAK, Springate JRC, Whitehead C (1984) The effects of constant photoperiods in timing of spawning in the rainbow trout. Aquaculture 43:213–223

Brown JL, Citino SB, Bush M, Lehnhardt J, Phillips LG (1991) Cyclic patterns of luteinizing hormone, follicle–stimulating hormone, inhibin and progesterone secretion in the Asian elephant (*Elephas maximus)* J Zoo Wildlife Med 22:49–57

Brown JL, Scott B, Shaw CJ, Miller Ch (1994) Endocrine profiles during the estrous cycle and pregnancy in the Baird's Tapir (*Tapirus bairdii).* Zoo Biol 13:107–117

Brüggemann H, Karg H (1960) Hormone in der Tiermast. In: Deutsche Gesellschaft für Endokrinologie (ed) 6. Symposium der Deutschen Gesellschaft für Endokrinologie. Springer Verlag, Berlin, S 195–206

Chapeau C, King GJ, Bamberg E (1993) Fecal estrogens in one primate and several ungulate species during various reproductive stages. Anim Reprod Sci 34:167–175

Choi HS (1987) Immunologische Bestimmung von Sexualsteroiden zur Fertilitätskontrolle bei Rind, Schwein und Pferd. Wien Tierärztl Monat 74:14–22

Claus R (1979) Pheromone bei Säugetieren unter besonderer Berücksichtigung des Ebergeruchsstoffes und seiner Beziehung zu anderen Hodensteroiden. Fortschr Tierphysiologie und Tierernährung, Beiheft 10, Paul Parey, Hamburg, S1–136

Claus R, Weiler U (1985) Influence of light and photoperiodicity on pig prolificacy. J Reprod Fertil Suppl 33:185–197

Claus R, Hoang–Vu C, Ellendorf E, Meyer HHD, Schopper D, Weiler U (1987) Seminal estrogens in the boar: origin and functions in the sow. J Steroid Biochem 27:331–335

Claus R (1990) Physiological role of seminal components in the reproductive tract of the female pig. J Reprod Fertil Suppl 40:117–131

Claus R, Over R, Dehnhard M (1990) Effect of male odour on LH–secretion and the induction of ovulation in seasonally anoestrous goats. Anim Reprod Sci 22:27–38

Claus R, Dimmick MA, Giménez T, Hudson LW (1992) Estrogens and prostaglandin $F_2\alpha$ in semen and blood plasma of stallions. Theriogenology 38:687–693

Cole PA, Robinson CH. (1988) A peroxide model reaction for placental aromatase. J Am Chem Soc 110:1284–1285

Concannon PW (1986) Canine pregnancy and parturition. Vet Clin N Am–Small 16:453–475

Corbin CJ, Khalil MW, Conley, AJ (1995) Functional ovarian and placental isoforms of porcine aromatase. Mol Cell Endocrinol 113:29–37

Czekala NM, Roocroft A, Bates M, Allen J, Lasley BL (1992) Estrogen metabolism in the Asian elephant (*Elephas maximus*). Zoo Biol 11:75–80
Curtis RF, Ballantine JA, Keverne EB, Bonsall RW, Michael RP (1971) Identification of primate sexual pheromones and the properties of synthetic attractants. Nature 232:396–398
Daels PF, Ammon DC, Stabenfeldt GH, Liu IKM, Hughes JP, Lasley B (1991) Urinary and plasma estrogen conjugates, estradiol and estrone concentrations in nonpregnant and early pregnant mares. Theriogenology 35:1001–1017
Davis SL, Ohlson DL, Klindt J, Anfinson NS (1977) Episodic growth hormone secretory patterns in sheep: relationship to gonadal steroids. Am J Physiol 233: E519–E523
Dawson A (1983) Plasma gonadal steroid levels in wild starlings (*Sturnus vulgaris*) during the annual cycle and in relation to the stages of breeding. Gen Comp Endocr 49:289–294
Dehennin L, Silberzahn P, Reiffsteck A, Zwain I (1984) 19–norandrostenedione and 19–nortestosterone in human and equine follicular fluid. Incidence on the accuracy of radioimmunoassay of some androgens. Pathol Biol 32:828–829
Dehnhard M, Claus R (1996) Attempts to purify and characterize the estrus–signalling pheromone from cow urine. Theriogenology 46:13–22
Desaulniers DM, Goff AK, Betteridge KJ, Rowell JE, Flood PF (1989) Reproductive hormone concentrations in faeces during the oestrous cycle and pregnancy in cattle (*Bos taurus*) and muskoxen (*Ovibos moschatus*). Can J Zool 67:1148–1154
Dhindsa DS, Dziuk PJ (1968) Effect on pregnancy in the pig after killing embryos of fetuses in one uterine horn in early gestation. J Anim Sci 27:122–126
Donham RS (1979) Annual cycle of plasma LH and sex hormones in male and female Mallards (*Anas platyrhynchos)* Biol Reprod 21:1273–1285
Dorfman RI, Ungar F (1965) Metabolism of Steroid Hormones. Academic Press, New York, London
Dunn TG, Kaltenbach, CC, Koritnik DR, Turner DL, Niswender GD (1977) Metabolites of estradiol–17β and estradiol benzoate in bovine tissues. J Anim Sci 46:659
Farber TM, Arcos M, Crawford L (1983) Safety evaluation standards in the United States. In: Meissonnier E (ed) Anabolics in Animal Production. Soregraph, Levallois, pp 509–513
Farner DS, Wingfield JC (1980) Reproductive endocrinology of birds. Annu Rev Physiol 42:457–472.
Farnsworth NR, Bingel AS, Cordell GA, Crane FA, Fong HHS (1975) Review article: Potential value of plants as sources of new antifertility agents I. J Pharm Sci 64:535–598
Feeny P (1976) Plant apparency and chemical defense. Recent Advances in Phytochemistry. 10:1–40
Fent K, Bucheli TD (1994) Inhibition of hepatic microsomal monooxygenase system by organotins in vitro in freshwater fish. Aquat Toxicol 28:107–126
Fischer HE, Bazer FW, Fields MJ (1985) Steroid metabolism by endometrial and conceptus tissues during early pregnancy and pseudopregnancy in gilts. J Reprod Fertil 75:69–78
Fieser LF, Fieser M (1967) Steroids. Reinhold Publishing Corporation, New York
Fürbass R, Vanselow J (1995) An aromatase pseudogene is transcribed in the bovine placenta. Gene 154:287–291
Fürbass R, Kalbe C, Vanselow J (1997) Tissue–specific expression of the bovine aromatase–encoding gene uses multiple transcriptional start sites and alternative first exons. Endocrinology 138:2813–2819
Gadsby JE, Balapure AK, Britt JH, Fitz FA (1990) Prostaglandin F2α receptors on enzyme–dissociated pig luteal cells throughout the estrous cycle. Endocrinology 12:787–795
Galli FE, Wassermann GF (1972) Steroid biosynthesis by testes and ovaries of 15 day–old chick embryos. Gen Comp Endocr 19:509–515

Gallien L, Le Fougoc MT (1957) Détection fluorimétrie et colorimétrie de stéroides sexuels et les gonades embryonnaires de poulet. C R Séanc Biol 151:1088–1089

Gaub MP, Bellard M, Scheuer I, Chambon P, Sossone–Corsi P (1990) Activation of the ovalbumin gene by the estrogen receptor involves the fos–jun complex. Cell 63:1267–1276

Geisert RD, Zavy MT, Moffatt RJ, Blair RM, Yellin T (1990) Embryonic steroids and the establishment of pregnancy in pigs. J Reprod Fertil Suppl 40:293–305

Goodwin M, Gooding KM, Regnier F (1979) Sex pheromones in the dog. Science 203:559–561

Graham–Lorence S, Khali MW, Lorence MC, Mendelson CR, Simpson ER (1991) Structure–function relationships of human aromatase cytochrome P–450 using molecular modeling and site directed mutagenesis. J Biol Chem 266:11939–11946

Greyling MD, van Aarde, RJ, Potgieter HC (1997) Ligand specificity of uterine estrogen and progesterone receptors in the subadult African elephant, *Loxodonta africana*. J Reprod Fertil 109:199–204

Gross TS (1992) Development and use of faecal steroid analyses in several carnivore species. In: Schaftenaar W, Buiter RM, Dielman SJ (eds) The First International Symposium on Faecal Steroid Monitoring in Zoo Animals, Rotterdam, pp 55–61

Hagen–Mann K (1994) Partielle Sequenzierung der bovinen Estrogen– und Androgenrezeptor mRNA, ihre Lokalisation und Quantifizierung über ein RT–PCR System in verschiedenen Geweben. Dissertation, Technische Universität München

Hara A, Hirai H (1978) Comparative studies on immunochemical properties of female–specific serum protein and egg yolk proteins in rainbow trout (*Salmo gairdneri*). Comp Biochem Phys 59B:339–343

Harada N (1993) Genetic analysis of human placental aromatase deficiency. J Steroid Biochem 44:331–340

Harvey S, Bedrak E Chadwick A (1981). Serum Concentrations of prolactin, LH, GH, corticosterone, progesterone, testosterone and oestradiol in relation to broodiness in domestic turkeys. J Endocrinol 89:187–195

Heap RB, Hamon M (1979) Oestrone sulphate in milk as an indicator of a viable conceptus in cows. Brit Vet J 135:355–363

Heard RDH, O'Donnell VJ (1954) Biogenesis of the estrogens: the failure of cholesterol–4–^{14}C to give rise to estrone in the pregnant mare. Endocrinology 54:209

Heard RDH, Jacobs R, O'Donnell V, Peron FG, Saffran JC, Solomon SS, Thompson LH, Willoughby H, Yates CH (1954) The application of ^{14}C to the study of the metabolism of the sterols and steroid hormones. Recent Prog Horm Res 9:383

Heistermann M, Tari S, Hodges JK (1993) Measurement of faecal steroids for monitoring ovarian function in New World primates, *Callitrichidae*. J Reprod Fertil 99:243–251

Hess DL, Schmidt AM, Schmidt MJ (1983) Reproductive cycle of the Asian elephant (*Elephas maximus*) in captivity. Biol Reprod 28:767–773

Hess RA, Bunick D, Lee K-H, Bahr J, Taylor JA, Korach KS, Lubahn DB (1997) A role for estrogens in the male reproductive system. Nature 390:509–512

Hesse R, Hoffmann B, Karg H, Vogt K (1981) Untersuchungen über den Nachweis von Phytoöstrogenen in Futterpflanzen und Hopfen mit Hilfe eines Rezeptortests. Zbl Vet Med A 28:442–454

Hickey GJ, Krasnow JS, Beattie WG, Richards JS (1990) Aromatase cytochrome P–450 in rat ovarian granulosa cells before and after luteinization: adenosin 3′,5′–monophosphate–dependent and independent regulation. Cloning and sequencing of rat aromatase cDNA and 5' genomic DNA. Mol Endocrinol 4:3–12

Hindle JE, Hodges JK (1990) Metabolism of oestradiol-17β and progesterone in the white rhinoceros (*Ceratotherium simum simum*). J Reprod Fertil 90:571–580

Hindle JE, Möstl E, Hodges JK (1992) Measurement of urinary estrogens and 20α-dihydroprogesterone during ovarian cycles of black (*Diceros bicornis*) and white (*Ceratotherium simum*) rhinoceros. J Reprod Fertil 94:237–249

Hirschenhauser K, Dittami J, Möstl E, Wallner B, Kotrschal K (1997) Testosterone and behaviour during different phases of reproduction in greylag geese (*Anser anser*) In: Taborsky M, Taborsky B (eds) Advances in Ethology, No. 32, Suppl. Ethology, Contrib. XXVth IEC, Vienna, Blackwell Wiss.-Verlag, New York, p 61

Hinshelwood MM, Zheng L, Conley AJ, Simpson ER (1995) Demonstration of tissue-specific promoters in nonprimate species that express aromatase P450 in placentae. Biol Reprod 53:1151–1159

Hodges JK, Brand HM, Henderson C, Kelly RW (1983) The levels of circulating and urinary estrogens during pregnancy in the marmoset monkey (*Callithrix jacchus)* J Reprod Fertil 67:3–82

Hodges JK, McNeilly AS, Hess DL (1987) Circulating hormones during pregnancy in the Asian and African elephants. Int Zoo Yb 26:285–289

Hoffmann B (1983) Untersuchungen zur Steroidhormonsynthese in der Plazenta des Rindes. Wien Tierärztl Monat 70:224–228

Hoffmann B, Höveler R, Nohr B, Hasan SH (1994) Investigations on hormonal changes around parturition in the dog and the occurrence of pregnancy–specific non conjugated estrogens. Exp Clin Endocrinol 102:185–189

Ing NH, Spencer TE, Bazer FW (1996) Estrogen enhances endometrial estrogen receptor gene expression by a posttranscriptional mechanism in the ovariectomized ewe. Biol Reprod 54:591–599

Jewgenow K, Meyer HHD (1998) Comparative binding affinity study of progestins to the cytosol progestin receptor of endometrium in different mammals. Gen Comp Endocr 110:118–124

Jobling S, Sumpter JP (1993) Detergent components in sewage effluent are weakly estrogenic to fish: An in vitro study using rainbow trout (*Oncorhynchus mykiss*) hepatocytes. Aquat Toxicol 27:361–372

Johnson AL, van Tienhoven A (1980) Plasma concentrations of six steroids and LH during the ovolatory cycle of the hen, gallus domesticus. Biol Reprod 23:386–393

Jung HJG (1977) Responses of mammalian herbivores to secondary plant compounds. The Biologist 59:123–136

Kapustin N, Critser, JK, Olson D, Malven PV (1996) Luteal estrous cycles of 3–week duration are initiated by anovulatory luteinizing hormone peaks in African Elephants. Biol Reprod 55:1147–1154

Karg H (1969) Hormonale Regulation. In: Lenkeit W, Breirem K, Crasemann E (Hrsg) Handbuch der Tierernährung, Band 1, Allgemeine Grundlagen. Paul Parey Verlag, Hamburg, S 362–384

Karg H, Meyer HHD, Hoffmann B (1990) Hormonale Leistungsförderer bei lebensmittelliefernden Tieren. Dtsch Ärtzeblatt 87:667–674

Kassam AAH, Lasley BL (1981) Estrogen excretory patterns in the Indian Rhinoceros (*Rhinoceros unicornis)*, determined by simplified urinary analysis. Am J Vet Res 42:251–255

Kasman LH, McCowan B, Lasley BH (1985) Pregnancy detection in tapirs by direct urinary estrone sulfate analysis. Zoo Biol 4:301–306

King GJ, Ackerly CA, (1985) Demonstration of estrogens in developing pig trophectoderm and yolk sac endoderm between days 10 and 16. J Reprod Fertil 73:361–367

Knight JW (1994) Aspects of placental estrogen synthesis in the pig. Exp Clin Endocrinol 102:175–184

Kossila VL (1983) The use of anabolic steroids in animal production – The FAO position. In: Meissonnier E (ed) Anabolics in Animal Production. Soregraph, Levallois, pp 497–503

Küst D (1934) über Sexualhormone bei den Haustieren. Klin Wochenschr 13:1782–1784

Lam TJ (1983) Environmental influences gonadal activity in fish. In: Hoar S, Randall DJ, Donaldson EM (eds) Fish Physiology. IXB. Academic Press, London, pp 65–101

Lamming GE, Ballarini G, Baulieu G, Brooks P, Elias PS, Ferrando R, Galli CL, Heitzmann RJ, Hoffmann B, Karg H, Meyer HHD, Michel G, Poulsen E (1987) Scientific report on anabolic agents in animal production. Vet Rec 24:389–392

Laudet V, Hani C, Coll J, Catzeflis F, Stehelin D (1992) Evolution of the nuclear receptor gene superfamuly. EMBO J 11:1003–1013

Lech JJ, Lewis SK, Ren L (1996) In vivo estrogenic activity of nonylphenol in rainbow trout. Fund Appl Toxicol 30:229–232

Legan SJ, Winans SS (1981) The photoneuroendocrine control of seasonal breeding in the ewe. Gen Comp Endocr 45:317–328

Lengwinat T, Meyer HHD (1996) Investigations of BrdU incorporation in roe deer blastocysts in vitro. Anim Reprod Sci 45:103–107

Lenkeit W (1969) Der mütterliche Stoffwechsel während der Gravidität. In: Lenkeit W, Breirem K, (Hrsg) Handbuch der Tierernährung, Band 2, Leistungen und Ernährung. Paul Parey Verlag, Hamburg, S 115–142

LeRoux MG, Theze N, Wolff J, Le Pennec JP (1993) Organization of the rainbow trout estrogen receptor gene. Biochim Biophys Acta 1172:226–230

Mahesh VB, Dhindsa DS, Anderson E, Kalra SP (1987) Regulation of ovarian and testicular function. Adv Med and Biol. Vol. 219, Plenum Press, New York

Maggi M, Fantoni G, Peri A, Giannini S, Brandi ML, Orlando C, Serio M (1991) Steroid modulation of oxytocin/vasopressin receptors in the uterus. J Steroid Biochem 40:481–491

Maghuin-Rogister G, Bosseloire A, Gaspar P, Dasnois C, Pelzer G (1988) Presence of 17β–19–nortestosterone in urine samples from boars. Ann Med Vet 132:437–40

Matsumine H, Herbst MA, Ignatius Ou S-H, Wilson JD, McPhaul MJ (1991) Aromatase mRNA in the extragonadal tissues of chickens with the Henny–feathering trait is derived from a distinct promoter structure that contains a segment of a retroviral long terminal repeat. J Biol Chem 266:19900–19907

McArthur C, Robbins CT, Hagerman AE, Hanley TA, (1993) Diet selection by a ruminant generalist browser in relation to plant chemistry. Can J Zool 71:2236–2243

McCracken JA, Schramm W, Okulicz WC (1984) Hormone receptor control of pulsatile secretion of $PGF_{2\alpha}$ from ovine uterus during luteolysis and its abrogation in early pregnancy. Anim Reprod Sci 7:31–56

McNeilly AS, Martin RD, Hodges JK, Smuts GL (1983) Blood concentrations of gonadotrophins, prolactin and gonadal steroids in males and non–pregnant and pregnant female African elephants (*Loxodonta africana)* J Reprod Fertil 67: 113–120

Means GD, Kilgore MW, Mahendroo MS, Mendelson CR, Simpson ER (1991) Tissue–specific promoters regulate aromatase cytochrome P450 gene expression in human ovary and fetal tissues. Mol Endocrinol 5:2005–2013

Meissonnier E (1983) Anabolics in Animal Production, Soregraph, Levallois

Messier F, Desaulniers DM, Goff AK, Nault R, Patenaude R, Crete M (1990) Caribou pregnancy diagnosis from immunoreactive progestins and estrogens excreted in feces. J Wildlife Manage 54:279–283

Meyer HHD, Rapp M (1985) Estrogen receptor in bovine skeletal muscle. J Anim Sci 60:94–300

Meyer HHD, Mittermeier T, Schams D (1986) Dynamics of oxytocin, estrogen, and progestin receptors in the bovine endometrium during the estrous cycle. Acta Endocrinol 118:96–104

Meyer HHD, Karg H (1989) Growth stimulators for farm animals: mode of action, effects on meat quality and potential risks originating from residues. Proc. FAO/CAAS Workshop on Biotechnology in Animal Production and Health, Beijing, pp 49–58

Meyer HHD, Sauerwein H, Mutayoba BM (1990) Immunoaffinity chromatography and a biotin–streptavidin amplified enzymeimmunoassay for sensitive and specific estimation of estradiol–17β. J Steroid Biochem 35:263–269

Meyer HHD (1991) The illegal practice and resulting risks versus the controlled use of licensed drugs: views on the present situation in Germany. Ann Rech Vet 22:299–304

Meyer HHD, Falckenberg D, Janowski T, Rapp M, Rösel EF, van Look L, Karg H (1992) Evidence for the presence of endogenous 19–nortestosterone in the cow peripartum and in the neonatal calf. Acta Endocrinol 126:369–73

Meyer HHD (1994) Luteal versus placental progesterone: the situation in the cow, pig and bitch. Exp Clin Endocrinol 102:190–192

Meyer HHD, Stoffel B, Hagen–Mann K (1995) β–Agonists, anabolic steroids and their receptors: New aspects in growth regulation. In: von Engelhardt W, Leonhard–Marek S, Breves G, Giesecke D (eds) Ruminant Physiology: Digestion, Metabolism, Growth and Reproduction, Proc. of the Eighth International Symposium on Ruminant Physiology. Ferdinand Enke Verlag, Stuttgart, pp 475–482

Meyer HHD, Jewgenow K, Hodges JK (1997a) Binding activity of 5α–reduced gestagens to the progestins receptor from African elephant (*Loxodonta africana*). Gen Comp Endocr 104:164–167

Meyer HHD, Rohleder M, Streich WJ, Göltenboth R, Ochs A (1997b) Sexualsteroidprofile und Ovaraktivitäten des Pandaweibchens YAN YAN im Berliner Zoo. Berl Münch Tierärztl 110:43–147

Möstl E, Meyer HHD, Bamberg E, von Hegel G (1988) Estrogen determination in faeces of mares by enzyme immunoassay on microtitre plates. In: Goeroeg S (ed) Advances in Steroid Analysis. Akademiai Kiado, Budapest, pp 219–224

O'Connell RJ, Singer AG, Pfaffmann C, Agosta WC (1979) Pheromones of hamster vaginal discharge. Attraction to femtogram amounts of dimethyldisulfide and to mixtures of volatile components. J Chem Ecol 5:575–585

Oehlmann J, Stroben E and Fioroni P (1992) The rough tingle Ocenebra erinacea (*Neogastropoda: Muricidae)* an exhibitor of imposex in comparison to *Nucella lapillus*. Helgoländer Meeresunter 46:311–328

Paech K, Webb P, Kuiper GGJM, Nilsson S Gustafsson J–A, Kushner PJ, Scanlan TS (1997) Differential ligand activation of estrogen receptors ERα and Erβ at AP1 sites. Science 277:1508–1510

Palme R, Holzmann A, Mitterer Th (1994) Measuring fecal estrogens for the diagnosis of cryptorchidism in horses. Theriogenology 42:1381–1387

Palme R, Fischer P, Schildorfer H, Ismail MN (1996) Excretion of infused ^{14}C–steroid hormones via faeces and urine in domestic livestock. Anim Reprod Sci 43:43–63

Paris A, Rao D (1989) Biosynthesis of estradiol–17beta fatty acyl esters by microsomes derived from bovine liver and adrenals. J Steroid Biochem 33:465–472

Pelissero C, Flouriot G, Foucher JL, Bennetau B, Dunogues J, Le Gac F, Sumpter JP (1993) Vitellogenin synthesis in cultured hepatocytes; an in vitro test for the estrogenic potency of chemicals. J Steroid Biochem 44:263–272

Perez LE, Czekala NM, Weisenseel KA, Lasley BL (1988) Excretion of radiolabeled estradiol metabolites in slow loris (*Nycticebus coucang*). Am J Primatol 16:321–330

Perry JS, Heap RB, Amoroso EC (1973) Steroid hormone production by pig blastocysts. Nature 245:45–47

Pitra C, Fürbass R, Seyfert HM (1997) Molecular phylogeny of the tribe Bovini (Mammalia: Artiodactyla): alternative placement of the Anoa. J Evol Biol 10:589–600

Pohl–Branscheid M, Holtz W (1990) Control of spawning activity in male and female rainbow trout (*Oncorhynchus mykiss*) by repeated foreshortened seasonal light cycles. Aquaculture 86:93–104

Pryce CR, Schwarzenberger F, Döbeli M (1994) Monitoring fecal samples for estrogen excretion across the ovarian cycle in Goeldi's monkey (*Callimico goeldii*). Zoo Biol 13:219–230

Purdom CE, Hardiman PA, Bye VJ, Eno NC, Tyler CR Sumpter JP (1994) Estrogenic effects of effluents from sewage treatment works. Chem Ecol 8:275–285

Raeside JL (1978) Seasonal changes in the concentration of estrogens and testosterone in the plasma of the stallion. Anim Reprod Sci 1:205–212

Raeside JI, Renaud RL, Friendship RM (1989) Aromatization of 19–norandrogens by porcine Leydig cells. J Steroid Biochem 32:729–735

Rasmussen LEL, Lee TD, Zhang, A, Roelofs WL, Daves GD (1997) Purification, identification, concentration and bioactivity of (Z)–7–dodecen–1–yl acetate: Sex pheromone of the female Asian elephant, *Elephas maximus*. Chem Senses 22:417–437

Reinboth R, Becker B (1984) In vitro studies on steroid metabolism by gonadal tissues from ambisexual teleosts. I. Conversion of C14-testosterone by males and females of the protogynous wrasse *Coris julis L*. Gen Comp Endocr 55:245–250

Reiter RJ (1978) Interaction of photoperiod, pineal and seasonal reproduction as exemplified by findings in the hamster. In: Reiter RJ (ed) Progr Reprod Biol 4. Karger Verlag, Basel, p 169

Ren L, Lewis SK, Lech JJ (1996) Effects of estrogen and nonylphenol on the post–transcriptional regulation of vitellogenin gene expression. Chem Biol Interact 100:67–76

Roach AH, Davies PL (1980) Catfish vitellogenin and its messenger RNA are smaller than their chicken and xenopus counterparts. Biochim Biophys Acta 610:400–412

Roberts RM, Cross JC, Leaman DW (1992) Interferons as hormones of pregnancy. Endocr Rev 13:432–445

Roche JF, Diskin MG (1995) Hormonal regulation of reproduction and interactions with nutrition in female ruminants. In: von Engelhardt W, Leonhard-Marek S, Breves G, Giesecke D (eds) Ruminant Physiology: Digestion, Metabolism, Growth and Reproduction, Proc. of the Eighth International Symposium on Ruminant Physiology. Ferdinand Enke Verlag, Stuttgart, pp 409–428

Safar-Hermann N, Ismail MN, Choi HS, Möstl E, Bamberg E (1987) Pregnancy diagnosis in zoo animals by estrogen determination in feces. Zoo Biol 6:189–193

Sauerwein H, Meyer HHD (1989) Androgen and estrogen receptors in bovine skeletal muscle: relation to steroid induced allometric muscle growth. J Anim Sci 67:206–212

Sauerwein H, Meyer HHD, Schams D (1992) Divergent effects of estrogens on the somatotropic axis in male and female calves. J Reprod Develop 38:271–278

Sauerwein H, Pfaffl M, Hagen–Mann K, Malucelli A, Meyer HHD (1995) Expression of estrogen and androgen receptor in the bovine gastrointestinal tract. Deut Tierärztl Woch 102:143–178

Schams D, Meyer HHD, Prokopp S, Chan JSD (1984) The role of steroid hormones, prolactin and placental lactogen on mammary gland development in ewes and heifers. J Endocrinol 102:121–130

Schopper D, Gaus J, Claus R, Bader H (1984) Seasonal changes of steroid concentrations in seminal plasma of an european wild boar. Acta endocr (Kbh) 107:425–427

Schubiger JL, Wahli W (1986) Linkage arrangement in the vitellinogenin gene family of *Xenopus laevis* as revealed by gene segregation analysis. Nucleic Acids Res 14:8723–8734

Schwarzenberger F, Möstl E, Palme R, Bamberg E (1996) Faecal steroid analysis for non–invasive monitoring of reproductive status in farm, wild and zoo animals. Anim Reprod Sci 42:515–526

Sharp DC, McDowell KJ, Weithenauer J, Thatcher WW (1989) The continuum of events leading to maternal recognition of pregnancy in mares. J Reprod Fertil Suppl 37:101–107

Sheahan DA, Bucke D, Matthiessen P, Sumpter JP, Kirby MF, Neall P, Waldock M (1994) The effects of low levels of 17α–ethynylestradiol upon plasma vitellogenin levels in male and female rainbow trout, *Oncorhynchus mykiss* held at two acclimation temperatures. In: Müller R, Lloyd R (eds) Sublethal and chronic effects of pollutants on freshwater fish. FAO, Fishing News Books, Oxford, pp 99–112

Shemesh M, Lindner HR, Ayalon N (1972) Affinity of rabbit uterine oestradiol receptor for phyto–estrogens and its use in a competitive protein-binding radioassay for plasma coumestrol J Reprod Fertil 29:1–9

Shideler SE, Ortuno AM, Moran FM, Moorman EA, Lasley BL (1993) Simple extraction and enzyme immunoassays for estrogen and progesterone metabolites in the feces of *Macaca fascicularis* during non–conceptive and conceptive ovarian cycles. Biol Reprod 48:1290–1298

Shille VM, Lundström KE, Stabenfeld GH (1979) Follicular function in the domestic cat as determined by estradiol–17β concentrations in plasma: relation to estrous behavior and cornification of exfoliated vaginal epithelium. Biol Reprod 21:953–963

Shille VW, Hagerty MA, Shackleton C, Lasley BL (1990) Metabolites of estradiol in serum, bile, intestine and feces of the domestic cat (*Felis catus*). Theriogenology 34:779–794

Short RV (1984) Oestrous and menstrual cycle. In: Austin CR, Short RV (eds) Reproduction in mammals. Book 3: Hormonal control of reproduction. 2nd edn. Cambridge University Press, Cambridge, pp 115–152

Silberzahn P, Dehennin L, Zwain I, Reiffsteck A (1985) Gas chromatography–mass spectrometry of androgens in equine ovarian follicles at ultrastructurally defined stages of development. Idenfication of 19-nortestosterone in follicular fluid. Endocrinology 17:2176–81

Simpson ER, Mahendroo MS, Means GD, Kilgore MW, Hinhselwood MM, Graham–Lorence S, Amarnreh B, ItoY, Fisher CR, Michael MD, Mendelson CR, Bulun SE (1994) Aromatase cytochrome P450, the enzyme responsible for estrogen biosynthesis. Endocr Rev 15:342–355

Simmons DL, Lalley PA, Kasper CB (1985) Chromosomal assignments of gene coding for components of the mixed function oxidase system in mice. J Biol Chem 260:515–521

Skidmore JA, Allen WR, Heap RB (1994) Estrogen synthesis by the peri–implantation conceptus of the one–humped camel (*Camelus dromedarius*). J Reprod Fertil 101:363–367

Smith SJ, Cox JE, Houghton E, Dumasia MC, Moss MS (1987) In vitro biosynthesis of C18 neutral steroids in horse testes. J Reprod Fertil 35:71–78

Taya K, Komura H, Kondoh M, Ogawa Y, Nakada K, Watanabe G, Sasomoto S, Tanabe K, Saito K, Tajima H, Narushima E (1991) Concentrations of progesterone, testosterone and estradiol–17β in the serum during the estrous cycle of Asian elephants (*Elephas maximus*). Zoo Biol 10:299–307

Tanabe Y, Nakamura T, Fujiioka K, Dori O (1979) Production and secretion of sex steroid hormones by the testes, ovary and the adrenal gonads of embryonic and young chickens (*Gallus domesticus*). Gen Comp Endocr 39:26–33

Tan N, Lam T, Ding J (1995) Molecular cloning and sequencing of the hormone–binding domain of oreochromis aureus estrogen receptor gene. DNA Seq 5:359–370

Terqui M, Palmer E (1979) Estrogen pattern during early pregnancy in the mare. J Reprod Fertil (Suppl 27):441–446

Theyerl–Abele M, Meyer HHD, Schams D, Karg H (1990) Konzentrationen von IGF–I und Östradiol–17β im Blutplasma des graviden Rindes. Deut Tierärztl Woch 97:382–385

Van Ginkel LA, Stephany RW, Zoontjes PW, van Rossum HJ, van Blitterswijk H, Zuijdendorp J (1989) The presence of nortestosterone in edible partes of non–castrated male pigs. Ned Tijdschr Diergeneesk 114:311–14

Van Lennep EW (1961) The histology of the placenta of the one-humped camel (*Camelus dromedarius)* during the first half of pregnancy. Acta Morphol Neerlando Scandinavica 4:180–193

Vanselow J, Fürbass R (1995) Novel aromatase transcripts from bovine placenta contain repeated sequence motifs. Gene 154:281–286

Wagner JF (1983) Estradiol controlled release implants. Efficacy and drug delivery. In: Meissonnier E (ed) Anabolics in Animal Production. Soregraph, Levallois, pp 129–142

Wagner R, Görlich L, Jungblut PW (1972) Multiple steroid hormone receptors in calf uterus. Binding specificities and distribution. Hoppe–Seyler's Z Physiol Chem 353:1654–1656

Wallace RA (1978) Oocyte growth in nonmammalian vertebrates. In: Jones RE (ed) The vertebrate ovary. Plenum Press, New York, pp 469–502

Wasser SK, Monfort SL, Wildt DE (1991) Rapid extraction of faecal steroids for measuring reproductive cyclicity and early pregnancy in free–ranging yellow baboons (*Papio cynocephalus cynocephalus*). J Repord Fertil 92:415–423

Wasser SK, Monfort SL, Southers J, Wildt DE (1994) Excretion rates and metabolites of oestradiol and progesterone in baboon (*Papio cynocephalus cynocephalus*) in faeces. J Repord Fertil 101:213–220

Wasser SK, Papageorge S, Foley C, Brown JL (1996) Excretory fate of estradiol and progesterone in the African elephant (*Loxodonta africana*) and patterns of fecal steroid concentrations throughout the estrous cycle. Gen Comp Encorinol 102: 255–262

Weiler U, Claus R, Schnoebelen–Combes S, Louveau I (1998) Influence of age and genotype on endocrine parameters and growth performance: a comparative study in wild boars, meishan and large white boars. Livest Prod Sci 54:21–31

White R, Jobling S, Hoare SA, Sumpter JP, Parker MG (1994) Environmentally persisten alkylphenolic compounds are estrogenic. Endocrinology 135:175–182

Whitehead C, Bromage N, Forster JRM (1978) Seasonal changes in reproductive function of female rainbow trout *Salmo gairdneri*. J Fish Biol 12:601–608

Wibbels T, Crews D (1992) Specificity of steroid hormone–induced sex determination in a turtle. J Endocrinol 133:121–129

Wiley HS, Wallace RA (1981) The structure of vitellogenin. Multiple vitellogenins in *Xenopus laevis* give rise to multiple forms of the yolk proteins. J Biol Chem 256:8626–8634

Ziegler TE, Sholl SA, Scheffler G, Haggerty MA, Lasley BL (1989) Excretion of estrone, estradiol, and progesterone in the urine and feces of the female cotton–top tamarin (*Sanguinus oedipus oedipus*). Am J Primatol 17:185–195

CHAPTER 47

Therapeutic Use of Estrogens in Veterinary Medicine

M. OETTEL

A. Introduction

Endogenous estrogens are mainly formed in the granulosa cells of the ovarian follicles, in the corpus luteum in primates, in the placenta, in the adrenal cortex, and in the testes. Compared with other male domestic animals, stallions produce extremely large amounts of estrogen in their testes (GAILLARD 1991). Due to the equine testicular aromatase, concentrations of estrone sulfate in peripheral blood plasma are about 100 times higher than those of testosterone. The concentrations of conjugated estrogens in the blood plasma of stallions range between 100 ng/ml and 400 ng/ml; the respective range for boars is "only" 1–5 ng/ml. Nevertheless, the boar also secretes a large amount of estrogens, considering the fact that concentrations secreted by other animals and humans are within the picogram range. Estrogens, while acting luteolytically in cattle, act luteotropically in sows. High concentrations of conjugated estrogens, including compounds with unsaturated rings A and B (equilin, equilenin etc.), are present in pregnant-mare's urine. The conjugated estrogens in pregnant-mare's urine (sodium salts of estrone-3-hydrogen-sulfate and equiline-3-hydrogen-sulfate) have been widely used as oral estrogen replacement in postmenopausal women (OETTEL 1996).

Phytoestrogens (zearalenone, β-zearalenol, coumestrol, genistein, daidzein, phloretin, formononetin, and biochanin A) found in certain plants may give rise to a hyperestrogenic state and associated diseases in ruminants (MAKAREVICH et al. 1997). Among synthetic estrogens are the stilbenes (diethyl stilbestrol or dienestrol), whose use is relatively widespread in veterinary medicine. However, stilbenes have not yet been sufficiently investigated with respect to specific veterinary medical angles as to their metabolic patterns or side effects, e.g., tumorigenesis in the F1-generation. Their use in food-producing animal species is prohibited.

There are other synthetic compounds that may accumulate in the environment, e.g., dichlorodiphenyl-trichloroethane 2′,3′-dideoxythymidine (DDT), alkyl phenols, or other pesticides, which show remarkable estrogenic activities in farm animals and may have a strong influence on the metabolic endocrine and reproductive endocrine systems (RAWLINGS et al. 1998; TYLER et al. 1998).

B. Pharmacokinetics

As estrogens are lipophilic substances, their absorption from the mucous membranes or skin is very good (even in ruminants!). In terms of professional safety, this good skin penetration should be taken into account by veterinarians when handling injection solutions. The plasma half-life of estrogens is characterized by the presence of a multicompartment system. So, like most other steroids, estrogens are deposited in the fatty tissue and released from it slowly. After an intravenous dose, the elimination half-life of 17β-estradiol (first phase without the fatty tissue compartment) is about 20–70 min. The next phase is characterized by an elimination half-life of around 5–6 h, while the half-life of the residual elimination exceeds 24 h. 17β-Estradiol is metabolized to estrone and estriol. In pigs, 17β-estradiol is directly glucuronized at more than 90% in the gastrointestinal mucosa before reaching the blood flow. Esterification of 17β-estradiol with specific fatty acids, as a function of the chain length of the respective fatty acid, gives effective depot preparations. For example, in cattle, the half-life of estradiol benzoate after a parenteral dose is 5 days, and the biological effect of estradiol valerate after subcutaneous application occurs after about 3 weeks. Stilbenes, such as diethylstilbestrol (DES), undergo a distinctly slower hepatic metabolism. Therefore, the residue issue must be viewed very critically. Exogenous estrogens can be secreted in the milk (GROSVENOR et al. 1992).

Withdrawal periods for intramuscular 17β-estradiol benzoate in cattle are 12 days for edible tissues and 5 days for milk. Injection into the fatty tissue must be avoided at any rate, since this will cause undesirable depot formation – at least at the injection site (LÖSCHER et al. 1994).

C. Undesirable Effects, Interactions, and Toxicity

Long-term use of estrogens in extremely large doses leads to cancerous tissue changes in a number of experimental animals. Most of these tumors develop in the target organs specific for estrogens and are independent of the estrogen type. However, they are characteristic of the respective laboratory animal species. The most common tumors in mice and rats include mammary cancer, chromophobe pituitary adenomas, carcinomas of the bladder, and tumors of the lymphatic system and of the adrenal cortex. Tumors typical of the Syrian hamster are clear cell carcinomas of the kidneys (IATROPOULOS 1994).

Specific side effects should be looked for in dogs, since strong disturbances of the hematopoietic system (anemia, thrombocytopenia up to fatal panmyelophthisis) may occur in the therapeutic dose range. In this context, it is interesting to note that there is no increase in the plasma 17β-estradiol concentrations during gestation in bitches, unlike the situation with all other species. This peculiarity probably enables bitches to become pregnant in spite of the canine hematopoietic system being extraordinarily sensitive to estrogens (bone-marrow suppression). In contrast, chronic-toxicity results have been reported from drug toxicology studies, in which varying doses of orally potent ethinyl estradiol had been used in dogs without producing such side effects. At any rate, if a dog is on repeat doses of estrogen, the blood count should be checked.

In the dog, one of the rare species, along with humans, to spontaneously develop benign prostatic hyperplasia (BPH) (Coffey and Walsh 1990), estrogens act synergically with androgens to promote basal cell squamous metaplasia (DeKlerk et al. 1979). In addition, when administered alone to castrated dogs, estrogens activate prostatic functions, resulting in increased fibromuscular stroma and basal cell metaplasia (Merk et al. 1982). In addition, estradiol activates $p60^{src}$, $p53/56^{lyn}$ and renatured p50/55 protein tyrosine kinases in the dog prostate (Allard et al. 1997). Chronic toxicity studies using different estrogens have shown that male dogs on estrogen develop strong BPH, which may give rise to cystitis due to ascending urinary tract infection.

It has also been discussed on various occasions that, in predisposed bitches, an estrogen dose may promote the development of endometritis and pyometra. Estrogen receptor expression was elevated in bitches suffering from cystic endometritis–pyometra complex, irrespective of the presence and severity of infection, an observation which supports the hypothesis that bacteria are not the primary cause of this disease (de Cock et al. 1997). As a result, contraindications for the administration of estrogens to dogs include disturbances of the hematopoietic system, acute and severe chronic diseases of the liver, endometritis, as well as young animals in the growth phase.

In cattle, gonadotropin blockage promotes the formation of ovarian cysts as well as hypogonadism, and there is a reduction in milk secretion or development of mastitis. Among the side effects are sodium retention and edema formation. Genital erythema, polydipsia, and polyuria are further signs of estrogen overdose. Excessive uptake of phytoestrogens (genistein and coumestrol) led to colpoptosis and balanoposthitis in sheep in Australia. Administration of sexual hormones may increase the incidence and intensity of nematosis.

D. Indications

For many of the claimed indications, other groups of hormones are preferred to estrogens due to selective effectiveness or specific legal restrictions. This particularly refers to the use of prostaglandins, gonadotropins, or gonadotropin-releasing hormone. Despite their major biological effects on animals, the use of estrogens, on their own, is limited (Roche 1991):

- Control of estrus and ovulation in the mare. If progesterone is used, it must be injected daily (100–150 mg) in conjunction with 10 mg estradiol benzoate. This treatment is effective if given for 14 days, and estrus and ovulation occur 3–10 days posttreatment. Fertility to natural mating appears to be normal.

- Misalliance in the bitch. Estrogens can be used as post-coital anti-nidatory agents in the bitch. Pregnancy can be prevented by estrogen administration during the first 5 days after mating. Estrogens exert their effect by interfering with the transport of the zygotes from the tube to the uterine cavity, probably by causing edema of the endosalpinx and thus a temporary tubal occlusion. Signs of estrus will be prolonged, and it is not advisable to repeat treatment if a mismating occurs again (NOAKES 1996).
- Misalliance in the cat. Estradiol cypionate by i.m. injection at a dose rate of 125–250 μg within 40 h of mating has been shown to be effective in preventing pregnancy. Similarly, injections of DES have been used. However, there is little data regarding the possible side effects, and such treatments should only be used in exceptional circumstances (NOAKES 1996).
- Induction of estrus and ovulation in anestrous animals. Estradiol may induce a gonadotropin surge (positive feedback), but not all animals will ovulate. The ovulatory response is dependent on the follicular status of the animal at the time of injection. Thus, estrogen is not recommended for induction of estrus because of the possibility of getting behavioral estrus without an ensuing ovulation. However, luteal function and estrus in peripubertal beef heifers treated with an intravaginal progesterone-releasing device is improved by adding 1 mg estradiol bezoate i.m. 24–30 h after device removal (RASBY et al. 1998).
- Control of the time of ovulation. Since estradiol is the positive feedback hormone responsible for induction of the gonadotropin surge, it is logical to consider it as a potential agent to synchronize precisely the time of ovulation following progestin treatment. However, results are not equivocal for this use, and it is not currently recommended because, in a number of studies, low fertility resulted, due presumably to adverse effects of estrogens on gamete transport in the female sexual tract.
- Control of the life span of the corpus luteum (estrus synchronization). In cattle, estradiol is partially luteolytic and is administered at the end of progesterone or progestin synchronization treatments (with or without prostaglandin $F_{2\alpha}$) (KASTELIC et al. 1996; HANLON et al. 1997; KESLER and FAVERO 1997; RATHBONE et al. 1998). Estradiol benzoate, when combined with progesterone, showed that estradiol had an effect on patterns of follicle development as reflected by the precision in the synchrony of estrus in heifers (MACMILLAN et al. 1993) and the posttreatment fertility of lactating cows (MACMILLAN and BURKE 1996).

These principles of estrus cycle control have been simplified by DAY et al. (1997). Estradiol benzoate was injected at the initiation of the synchronous treatment (2 mg) using progesterone administered per vaginam for 7 days. A luteolytic dose of prostglandin $F_{2\alpha}$ was administered at device removal and then estradiol benzoate (1 mg) was injected for a second time 48 h later. This sequence represents the integration of control principles relating to synchronized follicular development, synchronized luteolysis and synchronized follicular maturation to achieve synchronized estrus and ovulation with apparently normal and possibly improved fertility.

In sows, however, estrogens are luteotropic. This emphasizes the need to be careful when using estrogen due to the multitude of biological effects and species differences that are evident. They have also been used as abortifacients in cows and ewes.

- Companion animal uses. Estrogens in daily doses of 1–3mg by mouth or 0.5–1mg by injection can be used in bitches that are obese due to hypogonadism. They are also of value for urinary incontinence in old or spayed bitches. Estradiol benzoate (1mg injected daily for 3days followed by 1mg every third day as required) can be used. In the dog, estrogens are of value for the treatment of anal adenoma, especially where ulceration has occurred, and for causing a regression of hypertrophied prostate glands. For anal adenoma, long-acting estrogens can be used, but will cause increase in thirst and urination. For hypertrophied prostate glands, orally active estrogens can be used. Excessive libido in dogs, particularly young dogs, can be treated with estrogens, in daily oral doses of 1–3mg, by injection of daily doses of 0.5–1mg or, in persistent cases, by implants of 10mg, with which some feminization may occur. Lactation in bitches and cats can be diminished by the oral use of estrogens given in two doses 12h apart: 2 × 10mg for the bitch, and 2 × 3mg or 2 × 5mg for the cat. However, androgens are preferred.

In summary, the following indications are recommended for estrogens:

Dogs: Prevention of nidation, urinary incontinence after oophorectomy, benign prostatic hypertrophy (BPH), anal adenoma
Sheep: Induction of birth in cases of pregnancy toxemia, or when parturition is believed to be delayed
Cattle: Metritis/endometritis, pyometra, estrus cycle control
Horses: Too weak outer estrous symptoms

A significant and controversial pharmacological development has been the introduction of the various estrogenic growth promoters for use in beef cattle. The greatest benefits are seen in cows treated with androgens, bulls treated with estrogens, and castrates treated with combined preparations of androgens and estrogens. After estrogen treatment of ruminants, plasma concentrations of both insulin and growth hormones are increased, and elevated levels of these substances at the muscle cell result in increased protein accretion by increasing amino acid uptake. The major anabolic effects of estrogens occur via increased outpouring of growth hormone from the pituitary, and the raised plasma insulin concentrations are secondary to the changes in plasma glucose. The increase in secretion of growth hormone also results in a small increase in plasma glucose concentration. The combination of increased growth hormone and insulin in the muscle cell probably increases protein accretion (Barragry 1994).

Zeranol is a nonsteroidal compound with estrogenic activity, which is often used in the United States as a growth-promoting agent. Zeranol is produced by reduction of the natural product zearalenone, which is produced by fermentation from the mold *Giberalla zeae*. This mycotoxin from *fusarium* species was originally identified by its estrogenic effects on sows fed moldy grain (BARRAGRY 1994).

In most countries, it is forbidden to stimulate growth processes or protein synthesis and influence fat distribution using estrogens, especially in cattle (so-called estrogen fattening). For control, sensitive immunological and gas chromatographic/mass-spectrometric detection methods have been developed and validated (NASCIMENTO et al. 1996; HARTMANN and STEINHART 1997; HOFFMANN et al. 1997). In the European Union (EU), the use of sex hormones or sex-hormone-like compounds is closely restricted by the guideline 96/22/EC. Consequently, the therapeutic use of natural 17β-estradiol in food-producing animals is only allowed as an injection given by the veterinarian himself. Use is permitted only within the framework of zootechnical measures, such as the preparation of donor animals for embryo transfer or for estrus synchronization. Moreover, use is restricted to short-acting compounds, because under the above guidelines the use of products with a withdrawal period (time from last administration to slaughtering) exceeding 15 days is not permissible. In the EU, the use of these hormones to achieve anabolic effects is prohibited in food-producing animals. A respective treatment is only permitted in companion animals, e.g., after cachectic conditions (KROKER and SCHMÄDICKE 1998).

E. Dosage

Dogs (inhibition of nidation). Estradiol benzoate 0.1–0.3 mg/kg body weight i.m. or ethinyl estradiol or mestranol (prodrug of ethinyl estradiol) 0.2–0.5 mg/kg each three times within 5 days. To be started not later than 5 days after mating. SUTTON et al. (1997) recommend a much lower dose, namely estradiol benzoate 0.01 mg/kg on days 3, 5 and, sometimes, 7 after mating, which has proved very successful. It enables an essential reduction of unwanted side effects.

Dogs (BPH). Estradiol benzoate 1 mg i.m. weekly or diethylstilbestrol 0.5–1.0 mg orally per day over 5 days (HUTCHISON 1996).

Sheep (birth induction). Estradiol benzoate 15 mg given i.m. Birth normally follows about 48 h later (JACKSON 1995).

Cattle and horses. 17β-Estradiol 0.01–0.1 mg/kg twice within 24 h, estradiol cypionate up to 10 mg or estrone up to 20 mg/animal i.m. Estrogens can also be applied in conjunction with appropriate carriers as implants or vaginal rings. Estrogens even proved successful when used in various baits to reduce populations of pigeons and foxes (OETTEL 1981), while ethinyl estradiol preparations proved highly unpalatable to female rats (GAO and SHORT 1993).

F. Antiestrogens

The use of antiestrogens in veterinary medicine is rare. It may refer to mammary cancer or the postoperative treatment after mammary cancer surgery in bitches or cats. Aromatase inhibitors have been tried with great success in prostatic hypertrophy in male dogs.

References

Allard P, Atfi A, Landry F, Chapdelaine A, Chevalier S (1997) Estradiol activates $p60^{src}$, $p53/56^{lyn}$ and renatured p50/55 protein tyrosine kinases in the dog prostate. Mol Cell Endocrinol 126:25–34

Barragry TB (1994) Growth-promoting agents. In: Veterinary drug therapy. Lea & Febiger, Philadelphia, pp 619–621

Coffey DS, Walsh PC (1990) Clinical and experimental studies of benign prostatic hyperplasia. Urol Clin North Am 17:461–475

Day M, Burke CR, Nation DP, Rhodes FM, Macmillan KL (1997) Effect of hormonal environment at emergence on persistence of ovarian follicles in cattle. Proc NZ Soc Animal Production 57:239–240

de Cock H, Vermeirsch H, Ducatelle R, De Schepper J (1997) Immunohistochemical analysis of estrogen receptors in cystic-endometritis-pyometra complex in the bitch. Theriogenology 48:1035–1047

Deklerk DP, Coffey DS, Ewing LL, McDermott IR, Reiner WG, Robinson CH, Scott WW, Strandberg JD, Taladay P, Walsh PC, Wheaton LG, Zirbin BR (1979) Comparison of spontaneous and experimentally induced canine prostatic hyperplasia. J Clin Invest 64:842–849

Gaillard JL (1991) Equine testicular aromatase: substrates specificity and kinetic characteristics. Comp Biochem Physiol 100B:107–115

Gao Y, Short RV (1993) Use of an estrogen, androgen or gestagen as a potential chemosterilant for control of rat and mouse populations. J Reprod Fertil 97: 39–49

Grosvenor CE, Picciano MF, Baumbrucker CR (1992) Hormones and growth factors in milk. Endocr Rev 14:710–728

Hanlon DW, Williamson NB, Wichtel JJ, Steffert IJ, Craigie AL, Pfeiffer DU (1997) Ovulatory responses and plasma luteinizing hormone concentrations in dairy heifers after treatment with exogenous progesterone and estradiol benzoate. Theriogenology 47:963–975

Hartmann S, Steinhart H (1997) Simultaneous determination of anabolic and catabolic steroid hormones in meat by gas chromatography-mass spectrometry. J Chromatogr B Biomed Sci Appl 704:105–117

Hoffmann B, Goes de Pinho T, Schuler G (1997) Determination of free and conjugated estrogens in peripheral blood plasma, feces and urine of cattle throughout pregnancy. Exp Clin Endocrinol Diabetes 105:296–303

Hutchison M(1996) Manual of small animal endocrinology. Kingsley House, Shurdington/UK

Iatropoulos MJ (1994) Endocrine considerations in toxicological pathology. Exp Toxicol Pathol 45:391–410

Jackson PGG (1995) Handbook of veterinary obstetrics. Saunders, London, pp 208

Kastelic JP, McCartney DH, Olson WO, Barth DA, Garcia A, Mapletoft RJ (1996) Estrus synchronization in cattle using estradiol, melengestrol acetate and PGF. Theriogenology 46:1295–1304

Kesler DJ, Favero RJ (1997) Needleless implant delivery of gonadotropin-releasing hormone enhances the calving rate of beef cows synchronized with Norgestomet and estradiol valerate. Drug Dev Ind Pharm 23:607–610

Kroker R, Schmädicke I (1998) Stellt die veterinärmedizinische Anwendung von steroidalen Sexualhormonen und ihren Abkömmlingen ein Risiko für die Umwelt dar? Bundesgesundhbl 8:344–345

Löscher W, Ungemach FR, Kroker R (1994) Grundlagen der Pharmakotherapie bei Haus- und Nutztieren, 2nd edn. Verlag Paul Parey, Berlin

Macmillan KL, Burke CR (1996) Effects of oestrus cycle control on reproductive efficiency. Anim Reprod Sci 42:307–320

Macmillan KL, Taufa VJK, Day AM (1993) Principles of synchronising oestrus in dairy heifers. Proc NZ Soc Anim Production 53:267–270

Makarevich A, Sirotkin A, Taradajnik T, Chrenek P (1997) Effects of genistein and lavendustin on reproductive processes in domestic animals in vitro. J Steroid Biochem Mol Biol 63:329–337

Merk FB, Ofner P, Kwan PWL, Leav I, Vena RL (1982) Ultrastructural and biochemical expressions of divergent differentiation in prostates of castrated dogs treated with estrogen and androgen. Lab Invest 47:437–450

Nascimento ES, Salvadori MC, Ribeiro-Neto LM (1996) Determination of synthetic estrogens in illegal veterinary formulations by HPTLC and HPLC. J Chromatogr Sci 34:330–333

Noakes DE (1996) Pregnancy and its diagnosis. In: Arthur GH, Noakes DE, Pearson H, Parkinson TJ (eds) Veterinary reproduction & obstetrics. Saunders, London, pp 106–107

Oettel M (1981) Der Einsatz von Sexualhormonen beim Kleintier. In: Döcke F (ed) Veterinärmedizinische Endokrinologie, 2nd edn, Fischer Verlag Jena, pp 718–720

Oettel M (1996) Endokrinpharmakologie In: Frey HH, Löscher W (eds) Lehrbuch der Pharmakologie und Toxikologie für die Veterinärmedizin. Ferdinand Enke Verlag, Stuttgart, pp 388–391

Rasby RJ, Day ML, Johnson SK, Kinder JE, Lynch JM, Short RE, Wetteman RP, Hafs HD (1998) Luteal function and estrus in peripubertal beef heifers treated with an intravaginal progesterone releasing device with or without a subsequent injection of estradiol. Theriogenology 50:55–63

Rathbone MJ, Macmillan KL, Inskeep K, Burggraaf S, Bunt CR (1998) Fertility regulation in the cattle. J Controlled Release 54:117–148

Rawlings NC, Cook, SJ, Waldbillig D (1998) Effects of the pesticides carbofuran, chlorpyrifos, dimethoate, lindane, triallate, trifluralin,2,4-D, and pentachlorophenol on the metabolic endocrine and reproductive endocrine system in ewes. J Toxicol Environ Health 54:21–36

Roche JF (1991) The reproductive system. In: Brander GC, Pugh DM, Bywater RJ, Jenkins WL (eds) Veterinary applied pharmacology & therapeutics. Baillière Tindall, London, pp 302–301

Sutton DJ, Geary MR, Bergman JGHE (1997) Prevention of pregnancy in bitches following unwanted mating: a clinical trial using low dose oestradiol benzoate. J Reprod Fertil Suppl 51:239–243

Tyler CR, Jobling S, Sumpter JP (1998) Endocrine disruption in wildlife: a critical review of the evidence. Crit Rev Toxicol 28:319–361

Part 9
Estrogens, Antiestrogens, and the Environment

CHAPTER 48

Environmental Estrogens

S. MÄKELÄ, S.M. HYDER, and G.M. STANCEL

A. Introduction and Perspective

Chemicals present in the environment can mimic, antagonize or affect the actions of endogenous hormones. These compounds are collectively referred to as "endocrine disruptors", and have recently been defined by the European Union as any "exogenous agent that causes adverse health effects in an intact organism or its progeny, secondary to changes in endocrine function". A very large number of chemicals, both naturally occurring and synthetic, fall into this category. They cause endocrine-disrupting effects either directly by interaction with estrogen receptors (ERs), or indirectly by affecting the levels of endogenous hormones, the levels or function of hormone receptors or biological pathways regulated by the endocrine system. In this chapter, we have focused primarily on directly acting agents, since the focus of this monograph is on chemicals that act as estrogens.

Interest in this area was initiated by the identification of "environmental estrogens", i.e., chemicals in the environment that produce biological actions similar to those of ovarian estrogens (MCLACHLAN 1980), although it is now clear that compounds with other activities (e.g., androgenic or antiandrogenic) are also present in the biosphere (KELCE and WILSON 1997). There are three sources of environmental estrogens: phytoestrogens, which are compounds derived from plants; mycoestrogens, which are agents produced by fungi; and xenoestrogens, which are not naturally occurring, but are synthesized for commercial use or are formed as by-products of various manufacturing or combustion processes. Phytoestrogens fall into three major chemical classes, isoflavones, coumestans and lignans. All known mycoestrogens are β-resorcylic acid lactones. Xenoestrogens do not fit into one or a limited number of chemical classes, but tend to be very hydrophobic and have at least one relatively unhindered phenolic group with an *o*- or *p*-substituent (DUAX and WEEKS 1980; KATZENELLENBOGEN 1995).

Attention to these compounds has gone through a number of phases, each of which has evolved from new insights into their prevalence in the environment, their biological actions and their potential health effects on humans and animals. The existence of phytoestrogens and mycoestrogens was first recognized in the 1920s and 1930s by the effects of feeding moldy grain to domestic animals and from estrogenic bioassays of plant extracts in rodents. Concern about these compounds was further stimulated in the late 1940s by reports that sheep grazing on certain types of clover in western Australia became infertile ("clover disease") (BENNETS et al. 1946). In the 1950s, it was observed that the pesticide *o,p'*-dichlorodiphenyl-trichloroethane (DDT) had certain estrogenic actions in birds (BURLINGTON and LINDERMAN 1950) and, then, in the 1960s, the pesticide was found to have estrogenic effects in mammals (BITMAN et al. 1968; WELCH et al. 1969). Since the initial discovery of environmental estrogens, there has been potential concern that these agents might produce cancer, infertility, developmental defects or other diseases in humans and animals.

In the early 1970s, it was discovered that in utero exposure to the synthetic estrogen diethylstilbestrol (DES), which was prescribed for many pregnant women, increased the incidence of vaginal cancer in humans (HERBST and BERN 1981). This raised concerns that exposure to environmental estrogens in utero might cause serious developmental toxicities in humans and animals. Special mention should be made at this point of the leading contributions from the laboratories of JOHN MCLACHLAN at the National Institute of Environmental Health Sciences and HOWARD BERN at the University of California, who did many of the pioneering studies on animals models of DES exposure. Because of the human experience, they frequently used DES as the estrogen in many of their ground-breaking studies (for example see BERN and TALAMANTES 1981; NEWBOLD 1985; MORI et al. 1979). Since DES is not an "environmental" estrogen, much of their pioneering work is not explicitly discussed in this chapter, although they served to define the underlying mechanisms of the developmental toxicity of all types of estrogens, and were a major stimulus to many subsequent studies with environmental estrogens. While many of these concerns have persisted, a number of studies beginning in the 1980s have suggested that phytoestrogens present in certain foodstuffs may have important health benefits (ADLERCREUTZ and MAZUR 1997; KURZER and XU 1997). At present, it is clear that many chemicals in the environment can exhibit estrogen- or antiestrogen-like activity in one or more test systems, but their actual effects on public health remain to be firmly established.

B. Initial Identification of Environmental Estrogens

I. Phytoestrogens

Around 1920, the pioneering studies of PAPANICOLAOU in the guinea pig (STOCKARD and PAPANICOLAOU 1917) and EVANS in the rat (LONG and EVANS 1922) described the association of the cyclic pre-ovulatory swelling of the ovarian follicle, with uterine engorgement and vaginal cornification. In 1923, ALLEN and Doisy extracted *liquor folliculi* from the larger and more readily accessible hog follicles and showed that it produced behavioral changes typical of estrus, uterine growth and hyperemia, and vaginal cornification when injected into rats and mice. This latter effect was the basis for the Allen-Doisy test, which became the standard bioassay for identification of estrogenic substances and assessment of their potencies.

Using this assay, at least seven different groups had identified plant extracts that gave a positive test for estrogenic activity by 1930, and reviews of this topic appeared as early as 1933 (LOEWE 1933). In fact, so many plant extracts were found to contain activity when assayed by the Allen-Doisy test that DOHRN published a paper in 1927 with the self-explanatory title "Ist der Allen-Doisy-Test Spezifisch für das Weibliche Sexualhormon?". This was perhaps the first realization of the very large number of environmental chemicals with estrogenic activity. Indeed, by the mid 1950s, extracts from over 40 different plant species had been shown to possess estrogenic activity (BRADBURY and WHITE 1954) and, by the mid 1970s, this number had grown to over 300 (FARNESWORTH et al. 1975).

In 1931, it was discovered that plants such as soybeans contain especially high amounts of the glycosides of isoflavones, such as genistein (WALZ 1931). Since certain Oriental diets are very rich in soy products relative to Western diets, many epidemiological studies have investigated health effects in populations with a high soy diet. Observed effects in these studies have often been ascribed to the "phytoestrogens" present in soy, but it has not always been established that effects of Oriental diets are explicitly due to phytoestrogens rather than to other soy components or confounding factors.

Stimulated by the studies of BENNETS and COLLEAGUES (1946), BRADBURY and WHITE (1951) identified isoflavonoids (genistein and formononentin) in the 1950s as the estrogenic substances in clover, responsible for infertility in sheep ("clover disease"). It is now recognized that formononetin has only limited affinity for the ER, but is metabolized in vivo to daidzein which, in turn, is converted to equol; both of these metabolites have greater affinities for the receptor. This was the initial identification of the estrogenic activity of this class of compounds, and it is now recognized that isoflavones are one of the major types of phytoestrogens.

Several years later, BICKOFF et al. (1957) isolated a chemical with a coumarin structure from another type of clover (ladino clover). Since this agent had estrogenic activity in uterotrophic assays (BICKOFF et al. 1957), they proposed the name coumestrol. FOLMAN and POPE (1966, 1969) and SHUTT (1967) then showed that genistein, coumestrol and related phytoestrogens competed with estradiol for retention in target tissues, such as the uterus; SHUTT and COX (1972) later showed that they bound to the ER in cell-free uterine preparations. Collectively, these studies firmly established isoflavones and coumestans as two major classes of direct-acting estrogens.

Lignans are compounds possessing a 2,3-dibenzylbutane structure that are found in many plants where they play a role in the formation of plant cell walls. Mammalian lignans, such as enterolactone and enterodiol, were first identified in the urine of humans and animals in 1980 (SETCHELL et al. 1980; STITCH et al. 1980). These compounds are not found in plants, but are formed in the gut by bacterial biotransformation of plant lignans, such as matairesinol and secoisolariciresinol, which are ingested in the diet (AXELSON and SETCHELL 1981; SETCHELL et al. 1981). These compounds have NOT been convincingly shown to bind to ERs and they do NOT exhibit uterotrophic effects in mice (SETCHELL et al. 1981). However, there is one report demonstrating that they affect RNA synthesis in the rat uterus in vivo (WATERS and KNOWLER 1982) and several reports that they produce estrogenic responses in cultured cells (*vide infra*). Mammalian lignans are present in very high concentrations in human plasma and urine, and have attracted widespread attention as dietary components with potential anticancer activity (ADLERCREUTZ 1995; ADLERCREUTZ and MAZUR 1997; ADLERCREUTZ et al. 1982; SETCHELL and ADLERCREUTZ 1988). Lignans are particularly high in fiber-rich diets, and along with coumestans and isoflavones, they are considered to be the third major class of phytoestrogens, although it is not yet clear whether their biological effects are due to direct estrogen agonist or antagonist actions at ER sites.

II. Mycoestrogens

In the early 1900s, pig farmers in the US "cornbelt" complained of "false heat" and infertility in some of their swine herds. Anecdotal reports of similar effects were also noted in the early to mid part of the century in Australia, Ireland, France, Rumania, Hungary, Canada and Russia (MIROCHA et al. 1977; MOREAU and Moss 1979). In 1928, MCNUTT et al. (1928) first established a cause–effect relationship between feeding corn spoiled by fungal growth and what he termed "vulvovaginitis" in gilts. Working independently, STOB et al. (1962) and CHRISTENSEN et al. (1965) isolated the responsible mycotoxin, now known as zearalenone, from *Fusarium graminearum (Gibberella zeae)*, and its chemical structure was reported by URRY et al. (1966).

Zearalenone is a β-resorcylic acid-lactone derivative, and all mycoestrogens identified to date belong to this class of compounds. Many studies in the following decade demonstrated that zearalenone and related compounds bind directly to the ER and produce estrogenic responses in numerous tissues. These compounds are quite potent, relative to many other environmental estrogens and, in fact, when the use of DES to improve feeding efficiency in cattle was banned in the US, a compound closely related to zearalenone (zeranol or Ralgro®) was substituted in its place and is still in use today.

III. Xenoestrogens

The first report that a xenobiotic prevalent in the environment had estrogenic activity was by BURLINGTON and LINDERMAN (1950) who observed that DDT inhibited testicular growth and comb development in cockerels. They noted the structural similarity between DDT and DES and suggested that the pesticide might be producing these effects by acting as an estrogen. Actually, a physician in Arizona had noted oligospermia in young men exposed to DDT the year before, but had not suggested estrogenicity or any other specific cause as the basis of the observed oligospermia (SINGER 1949). The discovery that technical grade DDT had uterotrophic activity in mammals was accidentally made by WELCH et al. (1969). They attempted to use it to diminish the uterotrophic activity of estrogens by increasing their metabolism secondary to induction of mixed-function oxidases, but instead observed that "control" animals receiving the pesticide alone exhibited increased uterine weights. At about the same time, BITMAN et al. (1968) noted that DDT increased the glycogen content of rodent uteri and the weight of bird oviducts.

NELSON (1974) was the first to show that *o,p*'-DDT could bind to the ER in a cell-free system, and KUPFER and BULGER (1976b) were the first to show that it bound to the receptor and caused its nuclear localization in the uteri of intact animals. In another early serendipidous observation, TULLNER noted that the use of the insectide methoxychlor in the animal-care quarters at the National Cancer Institute increased the uterine weight of resident rodents, and went on to show that injections of the compound itself increased uterine weight (TULLNER 1961). These initial observations led many workers to investigate the potential estrogenic activity of a large number of pesticides, fertilizers, industrial chemicals and related products, and these studies have been outlined in a series of important monographs edited by MCLACHLAN, who is generally credited with coining the term "environmental estrogens" (MCLACHLAN 1980; MCLACHLAN 1985; MCLACHLAN and KORACH 1995).

Even though the estrogenic effects of DDT were first recognized in 1950, the identification of new environmental estrogens some 40 years later is often still accidental. For example, in 1991, SOTO and COLLEAGUES discovered the estrogenic activity of alkyl phenols, because these compounds leached out of polystyrene tubes used to store tissue culture reagents and caused an unexplained proliferation of estrogen-responsive cells grown in tissue culture. Similarly, FELDMAN and his colleagues (KRISHNAN et al. 1993) inadvertently discovered that autoclaving media in polycarbonate flasks released a substance that blocked the binding of estradiol to its receptor, and they went on to identify the competing compound as another phenolic chemical, bisphenol-A. Observations such as these led SOTO and her colleagues and others to extend their screening studies to determine whether chemicals likely to find their way into the food chain might have estrogenic activity. This led to the observation that toxaphene, endosulfane and dieldrin had measurable activity in cultured-cell test systems (SOTO et al. 1994).

As a historical aside, it is interesting to note that bisphenol-A and structurally related compounds were shown to have estrogenic activity by DODDS and LAWSON (1938), and MUELLER and KIM (1978) showed that many simple phenols could interact with the ER. However, these compounds were not thought of as major environmental pollutants at those times, and this is presumably the reason that these reports did not have a more immediate impact on endocrine disruptor research. Interest in these compounds has increased tremendously throughout this decade, with the realization that they can leach into the environment from a large number of manufactured products in which they are present. These and related observations have sparked an intense discussion about our research needs to develop practical, predictive approaches to screen the thousands of chemicals in commerce for estrogenic or other endocrine disrupting activities (GRAY 1997; KAVLOCK et al. 1996).

C. Structures and Sources of Major Prototype Environmental Estrogens

One of the most vexing issues in the study of environmental estrogens remains the inability to accurately predict hormone-like biological activity from a knowledge of chemical structure. Much is known about how modifications of ovarian estrogens, such as estradiol, alter their receptor-binding and biological activity, but it is not yet possible to accurately predict the estrogenic activity of the wide range of environmental chemicals that produce hormone-like actions. Thus, a bewildering array of structurally diverse chemicals derived from animal, plant, fungal and industrial sources can be shown to elicit estrogenic responses in one or more test systems. As noted by KATZENELLENBOGEN (1995), one reason for this situation is the tremendous sensitivity of the bioassays available to measure estrogenic activity, e.g., compounds with only 1/1,000,000th the potency of estradiol can be shown to have estrogenic activity in various assay systems.

In their 1938 paper, DODDS and LAWSON noted that "substances containing two phenol groups joined by a carbon chain are active (as estrogens)". This remains one of the most common general features of environmental estrogens, many of which are phenols that contain hydrophobic substituents in the *meta* or *para* positions. In other cases, e.g., that of methoxychlor, the parent compound undergoes activation via biotransformation to a phenol. In general, ER binding has a fairly rigid requirement for a phenolic structure, similar to the A ring of steroidal estrogens, but can tolerate more diversity in regions analogous to the D ring of steroids (DUAX and WEEKS 1980). There are exceptions and some xenoestrogens (e.g., kepone, endosulfane and toxaphene) are not even aromatic. However, compounds of this nature tend to be particularly hydrophobic, which contributes to estrogenicity in a general sense, and they tend to have the lowest estrogenic potency of the environmental estrogens. For a recent review of the relationship between the structure of estradiol-like ligands and ER binding, interested readers may consult several recent publications (ANSTEAD et al. 1997, BRZOZOWSKI et al. 1997); far less is known about the binding of nonsteroidal environmental estrogens.

Despite historical difficulty in predicting estrogenic activity from structural features, some progress is being made. For example, TONG et al. (1997) and WALLER et al. (1996) have recently used quantitative structure–activity relationship (QSAR) models based on comparative, molecular field analysis to predict the relative binding affinities of a series of compounds for ER-α. They were able to obtain good correlations between experimentally measured and predicted affinities for a number of compounds, including natural, synthetic and environmental estrogens. While the compounds analyzed were all phenols with hydrophobic substituents, QSAR approaches will undoubtedly continue to receive much attention because of the cost and difficulty of screening compounds solely with bioassays.

I. Phytoestrogens

Based on their chemical structures, phytoestrogens may be classified into one of three major groups: isoflavones, coumestans and lignans. The structures of the major representative compounds in each of these categories are shown in Figs. 1–3, respectively.

1. Isoflavones

The general structure of isoflavones and the specific structures of genistein and daidzein, two isoflavones found in human diets, are illustrated in Fig. 1, along with that of estradiol. In addition, the structure of equol, a related isoflavan derived from metabolism of daidzein, is also illustrated. Equol is actually the major circulating form of phytoestrogen present in sheep that graze on subterranean clover which causes infertility (BRADEN et al. 1971); it is also present in goats, cattle, horses, fowl and guinea pigs (SHUTT 1976). Equol is also found in humans and appreciable amounts may be present in the urine of people with typical Western diets, indicating that it may be derived from a variety of common dietary sources (ADLERCREUTZ et al. 1982; SETCHELL et al. 1984). Another isoflavanoid encountered in human diets is glycitein, which is the 6-*O*-methyl derivative of daidzein (ADLERCREUTZ and MAZUR 1997; KURZER and XU 1997). All these compounds are diphenols, with their phenolic A and B rings orientated in a similar fashion to the A and D rings of estradiol, and they exhibit estrogenic activity in various assay systems.

Isoflavones are present in high concentrations in leguminous plants, particularly red and subterranean clover and soy beans. The major isoflavones found in soybeans are genistein, daidzein and glycitein. The unconjugated forms of these compounds can be found in very small amounts (approximately 1% of the isoflavone content), but the major forms present are glycosides, including the 7-*O*-glucosides (daidzein, genistein and glycitein), acetyl glucosides and malonylglucosides (ELDRIDGE and KWOLEK; 1983, KUDOU et al. 1991; NAIM et al. 1974). Biochanin A and formononetin are the 4'-*O*-methyl ethers of genistein and daidzein, respectively, and are major isoflavonoids in clover (BRADBURY and WHITE 1954). Absorption of ingested isoflavones in sheep appears to involve deconjugation reactions, most likely catalyzed by β-glucosidases present in bacteria in the human gut and in food itself, as well as nonenzymatic hydrolysis by gastric hydrochloric acid (SHUTT et al. 1970). Urinary profiles after soya ingestion suggest that similar reactions may occur in humans (KELLY et al. 1993). Additional information about the isoflavonone

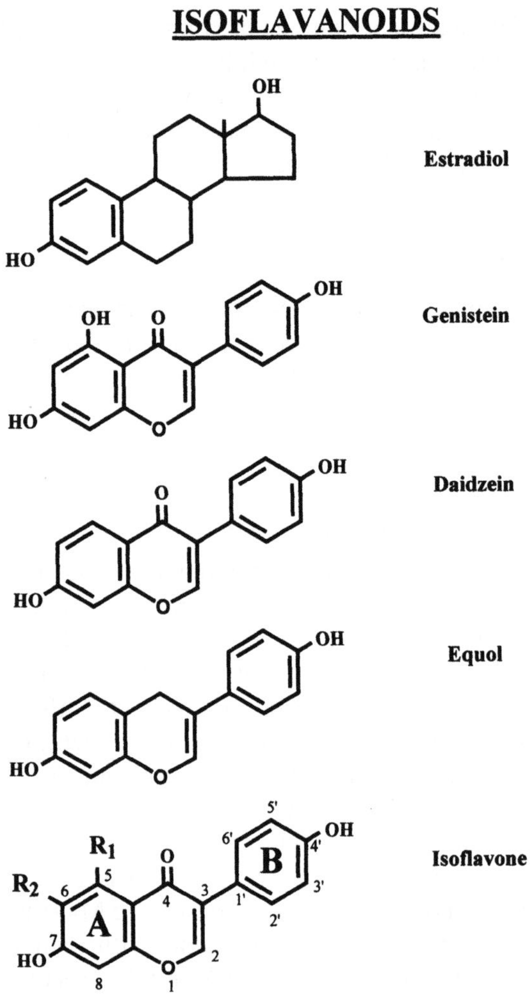

Fig. 1. Structures of major isoflavanoids. The structures of genistein, daidzein and equol are illustrated along with the general structure of isoflavones and the specific structure of estradiol for comparison

content of specific foodstuffs and their absorption can be found in excellent recent reviews by KURZER and XU (1997) and ADLERCREUTZ and MAZUR (1997).

In addition to the specific isoflavones noted above, several related types of bioflavonoids have been shown to possess estrogenic or antiestrogenic activity, including chalcones, flavones, flavonols, flavanones and isoflavans. Some specific compounds include phloretin, apigenin, kaempferol and naringenin (MIKSICEK 1993, 1994, 1995; RUH et al. 1995; SATHYAMOORTHY et al. 1994). MIKSICEK (1993, 1994, 1995) has performed the most thorough systematic study of the structural requirements of bioflavonoids required for ER activation and biological activity.

2. Coumestans

A large number of coumestans have been isolated from plants, including human foodstuffs and fodder crops, but few have been shown to possess estrogenic activity. Coumestrol and its 4'-methoxy derivative are the phytoestrogens found in alfalfa, ladino clover and other fodder crops that are the most significant source of this class of compounds. It has also been suggested that the content of coumestrol in either clover or alfalfa can increase in response to fungal disease, suggesting that the compound may be somehow associated with the defense mechanisms of plants to leaf pathogens (BICKOFF 1968). Split peas, kala-chana seeds, pinto-bean seeds, lima-bean seeds and soybean sprouts also contain some coumestrol (see KURZER and XU 1997; ADLERCREUTZ and MAZUR 1997 and references therein).

Coumestans are structurally similar to stilbenes, and the structures of diethylstilbestrol and coumestrol are illustrated in Fig. 2. The similarity is apparent if one traces the carbon backbone (shown by the asterisks in both structures) linking the phenolic rings in the two molecules.

COUMESTANS

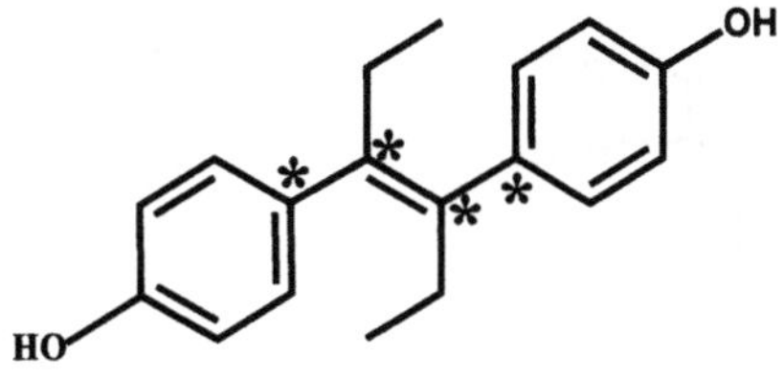

Diethylstilbestrol

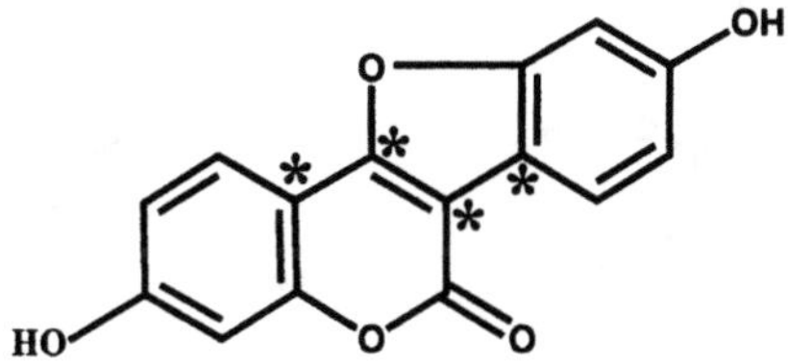

Coumestrol

Fig. 2. Coumestans. The structure of coumestrol, a prototype coumestan, is illustrated along with the synthetic estrogen Diethylstilbestrol for comparison. The *asterisks* are used to note the similarity of the carbon chain linking the phenolic rings of both compounds

3. Lignans

Lignans are compounds possessing a 2,3-dibenzylbutane structure which is illustrated in Fig. 3, along with the structures of several major mammalian and plant lignans. The major mammalian lignans are enterodiol and enterolactone which were originally identified in human urine (SETCHELL et al. 1980; STITCH et al. 1980). These compounds are not found in plants, but their precursors, secoisolariciresinol and matairesinol, are ingested in the diet. These plant lignans are found in foods such as flaxseed, unrefined grains (especially rye), fruits and berries (the seeds of which may be rich in lignans). In addition to these two compounds, the diglucoside of secoisolariciresinol, lariciresinol and isolariciresinol represent other dietary lignans (ADLERCREUTZ 1995; ADLERCREUTZ and MAZUR 1997; KURZER and XU 1997).

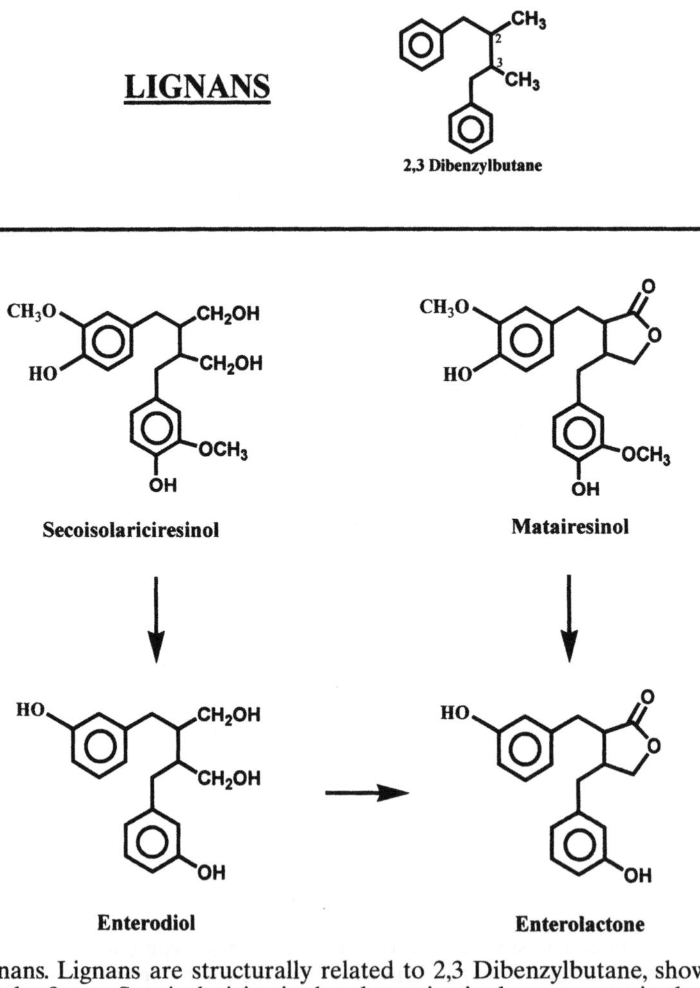

Fig. 3. Lignans. Lignans are structurally related to 2,3 Dibenzylbutane, shown in the *top half* of the figure. Secoisolariciresinol and matairesinol are present in the diet and converted to enterodiol and enterolactone by bacteria present in the gut

Initially, it was unclear whether mammalian lignans were of dietary origin or were produced by the ovary, since urinary lignan levels varied during the menstrual cycle and seemed to be altered in post-menopausal women (SETCHELL et al. 1980). However, subsequent experiments using germ-free animals (AXELSON and SETCHELL 1981) and antibiotic-treated humans (SETCHELL et al. 1981) established that enterodiol and enterolactone are formed in the gut by dehydroxylation and demethylation reactions, catalyzed by enzymes of the intestinal flora. Once formed, enterodiol can also undergo oxidation to enterolactone.

II. Mycoestrogens

Zearalenone is a nonsteroidal macrolide belonging to the class of natural products which, structurally, are β-resorcylic lactones (see Fig. 4). Zearalenone was isolated independently by both STOB et al. (1962) and CHRISTENSEN et al. (1965) on the basis of its uterotrophic activity in rodents; URRY et al. (1966) elucidated its structure shortly thereafter. Zearalenone is the major mycoestrogen found in most species of *Fusarium*, but other related compounds are also produced by fungi to a lesser extent, and hundreds of zearalenone derivatives have been synthesized. In mammals, the major metabolite of zearalenone is zearalenol, and both the parent compound and metabolite undergo further conjugation with glucuronic acid or sulfate (LONE 1997; MIGDALOF et al. 1983). Zearalenone and related compounds are some of the most potent environmental estrogens. Interestingly, α-zearalenol, which is a mammalian metabolite of zearalenone, is a more potent estrogen than the parent compound (EVERETT et al. 1987; WILSON and HAGLER 1985).

The very first report of the isolation of zearalenone (STOB et al. 1962) noted that the compound was also able to improve growth rate and feed efficiency in sheep. This prompted a number of studies to investigate the anabolic activity of this class of compounds which led to the extensive use of zeranol (also referred to as α-zearalanol) as an anabolic agent in sheep and cattle marketed under the proprietary name of Ralgro®. This compound was found to have approximately the same anabolic activity as DES in cattle and sheep, but less than 1% the estrogenic activity of DES in rodent uterotrophic assays. In addition, tests on the tissues of animals treated with either DES or zeranol showed far less residue of the mycoestrogen than the synthetic stilbene estrogen in their carcasses (HURD 1977; LONE 1997). Largely because of these properties, zeranol has been used extensively to promote growth of domestic animals in the US. In addition, zeranol is metabolized to zearalanone, which has less estrogenic activity than zearalenone which is metabolized to the more potent zearalenol. While zearalenone and its derivatives are anabolic in sheep and cattle, their effects are quite species specific; they cause too much reproductive toxicity in swine to be used for this purpose and they have little or no effect in laying hens and broilers (see PATHRE and MIROCHA 1980; RODRICKS et al. 1977 and references therein). It should also be noted that this

MYCOESTROGENS

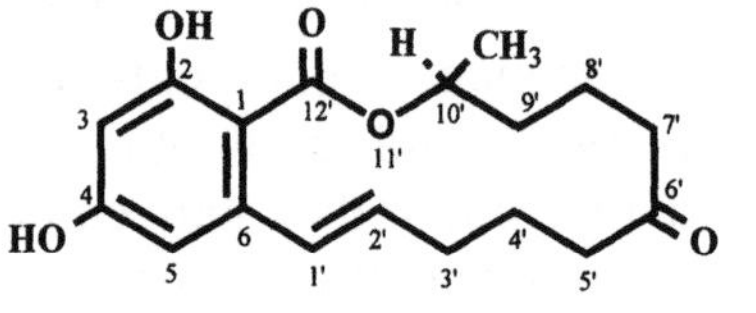

Zearalenone

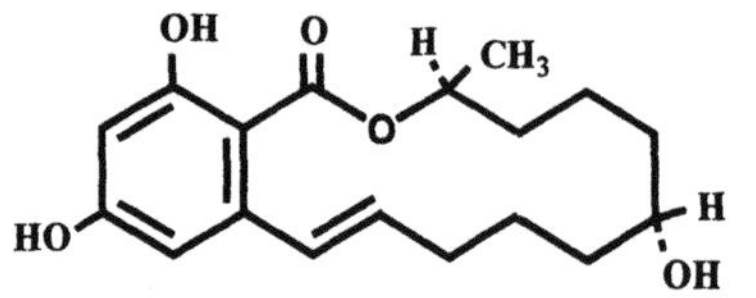

α - Zearalenol

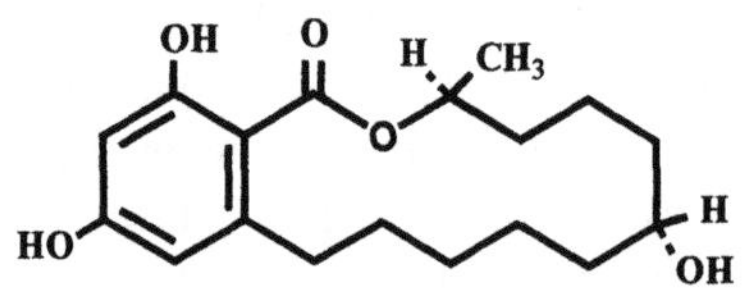

Zeranol
(α – Zearalanol)

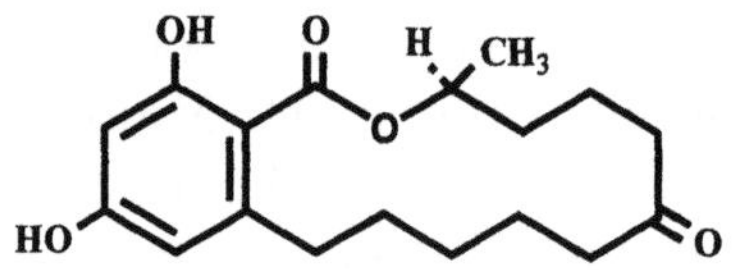

Zearalanone

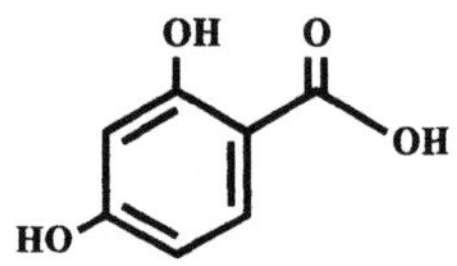

β - Resorcylic Acid

Fig. 4. Mycoestrogens. Mycoestrogens are structurally related to *β*-resorcyclic acid, illustrated at the *bottom* of the figure, and the structures of the four specific mycoestrogens, zearalenone, *α*-zearalenol, zeranol and zearalanone, are illustrated

class of compounds exhibits systemic estrogenic activity in humans (UTIAN 1973) and affects cultured human cells (MARTIN et al. 1978).

III. Xenoestrogens

Xenoestrogens do not fall into well-defined groups based on their chemical structures in a manner analogous to phytoestrogens and mycoestrogens. Rather, a bewildering array of man-made chemicals with widely different structures have been shown to exert estrogenic or antiestrogenic activity in a variety of test systems, albeit it at quite low potency in many cases. We have, thus, arbitrarily grouped xenoestrogens into functional classes, since this is the way they are often discussed. The three major groupings include pesticides and agricultural chemicals, phenolic compounds used in the manufacture of plastics and other synthetic materials, halogenated aromatic hydrocarbons used commercially and their by-products.

Most recently pthalates, such as butylbenzylphthalate and dibutylphthalate, have also been found to have estrogenic activity in cultured breast cancer cells (SOTO et al. 1995), to activate the ER of rainbow trout (JOBLING et al. 1995) and to decrease sperm production and testes size following in utero exposure (SHARPE et al. 1995). Massive amounts of these chemicals have been produced since the 1930s, as they are one of the major types of plasticisers used. Because relatively little information is available about the estrogenic actions of these compounds, they are not covered further in this chapter.

1. Pesticides and Agricultural Chemicals

As noted above, DDT was the first pesticide shown to have estrogenic activity. Technical grade DDT used commercially is primarily *p,p'*-DDT, but contains a mixture of isomers, of which *o,p'*-DDT (which comprises 15–20% of commercial preparations) is the most potent estrogen. This isomer binds to the ER and produces hormone-like effects in many systems (BULGER and KUPFER 1985; KUPFER and BULGER 1976a; NELSON et al. 1978; ROBISON et al. 1985a). Like other organochlorines, DDT exhibits marked bioaccumulation, and its use is now banned in the US and Europe. Methoxychlor is a structurally related compound without chlorine substituents on its aromatic rings that was developed to replace DDT and it continues to be used commercially at this time (CUMMINGS 1997). The structures of methoxychlor and DDT are shown in Fig. 5. Methoxychlor itself has little estrogenic activity, but as shown by KUPFER and BULGER (1979) in a landmark study, it is an environmental "proestrogen" which is *O*-demethylated by microsomal enzymes to the active estrogenic form 2,2-*bis*(*p*-hydroxyphenyl)-1,1,1-trichloroethane.

In addition to DDT and methoxychlor, a variety of other pesticides and agricultural chemicals have been reported to exhibit either estrogenic or antiestrogenic activity. These include chlordecone or Kepone® (EROSCHENKO and PALMITER 1980; GELLERT 1978a; GUZELIAN 1982; HAMMOND et al. 1979); endosulfane, toxaphene and dieldrin (SOTO et al. 1994); *β*-hexachlorocyclohexane, a component of the insecticide lindane (see STEINMETZ et al. 1996); and chloro-*S*-triazine-derived herbicides, such as atrazine (CONNOR et al. 1996; TRAN et al. 1996). The structures of chlordecone and endosulfane are also shown in Fig. 5 and illustrate the diversity of chemical structures capable of producing estrogenic responses.

2. Halogenated Aromatic Hydrocarbons

Polychlorinated biphenyls (PCBs) and their mixtures (e.g., Aroclors) have been extensively used for over 50 years as plasticizers, adhesives and dielectric

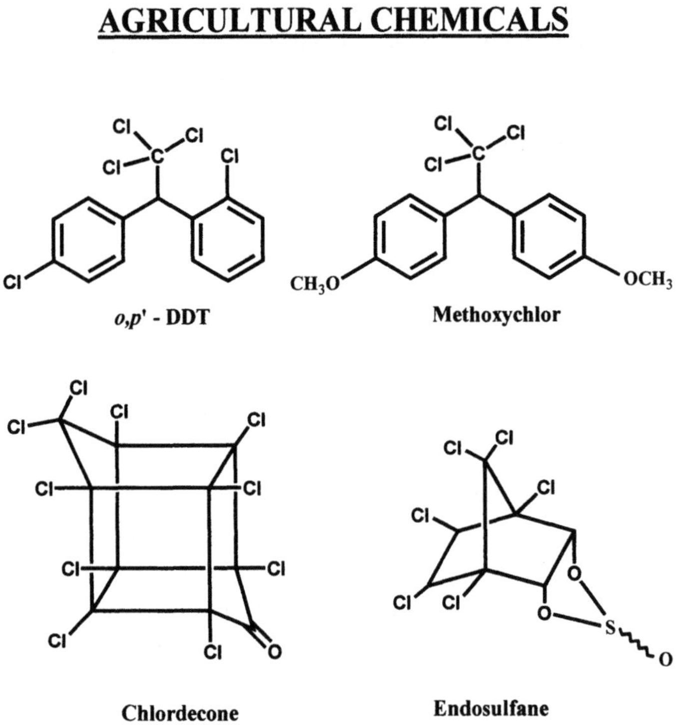

Fig. 5. Agricultural chemicals: the structures of *o,p′*-DDT, methoxychlor, chlordecone and endosulfane, four chemicals used in agriculture that possess estrogenic activity

fluids in capacitors and transformers. It is interesting that these endocrine disruptors were first found in the environment in the 1960s as unidentified chromatographic peaks (RISEBROUGH et al. 1968) in the analyses of environmental extracts for another endocrine disruptor – DDT! The related polychlorinated dibenzo-*p*-dioxins (PCDDs) and polychlorinated dibenzofurans (PCDFs) are not commercially produced as such, but are industrial by-products formed during the production of PCBs and herbicides such as Agent Orange. They are also formed during the burning of coal, wood, industrial and municipal wastes; from the automobile exhaust of leaded gasoline; and as by-products of the chlorine bleaching of pulp and paper products. These compounds are, thus, extremely widespread in the environment and are generally present as mixtures of many different isomers and congeners. This has made it extremely difficult to elucidate the estrogenic or other biological activities of these compounds, because it is difficult to obtain highly purified congeners for experimental testing, and different isomers and congeners may have estrogenic, antiestrogenic or no endocrine activity (BIRNBAUM 1994; GELLERT 1978b; JANSEN et al. 1993; KORACH et al. 1988; MCKINNEY and WALLER 1994; NESARETNAM et al. 1996; SAFE 1995). The general structures of PCBs, PCDDs and PCDFs, and the specific structure of 2,3,7,8-tetrachlorodibenzo-*p*-dioxin (TCDD) are illustrated in Fig. 6.

HALOGENATED AROMATIC HYDROCARBONS

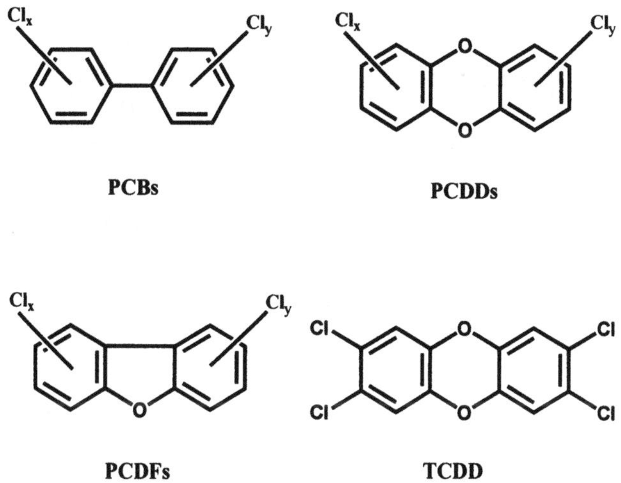

Fig. 6. Halogenated aromatic hydrocarbons: the general structures of polychlorinated biphenyls (PCBs), polychlorinated dibenzo-dioxins (PCDDs), polychlorinated dibenzofurans (PCDFs) and TCDD, a specific PCDD

One of the most toxic and widely studied compounds in this category is TCDD, which is often simply referred to as dioxin in the literature. This compound has little affinity for the ER, but exerts antiestrogenic effects in many biological systems. Therefore, these effects most likely occur by indirect mechanisms. One possibility is an inhibitory type of "cross-talk" between its cognate receptor (the aryl hydrocarbon or Ah receptor) and the ER (SAFE 1995). Also, TCDD, PCBs and other halogenated aromatic hydrocarbons are potent inducers of the cytochrome P450 family and other drug metabolizing enzymes and, thus, could act as indirect endocrine disruptors by affecting the synthesis or degradation of endogenous hormones.

3. Commercially Used Phenols and Related Compounds

MUELLER and KIM (1978) were the first to report that simple phenols containing either a fixed B ring (e.g., napthols), flexible 4- or 5-carbon side chains in the *o*- or *p*-positions, or other assorted phenols displace labeled estradiol from its receptor. These studies were not extensively pursued until it was serendipitously observed that alkyl phenols (SOTO et al. 1991) and bisphenol-A (KRISHNAN et al. 1993), which leach out of plasticware, possess estrogenic activity. Other work showed that alkyl phenol ethoxylates (e.g., *p*-Nonylphenol ethoxylates), which are very commonly used nonionic surfactants, also have estrogenic activity in fish (JOBLING and SUMPTER 1993). These compounds

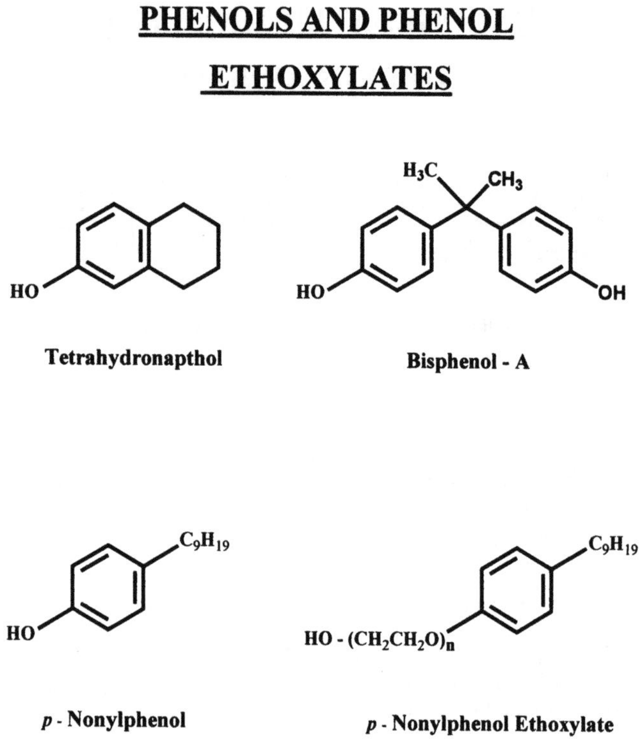

Fig. 7. Phenols and phenol ethoxylates: the structures of four phenol derivatives with estrogenic activity

are degraded to alkyl phenols, which have estrogenic activity, as noted above. The phenols are also more hydrophobic and stable in the environment than the parent ethoxylates (AHEL et al. 1987; GIGER et al. 1987). These were important observations because enormous amounts of these compounds are used in the production of polystyrene and polycarbonate materials (many of which are used in food storage containers and serving utensils), detergents, paints, herbicides, pesticides and many other formulated products. Large amounts of these compounds, thus, enter the biosphere as consumer and industrial wastes and as by-products of synthetic and manufacturing processes (NIMROD and BENSON 1996). The structures of 5,6,7,8-tetrahydronapthol-2, the prototype compound studied by MUELLER and KIM (1978), *p*-nonyl-phenol, *p*-nonylphenol-ethoxylate and bisphenol-A are shown in Fig. 7.

D. Mechanisms of Action

Chemicals may produce "estrogenic" or "antiestrogenic" responses directly, by interacting with hormone receptor sites or indirectly, by altering endogenous hormone levels, receptor levels or post-receptor steps in estrogen-regulated pathways. Because this monograph deals explicitly with effects of estrogens, this section will focus on direct mechanisms.

I. Receptor Interactions

Many of the environmental estrogens covered in this chapter are known to bind to the ER, albeit with a wide range of affinities. The major exceptions are the lignans, which have not been shown unequivocally to bind directly to the hormone binding site on the classical ER-α. Since the ER-β has only recently been discovered (KUIPER et al. 1996), its affinity for environmental estrogens largely remains to be established. Unless otherwise noted, receptor binding has been measured by competition of an unlabeled environmental estrogen with radiolabeled estradiol (or another ligand such as DES) for receptor binding sites in tissue or cell extracts, rather than to highly purified receptors. Most studies used endogenous receptors, but MIKSICEK (1993, 1995) used human ERs produced from expression plasmids in cell lines without endogenous receptor to study the binding of phyto- and mycoestrogens.

Because of the large number of such studies, we have not explicitly reviewed those that only demonstrated competitive binding without estimates of receptor affinity or binding curves from which a reader could obtain estimates. A number of studies in which the binding affinities (Kds) of ER-α for environmental estrogens have been determined, relative to estradiol or DES, and the few cases in which Kd values were directly determined are outlined in Table 1. One recent study found that the binding of most endogenous estrogens was similar for ER-α and ER-β, although the latter had higher affinities for genistein (7-fold) and coumestrol (2-fold) (KUIPER et al. 1997). There is tremendous interest in the ligand binding specificities of these two receptors, and additional studies of the interactions of environmental estrogens with these proteins can be expected.

While there is obviously a great range of relative binding activities of environmental estrogens to the ER, a few generalities are apparent. Zearalanol (i.e., zeranol) has consistently been found to have a very high receptor affinity. In fact, two reports that used radiolabeled compounds to directly obtain a Kd value found affinities essentially the same as that of estradiol (KATZENELLENBOGEN et al. 1979, MASTRI et al. 1985), and several other studies using competitive binding found affinities 10–33% that of the endogenous hormone. The relative affinity of coumestrol for the ER from various sources is approximately 10% that of estradiol, that of zearalenone is roughly 2% that of the hormone, and several flavonoids display relative binding affinities of 0.1–1% that of estradiol. Special note should be made, however, of a recent report using in vitro translated human ER-α that found the relative binding affinities of coumestrol and genistein to be 94% and 5%, respectively, of estradiol (KUIPER et al. 1997). This is much higher than has been found in previous studies, which most often used tissue or cell homogenates rather than highly purified receptor. As a group, the xenoestrogens have, by far, the lowest

Table 1. Relative binding activity of environmental estrogens to the estrogen receptor-α

Compound	Relative binding activity	ER Source	Reference
Estradiol	100%		
Phytoestrogens:			
Genistein	0.25%	Mouse uterus	Rosenblum et al. 1993
	0.4–1%	Sheep uterus	Shutt and Cox 1972
	2%	Human MCF-7	Martin et al. 1978
	0.4%	Human MCF-7	Miksicek 1994
	5%	Human ER	Kuiper et al. 1997
Equol	0.2–0.9%	Sheep uterus	Shutt and Cox 1972
Daidzein	0.015%	Mouse uterus	Rosenblum et al. 1993
	0.1%	Sheep uterus	Shutt and Cox 1972
	0.1%	Human MCF-7	Miksicek 1994
Biochanin A	0.0015%	Mouse uterus	Rosenblum et al. 1993
	<0.01%	Sheep uterus	Shutt and Cox 1972
	0.005%	Human MCF-7	Miksicek 1994
Formononetin	<0.01%	Sheep uterus	Shutt and Cox 1972
	<0.01%	Human MCF-7	Martin et al. 1978
Numerous Isoflavonoids	0.1–0.003%	Human ER	Miksicek 1993; Miksicek 1995
Coumestans:			
Coumestrol	19%	Rat uterus	Shutt and Cox 1972
	5%	Sheep uterus	Shutt and Cox 1972
	10%	Human MCF-7	Martin et al. 1978
	13%	Human MCF-7	Miksicek 1994
	94%	Human ER	Kuiper et al. 1997
4′-Methoxycoumestrol	<0.01%	Sheep uterus	Shutt and Cox 1972
Mycoestrogens:			
Zearalenone	18%	Rat uterus	Katzenellenbogen et al. 1979
	3.3%	Human MCF-7	Martin et al. 1978
	1.25%	Human MCF-7	Miksicek 1994
Zearalenol (α)	19%	Human ER	Kuiper et al. 1997
	1.3–10%	Human MCF-7	Martin et al. 1978
Zearalenol (β)	0.6%	Human MCF-7	Martin et al. 1978

Zearalanol (Zeranol)	100%[b]	Rat liver	MASTRI et al. 1985
	33%	Rat liver	POWELL-JONES et al. 1981
	13.6%[b]	Rat uterus	KATZENELLENBOGEN et al. 1979
	10%	Rat mammary	BOYD and WITLIFF 1978
Xenoestrogens:			
o,p′-DDT	0.1–0.004%	Rat uterus	STANCEL et al. 1980; NELSON 1974[a]; OUSTERHOUT et al. 1981[a]; KELCE et al. 1995 ; HAMMOND et al. 1979
	0.01–0.02%	Rat brain	BROWN and BLAUSTEIN 1984
	0.0003%	Human MCF-7	SOTO et al. 1995
p,p′-DDT,-DDE,-DDD	<0.0002%	Rat uterus	KELCE et al. 1995
mono-, bis-Hydroxy-methoxychlor	0.17–0.50%	Rat uterus	NELSON 1974; OUSTERHOUT et al. 1981
Chlordecone (Kepone)	0.02%	Chick oviduct	EROSCHENKO and PALMITER 1980
	0.04–0.1%	Rat uterus	HAMMOND et al. 1979; KELCE et al. 1995; WILLIAMS et al. 1989
	0.10%	Rat brain	WILLIAMS et al. 1989
	0.008%	Trout liver	THOMAS and SMITH 1993
Endosulfane	0.00024%	Human MCF-7	SOTO et al. 1995
Toxaphene	0.00032%	Human MCF-7	SOTO et al. 1995
Various PCBs	0.01–2.4%	Mouse uterus	CONNOR et al. 1997; KORACH et al. 1988
	0.1–0.0005%	Rat uterus	CONNOR et al. 1997
3,4,3′,4′-tetrachlorobiphenyl	5%	Human MCF-7	NESARETNAM et al. 1996[a]
Octylphenol	0.07%	Trout liver	WHITE et al. 1994
Nonylphenol	0.03%	Trout liver	WHITE et al. 1994
	0.021%	Human MCF-7	SOTO et al. 1995
Nonylphenol 1 Ethoxylate	0.005%	Trout liver	WHITE et al. 1994
Bisphenol A	0.1%	Rat uterus	KRISHNAN et al. 1993
	0.05%	Human MCF-7	KRISHNAN et al. 1993

ER, estrogen receptor; PCB, Polychlorinated biphenyls; DDT, dichlorodiphenyl-trichloroethane
[a] Relative binding affinities calculated relative to DES as 100%
[b] Binding affinities (Kd) determined directly by Scatchard analysis of binding of radiolabelled zearalanol to ER

relative binding, although several compounds in this class (e.g., phenolic metabolites of methoxychlor, bisphenol A, and the PCB 3,4,3′,4′-tetrachlorobiphenyl) are reported to have affinities in the range of some phytoestrogens.

II. Gene Expression

As described in preceding chapters of this volume, estrogens act in large part as transcriptional activators. In Table 2, we have thus tabulated reports of specific genes activated by the major prototype environmental estrogens in various systems. Some of these studies have monitored expression of endogenous genes in vivo or in cultured cells; others have measured activity of reporter constructs (referred to as "ERE-Reporters" in Table 2) which contained the vitellogenin estrogen response element(ERE) (GGTCAnnnTGACC) linked to a reporter, such as chloramphenicol acetyltransferase or luciferase. Because of the number of such studies, we cannot discuss each one in detail in the text, and readers are thus referred to the original articles for additional specifics.

While quite informative, it is important to note that many of these reports employed transfection studies that measured the activity of reporter constructs in plasmids rather than the response of endogenous genes in chromatin. This may be an important point, since recent studies on steroid action indicate that a major effect of this class of hormones is to recruit factors with histone acetyltransferase activity, which increase transcription by altering chromatin structure – an effect that obviously does not occur from plasmid templates. In addition, some studies have used reporters with multiple-consensus EREs, and such regulatory sequences are not generally present in endogenous genes. Also, it is now clear that the ability of a specific estrogen to stimulate transcription is both cell- and gene specific, and induction (or lack thereof) of one gene by a particular estrogen does not mean it will (or will not) induce other estrogen-responsive genes, or the same gene in a different context. The points are all discussed in detail in preceding chapters of this volume for interested readers.

With these considerations in mind, we believe that the regulation of endogenous estrogen-responsive genes by environmental estrogens is an area in which more research is badly needed, even though, at first glance, it appears that many such studies have been performed (Table 2). This is especially true since almost nothing is known about the regulation of growth factors, their receptors, cellular oncogenes and other regulatory molecules by environmental estrogens, and these genes are thought to be largely responsible for the growth-promoting effects of endogenous estrogens.

1. Phytoestrogens

Phytoestrogens induce expression of a number of estrogen-responsive genes and reporter constructs in vivo (MAKELA et al. 1995b; SANTELL et al. 1997) and

Table 2. Genes regulated by environmental estrogens

Compound	Gene / system	Reference
Phytoestrogens:		
Genistein	ERE-Reporter / Yeast	Collins et al. 1997
	" / HeLa Cells	Makela et al. 1994
	" / "	Miksicek 1993, 1994
	"/ LeC-9 cells	Mayr 1992
	c-*fos* / Rat uterus	Santell et al. 1997[b]
	pS2 / MCF-7 Breast cancer	Wang et al. 1996[a]
	"	Zava et al. 1997[a]
Coumestrol	ERE-Reporter / Yeast	Gaido et al. 1997
	" / HeLa cells	Miksicek 1993, 1994
	" / "	Makela et al. 1994
	" / LeC-9 cell	Mayr 1992
	pS2 / HeLa cells	Makela et al. 1994
	c-*fos* / Mouse prostate	Makela et al. 1995
Daidzein, Equol, Kaempferol, Enterolactone	pS2 / MCF-7 Breast cancer	Sathyamoorthy et al. 1994
	" / "	Zava et al. 1997
Enterodiol, Enterolactone	Progesterone receptor in MCF-7 Breast cancer cells and primary Rat uterine cells	Welshons et al. 1987
Equol	ERE-reporter / Yeast	Coldham et al. 1997
Mycoestrogens:		
Zearalenone and Zearalanol	ERE-reporter / Yeast	Collins et al. 1997
		Coldham et al. 1997
	ERE-reporter / MCF-7	Mayr 1988
	ERE-reporter / Le42	
	ERE-reporter / HeLa	Makela et al. 1994
	ERE-reporter / Yeast	Miksicek 1993, 1994
	Creatine kinase / Rat uterus	Katzenellenbogen et al. 1979[b]
	pS2 / MCF-7 breast cancer cells	Makela et al. 1994
Xenoestrogens:		
o,p'-DDT	ERE-reporter / Yeast	Chen et al. 1997
		Gaido et al. 1997
		Arnold et al. 1996
	ERE-reporter / GH4 Pituitary cells	Edmunds et al. 1997

Table 2. (*Continued*)

Compound	Gene / system	Reference
	pS2 / MCF-7 breast cancer cells	STEINMETZ et al. 1996[a]
	Progesterone R / MCF-7 cells	CHEN et al. 1997
	Estrogen R / MCF-7 cells	CHEN et al. 1997
	Creatine kinase / Rat uterus	ROBISON et al. 1984[b]
Methoxychlor	ERE-Reporter / Yeast	GAIDO et al. 1997
		ODUM et al. 1997[b]
	Creatine kinase / Rat uterus	CUMMINGS and METCALF 1995
Kepone	Progesterone R / Rat uterus	HAMMOND et al. 1979[b]
	Ovalbumin and conalbumin/Oviduct	PALMITER and MULVIHILL 1978
	Vitellogenin and estrogen R /Cultured trout hepatocytes	FLOURIOT et al. 1995
PCB (2′,4′,6′-Tri-chlorobiphenyl)	ERE-reporter / GH4 Pituitary Cells	EDMUNDS et al. 1997
PCB (3,4,3′,4′-Tetrachlorobiphenyl)	ERE-reporter / T47-D breast cancer	NESARETNAM et al. 1996
	pS2/ MCF-7 and ZR-75 breast cancer	
Aroclor-1245 (PCB Mixture)	Estrogen R and Vitellogenin / Cultured trout hepatocytes	FLOURINOT et al. 1995
Various PCBs	ERE-reporter / MCF-7 breast cancer	CONNOR et al. 1997[a,b]
Octylphenol	ERE-reporter / MCF-7 breast cancer	WHITE et al. 1994
	ERE-reporter / Chick Embryo Fibroblast	
	Vitellogenin/Cultured Trout Hepatocytes	
Nonylphenol	ERE-reporter / Yeast	GAIDO et al. 1997
	ERE-reporter / MCF-7 breast cancer	WHITE et al. 1997
	Progesterone R / MCF-7 cells	SOTO et al. 1991
	pS2, Estrogen R, and MUC-1 / MCF-7 breast cancer Cells	REN et al. 1997
	Estrogen R and vitellogenin / Cultured trout hepatocytes	WHITE et al. 1994, FLOURINOT et al. 1995
	Estrogen R and vitellogenin / Trout liver (in vivo)	REN et al. 1996, LECH et al. 1996
Nonyl- and Octyl-Phenolethoxylates	ERE-reporters/ MCF-7 and chick embryo fibroblasts	WHITE et al. 1994
Bisphenol-A	Progesterone R / MCF-7 cells	KRISHNAN et al. 1993
	ERE-reporter / Yeast	GAIDO et al. 1997
	ERE-reporter / Rat pituitary cells	STEINMETZ et al. 1997

PCB, Polychlorinated biphenyls; ERE, estrogen response element
[a] Study also measured effect of the indicated compound on MCF-7 cell proliferation or uterotrophic action[b]

in various cultured cell lines (see Table 2). In general, the dose ranges of compounds that produce these effects are those expected based on receptor affinities (Table 1); genistein and coumestrol are generally more potent than the other compounds in activating the genes shown in Table 2 and, in the few cases where lignans have been studied, they have the lowest potencies of the phytoestrogens. Several studies have shown that the effects of genistein and coumestrol in particular are blocked by antiestrogens, indicating that at least some of the reported actions are ER mediated.

2. Mycoestrogens

As illustrated in Table 2, the mycoestrogens induce a variety of estrogen-responsive genes, both in vivo and in cultured cells, and in most cases the magnitude of the responses is similar to those produced by endogenous hormones, such as estradiol. In keeping with the relatively high affinity of zeralanone and its derivatives for the ER relative to other environmental estrogens, these compounds are fairly potent and typically have potencies of the order of 1–10% those of estradiol or DES. Given the high affinity and potency of this class of compounds, it is worth noting that their levels in human diets are expected to be much lower than those of the isoflavonoids or lignans.

3. Xenoestrogens

Organochlorines have been shown to induce expression of a number of estrogen-responsive genes, both in vivo and in vitro. *o,p′*-DDT increases expression of pS2 (STEINMETZ et al. 1996) and the progesterone receptor (CHEN et al. 1997), decreases expression of the ER (CHEN et al. 1997) in human MCF-7 breast cancer cells, and increases expression of reporter constructs containing an ERE in yeast (ARNOLD et al. 1996; CHEN et al. 1997; GAIDO et al. 1997). In vivo, *o,p′*-DDT (ROBISON et al. 1984; STANCEL et al. 1980) and methoxychlor (CUMMINGS and METCALF 1995) induce the expression of creatine kinase B (referred to in the older literature as the estrogen "induced protein" or IP) in the rat uterus; Kepone® induces expression of the progesterone receptor in the rat uterus (HAMMOND et al. 1979) and ovalbumin and conalbumin in the chicken oviduct (EROSCHENKO and PALMITER 1980; PALMITER and MULVIHILL 1978). Kepone® also induces expression of both the ER and vitellogenin genes in cultured trout liver cells (FLOURIOT et al. 1995), and methoxychlor induces expression of ERE-reporter constructs in yeast (GAIDO et al. 1997; ODUM et al. 1997) and mammalian cells (FLOURIOT et al. 1995).

In general, the organochlorines produce responses that are similar in magnitude to those produced by estradiol and DES. However, in all cases, the potency of the organochlorines is very weak relative to estradiol, ranging from 1/1000 to 1/10,000th that of the hormone in most studies, although it has been reported to be as little as 1/8,000,000th that of estradiol in a yeast expression system (GAIDO et al. 1997). These potency values are, thus, in line with the relative binding affinities of these compounds to the ER (see Table 1). However, it should be kept in mind that these compounds may be relatively more potent in long-term studies because of their bioaccumulation (e.g., see HAMMOND et al. 1979) or because of pharmacokinetic properties, e.g., less serum binding than estradiol (ARNOLD et al. 1996), and the overall effects of many weak agonists may be additive (see SOTO et al. 1995).

PCBs are generally considered to have some of the lowest potencies and ER affinities reported for environmental estrogens. However, some specific congeners have recently been reported to be relatively more potent than most other PCBs tested to date. For example, 2′,4′,6′-trichlorobiphenyl (EDMUNDS et al. 1997) and 3,4,3′,4′-tetrachlorobiphenyl (NESARETNAM et al. 1996) have been reported to induce expression of ERE-reporter constructs in breast cancer and pituitary tumor cells at concentrations between 1 nM and 100 nM. A PCB mixture (Aroclor 1245) has also been reported to induce the expression of mRNAs for the ER and vitellogenin in cultured trout liver cells (FLOURIOT et al. 1995).

Nonylphenol, octylphenol, several of their ethoxylate derivatives and bisphenol-A have all been shown to induce the expression of estrogen-responsive genes, as indicated in Table 2. These have included studies in cultured cells from rat, human, fish and yeast cell systems for both endogenous genes (vitellogenin, ER, pS2, MUC-1 and progesterone receptor) and reporter constructs in yeast and mammalian cells. Other studies have monitored the in vivo expression of the ER and vitellogenin genes in trout liver (LECH et al. 1996; REN et al. 1996). Several of these studies have established that gene expression is blocked by antiestrogens, indicating involvement of the ER. In general, the potencies of these environmental estrogens are in the range of 1/1000th to 1/10,000th that of estradiol that is expected based on their receptor affinities (Table 1). However, as noted by NAGEL et al. (1997), the relative potencies in vivo may be either increased or decreased, depending on the degree to which specific compounds bind to serum proteins, and potencies in vivo can also be affected by other pharmacokinetic properties.

III. Cell Growth and Proliferation

Exogenous estrogens increase proliferation in many normal target tissues, tumors and tumor cell lines, and estrogens have been shown to cause cancer in humans and experimental animals, as discussed in previous chapters of this volume. Therefore, investigations on the effects of environmental estrogens on cell growth and proliferation have been of special interest as one possible mechanism of their toxicity. The two experimental systems which have been most widely used to study the proliferative effects of environmental estrogens have been the rodent uterotrophic assay and the growth of cultured human breast cancer cells, especially the MCF-7 cell line. The number of studies in these two experimental systems is so large that each could be the subject of a separate review itself. Therefore, we have only summarized the major findings and concepts here.

1. The In Vivo Uterotrophic Response to Environmental Estrogens

By 1980, a large number of studies on the uterotrophic actions of environmental estrogens in the rat or mouse uterus had already been performed. In some cases, a uterotrophic bioassay was used in the initial discovery of the estrogenic activity of the compounds as early as the 1950s and 1060s, e.g., coumestrol (BICKOFF et al. 1957), DDT (WELCH et al. 1969), methoxychlor (TULLNER 1961) and zearalenone (CHRISTENSEN et al. 1965; STOB et al. 1962). In other cases, uterotrophic actions in rodents were discovered shortly after the estrogenic activity of the compound was observed in other systems, e.g., genistein (NOTEBOOM and GORSKI 1963), Kepone® (GELLERT 1978a; HAMMOND et al. 1979) and equol (TANG and ADAMS 1980). Additional details of the large number of early studies on the uterotrophic activities of environmental estrogens may be found in early reviews (BULGER and KUPFER 1985; KUPFER and BULGER 1976a; NELSON et al. 1978; Robison et al. 1985a), articles in several monographs (MCLACHLAN 1980; MCLACHLAN 1985; MCLACHLAN and KORACH 1995) or more recent reviews (ADLERCREUTZ and MAZUR 1997; KURZER and XU 1997). In addition, some of the more recent references to primary studies that have monitored the uterotrophic effects of environmental estrogens are noted by a footnote in Table 2.

A number of major points have been consistently demonstrated in these studies.

1. Environmental estrogens effectively stimulate the true proliferation of uterine cells (ROBISON et al. 1985b; TANG and ADAMS 1980), not merely an increase in tissue edema, and the magnitude of stimulation at high doses is generally similar to that seen with estradiol or DES. There are scattered reports that environmental estrogens can "antagonize" the actions of estradiol in this system (FOLMAN and POPE 1966,1969; KITTS et al. 1983; TANG and ADAMS 1980), but it is not clear whether such effects represent true antagonism at the receptor site or stem from pharmacokinetic properties of the environmental estrogen being tested. Similarly, some of the halogenated aromatic hydrocarbons can antagonize the uterotrophic effect of estradiol, but this again is NOT thought to be a direct antagonism at the ER site (SAFE 1995).
2. Environmental estrogens bind to the uterine ER in cell-free systems, bind to the ER in the nuclei of uterine cells in vivo, and uterine proliferation correlates with occupancy of nuclear receptors in vivo (ROBISON et al. 1985b).
3. The overwhelming majority of these studies have monitored the acute effects (e.g., within 6–72 h) of one to three injections of the test compounds. Under these conditions, their potency generally correlates with their receptor binding affinity and they are, thus, far less potent than estradiol or DES. Obvious exceptions are compounds like methoxychlor that are relatively inactive prior to activation by biotransformation.

On the surface, environmental estrogens thus appear to be weak estrogens in this well-studied experimental test system, but several important considerations should be kept in mind. First, the low clearance and bioaccumulation of some compounds can have important effects. For example, the kinetics of uterine cell proliferation are different following treatment with DDT or estradiol (Robison et al. 1985b) and we do not know whether these differences have long-term consequences. Similarly, doses of Kepone® too low to produce acute effects may cause pronounced effects when administered for longer times (see Hammond et al. 1979). Second, environmental estrogens can have a variety of effects in vivo which might be opposing (e.g., direct ER binding might stimulate uterine growth, but induction of P450s might increase estrogen clearance), and this complicates interpretation of whole-animal studies with endpoints such as uterine growth. Finally, there may be species differences in the pharmacokinetic profiles of various environmental estrogens, and this may complicate extrapolations from uterotrophic effects in rats to effects in humans and domestic animals. In fact, there may even be significant differences among the effects observed in different strains of the same species, e.g., the differences in response of different rat strains to bisphenol-A (Steinmetz et al. 1997). These points emphasize that pharmacokinetic studies of plasma and body levels of environmental estrogens, coupled with measurements of receptor occupancy in vivo as a function of time and dose, are sorely lacking in most cases.

2. Proliferation of Estrogen-Sensitive Cancer Cell Lines

The above studies clearly established that many environmental estrogens can produce uterotrophic effects, which are a hallmark of estrogen action. However, chemicals may produce estrogenic responses by indirect mechanisms in vivo, and the results of in vivo studies are affected by the pharmacokinetics of the agent being investigated. Therefore, many groups have also studied the effects of environmental estrogens on the proliferation of estrogen sensitive cells in culture. MCF-7 human breast cancer cells have been used more than any other cell line to study the proliferative effects of environmental estrogens. Due to space limitations, we cannot review all these studies individually, but interested readers may find additional information in recent references (Adlercreutz and Mazur 1997; Kurzer and Xu 1997; Soto et al. 1995) or in those studies cited in Table 2.

To our knowledge Martin et al. (1978) were the first to perform a comprehensive study of environmental estrogens in MCF-7 cells. They measured the binding affinities of coumestrol, genistein, formononetin, zearalenone and zearalenol to the cytoplasmic ER from MCF-7 cells; the ability of these compounds to occupy the nuclear ER in intact cells; ER processing after exposure of the cells to these compounds; and the effect of the environmental estrogens on MCF-7 proliferation. Their results indicated that these compounds were all agonists and their effects correlated with their ability to interact with the ER.

Numerous other studies have used this system to study the effects of other environmental estrogens on MCF-7 cell proliferation and have, generally, come to the same conclusions. Special mention should be made of two particular studies: WELSHONS and COLLEAGUES (1990) used MCF-7 cells to devise an in vitro bioassay for the detection of estrogenic components in animal feeds. SOTO et al. (1995) developed the so-called "E-screen" assay to detect chemicals with estrogenic activity, and have screened a very large number of compounds using this assay over the last several years. Recent studies from her group thus contain information about a greater number of compounds than most other studies (SOTO et al. 1995).

While the great majority of studies have found that environmental estrogens stimulate proliferation of MCF-7 cells, a few studies have found additional effects. It has been reported that low concentrations of certain compounds can stimulate proliferation of MCF-7 cells, but that higher concentrations inhibit cell proliferation and are, thus, "antiestrogenic" (DEES et al. 1997; WANG and KURZER 1997; WANG et al. 1996; ZAVA and DUWE 1997). Several points should be made about these studies. First, the high concentrations that were used in these cultured-cell experiments are very unlikely to ever be reached in vivo and it, thus, seems unlikely that such effects would occur in most human or animal populations. Second, estradiol itself shows exactly the same type of "biphasic" effects of first stimulating and then inhibiting MCF-7 cell growth (SONNENSCHEIN et al. 1994). Thus, even the inhibitory effect of phytoestrogens seen at high concentrations can be viewed as an "estrogenic" rather than an antiestrogenic action. Finally, a number of studies have specifically reported that genistein inhibits the growth of human breast cancer cells (BARNES 1995; MONTI and SINHA 1994; PAGLIACCI et al. 1994). However, in these cases, it is clear that the phytoestrogen is not acting through the ER, since inhibition is observed in both ER-positive and ER-negative cells. It is, thus, probable that these specific actions are due to the inhibition of tyrosine kinase activity by genistein which has been extensively studied since its initial discovery by AKIYAMA et al. (1987). This is a rational explanation for genistein's effects, since many peptide growth factors known to stimulate proliferation of cultured cells act via stimulation of receptors with tyrosine kinase activity.

While many studies have investigated the proliferative effects of environmental estrogens on MCF-7 breast-cancer cell lines, there is surprisingly little data about other cell lines, and even less data about in vivo effects in animal models of estrogen-sensitive tumors. ROBISON et al. (1985c) demonstrated that *o,p'*-DDT supports the growth of an estrogen-sensitive rat mammary tumor, and SCRIBNER and MOTTET (1981) showed that DDT accelerates the induction of mammary tumors in male rats by 2-acetamidophenanthrene. Some, but not all, rat-feeding studies have demonstrated a protective effect of soy-rich diets on mammary tumors (MESSINA et al. 1994), but most such studies have not specifically addressed the role of purified phytoestrogens per se on cancer prevention or addressed the molecular mechanisms involved. Given the epidemiological interest in these compounds as potential chemopreventive agents against cancer (see Sect. E below), there is a great need for this type of data in controlled animal studies.

3. Angiogenesis

A sufficient nutrient supply is a basic requirement for cell growth and proliferation. The formation of blood vessels (angiogenesis) and regulation of their permeability are, thus, important for the proliferation of both normal and tumor cells. In fact, certain tumors may not grow beyond a certain size unless the density of micro-vessels in their vicinity increases (FOLKMAN 1994), and the density of blood vessels near a tumor also influences its ability to metastasize (CLAFFEY and ROBINSON 1996). To explore the possibility that dietary compounds might influence angiogenesis, FOTSIS et al. (1993, 1995) examined urinary extracts from human subjects consuming a diet rich in plant products. These studies established that the phytoestrogen genistein inhibits both the proliferation of vascular endothelial cells and angiogenesis in vitro. Subsequent studies from the same group led to the discovery of several flavonoids, including 3-hydroxyflavone, 3′,4′-dihydroyflavone, 2′,3′-dihydroxyflavone, fisetin, apigenin and luteolin as inhibitors of in vitro angiogenesis (FOTSIS et al. 1997).

The mechanism of this effect is unknown at present, but we mention it here because of several recent studies linking estrogens and antiestrogens to angiogenesis. First, it has recently been reported that the endogenous estrogen metabolite 2-methoxyoestradiol inhibits angiogenesis and tumor growth in vivo (FOTSIS et al. 1994). Other studies have shown that tamoxifen and other triphenylethylene antiestrogens can inhibit angiogenesis in vitro and block the growth of cultured endothelial cells (GAGLIARDI and COLLINS 1993; GAGLIARDI et al. 1996). In addition, estrogens and antiestrogens can induce the expression of vascular endothelial growth factor (CULLINAN-BOVE and KOOS 1993; HYDER et al. 1997; HYDER et al. 1996; NAKAMURA et al. 1996), which is the most potent angiogenic factor known. Whether the effects of genistein and other flavonoids on endothelial cell proliferation and angiogenesis are due to estrogenic or antiestrogenic activities, or other potential mechanisms (FOTSIS et al. 1995) remains to be established. However, given the importance of angiogenesis in many physiological and pathological processes, this appears to be an important area for further study.

IV. Development and Differentiation

Early developmental exposures to potent exogenous estrogens, such as DES, are known to have profound permanent adverse effects on the reproductive system [both on central nervous system (CNS) and reproductive organs] in many species, including humans (HERBST and BERN 1981). It has been suggested frequently that the less potent environmental estrogens may also affect the developing organism and induce permanent structural and/or functional changes. However, to date, the developmental effects of relatively few of the hundreds of estrogenic compounds present in our environment have been studied. At least some of these compounds exert effects similar, but not identical to those of 17β-estradiol or DES, but the doses of environmental estrogens required to produce effects in experimental animals are generally higher.

1. Effects on Female Sexual Differentiation

Levy et al. (1995) and Faber and Hughes (1993) studied the effects of perinatal genistein on the sexual differentiation of rats. Given on days 16–20 of gestation, genistein reduces anogenital distance, a typical estrogen effect. However, genistein, unlike DES or estradiol, did not increase the volume of the sexually dimorphic nucleus in the hypothalamus (SDN-POA), and delayed the onset of puberty (Levy et al. 1995). Neonatal exposures to high doses of genistein, 500 μg or 1000 μg per day on days 1–10 after birth, mimic the effect of DES and estradiol to increase SDN-POA volume and decrease LH secretion in response to GnRH in adulthood. Interestingly, a lower dose of genistein (10–100 μg per day) had an opposite, nonandrogenizing, pituitary-sensitizing effect, as indicated by an unaltered SDN-POA volume and increased pituitary responsiveness to GnRH (Faber and Hughes 1993). These results suggest that the developmental actions of genistein may not be identical to those of estrogens, such as estradiol or DES, and that the timing of genistein exposure is critical for the effects on the CNS.

2. Effects on Uterus, Vagina and Ovary

Several environmental compounds act as estrogens in the immature rat uterus, although the pattern of effects varies and is not necessarily identical to that of estradiol or DES. Given neonatally (on days 1–5 after birth) mycoestrogens, such as zearalenone and zearalanol, increase nuclear ER levels, induce premature growth and increase ornithine decarboxylase activity (Sheehan et al. 1984). Coumestrol acts very similarly to DES: given neonatally (on days 1–5 or days 1–10 after birth), it induces premature uterine growth and gland development; later, it reduces uterine weight and ER content (Medlock et al. 1995a,b). Equol given neonatally acts neither as a typical estrogen nor antiestrogen. It only increases uterine dry weight prematurely, but lowers the uterine weight at later ages, and does not induce luminal epithelial hypertrophy or affect the genesis of glands or ER content of the uterus (Medlock et al. 1995a,b). When given prepubertally, on days 10–14, which is the critical time for uterine gland development, both phytoestrogens act like DES to inhibit gland genesis (Medlock et al. 1995a, b).

Neonatal exposure to coumestrol by injection or via the dam's milk has permanent adverse effects similar to DES in the rodent ovary and vagina. Coumestrol-exposed mice and rats display premature vaginal opening, persistent estrus and an acyclic condition in young adults similar to premature anovulatory syndrome, i.e., persistent ovary-independent vaginal cornifcation, cervicovaginal pegs, downgrowths, cysts and adenosis, polyovular and hemorrhagic ovarian follicles (Burroughs et al. 1985, 1990; Whitten et al. 1993, 1995a, b; Whitten and Naftolin 1992).

The developmental toxicity of TCDD in the female reproductive system has been thoroughly reviewed by Brouwer et al. (1995). Perinatal exposure to TCDD induces urogenital malformations and altered estrus cyclicity (in the hamster, but not the rat), but does not affect sexual behavior or fertility.

3. Effects on Mammary-Gland Differentiation and Carcinogenesis

Lamartiniere et al. (1995a, b) and Murrill et al. (1996) have examined the effect of early genistein exposure on the differentiation of rat mammary gland and development of mammary cancer. Neonatal and prepubertal exposures to high doses of genistein (5 mg per day on day 2, day 4 and day 6, or 500 μg per g of body weight on day 16, day 18 and day 20 after birth) enhance mammary-gland differentiation (an estrogenic effect) and suppress the development of mammary cancers induced by dimethylbenz[a]anthracene. The protective effect of early genistein exposure against mammary carcinogenesis can be explained by the precocious gland maturation, which leads to a decreased number of undifferentiated glands which are the structures susceptible to the carcinogen administered at a later time. In addition, prepubertal genistein induces a slight change in the estrus cycle (an increase in time spent in the estrus phase), but had no other significant toxicity in the reproductive system (Murrill et al. 1996). Brown and Lamartiniere (1995) studied the effect of prepubertal exposure to genistein, *o,p'*-DDT, Aroclor 1221, Aroclor 1254, and TCDD on rat mammary-gland differentiation and cell proliferation. Only genistein and *o,p'*-DDT had typical estrogenic effects, i.e., enhanced gland differentiation and epithelial cell proliferation. TCDD had an opposite effect; reduction in cell proliferation and delayed gland differentiation, which is in keeping with its generally antiestrogenic activities (Safe 1995).

4. Effects in Males

Most of the studies in the developing male have investigated potential adverse effects in the male reproductive system. The developmental toxicity of polyhalogenated aromatic hydrocarbons in the male reproductive system has been thoroughly reviewed by Peterson et al. (1993) and Brouwer et al. (1995). One of the most studied compounds is TCDD. Perinatal exposure to very low doses of TCDD induces a wide spectrum of reproductive-tract abnormalities, including reduced anogenital distance; delayed testicular descent and increased plasma testosterone concentration after birth; decreased ventral prostate, seminal vesicle, epididymis, caudate epididymis, and testis weights; reduced responsiveness of ventral prostate to androgens; decreased sperm production; and demasculinization/feminization of sexual behavior. It is not clear, at present, if these actions result from the estrogenic, antiestrogenic or other effects of TCDD.

Xenoestrogens tested for male reproductive tract toxicity include alkyl phenols, estrogenic pesticides and phytoestrogens. In rats, maternal exposure (day 11.5 and day 15.5 post-coitum) to 4-octylphenol and DES, reduces the expression of cytochrome P450 17α-hydroxylase/C17–20 lyase (an enzyme critical for androgen biosynthesis) in fetal Leydig cells (MAJDIC et al. 1996). When added to the drinking water of female rats from 2 weeks prior to mating until 22 days after delivery, 4-octylphenol, octylphenol polyethoxylate (which is metabolized to octylphenol) and butyl benzyl phthalate reduced testicular weight and daily sperm production, similarly to DES, although the testicular morphology remained unaltered (SHARPE et al. 1995). In the same study, 4-octylphenol reduced the ventral prostate weight in a manner similar to DES. Prenatal exposure to bisphenol A (orally on days 11–17 of gestation), but not octyl phenol, increased the prostatic weight in adult mice. This is an effect similar to that of a low dose of oral DES (0.02–2.0 μg per day per kg of maternal body weight) or subcutaneous estradiol that results in a 50% increase in free estradiol in fetal serum (NAGEL et al. 1997). In male mice, prenatal exposure to the estrogenic pesticides *o,p′*-DDT and methoxychlor on days 11–17 of gestation induces a DES-like biphasic effect on the territorial behavior measured by the rate of urine marking in novel territory (VOM SAAL et al. 1995).

Phytoestrogens may also have adverse effects in the developing male reproductive system. Maternal exposure to genistein, DES or estradiol benzoate on days 16–20 of gestation reduces the anogenital distance in male offspring (LEVY et al. 1995). Lactational exposure of male rats to coumestrol results in altered sexual behavior; reductions in mount and ejaculation frequency and increased latency to mount and ejaculate, but induced no changes in testicular weights or plasma testosterone levels (WHITTEN et al. 1995a). Special mention should also be made of a recent study showing that very low doses of DES may produce opposite effects on the developing male reproductive tract than higher doses (VOM SAAL et al. 1997), yielding an inverted-U-shaped dose–response relationship. It is unknown, at present, whether environmental estrogens will display a similar dose–response profile; if they do, developmental defects may be produced in this system by exposure levels lower than previously recognized.

E. Human Exposures, Epidemiology and Potential Health Effects

The incidence of hormone-responsive cancers, such as those of the breast and prostate, is increasing and epidemiological studies generally indicate that environmental factors play a role in the etiology of this disease. Given that breast and endometrial cancer are estrogen responsive, there is great interest in the possible relationships between environmental estrogens and these diseases. In addition, it is clear that cardiovascular disease and osteoporosis in women are related to estrogens, and the prevalence of these diseases also varies in different populations, suggesting an environmental effect on their incidence. There have been a large number of studies of human exposures to some of the organochlorines, polyaromatic hydrocarbons and phytoestrogens covered in this chapter, but it is not practical to review them all here. Rather, we have focused on recent, selective studies that have explicitly investigated the potential association of levels of these chemicals with the most prevalent hormone-responsive cancers.

As a prelude to individual studies, it is interesting to consider the levels of environmental estrogens in human populations. A recent study measured levels of DDE, the major metabolite of the pesticide DDT, of approximately 20 nM in the plasma of US women from several geographical sites, and a roughly comparable level of PCBs (HUNTER et al. 1997) and levels of these lipophilic molecules are known to be much higher in human milk. While levels of organochlorines are expected to be somewhat variable, these numbers provide a crude frame of reference. Given the very long half-life for clearance of these compounds, their levels in humans are expected to change very slowly.

In contrast, levels of phytoestrogens are markedly dependent on specific diets and a variety of other factors. Adults on a typical Western diet devoid of soy generally have plasma genistein and daidzein levels below 15 nM. In Japanese adults eating a traditional diet or Westerners fed a soy-supplemented diet, plasma concentrations of 100–900 nM genistein and 50–500 nM daidzein are reached (ADLERCREUTZ et al. 1991, 1992, 1993; CASSIDY 1996; CASSIDY et al. 1994; FRANKE et al. 1995; GOODERHAM et al. 1997; MORTON et al. 1997) . In omnivorous adults on typical diets, plasma concentrations of enterodiol and enterolactone are <40 nM and <150 nM, respectively, but may approach 0.6 μM or 0.8 μM after flaxseed supplementation (ADLERCREUTZ et al. 1982, 1991; BRZEZINSKI et al. 1997; CASSIDY 1996; GOODERHAM et al. 1997; HORN-ROSS et al. 1997; LAMPE et al. 1994; MORTON et al. 1994; SCHULTZ et al. 1991).

Special mention should also be made of the levels of phytoestrogens of infants, since these compounds are found in cow and human milk, and since many infant formulas are soy based. In 4-month-old infants in the US fed human breast milk or cow-milk formula, plasma concentrations of genistein and daidzein are comparable to those of adults consuming a typical Western diet (i.e., <15 nM). However, infants fed typical commercially purchased soy-based formulas had plasma concentrations of 2.5 μM (genistein) and 1.2 μM (daidzein) of these phytoestrogens (SETCHELL et al. 1997). By way of comparison, these levels are approximately 10,000–20,000 times the circulating levels of endogenous estradiol in early life.

I. Organochlorines and the Incidence of Breast Cancer

Since DDT analogs and PCBs are the two most prevalent and persistent "estrogens in the environment", these compounds have been a major focus of study. Two small case-control studies reported higher levels of DDE (the major persistent metabolite of DDT) in breast cancer patients than in normal women (Falck et al. 1992; Wasserman et al. 1976), while one found an association with PCBs (Falck et al. 1992). Another study discovered an association of DDE levels only with ER-positive tumors (Dewailly et al. 1994). However, a European study actually found lower levels of DDE in adipose tissue of cancer patients than in normals (Van't Veer et al. 1997). Another small study found an association of β-hexachlorocyclohexane (a component of the pesticide lindane) with breast cancer patients in Finland (Mussalo-Rauhamaa et al. 1990). These were relatively small studies, however, and a number of appropropriate concerns have been expressed regarding the interpretation of these studies (for example, see Safe 1995).

More recently, there have been three prospective studies with larger numbers of women. A study with 58 cases in New York found a significantly higher level of DDE in the blood of breast cancer cases than controls, but did not find significantly higher levels of PCBs (Wolff et al. 1993). However, a larger study involving 150 cases in San Francisco did not find a significant increase in blood levels of these chemicals in cases when compared with controls (Krieger et al. 1994). Finally, the largest study to date, which involved 240 cases, did not find an increase in DDE or PCBs in women with the disease when compared with controls (Hunter et al. 1997). Furthermore, there is no evidence of increased breast or uterine cancer in women exposed to PCBs occupationally or from industrial accidents, although the incidence of other cancers may be increased in these populations (Bertazzi et al. 1993; Kogevinas et al. 1993, 1997; Pesatori et al. 1993). In fact, one study even suggested that exposure to TCDD was associated with a decrease in the incidence of breast cancer in the Seveso population (Bertazzi et al. 1993). While the mechanism behind this observation is unknown it may be in keeping with the antiestrogenic activity observed experimentally with this compound (Safe 1995). While one cannot exclude the possibility that other compounds might play a role in this disease or that its incidence would only be increased in certain subsets of people, the available human data does not support the concern that DDT or PCBs contribute significantly to the incidence of breast cancer in the general population.

II. Phytoestrogens and Prevention of Breast Cancer

It has been known for many years that environmental influences contribute significantly to the incidence of cancer, and many of the most common human cancers (e.g., breast cancer in women and prostate cancer in men) are influenced by endocrine factors. In 1980, two new compounds, the lignans enterodiol and enterolactone, were isolated from human urine (Setchell et al. 1980; Stitch et al. 1980). These compounds shared some structural similarities with steroid hormones; they were derived from diets (e.g., diets high in fiber content), generally associated with a decreased incidence of several human cancers, and other lignans (e.g., Podophyllotoxin) were known to possess antimitotic activity (Barclay and Perdue 1976; Hartwell 1976). This suggested to investigators in this area that consumption of these compounds might be inversely related to the incidence of certain cancers in humans (see Setchell and Adlercreutz 1988). Since the early 1980s a number of studies have investigated the relationship between phytoestrogens and cancer. There have been three major types of studies performed in this area: (1) epidemiological studies in humans, (2) long-term feeding studies in animals, and (3) basic investigations into potential anti-cancer mechanisms of phytoestrogens.

There have been two major types of epidemiological studies performed with phytoestrogens. The first type was observational studies where the intake of soy or high-fiber diets, rich in phytoestrogens, or the urinary excretion of phytoestrogens has been measured in women belonging to groups known to have different incidences of breast cancer (e.g., vegetarians vs omnivores). These studies have generally shown that women from groups with the lowest expected incidence of breast cancer have the highest intake of phytoestrogen-rich foods and the highest urinary excretion of phytoestrogens and vice versa (reviewed in Adlercreutz 1995; Adlercreutz and Mazur 1997; Messina et al. 1994; Setchell and Adlercreutz 1988). However, there were exceptions (e.g., Japanese women who emigrate to Hawaii (Adlercreutz et al. 1995).

The second type of epidemiological investigation has been case-control studies in which soy intake or urinary excretion of phytoestrogens is actually measured in patients with cancer and compared with that in matched controls. In case-control studies, the trend has generally been that a soy-rich diet is associated with a decreased incidence of breast cancer in premenopausal women (Adlercreutz 1995; Adlercreutz and Mazur 1997; Adlercreutz et al. 1995; Messina et al. 1994), but again there are exceptions (Yuan et al. 1995).

One of the strongest associations has been a recent case-control study demonstrating an association between high urinary excretion of equol and enterolactone over a 72-h period and a decreased incidence of breast cancer (Ingram et al. 1997). A similar result was obtained in an earlier study that examined fewer cases, but monitored urinary profiles for a much longer time (Adlercreutz et al. 1982). None of these studies has indicated that soy-rich diets are harmful. Similarly, we are not aware of any published studies of the effect of phytoestrogen-rich diets in women with known breast cancer, although a recent meeting report found that phytoestrogens stimulate proliferation in normal breast epithelial cells (McMichael-Phillips et al. 1996). On balance, the epidemiological studies conducted to date have not proven a protective effect of phytoestrogens, but are consistent with this possibility which seems to deserve further study.

A number of workers have also performed animal feeding studies (see Messina et al. 1994). In most cases, animals were fed diets rich in soy products or control diets, and were also treated with a cancer-inducing agent, e.g., dimethylbenzanthracene (DMBA), *N*-methyl-*N*-nitrosourea (NMU), or radiation. A variety of parameters, such as tumor incidence, tumor size, tumor latency period, etc, were then measured. The majority of such studies have found some beneficial effect of soy-rich diets on mammary cancer, but a sizeable fraction have not (MESSINA et al. 1994). Interpretation of these feeding studies and human epidemiological studies is complicated, however, because purified phytoestrogen preparations were not used. Rather, diets rich in soy or other foods were used, and these contain other ingredients besides phytoestrogens which have known anticarcinogenic activity, e.g., fiber, protease inhibitors, phytic acid and the sterol β-sitosterol (KENNEDY 1995). Thus, it is unknown whether it is specifically the phytoestrogens, or other components, that are responsible for the observed effects.

Experimental studies of phytoestrogens have not yielded a clear mechanistic basis for their potential cancer-protective effects. In a few cases, isoflavones (FOLMAN 1966; TANG and ADAMS 1980) and coumestrol (FOLMAN and POPE 1966) appear to decrease in vivo responses to estradiol, but this could be due more to pharmacokinetic profiles than true antiestrogenic effects at the receptor site. In other studies, genistein has been shown to inhibit the proliferation of breast cancer cells in vitro, but this inhibition does not appear to require that the cells have ERs, and inhibition occurs at concentrations unlikely to be achieved in vivo (BARNES 1995; MONTI and SINHA,1994; PAGLIACCI et al. 1994). Thus, despite the frequent implications in the literature that these compounds are antiestrogens, the available evidence overwhelmingly indicates that the flavonoids and coumestans (see Sects. D.II and D.III above), which bind to the ER, are hormone agonists in almost all test systems, including the ability to increase the proliferation of cultured human breast cancer cells.

Several other indirect or nonestrogen related mechanisms have been suggested as possible explanations of the protective effect of phytoestrogens on estrogen-dependent cancers.

1. Some studies have shown that soy diets decrease circulating LH and FSH levels and slightly lengthen the menstrual cycle, although this is still controversial. This is a plausible mechanism, since these changes would be expected to reduce the total exposure to unopposed estrogens in a woman's lifetime, but it is not clear on a cellular level how these changes would occur.
2. Another general type of mechanism proposed is the reduction in estrogen synthesis. This is, again, plausible because phytoestrogens inhibit aromatase and type-I 17β-steroid dehydrogenase present in certain cells, both of which are involved in the synthesis of estradiol.
3. Another potential mechanism is the induction of serum sex-steroid binding globulin, which is known to occur in response to phytoestrogens. In theory, this could cause a transient decrease in unbound plasma estrogens, but one would anticipate that this would be sensed by the hypothalamic/pituitary axis which would then re-establish higher circulating hormone levels.

4. Phytoestrogens are phenolic compounds with antioxidant properties and could act by this mechanism.
5. Genistein (but not other phytoestrogens) is a potent inhibitor of protein tyrosine kinases. Since many growth-factor receptors are ligand-activated tyrosine kinases and/or stimulate cascades involving tyrosine kinases, genistein could also act via this mechanism.

Since this volume is focused on estrogens, we have not reviewed these indirect mechanisms in detail, but felt it was important to note them because readers new to the field may assume that reported effects of phyto"estrogens", including their anticancer properties, are due solely to their "estrogenic" or "antiestrogenic" actions. Interested readers may find specific information about these potential mechanisms in a number of excellent reviews already mentioned (Adlercreutz 1995; Adlercreutz and Mazur 1997; Kurzer and Xu 1997; McLachlan and Korach 1995; Messina et al. 1994; Setchell and Adlercreutz 1988). These reviews also contain references to the effects of phytoestrogens on other types of cancers (e.g., liver and gastrointestinal). We have not explicitly reviewed these diseases since the bases for either environmental or endogenous estrogen effects in these diseases is not clear at present.

As noted previously, Lamartiniere and colleagues (1995a, b) have observed that neonatal or prepubertal (Murrill et al. 1996) exposure of animals to genistein suppresses DMBA-induced mammary adenocarcinomas later in life, and the proposed mechanism is that the phytoestrogen causes differentiation and reduced proliferation in the mammary gland. This is an intriguing idea, given that a similar concept has been proposed for the protective effect of elevated hormone levels early in life (i.e., associated with an early first pregnancy) on the development of breast cancer in women (Henderson et al. 1993).

III. Phytoestrogens and Prostate Cancer

It has also been suggested that phytoestrogens may offer protection against prostate cancer. Epidemiological studies indicate that Japanese men eating a diet rich in tofu have a low incidence of prostate cancer (Severson et al. 1989), and it has also been suggested that phytoestrogens may inhibit the growth of latent prostate cancer (Adlercreutz 1990). In experimental studies, it has also been found that soy has a protective effect on prostatitis in rats (Sharma et al. 1992), that soy protects against prostatic dysplasia in some animal models (Makela et al. 1995a) and that genistein can inhibit the growth of cultured prostate cancer cells (Peterson and Barnes 1993; Rokhlin and Cohen 1995). Potential mechanisms for this effect include estrogenic or antiestrogenic activities of phytoestrogens, an inhibition of growth factor-associated protein tyrosine kinase activity by genistein, or interference with androgen biosynthesis, although the underlying mechanisms or the specific compounds responsible for the epidemiological findings are unknown at present. Additional information about phytoestrogens and prostate cancer may be found in several recent reviews (ADLERCREUTZ 1995; ADLERCREUTZ et al. 1995; ADLERCREUTZ and MAZUR 1997; BARNES 1995; MESSINA 1994).

IV. Cardiovascular Disease and Osteoporosis

Ischemic heart disease and osteoporosis are two widespread diseases, the incidence of which has long been linked to decreasing estrogen levels in postmenopausal women, Thus it is not surprising that the potential role of environmental estrogens on these diseases is receiving increased attention. Various studies have investigated the effects of phytoestrogens on these diseases; these are reviewed briefly here. Readers may obtain additional details in recent reviews (ADLERCREUTZ and MAZUR 1997; CLARKSON et al. 1995; KURZER and XU 1997).

Because estrogen replacement therapy is known to have beneficial effects on plasma lipid profiles, the effects of diets rich in phytoestrogens on cholesterol and triglycerides have been a major focus in this area. It has been recognized for some time that people who consume diets rich in vegetable protein have less coronary heart disease and hypercholesterolemia than those eating diets rich in animal protein (STAMLER 1979) and a number of studies has specifically shown that diets rich in soy protein decrease total plasma cholesterol, low-density lipoprotein (LDL)-cholesterol and triglycerides (CARROLL 1991; CARROLL and KUROWSKA 1995; SIRTORI et al. 1993). While a number of components of soy-rich diets could be responsible for these beneficial effects (POTTER 1995), a recent study of rhesus monkeys indicates that much of this effect is probably due to the specific phytoestrogens in soy diets (ANTHONY et al. 1996). A recent *meta* analysis of controlled clinical trials involving the consumption of soy protein has documented the decrease in cholesterol and triglyceride levels, and also suggested that much of this effect is likely due to the phytoestrogen content of soy (ANDERSON et al. 1995b).

In addition to the studies on plasma lipids, a number of other effects of phytoestrogens and genistein in particular have been considered as possible mechanisms for the beneficial effect of plant-rich diets on coronary heart disease. These include an inhibition of the tyrosine kinase activity of growth factors involved in intimal smooth muscle proliferation, the anti-oxidant properties of phytoestrogens, and inhibition of thrombin formation and numerous thrombin mediated effects, platelet aggregation or atherosclerotic-plaque formation (see ADLERCREUTZ and MAZUR 1997; CLARKSON et al. 1995; KURZER and XU 1997; RAINES and ROSS 1995; WILCOX and BLUMENTHAL 1995; for additional details).

Osteoporosis is another serious disease linked to estrogen deficiency, and epidemiological studies have indicated that Asian women, who generally consume diets high in soy products, have historically had a lower incidence of osteoporosis than women in the US and Europe (COOPER et al. 1992; WHO 1994). A number of studies have shown that soy-based diets (ARJMANDI et al.

1996), coumestrol (Dodge et al. 1996; Tsutsumi 1995), genistein (Anderson et al. 1995a), methoxychlor (Dodge et al. 1996) and zeranol (Dodge et al. 1996) all have favorable effects on the maintenance of bone mass in animals. In addition, the synthetic isoflavone ipriflavone, which undergoes biotransformation to daidzein and other metabolites (Brandi 1992), maintains bone density in both rodents (Yamazaki and Kinoshita 1986) and women (Agnusdei et al. 1989; Gambacciani et al. 1994; Valente, et al. 1994). While the estrogenic activity of phytoestrogens provides a rational basis for these effects, the basis of these effects on bone are unknown and may involve additional mechanisms, e.g., maintenance of calcium levels, since some processed soy products have a high calcium content (Breslau et al. 1988; Haytowitz and Matthews 1976).

V. Fertility and Reproductive Health

There is an increasing concern about the possible impact of environmental estrogens, particularly xenoestrogens, on fertility and reproductive health in humans and wildlife. It is established that potent estrogens, such as estradiol and DES, can have adverse effects on male reproduction (e.g., decreased sperm production), and several previous reports have indicated that occupational exposure to pesticides, such as DDT (Singer 1949) and Kepone (Guzelian 1982), is associated with adverse reproductive outcomes. More recently, a number of epidemiological studies have reported an increase in the incidence of male reproductive-tract disorders, such as hypospadias and testicular cancer, and a decline in the sperm counts in some countries (Auger et al. 1995; Carlsen et al. 1992, 1995; Irvine et al. 1996; Sharpe and Skakkebaek 1993; Toppari et al. 1996). However, others have reported that sperm counts vary dramatically in different geographic locations (Fisch and Goluboff 1996; Fisch et al. 1996b) and did NOT observe a decline in sperm count in US men over a 25-year period in three georgraphical regions (Fisch et al. 1996a). In addition, others have discussed inconsistencies in published studies regarding the decline in sperm production and potential methodological concerns (Daston et al. 1997). It is, thus, not clear at present whether the incidence of male reproductive disorders and low sperm counts are increasing and, if so, whether or not environmental estrogens are primarily responsible for any adverse effects.

In other studies related to human reproductive health, Rogan and colleagues (Gladen and Rogan 1995; Rogan et al. 1987) reported that levels of PCBs and the DDT metabolite DDE in human milk are inversely related to the length of lactation in individual women in both the US and Mexico. Declines in both the initiation and duration of lactation have been reported throughout the world (Notzon 1984), and it is clearly established that potent estrogens, such as estradiol and DES, can suppress lactation. Rogan and colleagues (1987) have, thus, suggested that their findings are due to the estrogenic activity of these organochlorines.

It has long been recognized that xenobiotics can have adverse effects of fertility and reproduction in wildlife, which could be due to a great variety of mechanisms. In recent years, concern has been expressed that environmental estrogens may have serious adverse effects on the reproductive capacity of wildlife. This topic is beyond the scope of this review, but interested readers are referred to several recent publications and the references therein for further information (Colborn and Clement 1992; Guillette and Guillette 1996; McLachlan and Arnold 1996; McLachlan and Korach 1995; Toppari et al. 1996).

F. Conclusion

At present, it is unequivocally established that many chemicals found in the environment, the so-called "environmental estrogens", can bind to the ER and produce hormone like effects in the laboratory. These observations raise a number of very serious concerns, but it is not yet clear whether these agents are present at sufficient levels in the environment to pose health risks to humans or animals. In addition, there is also a solid rational for suggesting that some environmental estrogens, especially the phytoestrogens, may have beneficial health effects. Given this situation, it seems prudent to continue sound scientific approaches to: (1) investigating the basic mechanism of action of environmental estrogens; (2) assessing their levels in humans, animals and the biosphere at large; and (3) appropriate monitoring of their public-health effects. This is the approach most likely to improve the scientific basis for evaluating the potential risks and benefits of known environmental estrogens and to accurately predict the biological activities of new compounds before their introduction into the environment. In addition to answering public-health questions, an improved understanding of how the diverse chemical structures of environmental estrogens elicit biological responses will increase our basic knowledge of estrogen pharmacology.

References

Adlercreutz H (1990) Western diet and Western diseases: some hormonal and biochemical mechanisms and associations. Scand J Clin Lab Invest 50(suppl 201): 3–23

Adlercreutz H (1995) Phytoestrogens: epidemiology and possible role in cancer protection. Environ Health Perspect 103(suppl 7):103–112

Adlercreutz H, Fotsis T, Heikkinen R, Dwyer JT, Woods M, Goldin BR, Gorback SL (1982) Excretion of the lignans enterolactone and enterodiol and of equol in omnivorous and vegetarian post-menopausal women and in women with breast cancer. Lancet 2:1295–1299

Adlercreutz H, Hamalainen E, Gorback S, Goldin B (1992) Dietary phyto-estrogens and the menopause in Japan. Lancet 339:1233

Adlercreutz H, Honjo H, Higashi A, Fotsis T, Hamalainen E, Hasegawa T, Ikada H (1991) Urinary excretion of lignans and isoflavonoid phytoestrogens in Japanese men and women consuming a traditional Japanese diet. Am J Clin Nutrition 54:1093–1100

Adlercreutz H, Markkanen H, Watanabe S (1993) Plasma concentrations of phyto-estrogens in Japanese men. Lancet 342:1209–1210
Adlercreutz H, Mazur W (1997) Phyto-estrogens and western diseases. Ann Med 29:95–120
Adlercreutz HC, Goldin BR, Gorbach SL, Hockerstedt KAV, Watanabe S, Hamalainen EK, Markkanen MH, Makela RR, Wahala KT, Hase TA, Fotsis T (1995) Soybean phytoestrogen intake and cancer risk. J Nutrition 125:757S–770S
Agnusdei D, Zacchei F, Bigazzi S, Cepollaro C, Nardi P, et al. (1989) Metabolic and clinical effects of ipriflavone in established post-menopausal osteoporosis. Drugs Exp Clin Res 15:97–104
Ahel M, Conrad J, Giger W (1987) Persistant organic chemicals in sewage effluents. III. Determination of nonylphenoxy carboxylic acids by high resolution gas chromatography/mass spectrometry and high-performance liquid chromatography. Environ Sci Technol 21:697–703
Akiyama T, Ishida J, Nakagawa S, Ogawara H, Watanabe S, Itoh NM, Shibuya M, Fukami Y (1987) Genistein, a specific inhibitor of tyrosine-specific protein kinases. J Biol Chem 262:5592–5595
Allen E, Doisy EA (1923) An ovarian hormone. Preliminary report on its localization, extraction and partial purification, and action in test animals. J Am Med Assoc 81:819–821
Anderson JJ, Ambrose WW, Garner SC (1995a) Orally dosed genistein from soy and prevention of cancellous bone loss in two ovariectomised rat models. J Nutrition 125:799 S
Anderson JW, Johnstone BM, Cook-Newell ME (1995b) Meta-analysis of the effects of soy protein intake on serum lipids. N Engl J Med 333:276–282
Anstead GM, Carlson KE, Katzenellenbogen JA (1997) The estradiol pharmacophore: ligand structure-estrogen receptor binding affinity relationships and a model for the receptor site. Steroids 62:268–303
Anthony MS, Clarkson TB, Hughes CL, Morgan TM, Burke GL (1996) Soybean isoflavones improve cardiovascular risk factors without affecting the reproductive system of peripubertal rhesus monkeys. J Nutrition 126:43–50
Arjmandi BH, Alekel L, Hollis BW, Amin D, Stacewicz-Sapuntzakis M (1996) Dietary soybean protein prevents bone loss in an ovariectomized rat model of osteoporosis. J Nutrition 126:161–167
Arnold SF, Robinson MK, Notides AC, Guillette LJ J, McLachlan JA (1996) A yeast estrogen screen for examining the relative exposure of cells to natural and xeno-estrogens. Environ Health Perspect 104:544–548
Auger J, Kunstmann JM, Czyglik F, Jouannet P (1995) Decline in semen quality among fertile men in Paris during the past 20 years. N Eng J Med 332:281–285
Axelson M, Setchell KDR (1981) The excretion of lignans in rats-evidence for an intestinal bacterial source for this new group of compounds. FEBS Lett 123:337–342
Barclay AS, Perdue RE Jr (1976) Distribution of anticancer activity in higher plants. Canc Treat Rep 60:1081–1113
Barnes S (1995) Effect of genistein on in vitro and in vivo models of cancer. J Nutrition 125:777S–783 S
Bennets H, Underwood EJ, Shier FL (1946) A specific breeding problem of sheep on subterranean clover pasture in Western Australia. Aust Vet J 22:2–12
Bern HA, Talamantes FJ (1981) Neonatal mouse models and their relation to disease in the human female. In: Herbst AL and Bern HA (eds.) The Developmental Effects of Diethylstilbestrol (DES) in Pregnancy. Thieme-Stratton, New York, pp 129–147
Bertazzi PA, Pesatori AC, Consonni D, Tironi A, Landi MT, Zocchetti C (1993) Cancer incidence in a population accidentally exposed to 2,3,7,8-tetrachlorodibenzo-*para*-dioxin. Epidemiol 4:398–406
Bickoff EM (1968) *Rev Ser Commonw Bur Past Fiel*d Crops No. 1

Bickoff EM, Booth AN, Lyman RL, Livingston AL, Thompson CR, De Eds F (1957) Coumestrol, a new estrogen isolated from forage crops. Science 126:969–970
Birnbaum LS (1994) Endocrine effects of prenatal exposure to PCBs, dioxins and other xenobiotics: implications for policy and future directions. Environ Health Perspect 102:676–679
Bitman J, Cecil HC, Harris SJ, Fries GF (1968) Estrogenic activity of o,p′-DDT in the mammalian uterus and avian oviduct. Science 162:371–372
Bradbury RB, White DE (1951) The chemistry of subterranean clover. Part I. Isolation of formononetin and genistein. J Chem Soc 3447–3449
Bradbury RB, White DE (1954) Estrogens and related substances in plants. Vitam Horm 12:207–233
Braden AWH, Thain RI, Shutt DA (1971) Comparison of plasma phyto-estrogen levels in sheep and cattle after feeding on fresh clover. Aust J Agric Res 22:663–670
Brandi ML (1992) Flavonoids: biochemical effects and therapeutic applications. Bone Miner 19:3S–14S
Breslau NA, Brinkley L, Hill KD, Pak CYC (1988) Relationship of animal protein-rich diet to kidney stone formation and calcium metabolism. J Clin Endocrinol Metab 66:140–146
Brouwer A, Ahlborg UG, Van den Berg M (1995) Functional aspects of developmental toxicity of polyhalogenated aromatic hydrocarbons in experimental animals and human infants. Eur J Pharmacol 293:1–40
Brown NM, Lamartiniere CA (1995) Xenoestrogens alter mammary gland differentiation and cell proliferation in the rat. Environ Health Perspect 103:708–713
Brzezinski A, Adlercreutz H, Shaoul R, Rosler A, Shmueli A, Tanos V, Schenker JG (1997) Short-term effects of phytoestrogen-rich diet on postmenopausal women. J N Amer Menopause Soc 4:89–94
Brzozowski AM, Pike ACW, Dauter Z, E. HR, Bonn T, Engstrom O, Ohman L, Greene GL, Gustafsson J-A, Carlquist M (1997) Molecular basis of agonism and antagonism in the estrogen receptor. Nature 389:753–758
Bulger WH, Kupfer D (1985) Estrogenic activity of pesticides and other xenobiotics on the uterus and male reproductive tract. In: (Thomas JA, ed.) Endocrine Toxicology. Raven Press, New York, pp 1–33
Burlington H, Linderman VF (1950) Effect of DDT on testes and secondary sex characteristics of white leghorn cockerels. Proceed Soc Exp Biol Med 74:48–51
Burroughs CD, Bern HA, Stokstad ELR (1985) Prolonged vaginal cornification and other changes in mice treated neonatally with coumestrol, a plant estrogen. J Toxicol Environ Hlth 15:51–61
Burroughs CD, Mills KT, Bern HA (1990) Reproductive abnormalities in female mice exposed neonatally to various doses of coumestrol. J Toxicol Environ Hlth 30:105–122
Carlsen E, Giwercman A, Keiding N, Skakkebaek NE (1992) Evidence for decreasing quality of semen during the past 50 years. Br Med J 305:609–613
Carlsen E, Giwercman A, Keiding N, Skakkebaek NE (1995) Declining semen quality and increasing incidence of testicular cancer: is there a common cause. Environ Health Perspect 103(suppl 7):137–139
Carroll KK (1991) Review of clinical studies on cholesterol-lowering response to soy protein. J Am Diet Assoc 91:820–827
Carroll KK, Kurowska EM (1995) Soy consumption and cholesterol reduction: review of animal and human studies. J Nutrition 125:594S–597S
Cassidy A (1996) Physiological effects of phyto-estrogens in relation to cancer and other human health risks. Proceed Nutr Soc 55:399–417
Cassidy A, Bingham S, Setchell KDR (1994) Biological effects of a diet of soy protein rich in isoflavones on the menstrual cycle of premenopausal women. Am J Clin Nutrition 60:333–340

Chen CW, Hurd C, Vorojeikina DP, Arnold SF, Notides AC (1997) Transcriptional activation of the human estrogen receptor by DDT isomers and metabolites in yeast and MCF-7 cells. Biochem Pharmacol 53:1161–1172
Christensen CM, Nelson GH, Mirocha CJ (1965) Effect on the white rat uterus of a toxic substance isolated from *Fusarium*. Appl Microbiol 13:653–659
Claffey KP, Robinson GS (1996) Regulation of vegf/vpf expression in tumor cells – consequences for tumor growth and metastasis. Canc Metast Rev 15:165–176
Clarkson TB, Anthony MS, Hughes CL Jr (1995) Estrogenic soybean isoflavones and chronic disease. Risks and benefits. Trend Endocrinol Metab 6:11–16
Colborn T, Clement C (1992) Chemically-induced alterations in sexual and functional development: the wildlife/human connection (Advances in Modern Experimental Toxicology, Volume 21), Princeton Scientific Publishing, Princeton.
Connor K, Howell J, Chen I, Liu H, Berhane K, Sciarretta C, Safe S, Zacharewski T (1996) Failure of chloro-S-triazine-derived compounds to induce estrogen receptor-mediated responses in vivo and in vitro. Fund Appl Toxicol 30:93–101
Cooper C, Campion G, Melton LJ III (1992) Hip fractures in the elderly: a world-wide projection. Osteopor Intern 2:285–289
Cullinan-Bove K, Koos RD (1993) Vascular endothelial growth factor/vascular permeability factor expression in the rat uterus: rapid stimulation by estrogen correlates with estrogen induced increases in uterine capillary permeability and growth. Endocrinol 133:829–837
Cummings AM (1997) Methoxychlor as a model for environmental estrogens. Crit Rev Toxicol 27:367–379
Cummings AM, Metcalf JL (1995) Methoxychlor regulates rat uterine estrogen-induced protein. Tox Appl Pharmacol130:154–160
Daston GP, Gooch JW, Breslin WJ, Shuey DL, Nikiforov AI, Fico TA, Gorsuch JW (1997) Environmental estrogens and reproductive health – a discussion of the human and environmental data. Reprod Toxicol 11:465–481
Dees C, Foster JS, Ahamed S, Wimalasena J (1997) Dietary estrogens stimulate human breast cells to enter the cell cycle. Environ Health Perspect 105(suppl 3):633–636
Dewailly E, Dodin S, Verreault R, Ayotte P, Sauve L, Morin J (1994) High organochlorine body burden in women with estrogen receptor positive breast cancer. J Natl Canc Inst 86:232–234
Dodds EC, Lawson W (1938) Molecular structure in relation to estrogenic activity. Compounds with a phenanthrene nucleus. Proceed Roy Soc Lond (B) 125:222–232
Dodge JA, Glasebrook AL, Magee DE, Phillips DL, Sato M, Short LL, Bryant HU (1996) Environmental estrogens: effects on cholesterol lowering and bone in the ovariectomized rat. J Steroid Biochem Mol Biol 59:155–161
Dohrn M (1927) Ist der Allen-Doisy-Test spezifisch fur das weibliche sexualhormon? Klin Wochensch 6:359–360
Duax WL, Weeks CM (1980) Molecular basis of estrogenicity: x-ray crystallographic studies. In: McLachlan JA (ed.) Estrogens in the environment. Elsevier/North Holland, New York, pp 11–31
Edmunds JS, Fairey ER, Ramsdell JS (1997) A rapid and sensitive high throughput reporter gene assay for estrogenic effects of environmental contaminants. Neurotoxicol 18:525–32
Eldridge AC, Kwolek WF (1983) Soybean isoflavones: effect of environment and variety on composition. J Agric Food Chem 31:394–396
Eroschenko VP, Palmiter RD (1980) Estrogenicity of kepone in birds and mammals. In: (McLachlan JA, ed.) Estrogens in the Environment. Elsevier, New York, pp 305–326
Everett DJ, Perry CJ, Scott KA, Martin BW, Terry MK (1987) Estrogenic potencies of resorcyclic acid lactones and 17beta-estradiol in female rats. J Toxicol Environ Hlth 20:435–443
Faber KA, Hughes CL (1993) Dose-response characteristics of neonatal exposure to genistein on pituitary responsiveness to gonadotropin releasing hormone and

volume of sexually dimorphic nucleus of the preoptic area (SDN-POA) in post-pubertal castrated female rats. Reprod Toxicol 7:35–39
Falck F Jr, Ricci A Jr, Wolff MS, Godbold J, Deckers P (1992) Pesticides and polychlorinated biphenyl residues in human breast lipids and their relation to breast cancer. Arch Environ Hlth 47:143–146
Farnesworth NR, Bingel AS, Cordell GA, Crane FA, Fong HHS (1975) Potential value of plants as antifertility agents. II. J Pharm Sci 64:717–754
Fisch H, Goluboff ET (1996) Geographic variations in sperm counts: a potential cause of bias in studies of semen quality. Fert Steril 65:1044–1046
Fisch H, Goluboff ET, Olson JH, Feldshuh J, Broder SJ, Barad DH (1996a) Semen analyses in 1,283 men from the United States over a 25-year period: no decline in quality. Fert Steril 65:1009–1014
Fisch H, Ikeguchi EF, Goluboff ET (1996b) Worldwide variations in sperm counts. Urology 48:909–911
Flouriot G, Pakdel F, Ducouret B, Valotaire Y (1995) Influence of xenobiotics on rainbow trout liver estrogen receptor and vitellogenin gene expression. J Mol Endocrinol 15:143–151
Folkman J (1994) Angiogenesis and breast cancer. J Clin Oncol 12:441–443
Folman Y, Pope GS (1966) The interaction in the immature mouse of potent estrogens with coumestrol, genistein and other utero-vaginotrophic compounds of low potency. J Endocrinol 34:215–225
Folman Y, Pope GS (1969) Effect of norethisterone acetate, dimethylstilboestrol, genistein, and coumestrol on uptake of [H]oestradiol by uterus, vagina and skeletal muscle of immature rats. J Endocrinol 44:213–218
Fotsis T, Pepper M, Adlercreutz H, Fleischmann G, Hase T, Montesano R, Schweigerer L (1993) Genistein, a dietary-derived inhibitor of in vitro angiogenesis. Proc Natl Acad Sci USA 90:2690–2694
Fotsis T, Pepper M, Adlercreutz H, Hase T, Montesano R, Schweigerer L (1995) Genistein, a dietary ingested isoflavonoid, inhibits cell proliferation and in vitro angiogenesis. J Nutrition 125:790S–797 S
Fotsis T, Pepper MS, Aktas E, Breit S, Rasku S, Adlercreutz H, Wahala K, Montesano R, Schweigerer L (1997) Flavonoids, dietary-derived inhibitors of cell proliferation and in vitro angiogenesis. Canc Res 57:2916–2921
Fotsis T, Zhang Y, Pepper MS, Adlercreutz H, Montesano R, Nawroth PP, Schweigerer L (1994) The endogenous estrogen metabolite 2-methoxyoestradiol inhibits angiogenesis and suppresses tumour growth. Nature 368:237–239
Franke AA, Custer LJ, Cerna CM, Narala K (1995) Rapid HPLC analysis of dietary phytoestrogens from legumes and from human urine. Proceed Soc Exp Biol Med 208:18–26
Gagliardi A, Collins DC (1993) Inhibition of angiogenesis by antiestrogens. Canc Res 53:533–535
Gagliardi AR, Hennig B, Collins DC (1996) Antiestrogens inhibit endothelial cell growth stimulated by angiogenic-growth factors. Anticanc Res 16:1101–1106
Gaido KW, Leonard LS, Lovell S, Gould JC, Babai D, Portier CJ, McDonnell DP (1997) Evaluation of chemicals with endocrine modulating activity in a yeast-based steroid hormone receptor gene transcription assay. Tox Appl Pharmacol 143: 205–212
Gambacciani M, Spinetti A, Piaggesi L (1994) Ipriflavone prevents the bone mass reduction in premenopausal women treated with gonadotropin hormone-releasing hormone agonists. Bone Miner 26:19–26
Gellert RJ (1978a) Kepone, mirex, dieldrin, and aldrin: estrogenic activity and the induction of persistent vaginal estrus and anovulation in rats following neonatal treatment. Environ Res 16:131–138
Gellert RJ (1978b) Uterotrophic activity of polychlorinated biphenyls (PCB) and induction of precocious reproductive aging in neonatally treated female rats. Environ Res 16:123–130

Giger W, AHEL M, Koch M, Laubscher HU, Schaffner C, Schneider J (1987) Behaviour of alkylphenol-polyethoxylate surfactants and of nitriloacetate in sewage treatment. Water Sci Technol 19:449–460

Gladen BC, Rogan WJ (1995) DDE and shortened duration of lactation in a northern Mexican town. Am J Publ Hlth 85:504–508

Gooderham MJ, Adlercreutz H, Ojala S, Wahala K, Holub BJ (1997) A soy protein isolate rich in genistein and daidzein and its effects on plasma isoflavone concentrations, platelet aggregation, blood lipids and fatty acid composition of plasma phospholipid in normal men. J Nutrition 126:2000–2006

Gray LE et al. (1997) Endocrine screening methods workshop report – detection of estrogenic and androgenic hormonal and antihormonal activity for chemicals that act via receptor or steroidogenic enzyme mechanisms. Reprod Toxicol 11: 719–750

Guillette LJ Jr, Guillette EA (1996) Environmental contaminants and reproductive abnormalities in wildlife: implications for public health? Toxicol Indust Hlth 12:537–550

Guzelian PS (1982) Comparative toxicology of chlordecone (Kepone) in humans and experimental animals. Ann Rev Pharmacol Toxicol 22:89–113

Hammond B, Katzenellenbogen BS, Krauthammer N, McConnell J (1979) Estrogenic activity of the insecticide chlordecone (Kepone) and interaction with uterine estrogen receptors. Proc Natl Acad Sci USA 76:6641–6645

Hartwell JL (1976) Types of anticancer agents isolated from plants. Canc Treat Rep 60:1031–1067

Haytowitz DB, Matthews RH (1976) Composition of foods: legumes and legume products. In: Agriculture Handbook, No. 8. Human Nutrition Information Service USDA, Washington DC

Henderson BE, Ross RK, Pike MC (1993) Hormonal chemoprevention of cancer in women. Science 259:633–638

Herbst AL, Bern HA (1981) The Developmental Effects of Diethylstilbestrol (DES) in Pregnancy Thieme-Stratton, New York.

Horn-Ross PL, Barnes S, Kirk M, Coward L, Parsonnet J, Hiatt RA (1997) Urinary phytoestrogen levels in young women from a multiethnic population. Canc Epid Biom Prev 6:339–345

Hunter DJ, Hankinson SE, Laden F, Colditz GA, Manson JE, Willett WC, Speizer FE, Wolff MS (1997) Plasma organochlorine levels and the risk of breast cancer. The N Engl J Med 337:1253–1258

Hurd RN (1977) Structure activity relationships in zearalenones. In: (Rodricks JV et al. eds.) Mycotoxins in human and animal health. Pathotox Publishers, Inc., Park Forest South (IL)

Hyder SM, Chiappetta C, Stancel GM (1997) Triphenylethylene antiestrogens induce uterine vascular endothelial growth factor expression via their partial estrogen agonist activity. Canc Lett 120:165–171

Hyder SM, Stancel GM, Chiappetta C, Murthy L, Boettger-Tong HL, Makela S (1996) Uterine expression of vascular endothelial growth factor is increased by estradiol and tamoxifen. Canc Res 56:3954–3960

Ingram D, Sanders K, Kolybaba M, Lopez D (1997) Case-control study of phytooestrogens and breast cancer. Lancet 350:990–994

Irvine S, Cawood E, Richardson D, MacDonald E, Aitken J (1996) Evidence of deteriorating semen quality in the United Kingdom: birth cohort study in 577 men in Scotland over 11 years. Br Med J 312:467–471

Jansen HT, Cooke PS, Porcelli J, Liu TC, Hansen LG (1993) Estrogenic and antiestrogenic actions of PCBs in the female rat: in vitro and in vivo studies. Reprod Toxicol 7:237–248

Jobling S, Reynolds T, White R, Parker MG, Sumpter JP (1995) A variety of environmentally persistent chemicals, including some phthalate plasticizers, are weakly estrogenic. Environ Health Perspect 103(suppl 7):582–587

Jobling S, Sumpter JP (1993) Detergent components in sewage effluent are weakly estrogenic to fish: an in vitro study using rainbow trout (*Oncorhynchus mykiss*) hepatocytes. Aquat Toxicol 27:361–372

Katzenellenbogen JA (1995) The structural pervasiveness of estrogenic activity. Environ Health Perspect 103(suppl 7):99–101

Katzenellenbogen BS, Katzenellenbogen JA, Mordecai D (1979) Zearalenones: characterization of the estrogenic potencies and receptor interactions of a series of fungal beta-resorcylic acid lactones. Endocrinol 105:33–40

Kavlock RJ et al. (1996) Research needs for the risk assessment of health and environmental effects of endocrine disruptors: a report of the U.S. EPA-sponsored workshop. Environ Health Perspect 104 (supplement 4):715–740

Kelce WR, Wilson EM (1997) Environmental antiandrogens – developmental effects, molecular mechanisms, and clinical implications. J Molec Med 75:198–207

Kelly GE, Nelson C, Waring MA, Joannou GE, Reeder AY (1993) Metabolites of dietary (soya) isoflavones in human urine. Clin Chim Acta 223:9–22

Kennedy AR (1995) The evidence for soybean products as cancer preventive agents. J Nutrition 125:733S–743S

Kitts WD, Newsome FE, Runeckles VC (1983) The estrogenic and antiestrogenic effects of coumestrol and zearalenol on the immature rat uterus. Canad J Anim Sci 63:823–834

Kogevinas M, Becher H, Benn T, Bertazzi PA, Boffetta P, Bueno-de-Mesquita HB, Coggon D, Colin D, Flesch-Janys D, Fingerhut M, Green L, Kauppinen T, Littorin M, Lynge E, Mathews JD, Neuberger M, Pearce N, Saracci R (1997) Cancer mortality in workers exposed to phenoxy herbicides, chlorophenols, and dioxins. An expanded and updated international cohort study. Am J Epidemiol 145:1061–1075

Kogevinas M, Saracci R, Winkelmann R, Johnson ES, Bertazzi PA, Bueno de Mesquita BH, Kauppinen T, Littorin M, Lynge E, Neuberger M (1993) Cancer incidence and mortality in women occupationally exposed to chlorophenoxy herbicides, chlorophenols, and dioxins. Canc Causes Control 4:547–553

Korach KS, Sarver P, Chae K, McLachlan JA, McKinney JD (1988) Estrogen receptor-binding activity of polychlorinated hydroxybiphenyls: conformationally restricted structural probes. Mol Pharmacol 33:120–126

Krieger N, Wolff MS, Hiatt RA, Rivera M, Vogelman J, Orentreich N (1994) Breast cancer and serum organochlorines: a prospective study among white, black, and Asian women. J Natl Canc Inst 86:589–599

Krishnan AV, Stathis P, Permuth SF, Tokes L, Feldman D (1993) Bisphenol-A: an estrogenic substance is released from polycarbonate flasks during autoclaving. Endocrinol 132:2278–2286

Kudou S, Fleury Y, Welti D, Magnolato D, Uchida T (1991) Malonyl isoflavone glycosides in soybean seeds (Glycine max MERRILL). Agric Biol Chem 55:2227–2233

Kuiper GGJM, Carlsson B, Grandien K, Enmark E, Haggblad J, Nilsson S, Gustafsson JA (1997) Comparison of the ligand binding specificity and transcript tissue distribution of estrogen receptors alpha and beta. Endocrinol 138:863–870

Kuiper GGJM, Enmark E, Pelto-Huikko M, Nilsson S, Gustafsson JA (1996) Cloning a novel estrogen receptor expressed in rat prostate and ovary. Proc Natl Acad Sci USA 93:5925–5930

Kupfer D, Bulger WH (1976a) Interactions of chlorinated hydrocarbons with steroid hormones. Fed Proceed 35:2603–2608

Kupfer D, Bulger WH (1976b) Studies on the mechanism of estrogenic actions of o,p′-DDT: interactions with the estrogen receptor. Pesticide Biochem Physiol 6:561–570

Kupfer D, Bulger WH (1979) A novel in vitro method for demonstrating proestrogens. Metabolism of methoxychlor and o,p'-DDT by liver microsomes in the presence of uteri and effects on intracellular distribution of estrogen receptors. Life Sci 25:975–983

Kurzer MS, Xu X (1997) Dietary phytoestrogens. Ann Rev Nutrition 17:353–381

Lamartiniere CA, Moore J, Holland M, Barnes S (1995a) Neonatal genistein chemoprevents mammary cancer. Proceed Soc Exp Biol Med 208:120–123
Lamartiniere CA, Moore JB, Brown NM, Thompson R, Hardin MJ, Barnes S (1995b) Genistein suppresses mammary cancer in rats. Carcinogen 16:2833–2840
Lampe JW, Martini MC, Kurzer MS, Adlercreutz H, Slavin JL (1994) Urinary lignan and isoflavonoid excretion in premenopausal women consuming flaxseed powder. Am J Clin Nutrition 60:122–129
Lech JJ, Lewis SK, Ren L (1996) In vivo estrogenic activity of nonylphenol in rainbow trout. Fund Appl Toxicol 30:229–232
Levy JR, Faber KA, Ayyash L, Hughes CL (1995) The effect of prenatal exposure to the phytoestrogen genistein on sexual differentiation in rats. Proceed Soc Exp Biol Med 208:60–66
Loewe S (1933) Klein's Handbuch der Pflanzenanalyse. Springer, 4:1034–1041
Lone KP (1997) Natural sex steroids and their xenobiotic analogs in animal production – growth, carcass quality, pharmacokinetics, metabolism, mode of action, residues, methods, and epidemiology. Crit Rev Food Sci Technol 37:93–209
Long JA, Evans HM (1922) The oestrous cycle in the rat and its associated phenomena. Mem Univ California 6:1–6
Majdic G, Sharpe RM, O'Shaugnessy PJ, Saunders PT (1996) Expression of cytochrome P450 17alpha-hydroxylase/C17–20 lyase in the fetal rat testis is reduced by maternal exposure to exogenous estrogens. Endocrinol 137:1063–1070
Makela S, Pylkkanen LH, Santti RS, Adlercreutz H (1995a) Dietary soybean may be antiestrogenic in male mice. J Nutrition 125:437–445
Makela S, Santti R, Salo L, McLachlan JA (1995b) Phytoestrogens are partial estrogen agonists in the adult male mouse. Environ Health Perspect 103:123–7
Martin PM, Horwitz KB, Ryan DS, L MW (1978) Phytoestrogen interaction with estrogen receptors in human breast cancer cells. Endocrinol 103:1860–1867
Mastri C, Mistry P, Lucier GW (1985) In vivo estrogenicity and binding characteristics of alpha-zearalanol (P-1496) to different classes of estrogen binding proteins in rat liver. J Steroid Biochem 23:279–289
McKinney JD, Waller CL (1994) Polychlorinated biphenyls as hormonally active structural analogs. Environ Health Perspect 102:290–297
McLachlan JA (1980) Estrogens in the environment Elsevier/North-Holland, New York.
McLachlan JA (1985) Estrogens in the environment II. Influences on development Elsevier, New York.
McLachlan JA, Arnold SF (1996) Environmental Estrogens. Am Scientist 84:452–461
McLachlan JA, Korach KS (1995) Estrogens in the environment, III. Global health implications. Environ Health Perspect 103(supple 7)
McMichael-Phillips DF, Harding C, Morton M, Potten CS, Burdred NJ (1996) The effects of soy supplementation on epithelial proliferation in the normal human breast. In: Abstracts of the Second International Symposium on the Role of Soy in Preventing and Treating Chronic Disease. Brussels, p 35
McNutt SH, Purwin P, Murray C (1928) Vulvo-vaginitis in swine; preliminary report. J Am Vet Med Assoc 73:484
Medlock KL, Branham WS, Sheehan DM (1995a) Effects of coumestrol and equol on the developing reproductive tract of the rat. Proceed Soc Exp Biol Med 208:67–71
Medlock KL, Branham WS, Sheehan DM (1995b) The effects of phytoestrogens on neonatal rat uterine growth and development. Proceed Soc Exp Biol Med 208:307–313
Messina MJ, Persky V, Setchell KDR, Barnes S (1994) Soy intake and cancer risk: a review of the in vitro and in vivo data. Nutr Canc 21:113–131
Migdalof BH, Dugger HA, Heider JG, Coombs RA, Terry MK (1983) Biotransformation of zeranol: disposition and metabolism in the female rat, rabbit, dog, monkey and man. Xenobiotica 13:209–221

Miksicek RJ (1993) Commonly occurring plant flavonoids have estrogenic activity. Mol Pharmacol 44:37–43
Miksicek RJ (1994) Interaction of naturally occurring nosteroidal estrogens with expressed recombinant human estrogen receptor. J Steroid Biochem Mol Biol 49:153–160
Miksicek RJ (1995) Estrogenic flavonoids: structural requirements for biological activity. Proceed Soc Exp Biol Med 208:44–50
Mirocha CJ, Pathre SV, Christensen CM (1977) Zearalenone. In: Rodricks JV et al. (eds.) Mycotoxins in human and animal health. Pathotox Publishers, Inc., Park Forest South, Illinois, pp 345–364
Monti E, Sinha BK (1994) Antiproliferative effect of genistein and adriamycin against estrogen-dependent and -independent human breast carcinoma cell lines. Anticanc Res 14:1221–1226
Moreau C, Moss M (1979) Estrogenic effects:zearalenone. In: Moulds, Toxins and Food. John Wiley & Sons, Chicchester, pp 233–235
Mori T, Nagasawa H, Bern HA (1979) Long-term effects of perinatal exposure to hormones on normal and neoplastic mammary growth in rodents: a review. J Environ Pathol Toxicol 3:191–205
Morton MS, Chan PSF, Cheng C, Blacklock N, Matos-Ferreira A, Abranches-Monteiro L, Correia R, Lloyd S, Griffiths K (1997) Lignans and isoflavonoids in plasma and prostatic fluid in men: samples from Portugal, Hong Kong, and the United Kingdom. Prostate 22:122–128
Morton MS, Wilcox G, Wahlqvist ML, Griffiths K (1994) Determination of lignans and isoflavonoids in human female plasma following dietary supplementation. J Endocrinol 142:251–259
Mueller GC, Kim UH (1978) Displacement of estradiol from estrogen receptors by simple alkylphenols. Endocrinol 102:1429–1435
Murrill WB, Brown NM, Zhang JX, Manzolillo PA, Barnes S, Lamartiniere CA (1996) Prepubertal genistein exposure suppresses mammary cancer and enhances gland differentiation in rats. Carcinogen 17:1451–1457
Mussalo-Rauhamaa H, Hasanen E, Pyysalo H, Antervo K, Kauppila R, Pantzar P (1990) Occurrence of beta-hexachlorocyclohexane in breast cancer patients. Cancer 66:2124–2128
Nagel SC, vom Saal FS, Thayer KA, Dhar MG, Boechler M, Welshons WV (1997) Relative binding affinity-serum modified access (RBA-SMA) assay predicts the relative in vivo bioactivity of the xenoestrogens bisphenol A and octylphenol. Environ Health Perspect 105:70–76
Naim M, Gestetner B, Zilkah S, Birk Y, Bondi A (1974) Soybean isoflavones. Characterization, determination and fungal activity. J Agricult Food Chem 22:806–810
Nakamura J, Savinov A, Lu Q, Brodie A (1996) Estrogen regulates vascular endothelial growth factor -permeability factor expression in 7,12-dimethylbenz(a) anthracene-induced mammary tumors. Endocrinol 137:5589–5596
Nelson JA (1974) Effects of dichloro-diphenyltrichloroethane (DDT) analogs and polychlorinated biphenyl (PCB) mixtures on 17beta-[H]-estradiol binding to rat uterine receptor. Biochem Pharmacol 23:447–451
Nelson JA, Struck RF, James R (1978) Estrogenic activities of chlorinated hydrocarbons. J Toxicol Environ Hlth 4:325–339
Nesaretnam K, Corcoran D, Dils RR, Darbre P (1996) 3,4,3′,4′-Tetrachlorobiphenyl acts as an estrogen in vitro and in vivo. Mol Endocrinol 10:923–936
Newbold R, McLachlan JA (1985) Diethylstilbestrol associated defects in murine genital tract development. In: McLachlan JA (ed.) Estrogens in the Environment II. Influences on development, Elsevier, New York, pp 288–318
Nimrod AC, Benson WH (1996) Environmental estrogenic effects of alkylphenol ethoxylates. Crit Rev Toxicol 26:335–364
Noteboom WD, Gorski J (1963) Estrogenic effect of genistein and coumestrol diacetate. Endocrinol 73:736–739

Notzon F (1984) Trends in infant feeding in developing countries. Pediatrics 74:648–666
Odum J, Lefevre PA, Tittensor S, Paton D, Routledge EJ, Beresford NA, Sumpter JP, Ashby J (1997) The rodent uterotrophic assay: critical protocol features studies with nonyl phenols, and comparison with a yeast estrogenicity assay. Reg Toxicol Pharmacol 25:176–188
Pagliacci MC, Smacchia M, Migliorati G, Grignani F, Riccardi C, Nicoletti I (1994) Growth-inhibitory effects of the natural phytoestrogen genistein in MCF-7 human breast cancer cells. Eur J Canc 30 A:1675–1682
Palmiter RD, Mulvihill ER (1978) Estrogenic activity of the insecticide kepone on the chicken oviduct. Science 201:356–358
Pathre SV, Mirocha CJ (1980) Mycotoxins as estrogens. In: McLachlan JA (ed.) Estrogens in the environment. Elsevier/North-Holland, New York, pp 265–278
Pesatori AC, Consonni D, Tironi A, Zocchetti C, Fini A, Bertazzi PA (1993) Cancer in a young population in a dioxin-contaminated area. Int J Epidemiol 22:1010–1013
Peterson G, Barnes S (1993) Genistein and biochanin-A inhibit the growth of human prostate cancer cells but not epidermal growth factor receptor tyrosine autophosphorylation. Prostate 22:335–345
Peterson RE, Theobald HM, Kimmel GL (1993) Developmental and reproductive toxicity of dioxins and related compounds: cross-species comparisons. Crit Rev Toxicol 23:283–335
Potter SM (1995) Overview of proposed mechanisms for the hypocholesterolemic effect of soy. J Nutrition 125:606S–611S
Raines EW, Ross R (1995) Biology of atherosclerotic plaque formation: possible role of growth factors in lesion development and the potential impact of soy. J Nutrition 125:624S–630S
Ren L, Lewis SK, Lech JJ (1996) Effects of estrogen and nonylphenol on the post-transcriptional regulation of vitellogenin gene expression. Chem-Biol Interact 100:67–76
Risebrough RW, Rieche P, Herman SG, Peakall DB, Kirven MN (1968) Polychlorinated biphenyls in the global ecosystem. Nature 220:1098–1102
Robison AK, Mukku VR, Spalding DM, Stancel GM (1984) The estrogenic activity of DDT: the in vitro induction of an estrogen-inducible protein by o,p′-DDT. Tox Appl Pharmacol 76:537–543
Robison AK, Mukku VR, Stancel GM (1985a) Analysis and characterization of estrogenic xenobiotics and natural products. In: McLachlan JA (ed.) Estrogens in the Environment II. Influences on Development. Elsevier, New York, pp 107–115
Robison AK, Schmidt WA, Stancel GM (1985b) Estrogenic activity of DDT: estrogen-receptor profiles and the responses of individual uterine cell types following o,p′-DDT administration. J Toxicol Environ Hlth 16:493–508
Robison AK, Sirbasku DA, Stancel GM (1985c) DDT supports the growth of an estrogen responsive tumor. Toxicol Lett 27:109–113
Rodricks JV, Hesseltine CW, Mehlman MA (1977) Mycotoxins in human and animal health Pathotox Publishers, Inc., Park Forest South (IL)
Rogan WJ, Gladen BC, McKinney JD, Carreras N, Hardy P, Thullen J, Tingelstad J, Tully M (1987) Polychlorinated biphenyls (PCBs) and dichlorodiphenyl dichloroethene (DDE) in human milk: effects on growth, morbidity, and duration of lactation. Am J Publ Hlth 77:1294–1297
Rokhlin OW, Cohen MB (1995) Differential sensitivity of human prostatic cancer cell lines to the effects of protein kinase and phosphatase inhibitors. Canc Lett 98:103–110
Ruh MF, Zacharewski T, Connor K, Howell J, Chen I, Safe S (1995) Naringenin: a weakly estrogenic bioflavonoid that exhibits antiestrogenic activity. Biochem Pharmacol 50:1485–1493
Safe SF (1995) Modulation of gene expression and endocrine response pathways by 2,3,7,8-tetrachlorodibenzo-*p*-dioxin and related compounds. Pharmacol Therap 67:247–281

Santell RC, Chang YC, Nair MG, Helferich WG (1997) Dietary genistein exerts estrogenic effects upon the uterus, mammary gland and the hypothalamic/pituitary axis in rats. J Nutrition 127:263–269
Sathyamoorthy N, Wang TT, Phang JM (1994) Stimulation of pS2 expression by diet-derived compounds. Canc Res 54:957–961
Schultz TD, Bonorden WR, Seaman MS (1991) Effect of short-term flaxseed consumption on lignan and sex hormone metabolism in men. Nutr Res 11:1089–1100
Scribner JD, Mottet NK (1981) DDT acceleration of mammary gland tumors induced in the male Sprague-Dawley rat by 2-acetamidophenanthrene. Carcinogen 2: 1235–1239
Setchell KD, Borrielo SP, Hulme P, Kirk DN, Axelson M (1984) Non-steroidal estrogens of dietary origin: possible roles in hormone dependent disease. Am J Clin Nutrition 40:569–578
Setchell KDR, Adlercreutz H (1988) Mammalian lignans and phyto-estrogens. Recent studies on their formation, metabolism and biological role in health and disease. In: Rowland IR (ed.) Role of the Gut Flora in Toxicity and Cancer. Academic Press, London, pp 315–345
Setchell KDR, Lawson AM, Borriello SP, Harkness R, Gordon H, Morgan DML, Kirk DN, Adlercreutz H, Anderson LC, Axelson M (1981) Lignan formation in man-microbial involvement and possible roles in relation to cancer. Lancet 2:4–8
Setchell KDR, Lawson AM, Mitchell FL, Adlercreutz H, Kirk DN, Axelson M (1980) Lignans in man and in animal species. Nature 287:740–742
Setchell KDR, Zimmer-Nechemias L, Cai J, Heubi JE (1997) Exposure of infants to phyto-estrogens from soy-based infant formula. Lancet 350:23–27
Severson RK, Nomura AMY, Grove JS, Stemmermann GN (1989) A prospective study of demographics, diet, and prostate cancer among men of Japanese ancestry in Hawaii. Canc Res 49:1857–1860
Sharma OP, Adlercreutz H, Strandberg JD, Zirkin RR, Coffey DS, Ewing LL (1992) Soy of dietary source plays a preventive role against pathogenesis of prostatitis in rats. J Steroid Biochem Mol Biol 43:557–564
Sharpe RM, Fisher JS, Millar MM, Jobling S, Sumpter JP (1995) Gestational and lactational exposure of rats to xenoestrogens results in reduced testicular size and sperm production. Environ Health Perspect 103:1136–1143
Sharpe RM, Skakkebaek NE (1993) Are estrogens involved in falling sperm counts and disorders of the male reproductive tract? Lancet 341:1392–1395
Sheehan DM, Branham WS, Medlock KL, Shanmugasundaram ER (1984) Estrogenic activity of zearalenone and zearalanol in the neonatal rat uterus. Teratol 29:383–92
Shutt DA (1967) Interaction of genistein with oestradiol in the reproductive tract of the ovariectomized mouse. J Endocrinol 37:231–232
Shutt DA (1976) The effect of plant estrogens on animal reproduction. Endeavor 35:110–113
Shutt DA, Cox RI (1972) Steroid and phyto-estrogen binding to sheep uterine receptors in vitro. J Endocrinol 52:299–310
Shutt DA, Weston RA, Hogan RJ (1970) Quantitative aspects of phyto-estrogen metabolism in sheep fed on subterranean clover (*Trifolium subterranean*, cultivar Clare) or red clover (*Trifolium pratense*). Aust J Agric Res 21:713–722
Singer PL (1949) Occupational oligospermia. J Am Med Assoc 140:1249
Sirtori CR, Even R, Lovati MR (1993) Soybean protein diet and plasma cholesterol: from therapy to molecular mechanisms. Ann NY Acad Sci 676:188–201
Sonnenschein C, Szelei J, Nye TL, Soto AM (1994) Control of cell proliferation of human breast MCF7 cells; serum and estrogen resistant variants. Oncol Res 6:373–381
Soto AM, Chung KL, Sonnenschein C (1994) The pesticides endosulfan, toxaphene, and dieldrin have estrogenic effects on human estrogen-sensitive cells. Environ Health Perspect 102:380–383

Soto AM, Justicia H, Wray JW, Sonnenschein C (1991) *p*-Nonyl-phenol: an estrogenic xenobiotic released from "modified" polystyrene. Environ Health Perspect 92:167–173

Soto AM, Sonnenschein C, Chung KL, Fernandez MF, Olea N, Serrano FO (1995) The E-SCREEN assay as a tool to identify estrogens: an update on estrogenic environmental pollutants. Environ Health Perspect 103:113–22

Stamler J (1979) Population studies. In: Levy RI et al. (eds.) Nutrition and Lipids and Coronary Heart Disease. Raven Press, New York, pp 25–88

Stancel GM, Ireland JS, Mukku VR, Robison AK (1980) The estrogenic activity of DDT: in vivo and in vitro induction of a specific estrogen inducible uterine protein by o,p′-DDT. Life Sci 27:1111–1117

Steinmetz R, Brown NG, Allen DL, Bigsby RM, Ben-Jonathan N (1997) The environmental estrogen bisphenol A stimulates prolactin release in vitro and in vivo. Endocrinol 138:1780–1786

Steinmetz R, Young PC, Caperell-Grant A, Gize EA, Madhukar BV, Ben-Jonathan N, Bigsby RM (1996) Novel estrogenic action of the pesticide residue beta-hexachlorocyclohexane in human breast cancer cells. Canc Res 56:5403–5409

Stitch SR, Toumba JK, Groen MB, Funke CW, Leemhuis J, Vink J, Woods GF (1980) Excretion, isolation and structure of a new phenolic constituent of female urine. Nature 287:738–740

Stob M, Baldwin RS, Tuite J, Andrews FN, Gillette KG (1962) Isolation of an anabolic, uterotrophic compound from corn infected with *Gibberella zeae*. Nature 196:1318

Stockard CR, Papanicolaou GN (1917) Existence of a typical oestrous cycle in the guinea pig, with a study of its histological and physiological changes. Am J Anat 22:225–283

Tang BY, Adams NR (1980) Effect of equol on estrogen receptors and on synthesis of DNA and protein in the immature rat uterus. J Endocrinol 85:291–297

Tong W, Perkins R, Xing L, Welsh WJ, Sheehan DM (1997) QSAR models for binding of estrogenic compounds to estrogen receptor alpha and beta subtypes. Endocrinol 138:4022–4025

Toppari J, Larsen JC, Christiansen P et al. (1996) Male reproductive health and environmental xenoestrogens. Environ Health Perspect 104(suppl 4):741–803

Tran DQ, Kow KY, McLachlan JA, Arnold SF (1996) The inhibition of estrogen receptor-mediated responses by chloro-S-triazine-derived compounds is dependent on estradiol concentration in yeast. Biochem Biophys Res Commun 227: 140–146

Tsutsumi N (1995) Effect of coumestrol on bone metabolism in organ culture. Biol Pharm Bull 18:1012–1015

Tullner WW (1961) Uterotrophic action of the insecticide methoxychlor. Science 133:647–648

Urry WH, Wehrmeister HL, Hodge EB, Hidy PH (1966) The structure of zearalenone. Tetrahedron Lett No. 27:3109–3114

Utian WH (1973) Comparative trial of P-1496, a new nonsteroidal estrogen analog. Br Med J 1:579–581

Valente M, Bufalino L, Castiglione GN (1994) Effects of 1-year treatment with ipriflavone on bone mass in postmenopausal women with low bone mass. Calcified Tissue Int 54:377–380

van't Veer P, Lobbezoo IE, Martin-Moreno JM (1997) DDT (dicophane) and postmenopausal breast cancer in Europe: case-control study. Br Med J 315:81–85

vom Saal FS, Nagel S, Palanza P, Boechler M, Parmigiani S, Welshons WW (1995) Estrogenic pesticides: binding relative to estradiol in MCF-7 cells and effects of exposure during fetal life on subsequent territorial behavior in male mice. Toxicol Lett 77:343–350

vom Saal FS, Timms BG, Montano MM, Palanza P, Thayer KA, Nagel SC, Dhar MD, Ganjam VK, Parmigiani S, Welshons WW (1997) Prostate enlargement in mice due

to fetal exposure to low doses of estradiol or diethylstilbestrol and opposite effects at high doses. Proc Natl Acad Sci USA 94:2056–2061
Waller CL, Oprea TI, Chae K, Park HK, Korach KS, Laws SC, Wiese TE, Kelce WR, Gray LE (1996) Ligand-based identification of environmental estrogens. Chem Res Toxicol 9:1240–1248
Walz E (1931) Isoflavon- and saponin-glucoside in *Soja hispida*. Justus Liebigs Annal Chem 498:118–155
Wang CF, Kurzer MS (1997) Phytoestrogen concentration determines effects on DNA synthesis in human breast cancer cells. Nutr Canc 28:236–247
Wang TY, Sathyamoorthy N, Phang JM (1996) Molecular effects of genistein on estrogen receptor mediated pathways. Carcinogen 17:271–275
Wasserman M, Nogueira DP, Thomatiis L, Mirra AP, Shibata H, Arie G, Cucos S, Wasserman D (1976) Organochlorine compounds in neoplastic and adjacent apparently normal breast tissue. Bull Environ Contam Toxicol 15:478–484
Waters AP, Knowler JT (1982) Effect of a lignan (HPMF) on RNA synthesis in the rat uterus. J Reprod Fertil 66:379–381
Welch RM, Levin W, Conney AH (1969) Estrogenic action of DDT and its analogs. Tox Appl Pharmacol14:358–367
Welshons WV, Rottinghaus GE, Nonneman DJ, Dolan-Timpe M, Ross PF (1990) A sensitive bioassay for detection of dietary estrogens in animal feeds. J Vet Diagn Invest 2:268–273
Whitten PL, Lewis C, Naftolin F (1993) A phytoestrogen diet induces the premature anovulatory syndrome in lactationally exposed female rats. Biol Reprod 49:1117–1121
Whitten PL, Lewis C, Russell E, Naftolin F (1995a) Phytoestrogen influences on the development of behavior and gonadotropin function. Proceed Soc Exp Biol Med 208:82–86
Whitten PL, Lewis C, Russell E, Naftolin F (1995b) Potential adverse effects of phytoestrogens. J Nutrition 125(suppl 3):771S–776S
Whitten PL, Naftolin F (1992) Effects of phytoestrogen diet on estrogen-dependent reproductive processes in immature female rats. Steroids 57:56–61
WHO SG (1994) Assessment of fracture risk and its application to screening for postmenopausal osteoporosis. WHO Technical Series 843:11–13
Wilcox JN, Blumenthal BF (1995) Thrombotic mechanisms in atherosclerosis: potential impact of soy proteins. J Nutrition 125:631S-638S
Wilson ME, Hagler WM Jr (1985) Metabolism of zearalenone to a more estrogenically active form. In: McLachlan JA (ed.) Estrogens in the environment. II. Influences on development. Elsevier, New York, pp 238–250
Wolff MS, Toniolo PG, Lee EW, Rivera M, Dubin N (1993) Blood levels of organochlorine residues and risk of breast cancer. J Natl Canc Inst 85:648–652
Yamazaki I, Kinoshita M (1986) Calcitonin-secreting property of ipriflavone in the presence of estrogen. Life Sci 38:1535–1541
Yuan J-M, Want Q-S, Ross RK, Henderson BE, Yu MC (1995) Diet and breast cancer in Shanghai and Tianjin, China. Br J Canc 71:1353–1358
Zava DT, Duwe G (1997) Estrogenic and antiproliferative properties of genistein and other flavonoids in human breast cancer cells in vitro. Nutr Canc 27:31–40

Subject Index

C

MIX
Papier aus verantwortungsvollen Quellen
Paper from responsible sources
FSC® C105338

If you have any concerns about our products,
you can contact us on
ProductSafety@springernature.com

In case Publisher is established outside the EU,
the EU authorized representative is:
Springer Nature Customer Service Center GmbH
Europaplatz 3, 69115 Heidelberg, Germany

Printed by Libri Plureos GmbH
in Hamburg, Germany